# CAMBRIDGE LIBRARY COLLECTION

*Books of enduring scholarly value*

## Technology

The focus of this series is engineering, broadly construed. It covers technological innovation from a range of periods and cultures, but centres on the technological achievements of the industrial era in the West, particularly in the nineteenth century, as understood by their contemporaries. Infrastructure is one major focus, covering the building of railways and canals, bridges and tunnels, land drainage, the laying of submarine cables, and the construction of docks and lighthouses. Other key topics include developments in industrial and manufacturing fields such as mining technology, the production of iron and steel, the use of steam power, and chemical processes such as photography and textile dyes.

## Submarine Telegraphs

An accomplished telegraph engineer in his own right, Sir Charles Bright (1863–1937) was the son of Sir Charles Tilston Bright (1832–88), who had achieved greatness in laying the first transatlantic cable in 1858. The younger Bright worked alongside his father for a time, continued his research, and became an authority on the subject. Examining the history, construction and working of submarine telegraphs, this 1898 treatise traces both technical and commercial developments, looking also at the labour involved. Bright addresses the laying of cables across the globe, giving accounts of projects in India, South America and beyond. Illuminating the many commercial uses for submarine cables, Bright provides an informed survey of the early standardisation of telegraphy systems. Replete with detailed illustrations and technical drawings, this work remains an indispensable resource in the history of telecommunications and electrical engineering.

Cambridge University Press has long been a pioneer in the reissuing of out-of-print titles from its own backlist, producing digital reprints of books that are still sought after by scholars and students but could not be reprinted economically using traditional technology. The Cambridge Library Collection extends this activity to a wider range of books which are still of importance to researchers and professionals, either for the source material they contain, or as landmarks in the history of their academic discipline.

Drawing from the world-renowned collections in the Cambridge University Library and other partner libraries, and guided by the advice of experts in each subject area, Cambridge University Press is using state-of-the-art scanning machines in its own Printing House to capture the content of each book selected for inclusion. The files are processed to give a consistently clear, crisp image, and the books finished to the high quality standard for which the Press is recognised around the world. The latest print-on-demand technology ensures that the books will remain available indefinitely, and that orders for single or multiple copies can quickly be supplied.

The Cambridge Library Collection brings back to life books of enduring scholarly value (including out-of-copyright works originally issued by other publishers) across a wide range of disciplines in the humanities and social sciences and in science and technology.

# Submarine Telegraphs

*Their History, Construction, and Working*

CHARLES BRIGHT

# CAMBRIDGE
## UNIVERSITY PRESS

University Printing House, Cambridge, CB2 8BS, United Kingdom

Published in the United States of America by Cambridge University Press, New York

Cambridge University Press is part of the University of Cambridge.
It furthers the University's mission by disseminating knowledge in the pursuit of
education, learning and research at the highest international levels of excellence.

www.cambridge.org
Information on this title: www.cambridge.org/9781108069489

© in this compilation Cambridge University Press 2014

This edition first published 1898
This digitally printed version 2014

ISBN 978-1-108-06948-9 Paperback

# SUBMARINE TELEGRAPHS

## THEIR HISTORY, CONSTRUCTION, AND WORKING

H.M.S. "Agamemnon" in Lat. 54° N., Long. 30° W., on 21st June 1858 (*see p.* 44).
(After the painting by Henry Clifford.)

H.M.S. "Agamemnon" entering Doulus Bay, Valentia, Ireland, at 5 A.M., 5th August 1858 (*see pp.* 47, 48),
thus completing the successful laying of the First Atlantic Cable. (After the painting by Henry Clifford.)

# SUBMARINE TELEGRAPHS

## Their History, Construction, and Working

FOUNDED IN PART ON WÜNSCHENDORFF'S 'TRAITÉ DE TÉLÉGRAPHIE SOUS-MARINE'
AND
COMPILED FROM AUTHORITATIVE AND EXCLUSIVE SOURCES

BY

## CHARLES BRIGHT, F.R.S.E.

ASSOCIATE MEMBER OF THE INSTITUTION OF CIVIL ENGINEERS
MEMBER OF THE INSTITUTION OF MECHANICAL ENGINEERS
MEMBER OF THE INSTITUTION OF ELECTRICAL ENGINEERS

LONDON
CROSBY LOCKWOOD AND SON
7, STATIONERS' HALL COURT, LUDGATE HILL
1898

Printed at THE DARIEN PRESS, *Bristo Place, Edinburgh.*

For many a fathom gleams and moves and moans
The tide that sweeps above.   .    .    .    .

Nor where they sleep shall moon or sunlight shine,
Nor man look down for ever.   .    .    .    .
And over them, while death and life shall be,
The light and sound and darkness of the sea.

SWINBURNE : Tristram of Lyonnesse.

# PREFACE.

———◆———

THE present volume had its origin in a scheme for an English revision of Monsieur Wünschendorff's famous, and almost classic, work on Submarine Telegraphy.*

I soon found, however, that—owing to the lapse of time and other circumstances—it would be impossible to deal with the entire treatise, and bring it up to date, within the confines of a single cover. The course I have adopted, therefore, has been to prepare a fresh treatise—partly based on a portion of M. Wünschendorff's. The scope—though of equal, if not greater, length—is restricted to three of the five Parts which, in touching on the fringe of the subject generally, were so admirably and systematically put together by that author, from the available literature and data—the former mainly consisting of early papers dealing with the different aspects of Submarine Telegraphy. It is thought that this is as much as could be done justice to at all completely on the lines adopted here.

The three titles selected for our present theme are — (1)

---

* "Traité de Télégraphie Sous-Marine," by E. Wünschendorff, M.I.E.E., Directeur-Ingéniéur de Télégraphes de la Region de Paris (Baudry et Cie, Paris, 1888).

History, (2) Construction, and (3) Working, of Submarine Telegraphs. This choice was not made without reason. In the first
place—strange as it may appear—notwithstanding the vast sums
at present invested in Submarine Telegraphy (far more than in
any other branch of electrical industry), nobody has yet attempted
to present to the reader in book form an account of the various
enterprises of this nature from the days of inception up to the
present advanced period.* I venture to hope that this omission is,
in some measure. made good in PART I.—to a great extent based
on original and official documents hitherto unpublished. Here,
the reader may observe that Submarine. Telegraphy, like everything else, is the work of many hands.

---

* Shortly before his death, the late Mr Willoughby Smith gave us an interesting
narrative of his extensive experiences, under title "The Rise and Extension of
Submarine Telegraphy" (London: J. S. Virtue and Co.), but this is more in the
nature of an autobiography of that distinguished pioneer.

Again, within the same month of the same year (1867), the late Mr Robert
Sabine and Mr Edward Bright, M.Inst.C.E., brought out works on "The Electric
Telegraph," treating the subject in its more general sense—the work of the latter
author being a revision of one of Dr Lardner's now classic volumes entitled "The
Museum of Science and Art." Both works are published from the same office as
this book.

Finally, Mr R. S. Culley has produced periodic editions of his excellent
"Handbook of Practical Telegraphy" (London: Longmans, Green, and Co.); and
Mr George Prescott's "Electricity and the Electric Telegraph" (Spon) corresponds
to this in the States. Preece and Sivewright's text-book of "Telegraphy"
(Longmans) then completes the available book literature on the subject in its
broadest phase; but none of the latter can be said to be historical in the ordinary
sense, except perhaps "Lardner and Bright"—and that only in a very general way
up to the end of 1866.

To turn to the future, it may be stated that a more detailed history of matters
connected with the First Trans-Atlantic Cable than was possible here will be
published very shortly in the "Life-Story of Sir Charles Tilston Bright, C.E.,
M.P." (London: Archibald Constable and Co.), wherein also will be found a fuller
account of the Telegraph to India.

Of late it has been rather the custom to question the utility of history with reference to engineering in its various aspects; but those who are inclined that way should remember that it is only by a study of what has been done before that the repetition of failures and wrong methods can be avoided.* In the historical digest here furnished, a reference will be found to the commercial and financial aspects of Submarine Telegraphy, as well as to the part played by those men of business who were so largely responsible for the success ultimately arrived at. It would be well if Science, Capital, and Labour were equally ready to admit each other's claims for recognition in all such great achievements. In this connection it may be stated here that there is a project afoot to commemorate the Jubilee of Submarine Telegraphy when it falls due in 1901.†

In the preparation of this book, no pains have been spared to

---

* A somewhat striking example of the way in which history is liable to repeat itself—especially when unrecorded—is the Statham and Whitehouse patent of 1856 for a return wire. Apparently this was taken out regardless of the fact that, previous to Steinheil's 1836 apparatus, all telegraphs were worked by a metallic circuit, Dr Watson's 1747 demonstration of the use of the earth having, it would seem, been forgotten.

† The Executive Committee of the International Submarine Telegraph Memorial—formed in part to carry out this object, as well as to memorialise the leading part taken by the late Sir John Pender, G.C.M.G., in the commercial development and extension of submarine telegraphy throughout the world — is composed as follows :—The Marquis of Tweeddale (Chairman), The Right Hon. Viscount Peel, Lord Kelvin, G.C.V.O., F.R.S., Lord Sackville Cecil, Sir Robert Herbert, G.C.B., Sir Eyre Shaw, K.C.B., Sir Albert J. Leppoc Cappel, K.C.I.E., Mr J. C. Lamb, C.B., C.M.G. (H.M. Post Office), Mr W. S. Andrews, Mr W. H. Baines, Mr F. A. Bevan, Mr G. von Chauvin, Dr J. A. Fleming, F.R.S., Mr R. Kaye Gray, Dr John Hopkinson, F.R.S., Dr Alexander Muirhead, Mr John Newton, Mr F. C. C. Nielsen, Mr J. Denison Pender, Mr J. W. Swan, F.R.S., Mr J. H. Tritton, Mr E. M. Underdown, Q.C., The Editor of the *Electrician*, The President of the Institution of Civil Engineers (Sir J. Wolfe Barry, K.C.B., F.R.S.),

give an accurate idea of the share taken by each in contributing to the foundation and perfection of this branch of electrical industry and engineering.

According to my lights and to the best of my ability, strict impartiality has been studied by me in all personal matters, whether affecting the reputations and interests of private individuals, or those of business firms and corporations. The same impartial attitude has been applied to nationality as to personality. This, I believe, will be admitted by any foreigner who peruses the book carefully, notwithstanding the obvious and necessary fact that it is addressed chiefly to English-speaking engineers and electricians, and deals with an industry which is at present almost entirely in the hands of British companies.* However patriotic the writer may be in his private sentiments, national prejudices are quite out of place in a work of this kind. Two of the accompanying maps are purposely drawn up in French, partly because that is still the *lingua franca* of modern nations, and also as being 'official publications of the International Telegraph Bureau at Bern.

A further point remains upon which it seems to me that I owe

---

The President of the Institution of Electrical Engineers (Sir Henry Mance, C.I.E.), and the author.    Mr G. R. Neilson acts as Honorary Secretary.

It may be mentioned here in passing that one of the Commemorative Honours conferred during the present jubilee year was the baronetcy of Sir James Pender, M.P.,—in part a fitting tribute to the energies of his father.

* For example, it has, of course, been impossible to include detailed descriptions of all the various Relay devices used on the extensive French land lines, which the English electrician in connection with submarine cable work is never likely to have occasion to use.    Had space, however, permitted, it would have been interesting to describe and illustrate these at length—if only on account of their entirely novel and ingenious character.    They almost demand a booklet to themselves.

some explanation to my readers.  I refer to the matter of repetitions in different parts of the book.  Although from the purely literary standpoint these repetitions are certainly blemishes, I am inclined to think that their advantage for purposes of reference will be found more than counterbalancing.  Research is considerably facilitated by enabling the student to readily find information on the special point he desires under various heads—possibly in one part of the book, possibly in another.*

Again, an additional matter calling for explanation is the free use of the footnote system.  This is partly to maintain a more easy sequence of sentences, and partly to avoid thrusting items of pure detail into the teeth of the not too inquiring reader.

No attempt is made here to give absolute recipes or up-to-date constants.  Anything of the sort is purposely eschewed, the technical portion of the book being only intended to give an insight into the leading principles involved in the art of submarine cable construction and working.  Moreover, formulæ, to be of any practical value, should be obtained at the individual factories at the time.  Again, all reference tables, such as, strictly, only apply for a short period and under certain conditions, are conspicuous by their absence, on the ground of being misleading if not absolutely up-to-date, and unless they can be made to refer generally to all the materials known under the same name.  A further reason for this is that there are already several pocket, and hand, books which profess to supply this class of information : appearing as they do at frequent intervals, there is some chance of the data being con-

---

* Moreover, it is partly to meet the possible separation of these parts in any subsequent publication.

temporaneous, where the same could not possibly be hoped with a work of this character. For books of pure reference with tables, I commend the reader to the now somewhat classic "Clark and Sabine,"* besides "Munro and Jamieson"† of more recent date, and also Kempe's‡ "Pocket-Book." For many useful and instructive articles connected with different phases of cable work, it is additionally suggested that he should peruse the early volumes of *The Electrician* and the *Electrical Review* (formerly the *Telegraphic Journal*), as well as the *Electrical Engineer* of a later date. His attention is further called to the columns of the *Engineer* (at one time the only applied science periodical), and, moreover, to those of *Engineering*, more especially of a few years subsequently. Some of these are alluded to further in the body of this book.

The parts of the subject which, for the present, are taken up here are those which have been less, if at all, dealt with by Mr Wilkinson and other recent authors, to whom I have referred the student for full particulars regarding the laying,§ repairing, and testing of cables, with which branches of Submarine Telegraphy this volume only deals incidentally—and mainly, in PART I., from an historical standpoint — in quite a general way, as occasion requires in connection with the subjects proper of the book.

I hope that the Index at the end will meet every requirement

---

* "Electrical Tables and Formulæ," by Latimer Clark and Robert Sabine (London : E. and F. N. Spon), 1871.

† "Electrical Rules and Tables," by John Munro and Andrew Jamieson, F.R.S.E., M.Inst.C.E. (London : Charles Griffin and Co., 12th edition, 1897).

‡ "The Electrical Engineer's Pocket-Book of Modern Rules, Formulæ, Tables, and Data," by H. R. Kempe, A.M.Inst.C.E. (Crosby Lockwood and Son, London), 2nd edition, 1892.

§ Besides particulars regarding submarine survey and sounding work.

for reference, no pains having been spared to render it in every sense complete and a fitting companion to a treatise of this character.

Though naturally a more or less technical work, intended for the "cable-man," I have attempted in the course of it to present matter of general interest—*i.e.*, of interest to a large reading public outside the confined circle of telegraph engineers and electricians. There has been no want of material to work on. The difficulty has rather been a superabundance, and the necessity of a severe sifting of evidence.

The commencement and rapid development of the world's electric nerve system constitutes, without a doubt, one of the most marvellous characteristics of the Victorian Era. The laying of the first effective submarine cable (1851) was almost exactly contemporaneous with the first great International Exhibition; and telegraphy on Cooke and Wheatstone's electro-magnetic system dates back to the first few years of Queen Victoria's reign.* Thus the present year, in which we have been celebrating the sixtieth anniversary of Her Majesty's "Record Reign," seems to me a particularly suitable one for the appearance of a book like this.

In conclusion, I desire to express my cordial thanks for the kind assistance I have received from many quarters in its preparation. As already observed, a part of M. Wünschendorff's treatise —so carefully compiled and edited in French, albeit mainly from

---

* "Introduction to the Catalogue of the Victorian Era Exhibition on Science and Engineering, 1837-97," by Charles Bright, F.R.S.E.

English sources—constitutes the original "stock" upon which I worked a large part of mine. I have, moreover, availed myself of a great many of the same blocks—some actually of French origin—for my illustrations. To M. Wünschendorff, therefore, I owe a debt of more than ordinary gratitude, and his connection with this volume is too intimate for me to include him in the ordinary catalogue of contributors of facts or ideas.

Next to M. Wünschendorff, my special thanks are due to Lieut. Anthony Thomson, R.N.R., C.B., in regard to the original scheme already referred to.

The following gentlemen have been good enough to render me material assistance in various ways and in various departments of the subject :—The Right Hon. Lord Kelvin, G.C.V.O., LL.D., F.R.SS. (L. & E.); Sir Samuel Canning, M.Inst.C.E.; Mr W. H. Preece, C.B., F.R.S., M.Inst.C.E.; Mr Latimer Clark, F.R.S., M.Inst.C.E.; Mr Edward Bright, M.Inst.C.E.; Mr F. C. Webb, M.Inst.C.E.; Mr Henry Clifford, Mr H. A. C. Saunders, Prof. Andrew Jamieson, F.R.S.E., M.Inst.C.E.; Dr Alexander Muirhead, Mr Herbert Taylor, M.Inst.C.E.; Mr Alexander Siemens, M.Inst.C.E.; Mr R. Kaye Gray, M.Inst.C.E.; Mr W. Claude Johnson, M.Inst.C.E.; Mr F. R. Lucas, Mr Willoughby S. Smith, Mr Frank Jacob, Mr P. B. Delany, Mr Thomas Bolas, Mr W. R. Culley, and Mr Edward Stallibrass, Assoc.M.Inst.C.E.

For information or assistance of another character I have also to thank :—Lord Sackville Cecil, Rear-Admiral Sir W. J. L. Wharton, K.C.B., F.R.S. (Hydrographer to the Admiralty); Sir Douglas Galton, K.C.B., D.C.L., LL.D., F.R.S.; Sir C. L. Peel, K.C.B.; Major General Sir F. J. Goldsmid, K.C.S.I., C.B.;

Mr B. T. Ffinch, C.I.E.; Captain A. W. Stiffe, R.I.M.; Mr S. W. Silver, Mr W. S. Andrews, Mr W. Shuter, Mr J. Denison Pender, and Mr William O. Smith; besides Mr Charlton J. Wollaston —one of the very earliest workers in the field—and Mr T. Hillas Crampton, the son of another.

There are also others to whom I am indebted—mostly in connection with matters to which their names are attached in the body of the book.

I desire, moreover, to express my gratitude for assistance rendered by the following Government Departments, Firms, and Companies:—H.M. Post Office, Admiralty, War Office, Foreign Office, Colonial Office, The Agents-General for the Colonies, India Office (Indo-European Telegraph Department), Patent Office, Messrs Clark, Forde, and Taylor, The Telegraph Construction and Maintenance Company, Messrs Siemens Brothers and Co., The India-rubber, Gutta-percha, and Telegraph Works Company, of Silvertown, W. T. Henley's Telegraph Works Company, Hooper's Telegraph and India-rubber Works, Messrs Dixon and Corbitt and R. S. Newall and Co., Messrs Johnson and Phillips, Messrs Easton, Anderson, and Goolden, Messrs Frost Brothers, Messrs Brown, Lenox, and Co., Messrs Elliott Brothers, and Mr James White; also to the Anglo-American Telegraph Company, The Eastern Telegraph Company, The "Eastern-Extension" Telegraph Company, The Eastern and South African Telegraph Company, The Direct United States Cable Company, The Western Union Telegraph Company, The Great Northern Telegraph Company, The Brazilian Submarine Telegraph Company, The Western and Brazilian Telegraph Company, The

Central and South American Telegraph Company, The West India and Panama Telegraph Company, and the South American Cable Company.

Further, I am especially indebted to my brother, Mr J. Brailsford Bright, M.A., of the Inner Temple, for legal and general advice, assistance, etc., partly in regard to the various patents connected with the subject.

I also take this opportunity of thanking the Publishers, and Mr R. M. Johns, of the Middle Temple, for their valuable and earnest co-operation.

Last—but by no means least—I wish to record my thanks to Mr C. R. Wylie for the design on the cover of this volume.

In fully acknowledging assistance rendered to me in divers ways with willing and unstinting hands, I desire to state that the entire responsibility for all defects or errors must rest with myself.

CHARLES BRIGHT.

53 West Cromwell Road,
    London, S.W.  *December* 1897.

*P.S.*—It should be understood that the term "projected" in the Map facing page 208 does not include lines which have been discussed, but which, so far, have not taken definite shape. Recent events seem, however, to point towards the principal Powers being brought into direct and independent telegraphic communication with their individual Colonies.

# CONTENTS.

——◆——

## PART I.—THE HISTORY OF SUBMARINE TELEGRAPHS.

————

### CHAPTER I.

### EARLY SUBAQUEOUS TELEGRAPHY.

### CHAPTER II.

### THE DAWN OF OCEAN TELEGRAPHY.

## CHAPTER III.

## DEVELOPMENTS.

# CHAPTER IV.

## COMMERCIAL AND MISCELLANEOUS RÉSUMÉ.

# PART II.—THE CONSTRUCTION OF SUBMARINE TELEGRAPHS.

## CHAPTER I.

### THE CONDUCTOR.

## CHAPTER II.

### THE INSULATING ENVELOPE.

## CHAPTER III.

### JOINTING.

CHAPTER IV.

## MECHANICAL PROTECTION AND STRENGTH.

CHAPTER V.

## COMPLETED CABLE.

APPENDICES.

# PART III.—THE WORKING OF SUBMARINE TELEGRAPHS.

## CHAPTER I.

## THEORY OF THE TRANSMISSION OF SIGNALS THROUGH CABLES.

## CHAPTER II.

## SIGNALLING APPARATUS.

# CHAPTER III.

## DUPLEX TELEGRAPHY.

# CHAPTER IV.

## AUTOMATIC MACHINE TRANSMISSION.

# CHAPTER V.

## RECENT DEVELOPMENTS.

# APPENDIX.

# LIST OF CHARTS AND MAPS

ILLUSTRATING

## SYSTEMS OF SUBMARINE TELEGRAPH COMMUNICATION.

# LIST OF ILLUSTRATIONS

## (EXCLUDING MINOR AND THEORETICAL DIAGRAMS).

# INTRODUCTION.

———◆———

THE idea of transmitting thought across the seas by means of the electric fluid—that wonderful agent* which reserves fresh surprises for us every day—dates back to the first years of this century. It was, however, only realised practically a good deal later, when the application of steam to industry, and the construction of the first electric telegraphs, had increased the intercourse of different nations. Still, many disappointments awaited the pioneers of this useful work, and the disasters (financial and otherwise) which accumulated for several years reached such proportions that—but for the persevering energy of a few men, resolved to triumph over all obstacles, and the genius of those eminent men of science who, for many years, pursued, without ceasing, the solution of the very troublesome questions which were presented by this new application of science—submarine telegraphy would certainly never have survived its first reverses. By an admirable interweaving of events, the industry which had been saved by science, returned to her the services which she had received from her, and contributed to secure for her a progress which, left to herself, she never could have achieved for many years to come.

To-day, submarine telegraphy—although still susceptible of numerous improvements—has, like other applied sciences, its laws and its *routine*. If the enterprises necessitated by it are complicated by the formidable element

---

* Wonderful, indeed, were the things expected of electricity in those days—though not very different, perhaps, from what is even now sometimes looked for. For instance, there was actually a patent taken out by one Wagner (Specification No. 173 of 1854) for "indicating a person's thoughts by the agency of *nervous* electricity"! The wording here is, at any rate, a curiosity in expression.

in the midst of which they have to be carried out, and if they still present a good many difficulties—and sometimes even a good many dangers*— these difficulties are not insurmountable; and the millions which are invested in enterprises of this nature testify alike to its success, to the confidence which it inspires, and to the services which it renders to humanity.

Sir Henry Mance, C.I.E., M.Inst.C.E., when, as President of the Institution of Electrical Engineers, he delivered his Inaugural Address at the beginning of this year, sketched out for us some of the more notable features and landmarks in the step-by-step progress of submarine telegraphy from its birth to the present time.† This luminous address positively bristled with interesting facts. It also touched on what may be termed the romance of the subject, à propos of the various travels involved in every corner of the globe.

The author had occasion to refer more particularly to PART I. of this work in his Preface. As to PARTS II. and III., on Construction and Working respectively, they may appear to enter somewhat closely into detail at times; but it was felt that having at hand the various patent specifications connected with the subject, it would be well to shew, as far as possible, everything that led up to the practical results arrived at to-day.

---

* The Telegraph Ship "La Plata," on her way (from London to South America) to lay a cable on behalf of Messrs Siemens Brothers, along the Brazilian coast, was lost with all hands on board, off the isle of Ushant, in November 1874—the peculiar nature of her cargo having prevented her from being lightened quickly enough—during a hurricane which caused a leak in her hull. A few months previously, the T.S. "Gomos," which was destined for the very same work, had foundered on the Brazilian coast. The Anglo-American Telegraph Company's repairing steamer "Robert Lowe," in 1873, met a similar fate near Newfoundland; and so did the "Volta" belonging to the Eastern Telegraph Company (1887), in the Archipelago.

But perhaps saddest and strangest of all was the fate which befell T.S. "Magneta." This vessel, having just been built for the "Eastern Extension" Telegraph Company, sailed from London on 8th March 1885, and was never heard of again. It is supposed that she capsized; but the mystery remains as profound as the depths of the sea which hold it.

† Since the above, the author has endeavoured to summarise the advances made in submarine telegraphy during the last quarter of a century in the Jubilee number of the *Electrical Review* (see its issue of 12th November 1897, "Twenty-Five Years of Submarine Telegraphy," by Charles Bright, F.R.S.E.).

Notwithstanding these years of submarine telegraphy — and that all the information thereon is naturally of English origin—there was no book in our language* touching on the construction of cables† until an able production of Mr Wilkinson's‡ appeared last year. The subject of manufacture was partially dealt with there, though perhaps more fully in the present volume.

With further reference to PART II., it is hoped that the chapters relative to insulating materials may also prove interesting—if not actually useful—to those engaged in electric light, power and traction work, more especially as they touch on subjects only treated in a fragmentary manner in previous publications of whatever description.

PART III.—on the *Working* of submarine telegraphs—is, as may be imagined, that which contains the greatest amount of novelty.§ Here, it has been found necessary to pre-suppose a knowledge and understanding of electrical phenomena and terms, particularly in regard to telegraphic problems. In any case, these can be readily gathered from such sources as the work on Testing by Mr Kempe,‖ and Mr Young's forthcoming

---

* Since Monsieur Wünschendorff's French work (first published as a series of articles in *La Lumière Electrique*), Signor E. Jona has produced (1895) an admirable little Italian treatise entitled "Covi Telegraphici Sottomarini."

† It is scarcely surprising that those who practically had the field to themselves—or nearly so—in early days (when the science and practice was still in an unsettled state), should have been reluctant to open their professional treasures to the public gaze. Now, however, the question rests on a different footing ; and it is doubtful whether any individual or firm can lose as much as they can gain by free intercourse with their neighbours, and by reasonable and honourably maintained publicity : the more so when we remember that most of the mechanism involved has already been patented at some time or another, and that patenting entails publication. Let us recognise too that a man does not learn his business from reading books : that is merely ancillary to his practical instruction and experience.

‡ "Submarine Cable-Laying and Repairing," by H. D. Wilkinson, M.I.E.E. (London : *The Electrician* Printing and Publishing Company).

§ Partly owing to nothing having been hitherto published thereon, except in a disjointed form, from time to time, in the course of papers and articles. It may also be explained by the comparatively recent application of automatic transmission to the exigencies of submarine cables, as well as to other improvements and suggestions regarding the apparatus generally.

‖ "A Handbook of Electrical Testing," by H. R. Kempe, A.M.Inst.C.E., M.I.E.E. (London : E. and F. N. Spon), 5th edition, 1892.

treatise on the same subject, but more particularly with reference to the *routine* of cable-work,* besides Mr Wilkinson's, already alluded to.

Method and consistency have been observed as far as possible in the electrical symbols and nomenclature adopted. In selecting these, consideration has been given to what is likely—by wont and usage—to be most familiar to the cable electrician, and what is least likely to lead to doubt or confusion. It is, perhaps, to be regretted—if only for the sake of uniformity—that Professor Jamieson's suggestions in this direction have not so far received definite recognition.† In the present disorganised state many eminent authors use "F" for "Capacity" and "V" for "Potential." On that principle, however, surely they should also employ *w* for "Resistance," and "A" (or *a*) for "Current," thereby introducing yet another representation of Ohm's law, which most of us prefer to remember as $C = \dfrac{E}{R}$. To be consistent, they should further speak of the "F *w* law" for speed, instead of the now almost classic KR law!

In the references to various patents it is hoped that some interest and use will be discovered, this being the outcome of an exhaustive search through all specifications relating, in any shape or form, to submarine telegraphy from its inception up to the present time. A study of patent specifications will be found eminently instructive. It is of great service if only in revealing to the would-be inventor that which has already been published and patented (in some instances several times over)—thereby chastening his too impetuous zeal, or, at any rate, focussing his ingenuity upon more original tasks and problems. In these pages certain comparative novelties will be found described somewhat in detail—usually for the very reason that they *are* novelties, and have not hitherto received public notice. In particular instances again—as in that of the siphon recorder—

---

* "Electrical Testing for Telegraph Engineers," by J. Elton Young (*The Electrician* Printing and Publishing Company).

Another forthcoming work on somewhat similar lines is "Electrical Instruments and Measurements," by Charles H. Yeaman (Griffin).

For several years past the French have had the benefit of Mons. H. Thomas' "Traité de Télégraphie Electrique" (Baudry et Cie, Paris), which deals largely with electrical instruments and methods of testing.

† "Electrical Definitions, Nomenclature, and Notation," by Prof. A. Jamieson, F.R.S.E. (*Jour. Inst. E.E.*, vol. xiv., p. 297).

some of the older forms of apparatus are more fully described than would otherwise appear necessary, both because it is found that they are still in use, and also on account of the instructive illustrations which they afford of the manner in which most great inventions have grown up. On the other hand, where an old device has no practical application, and is of no special interest, silence—or, at any rate, quite a brief reference—is the order of the day.

The person responsible for the construction of a submarine cable should be one with an efficient mechanical training: similarly in regard to the engineer attending to the design and setting up of the gear fitted to the vessel for its proper submersion and recovery. The former should also be an electrician, besides having a knowledge of the materials involved.

The person in charge of a cable (laying or repairing) expedition should be largely a sailor, with plenty of self-possession and a power over men. To do the work satisfactorily, on scientific principles, he should also be an engineer with sufficient technical training, besides being a bit of an electrician. But, above all, perhaps, he should be a man possessed of the less common feature—a power for administration—*i.e.*, a capacity for ruling others, and maintaining complete discipline amongst a more or less large staff and body of cable-hands, sailors, etc.

An electrician with a proper training and a cool head is, under all circumstances—especially during repairs—absolutely necessary at the head of the Testing department.

Lastly, the superintendents of the stations from which the cable is worked require to be men who, besides the necessary knowledge of the instruments and appliances they have in their command for turning to account, must also have a capacity for organisation and business—not only as regards staff and telegraph office, but also as representing the company in dealing with others in the town and locality.

In conclusion, the author desires to call attention to the importance of studying the specifications or shorter descriptions of patents—as much in this as in the case of any other branch of engineering the student may desire to become a master of. Plenty of food for reflection is furnished in the early

"Abridgments of Specifications relating to Electricity and Magnetism." These volumes are also provided with an introduction, which contains a convenient and brief *résumé* of the leading early discoveries and inventions connected with electro-telegraphy generally.

Quite recently a good deal has been heard about the "New Telegraphy," and it may be observed with reference to this that a company has been registered as the "Wireless Telegraph and Signal Company Limited," which proposes to acquire from Signor Guglielmo Marconi certain letters patent, etc. etc. Though the Marconi system is certainly the most promising at present before the public for solving the problem of telegraphic communication between outlying lighthouses or lightships and the shore—and a description thereof will, on this account, be found at the end of the book—yet shareholders in cable-making companies need by no means tremble for their immediate future. We have probably many a long day to wait for the advent of that wonderful electrician who shall succeed (without cables) in sufficiently centralising the Hertzian waves upon a given spot a thousand —or, say, even a hundred — miles distant, for purposes of transmarine telegraphy.

Meanwhile, vast improvements are still conceivable in the type of submarine line (electrically speaking), and even in the signalling apparatus, whilst oceanic telephony is yet in the making.

# PART I

## THE HISTORY OF SUBMARINE TELEGRAPHS

# CONTENTS OF PART I.

# SUBMARINE TELEGRAPHS.

## PART I.—HISTORY.

### CHAPTER I.

#### EARLY SUBAQUEOUS AND SUBMARINE TELEGRAPHY.

#### SECTION 1.—THE EVOLUTION OF THE ART.*

A SPANIARD named Salvá appears to have suggested the feasibility of Submarine Telegraphy as far back as 1795—a full century ago!—in the course of a paper read at the Barcelona Academy of Sciences.

In 1803 Aldini, a nephew of Galvani, is said to have performed experiments of this nature in the sea off Calais, and also across the river Marne, near Charenton. It is believed, however, that no authentic records of these are in existence.

Similar investigations were conducted jointly by Sömmering and Schilling in 1811 across the river Isar, near Munich. This was probably the earliest instance of a soluble insulating material—sometimes described as

---

\* For full particulars regarding pre-electrical telegraphy, and the first attempts at electro-telegraphy, the reader curious upon those subjects may be referred to " A History of Electric Telegraphy to the Year 1837," by J. J. Fahie, M.I.E.E. (London, 1884 : E. and F. N. Spon) ; also to "Ancient Methods of Signalling" (*Cornhill Magazine*, December 1897), by Charles Bright, F.R.S.E. ; and "The Evolution of Telegraphy during the Victorian Era" (*Gentleman's Magazine*, of the same month), by Charles Bright, F.R.S.E.

india-rubber—being applied to a conducting wire.  The following year
Schilling executed the first attempt to ignite gunpowder at a distance, and
to explode mines across the river Neva, near St Petersburg, by means of
a subaqueous conducting wire, insulated as in the previous experiment.

In 1813 John Robert Sharpe transmitted signals through seven miles
of insulated wire submerged in a pond.

The earliest record of practical telegraphy under water, of which there
appears to be any particulars, relates to experiments made by Colonel
Pasley, of the Royal Engineers, at Chatham, in 1838.  His method of
insulation—gutta-percha being then unknown—was to surround a conducting wire with strands of tarred rope, and then again with pitched yarn.*

In the following summer, Dr O'Shaughnessy (afterwards Sir William
O'S. Brooke, F.R.S.), the Director of the East India Company's Telegraphs, made a series of experiments across the Hugli—a very broad
river—which he thus described in the *Journal of the Asiatic Society*,
September 1839:—"Insulation, according to my experiments, is best

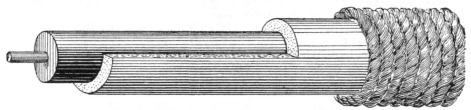

FIG. 1.—O'Shaughnessy's Line across the River Hugli.

accomplished (Fig. 1) by enclosing the wire (previously pitched) in a split
ratan, and then paying the ratan round with tarred yarn ; or the wire may,
as in some experiments made by Colonel Pasley, at Chatham, be surrounded
by strands of tarred rope, and this by pitched yarn.  An insulated rope
of this kind may be spread across a wet field, nay, even led through a
river, and will still conduct, without any appreciable loss, the electrical
signals above described."

In February 1840, Professor Wheatstone (afterwards Sir Charles
Wheatstone, F.R.S.) exhibited, at a Committee of the House of Commons,
the methods by which he thought it possible to establish telegraphic
communication between Dover and Calais.  He seems to have been
unaware of Colonel Pasley's and Dr O'Shaughnessy's prior experiments

---

* Colonel (afterwards Major-General Sir F. C.) Pasley, R.E., employed this system of
insulation, both experimentally, in the Medway at Chatham, and during the diving operations he directed in 1838 in connection with the wreck of the "Royal George," off
Spithead.

referred to above, for his method of insulating with tarred hemp was similar to theirs, with the exception that he omitted the pitching, which tended to further increase the insulation, as well as the split ratan covering, which in itself afforded some protection when laid.*

Wheatstone took, however, considerable trouble in the details of his plan, which is worthy, therefore, of a full record. His cable was to be composed of seven copper conductors, each of which formed the heart of a strand of hemp thoroughly saturated with boiled tar; the seven strands were then to be laid up together, and the whole covered with hemp treated in a similar manner.† He prepared two sets of drawings, as in Figs. 2 and 3.‡

The first set of drawings (Fig. 2, Plate I.) shews elevations and sections of the following machines :—

    *a.* Machine for covering the conductor with the insulating material.
    *b.* Machine for simultaneously covering with hemp and laying up together seven such similar conductors at one operation, thus forming a multiple-cored cable.
    *c.* Machine for binding together the sheaf of insulated conductors with hemp so as to combine them in a single cable. On emerging from each of the above machines the cable so far made was to be passed through a bath of insulating substance, and the completed cable was to be finally taken on board the ship.
    The various stages above described, with the cable passing through the insulating fluid and then drawn off aboard the ship, are shewn in section *d* of this figure.

The second set of drawings (Fig. 3, Plate II.) shews—

    *a.* The profile and plan of Dover Straits between the South Foreland and Cape Grisnez, with the proposed landing-places and route for the projected cable.
    *b.* A view of the vessel, in tow of a tug, laying the cable.
    *c.* An elevation and plan of the drums on which the cable was to be wound, and of the paying-out pulley mounted at the stern of the vessel.
    *d.* An illustration of the method of soldering together the ends of the sections wound on the several drums.
    *e.* A view of the ship underrunning the cable in order to recover a fault.
    *f.* A section and a perspective view of the cable.

In 1844 Wheatstone made several trials of this method of communication in Swansea Bay, and succeeded in exchanging signals with a neighbouring lighthouse from a boat.

As soon as gutta-percha made its appearance in England, Wheatstone thought of using it for his cable; he proposed to enclose the wire so insulated in a leaden pipe.

---

* Some years later it was asserted that Professor Wheatstone had been engaged in correspondence on this subject as early as 1837.

† It is stated that in 1845 Professor Wheatstone contemplated employing gutta-percha as a means of insulation, but no particulars are to hand as to how he intended to apply it.

‡ These drawings are reproduced from photographs, published in 1876, by the late Mr Robert Sabine, in the *Journal of the Society of Telegraph Engineers.*

The experiments he made with gutta-percha were, however, insufficient; and he failed to discover a method of coating his conducting wires with this insulating substance.

Professor S. F. B. Morse, the well-known inventor of the telegraph apparatus which bears his name, also studied this question. Morse relates that in 1842 he laid down in New York Harbour an insulated copper wire through which he sent electric currents.* The following year he is said to have submitted to the Government of the United States a proposal to establish electric communication between Europe and America.

In 1845 Ezra Cornell laid a cable, 12 miles long, in the Hudson River, to connect Fort Lee and New York. The cable consisted of two cotton-covered copper wires, insulated with india-rubber, and enclosed in a lead pipe. This cable worked well for several months, but was broken by ice in 1846.

Mr Charles West, associated with Messrs S. W. Silver and Co.—having obtained permission, in 1846, from the British Government, to establish telegraphic communication between Dover and Calais—paid out a wire, by hand, in Portsmouth Harbour, insulated with india-rubber, through which he was able to telegraph from his boat to the shore.† This experiment was witnessed by a large number of spectators. Mr West, however, was unable to get the necessary pecuniary assistance he applied for from the Electric Telegraph Company, such as would have enabled him to comply with the stipulations of the French Government, and the project fell through.‡

Two trials of wire insulated with gutta-percha were made in the course of the year 1848 in different parts of the world—one in the Hudson River, U.S. America, by Armstrong, who, emboldened by success, wrote to the *Journal of Commerce of New York* proposing a similar cable to be laid along the bottom of the Atlantic; the second was carried out by the late Dr Werner von (then Mr) Siemens, in the harbour of Kiel, the wire being used to fire submarine mines.§

---

* The method of insulation used here was described by Morse, later on, as hemp soaked in tar and pitch, surrounded with a layer of india-rubber. Morse was a great letter-writer, and records of this early work are solely based on his own statements, at a time when he noted in his diary : " I am crushed for want of means. My stockings all want to see my mother, and my hat is hoary from age."

† Sir Joseph Paxton, original designer of the 1851 exhibition building (now the Crystal Palace), and Charles Dickens, the novelist, are said to have been jointly interested in this experimental line.

‡ In these matters, whereas the English Government invariably grant leave for landing a cable at once without demur, the French Government have—perhaps wisely—always demanded some definite assurance of the undertaking being backed up by the required funds.

§ In 1846 to 1849, great lengths of gutta-percha covered wires were laid as *subterranean* lines for telegraphic purposes in Germany and Prussia, at the instance of the brothers Siemens. Werner Siemens, in 1847, invented a machine for applying

[PLATE I.

[To face p. 4.

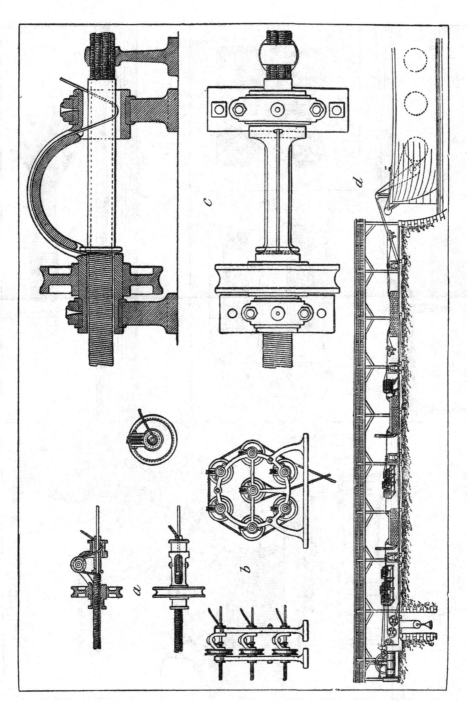

FIG. 2.—Wheatstone's Proposed Telegraph Line between England and France, 1840 (Construction and Shipment).

The material originally positioned here is too large for reproduction in this reissue. A PDF can be downloaded from the web address given on page iv of this book, by clicking on 'Resources Available'.

[PLATE II.

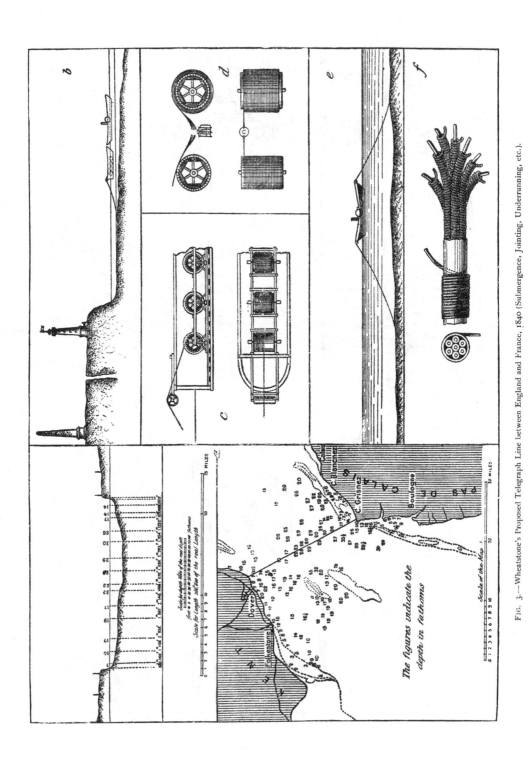

FIG. 3.—Wheatstone's Proposed Telegraph Line between England and France, 1840 (Submergence, Jointing, Underrunning, etc.).

[To face p. 4.

SECTION 2.—THE FIRST SUBMARINE TELEGRAPH CABLES.

On the 10th day of January 1849, Mr C. V. Walker, F.R.S., telegraphic superintendent and electrician to the South-Eastern Railway Company, laid a gutta-percha covered cable, two miles long, in the Channel. Starting from the beach at Folkestone, the cable was joined up to an aerial line 83 miles in length, carried along the South-Eastern Railway, and Mr Walker, on board the " Princess Clementine," at anchor, succeeded in exchanging telegrams with London.

**The First Dover-Calais Line.**—On 23rd July 1845, the brothers Brett addressed themselves to Sir Robert Peel, as Prime Minister and First Lord of the Treasury, relative to a proposal of theirs for establishing a general system of telegraphic communication — oceanic and otherwise. Their original copy of the letter is reproduced here (Plate III.). They were referred to the Admiralty, Foreign Office, etc., and gradually became involved in a departmental correspondence—more academic than useful— in which they were passed backwards-and forwards from one Government office to another.

In 1847, after considerable negotiations, the Messrs Brett * obtained a concession from the French Government to establish a submarine telegraph between France and England. However, not even the necessary preliminaries being effected within the stipulated limits of time, the

---

*seamless* gutta-percha to a wire, similar in principle to the machine for making macaroni, and practically the same as those now in use. Previously, Siemens had covered his wire with strips of gutta-percha (probably the first attempt made public), but much trouble was experienced by moisture penetrating at the seams where the strips met.

Following this, in 1851, came the extensive system of underground lines laid down in England for the British and Irish Magnetic Telegraph Company, by their engineer, Mr (afterwards Sir Charles) Bright, according to a patent of the Messrs Bright—besides comparatively short lengths through tunnels and under towns by the " Electric " Company.

* Of the two brothers, Jacob Brett (whose death we have more recently lamented) appears to have first become enthused with the idea of submarine telegraphy. On 16th June 1845, he registered the first company having the above for its object, and the original certificate of registration is here reproduced (Plate IV.), though the proposed company was not ever formed. He became embued with the idea of adapting House's printing telegraph instrument to the purposes of cable signalling. These notions took shape in a patent taken out on 13th November of the same year (specification No. 10,939, A.D. 1845), being mainly a communication from Professor Royal E. House, of America.

Mr Brett then induced his elder brother—a man of some wealth and business capacity, formerly a dealer in curiosities—to become interested in the scheme, and from that date John Watkins Brett always "held the reins." However, several of the actual concessions— especially in France, where more was required—were granted in the name of Jacob Brett alone. In France these concessions ensured sole landing rights for ten years. In England it was merely a matter of permission, without exclusion to others.

agreement was cancelled.  It was again renewed on the 10th August 1849, for the period of ten years, on the understanding that communication was to be established by the 1st of September 1850 at the latest.*  A company was formed under the title of the "English Channel Submarine Telegraph Company," † and 25 nautical miles of line were manufactured at the works of the Gutta-percha Company, having a central copper conductor, No. 14 Birmingham wire gauge, covered with gutta-percha half an inch in thickness.  A single coating of gutta-percha was applied, in 100-yards lengths, to wire which had not been annealed.  These lengths were united by twisting the ends of the copper wires together, and soldering over the joint; ‡ a covering of gutta-percha was then applied, the gutta being softened by heat and compressed in a wooden mould, through which the copper conductor was drawn.  The cumbersome character of the joints may be gathered from Fig. 4. §

The sections thus formed in convenient lengths were coiled up, placed under water, and tested with a battery of twenty-four ordinary cells : if no decided leakage was indicated, by a very crude form of galvanometer,‖ the

FIG. 4.—Type of Joint in First Channel Line (1850), ⅛th of actual size.

coil was accepted, placed on board, jointed to the length already manufactured, and wound on a large wood and iron drum.  This drum, which revolved on a horizontal axis, was placed on the deck of the "Goliath,"

---

* It may be remarked that this period of grace was none too long in the absence of any data to go upon.

† This was brought about by personal subscriptions of £500 from Mr (afterwards Sir Charles) Fox, C.E., Mr Francis Edwards, Mr J. W. Brett, and Mr Charlton J. Wollaston, making in all £2,000, enough to cover the estimated cost for carrying out the enterprise. Ultimately these gentlemen banded themselves together to be represented by a *Société en Commandite* in Paris, partly with a view to limiting their liability.

‡ In those days the copper conductor used to be joined by what is termed a bell-hanger's twist, and then further secured with soft solder.  Gutta-percha in a plastic state was then applied in a single coating and pressed into shape in a wooden mould, the diameter of the finished joint being usually about two inches.  Of course the joints were not tested separately at that time, neither was any notice taken of the resistance or conductivity of the copper, or of the resistance or electro-static capacity of the gutta-percha.—"*Résumé* of the Earlier Days of Electric Telegraphy," by Willoughby Smith, 1891 (*Jour. Soc. Tel. Eng.*, vol. x., p. 312).

§ Taken from originals in the possession of Mr H. Marsh, of H.M. Postal Telegraphs, formerly an officer of the Submarine Telegraph Company, and connected with cable matters almost from the first.

‖ The insulation of this line is said to have been, on occasions, tested by Mr Wollaston's tongue, when no other apparatus was at hand or in working order.

[PLATE III.

COPY OF A LETTER SUBMITTED TO THE GOVERNMENT IN JULY, 1845

PRINTED BY BRETT'S ELECTRIC TELEGRAPH.

TO THE RIGHT HONOURABLE SIR ROBERT PEEL, BART.

LONDON. JULY, MDCCCXLV.

WE BEG THE HONOUR TO SUBMIT A PLAN FOR GENERAL COMMUNICATION BY MEANS OF OCEANIC. AND SUBTERRANEAN, INLAND, ELECTRIC TELEGRAPHS. FOR WHICH PATENTS HAVE BEEN SECURED BY THE UNDERSIGNED AND FOR THEIR CONS -STRUCTION, ON CHEAP AND EFFICIENT PLANS

BY MEANS OF THIS TELEGRAPH, ANY COMMUNICATION MAY BE INSTANTLY TRANSMITTED FROM LONDON. OR ANY OTHER PLACE AND DELIVERED IN A PRINTED FORM ALMOST AT THE SAME INSTANT OF TIME, AT THE MOST DISTANT PARTS OF THE UNITED KINGDOM OR OF THE COLONIES.

THE ADVANTAGE AND POWER OFFERED TO THE GOVERNMENT BY THIS INVENTION. RENDER IT OF THE GREATEST IMPORTANCE THAT THEY SHOULD HAVE IT UNDER THEIR OWN CONTROUL, AND ARRANGE AND CONDUCT THIS PLAN OF GENERAL TELEGRAPHIC COMMUNICATION.

THE FOLLOWING ARE A FEW OF THE ADVANTAGES OFFERED BY THIS PATENT.

I. THE IMMEDIATE COMMUNICATION OF GOVERNMENT ORDERS AND DESPATCHES TO ALL PARTS OF THE EMPIRE AND THE INSTANT RETURN OF ANSWERS TO THE SAME, FROM THE SEATS OF LOCAL GOVERNMENT. ETC. ALL DELIVERED IN AN UNERRING AND PRINTED FORM.

II. A GENERAL TELEGRAPHIC POST OFFICE SYSTEM UNITING THE CHIEF AND BRANCH OFFICES IN LONDON IN CONNECTION WITH ALL THE OFFICES THRO--UGHOUT THE KINGDOM. FOR TRANSMITTING MESSAGES OF BUSINESS, ETC. FROM MERCHANTS, BROKERS, TRADESMEN, AND PRIVATE PERSONS. AT A FIXED RATE OF CHARGE. THESE COMMUNICATIONS WOULD BE PRINTED ON PAPER AND ALL EN-CLOSED IN SEALED ENVELOPES, AND ADDRESSED BY CONFIDENTIAL CLERKS. AND ISSUED BY SPECIAL MESSENGERS OR THE USUAL POST OFFICE DELIVERY.

III. THE ADVANTAGES OF THIS PLAN, APPLIED TO POLICE ARRANGEMENTS THROUGHOUT THE UNITED KINGDOM, AND TO THE ARMY, AND NAVY DEPARTMENTS, MUST BE AT ONCE OBVIOUS TO THE GOVERNMENT. BY IT INSTRUCTIONS MIGHT BE CONVEY-ED INSTANTANEOUSLY. AND THE MOVEMENTS OF THE FORCES SO REGULATED THAT ANY AVAILABLE NUMBER OF THEM MAY BE BROUGHT TOGETHER AT ANY GIVEN POINT. IN THE SHORTEST POSSIBLE TIME NECESSARY FOR THEIR CONVEYANCE.

THESE ARE SOME OF THE ADVANTAGES. OTHERS READILY SUGGEST THEMSELVES. NAMELY, GENERAL COMMUNICATION BETWEEN STATIONS ON THE COAST. SUCH AS LIGHT-HOUSES CHANNEL ISLANDS. ETC. SO THAT A GENERAL SUPERVISION OF THE COAST MIGHT BE OBTAINED FOR THE USE OF THE NAVY, LLOYDS, AND FOR THE PREVENTION OF SMUGGLING. ETC

SIGNED. . . . J. AND J.W. BRETT.
NO 2. HANOVER SQUARE

THE SUBTERRANEAN. AND OCEANIC. ELECTRIC. PRINTING TELEGRAPH. OF WHICH J. BRETT. IS THE SOLE PATENTEE, AND ORIGINATOR IS THE ONLY ONE THAT PRINTS AND REGISTERS. IT SERVES AT THE SAME TIME. BY ONE WIRE ONLY. ALL THE OBJECTS REQUIRED. VIZ. ALARUM BELLS, ETC. AND PRINTS PERMANENTLY ON PAPER EIGHTY SEVEN, LETTERS AND CHARACTERS A MINUTE.

FACSIMILE COPY (REDUCED) OF ORIGINAL LETTER.

[To face p. 6.

[PLATE IV.

No. *384* CERTIFICATE OF PROVISIONAL REGISTRATION

of the *General Oceanic Telegraphic* Company

Pursuant to the Act 7 & 8 Vict., c. 110.

---

I *Frederic Rogers Esquire*,

**Registrar of Joint Stock Companies, do hereby Certify that the**
*General Oceanic Telegraphic*
**Company is PROVISIONALLY Registered according to Law.**

*Given under my Hand, and Sealed with my Seal of Office,*
*this* *sixteenth* *day of* *June* *eighteen*
*hundred and* *forty-five*

*Frederic Rogers*

Registrar of Joint Stock Companies

*⁎⁎* This Certificate remains in force until the Company be completely Registered, or until the expiration
of 12 months from the date hereof; and if the Company be not then completely Registered, this
Certificate must be renewed

FACSIMILE COPY (REDUCED) OF CERTIFICATE OF REGISTRATION OF THE FIRST PROPOSED
SUBMARINE TELEGRAPH COMPANY.

[*To face p.* 6.

a small steam-tug on the Thames, and taken to the port of Dover. The number and size of the joints prevented the cable coiling evenly on the drum, so the spaces were filled in with wadding; small laths of wood, adapted to the curvature of the drum, being used to separate the successive flakes.

The shore ends were laid first, extending on the English side from a horse-box in the yard of the South-Eastern Railway terminus out to a structure of piles for the new Admiralty Pier in Dover Harbour; and on the French side to just beyond the rocky ledge which stretches out a considerable distance from Cape Grisnez, a headland in the vicinity of Calais—but away from the anchorage ground of that town. This shore-end cable consisted of a copper conductor, No. 16 B.W.G., covered with cotton soaked in india-rubber solution, the whole being encased in a very thick lead tube.

On the 23rd of August 1850, the "Goliath" left Dover at 10 A.M., escorted by the "Widgeon" (a Government surveying-vessel) to shew her the way, previously marked out by a line of buoys.* The seaward extremity of the shore end was picked up in a boat, and then joined on to the main portion of the line, the paying

FIG. 5.—Leaden Weight fastened to First Channel Line (1850) at 100-yards intervals.

Scale, ¼ actual size.

out of which was then commenced, Mr Charlton Wollaston † being the responsible engineer of this undertaking. At intervals of a hundred yards a leaden weight (Fig. 5), weighing from 10 lbs. to 30 lbs., according to the depth,‡ was fastened to the cable§ to ensure it going to the bottom.‖ They were made in two flat halves, secured together by bolts, and the vessel had to be stopped each time one was put on.

---

* Mr F. C. Webb was probably the first to actively demonstrate the great virtue attached to a system of buoying out the route for the line of a short cable beforehand. This particularly applies where a heavy and expensive type is involved near an irregular bottom; or where it is of special importance that the conductor resistance and electro-static capacity of the circuit should, for working speed purposes, be as low as possible.

† A nephew of Wollaston, the illustrious philosopher (who introduced the famous electro-chemical cell bearing that name), and a former pupil of Brunel.

‡ The maximum depth in the English Channel is about 30 fathoms. 1 fathom = 6 feet.

§ It was gravely suggested by a prominent naval officer to thread the line through old cannonades lying idle at Portsmouth and other dockyards.

‖ This was considered necessary, owing, presumably, to the large proportion of gutta-percha which went to complete the entire line, the comparatively small conductor being

The weather being fine, no accidents of a serious nature occurred during the submersion (an idea of which may be gained from Fig. 6), but some trouble was caused by the laths, which sprung up from the drum when held momentarily by the cable at one end only. The "Goliath" anchored at six o'clock in the evening close to the buoy marking the seaward end of the line to the shore on the French side. Before making the joint, they tried to communicate with Dover, using Mr Jacob Brett's modification of the House printing instrument (Fig. 7), supposed to print, in Roman characters, sixty messages of fifteen words per hour.* As a

FIG. 6.†—Laying of the First Channel Line, 1850.

matter of fact, some few, more or less incoherent, letters appeared here

---

covered to half-an-inch with gutta-percha, which—with the purity of those days—usually had a slightly lower specific gravity than salt-water. A light chain twined round the insulated conductor throughout its length would no doubt have served the purpose better, inasmuch as it would have also protected it from chafing, besides being less liable to damage the core.

* The transmitting apparatus here was constituted by a keyboard in front of the table (see figure), on which the receiving instrument stood. On the face of the latter was an indicating dial (as shewn), the hand of which pointed successively to the letters printed upon the scroll of paper by the apparatus behind the dial.

† From a drawing originally published in the *Illustrated London News* for September 1850. About that time this journal also contained several other interesting illustrated articles concerning the manufacture (as well as the laying) of the above line and those which followed it.

and there on the slip of paper, but intelligible words were conspicuous by their absence.

As night was approaching, these experiments were suspended, and the joint between the two ends was made in a boat. A Cooke and Wheatstone needle instrument and signalling apparatus was connected up with the line in a bathing-machine on the beach at Cape Grisnez, and afterwards at the lighthouse, where for several hours signals from Dover were anxiously awaited. Ultimately signals were exchanged ; one being sent to Louis Napoleon Buonaparte.* A few hours later communication entirely failed. The situation remaining unchanged, it became evident on 31st August that the cable was broken.† Attempts were now made to pick

FIG. 7.—House and Brett's Type-printing Telegraph.

it up near the position of the final splice, but the weight of the leaden pipe prevented this, and the line had to be abandoned.‡

---

\* Shortly afterwards Napoleon III., Emperor of the French.

† It subsequently transpired that a Boulogne fisherman had—accidentally or otherwise —raised it to the surface with his trawl. Imagining that he had discovered a new kind of brown kelp seaweed, snake, or coral " with gold in its centre," he cut out a considerable length. As has since been abundantly proved, there was, indeed, gold in its heart, though not in the literal sense. This fisherman has been aptly referred to by Dr W. H. Russell as a *piscatore ignobile*, though probably only anticipating by a few hours the fate to which such a line was surely doomed.

‡ The carrying out of this enterprise excited little or no attention at the time. It was, in fact, looked upon as a mad freak—and even as a gigantic swindle—indulged in only by wild minds. When accomplished, *The Times* remarked, in the words of Shakespeare, " The jest of yesterday has become the fact of to-day." But a few hours afterwards it might with equal truth have been said, " The fact of yesterday has become the jest of to-day."

Nevertheless, the labours of Messrs Brett, Wollaston, and Reid,* in endeavouring to facilitate intercourse between the Continent and their own country, were not entirely thrown away. The feasibility of laying a cable, and of transmitting electric signals across the Channel, for a distance of over 20 miles, had been proved. This experimental line and the signals obtained through it had, at any rate, the effect of eradicating the (at that time) very prevalent belief that, if successfully submerged, the current would become dissipated in the water, notwithstanding the insulating covering to the conductor.† Although the above line was, practically speaking, never turned to account, it was not by any means abortive, for it served to maintain the concession granted to the Messrs Brett in virtue of the signals that had been conveyed.‡ It only remained to find a satisfactory method of protecting the insulated conductor from injury during and after the laying. The problem, once embarked on in a practical way, could not be very difficult to solve.

On the 19th December 1850, the Bretts obtained a new concession from the French Government, under the same conditions as before, which was to be in force for ten years counting from the 1st of October 1851, and to hold good only in case the submarine line was in working order on that day. But the unsuccessful attempt in the preceding August had made capitalists distrustful, and only seven weeks before the expiration of the time limit the necessary funds had not been subscribed. The undertaking was saved by the energy and talent of one man, Mr T. R. Crampton, a well-known railway engineer and inventor, whose name should always be recorded in every history of the early days of submarine telegraphy. He raised the necessary capital of 375,000 francs (£15,000), putting his own name down for half this amount, and being joined by Lord de Mauley and

---

* Mr William Reid, sen., acted in the capacity of contractor, undertaking the equipment of the "Goliath" with the necessary gear, and attending to the laying operations, to the designs and instructions of Mr C. J. Wollaston, as engineer of the whole work on behalf of the company he represented. Mr Reid additionally attended to the electrical testing of the cable. He had previously contracted for the laying down of the Electric Telegraph Company's underground lines, and was skilled in the rough-and-ready tests of those days. Mr Willoughby Smith was also aboard, for superintending the jointing of the line as made by the Gutta-percha Company with which he was associated.

† Some critics had actually supposed that the method of signalling was that of pulling the wire after the manner of mechanical house bells, and pointed out that the bottom of the Channel was far too rough and uneven to permit of this being possible for any length of time !

‡ A series of electrical tests conducted by Mr Marsh on a short length of this line, when picked up after twenty-five years' submersion, served to shew what excellent stuff it was composed of.

Sir James Carmichael.* Mr Crampton then settled the type of cable to be laid—a type which has since survived, in all essential features.†

The construction of the cable was started by Messrs Wilkins and Weatherly, at Wapping, but owing to law proceedings with regard to a manufacturing patent it was completed there by Messrs Newall and Co.,‡ and on the 25th September 1851 successfully laid across the Channel, under the supervision of Mr Crampton, by contract with the company.

The cable (Fig. 8) contained four copper wires of No. 16 Birmingham wire gauge, each one covered with two layers of gutta-percha to No. 1 gauge; these four insulated conductors, or cores,§ were laid together, and

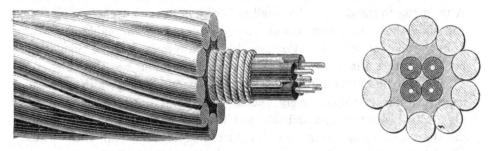

FIG. 8.‖—Dover-Calais Cable, 1851.

---

* Later on, in 1852, they formed a *Société en Commandite*, entitled "La Compagnie du Télégraphe Sous-Marin," with a domicile in Paris, in order to secure concessions from the French Government. This was merged into the once famous Submarine Telegraph Company, Messrs Crampton and Wollaston being the engineers, and later on Mr J. Bordeaux, followed by Mr J. R. France.

† The general type of iron-wire sheathed cable, for submarine purposes, is said to have been originally suggested to Mr Crampton—and formerly to Mr Wollaston—for the purposes of the proposed Dover-Calais cable, by Mr William Küper, a colliery rope manufacturer, the heart (or core) of hemp in a wire pit rope being merely substituted by the insulated conductor. Mr Willoughby Smith is accredited with having made a similar suggestion ; also Mr Edward Highton, Mr R. S. Newall, and Mr W. Reid.

‡ With reference to this the late Mr Willoughby Smith once wrote :—"Several miles of the core and a few knots of the cable had been manufactured when everything was suddenly stopped. A silence that might be felt held sway. Reports circulated that Messrs Newall and Co., wire-rope makers of Gateshead, had a patent for inserting a core of some soft material into wire ropes, with a view to rendering them more pliable and more manageable. They, therefore, considered that this submarine telegraph cable was an infringement of their patent, and obtained an injunction to stop its manufacture.

"It appeared strange—especially to the initiated—that this submarine telegraph cable should be looked upon as belonging to the same category as a wire rope, considering how different were their respective purposes !"

§ The term "core" will, in future, be employed throughout to indicate the insulated conductor forming the heart (or core) of an electric cable.

‖ Unless otherwise stated, all the illustrations of cables in this book represent the actual size of the type in question, and are, as nearly as possible, to scale—in detail as well as collectively.

the interstices filled up with strands of tarred Russian hemp; tarred spun-yarn was then wound round outside the several cores thus laid up.* The above work was carried out by the Gutta-percha Company, under the superintendence of Mr Samuel Statham.† The outer covering consisted of ten galvanised iron wires of No. 1 gauge, wound spirally round the bundle of cores with a long lay, and was intended to protect the conductors from the strains and chafing which had so quickly destroyed the cable laid the year before.‡

This cable weighed about seven tons to the mile. It was put (in an oblong coil) in the hold of the "Blazer"—an old pontoon hulk belonging to Her Majesty's Government—which was taken in tow by two steamers. A third tug to stand by, and a small man-of-war steamer to act as pioneer and shew the course, accompanied the expedition. The cable was laid from the foot of the South Foreland Lighthouse towards Cape Sangatte,§ but the weather was not so favourable as on the previous occasion.

The weight of the cable caused it to pay out rapidly, although the depths in this locality scarcely exceed 30 fathoms. Added to this, the tugs drifted with the wind and tide, and being held back by the cable at the stern, were unable to make a straight course for Cape Sangatte. Thus when the vessels arrived within about a mile of the French coast, there was no more cable left on board.

Communication was temporarily established with three gutta-percha covered wires twisted up together, and on the 19th of October this was replaced by armoured cable similar to that already laid. On the 13th of November following the line was opened to the public.‖

---

* The helical method of laying up the several insulated conductors with the hemp worming was, it is believed, at the instigation of Mr Wollaston. Strange as it may appear nowadays, they would otherwise have been drawn straight!

† This was tested by Mr Wollaston, and its manufacture subject to his supervision, on behalf of the Submarine Telegraph Company and Mr Crampton.

‡ It will be readily understood that the conductor and insulation joints in vogue at that time, already described in connection with the previous line, frequently became damaged in passing through the lay plate of the sheathing machine, or at any rate often required spoke-shaving to enable them to pass through.

§ Though involving a greater length of cable by five miles, this was decided on for the landing-place instead of Cape Grisnez, owing to the latter having been found to be seriously pervaded with rocks. Whereas Cape Grisnez is some 13 miles from Calais (in a southerly direction), Cape Sangatte is only about three miles distant. Several gutta-percha covered wires had been laid underground between Sangatte and the Calais Telegraph Office. Thus, when these were afterwards joined up to the several conductors of the Channel Cable, direct communication was established between the South Foreland and Calais town.

‖ On 23rd October 1851, another (third) ten-year concession was obtained from the French Government in favour of Lord de Mauley, the Honourable F. W. Cadogan, Sir James Carmichael, Bart., and Mr J. W. Brett. It was under this that the Submarine Telegraph Company proceeded to work the line.

This 25-mile cable has since undergone numerous and extensive repairs, and was not entirely renewed for many years.*    Fig. 9 represents a piece picked up in 1859 after being eight years under water. The outside iron wires were partially eaten away, but the gutta-percha—of so highly durable a quality, in those days of practically no competition—was in excellent preservation.

**Other Anglo-Continental Cables.**—The success of Crampton's cable gave considerable impetus to submarine telegraphy.    On all sides similar enterprises started up, but many failures occurred before these operations came to be looked upon as ordinary industrial undertakings.

In the course of the following year three unsuccessful attempts were made to establish telegraphic communication between England and Ireland. Firstly, by a line between Holyhead and Howth, near Dublin.    We know now that this cable was not heavy enough for contending with the rough

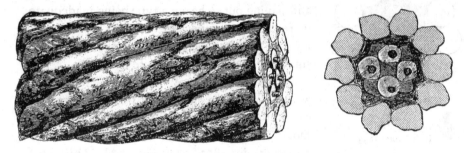

FIG. 9.--Piece of 1851 Channel Cable picked up in 1859.

bottom, strong currents, and disturbance from anchors experienced in these waters.    This undertaking is mainly remarkable, on account of it being the only attempt to do without any intermediate serving between the insulated conductor (made by the Gutta-percha Company) and the sheathing.    The conductor was composed of a single wire of No. 16 gauge. This was covered with gutta-percha to gauge No. 2, the completed core being closed with twelve No. 12 galvanised iron wires.    A few miles from each shore the cable had larger iron wires—about No. 6—so that this was the first cable with shore ends in the ordinary sense as now adopted.    Messrs R. S. Newall and Co. undertook the carrying out of the above work.    On it was also engaged Mr Henry Woodhouse.    At that time they did not test

---

* The vitality displayed by this early line was very remarkable.    The last piece in circuit was a 40-yards length on the beach at Sangatte.    This was only cut out quite a short time ago, and a new shore end laid.

the cable *under water* prior to being laid. Thus, when it had been submerged but a few days, the insulation proved to be so bad that signals could not be made to pass, and all efforts to pick up and repair the cable were unavailing.

A second attempt was made between Portpatrick and Donaghadee, the cable consisting of a central copper conductor, covered, first with india-rubber, then with gutta-percha, and hemp outside all. The cable, being far too light, was actually carried away by the strong tidal currents and even broken into pieces during laying. The undertaking had, therefore, to be abandoned.

The third endeavour was made between the same two points for the Magnetic Telegraph Company, with a cable of a similar character to that successfully adopted by Mr Crampton in the cable he had recently laid from England to France, but containing six conductors. Unfavourable weather was experienced, and the arrangements for checking and paying out so heavy a type being inadequate, more cable was expended than had been allowed for by the engineers, so that there was not sufficient to reach the opposite coast.

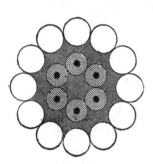

FIG. 10.—First Effective Anglo-Irish Cable, 1853.

What the fate of submarine telegraphy would have been had these three failures preceded the establishment of the line between Dover and Calais may be readily surmised. However, in 1853, a heavy cable (Fig. 10), weighing seven tons per mile, with six conductors, was successfully laid in upwards of 180 fathoms* across the Irish Channel, between Portpatrick (Scotland) and Donaghadee (Ireland), for the Magnetic Company, of which the late Sir Charles Bright was engineer, with permanent success, establishing communication with Ireland for the first time telegraphically.

Only a year elapsed before it became evident that another cable was required to meet the traffic between England and the Continent, and in 1853 an additional line was laid—this time to connect us with Belgium—from Dover to Ostend. This was a heavy multiple cable, similar to that laid between Dover and Calais in 1851, but embracing six conductors instead of four.†

---

* This was the deepest water in which a cable was laid for some time—*i.e.*, up till the date of the Spezia-Corsica line in 300 fathoms and over.

† In his "Museum of Science and Art," published in 1854, Dr Lardner says, with reference to this undertaking :—"The next great enterprise of this kind, of which the accomplishment must render for ever memorable the age we have the good fortune to

Anglo-Dutch and Anglo-German cables followed in due course; and in less than ten years from the commencement of its operations, the Sub-marine Telegraph Company was working at least half-a-dozen excellent

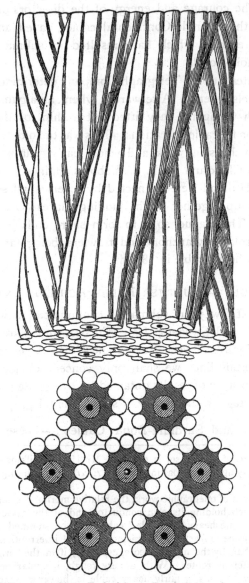

FIG. 11.—Shore End of Anglo-Dutch Cable, 1853.

cables ranging from 25 to 117 miles in length, connecting England with

---

live in, was the deposition in the bed of the Channel of a like cable connecting the coasts of England and Belgium, *measuring 70 miles in one unbroken length!* This colossal rope of metal and gutta-percha was also constructed at the works of Messrs Newall."

the Continent.   That the business was highly profitable there is no doubt whatever; and knowing what we now do about cable work the risk was inconsiderable.   But it appeared to be far greater in those days, and much praise is due to the courage and energy of the directors of the company, who certainly, by these Anglo-Continental cables, paved and led the way to the greater enterprises which have resulted in submarine telegraphy throughout the whole world.

In 1853 the International Telegraph Company determined to lay cables to Holland and elsewhere.   Afterwards, when . amalgamated with the Electric Telegraph Company, they fitted out a ship for the first time in a permanent fashion for cable operations.   This was the old "Monarch,"* 500 tons, which did much useful work, and whose early gear was designed by the late Mr Edwin Clark and Mr F. C. Webb as engineers to the company.

England and Holland were connected by seven small separate cables, which were bound up together at the landing places (Orfordness, on the Suffolk coast, and The Hague) so as to form shore ends (Fig. 11).   These were repeatedly broken by anchors and trawlers, being far too light, and were replaced by a heavy cable in 1858.†

SECTION 3.—FURTHER EARLY ACHIEVEMENTS.

During the next few years submarine communication was established between Denmark and Sweden; Italy, Corsica, and Sardinia; Sardinia and Africa; and, finally, between Ireland and North America.

Of these numerous lines we shall only concern ourselves with those which, by the importance of results obtained, or by the degree of difficulty overcome, mark a stage in the history of submarine telegraphy.

**Spezia-Corsica and Sardinia-Bona Cables.‡**—These two cables, in

---

* A striking picture of this vessel is given in Mr H. D. Wilkinson's recent book, "Submarine Cable Laying and Repairing" (*The Electrician* Printing and Publishing Company, London).

† For a given amount of telegraphic work, the relative merits of a number of separate cables of light type, each holding its own conductor, and of the same total number of conductors encased in one heavy "multiple" cable, may be summed up as follows :— The former are more prone to fracture or abrasion, but total interruption of traffic is less liable to occur here than by the multiple core, especially if in the former the separate cables are laid on different routes.   With a very rough, irregular bottom subject to anchorage and stormy currents, a fairly heavy cable is, however, practically essential, and this being so, it may as well contain more than one conductor.   Some form of multiple-conductor cable is, in fact, the invariable custom for short distances involving telegraphic inter-communication of the "heavy traffic" order.

‡ This was the first undertaking with which Messrs Glass, Elliot, and Co. were concerned, as a firm, after taking over the business of Messrs W. Küper and Co.   They had been more or less associated since 1851, when Mr Küper came over from Germany, and manufactured several of the early cables, more especially those just referred to.

conjunction with a short submarine line between Corsica and Sardinia, and the aerial lines across the two islands, were to connect France with its colony Algeria *via* Italy.   The cable laid from Spezia, and also the one across the Strait of Bonifacio, contained six conductors each, and were sheathed with twelve (ungalvanised) iron wires of No. 1 gauge, bringing the total weight of each to as much as eight tons per nautical mile.   The depth of water between Corsica and Genoa being over 300 fathoms, Mr J. W. Brett was provided with two drums (Fig. 12), furnished with double sets of flanges, and placed in a line, one in front of the other, and round each of which the cable was given five turns.   With this description of holding-back gear, and with the assistance of Mr John Thompson, Mr Brett was able to lay the first of the above two cables—100 N.M.*—in the course of the year 1854.   With this, however, the cable took charge in some 300 fathoms, about midway; running out with such velocity under the strain that it was bent nearly flat as it passed round the brake drum.   After anchoring by it for a day, the injured part was hauled back and made good.

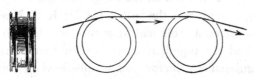

FIG. 12.—Paying-out Apparatus for Mediterranean Cables, 1854.

Shortly afterwards, an attempt was made to pay out a similar type of cable between Bona and Chia, near Cagliari.   But between these points —distant about 150 N.M.—there are depths of 1,600 to 1,800 fathoms, and the cable, which they were unable to restrain, ran out at a terrific speed. Added to this, the small steamer "Tartar," towing the vessel which had the cable on board, could go no faster, so that the impossibility of reaching Sardinia with the remaining cable soon became apparent.   The paying out had been suspended, as was then customary, during the night, and they were deliberating as to what should be done, when the cable suddenly parted.   It was passed out through a hawse pipe, sustaining a sharp nip as it hung down at right angles, and, being unprotected against chafing,

---

* Throughout this work N.M. and "naut" are the abbreviations adopted for nautical miles.   By some the expression "knot" is used in this sense, but inaccurately so.   A knot is a velocity, or rate, rather than a measure of distance.   1 knot = 1 nautical mile (N.M.) per hour; 1 nautical mile (N.M.) = 2,029 yards—or approximately 1,000 fathoms—for telegraph work; 1 statute land mile (S.M.) = 1,760 yards.   Unfortunately, such high authorities as Raper and Norrie have given currency to this misnomer by speaking of "knots" for nautical miles.

the pitching motion of the ship had caused the wires to gradually wear away at a point.

In 1858, under the direction of Mr Charles Liddell,* the "Elba" succeeded in picking up about 60 miles of this cable from Bona seawards, and thus the possibility of recovering and repairing a somewhat heavy cable, laid in comparatively deep water, was practically demonstrated.

On starting to make a second attempt, the expedition steered more in the direction of Galita Island, following a line where previous soundings shewed more moderate depths. During the night, however, they deviated considerably from the proper track, thus wasting cable, and at daylight found themselves off Galita, in 400 fathoms, unable to reach the island for want of cable. The end was passed forward outside the ship and properly secured, a vessel being immediately despatched to Bona for a buoy. At the same time they telegraphed through to London asking that sufficient cable to reach the shore might be sent out. Unfortunately, however, on the fifth day, the cable parted, the continuous pitching and sheering of the vessel causing it to chafe through on the rocky bottom.

Mr Brett having retired from the enterprise, Mr R. S. Newall took it up in 1857, and manufactured a cable with four conductors, each consisting of four copper wires laid up together, covered with two coatings of gutta-percha. These insulated conductors were embedded in hemp,† and the whole sheathed with iron wires of No. 11 B.W.G. for the deep-sea type, and with twelve iron wires of No. 6 B.W.G. for those portions of the cable which were to be laid near the land. The weight of the deep-sea cable was 1.85 tons, and that of the heavier type 3 tons, to the mile.

The cable was coiled below the deck in a cylindrical iron tank, from the centre of which rose a sort of cone (Fig. 13), secured to the bottom of the tank, but free all round at its upper end. Four rings of iron piping were suspended by ropes in a horizontal position round the cone. These rings were of different diameters, largest at the bottom and successively smaller towards the top. The cable was guided and controlled (against flying out by centrifugal force) by passing up between these rings and the central cone, being thus forced to uncoil uniformly and horizontally without kinking. The two lower rings—one of which is shewn—were lowered from time to time as the quantity of cable in the tank diminished, the bottom one being kept pretty close to the coil to keep the outer turns in paying out from

---

* This gentleman and Professor Lewis Gordon were partners with Mr R. S. Newall.

† In those days the description "hemp" or "ordinary hemp" seems to have been commonly applied to what, at the present time, we should term *jute*.

rising in too direct a line to the eye of the tank. From the preceding, it will be seen that Mr Newall was the originator of the rings, or so-called "crinolines," forming a part, in some shape or another, of the fittings of all cable tanks, provided for the purpose of regulating and directing the cable's egress from its resting-place.*

The cable passed up out of the tank through a small horizontal ring placed centrally over the axis of the cone, and then through a casting screwed above, whence it ran along a wooden trough † which was supported on trestles. Battens were placed across the trough at intervals, to prevent the cable jumping up when being paid out rapidly. The cable then passed between two flanged rollers, placed one over the other, and again between two pieces of wood shod on the inside with iron, the upper piece being hinged at one end. A long lever

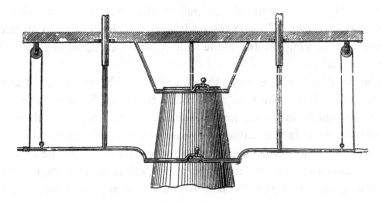

FIG. 13.—Newall's Cone and Rings for Cable Tank.

attached to the hinged piece enabled pressure to be exerted on the cable, which could thus be held as in the jaws of a vice. Thence the cable passed over a V-wheel ‡ to the paying-out drum, round which it took seven turns—the first turn being kept over towards the outer flange of the drum by passing over a conical roller, and a knife § attached to the framework forced the turns inwards and prevented them riding. After leaving the

---

* This device had formed the subject of a patent in 1855.

† The friction which the cable exerts against the sides and bottom of this open trough assists the brakes in preventing too rapid egress from the ship.

‡ A V-wheel, or sheave, is an article peculiar to cable work, having a deeper groove than an ordinary U-pulley, so as to form an extra safe bed and guide for the cable.

§ The term knife, as invariably adopted, is rather misapplied here. It would be more accurate to describe it as a *fleet* or *skid*, the object being to "fleet" the cable on the drum—*i.e.*, to press the cable over, so that it makes way for the fresh turn, to prevent the latter riding immediately over the previous one, as it would tend to otherwise.

drum, the cable passed over another V-sheave and under a jockey wheel forming a sort of dynamometer into the sea, over a cast-iron flanged pulley at the stern of the vessel.

The brake, intended to check the speed of the cable when paying out, consisted of a strong strap of sheet-iron four inches wide, which surrounded the entire circumference of a brake wheel bolted on one side of the drum. By means of a bent lever this iron band could be tightened up or slackened at will, so as to cause more or less friction between it and the revolving drum.

At the last moment, on the advice of Mr (afterwards Dr) Werner Siemens, the above form of dynamometer was fitted up and placed between the paying-out drum and the stern of the vessel. It consisted of a flanged pulley resting on the cable and revolving at the after end of a wooden lever, the other end of which was pivoted. Noting the deflection of the cable caused by the weight of the pulley, the strain could be deduced by a single calculation. To prevent undue heating by friction, a stream of water was kept playing on the cable from a tank placed above the drum and fed by a pump.

On the 7th of September 1857, the end of the cable was brought ashore at Fort Genoa. Here it was firmly secured to two posts fixed in the bottom of the trench which had been previously dug out to a depth of three feet. At one of these posts the cable end was joined up direct to the aerial line in communication with the fort. Up to this date the necessity of lightning guards between an aerial line and a cable, to protect the latter from lightning conveyed from the land line, does not appear to have forcibly commended itself. At 8.30 P.M., the "Elba," in tow of the "Mozambano," and escorted by two French men-of-war, left Cape de Garde, near Bona, paying out the cable. From time to time wires in the iron sheathing broke and sprung up out of the lay; becoming entangled, they soon formed a kind of skein round the cable, which then passed through the machinery with difficulty. Once, indeed, a sudden stoppage was caused in this way, and the cable would certainly have parted if the ship had not been going very slowly at the time. The expenditure of cable when passing over the deep water having exceeded the distance actually covered by more than a third, a change of the ship's course was decided on in order to make Cape Teulada, and the speed increased to six knots. In spite of these precautions, the cable had all run out when the ship was still 12 miles from the land; but, the depth here being only 80 fathoms, the laying was continued the next day by joining on a small single conductor cable, which, however, broke, unfortunately, about two miles' distance from the shore. In the course of the following October the end of the main cable was picked up;

and a new piece of similar type spliced on, with which the shore was reached, near Cape Spartivento.*

Electrical apparatus had been set up on board the "Elba," in order to enable a constant test to be kept on the cable for continuity during the laying operations.   Mr Siemens also made some approximate measurements of the time occupied by the current in passing from one end of the cable to the other, as well as on the inductive influence of a wire on neighbouring conductors.   These effects were found to disappear when a return wire was used so as to effect a metallic circuit †   By only two of the conductors were good signals ever transmitted, and these two failed in as many years' time. ‡   In 1860 Mr Jenkin, afterwards Professor Fleeming Jenkin, F.R.S. (L. & E.), picked up altogether about 57 miles of this cable, half on the African and half on the Sardinian side.   A portion of the recovered cable came up full of kinks, due to the excess of slack paid out.   He was, however, unable to reach the fault, and the work had to be abandoned.

**Black Sea Cable.**—Following the Spezia cable in 1855, a gutta-percha covered wire, No. 16 copper covered to No. 1 gauge, was laid§ in the Black Sea, by Mr Charles Liddell, in engineering charge on behalf of Messrs Newall, between Varna and Balaclava, for use during the Crimean War.   It was, however, sheathed with iron wires for about 10 miles out from each landing-place.   It was when fitting out for this expedition that Mr Newall had first introduced the arrangement of central cone and iron rings, which has already been described.   This line, some 300 miles in length, was laid with very little slack.   Communication through it ceased a short while after the taking of Sebastopol, and no efforts were made to restore it.

---

* A land line was afterwards established from Cape Spartivento to Cagliari, partly in order to bring this cable into direct communication with that to Malta laid subsequently.

† In 1856 Mr Samuel Statham and Mr Wildman Whitehouse had actually secured provisional protection (specification No. 1,726 of that year) for the use of a return wire, though this was originally the custom until Steinheil put Dr Watson's "metallic circuit" into practice !   No doubt Messrs Statham and Whitehouse had these inductive influences and earth currents in mind when they abandoned using the earth as a "return."

‡ Mr Siemens estimated the speed of the current in passing from Cape Spartivento to Bona as being at the rate of 0.25 seconds, *i.e.*, he found it took each signal impulse that time to reach to the further end.

§ Great trouble was experienced in this owing to the length falling short, the comparative lightness causing it to be drawn away out of its direct course.   In the open sea (subject to strong currents and boisterous water), it would probably have been absolutely impossible to lay a line of this character except by using a length far beyond that represented by the distance covered.

**The Sardinia-Malta and Malta-Corfu Cables.**—These two cables, each about 500 miles long, and weighing 18 cwt. per naut, were laid in 1857, only lasting for a very short time, being far too light for the locality. The Sardinia-Malta section, repaired for the first time in 1859, failed about six weeks afterwards in much the same place. Further attempts were made to repair it, but when five months had elapsed the operations were abandoned. The bottom was very uneven, and both breakdowns were attributed to volcanic action; they were, however, more probably caused by fishermen's trawls.

The Malta-Corfu section ceased working in 1861, during a heavy thunderstorm, although protected by a lightning guard. The tests shewed want of continuity in the conductor, and the fault seemed to be about 20 miles from Corfu. Some parts of the cable were afterwards picked up, and found to be in a good state of preservation.

*From* PUNCH; OR, THE LONDON CHARIVARI, 14*th September* 1850 (reduced *facsimile* of cartoon).

EFFECT OF THE SUBMARINE TELEGRAPH; OR, PEACE AND GOOD-WILL
BETWEEN ENGLAND AND FRANCE.

## THE ATLANTIC TELEGRAPH CABLE WAS SUCCESSFULLY LAID,
### 5TH AUGUST 1858.

Two mighty lands have shaken hands
    Across the deep wide sea ;
The world looks forward with new hope
    Of better times to be ;
For, from our rocky headlands,
    Unto the distant West,
Have sped the messages of love
    From kind Old England's breast.

And from America to us
    Hath come the glad reply,
"We greet you from our heart of hearts,
    We hail the new-made tie ;
We pledge again our loving troth
    Which under Heaven shall be
As steadfast as Monadnoc's cliffs,
    And deep as is the sea !"

Henceforth the East and West are bound
    By a new link of love,
And as to Noah's ark there came
    The olive-bearing dove,
So doth this ocean telegraph,
    This marvel of our day,
Give hopeful promise that the tide
    Of war shall ebb away.

No more, as in the days of yore,
    Shall mountains keep apart,
No longer oceans sunder wide
    The human heart from heart,
For man hath grasped the thunderbolt,
    And made of it a slave
To do its errands o'er the land,
    And underneath the wave.

Stretch on, thou wonder-working wire !
    Stretch North, South, East and West,
Deep down beneath the surging sea,
    High o'er the mountain's crest.
Stretch onwards without stop or stay,
    All lands and oceans span,
Knitting with firmer, closer bonds
    Man to his brother man.

Stretch on, still on, thou wondrous wire !
    Defying space and time,
Of all the mighty works of man
    Thou art the most sublime.
On thee, bright-eyed and joyous Peace,
    Her sweetest smile hath smiled,
For, side by side, thou bring'st again
    The mother and the child.

Stretch on ! O may a blessing rest
    Upon this wondrous deed,
This conquest where no tears are shed,
    In which no victims bleed !
May no rude storm disturb thy rest
    Nor quench the swift-winged fire
That comes and goes at our command
    Along thy wondrous wire.

Long may'st thou bear the messages
    Of love from shore to shore,
And aid all good men in the cause
    Of Him whom we adore :
For thou art truly but a gift
    By the All-bounteous given ;
The minds that thought, the hands that wrought,
    Were all bestowed by Heaven.

*The British Workman.*

Deck of H.M.S. "Agamemnon" (see pp. 47, 48), shewing Paying-out Apparatus, 1858
(from the *Illustrated London News*).

# CHAPTER II.

## THE DAWN OF OCEAN TELEGRAPHY.

### SECTION 1.—THE FIRST TRANS-ATLANTIC LINE.

HITHERTO the efforts of the early projectors of submarine telegraphy had been confined to the work of connecting countries divided only by narrow seas, or establishing communication between points on the same seaboard.

The next step forward in the science of submarine telegraphy was a gigantic one—no less, in fact, than that of spanning the Atlantic Ocean between Europe and America. This was aptly characterised at the time, by Professor Morse, as "the great feat of the century." It was the first venture in the direction of trans-oceanic telegraphy, so that there were no applicable data to go upon. The vast difference between laying comparatively short lengths of cable across rivers and bays, or in shallow water, and that of laying a long length of cable in depths of over two miles across an open ocean, will be evident alike to the sailor and the engineer.*

During the year 1855, the North American telegraph lines had been extended as far as Newfoundland, while in Europe the wires of the "Magnetic" Company had, several years previously, been carried by their engineer, Charles Bright, to various points on the west and south-west of Ireland, including Portrush, Sligo, Galway, Limerick, Tralee, and Cape Clear.

The feasibility of uniting the two vast systems of telegraphs had engaged the consideration of some of the most enterprising of those occupied in their development on both sides. It had been already proved that cables could be successfully laid in comparatively deep water; but the nearest points between the British Islands and Newfoundland are nearly 2,000 miles apart, and the greatest length of submarine line which had been successfully submerged prior to 1856 would form but an insignificant part of such an enormous distance, and that too embracing depths of nearly three miles.

Apart from the engineering difficulties entailed by this vast distance and depth, the question was then undetermined as to the possibility of conveying electric currents through such a length in an unbroken circuit, and at a speed that would enable messages to be passed quickly enough in succession to prove remunerative.

---

* The *greatest depths* in which any cables had been laid previous to this were those in the Mediterranean and Black Seas, during 1854, 1855, and 1856. Some of these were, temporarily speaking, partial successes, others absolute failures ; indeed, several of them had met with disaster even in the process of laying. An example of this was the cable between Sardinia and Algeria of 1854 (600 N.M.) in a maximum depth of about 800 fathoms. Again, the unprotected gutta-percha core laid from Varna to Balaclava in 1855 failed, as already shewn, not long after being submerged. In 1855 a cable was successfully laid between Spezia and Corsica in water running into an outside depth of 325 fathoms, the length being about 110 miles.

The *greatest length* of line which had been submerged in a satisfactory manner previous to the first Atlantic cable was that between Varna and Constantinople—171 miles—along the more or less protected Black Sea shore in 1855, the depth being upwards of 100 fathoms.

In the early experiments made by Professor Wheatstone with frictional electricity on a short length of bare wire in a room, the subtle influence was shewn to pass at the rate of nearly 300,000 miles in a second of time. Later experiments with voltaic electricity (*i.e.*, that obtained from a voltaic pile or battery), established the speed of transmission on bare overhead telegraph wires to be about 16,000 miles per second.

When, however, the underground gutta-percha insulated wires laid by the "Magnetic" Company in 1851, which extended between London and Dublin *via* Manchester and Liverpool, were tested on this point, with variously increased length, it was found that a far lower rate only was attainable.

In a paper read by Mr Edward Brailsford Bright, accompanied by experiments, at the meeting of the British Association in 1854, the velocity of ordinary telegraphic currents* in subterranean conductors was given as not exceeding 1,000 miles per second,† and he shewed that the gutta-percha insulated wire tended to retain part of the charge passed into it. In fact, the similarity between the conductor and a Leyden jar was on this occasion clearly set forth, in the sense of the induced charge on the outside of the dielectric holding back as a static charge some of the electricity flowing as current through the conductor, in the same way that the charge induced on the outside plate of a Leyden jar statically holds the primary charge on the inner plate, until either are neutralised.

The doubts as to working across the Atlantic were, however, very much modified by a series of experiments, instituted by Mr (afterwards Sir Charles Tilston) Bright, and also by others independently carried out on a larger scale by Mr Edward Orange Wildman Whitehouse.‡ Both of these sets of experiments were upon the underground wires of the "Magnetic" Company's system,§ which were so connected on various occasions as to afford a length of upwards of 2,000 miles in one con-

---

* It soon became evident that the rate of working was—on broad principles and within certain limits—entirely independent of the nature of the generating agent or of its power—*i.e.*, was practically uninfluenced by the number of cells employed. However, nowadays, with automatic transmission, a somewhat higher electro-motive force often renders signals readable which, owing to the ultra high speed at which the impulses follow one another, would not otherwise be.

† Later on, it was demonstrated by Professor Thomson that electricity could not be said to possess velocity, in the ordinary sense, at all.

‡ Previously a surgeon at Brighton.

§ Mr Whitehouse's conclusions were considerably more favourable than those arrived at by Mr Bright, or any one else, for the proposed cable.

tinuous circuit. According to Mr Whitehouse, "signals* were clearly and satisfactorily transmitted over this vast length at the rate of 210, 241, and 270 per minute, with a facility that would answer every commercial requirement."†

The difficulty in working that had been found to arise from the retardation of the electric current, due to induction, was overcome, as it had previously been in the magneto-electric instruments of the "Magnetic" Company,‡ by using a succession of opposite currents. By this means the latter or retarded portion of each current was wiped out by the opposite current immediately following it; and thus a series of electric waves could be made to traverse the wire, one after the other, several being in the act of passing onward at different points along the conductor at the same time.

While prior to, and during, 1855 the essential conditions and methods of signalling through an Atlantic cable had been independently investigated by Mr Charles Bright and Mr Whitehouse in England, the connection of the United States and Canada with Newfoundland had at last been brought to a successful issue.

In 1852 Mr F. N. Gisborne, a very able English engineer (previously engaged in constructing the Nova Scotia telegraph lines), in concert with an American syndicate, headed by Mr Tebbets of New York, obtained an exclusive concession for connecting St John's, Newfoundland, with Cape Ray, in the Gulf of St Lawrence, by an overhead line. The idea was to "tap" steamers coming from London at Cape Race, St John's, and pass messages between that point and Cape Breton, on the other side of the gulf, by carrier pigeons. A few miles of cable were made in England, and laid between Prince Edward Island and New Brunswick. Mr Gisborne then surveyed the route for the Newfoundland land line, and had erected some 40 miles of it, when the work was stopped for want of funds.

When in New York in 1854, Gisborne was introduced to Mr Cyrus West Field, a retired merchant, who became enthusiastic on the subject, and formed a small but strong syndicate. Better terms were then obtained

---

* *I.e.*, battery key contacts, producing single impulses.

† Professor S. F. B. Morse was also present at these experiments. He, moreover, accompanied the first expedition, aboard the "Niagara," as electrician to the New York, Newfoundland, and London Telegraph Company, which was naturally largely interested in the proposed telegraphic connection between Europe and Newfoundland. Professor Morse gave the scheme his entire support and scientific approval.

‡ Devised by Mr W. T. Henley, and improved on by Messrs Bright.

from Newfoundland, covering (with the idea of an Atlantic cable, suggested many years before both in England and the States) exclusive rights to land cables for fifty years, which monopoly they were able to get extended to New Brunswick, Cape Breton Island, Prince Edward Island, Nova Scotia, and the shores of the State of Maine.

In the meantime, Mr John W. Brett* had joined Mr Field's syndicate, to which he subscribed a considerable sum. A single-conductor cable of 85 miles was made† in England by Messrs Glass, Elliot, and Co.,‡ to be laid between Cape Breton and Newfoundland, from a sailing ship to be towed by an American steamer, " James Adger," with Mr Field, Professor Morse, and many friends on board. Mr (now Sir Samuel) Canning§ was in engineering charge on behalf of Messrs Glass and Elliot, but after 40 miles were laid, in August 1855, rough weather ensued, and the captain of the barque had to cut the cable to save his vessel.

A fresh instalment was sent out in 1856, and laid successfully across the gulf, thus connecting St John's with Canada and the American lines. It had one conductor, into which a great improvement was introduced for the first time. It was made of seven small copper wires laid up in the form of a strand, with a view to preventing a flaw in one of the wires at any point entirely stopping the conductivity. ‖ The insulated conductor was covered with tarred yarn, and protected by a sheathing of twelve outer iron wires. The weight was 2½ tons per N.M., and it lasted a long time, being successfully repaired ten years later.

---

* This gentleman, together with his brother, Mr Jacob Brett, had registered a company as early as 1845 for uniting Europe with America by telegraphic communication under the title of a " General Oceanic Telegraph Company."

† This was probably the first occasion on which the cable, as manufactured, was properly coiled in specially constructed tanks at the factory previous to shipment. Hitherto, the floor had been used for this purpose, and the coiling was of a somewhat rough-and-ready order, leading sometimes to entanglements between the turns and flakes.

‡ This firm had lately taken over the business of Messrs Küper and Co., one of the leading wire-rope makers (for colliery and other purposes), who had been associated with several of the previous early submarine cables.

§ Formerly a railway engineer under the late Mr Joseph Locke, M.P., F.R.S.

‖ This substitution for the single solid conductor not only obviated the objection of an imperfection in the copper at any one spot having serious results, but also lessened the risk of complete discontinuity due to any mechanical tension. It, moreover, gave greater pliability, thereby reducing the chance of breakage under an undue lateral strain, and also of injury to the insulating envelope.

Since the above occasion the conductor of a submarine cable has been invariably built up by several wires stranded together.

This type of conductor was provisionally protected in 1854 (see specification No. 2,547 of that year) by Professors William and John Thomson with Professor W. J.

Mr Cyrus W. Field, the Vice-President of the New York, Newfoundland, and London Telegraph Company (formed by the above-mentioned syndicate), then came over to England in July 1856, empowered by his associates to deal with exclusive concessions; and on the 29th September 1856, an agreement was entered into between Mr J. W. Brett, Mr Charles Bright, and Mr Field,* in which, on an equal footing, they mutually agreed

Original Station of the New York and Newfoundland Telegraph Company, 1855,
previous to erection of permanent building.

(as the "projectors") to exert themselves "with the view and for the purpose of forming a company for establishing and working electric telegraphic communication between Newfoundland and Ireland, to be called the 'Atlantic Telegraph Company,' or by such other name as the parties hereto shall jointly decide upon."

The nature of the ocean's bed had by this time become ascertained by several series of soundings taken by Lieut. O. H. Berryman, U.S.N., from

---

Macquorn Rankine, though never completed as a patent. The main object with the devisers was to enable a larger conductor to be used than was possible, for mechanical reasons, with a single wire.

It should be remarked, however, that the mechanical advantages of a strand had been previously appreciated and adopted in practice in various directions, including that of lightning conductors, by Newall and others.

* Joined later by Mr Whitehouse.

U.S.S. "Arctic," and also by Commander Joseph Dayman, R.N. (H.M.S. "Cyclops"), shewing that a gently undulating plateau of great breadth, at a depth varying gradually from 1,700 to 2,400 fathoms, extended nearly the whole distance between Ireland and British North America. These depths, though great for the purpose in view, compared favourably with the soundings of 6,000 to 7,000 fathoms that had presented themselves further southward.

The soundings were taken with the ingenious apparatus of Lieut. J. M. Brooke, U.S.N., by which the weight was automatically detached on reaching the bottom, while a small tube still retained by the line brought up specimens of the bottom, consisting of a soft ooze formed by the tiny shells of microscopic "infusoria," borne along, while alive, by the warm water of that "river in the ocean" the Gulf Stream, and dying in countless myriads on the temperature being lowered by contact with the more northern seas.*

FIG. 14.—Infusoria at the Bed of the Atlantic Ocean: Magnified 10,000 times.

This table-land—thus seemingly raised at the bottom of the sea—was christened "Telegraph Plateau" † by Lieut. M. F. Maury, U.S.N., Chief of the U.S. National Observatory.

---

* In those days soundings were effected by a hempen line offering enormous surface resistance. Soundings by such means were unreliable on account of the slow rate rendering it difficult to ascertain when bottom was reached. For the same reason the process was a lengthy one. Sir William Thomson introduced his pianoforte wire machine in 1872, and this has since been considerably improved on in the sounding machines of Mr F. R. Lucas, Messrs Johnson and Phillips, and the Silvertown Company, the latter having made a speciality of submarine survey.

† The specimens which were brought up from depths ranging between 1,700 to 2,400 fathoms, in the region of the so-called "telegraph plateau," bore a very strong resemblance to exceedingly finely powdered chalk. Their appearance at the bottom of a glass vessel was that of a light brown muddy sediment, in which were observed minute hard particles, hardly any of which exceeded one-fiftieth of an inch in diameter.

To quote from the report made by Prof. T. H. Huxley, F.R.S., on the specimens of the bottom obtained, and illustrated in Fig. 14 :—"Fully nine-tenths consist of minute animal organisms, called *foraminiferæ*, provided with thick skeletons composed of carbonate of lime. The species of the 'foraminiferæ,' of which 85 per cent. of the specimens consist, is called *globigerinæ*."

The specimens shewn have been magnified about 250 times their natural size, and were, as may be seen, of various dimensions and shapes, yet with general points of resemblance.

The actual words used, in this connection, by Lieut. Maury, in the course of his report on the proposed trans-Atlantic cable, were :—"The bed of the sea between

The question of suitable landing-places for the cable ends necessarily arose, and Trinity Bay, Newfoundland, was considered eminently adapted for the purpose, being about the nearest point, and having a very deep entry guarded by banks on each side. Thus, if icebergs grounded on either bank—as frequently happened—they could not, when reduced in their size by melting, touch, while drifting, the greater depth of the main channel, in the deepest part of which the cable would be submerged.

On the Irish side, Mr Edward Bright, the manager of the "Magnetic" Telegraph Company, with some of his staff, examined the various harbours and beaches between Dingle and Bantry Bays, on the extreme south-west promontory of Ireland, the nearest land towards America. Mr Bright chartered a fishing-smack for the purpose ; and after considering the question of freedom from anchorage, rocks, soft protected landing, together with the available Admiralty soundings, he pronounced Valentia* Bay as the most suitable locality. This selection has since been well justified, judging by the number of cables subsequently landed there, or in the immediate vicinity.

The actual route decided on for the entire line is shewn in the chart facing this page, with the soundings referred to above.

During the above preliminaries a large number of specimens of types of cables were made by Messrs Glass, Elliot, and Co., and tested for strength, in connection with their weight and other conditions.

On the 3rd of October 1856, the results of Mr Whitehouse's experimental researches—extending over several years—were shewn to Professor Samuel Morse, LL.D., the electrician of the New York and Newfoundland Company, at the central office of the "Magnetic" Company, in London, and on the 20th of that month the Atlantic Telegraph Company was registered.

---

Ireland and Newfoundland is a plateau, which seems to have been placed there especially for the purpose of holding a submarine telegraph, and of keeping it out of the way."

In later years, the necessity of soundings being taken in greater proximity to one another having forced itself upon the minds of those engaged in cable work, subsequent experience has shewn this bed not to be so plateau-like as was then supposed, though no very serious irregularities occur on this route as against what may be found a hundred miles or so south, where the "Faraday Hills" were discovered later on by S.S. "Faraday," when cable-laying for Messrs Siemens Brothers.

The above-mentioned supposed "plateau" was certainly made the most of at the time in various diagrams and maps, purporting to shew the extreme evenness of the proposed course in comparison with neighbouring beds of the Atlantic.

* Sometimes still spelt Valencia—*i.e.*, in the same way as the place on the east coast of Spain.

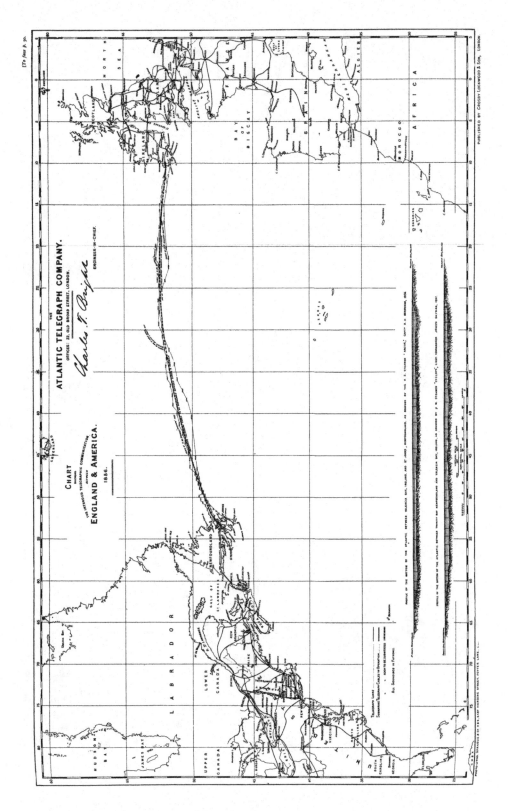

The material originally positioned here is too large for reproduction in this reissue. A PDF can be downloaded from the web address given on page iv of this book, by clicking on 'Resources Available'.

The British Government encouraged the project by a guarantee of £14,000 per annum (during the working of the cable), and promised vessels to assist in laying it.

The £350,000 necessary for the work was then obtained in an absolutely unprecedented manner. There was no promotion money, no prospectus published, no advertisements, no brokers, and no commissions; neither was there at that time any Board of Directors, or executive officers.

The election of a board was left to a meeting of shareholders, to be held after allotment by the provisional committee, consisting of the subscribers to the memorandum of association. Any remuneration to the projectors was left wholly dependent upon, and subsequent to, the shareholders' profits being over 10 per cent. per annum, after which the projectors divided the surplus.

The campaign was opened in Liverpool, the headquarters of the "Magnetic" Company, the greater proportion of that company's shareholders being merchants and shipowners there, who foresaw the value of the United States being connected telegraphically with this country and Europe through their Irish lines.

A notice, issued by Mr Edward Bright, on a half-sheet of notepaper, brought together a crowded meeting of the foremost people of Liverpool, at the Underwriters'-Rooms, on the 12th November 1856. The inspiriting addresses of Messrs Field and Brett, accompanied by the scientific explanations and answers of Mr Charles Bright, were exceedingly well received.

The "Magnetic" Company had prospered under Edward Bright as manager and Charles Bright, their engineer; and the very grandeur of the enterprise, coupled with Charles Bright's convincing experiments and the facts so lucidly laid before them by Messrs Brett and Field (the latter acting subsequently as general manager), commended it to the company's shareholders and friends, who may be said to have represented the adventurous spirit of the mercantile and manufacturing pioneers in the chief centres of Lancashire, Yorkshire, and Scotland.

In the matter of the Atlantic cable, Liverpool led the way. Charles Bright, with the then Mayor and Mr Charles Pickering (of Messrs Schroder), headed the public subscription list, their example being soon followed by many others. Similar meetings were at once held in Manchester and Glasgow, subscription lists being opened at the Magnetic Company's offices there, and at London, etc. In a few days the £350,000 *

---

* To meet the cost of additional cable and expenses, this capital was increased the following year to £465,000 by the issue of "Debentures."

was raised by the issue of 350 ordinary shares of £1,000 each!—chiefly taken by the shareholders of the "Magnetic" Company and their friends.*

The board was then formed, including such names as the late George Peabody, Samuel Gurney, T. A. Hankey, C. M. (afterwards Sir Curtis) Lampson, and Mr (afterwards Sir William) Brown, of Liverpool. Of the number elected, nine were directors or shareholders of the "Magnetic" Company, including Mr J. W. Brett. Two names may be specially referred to as destined, in different ways, to have the greatest possible influence subsequently in the development of submarine telegraphy. Mr (afterwards Sir John) Pender, G.C.M.G., M.P., was then a director of the "Magnetic" Company, and has since taken the most active and foremost part in the vast extensions—linking up the whole of the world—that have followed, including the Mediterranean *via* Gibraltar, Malta, and Alexandria; India *via* the Red Sea; China, Australia, Brazil, Africa (east and west to the Cape), and a large proportion of the many existing trans-Atlantic cables. He was the first chairman of the Telegraph Construction and Maintenance Company; and is at present chairman of nine great telegraph companies, representing some £15,000,000 of capital, and mainly carried through prosperously by his foresight, influence, and indomitable business energy.†

Another director was Professor William Thomson, F.R.S. (L. & E.), of Glasgow University (now Lord Kelvin), who was not only a tower of scientific strength on the board, but was an ardent believer in the Atlantic cable—so much so that he had communicated his views as to its practicability to the Royal Society in 1854. His accession was destined to prove of vast importance in influencing the development of the trans-oceanic system, for his subsequent experiments on the Atlantic cable, during 1857-58, led up to his invention of the marine galvanometer, whereby the most attenuated currents of electricity, incapable of producing visible signals on other known electrical instruments, were, by the use of a reflected beam of light, so magnified in their effect as to be readily legible.

---

* Ultimately by far the largest individual subscribers were Mr John Watkins Brett and Mr Cyrus Field, both men of fortune. They each took up twenty-five shares in the first instance, but both reduced their personal interest very considerably before allotment; indeed, Mr Brett reduced his by half, and Mr Field endeavoured to get the same proportion of his interest placed elsewhere.

† Since these lines were written Sir John Pender's lamented death has occurred. A memorial is to be raised in his honour to commemorate his services to submarine telegraphy. Commercially speaking, probably no man has done so much for the cause as Sir John Pender, and he has been not inaptly termed the "Cable King." Largely owing to Sir John's conspicuous connection with submarine telegraphy, his eldest son — Mr James Pender, M.P.—has recently been made a baronet.

Mr (afterwards Sir Charles) Bright was appointed chief engineer by the board, with Mr Wildman Whitehouse as electrician, and Mr George Saward, secretary.

The construction of the cable was taken in hand the following February. The distance from Valentia, on the Irish coast, to Trinity Bay in New-foundland, the two landing points selected, being 1,640 N.M., it was estimated that a cable length of 2,500 N.M. would be sufficient to meet all requirements. The Gutta-percha Company of London were entrusted with the manufacture of the core, consisting of a strand of seven No. 22 copper wires (total diameter No. 14 gauge) weighing 107 lbs. per N.M., insulated with three coatings of gutta-percha (to ⅜ inch diameter) weighing 261 lbs. per N.M., the conductor being, in fact, covered to No. 00 B.W.G.*

After various experiments† with sample-lengths of different iron wires made up into cable, the contract for the outer sheathings was divided equally between Messrs Glass, Elliot, and Co., of Greenwich,‡ and Messrs R. S. Newall and Co., of Birkenhead.§ The core was first sur-rounded with a serving of hemp saturated with a mixture of Norwegian

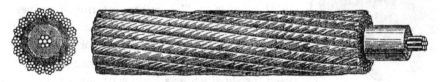

FIG. 15.—First Atlantic Cable (1857-58).

(Stockholm) tar, pitch, linseed oil, and wax, and then sheathed with eighteen strands, each containing seven Best (bright) iron wires of No. 22 gauge,‖ the completed strand being No. 14 gauge in diameter. The cable (Fig. 15)

---

* This formed a much heavier core than had ever been previously adopted, and the difficulties of manufacture were proportionately greater.

† These experiments were superintended at the works of Messrs˙ Brown, Lenox, and Co., the famous engineers of cable work appliances.

‡ At Morden and Enderby's wharves, the latter being rope-works just taken over by this firm.

§ This subdivision of labour—by half the contract being eventually assigned to Messrs Newall—was decided on in order to complete the work in the time, and with a view to meeting threatened opposition.

‖ This form of sheathing was the suggestion of that distinguished engineer, the late Mr I. K. Brunel, and was, no doubt, to a great extent adopted as being a "set off" to the heavy types which had just before proved so unsuitable for laying (or recovering) in the deep waters of the Mediterranean. In the same way, some authorities even advocated a hempen cable without any iron sheathing whatever. Brunel's strand armour of comparatively small wires proved beautifully pliable, and, when new, had many points in its favour for cable operations, though the individual wires were somewhat subject to breaking and getting loose from the rest. A large number of such small wires was then passed through (or received a coating of) a mixture of tar, pitch,* and linseed oil.† Its weight in air was 1 ton per N.M., and in water only 13.4 cwt., with a breaking strain of 3 tons 5 cwt., equivalent to nearly five miles of its weight in water. It was found that the process of galvanising could not be applied to so fine a wire as that employed in the strand-sheathing of this type.

For a length of 10 miles at the Valentia end, and for 15 miles out from Trinity Bay, the sheathing consisted of twelve iron wires of No. 0 gauge, making the weight of the shore-end cable 8.1 tons to the N.M. (Fig. 16).

A part of this was furnished with an increased thickness of insulation so as to cope with the rough work it was likely to be subjected to, and thus imbue it with a longer life. Outside the previously constructed core it was further covered up to ¾ inch with an additional two coats of gutta-percha compound, being a mixture of common gutta with mahogany and wood-dust.‡ This part of the shore-end type, with the core as above, is shewn in the figure. The greater portion, however (and all that subsequently laid at the New-foundland end), had the ordinary core, the same as the deep sea, and a lighter shore-end type used for shoal water a little way out.

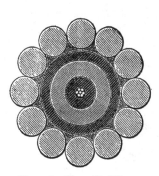

FIG. 16.—Shore-End Type.

The cable had to be delivered in June 1857, which only allowed four months for the entire completion of its manufacture! To give an idea of what this meant, it will suffice to say that no less than 119½ tons of copper

---

do not, however, make up a *durable* type of cable, for the following reasons :—(1) Increased total area exposed to rust ; (2) greater seriousness in same degree of rusting of each wire ; (3) difficulty of galvanising so small a section. Moreover, on account of the length of time involved in laying up the strands beforehand, this may be said to be an expensive form of armour. It was certainly, however, a wonderfully good type, mechanically speaking, for that time ; and it is quite a question now whether it was not preferable to that which followed over the cables of 1865, 1866, 1869, etc., in the same waters. In 1874 a portion of this cable was picked up from a depth of 2,000 fathoms in very good condition, on the occasion of other repairs being effected ; also again in 1880 in 212 fathoms.

* New specimens of this cable are usually made up without any external compound for reasons of convenience. This compound was again applied by brushes after the cable had been coiled in the tank.

† A series of interesting articles, descriptive and illustrative of the various stages of constructing this line, appeared in the *Illustrated London News* at the time.

‡ This was mainly intended as a further mechanical protection to the inner core from the weight and stiffness of the heavy wire sheathing.

had to be provided for the construction alone. This copper had to be drawn out into 20,500 miles of wire (providing for the lay); and seven parts of this wire had to be laid up into a strand 2,500 miles long. For the insulation, moreover, nearly 300 tons of gutta-percha were required to be prepared, and applied to the conductor in three separate coatings. Lastly, and with a due allowance for lay, 367,500 miles of wire had to be drawn from 1,687 tons of charcoal iron, and laid up into about 50,000 miles of strand* for the outer sheathing.†

Added to this, the ships had to be selected and got ready to receive the cable. Moreover, machines both for manufacturing and laying had to be constructed as well as designed. ‡

This race against time was the outcome of an unfortunate engagement (insisted upon by Mr Field in connection with his American arrangements) on the part of the company towards its shareholders and the public. Messrs Bright and Whitehouse urged that more time should be given, to ensure greater care in manufacture, and the former advocated a different type of cable, with a conductor more than three times as large, and a much greater thickness of insulation. §   But the contracts || had been given out by the provisional committee before he was appointed to the post of engineer.

---

* The entire length of wire used in the manufacture of this cable was, in fact, enough to girdle the earth thirteen times.

† The two firms engaged in the construction of this cable unfortunately applied the sheathing wires with opposite lays. Messrs Newall adopted the more ordinary (right-handed) lay, as had previously been the custom for telegraph cables as for ropes. Messrs Glass and Elliot discovered, however, that fresh turns would be put into a cable with such a lay in the act of coiling (right-handedly) into the tank of the laying vessel. They, therefore, laid up the wires the opposite way, so that in coiling down, the turns set up in manufacture would be taken out again. They did not, however, advise the other parties concerned of this change, and thus it was not known till afterwards. The left-handed lay of Messrs Glass and Elliot is now invariably adopted in the construction of telegraph cables.

‡ The late Mr Willoughby Smith bore suitable and independent testimony to these abnormally hurried conditions for the amount of work to be done in his "*Résumé* of the Earlier Days of Electric Telegraphy," delivered to the Society of Telegraph Engineers during their meetings at Paris on the occasion of the Electrical Exhibition in 1881. (See *Journal Soc. Tel. Engrs.*, vol. x. No. 38.)

§ The type of core actually recommended by Sir Charles Bright was "a copper conductor composed of seven equal wires of maximum purity stranded together, of such a gauge as is equivalent to a weight of 392 lbs. (3½ cwt.) per N.M. This conductor to be covered with three coatings of gutta-percha of a thickness represented by the same weight per N.M. as the conductor."

|| The contract price for the entire length of cable manufactured for the First Atlantic Line was £225,000, the core costing £40, and the armour £50, per mile.

An Atlantic cable of the present day runs into about half a million sterling. Gutta-percha in those days was less scarce ; on the other hand, its manufacture was more of a novelty, and there was less competition in the whole practice of cable-making.

There was not time to provide proper buildings or tanks on shore, and the cable consequently was laid dry, and exposed to the sun's heat, which injured some upper layers that had to be subsequently cut out. Unfortunately no experience in these matters was at hand, or no doubt the extreme importance of such questions would have been better appreciated.*

To carry through the then unprecedentedly heavy work devolving upon the engineering department, Charles Bright associated with himself Messrs Samuel Canning, W. Henry Woodhouse,† F. C. Webb,‡ and Henry Clifford. The three former had been prominently connected with cable-laying for some years, and the fourth was a most able mechanical engineer.

The British Government placed H.M. battle-ship "Agamemnon" (ninety-one guns) at the Company's service. She had been Admiral Lyons' flag-

U.S. Frigate "Niagara," used for laying the Atlantic Cable of 1858.

ship at the bombardment of Sebastopol, and was well suited for the purpose, her tonnage being 3,200, and her two screw engines well aft, while amidships she had a magnificent hold 45 feet square and about 20 feet deep. In this capacious receptacle nearly half the cable was stowed away, from the works at Greenwich, the balance being divided in two other small coils. She was in charge of Commander C. T. A. Noddall, R.N., and was

---

* Iron tanks were strongly urged by Mr Bright for stowing the cable in to permit of it being kept constantly covered with water after manufacture, and until being submerged. This recommendation was, however, only followed to the extent of keeping the core in water during testing operations.

† Like Mr Canning, this gentleman was originally a railway engineer.

‡ Mr Webb has probably had a greater experience in all phases of cable work than any one ; moreover, he has taken prominent part in more of the early expeditions than most of the pioneers.

one of the finest of the line of battle-ships in our navy at the time. Mr H. A. Moriarty, R.N., was to serve as navigating master.

The other half of the cable, made at Birkenhead, was coiled on board the U.S. steam frigate "Niagara" (see p. 36), Captain W. L. Hudson, lent by the United States Government. She was the finest vessel in their navy, of 5,000 tons, and had only recently been modelled by Mr Steers, the great yacht builder of America. Her lines were in fact those of a yacht; and it cost Captain Hudson a great pang when he discovered how much she had to be chopped about to take the cable.

During the brief period available, Mr Bright, in conjunction with Mr C. De Bergue, of Manchester (a mechanical engineer of some note), and with the able assistance of Mr Clifford, devised the machinery for paying out the cable, the main principle of which is shewn in Fig. 17. It included a Salter's balance arrangement, intended to indicate the strain on the cable.*

The above application of a Salter's balance as a dynamometer for cable-laying did not prove satisfactory. The vibration of the levers and rigidity

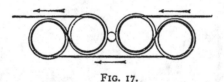

FIG. 17.

of the friction-brake prevented it giving a steady indication of the strain, such as could be read off with any sense of accuracy. The same plan had, however, previously been adopted with some success when cable-laying in the comparatively shoal waters of earlier cables.

Mr Bright also had what was aptly termed a "crinoline,"† or cage of iron bars, fixed round each ship's stern as an external guard to prevent the cable from fouling the screw, if necessity arose for backing the vessel.‡

---

* This machine was constructed and set up on both ships by Messrs De Bergue and Co. It was afterwards substituted by another, in the following expedition, designed by Mr Charles Bright, with the co-operation of Mr H. Clifford, and Messrs Easton and Amos, the makers.

† Nowadays known as a screw-guard, and fitted to the telegraph ship "Faraday."

‡ This was a particularly suitable precaution in this instance, owing to the fact that on both ships the picking-up apparatus (as well as that for paying out) was at the stern. On more than one occasion the cable was thus spared from coming into contact with the ship's propeller.

The question of recovering and repairing a cable in deep water had not been much gone into at that time. Thus, the picking-up gear, placed alongside that for paying out, and worked by steam, was merely intended as an auxiliary for hanging on to, or hauling inboard again, any short length in the event of a fault or any mishap occurring whilst laying the cable, the paying-out apparatus not being fitted with any steam-engine.

In addition, he devised an electrical log, completing and breaking the circuit at every revolution. Here, by means of a gutta-percha covered wire passed up the line, the log itself continuously signalled and recorded the speed through the water.

It had been intended to start laying from mid-ocean from both ships, one towards Ireland and the other to Newfoundland, so as to ensure a quiet time for making the splice (for thus they could wait in the middle and choose their weather), and also thereby reducing the period of laying by one-half. This course was strongly urged by Mr Bright and his experienced staff of engineers. However, the electrician, Mr Whitehouse—

The Lord-Lieutenant of Ireland making a Speech on the Starting of the First
Expedition from Valentia Bay, 6th August 1857.

whose health did not permit him to go out on the expedition—urged starting from Ireland. Owing largely to the anxiety of the board to have continued reports of progress, Mr Whitehouse's views prevailed.

After valedictory speeches from the Lord-Lieutenant (the Earl of Carlisle) and others (see illustration from *Illustrated London News*), the expedition started from Valentia on the 6th August 1857. It consisted of the " Niagara," from which the first half of the cable was to be laid, and the " Agamemnon "; and, as escorts, H.M. paddle frigate " Leopard," Captain J. F. B. Wainwright, R.N., and the U.S. paddle frigate "Susque-hanna,' Captain J. R. Sands, U.S.N., whilst H.M. sounding vessel "Cyclops," Commander Dayman, R.N., preceded what became known as the " Wire Squadron " to shew the way.

The greater part of the heavy "shore-end" cable had been laid just before (see illustration below), by a small steamer, the "Willing Mind," assisted by H.M. tender "Advice" and several launches, from Ballycarberry, a little cove off Caherciveen, Valentia Harbour.*

The "Niagara," having three miles of the shore end on board joined to the main deep-sea cable, spliced on to what was already laid, and then started paying out, at the rate of two knots, the rest of the heavy shore end. Three-quarters of an hour later, the shore end, owing to its extreme rigidity, slipped off the paying-out sheaves, jammed on the axle, and parted. It was at once underrun from shore, and spliced up again; and, on the 7th, the "Niagara" resumed paying out.

Landing the Shore End of the First Atlantic Cable by S.S. "Willing Mind,"
at Ballycarberry, Caherciveen, Valentia Harbour, 5th August 1857.

"For three days everything proceeded as satisfactorily as could be wished. The paying-out machinery worked perfectly in shallow, as well as in the deepest, water; and, moreover, during sudden transitions from one to the other.

"At four o'clock in the morning of the 10th, the depth of water began to increase rapidly, from 550 to 1,750 fathoms in a distance of eight miles. Up to this time, a 7-cwt. stress sufficed to keep the rate of paying out near enough to that of the ship; but, as the water deepened, the proportionate speed of the cable advanced, and it was necessary to augment

---

* A branch cable was also eventually laid from Ballycarberry (on opposite side of river Caher) to Knightstown, Valentia Island, the distance being about $1\frac{1}{2}$ miles.

the holding-back pressure by degrees, until, in the depth of 1,700 fathoms the indicator shewed a strain of 15 cwt., while the cable and ship were running 5½ and 5 knots respectively."

Mr Bright (as engineer-in-chief) wrote afterwards :—

"At noon on the 10th we had paid out 255 miles of cable, the vessel having made 214 miles from the shore. From this period, having reached 2,000 fathoms of water, it was necessary to increase the strain to a ton, by which the rate of the cable's egress was maintained at the required proportion (to yield the required slack) to that of the ship's speed. At six o'clock in the evening some difficulty arose through the cable getting out of the sheaves of the paying-out machine, owing to the tar and pitch* hardening in the groove† and a splice of large dimensions passing over them. This was rectified by fixing additional guards (or scrapers), and softening the tar with oil. It was necessary to bring up the ship, holding the cable by stoppers until it was again properly disposed round the pulleys. This event was of some importance, as shewing that it is possible to 'lay to' in deep water without continuing to pay out the cable, a point upon which doubts have frequently been expressed."‡

The laying proceeded with regularity till daylight of the 11th. Then, during the temporary absence of Mr Bright, the brakes were put down by the mechanic in charge at the wrong moment. There was a heavy head sea at the time, causing the ship to pitch very much, and a sudden scud broke the cable 20 fathoms below the surface. The depth where the accident occurred was 2,050 fathoms. The distance made good was 274 N.M., and the cable expended 334 miles.

Only 916 miles of cable now remained on board the "Niagara," making, with 1,250 miles in the "Agamemnon," a total of 2,166 miles. This not being considered sufficient for the distance to be covered, the expedition returned to Plymouth. The cable was taken ashore at Keyham,§ well

---

* This is, to a great extent, explained by the fact that the inner serving compound oozed out of the hemp through the stranded armour, under pressure during paying out. It was the hottest time of the year, it must be remembered, and the cable had not been cooled in water at all. This compound consisted of an undue allowance of pitch (in proportion to the quantity of tar), with the result that, being somewhat heavy and sticky, it gave much trouble both in the tank and on the drum during paying-out operations. It was responsible, indeed, for many an anxious moment.

† The unequal accumulation of compound on the grooves of a part of some of the sheaves, in an irregular fashion—causing them practically to present varying diameters—had, moreover, the effect of introducing continually unequal rates for the cable at different parts of the apparatus ; and this alone would be enough to result in the cable jumping off the sheaves.

‡ Mr Charles Bright's Report to Directors, based on Engineer's log.

§ At that time forming a part of the present Devonport Dockyard, near Plymouth.

tarred, and left dry for fear of rusting the iron sheathing wires, which, for reasons already mentioned, were ungalvanised. The faulty portions were cut out, and a new length of 700 N.M. manufactured by Messrs Glass, Elliot, and Co., which, with 39 miles recovered of the 1857 cable, brought up the total amount of available cable to 2,905 N.M.

The paying-out machinery was entirely altered by Mr Charles Bright. The four wheels over and under which the cable passed in the form of a figure eight, as shewn in Fig. 17 (p. 37), were dismounted and replaced by a brake gear, constructed to the orders of Mr Bright on lines agreed on with a committee called together by him on behalf of the Atlantic Company.* Figs. 18, 19, 20, 21 (Plate V.), shew the details of the machinery which was eventually mounted on board both ships in substitution for the previous gear. Two drums of large diameter, A and B, each having four deep grooves, were placed tandem fashion, one in front of the other.† Two smooth brake wheels, placed side by side, were "keyed" on to each drum-shaft and partially surrounded with wooden blocks, which were held in place by strong bands of sheet-iron.

One end of each band was secured to the framework, and the other connected up to a bent lever N. The lever arms, to which weights could be attached, were raised or lowered as required, to regulate the pressure of the blocks on the brake wheels, and to control the speed of revolution. The two pairs of brake wheels were coupled together by large spur wheels on the outer ends of their shafting, both of which geared into an intermediate pinion wheel E. The cable coming from the tanks and passing under a lightly weighted jockey,‡ was guided by a grooved pulley or V-sheave L, along the tops of both drums in the outer groove, then three times round the two drums, passing finally along· the top of both, in the inner groove, over a pulley F, and thence to the dynamometer. By this arrangement the cable only touched the outer half circumference of each drum. The dynamometer consisted of a grooved pulley G and crosshead O sliding between two upright guides on the framework. From the cross-head O a rod was suspended, to which weights could be attached, and which

---

* This committee was constituted as follows :— Thomas Lloyd, Chief of Steam Department, H.M. Navy ; Joshua Field, of Messrs Maudsley and Field; and John Penn, of Greenwich. Mr Bright also conferred with Mr Appold and Mr Amos (of Easton and Amos), as well as with Mr Everett and Mr Clifford, on the subject.

† The V-sheaves of the 1857 machine were replaced by grooved drums, as giving more control over the cable, thereby rendering them especially useful in an emergency. The same principle had previously been turned to account in Bright's picking-up apparatus of 1857, which was now practically modified to the requirements of a paying-out gear.

‡ The cable here was led through a guide (with a jockey-weight as a compressor). This arrangement, whilst leading the line on to the drums, at the same time checked it slightly. It is shewn at J in the figure.

terminated in a piston, working up and down in a cylinder filled with oil to lessen the jar of the descending weights under sudden variations of strain. From the top of the pulley F, the cable passed under the movable sheave of the dynamometer, then over a second sheave fixed at the same height as F, and so into the sea. Vertical scales, graduated for the different weights in use, were set up on the dynamometer frame, a pointer attached to the moving crosshead, or carriage, indicating by its rise and fall the varying tension of the cable. This species of dynamometer was designed by Mr Charles Bright, and is to be seen on all telegraph ships at the present time.*

Near the base of the dynamometer was placed a hand wheel W, similar to those in use for steering purposes. On the barrel of this wheel was wound one end of a chain, which passed over two pulleys, R and Q, secured to the outer arms of the bent levers N. The assistant stationed at the wheel (see illustration at head of chapter) easily followed the indications of the dynamometer, and, by turning the wheel in the required direction, could regulate the strain by lifting or lowering the weights attached to the bent levers, and so opening or closing the brakes. In Fig. 18, S is a water (or oil) cylinder with piston-rod and weights V regulating the pressure on brakes attached to paying-out drums. T is a water cylinder with piston-rod and small weight attached to brake of sheave guiding cable with one turn from the jockey lever J to the drums.

This friction gear for controlling the speed of egress is usually spoken of as "Appold's brake."† As may be seen from the foregoing description, and still better in detail from Figs. 22 and 23, the above ingenious form

---

* A dynamometrical arrangement of the same principle, but of a somewhat heavier character for the greater strains experienced, is also employed in the operations connected with the recovery of submarine lines. Here, in dragging for a cable, the dynamometer apparatus may be said to take the same part as the float does in fishing—by giving warning of any nibble, so to speak. In the case of cable work, a bite, during grappling, is indicated by a tug on the line to the extent of something like three tons additional strain.

† The late Mr John George Appold had previously invented a brake on this principle (Patent No. 13,586 of 1851), for "regulating and ascertaining the labour performed by manual or other power." This was intended more especially for application to the work done by prisoners on the crank, by means of which the exertion of turning it could be absolutely regulated from outside the wall to meet the varying strengths of prisoners. It was also used later for measuring the power of agricultural and other portable engines. The actual brake apparatus, described in the text above, and shewn in detail, with dynamometric connection, in Figs. 22 and 23, was an improved modification of that embodied in the patents of Mr Charles Bright (Nos. 990 and 1,294 of 1857), adapted for cable operations, and referred to earlier. Fig. 24 shews the brake-drum, in section, running through the tank of water. This modification was mainly due to the late Mr J. C. Amos, an engineer of great ability, of that eminent firm, Messrs Easton and Amos (now Messrs Easton, Anderson, and Goolden), who not only constructed and fitted up all the cable gear for this undertaking, but also took an active part in the arrangements.

[PLATE V.

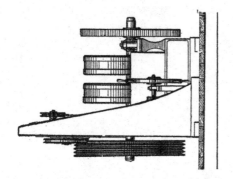

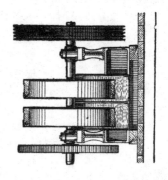

[To face p. 42.

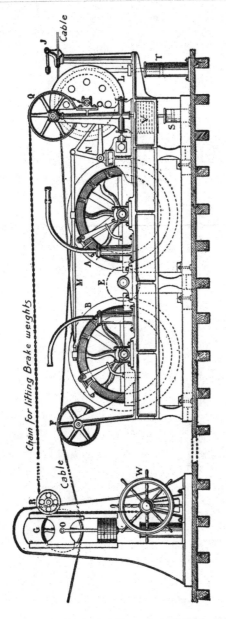

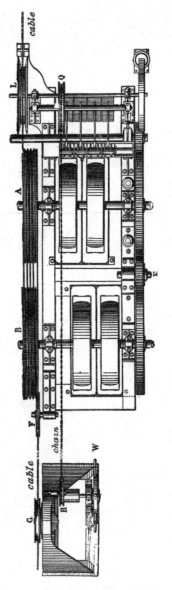

FIGS. 18, 19, 20, 21.—Paying-out Machinery of First Atlantic Cable, 1858.

of brake is self-relieving—indeed, in certain senses, self-adjusting. By its means, though the strain could be *reduced* at a moment's notice, it could not be *increased* by the man at the wheel.

This entire apparatus has since been universally adopted for submarine cable work, with the exception that a single flanged drum fitted with knives (end view of Fig. 22), takes the place of the grooved drums as a rule.*

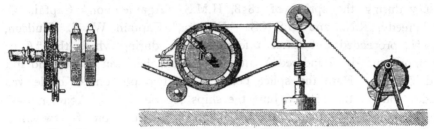

End Elevation.  Side Elevation.

FIGS. 22 and 23.—Details of Appold Friction Brake, Dynamometer, etc., as applied to Cable Work by Bright and Amos.

Mr W. E. Everett, U.S.N., chief engineer of the "Niagara," who had joined Mr Bright's staff, and Mr Clifford, superintended the construction and installation of all the cable machinery.

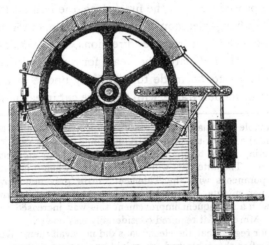

FIG. 24.—Appold Brake running through Tank of Water to Prevent Undue Heating.

In April 1858, Professor William Thomson (afterwards Sir William, and now Lord Kelvin) had designed his first marine galvanometer † (Letters

---

* Besides the fact that the grooves introduce extra friction to the cable, there are, even nowadays, several reasons for preferring this form of gear to any apparatus in which a knife is introduced as a guide to "fleet" the incoming turn.

† A highly sensitive modification of Gauss and Weber's very heavily constructed reflecting telegraph of 1837. In virtue of its extreme sensitiveness, it had the effect of materially reducing the length of time taken by a sufficient force of electricity reaching the further

Patent No. 328), an instrument of extreme delicacy, adapted both for testing purposes—especially aboard ship during cable expeditions—and for receiving signals through long submarine cables.*

**Successful Expedition of 1858.**—After some preliminary trials of paying out and picking up cables in 1,800 fathoms in the Bay of Biscay during the spring of 1858, H.M.S. "Agamemnon," Captain G. W. Preedy, R.N., and U.S.N.S. "Niagara," Captain W. L. Hudson, U.S.N., proceeded out (after a fearful storm, during which the "Agamemnon" nearly foundered†) into mid-ocean, between Newfoundland and Valentia. Here the splice between the two portions of cable was made; and on the 16th of June the ships separated, the "Agamemnon" laying towards Valentia, and the "Niagara" in the direction of Newfoundland. Whilst paying out, the cable parted three times in succession, and each time the operation had to be commenced over again, 540 miles of cable being lost in this way. After putting into Queenstown for supplies, the expedition sailed again for mid-ocean, where once more work was begun in the same way. To effect the splice aboard the "Agamemnon" between the cable ends on the two ships, in the middle of the Atlantic, the following was the course of procedure. In the first place, the end on the "Niagara" was passed to the "Agamemnon." Owing to the outer wire covering of each cable being laid in reverse directions, the ordinary splice was impossible. A special, and naturally weak, form was involved; and provision had accordingly to be made to avoid any undue strain coming on it. Two halves of a wooden frame, with a groove cut in each, were employed

---

end, such as was capable of actuating the indicating apparatus. This was the forerunner of what we now term the mirror-speaking instrument, and may be said to have been the means of first rendering ocean telegraphy a *fait accompli* from an electrical and commercial point of view. The latter fact will be appreciated, when it is stated that the best instrument, contemporaneous with the Thomson mirror galvanometer, could scarcely receive two words per minute, where the working rate of the "mirror" was ten to twelve words, and with a subsequent improvement this was increased to a capability of twenty per minute. Moreover, it required considerably less power.

It was a matter for regret that the electricians did not avail themselves of this beautiful instrument until after the cable had been laid some time, and all other signalling apparatus had failed, following on a few weeks' use. The *astatic* reflecting galvanometer was not invented by Professor Thomson till some years later.

\* This was an entirely different form of marine galvanometer to what is at present distinguished by that name, which was brought out by Professor Thomson in 1863—about the same time as the astatic reflecting galvanometer.

† See Frontispiece, upper portion, from an original drawing by Mr Henry Clifford, which was reproduced at the time by the *Illustrated London News*. A life-like description of this event also appeared in *The Times* shortly after the occasion. During this memorable storm the ship rolled to an angle of 45° at times, occupying over ten seconds. The height of some of the waves from crest to hollow was said to be over 40 feet.

for the two spliced ends to lay in, as illustrated in Fig. 25. The union of the ends—by what is sometimes described as 'a "ball splice"—was the first operation. This merely consisted of the wires overlapping each other bent backwards and forwards — the only possible form for two cables with opposite lays. This splice was then laid in the lower half of the frame, the cable on each side resting in the groove. The arrangement of each end in the frame is shewn here. The part forming the loops (served with spun yarn and fitted securely in the groove) tended to prevent any strain coming upon the splice, intermediate between them, by themselves taking all the stress. The two halves of the frame (covered over with iron boiler plate) were then bolted together, and thus formed a solid protec-

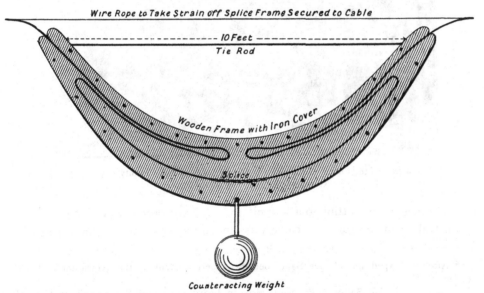

FIG. 25.—Frame for Mid-Ocean Splice of First Atlantic Cable, 1858.

tion to the splice. A wire-rope stay was secured to the cable, and this prevented any undue strain coming on the splice frame. The weight attached to the frame was to prevent the frame and splice turning over under tension. The whole arrangement has been characterised generally as extremely ingenious. The "Agamemnon" started paying out as soon as the splice was lowered into the sea. After paying out a certain length a signal was made to the "Niagara" to do likewise. Then both ships continued to pay out until the splice was supposed to have reached the bottom, when each vessel proceeded on her course at 1 P.M. on 29th July 1858, paying out cable in lat. 52° 9′ N. and long. 32° 27′ W. The "Niagara" steered towards America, with Mr Bright's assistants, Messrs Woodhouse and Everett, in charge respectively of the cable and machinery,

whilst Mr C. V. de Sauty * was in control of the electrical department.†
Mr Cyrus Field was also on board;‡ and H.M.S. "Gorgon," Commander
Joseph Dayman, R.N., escorted the "Niagara." She had an uneventful
voyage, in fine weather, and was met by H.M.S. "Porcupine," Captain
Henry Otter, at the entrance to Trinity Bay, to pilot her to the landing-

Landing of the First Atlantic Cable in the Bay of Bull Arm, Trinity Bay,
Newfoundland, 5th August 1858.

place. Except for cutting out a fault in the wardroom coil on the 2nd
August, all went well, and the cable was landed (as shewn above) in a little

* After the expedition Mr De Sauty became superintendent at the Newfoundland
station.

† Besides Mr De Sauty, there were also on board either the "Niagara" or the
"Agamemnon," as assistant electricians representing Mr Whitehouse and working for
Professor Thomson, Mr J. C. Laws, Mr Whitehouse's chief; Mr E. G. Bartholomew,
representing Professor Thomson on the "Agamemnon"; Mr F. Lambert, who afterwards
became a prominent electrician and member of Messrs Bright and Clark's staff, and later
attached to Messrs Clark, Forde, and Taylor; Mr H. A. C. Saunders; Mr Benjamin Smith,
now an authority on all electrical matters connected with cables, and superintendent
at the Eastern Company's Alexandria station; Mr Richard Collett; and Mr Charles
Gerhardi. Mr Whitehouse himself was not able to go on either expeditions on account
of his health. Mr Samuel Phillips, sen., was closely associated with Mr Whitehouse in
most of his early researches ashore. Mr Saunders has since taken a prominent part in
the extension of submarine telegraphs in the capacity of chief electrician to the Eastern
and Eastern Extension Companies, besides acting as consulting electrician to other of
the allied companies. Mr Gerhardi has for many years represented the Direct Spanish
Company as its manager; and Mr Collett is the secretary of the Brazilian Submarine
Company.

‡ The expedition was additionally accompanied by Mr Nicholas Woods, representing
*The Times* newspaper. Mr Woods wrote a number of articles for *The Times* during the

bay, Bull Arm,* at the head of Trinity Bay, Newfoundland, at 5.15 A.M. on the 5th August, when they "received very strong currents of electricity through the whole cable from the other side of the Atlantic."† The telegraph house at the Newfoundland end (see illustration at foot of this section, p. 56), was some two miles from the beach, and connected to the cable end by a land line, as shewn.

The voyage of the "Agamemnon," laying the cable towards Ireland, piloted by H.M.S. "Valorous," was by no means so prosperous, as she experienced very rough weather and heavy head winds nearly all the way. It was only by the constant watchfulness of Mr Bright and Messrs Canning and Clifford that accident was avoided during the violent pitching of the ship. On one occasion a fault was discovered only a very short distance from the paying-out machine, Professor Thomson reporting that continuity had ceased. "The ship was stopped, and the splice was worked at as men only could who felt that the life and death of the expedition depended upon their rapidity. All their zeal was, however, to no purpose. As a last and desperate resource the cable was stopped altogether, and for a few minutes the ship hung on by the cable. The strain was continually rising above two tons, and it would not hold out much longer. Fortunately it was only for a few minutes; so as soon as the splice was finished, the signal was made to loose the stopper, and the cable passed overboard safely enough."‡ The rough weather continued until the "Agamemnon" got into shoaler water (off Doulus Head) on the 4th August; and on the following morning, the "Agamemnon," ably guided (see Frontispiece, lower portion) by Mr H. A. Moriarty, the navigating master, entered Doulus Bay, Valentia, and anchored at 6 A.M.§ She had been under the escort of H.M.S. "Valorous," Captain W. C. Aldham, R.N.

The total length submerged by both ships was 2,050 N.M., the average slack paid out being about 17 per cent. The end was landed

---

expedition, which were much appreciated by its readers, giving detailed accounts of the expedition as it progressed, somewhat in the form of a narrative. Similarly, Mr John Mullaly was aboard the U.S.S. "Niagara," reporting on behalf of the *New York Herald*. This gentleman afterwards produced a book on the subject, treated from an American standpoint.

* This spot was selected on account of its seclusion from prevailing winds, and owing to its shelter from drifting icebergs.

† Engineer's log, U.S.N.S. "Niagara."

‡ Engineer's log, H.M.S. "Agamemnon."

§ The successful laying of the first Atlantic cable being completed on 5th August 1858, it was exactly one hundred and eleven years after Dr (later Sir William) Watson had astonished the scientific world by sending an electric current through a wire over two miles long, using the earth as the return part of the circuit.

(see illustration below) at Knightstown,* where Mr Bright, Professor
Thomson, and the other greatly tried members of the expedition were
heartily welcomed by the Knight of Kerry, Mr (afterwards Sir Peter)
Fitzgerald.  It was at once taken into the cable-rooms by Mr Whitehouse, the electrician, and attached to a galvanometer, when the first
message was received through the entire length.

Thus was this remarkable achievement carried out by Mr Bright and
his able assistants, the success being attained in the face of the greatest
difficulties, and after such disappointments and failures as might have
daunted many.  All the world was taken by surprise, and applauded not
only the triumph of such determined perseverance, but also the engineering

Landing the Cable from H.M.S. "Agamemnon" by boat at Knightstown, Valentia Island,
on 5th August 1858, thus Completing the First Trans-Atlantic Telegraph.

and nautical skill displayed in this triumph over the elements.  In the
course of a leading article on the successful completion of the work,
*The Times* remarked that, "since the discovery of Columbus, nothing has
been done in any degree comparable to the vast enlargement which has
thus been given to the sphere of human activity."  The Atlantic telegraph
had been justly characterised by Professor Morse, the American electrician,

---

* Partly by the desire of the Knight of Kerry, and partly owing to the local importance of the place, it was decided ultimately that Knightstown was to be the main station.
Communication was, however, temporarily established by the main cable being laid by
boats from the "Agamemnon" to the newly selected terminus.  A branch cable (shore-end
type) was then laid across the harbour, between Ballycarberry and Knightstown.  A
few days later this was underrun out to the buoyed shore end from Ballycarberry of the
year before, where it was cut, and a splice effected between the seaward side and the
heavy shore end.

as "the great feat of the century," and this was re-echoed by all the press on its realisation.

A knighthood honoured Charles Bright, when only twenty-six years old, being the youngest man who had received the distinction for generations past. It was the first title conferred in the telegraph profession, and remained so for many years. Captains G. W. Preedy and W. C. Aldham were both made Companions of the Bath, and other officers received promotion. In America, Mr Cyrus Field and Captain W. L. Hudson, with Messrs Everett and Woodhouse, and Messrs De Sauty and

Facsimile of First Message "cabled" across the Atlantic.

Laws, of the engineering and electrical staffs, received a perfect ovation both in Newfoundland and the States. In England there was much enthusiasm, and a great banquet was given to Sir Charles Bright and his coadjutors in Dublin; but in the United States the excitement was almost without bounds.

Congratulatory messages were exchanged between the Queen of England and the President of the United States. Noisy rejoicings, accompanied by illuminations, torchlight processions (which in New York caused the Town Hall to be set on fire), and salvoes of artillery, were universal throughout America. England was also about to celebrate in turn the realisation of her proud dream of union between two worlds, when the signals became confused.

A new and very serious fault of insulation was found to exist, which appeared to be located about 300 miles from Valentia. For a few more days communication was maintained, with the help of Professor Thomson's new mirror receiving instruments; but on the 1st of September 1858 signals became unintelligible. A few more words were transmitted at intervals up to the 20th October, when a total of 732 messages (some of great length) had been conveyed by this cable.*

The Whitehouse induction coils had only been used at Valentia for a few days, Daniell's cells being afterwards employed at this station, where the mirror instruments were also installed almost as soon as the line was opened. At Newfoundland, on the other hand, electro-magnetic instruments and a relay were adopted, and here the signals were always more difficult to read than at Valentia.

For reasons which remain unexplained, the sending instruments at Newfoundland were not ready for working till the 10th of August, so that communication actually lasted only twenty days, and the line was never opened for public traffic at this end. Nevertheless, the English Government had time to countermand the departure of two regiments about to leave Canada for England, which resulted in a saving of about £50,000.† This circumstance served to demonstrate the advantages to be derived from telegraphic communication between distant lands, and largely helped the starting of other kindred undertakings.

In 1860 attempts were set afoot to repair the cable, but were soon abandoned, owing to the bad condition of the sheathing wires, the shore end only being recovered.‡ The immediate cause of the interruption

---

* Amongst other services accomplished by this cable during the short time it was in operation, may be mentioned a message of peace and congratulation between Her Majesty and the President of the United States, besides similar messages between the corporations of London and New York. Through its instrumentality, intelligence was also conveyed (see illustration on p. 49) of the collision of two steamers of the Cunard Line—the "Europa" and the "Arabia"; and the same information transmitted to the relatives of all passengers, with an assurance of the safety of all on board.

† "The Electric Telegraph" (1867), by E. B. Bright, F.R.A.S., p. 115.

‡ Efforts were made to recover and make good the Valentia shore end, which, as the result of tests by Mr C. Bright, Professor Thomson, and Mr C. F. Varley, had been supposed to be especially faulty. It was underrun from a catamaran raft (see illustration, next page) for a distance of some three miles, but, on being cut, no fault could he found. The idea of repairs had, therefore, to be put aside, and the cable was spliced up again. Again, in 1860, an attempt was made to renew a portion of the cable near the Newfoundland end. Five miles were underrun before the cable got jammed, the bottom being very rocky. Beyond this, the expedition only revealed how enormously the gutta had improved by submersion.

could not, therefore, be precisely ascertained.  Blame has sometimes been attached to the electricians for applying to the cable, for signalling purposes, after it was laid, a potential equivalent to 500 volts, as well as alternating magneto-electric currents, actuated by a potential even five times greater— from induction coils five feet in length!  The maximum working speed gave 105 impulses per minute with alternating currents, implying by impulses a succession of equidistant dots.  This speed would give about 1.85 (five-letter) words per minute.*  With the Thomson galvanometer, however, a speed of quite three words per minute was obtained.

Notwithstanding the final failure of the undertaking, three facts were conclusively demonstrated—first, the possibility of laying 2,050 nauts of cable in ocean depths of two to three miles ; secondly, that by means of an electric current, distinct and regular signals could be transmitted

Underrunning Irish Shore End of First Atlantic Cable, 1860.

and received through an insulated conductor, even when extended beneath the sea, across this vast distance separating Ireland and America ; and thirdly, that the paying-out ships could be hove to in deep water without necessarily parting the cable.

Thus, regarded as an engineering work, it may be said that the great project (about the possibility of which so many eminent engineers and scientists had expressed their absolute disbelief †) was successfully accomplished as soon as the cable was laid ; and it has never been suggested

---

* More than thirty hours were required to transmit President Buchanan's telegram.
† That eminent scientist, Professor G. B. Airy, F.R.S. (Astronomer-Royal), had very forcibly stated that it would be impossible to deposit the cable at so great a depth ; and that in any case it was mathematically out of the question to transmit electrical signals through such a length.

that it actually broke after submergence, the cessation of signals being too gradual for this.*   It is true that there occurred, during the laying, two interruptions of signals between the ships.   These were much commented upon, and generally assigned to faults left by the electricians in the cable already paid out, which shortly afterwards unaccountably corrected themselves at the bottom of the sea!

When the cable passed out of the hands of Charles Bright and his assistant engineers, and from Professor Thomson and the electricians on board, to Mr Whitehouse on shore, it was in excellent order—far better (on account of the low temperature and pressure at the bottom) than before laying.   It had only been subjected to battery currents, derived from about seventy ordinary Daniell's cells, by which all the signals were interchanged between the ships.

Unfortunately for the life of the cable, Mr Whitehouse, the chief electrician—a man highly respected for his scientific attainments—was imbued with a belief that currents of very high intensity (*i.e.*, a high potential) were the best for signalling; and he had enormous induction coils constructed, *five feet long*, excited by a series of very large cells of the potent Smee type, and yielding currents estimated at about 2,000 volts potential.†   Instead of being satisfied with the comparatively mild and innocuous currents from the cells used on board the ships, he brought these induction coils into operation, with the result that faults were developed. The insulation was, in fact, unable to bear the electrical strain, and thus the signals began to fail gradually.   Ordinary battery power was resumed, but too late, for then the number of Daniell's cells had to be continually increased, till they actually numbered 480!   On this, the cable soon ceased to speak, even when using Professor Thomson's mirror galvanometer. ‡

As a proof of the intense and destructive power of the induction coil currents employed by Mr Whitehouse, the following experiment may be

---

* In reference to the above line, the late Mr Robert Sabine said (in his work on "The Electric Telegraph," Crosby Lockwood and Son) :—"At the date of the first Atlantic cable, the mechanical department was far ahead of the electrical.   The cable was successfully laid—mechanically good, but electrically bad."

† Some of the apparatus employed by Mr Whitehouse for working this line may be seen at Messrs Elliott Brothers, the famous, and now classic, instrument makers of St Martin's Lane, London.   Amongst other things, are some large condensers used in connection with the earth-plate carried out into the harbour at each terminus, for establishing efficient earth connection.

‡ An unusually violent lightning storm occurred shortly after the cable had been laid. This has been spoken of as a possible part-cause of the gradual failure of the line ; also a supposed "factory fault," masked by the tar in the hemp.

mentioned. It was carried out by Mr E. B. Bright and Mr C. F. Varley,* at Valentia, shortly after the cable broke down. A very fine prick was made with a needle in a spare length of the core of the cable, sufficiently deep just to touch the conductor, which was then placed in a large earthenware jar filled with water and connected to earth, as was one end of the great induction coil. On the other end of the secondary coil being joined to the conductor of the piece of cable, and the current applied, the interior of the jar was immediately lit up as if it were a lantern; and on withdrawing the specimen of core, the gutta-percha was found to be melted away to about half an inch around the tiny needle puncture.

The consensus of opinion was, that the fearful electric charges applied were the cause of the failure; and, notwithstanding the enormous improvements made in insulation (especially as regards the joints) up to the present time, it is doubtful whether any cable now in existence would long stand a trial with currents generated from such apparatus. Professor Thomson subsequently expressed his belief that if proper methods of handling the cable electrically had been in use from the beginning, its performance would have been lasting—and, in the main, satisfactory. It may be added that Professor Thomson, in 1856, previously † suggested what proved to be the more accurate requirements for working an Atlantic cable, in the course of a somewhat protracted correspondence in the columns of the *Athenæum*. Mr C. F. Varley gave expression to similar opinions, when giving evidence before the Board of Trade Commission of 1860 on the Construction of Submarine Telegraph Cables; as did also Professor D. E. Hughes, F.R.S., who had made numerous experiments on the cable with his type-printing telegraph instrument. Professor Hughes, indeed, expressed the firm conviction, that a current of that intensity from induction coils was quite sufficient to burst through the gutta-percha. No doubt the *primary* cause of the failure of the first Atlantic cable was the fact of the

---

* Mr Varley became electrician to the Atlantic Telegraph Company in succession to Mr Whitehouse, besides being associated with Professor W. Thomson.

† Mr E. O. W. Whitehouse had read a paper at the British Association Meeting of 1856 on the "Atlantic Telegraph." Further, commenting on this in a letter to the *Athenæum* of 1st November of that year, Professor W. Thomson pointed out, in opposition to Mr Whitehouse, that the number of words which could be sent through a long submarine cable varied inversely as the *square* of the length of that cable; and thus that, when the length of that cable was doubled, only one quarter the number of messages per diem could be sent through it.

At another time (see Part III.), in a paper published in the Proceedings of the Royal Society and in the *Philosophical Magazine*, he gave the complete theory, shewing that on all telegraph lines a limit existed to the speed of transmission. This important paper brought to light, in fact, the now famous KR law, which has ever since prevailed.

core, especially the conductor, being insufficiently large,* coupled with its low specific conductivity.†

When scientifically criticising the first Atlantic cable, regard should be had to the fact that it was made at the beginning of 1857—only about five years after the laying of the first Dover-Calais line. The next Atlantic cable was not constructed till a further eight years had passed, during which electrical knowledge and invention, especially in the manufacture and testing of the insulation, had advanced by leaps and bounds—not to mention the exhaustive evidence taken on the subject throughout a whole year (1860) at the afore-mentioned Board of Trade Commission. Any comparison, therefore, of these two enterprises is really out of the question, beyond that of passing in review the very different conditions under which the two were undertaken—the first being without any previous experience of the sort in deep water, the second after the actual laying was known to be a possibility, provided the required engineering skill were at hand.

---

* As already stated, the type of core (besides its protective armour) was settled by contract before Mr Charles Bright became the engineer, and though he strongly urged the adoption of a larger conductor with thicker insulating covering, the change was not considered practicable, involving, as it would, the raising of a considerable amount of further capital.

It may be remarked, however, that Mr Whitehouse's mistaken views of a low capacity being of much greater moment than a low conductor resistance, for signalling purposes, received the entire support of Michael Faraday, the greatest electrical *savant* of the age.

In any case, it was a core of far larger proportions than had ever previously been adopted, with the result that the difficulties of, manufacture were correspondingly increased. Moreover, we must remember that the enormously increased depth tended to bring to light faults which might otherwise have remained dormant.

It is not improbable that weak joints were the real cause of ultimate failure. A bad metallic joint—*i.e.,* a case of the two ends of the conductor not being properly united—was not a very uncommon occurrence in those days, when no method of joint-testing was in vogue beyond a comparison of the leakage from the cable (of, perhaps, several hundred miles) before and after the union had been made. A carelessly effected metallic joint is liable to draw apart on being subjected to a strain during the operation of laying. On the strain being taken off, the ends may join together again temporarily ; but, under such circumstances, the points of contact soon become oxidised, and, thus, all communication gradually ceases. This—together with a gradual percolation of water—is, perhaps, more likely what took place in the case of the first Atlantic line than any other of the explanations for the subsequent cessation of signals. Faults of insulation are scarcely ever of such a character as to account for signals being *entirely* stopped, and it is never suggested that the cable actually broke.

† The purity of copper in those days was so low that an electrical conductivity of 40 per cent. was as much as was ordinarily obtained for telegraphic purposes. However, between the expedition of 1857 and 1858, Professor Thomson drew attention to the importance of this matter, whilst pointing out (for the first time) that all copper had not the same conducting power. He also established a system of testing samples such as had a beneficial effect in the direction of increasing the electrical value of the wire obtained, by the last 400 N.M. being composed of copper of a higher degree of purity.

The first Atlantic cable has been dealt with here in a complete form, not only in the light of an engineering feat and the first successful attempt of its sort, but also as a great commercial undertaking, which it might be well supposed—considering the doubts expressed by various eminent engineers—would be a tremendous task to obtain supporters for. This work paved the way, by demonstrating the possibility of trans-Atlantic telegraphy, an idea almost universally scouted at the time.

Just as the notion of an Atlantic cable had met with great opposition from the incredulous, so also it formed the subject of much enthusiasm at the hands of the amateur engineer. No sooner was the project started than would-be inventors were all agate airing their various—more or less fantastic—notions. One suggested suspending the cable from the bottom (a little below the surface) by buoys or floats at regular intervals, so that ships might telegraph from them *en passant*.* Another gravely proposed to festoon it across to America by balloons. Not a few of these schemes were based on a somewhat prevalent belief of that time—*i.e.*, that no cable would properly sink to the bottom, and that it would in consequence be liable to impede the passage of ships! Then, one enthusiast wished to see parachutes attached to the cable during paying out "to avoid too rapid sinking through the water." A naval officer of eminence (now Admiral J. H. Selwyn, C.B.) devised a huge iron cylinder, around which 2,500 tons

---

* By this plan, the buoys were to be held entirely under water (Fig. 26) by moorings from the bottom, their position being indicated by small "watch-buoys" at the surface, fitted with flag, night light, and telegraph apparatus. The idea was fascinating, but its

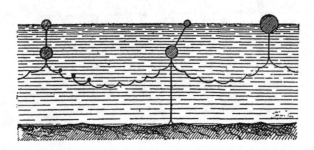

FIG. 26.—Proposal to buoy the Atlantic Cable at Interval Stations, for Ships to communicate from.

realisation impracticable. Putting aside the enormous difficulties which would have attended the laying and working of such a cable, the buoys (as experience has since amply shewn) would soon have got adrift. Moreover, the cable would not take long in chafing away at the points of suspension. "Monsters of the deep" would also probably find a line of this character more irksome to them than one lying continuously at the bottom.

of cable were to be coiled. It was then to be towed across the ocean paying out the cable as it went. Some, again (several times over) absolutely went so far as to patent the conversion of the laying vessel into a huge factory, so as to make the cable on board in a continuous length and submerge it during the process!* These were the kind of "wild notions" (as we now naturally characterise them) which the responsible engineers of that period had to deal with—and deferentially too.

---

* The main object here was to avoid joints, as well as to obviate the risk of damage between the operations of manufacture and laying.

Telegraph House at Trinity Bay, Newfoundland, 1858.

## SECTION 2.—THE RED SEA AND EAST INDIAN TELEGRAPH.

In 1857 Mr Lionel Gisborne, having obtained powers from the Turkish Government to carry a telegraph line across Egypt and lay a cable in the Red Sea, promoted the formation of the "Red Sea and India Telegraph Company," with a view to establishing communication between England and her East Indian possessions. The importance of this line, and the reverses which the Atlantic Telegraph Company at first met with in this same year, weighed with the British Government, which now decided to give its assistance. In 1858 a somewhat similar guarantee* of dividends was granted to the new association as had been (in the form of a subsidy) for the "Atlantic" Company.

The proposed line, 3,043 miles in length, between Suez and Kurrachee, was divided into two parts. The first portion, from Suez to Aden (1,358 N.M.), with intermediate landings at Kósseir and Suakin, was laid in 1859. The different sections of this cable broke down one after the other. They were all laid very taut, the slack in some cases being of less than unit value.† This though the bottom was, in certain parts, extremely uneven; moreover, the type of cable adopted was of a distinctly fragile character for some of its rough resting-places. The Suakin-Aden section was first repaired in 1860; but another interruption occurred a few days afterwards. A portion of the cable picked up was found to be covered with shells and marine growths‡ which, in some instances, had preserved the iron from rust; in many other cases the outer sheathing was completely worn through by the rocks on which the cable had rested.

The second portion of the line, from Aden to Kurrachee (1,685 miles), with intermediate landings at Hallania Island and at Muscat, was laid during the year 1860, at certain points in depths of 2,000 fathoms, the slack working out at 0.1 per cent. over the entire length. Faults developed very quickly in all three sections; and the company, having neither specially

---

* Eventually a subsidy was granted, which cost British taxpayers £36,000 per annum.

† By the agreement of this undertaking, the surplus cable belonged to the contractors: no wonder that faults soon developed which it was impossible to repair! In some places, owing to the tightness and high speed of laying, the elongation of the iron wires had "nipped" the gutta-percha; and, in others, the sheathing was much stretched and broken.

‡ See illustrations at foot of page 190. These figures may be taken also as representing the state in which cables are very generally found after submergence for a certain length of time in those localities which are pervaded by marine growths of various descriptions.

qualified men, nor the necessary materials, for carrying out repairs, was obliged to abandon the line, before any commercial use had been made of it.*

The cable was manufactured and laid by Messrs Newall, the conductor being a strand of seven copper wires equivalent to 180 lbs. per N.M. The insulation consisted of two layers of gutta-percha, alternating with two coatings of Chatterton's compound,† a substance (of high insulating qualities) more plastic than gutta itself, and intended to ensure better adherence between the gutta and the copper, as well as between the layers of gutta-percha themselves. The total weight of dielectric was represented by 212 lbs. per N.M.

The core was encased in tarred hemp and eighteen iron sheathing wires of No. 16 gauge. The outer sheathing of the shore end was composed of nine iron wires of No. 2 gauge. The working speed—during the thirty days' trial‡—between Aden and Hallania, the longest section (718 N.M.), was five words per minute. Messrs Gisborne and Forde were the engineers,§ and Messrs Siemens and Halske the electricians of this enterprise,‖ Mr Werner Siemens being out on the expedition.¶

---

* This was a most unfortunate line in every way. Report has it that a complete message was never got through the entire length, but only through each section separately. It ultimately failed altogether.

† This compound was the result of exhaustive experiments by Mr Willoughby Smith. Coal-tar naphtha had been previously used for adhering the successive coats of gutta-percha, with prejudicial effects owing to being a rapid solvent of that material. This compound, which has been almost invariably employed in all successive cables, is composed of certain proportions of Stockholm tar, resin, and gutta-percha.

‡ This was the first instance of a stipulated length of time for testing the cable after submergence, previous to taking it over from the contractors. This particular period has been almost invariably adhered to for all subsequent submarine telegraph systems, but nowadays the time is employed in applying definite electrical tests rather than in "working" through it.

§ They, however, took no part in the designing, manufacturing, or laying of the cable.

‖ An agreement existed for some time between Messrs R. S. Newall and Co. and Messrs Siemens and Halske, by which the latter acted as electricians and as consulting engineers to the former firm.

¶ For further particulars regarding this undertaking, see Mr F. C. Webb's "Old Cable Stories Re-told" in *The Electrician* of 1885.

SECTION 3.—COMMITTEE OF INQUIRY ON THE CONSTRUCTION OF
SUBMARINE TELEGRAPH CABLES.

Aroused by the successive failures of the "Atlantic" Company and the
"Red Sea and India" Company—the joint losses of which amounted to more
than a million sterling, and to the latter of which a continuous Treasury
guarantee had been given—the Government, in 1859, before undertaking
further responsibility, resolved to thoroughly investigate the entire question
of submarine telegraphy. A Joint-Committee of eight members was
appointed, half of whom were nominated by the Board of Trade, and the
remainder by the Atlantic Telegraph Company. The representatives of
the Board of Trade were Robert Stephenson, Douglas Galton, Charles
Wheatstone, William Fairbairn,* and George Bidder; and on behalf of the
"Atlantic" Company, Edwin Clark, Cromwell F. Varley, Latimer Clark,
and George Saward.† Mr Stephenson died soon after the Committee
commenced its work. Besides the question of construction, they were
instructed to discuss the best methods of "laying and maintaining submarine telegraph cables."‡

This Committee, with Captain Galton, R.E.,§ in the chair, representing
the Government, devoted twenty-two sittings, from the 1st of December
1859 to the 4th of September 1860, to questioning engineers, electricians,
professors, physicists, seamen, and manufacturers, who had taken part
in the various branches of submarine work, and whose knowledge or
experience might throw light on the subject. Investigations were
instituted concerning the structure of all cables previously made or in
course of manufacture, and the quality of the different materials used,
as to special points arising during manufacture and laying, on the
routes taken, electrical testing, and on sending and receiving instruments, speed of signalling, etc. etc. Experiments were also made
under the direction of the Committee to ascertain (1) the effects of
temperature and pressure on the insulating substances employed; (2) the
elongation and breaking strain of copper wires; of iron, steel, and tarred

---

* President of the British Association for that year, and afterwards Sir William
Fairbairn, F.R.S.

† The Right Hon. J. Stuart-Wortley was also a member of the Commission originally, but, owing to illness, only attended the two first meetings.

‡ In forming this Committee, regard was paid to the fact that it would be unsuitable
to have any engineer or electrician thereon who had taken a leading part, practically,
in the various larger undertakings which had already taken place, especially as their
evidence would be of special value, and should be reviewed impartially by others, rather
than by themselves.

§ Then of the War Office, and now Sir Douglas Galton, K.C.B., F.R.S.

hemp, separately and combined. Eminent scientists and engineers, including Professor Wheatstone, Professor William Thomson, Sir Charles Bright, Mr R. S. Newall, Mr R. A. Glass, Mr Wildman Whitehouse, Mr Samuel Canning, Mr C. W. Siemens, Mr Willoughby Smith, Mr C. F. Varley, Mr F. C. Webb, and Mr Latimer Clark,* made known to the Committee the science and practice of cable making and laying; the results of their investigations on the electrical properties of copper, pure and alloyed, and of gutta-percha; the permeability by water of the various insulating substances, and the chemical reasons for their change of condition; the electrical phenomena connected with charging and discharging conductors; the inductive action to which a cable is subjected; methods of testing conductors and of locating faults, as well as concerning other important points.†

The report of the Committee—condensing more such information in a given space than can be found elsewhere—was received in April 1861. It stated finally, in general terms, that, in the opinion of the Committee, based on the evidence adduced, the failures of the submarine cables submitted for investigation were due to causes which might have been avoided had the conditions been sufficiently understood beforehand. Further, the Committee expressed their conviction that submarine telegraphy might be as sure and remunerative in the future as it had been speculative in the past, provided that the specification, manufacture, laying, and maintenance of the cable were proceeded with on the lines laid down in their report.

Events have entirely realised their presages, and the fundamental principles indicated by the Committee still hold good in submarine telegraphy.

---

* This gentleman supplemented his evidence in chief by a report on the electrical phenomena in connection with the problem of submarine telegraphy. This (along with other reports from other experts) forms one of the appendices of the Bluebook, and was evidently the result of much work and experiment. As in the case of Faraday's "Researches," it may be said to have formed the basis of much that has been written since by way of enlightenment on the conditions met with in electric telegraphy, submarine and otherwise, and has no doubt led to further discovery. The principles which Mr Clark here enunciated are, to a great extent, clearly set forth in that now almost classic treatise on "Electrical Measurement" (E. and F. N. Spon), which he produced in 1867.

† Messrs Gisborne and Forde, associated with Mr C. W. Siemens, sent in some extremely valuable tables, which were the result of exhaustive experiments on the breaking strains of the various materials composing different classes of iron-sheathed cables. These experiments had for their object the determination of the relative strength, etc., of different forms of outer covering for submarine telegraphs, and were, in fact, conducted with a view to arriving at the best form. Mr H. C. Forde's evidence in this connection was especially to the point, as was also that of Mr Siemens.

This finding of the Committee was published by order of the Government, as also the reports of the meetings and descriptions of the experiments, together with papers and drawings sent in by the experts who were consulted, the whole being included in the form of a Parliamentary Bluebook, a memorial of honour to the Committee, whose work will ever be considered a model of scientific investigation.*

## SECTION 4.—THE FORMULATION OF ELECTRICAL STANDARDS AND UNITS.

In this same year (1861) a very important paper was given by Sir Charles Bright and Mr Latimer Clark, on electrical units and measurements, to the British Association for the Advancement of Science. Upon this being read, Professor William Thomson obtained the appointment of a committee with the object of determining a rational system of electrical units, and to construct an equivalent standard of measurement. The members were—Professors Williamson, Wheatstone, Thomson, Miller, Clerk Maxwell, and G. C. Foster, who were joined by Sir Charles Bright, Dr J. P. Joule, Dr A. Matthiessen, Messrs Balfour Stewart, David Forbes, C. W. Siemens, C. F. Varley, Latimer Clark, Charles Hockin, and Fleeming Jenkin.†

The work of this committee lasted eight years, and was not entirely finished until the close of the year 1869. As the result of its labours, we have the system of electro-magnetic absolute units, from which are derived the ohm, ampère, farad, volt, and coulomb, being a system of nomenclature suggested by Messrs Bright and Clark in their paper of 1861. This system was confirmed by an International Congress, in 1881, at which every civilised nation was represented. The creation of these standards has substituted perfectly definite and identical quantities for the many arbitrary units formerly in general use among electricians, has introduced precise definitions in all questions of electrical measurements, and has, indeed, rendered immense service both to the electrical industry and to science generally.

---

* Referring to this Bluebook, Sir Charles Bright, in his Presidential Address (of 1887) to the Institution of Electrical Engineers, remarked:—"I consider it to be the most valuable collection of facts, warnings, and evidence which has ever been compiled concerning submarine cables, and that no telegraph engineer or electrician should be without it, or a study of it. It is like the boards on ice marked 'Dangerous' as a caution to skaters. The succinct report of the Committee at the beginning of the book, which is, of course, based on the evidence obtained, should especially commend itself."

† See "Reports of Electrical Standards." Edited by Fleeming Jenkin, F.R.S.

SECTION 5.—OTHER ENTERPRISES OF THE PERIOD.

**Malta to Alexandria Line.**—This cable was manufactured in 1859, on the specification of Sir Charles Bright, supplied, together with a report,* to the British Government, who then intended it to be laid between Falmouth and Gibraltar, where depths of 2,500 fathoms are to be met with. Subsequently it was decided to put it down between Rangoon and Singapore, in depths not exceeding 100 fathoms; but the war with China having come to an end before the cable was ready, its destination changed a third time, and it finally came into use as a means of communication with Egypt, one of the stages on the road to India. The core, calculated to transmit at least five words per minute through a 1,200-mile circuit, was composed of seven copper wires stranded together, weighing 400 lbs. per N.M., covered with three coatings of gutta-percha, also weighing 400 lbs.† per N.M., including Chatterton's compound between

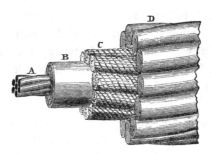

FIG. 27.—Malta-Alexandria Main Cable.      FIG. 28.—Longitudinal View of Shore End.

the layers. As only the core was finished when the destination of the cable was first changed, the heavy covered steel wires, with which it was to be sheathed, were replaced by eighteen iron wires of No. 11 gauge. The cable (Figs. 27, 28) thus weighed 2 tons per N.M.

Special care had been given to the manufacture of this line.‡ The

---

* This resulted from Mr Robert Stephenson and Sir C. Bright being consulted by the President of the Board of Trade (the late Sir Stafford Northcote, Bart., M.P., afterwards Lord Iddesleigh) on the subject, who eventually requested the latter to send in a detailed report, estimate, and specification to the Treasury. See Parliamentary Bluebook respecting "The Establishment of Telegraphic Communication in the Mediterranean and with India," 1859.

† This was the first instance in which the relative dimensions of the conductor and insulator were actually specified in weights (instead of merely in gauges) as a check on the qualities of materials employed.

‡ For full particulars see "The Malta-Alexandria Telegraph Cable," by Henry Charles Forde, M.Inst.C.E., *Minutes of Proceedings of the Institution of Civil Engineers*, vol. xxi.

Government placed the contract for laying in the same hands as its manufacture had originally been entrusted to, namely, with Messrs Glass, Elliot, and Co.*   Mr H. C. Forde, M.Inst.C.E., was the engineer,† and Messrs Siemens, Halske, and Co. were the electricians connected with the enterprise. This firm devised a rational method of testing during manufacture and laying.‡ They constructed for the purpose extremely delicate instruments, and finally brought out a standard of resistance—the column of mercury unit—which came into general use among electricians until the actual appearance of the British Association's new unit, the "ohm." The coils of core were tested after twenty-four immersions in water maintained at a temperature of 24° C. (75° F.), and were further submitted to alternate vacuum and pressure trials (the air being exhausted from the tank before the water was turned in§) so as to bring to light any air-bubbles in the dielectric. The application of a pressure test came about thus-wise : One day the gutta-percha was accidentally pierced, and it was then noticed that the water under pressure had forced a passage along the conductor between it and the insulation.

It was in the first instance to remedy this that Mr Willoughby Smith proposed to coat the central wire of the copper conductor with what is now generally known as Chatterton's compound,‖ before laying up the strand, so as to fill up all interstices, and thus, therefore, to

---

* Mr Samuel Canning and Mr Henry Clifford acted in an engineering capacity for the contractors, and Mr C. V. de Sauty as their electrician, Mr Whitehouse being in a consulting position.

† Messrs Gisborne and Forde were the engineers for this work; but Mr Lionel Gisborne died before the expedition took place, leaving Mr Forde to act alone in the position of engineer, with control of the expedition. Quite recently we have also had occasion to deplore the latter's death.

‡ See "The Electrical Tests Employed during the Construction of the Malta-Alexandria Telegraph," by Charles William Siemens, *Mins. Inst. C.E.*, vol. xxi.

§ The water pressure employed was equivalent to 600 lbs. per square inch, and was applied by apparatus specially designed by Mr W. Reid. This consisted of a large iron cylinder, from which the air was first exhausted by means of an air pump. The cylinder was then filled with water, and a hydraulic press put in action capable of producing the above pressure. After some hours the air-bubbles, if any existed, pierced minute holes in the gutta-percha ; these immediately enlarged under the effect of the negative current which then became discernible on the galvanometer.

‖ This compound (fully described in Part II.), though always referred to as Chatterton's compound, appears in reality to have originally emanated from the late Mr Willoughby Smith in 1858—Patent No. 1,811 of that year, "for insulating electric telegraph wires." The late Mr John Chatterton (at one time manager of the Gutta-percha Company's works, in succession to Mr Statham) took out a patent in the following year for "a method of applying Willoughby Smith's or other compounds for similar purposes." No doubt this was to place the special application of the patent on a firmer footing, these two gentlemen working together in the same interests. The manner in which this slight misnomer has occurred is only a repetition of many other such instances.

avoid the occurrence of air-bubbles, with the consequent presence of minute holes in the insulation when under the pressure of the sea. The core was made by the Gutta-percha Company, and the serving and sheathing by Messrs Glass, Elliot, and Co. As made, the cable was coiled in brick and cement tanks, and covered with water. Unfortunately, however, the foundations gave way. This defect in the construction of the tanks prevented the cable being always entirely covered with water, after manufacture. It was, in fact, exposed to alternations of wet and dry conditions, which would especially favour the rapid oxidation of the iron sheathing wires.* Heat was evolved from the enormous mass of metal, confined in a comparatively small space, and the temperature rose to 85° F. Water had to be continuously poured on the cable, and pumped out again. After the seriousness of spontaneous combustion had been thus experienced, iron tanks were ultimately fitted to the laying vessels.† This cable was successfully laid during the year 1861, by Mr Canning, assisted by Mr Clifford, acting for Messrs Glass, Elliot, and Co., in three sections, i.e., Malta-Tripoli, Tripoli-Benghazi, Benghazi-Alexandria, the total length being 1,331 miles. Notwithstanding the dimensions of the core adopted through each section separately, the working speed, with the Morse system,‡ was up to ten words per minute at the outside, and with the entire length in circuit, only three (five-letter) words could be obtained.§ On all three sections frequent interruptions occurred,|| but the depth where they occurred being always under 100 fathoms¶ repairs were always easily accomplished. The above cable proved a perfect and permanent success.

---

* No doubt this trouble was largely due to the iron wires having no outer covering, the result being that they were liable to be entirely exposed to the air, and when wet under such conditions would rapidly oxidise. The hemp (or jute) and asphalte casing introduced later on by Messrs Bright and Clark met this difficulty to a great extent.

† This example was subsequently followed at all the *factories*, as well as on other ships.

‡ The instrument used was an improved form of Morse recorder, with polarised relay, due to Messrs Siemens and Halske. A system of automatic translation was also applied at the intermediate stations.

§ It should be stated, however, that it was never intended to work this line right through. Though the Thomson mirror instrument was available, clerks to work it efficiently were scarce at that time. Moreover, the absence of record was regarded in those days as a serious objection to the latter system.

|| This cable was replaced in 1868 by a direct line from Malta to Alexandria. Sir Charles Bright, M.P., acted both as engineer and electrician. Mr Samuel Canning and Mr Willoughby Smith represented the Telegraph Construction and Maintenance Company (the contractors) in an engineering and electrical capacity respectively.

¶ Owing to this cable having been made originally for essentially shallow water, special care had to be taken to lay it quite close to the northern coast of Africa, adhering, in fact, to the 100-fathom line quite irrespective of the extra length entailed thereby. The average depth for the two sections along the coast was some 50 fathoms or so, and of that between Malta and Tripoli some 200, with a maximum of 420 fathoms.

The first use of resistance coils for testing submarine cables was with this undertaking. They had been employed before for testing *subterranean* wires\*—first of all by the patentees (for detecting faults) in November 1851.

**The Balearic Islands Cables** † (*Barcelona-Minorca, Majorca-Ivica, ana San Antonio, Spain*).—For a number of years from 1855, as will be gathered from this history, the deep waters of the Mediterranean had proved a sort of *bête noir* to cable-layers, commencing with Mr Brett's unsuccessful efforts between Sardinia and Bona in Algeria; continued by three failures in 1858 and 1859 to connect Candia with Alexandria; followed by two successive mishaps in 1860 and 1861 to lay a cable between Algiers and Toulon; and culminating in the untoward essays, thrice repeated, to lay a short cable of 113 miles between Oran and Cartagena in 1864.

In 1860, however, Sir Charles Bright broke the spell for a time, by laying, with success, an important series of four cables for the Spanish Government‡—viz., between Barcelona and Port Mahon, Minorca, 180 miles; Minorca to Majorca, 35 miles; Majorca to Ivica, 74 miles; and Ivica to San Antonio, Spain, 76 miles—in all 365 nauts.

These cables were submerged in great depths, that between Barcelona and Port Mahon being in 1,400 fathoms.

They were manufactured by Mr W. T. Henley. The sections between the three islands contained two conductors each, protected by eighteen outer wires, and weighed 1 ton 18 cwt. to the N.M.; and the two to the mainland were single wire cables, cased with sixteen wires, and weighing a ton and a quarter per N.M.

This work was carried out with great expedition. On the 29th August 1860, Sir C. Bright laid the Minorca to Majorca section, completing the shore end and connections next day. The 31st saw the shore end and connections made at the opposite end of the island; and the following day the cable was laid between Majorca and Ivica, the landing portion being carried out on the 2nd September. Rough weather delayed operations for two days; but on the 5th, Ivica island was put into telegraphic communi-

---

\* See specification of patent granted to C. T. and E. B. Bright, No. 14,331 of 1852. The part referring to this mode of testing with known resistances is published in the "Submarine Telegraph Committee's Evidence," p. 53, with drawings; also the "Museum of Science and Arts," 1854, and "The Electric Telegraph," by Lardner and Bright, p. 76, edit. 1867.

† From "The Life-Story of Sir Charles Tilston Bright, C.E., M.P.," by Edward Brailsford Bright, C.E., and Charles Bright, C.E., F.R.S.E. (Archibald Constable and Co., London).

‡ These cables were for the purpose of connecting up Spain with her island possessions —the Balearic group—from Barcelona to San Antonio at different parts of the mainland.

cation with the Spanish mainland at Javea Bay, alongside Cape San Antonio. The remaining section to be laid was that between Barcelona and Minorca, a distance of about 150 N.M. This gave some trouble, owing to broken wires in deep water, but was before long successfully carried out.

**Toulon to Algiers and Algiers to Port Vendres Cable.**—As the first concession, granted by the French Government in 1859, to establish a direct cable between France and Algeria, remained in abeyance for want of funds, Messrs Glass, Elliot, and Co., in 1860, engaged to undertake the work, which was to be completed by the 31st August of the same year. The cable—manufactured about the same time as the one laid between Falmouth and Gibraltar, and under similar conditions—was composed as follows:—1. A strand of seven copper wires, weighing 107 lbs. per N.M.; 2. Four coverings of gutta-percha, laid on alternately with four coatings of Chatterton's compound, the insulation weighing 197 lbs. per N.M., being, in fact, covered to a diameter of No. 0 B.W.G.; 3. A serving of tarred yarn; 4. An outer sheathing of ten iron wires, of No. 14 B.W.G. in diameter, each one separately covered with tarred yarn to prevent corrosion.* The cable weighed 2 tons per N.M., and the breaking strain was 6 tons. For the shallow-water portions, the sheathing was composed of eighteen bare iron wires close together, and of sufficient thickness to give the cable a weight of 2 tons 10 cwt. per N.M., for depths between 110 and 44 fathoms. It is needlesss to say the shore end was much heavier. The core, during manufacture, was tested under water at a uniform temperature of 75° F.

The cable was shipped on board the "William Cory,"† which arrived at Algiers on the 9th of September 1860. The landing-places chosen were Salpêtrière Bay near Algiers, and Sablettes Bight in the neighbourhood of Toulon. The cable would also cross the shallower depths off the island of Minorca, which is almost in a direct line between these two points.

Starting from the African coast, the laying at first proceeded smoothly, but the next day a kink was paid out, and loss of communication with the

---

\* This arrangement of enveloping each iron wire in a separate protective serving, such as tended to advantageously reduce the specific gravity for recovery purposes, had already been covered in a patent (No. 2950 of 1856), taken out by Messrs John and Edwin Wright. This patent was really intended merely for ordinary ropes, the novelty consisting in the introduction of iron with a view to extra strength. The above species of cable was afterwards adopted for the Atlantic cables of 1865, 1866, and 1869, as well as in a few subsequent lines, since which it has been entirely abandoned, being found to involve decay to a much greater extent than where each wire abuts immediately against its neighbour.

† Commonly known in those days as the "Dirty Billy."

shore speedily ensued. Although the depth was on the average between 1,300 and 1,400 fathoms, about three miles of cable were successfully picked up, and the defective portion cut out.* The operation of laying was then carried out without incident till—when about 45 miles from Cape Sicié, owing to rough weather—the cable parted in 1,300 fathoms, a depth such as under existing circumstances left no hope of recovery.

The Balearic Islands having, as already described, been joined up to Spain by two submarine lines, advantage was taken of the fact to obtain communication with Algeria by landing the cable at Minorca. It was grappled in a depth of 70 fathoms, and both the ends taken ashore near Mahon, using a deep-sea type of cable for this purpose, to be replaced in a few months' time. In this way messages were exchanged with Algeria *via* Barcelona and Mahon, whilst awaiting the completion of the direct line.

The "William Cory" returned to Toulon with a new stock of cable, on the 14th September 1860, and started from Sablettes to lay the Toulon-Mahon section, which was to be joined up with the Algiers cable off Minorca. Unfortunately, when about 90 miles from the land, she was run into by the "Gomer," a French Government vessel, acting as escort. The cable had to be cut and buoyed, and the injured vessel put into Toulon for repairs.

The "William Cory" was not ready to resume work till 13th January 1861. The buoy was easily found, and the rope picked up, but the chain parted on the drum, and the cable sank to the bottom again. As grappling seemed impracticable in 1,200 or 1,300 fathoms, the enterprise, as already stated, was abandoned.

The "Brunswick," another ship belonging to the same company, which had come to lay a cable between Toulon and Corsica, was ordered to attempt the recovery of the two sections now lying useless between Port Mahon and Toulon. She was only able to pick up eight miles from Sablettes, and twenty miles out from Mahon, the cable on both occasions breaking in deep water.

A new contract was entered into with Messrs Glass, Elliot, and Co., for completing the direct line, to which the French Government attached much importance. This time Port Vendres was settled on as the starting-point, which, being nearer to Minorca than Toulon, permitted of a route being taken in somewhat shallower water. The work, commenced by the "Brunswick" on the 31st August 1861, was entirely finished by the 19th of

---

* At this time, and during the subsequent expedition, Mr (afterwards Sir Samuel) Canning was acting as engineer-in-chief to Messrs Glass, Elliot, and Co.

the following September.   A fault which declared itself during the laying
necessitated about fifteen miles of cable being picked up ; other faults
appeared, but the sole condition exacted before taking over the cable being
that a twenty-word message should be sent both ways after fifteen days'
interval from laying, they were allowed to pass.

The length of cable laid between Port Vendres and Minorca was 226
miles, and that of the cable laid the year before between Minorca and
Algiers 230 miles.   Signals were sent through the whole length with the
help of Siemens' polarised relay, setting into action a local battery for
influencing the Morse recording coils.   The working speed was eight words
per minute, and on "trials" thirteen words.   This cable broke down, between
Minorca and Algiers, on the 25th of September 1862, on the occasion of a
violent storm.   A few attempts with insufficient appliances were made to
raise it, but were soon given up.

**Cable from Oran to Cartagena.**—Nothing daunted by the failures
of the cables to Corsica and Algeria, the French Government persisted

FIG. 29.

in the idea of telegraphically connecting, by a submarine line, their great
African colony with the mother country.   In order to lessen the expense,
the advantage of direct and independent communication was given up,
and it was decided to land the cable on the southern coast of Spain.   After
soundings had been taken during July 1863, Oran in Algeria, and Cartagena
in Spain, were chosen as the terminal points of the new cable, the distance
between them being only about 113 N.M.   Outwards from both shores
the depths increased rapidly to about 1,000 fathoms, after which the
soundings kept tolerably regular, the greatest depth met with on the
proposed track being about 1,400 fathoms.   At this depth there was
found to be an excellent bottom of soft ooze.   The French Government
entered into negotiations with Messrs Siemens and Halske to manufacture*
and lay between the places mentioned a cable of a type made and exhibited by them in London in 1862.   The core (Fig. 29) was a strand
of three copper wires, No. 16 gauge, covered with two layers of gutta-

---

* This was the first cable made at the newly-established factory of Messrs Siemens
Brothers on behalf of Messrs Siemens and Halske.

percha to No. 2 gauge. The deep-sea portion of this cable was constituted by the core being surrounded with two coverings of stout hemp line wound in opposite directions, with a very long lay, almost parallel to the axis of the cable; round the whole, flexible strips of copper were laid, the spirals of which overlapped like the scales of a fish. For this metal taping, copper containing phosphorus was used, which has the advantage of not being prejudicially attacked by sea-water. The total diameter of the cable was only 0.43. For the shallow-water portions, the hempen covering was increased in thickness, so as to bring the diameter up to a little over half an inch. In the case of the shore ends, the copper ribbon was replaced by the ordinary iron wire sheathing.

It should be mentioned here that (although Messrs Siemens were probably unaware of it at the time) this method of protecting submarine

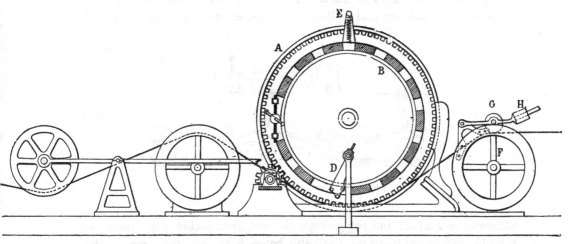

FIG. 30.—Paying-out Apparatus for Oran-Cartagena Cable, 1864.

cables by wrapping them spirally with overlapping strips or ribands of metal was invented and patented by Messrs E. B. and C. T. Bright eleven years before—No. 14,331, 21st October 1852—with full descriptions and drawing, coupled with a claim for the use of such spiral overlapping metallic ribands, and further improved by the addition of a washer to give flexibility.*

---

* This fact was referred to subsequently at page 282 of M. Wünschendorff's "Traité de Télégraphie Sous-Marine," on which this work is partly based. M. Wünschendorff gave credit to the Messrs Bright as the actual inventors of this system, which has been, and is, so extensively used, though Messrs Siemens were the first to apply it, in the above case, as the sole protective armour, forming, with the hemp bedding, a light cable for deep-water recovery purposes.

It may also be mentioned that subsequently Mr Henry Clifford adapted a form of metal riband (or tape) outside the core of cables to the special purpose of protecting the

The cable was shipped on board the "Dix-Décembre" * a small steamship, bought by the French Government the year before, and fitted up in England for submarine cable work. Her deck machinery (Fig. 30) consisted of a large drum A keyed on to a shaft which also carried two brake wheels B; each of these wheels was surrounded by a series of wooden blocks, kept in place by encircling bands of iron, which could be tightened or eased up by hand with a double-threaded screw C. These bands were supported below by an arm pivoted at one end to a strong block, and hung at the top by a powerful spring E which could be made to lift the blocks off the drum, allowing it to revolve freely when the brake straps were slackened.

Forward by the drum was the V-wheel F, surmounted by a jockey wheel which was loaded with a detachable weight H. The cable coming out of the tanks in the fore part of the vessel was guided, where necessary, by sheaves and rollers, and passed through the space left

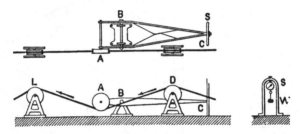

FIG. 31.—Salter's Balance Dynamometer.

between the V-wheel and the jockey; the weighted jockey wheel acted as a brake, and straightened out the cable before it reached the drum, so that the turns could not ride one on the top of the other. A knife arrangement, fitted flush to the drum on the fore side, continually forced the turns towards the centre of the drum, thus making room for the new turn coming on.

Abaft the drum came the dynamometer (Fig. 31), of a special design, due to Sir Charles Bright, and fitted for the first time, in 1867-68, by Mr F. C. Webb, M.Inst.C.E., on board the "Narva." It consisted of a flanged

---

same from the ravages of the teredo and other marine insects in their attacks on the jute or hemp serving. This precaution has met with complete success, and is universally resorted to for cables in waters frequented by boring insects of any description, the first example being the Penang-Malacca cable laid in 1879.

 * This vessel, still employed in repairing cables in the Channel, and along the western seaboard of France, was renamed the "Ampère" in 1870.

frilling A, turning on an axis supported by 1 inch of a braced lever (with latticed framework) pivoted at B. The other end of the lever was connected to a Salter's balance, suspended from a frame W. The length of the lever arm B C was five times that of A B, and the apparatus was adjusted so as to remain horizontal when in normal equilibrium.

At equal heights and distances, before and abaft, two fixed wheels with flanges were fitted, D and L; the cable passed over each of these, and under the dynamometer wheel A, between them. The wheels D and L were so placed that the cable made an angle with the horizontal of about 14° 31′, the sine of which being $\frac{1}{4}$ when the dynamometer was in its normal position (Fig. 32). Resolving by parallelogram of forces, the tensions OM and ON, equal and oppositely inclined at the point O, the length of the resultant was found as follows :—

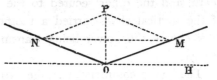

FIG. 32.—Dynamometric Principle.

$$\frac{OP}{2} = OM \times \sin. MOH = OM \times \tfrac{1}{4}OP = \tfrac{1}{2}OM.$$

Therefore, the vertical force tending to lift the dynamometer wheel was always exactly half the strain on the cable. The length of the lever arms A B and B C being as 1 to 5, the weight shewn on the Salter's scale was only a fifth of the upward thrust, which it balanced at the other end of the lever, and consequently only a tenth of the strain on the cable.*

The sheave L (Fig. 31) guided the cable over the ship's stern, and was furnished (as nowadays) with two check pieces, to prevent injury by the sharp edges of the sheave when the cable left the ship in a direction not parallel to the fore-and-aft line.

Another similar sheave was placed on the same platform, immediately above the stern of the vessel, and was used for picking-up purposes. The cable—carried aft over portable sheaves suspended from the rigging—took three turns round the drums, and then went to one or other of the tanks, following the same leads, but in reverse order, as the cable when paying out. By means of a connecting clutch, a pinion wheel, driven by a small deck steam-engine, could be geared into a large toothed wheel keyed on the same shaft as the drum, the paying-out machinery being thus transformed into a kind of huge steam winch. The cable was hauled away from the drum to the tanks by manual labour. Whilst paying out, the steam-

---

* The principle of the dynamometer used at the present time in cable operations is the same as here stated, and the form of it is, practically speaking, also the same.

engine was disconnected, and used to work a pump for the purpose of throwing a continuous stream of water over the cable and the brakes.

In order to obviate as far as possible the formation of kinks, and the difficulties experienced in the tanks when the cable passes rapidly from the outer turn of one flake to the smaller inside coils of the next, Mr C. W Siemens wound his cable on a large drum with a vertical axis, having a platform at each end.*   The drum revolved on cast-iron rollers placed between two circular rails, one of which was bolted to the lower platform underneath, and the other secured to the hull of the vessel.   The upper end of the vertical shaft carried a toothed wheel, which was turned by the deck engine, the power being transmitted by a system of cogs and chain pulleys.

The laying commenced on the 12th of January 1864.   After proceeding a few hours the cable broke, on the occasion of a stoppage caused by a derangement of the great reel.   The tremendous weight of the reel soon bore down the rollers, which became flattened in places, offering great resistance to turning.   Mr Siemens then gave up this method of paying out, and the cable was coiled in a tank.†

On the 28th of January a second start was made, paying out at first very slowly; afterwards the speed was increased, little by little, till it reached six knots, and even more.   When twelve hours had elapsed the cable parted in deep water, and was abandoned.   At the moment of the accident the dynamometer shewed a strain of only 6½ cwt.

A third attempt took place in the following September.   All the deep-sea cable was successfully laid, but a break occurred about ten miles from Cartagena just when the end was being landed.   It appears that, in order to avoid running short of cable, too little slack had been paid out on approaching the Spanish coast, where the water "shallows" very rapidly. Thus a considerable portion of the cable may have been suspended over sharp ledges of rock, which would account for the sudden breakage.   More probably, however, this was due to too great a strain being applied in tautening what was an exceedingly weak cable to start with.   Some miles were picked up, but without reaching the break, for the cable again parted in 900 fathoms.   After twenty days or so had been passed in unsuccessful dragging, further operations were abandoned.

---

* Similar measures had previously been taken, with the same end in view, over the first (1850) Dover-Calais cable.

† It is for the above reason that cables are now invariably stowed in tanks, instead of on drums, aboard a cable-laying ship, as well as on account of the obvious economy of space, where any considerable length is in question.

## THE TELEGRAPH TO INDIA.

[*From the "Illustrated London News."*

The Indo-European Telegraph : Landing the Cable in the Mud at Fāo, Persian Gulf.

The cables laid by Messrs Newall and Co. in the Red Sea, in 1859, and remaining unrepaired and abandoned,* the Indian Government, in 1863, determined to undertake on its own account the establishment of telegraphic communication with Europe.† However, with a view to make the submarine portion as short as possible, and to avoid deep water, they decided to limit themselves to laying a cable in the Persian Gulf, along the coast of Beloochistan, thus guarding the line of communication from the vandalism of the barbarous and unconquered natives of these parts. The landing-places selected were Kurrachee, on the frontier between India and Mekran, and Fāo, at the mouth of the Shatt-el-Arab River, formed by the union of the Tigris and Euphrates. The cable, with a total length of 1,450 miles,‡ was divided up into four sections, the intermediate landing-places being Gwadur, Mussendom, and Bushire. A land line, passing through Bussorah, Baghdad, Mosul, and Diarbekr, would convey the messages from Fāo to Constantinople,§ and thence to the already established Turkish land-line

---

* In spite of the efforts of a company formed to put the line through.

† The late, and highly esteemed, Colonel Patrick Stewart, R.E., C.B., was the Director-General of Indian Telegraphs at the time. He was assisted by Colonel Goldsmid, now Major-General Sir F. J. Goldsmid, C.B., K.C.S.I., and by Major (afterwards Sir J. U.) Bateman-Champain, R.E. Sir Frederick Goldsmid wrote an interesting book on this subject, entitled " Telegraph and Travel " (Macmillan and Co.).

‡ Weighing no less than 5,028 tons, and representing by far the heaviest length previously despatched on one expedition.

§ This cable system being only connected with Europe by land lines through Persia on the one hand and Arabia and Turkey on the other, the traffic was at first very slow, but better than taking a month to communicate with India by steamer.

system, as well as of the respective European Governments.  The form of cable decided on by Sir Charles Bright and Mr Latimer Clark (as engineers to the Government), and manufactured by Mr W. T. Henley, differed in certain important and novel particulars from all previous types.  With a view to combining the mechanical properties of a strand with the electrical advantages of a single solid wire,* the copper conductor was formed of four segments or segmental bars—*i.e.,* a quadrant of a circle in section (Fig. 33)—fitted into one another and drawn into a copper tube, to hold them together.  The whole was then rolled and drawn down to a straight and absolutely circular wire (or rod) of the requisite size—apparently solid but in reality multiple in construction, and consisting of five distinct pieces.  Sir Charles Bright had previously devised a wormed conductor to meet the same ends ; but Mr Clark's, here described, was at that time considered preferable.†  The weight of the conductor so formed

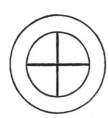

was 225 lbs. per naut.  The copper was specially tested, and the mean conductivity was as high as 89.14 per cent., whereas the maximum conductivity at the time of the first Atlantic cable was only 45 per cent. The conductor was then taken in hand by the Gutta-percha Company, who covered it with four coatings of gutta-percha, alternated by the application of Chatterton's compound, weighing 275 lbs. per N.M.  The finished core was tested in three-mile lengths, under water, at 75° F.; and also under pressure of 600 lbs, to the square inch.  Mr Latimer Clark's well-known "accumulation method"‡ was used for the first time in electrically testing the joints.  Not only is this the only joint test of any value, but it was the first occasion on which the joints in the core of a cable were tested at all.  Indeed, the joints themselves had hitherto been of so rough a character that they would scarcely bear testing, and had the effect of seriously lowering the insulation resistance of the early cables, which would otherwise—owing to the excellence of the manufacture of gutta-percha at that time—have been abnormally high.  For the covering of tarred hemp, which had been found to temporarily and partially conceal defects of insulation, Mr Willoughby Smith's plan·was adopted of substituting tanned hemp,

FIG. 33.—Diagram to shew Building-up of Segmental Conductor, 1863.

---

* By the smaller surface, etc., for a given quantity of copper, of a truly circular conductor.

† The wormed conductor of Sir C. Bright consisted of several wires stranded together, some of which were of a small gauge for the purpose of fitting into the interstices of the larger wires, the whole being drawn down to a tubular form.

‡ See " The Telegraph to India, and its Extension to Australia and China," by Sir Charles Tilston Bright, C.E., M.P.

which would retain moisture, and thus, being a ready conductor, would tend to show up any incipient faults at an early stage during manufacture.* This was the first occasion, moreover, on which the served core was kept continuously in water-tanks until required for sheathing ; thus securing, by frequent electrical tests taken under proper conditions, the detection of any faults up to the last moment before the application of the armour. The outer sheathing consisted of twelve galvanised iron wires of No. 2 B.W.G. To help to preserve the wires from rust, and to prevent the accidents so frequently occasioned by broken wires when passing through the paying-out machinery, the sheathing was covered with two servings of hempen yarn, wound on in reverse ways. This outer covering was impregnated with a bituminous preservative compound,† just previously introduced by Sir Charles Bright and Mr Latimer Clark,‡ which was applied warm whilst thoroughly plastic, and was a mixture§ of certain proportions—to be varied according to circumstances—of mineral pitch, tar, and powdered silica, the last ingredient being incorporated as an extra safeguard against teredoes and such like. Finally, the cable was led through semicircular rollers, by which the coating of compound was thoroughly pressed into all the interstices of the wires.

Fig. 34 represents the deep-sea type of the finished cable. To prevent the adjacent turns and flakes in the tanks sticking together,‖ the cable was "whitewashed."¶ Among other advantages derived from this outer yarn

---

* This process consisted of impregnating the jute (or hemp) with a solution of catechu (or "cutch," as it is often termed), a bark containing a large quantity of tannin ; hence the jute so prepared, by immersion in "cutch" solution, is nowadays very often described briefly as *tanned* jute. Though an excellent antiseptic and preservative against decay under some conditions, the tarring of the yarn had been found also to have the effect of lowering the insulation, besides sometimes having a detrimental action in the case of jointing.

† The outer yarn serving may be also regarded as a vehicle for the compound, some mixtures of which (under certain conditions) would not "take" on the iron wire direct.

‡ This was under a patent of Sir C. Bright—No. 538 of 1862—as a further preservative of the sheathing wires beyond that secured by previously galvanising them ; and a full reference to this cable compound, and the method of applying it, will be found in Part II. of this work. Its first application to the above cable was dwelt on at length by the present writer, in the course of some remarks in the *Journal of the Society of Arts* for 16th February and 9th March 1894. (See also "The Telegraph to India," by Sir Charles Bright, vol. xxv., *Proceedings of the Institution of Civil Engineers.*)

§ Being to some extent waterproof—and partially air-tight—a compound such as this is supposed to prevent the free circulation of water, with the result that but little oxygen is present : the iron wires cannot, therefore, be materially exposed to oxidation.

‖ Thus causing a foul during paying out.

¶ This external coat of lime—or chalk—and water was involved by the introduction of the compounded yarn (or canvas) outer covering. Previously, where the external wires had received an outer coat of tar, the turns had no inclination to stick—indeed, for paying-out purposes, the cable was too slippery rather than too sticky.

serving, the cable was found to be more flexible and to coil better in the tanks, facilitating the paying out in laying. The total weight of the cable was 4 tons per naut.. For the shore-end portions the sheathing wires were

materially increased in thickness, bringing the weight up to 8 tons per N.M. The cable when finished was coiled in tanks filled with water, and a series of electrical tests made for conductivity, insulation, and electro-static capacity. Finally it was placed in water-tight tanks fitted on board five large ships, and taken round the Cape to India by Sir Charles Bright, with his able staff, including Messrs J. C. Laws, F. C. Webb, F. Lambert, and others.* This cable was very successfully laid, but not without difficulties. At Khor Abdullah, for instance, at the mouth of the Shatt-el-Arab, the flotilla was anchored with six to eight miles of mud banks between the ship and the shore. As the cable could only be landed in a flat-bottomed vessel, a small craft named the "Comet" was requisitioned, belonging to the Bombay Marine Service, and employed in running between Bussorah and Bombay. In order to make room for the cable, she had to disembark her guns and coal, the operation occupying as much as fourteen days. On the banks of the Shatt-el-Arab the mud was so soft that Sir Charles Bright and his men had to drag the cable to the shore partly crawling on their stomachs (see illustration on page 73). Even when the solid part of the bank was reached which separates Khor Abdullah from Fāo, the cable required to be cut into mile and a half lengths, carried on the backs of several hundred Arabs engaged for the purpose, and then joined up again. These difficulties of landing the cable had scarcely been overcome when a fault occurred in

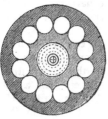

Fig. 34.—Persian Gulf Cable (Main Type).

the Fāo-Bushire † section. The necessary repairs were, however, quickly executed.

---

* Mr F. C. Webb (who subsequently remained to carry out deviations and completions as well as station arrangements) was Sir C. Bright's chief assistant engineer. Mr J. C. Laws acted as chief electrician, Mr F. Lambert being his principal assistant. Dr Esselbach was also engaged on the electrical staff up to the time of his lamented death.

† At the latter a specially heavy shore end was subsequently laid containing two cores (each jointed to that in the cable on each side), so as to make one cable do service

This was the first cable expedition in which Morse flag and lamp signalling were made use of by day and night respectively.*

On no section of the cable was the insulation after immersion less than 300,000,000 ohms (300 megohms) per naut, and the conductor resistance nowhere exceeded seven units per naut. Such satisfactory results were solely due to the exercise of minute and incessant care and testing during construction and laying of the cable; indeed, it may be said that a systematic course of electrical testing whilst in water-tight tanks originated with this undertaking.† The whole of this work was carried out by Sir Charles Bright personally. ‡ This was the first instance of any great length of cable being a complete and lasting success; and it was certainly due, in a large measure, to the many novel and improved methods introduced in its manufacture, and to the complete system of electrical and mechanical testing here applied.

By the date of laying the above cable—forming the first telegraphic connection between the United Kingdom, Europe, and India—the science of constructing and laying submarine telegraphs was pretty definitely worked out, and no very striking departure in general principles has since been introduced; indeed, the pioneer stage may be said at this juncture to have reached its termination.

It has been thought well, however, to continue the history up to the present date.

---

for the circuit in each direction, thus reducing the chances of interruption due to anchorage. This was eventually shifted to Jask, but this form of T-piece (compared with that of more recent practice) could never be regarded as very satisfactory, for in the event of a breakdown, and during the necessary repairs, it involved the interruption of traffic on the two sections immediately on each side, instead of on one only.

* Captain (now Vice-Admiral) P. H. Colomb, R.N., had a short time previously suggested the application of the Morse code of telegraphy to flag and lamp signalling for the navy.

† This was, no doubt, largely due to the fact that the perfected reflecting galvanometer for testing purposes had only just been devised by Professor Thomson, also his marine galvanometer with cast-iron magnetic screen. Both of these—now in everyday use— were employed on this occasion for the first time in practice.

‡ See "Bright on the Telegraph to India," *ibid.*, and "Old Cable Stories Retold," in *The Electrician*, 1896, by F. C. Webb.

# CHAPTER III.

## DEVELOPMENTS.

## SECTION I.—THE SECOND AND THIRD ATLANTIC CABLES OF 1865 AND 1866.

SINCE the 1858 cable ceased to work, the promoters of Atlantic telegraphy, with Cyrus Field at their head, had never for an instant relinquished hopes of establishing a more permanent line.* The Atlantic Telegraph Company,† whilst busy endeavouring to raise fresh capital, had

S.S. "Great Eastern."

prevailed on the British Government to despatch two vessels to further examine the ocean floor 300 miles out from the coasts of Ireland and

---

* After the failing of the 1858 cable, several schemes for fresh lines on new routes were brought forward. One of these was known in 1860 as the "Grand North Atlantic Telegraph," its object being to reduce the continuous length by laying a cable in four sections, bringing into use the inhospitable regions of Iceland and Greenland. The cable was to go from the extreme north of Scotland to the Faroe Islands, hence to Iceland, from there to Greenland, and so to Labrador. A project of this character had been originally brought to the notice of the Danish Government by Mr Wyld, the geographer, some years previously, and again, in a different form, by Colonel T. P. Shaffner, an American electrician of some note. This received favourable support from Admiral Sir Leopold M'Clintock, K.C.B.; Captain (now Sir Allen) Young, C.B.; Captain (afterwards Rear-Admiral) Sherard Osborn, R.N., C.B.; Dr John Rae, Dr Wallich, and others of Arctic expedition fame, who had explored the route; as well as partially by Sir Charles Bright, who drew up a report on the subject. In 1861 various papers were read at the Royal Geographical Society, calling further attention to the matter, and bringing to light the result of additional investigations. Ultimately—from the result of surveys, etc.—it appeared that the prevalence of ice, and the unsuitability of the proposed landing-places as habitable stations for the operative clerks, were sufficiently serious objections to render the scheme impracticable.

Another project which attracted some attention about the same time was described as the South Atlantic Telegraph. This was for a cable between the south of Spain and the coast of Brazil and British Guiana, touching at various islands on the way, and stretching on to the West Indies and the United States. Being, however, to a great extent foreign in its scope, this scheme found little favour with those who promoted such enterprises; and this remark equally applies to similar projects—apart from other reasons.

† This company had been kept alive mainly by the efforts of Mr (afterwards Sir Curtis) Lampson, the vice-chairman, aided by the secretary, Mr George Saward, as well as by Mr Field (the general manager) in both countries.

Newfoundland respectively. The expedition of H.M.S. "Porcupine," in 1862, to explore the western end of the Atlantic, in about the same latitude as Ireland, is still famous. It was found that the sea-bottom in this vicinity, instead of forming a precipice, as had been supposed by many, had in reality a very gentle slope.

It took, however, a considerable time to raise the full amount of capital required for another Atlantic cable, and this could only be done gradually. The great Civil War in America stimulated capitalists to renew the attempt. Mr Cyrus Field, who compassed land and sea incessantly,* pressed his friends on both sides of the Atlantic for aid, and constantly agitated the question in London and New York.†

When it was estimated that a sufficient advance had been made, the directors of the "Atlantic" Company (Chairman, the Right Honourable James Stuart Wortley) sought the assistance of certain independent scientific gentlemen as a consultative committee to advise them upon the electrical and mechanical questions involved in the proposed undertaking, with especial reference to the form of cable to be adopted. This committee (partly drawn from the members of the Government Commission already alluded to) was composed as follows:—Captain Douglas Galton, R.E., F.R.S.; William Fairbairn, F.R.S.; Professor W. Thomson, F.R.S.; Professor C. Wheatstone, F.R.S.; and Joseph Whitworth, F.R.S.

These gentlemen had under their consideration a number of proposals, and in the end the somewhat considerable experience of Messrs Glass, Elliot, and Co was thought to be a strong point in their favour as proposed contractors, especially as they had tendered a very large number of samples for the benefit of the committee's consideration of the subject.

The previous Atlantic cable had forwarded the possibility of communicating intelligence at a great distance and at a fair rate of speed by disclosing, in a practical way, the laws of resistance and electro-static capacity as affecting the speed of transmission in working submarine cables of great length. It also suggested the best means of reducing to a minimum the effect of retardation of signals due to those laws by a mathematical adjustment of the conducting and insulating constituents

---

* Mr Field is said, indeed, to have crossed the Atlantic sixty-four times—suffering from sea-sickness on each occasion—in connection with the various Atlantic cable enterprises in which he formed so prominent a figure.

† One of the main advantages pointed to on this occasion was the avoidance of misunderstandings between the two countries. Another (intended as a special attraction to Americans) was the improvement of the agricultural position of the United States, by extending to it the facilities, already enjoyed by France, of commanding the foreign grain markets. On this latter account, the project was warmly supported by the late Right Hon. John Bright, M.P., and other eminent Free Traders.

of a cable, both in a given ratio to themselves and relatively to the length of the line to be constructed.

Again, Professor Thomson had pointed out in the columns of *The Athenæum* the mistaken notion which had previously found favour (supported by Faraday, amongst others) regarding the requirements for obtaining a given working speed for submarine cables of certain length. It was partly, however, left for Mr S. A. Varley to put the matter correctly in a complete, yet non-mathematical, form; and this he had very ably done in the course of a paper read before the Institution of Civil Engineers,* and also in another before the Society of Arts,† some years previous to this project. The main point of Mr Varley's paper (as well as of Professor Thomson's contributions) was to shew that the signalling speed attained on a cable depended not only on the electro-static inductive capacity (as seemed to have been imagined by many), but also equally on its conductor resistance. ‡

It was now absolutely manifest that the circumference of the conductor and the thickness of the dielectric should not only bear a certain relation to one another for purposes of low capacity, but also that—in order to put a limit on its resistance—the conductor must not be smaller than a given diameter according to the length. Mr Cromwell Varley had also spoken before the Government Commission§ to the same effect; and Professor Thomson eventually laid down definitely the well-known law ‖ for speed through a cable as depending, on the one hand, inversely on the conductor resistance, and also inversely on the electro-static capacity of the system. ¶

Thus it came about that the type of core recommended by the Telegraph Construction and Maintenance Company (formerly Messrs Glass, Elliot, and

---

* "On the Electrical Qualifications requisite in Long Submarine Telegraph Cables," *Minutes of Proceedings, Institution of Civil Engineers*, vol. xvii.

† *Journal of the Society of Arts*, vol. vii.

‡ Mr Varley went further than this, indeed. He shewed that by doubling the diameter of a wire, whereas the conductor resistance became a fourth of what it had been (by the sectional area being quadrupled), the circumference being only doubled, the electro-static capacity would be but twice what it was before. The result is that, as these two are equal factors in the speed, by doubling the diameter of the conductor and providing for the same thickness of dielectric, the working speed admissible on a given cable is twice as much as previously.

§ This Commission had been arranged more especially with a view to considering the best type of cable for Atlantic telegraphy.

‖ Stated in a complete mathematical form in Part III.

¶ Signs of this fact being appreciated occurred previously, when Mr (afterwards Sir Charles) Bright, on becoming engineer to the Atlantic Telegraph Company in 1857, immediately recommended a much heavier core—392 lbs. copper to the same weight of gutta-percha—than had been arranged for the first Atlantic cable. It was, however, for financial and technical reasons, found impossible to effect the change.

Co.),* and adopted by the Scientific Committee for the new cable,† was to consist of a strand of seven copper wires, each of No. 18 B.W.G., weighing 300 lbs. per N.M., and coated with four layers of gutta-percha (each shewn in Fig. 40) alternating with four of Chatterton's compound, weighing 400 lbs. per N.M.‡ The central wire was covered with Chatterton's compound, to prevent small air-bubbles remaining in the interstices of the wires when the gutta-percha was put on, as well as to prevent the percolation of water along the cable from a loose, buoyed, or lost, end. The conductivity of the copper used was to be at least 85 per cent. of that of chemically pure copper. The sectional area of the conductor being nearly three times that of the 1858 cable, and the resistance being much less by reason of the greater purity obtained for the copper, it was estimated that a working speed of about seven words per minute would be afforded by this cable when laid.

The core was to be covered with a cushion of tanned jute,§ and over

---

* The Gutta-percha Company had just been (in April 1864) incorporated with Messrs Glass, Elliot, and Co., forming a limited liability company under the above title. Mr Pender (afterwards Sir John Pender, G.C.M.G., M.P.) was largely instrumental in effecting this, and he accordingly became the first chairman ; whilst Mr (afterwards Sir Richard) Glass was the managing director.

† Payment to the contractors for the manufacture of this cable was made partly by £250,000 in cash ; partly by £250,000 in 8 per cent. Preference shares of the company ; and partly by £100,000 in Mortgage Debentures ; the contractors in this instance being prepared to stake these two latter amounts on the success of the enterprise, the parties concerned being to a large extent financially interested therein. In the case of the successful submersion and working of the cable, a further sum of £137,140, in the original (unguaranteed) stock of the old "Atlantic" Company, was to be paid in instalments as long as the cable continued to work. The capital of the newly-formed "Atlantic" Company was £600,000, of which it will be seen the Telegraph Construction Company had practically found more than half by taking the enterprise on their shoulders in the way stated. Mr Field had been only able to raise £70,000 in the United States. The late Mr Thomas Brassey was the first individual he appealed to in this country, from whom he received much encouragement and noble support, as well as from Mr Pender, the second to be applied to individually. It will be seen that this was an expensive cable, as was but natural, considering the large proportion not paid for in cash, and the corresponding risk incurred by the contractors in accepting shares and bonds as part payment.

‡ This was made up as follows :—

|  |  |  |  |
|---|---|---|---|
| 1st coat | = 62 lbs. | 3rd coat | = 119 lbs. |
| 2nd „ | = 91 „ | 4th „ | = 128 „ |

being as near to the same *thickness* of gutta-percha for each individual covering as the machine would well permit of. The first coat was of purer quality than that of the remaining coats, and made according to the process of Mr Edwin T. Truman.

§ In the earlier days it appears to have been very much the custom to employ what was called hemp (where jute is now always used) as the protective serving, or padding, applied round the core. Possibly no very accurate distinction could at that time be drawn between the materials, or else owing to the iron sheathing employed not affording so much strength, it was considered that this bed for the core must also sensibly contribute to the strength of the finished cable. This seems to be the first occasion where ordinary jute is actually stated to have been used.

this, wound spirally, ten homogeneous iron wires (drawn from Bessemer steel*) of No. 13 B.W.G. Each wire (Fig. 35) was to be separately covered with yarns† of Manilla hemp previously soaked in a mixture of tar, india-rubber, and pitch ; the object of this serving‡ to each wire was to protect them from rust due to complete exposure to air and water,§ as well as with a view to lessening the specific gravity of the cable. There had been evidence to the effect that the tensile strength of a hemp-covered wire was greater than the sum of the strengths of the iron and hemp taken separately. It was calculated, therefore, that the deep-sea cable would resist a strain of 7 tons 15 cwt., *i.e.*, would be capable of suspending in water 11 miles of its own weight, for it was to weigh 1 ton 15¾ cwt. per N.M. in air, and only 14 cwt. in water, the total diameter being 1.1 inch. The above form of cable—adopted by the said committee—was

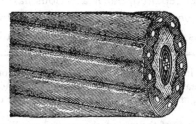 

FIG. 35.—Second Atlantic Cable (1865), Main Type.

considered by many leading authorities at that time to almost perfectly fulfil the conditions required for deep-sea work. ‖

---

* The Bessemer process had been introduced some years previously—*i.e.*, in 1855. That which comes under the description of *homogeneous* iron wire is drawn from charcoal puddled bars, often mixed with mild Bessemer steel. Messrs Webster and Horsfall (the famous pianoforte wire makers) were the originators of this class of iron wire, and this was the first occasion on which it was applied for submarine cables. It is advocated for homogeneous iron that it possesses almost the same strength as steel without the same springiness. In this instance, however, it was probably mixed with rather too much steel.

† Messrs John and Edwin Wright and Messrs Frost Brothers (ropemakers) supplied the whole of the Manilla yarn for this purpose. As previously stated, the former were the patentees of this type of cable.

‡ In laying up this form of cable the force exerted at the lay plate pressed the serving of each wire together into the shape shewn in the figure. This flattened appearance became still more conspicuous after stowage in the tanks for any considerable period.

§ It seems to have been thought that each wire being enclosed in the above serving did away with the necessity of galvanising. It was also considered that the latter process tended to weaken the wire.

‖ It appears perhaps strange nowadays that the possibility of laying the cable with success should ever have been a subject of serious doubt. Nevertheless considerable difficulty had been experienced on several occasions in keeping the egress of the cable in subjection when paying out in deep water : much trouble had often been met with owing to the brakes not working properly, and the cable not only running out too fast, but also

The shore-end cable (Fig. 36) was to have a second outer sheathing of twelve strands, each strand containing three galvanised iron wires of No. 2 B.W.G.; its weight was to be 20 tons per naut. It was to be joined on to the deep-sea type by a gradually tapering length of 25 fathoms.

The core was manufactured at the Wharf Road Gutta-percha Works in 4,500-yard lengths, equivalent to 2¼ N.M., each of which was kept twenty-four hours under water at 75° F., and was rejected if the insulation resistance per N.M. fell below 150,000,000 units. The measured resistance of the dielectric, as a matter of fact, proved to be nearly always more than double the required minimum. The core was again tested electrically whilst under pressure, then unwound from the reels, and examined with the greatest possible care, by passing through the fingers of a skilled workman capable of detecting any flaw by touch. In cases of slight injury the gutta was repaired, but if a serious flaw occurred the faulty piece was cut out. Eventually the apparatus employed for testing the core under pressure was that of Mr W. Reid, already alluded to

---

getting jammed in consequence. It was indeed in order to meet these objections that a cable of lower specific gravity and of greater surface was devised for this undertaking. Such a cable would sink more gradually, and would require less brake power to hold it back.

Though seeming to meet the difficulties of the period, the reputed excellence of this type has since proved fallacious, and is never now adopted. It is known to possess a comparatively short life on account of the rapid decay of the hemp—in close juxtaposition to the iron and the salt-water—and when the hemp has decayed, the wires, on being left somewhat in the form of an open cage, soon follow suit. The salt-water on coming into direct contact—and entirely, with each one separately—causes the section of each to be rapidly reduced by oxidation. Thus the strength of such a cable soon becomes too low for lifting purposes, though often comparatively light specifically.

On account of its pliability, a cable of this description renders itself easy of manipulation during cable operations generally. From an owner's point of view, however—i.e., for purposes of durability—such a type is considered nowadays most undesirable. Any preservation of the iron wires by the hemp laid up round each is at the best of very short duration, even when the latter has been saturated with a tar compound. The real strength for repairs is in the iron wires, and the hemp serving seldom lasts long enough to save these appreciably.

Moreover, much trouble is frequently experienced with this type owing to any given wire breaking away and getting jammed into the core, an event which can scarcely happen with an ordinary close-sheathed cable, having a more or less complete arch, or tube, of wires.

The cable above described is often aptly referred to as the "open-jawed" type, being a ready prey to the teredo—et hoc genus omne—which can the more readily burrow into it, and, in so doing, damage the core.

Again, having a low specific gravity, and offering a great deal of friction (both on account of its large circumference and rough surface), though comparatively easy to lay in some fashion without mishap, it is liable not to sink properly into all the irregularities, and, in fact, to be paid out with too little slack. Similarly—owing to skin friction and resistance—a cable with an exterior of these dimensions and character is undesirable for purposes of recovery.

with reference to the Malta-Alexandria cable—the first to be so tested. The degree of pressure adopted was the same in this instance, *i.e.*, equivalent to 640 lbs. per square inch, this being the limit of pressure which was considered at that time necessary or desirable as the nearest safe approximation to the pressure to be experienced by the cable on submersion. Such a pressure was, of course, a very long way from that which this cable subsequently had to contend with, though perhaps sufficiently representative in the case of the previous cables to which this system had been applied.*

The sheathing was undertaken at the Telegraph Construction Company's Works at Morden Wharf, Greenwich, under the personal direction and superintendence of Mr (afterwards Sir Richard Atwood) Glass. The reels of core were kept under water, in locked receptacles, until required, and the cable when finished was coiled into light iron tanks. These tanks were kept constantly filled with water, and the cable tested daily, both by the contractors and by the "Atlantic" Company's electrician, Mr C. F. Varley.

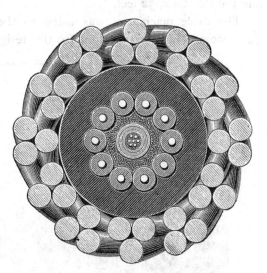

FIG. 36.—Shore End of the 1865 Atlantic Cable.

The "Great Eastern,' an enormous vessel of 22,500 tons burden,† which

---

* Nowadays a pressure of this amount would be considered utterly inadequate. Where any such test is applied at all, a degree of pressure is used which is fully equivalent to that at the bottom of the ocean where the cable is to be laid. Messrs Siemens Brothers are the only contractors who adopt this plan. At their factory very elaborate apparatus is used, and this is referred to in Part II. As the result of experience, other authorities consider that the pressure test system, whilst distinctly costly, is as liable to temporarily seal up and screen small faults of manufacture as to bring them to light.

† On account of her dimensions, originally christened the "Leviathan," and the conception of that distinguished engineer, Isambard Kingdom Brunel, F.R.S. (see p. 79). This vessel (built by the late Mr Scott Russell) was but a bit before her time. In the present day, with improved engines, she could be usefully and profitably employed, had she not been broken up. At the time referred to above, she could not get a suitable cargo, and laying the Atlantic cable was the first piece of work she did which was at all worthy of her, after lying more or less idle for nearly ten years. Mr Brunel drew attention to her as being the only vessel that could do the work. This was a few years after the "Eastern Steam Navigation Company" had failed over her, and a new company formed by Mr Gooch (afterwards Sir Daniel Gooch, Bart., M.P.), under the title of the "Great Ship Company," subsequently (in 1864) converted into the "Great

had been lying up idle for some years, was secured for the work, she being
the only ship that could take the entire length of cable.  As it was not
practicable to moor her off the works at East Greenwich, the cable had to
be cut into lengths and coiled on two pontoons, and thence transferred
to the big ship (see Fig. 37 opposite, and also Fig. 48, p. 108).*  It was
coiled into three large water-tight iron tanks built up in the hold of the
vessel.  The main tank measured 75 feet across; the after tank, 58 feet;
and the fore tank, 52 feet.

The cable machinery was fitted to the "Great Eastern" under the
guidance of Mr Henry Clifford (to the design of Messrs Canning and Clif-

FIG. 38. —Paying-Out Machine, Atlantic Cables, 1865-66.

ford), assisted by Mr S. Griffith.  Mr Appold's type of brake formed a part
of the paying-out apparatus—in fact the general character of the gear (Fig.
38) did not differ largely from that adopted in the second (1858) expedition
of the "Agamemnon" and "Niagara."  The main point of difference was
the further application of "jockeys" † (Fig. 39)—of a much more com-

Eastern Steamship Company."  As the chairman of the latter company, Mr Gooch
was an active spirit in arranging for the charter of the "Great Eastern" over this
work.  In this way he became an ardent supporter of the enterprise, and occupied
a seat on the board of the Telegraph Construction Company—assuming the chair later
on—up to the time of his lamented death in 1889.

* Taken from the original drawings of Mr S. Griffith on behalf of the Telegraph
Construction Company.

† So-called because they ride on the cable.

[PLATE VI.

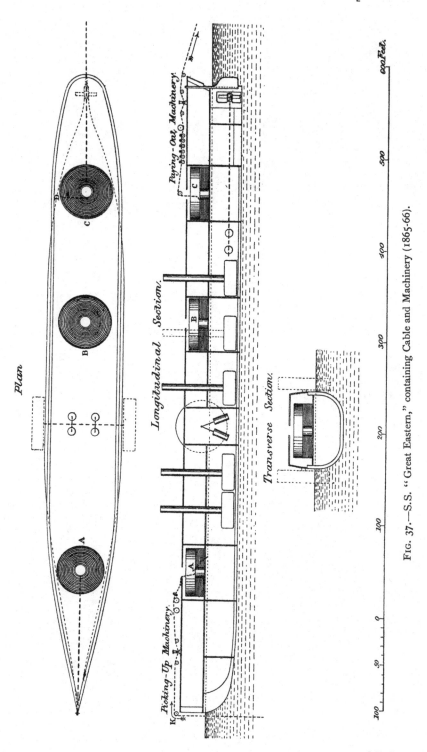

Plan

Longitudinal Section.

Paying-Out Machinery.

Picking-Up Machinery.

Transverse Section.

FIG. 37.—S.S. "Great Eastern," containing Cable and Machinery (1865-66).

[To face p. 86.

plete form—to the "Great Eastern" gear, which was entirely constructed and set up by that famous firm of engineers, Messrs Penn, of Greenwich.* The cable, in coming up from the tank in the hold of the vessel, passed along a conducting trough to the first of the six leading V-wheels of the machine (Fig. 39). It did not take a turn round this wheel B, but merely passed over the top of it and the five other wheels consecutively, being pressed down into their grooved rims by a small weighted roller or jockey pulley A, around the circumference of which there was a band of india-rubber so as to produce a retarding effect upon the cable when necessary. Each of the wheels over which the cable passed had a light brake fitted, so as to introduce a slight retardation for giving the cable a straightened lead to the drum. For the sake of clearness this brake is not shewn in the figure. The "jockey" turned upon an axle at the end of an arm centred at *a*, and the weights of the "jockeys" could be released at once by turning a hand-wheel. After leaving the last of these pulleys, the cable took several turns round a large drum, the axle of which was connected to a brake arrangement similar to that of 1858, by means of which the speed of the drum with a given strain was checked or accelerated, according to the increase or reduction of

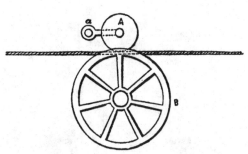

FIG. 39.—Diagram of One of the Six Leading Wheels of the 1865-66 Atlantic Cable Machinery.

a series of hand-weights, that could be attached or taken off as required. The cable was then led through the dynamometer system and hence outboard. As may be seen in Fig. 37 (and still better in Fig. 42 further on), there was a second brake drum in tandem with the former, but this was only intended as a reserve in case of accident.

The shore-end cable was manufactured by Mr W. T. Henley, of North Woolwich. It was some 30 miles in length, of which 26 N.M. were put on board S.S. "Caroline," fitted up for the purpose. This part was laid by her (see illustration on next page) from Foilhommerum Bay, Valentia Island, on the 22nd of July 1865.†

---

* The late Mr John Penn used also to act as referee in the affairs of the Telegraph Construction Company.

† The cable was landed here at the lower end of the island, and about six miles from Knightstown, instead of at Ballycarberry Strand, where the previous (1858) cable had been landed, partly because it was the more direct point, and partly to avoid the line of the old cable. It was, however, afterwards found to be more exposed, and to involve a

The next day the "Great Eastern," under the command of Captain (afterwards Sir James) Anderson,* with a total dead weight of 21,000 tons, joined up her cable to the shore end, and started paying out towards America, at a speed of six knots. She was escorted by two British men-of-war, the "Terrible" (Captain Napier) and the "Sphinx" (Captain Hamilton).

On behalf of the contractors, the Telegraph Construction and Maintenance Company, Mr (now Sir Samuel) Canning was the engineer in charge of the expedition, with Mr Henry Clifford as his chief assistant.† Both

Atlantic Cable of 1865. Landing the Shore End in Boats from S.S. " Caroline,"
off Foilhommerum Bay, Valentia Island.

these gentlemen had been engaged with Sir Charles Bright on the first Atlantic expedition, and had had much experience alike in cable work

---

rough bottom, besides the disadvantage of it not being on the mainland. This spot has therefore since been given up as a landing-place for Atlantic cables. Nowadays they are invariably taken direct to the mainland, so as to avoid dependence on a branch cable.

* Mr R. C. Halpin being the chief officer. Both these gentlemen had the reputation of possessing great skill in the handling of a ship. Captain Anderson was at the time captain in the Cunard Steamship Company's service, by whose permission, therefore, he accompanied the expedition.

† Since the first Atlantic expedition Messrs Glass, Elliot, and Co. had gathered together a full and permanent staff of engineers and electricians (of which Mr Canning became the chief and Mr Clifford the second engineer) in such a way as enabled them to undertake direct contracts for laying—as well as manufacturing—submarine telegraph cables. When this firm, in 1864, merged itself into the Telegraph Construction Company (as already described), position in the above respect was still further established.

and mechanical engineering. There were also on the engineering staff of the contractors, Mr John Temple and Mr Robert London. Mr C. V. de Sauty acted for them as chief electrician, assisted by Mr H. A. C. Saunders, and several others. Mr Willoughby Smith, the electrician to the Gutta-percha Company,* was also out on the expedition at the request of the contractors, though holding no exact official position. By arrangement with the Admiralty, Staff-Commander H. A. Moriarty, R.N., acted as the navigator of the expedition. This officer was well known to be possessed of great skill in this direction, and had previously served in a like position on the first Atlantic cable expedition, when his services were similarly lent by the Admiralty.

The Atlantic Telegraph Company was represented on board by Professor William Thomson (now Lord Kelvin), LL.D., F.R.S., and Mr C. F. Varley, as electricians, the former acting mainly as scientific expert in a consultative sense. Both Mr Field and Mr Gooch accompanied the expedition, the former as the chief promoter of the enterprise, and the latter on behalf of the "Great Eastern" Company. Representing the Press, there were also on board Dr W. H. Russell, the well-known journalist, of *The Times*, acting as the historian of the enterprise;† and Mr Robert Dudley, a famous artist, who produced several excellent pictures of the work in its different stages, as well as articles for the *Illustrated London News*, and for Dr Russell's book, "The Atlantic Telegraph," now very scarce.‡

Unfortunately trouble soon arose. The first fault declared itself on the 24th, when 84 miles had been paid out. It was decided to pick up back to the fault, which was found after 10½ miles of cable had been brought on board. A piece of iron wire was found to have pierced the cable diametrically, so as to- make contact between the sea and the con-

---

* Commercially speaking, incorporated with the newly-formed Telegraph Construction Company ; Mr John Chatterton then becoming manager of the Gutta-percha Company's works in succession to the late Mr Samuel Statham, who had just died.

† This famous war correspondent, author, and general newspaper writer, has since become Sir William Howard Russell, LL.D.

‡ Detailed and stirring accounts of the events of this expedition also appeared subsequently in *Blackwood's Magazine*, in *Cornhill*, and in *Macmillan's*. The former was written by Mr Henry O'Neil, A.R.A., and the latter by Mr John C. Deane, both of whom were eye-witnesses aboard the "Great Eastern." Mr O'Neil also brought out an illustrated comic journal during this and the following expedition, issued at periodic intervals, which was a source of much amusement to those who had time for perusing it. Still more was an "extravaganza"—written by Nicholas Woods and J. C. Parkinson—on the subject, performed on board on the completion of all the work in 1866. Both these were afterwards published in booklet form, and are much treasured by the parties caricatured therein.

ductor. The faulty portion was cut out, and the paying out resumed as soon as the cable was spliced up again. On 29th July, when 716 miles had been laid, another and more serious fault appeared. The arduous operation of picking up again commenced, and after nine hours' work the fault was safe inboard, and the necessary repair effected. On stripping the cable another piece of iron wire was discovered, sticking right through the core. Anxiety and misgivings were now felt by all on board, for it seemed that such reverses could only be attributed to malevolence. On the 2nd of August a further fault was reported; they were now two-thirds of the way across, 1,186 miles of cable being already laid. Again they had to pick up, and this time in a depth of 2,000 fathoms. One mile only had been recovered, when an accident of some kind happened to the picking-up machinery.* The great ship, having stopped, was at the mercy of the wind and swell, and heavy strains were brought on the cable, which consequently suffered badly in two places. Before the two injured portions could be secured on board, the cable parted and sank. Mr Canning at once decided to endeavour to recover the cable, notwithstanding the fact that it lay in 2,000 fathoms. An iron grapnel was lowered by means of a wire rope which had been brought in case a mark buoy had to be put down, or the cable had to be cut, owing to some unforeseen accident, and temporarily buoyed.† This rope was in lengths of 100 fathoms, and measured in all about 5,000 fathoms. The vessel then proceeded to make a series of short runs crossing and recrossing the supposed line of cable. After manœuvring in this way for about fifteen hours, picking up was started on the supposition of the cable having been hooked by the grapnel; 700 fathoms of rope had been hove in, when one of the connecting links gave way, and all beyond it sank to the bottom. The work was recommenced with hempen ropes, two miles further west, in a depth of 2,300 fathoms, and on the 8th of August the cable was again hooked; but when raised to within 1,500 fathoms of the surface another connecting link parted, the strain being about nine tons. Two more attempts were made. The first time the grapnel fouled the rope, and had to be hove up again; the second time the rope parted close to the capstan when they had picked up 765 fathoms with a gradually increasing strain. The store of grappling rope being now quite exhausted, the work had to be given up, and on 11th August 1865 the fleet of ships parted company—shattered in hopes as well as in ropes.

---

* This constantly failed for want of sufficient strength and an adequate supply of steam.
† But with no special idea at the time of recovery work in such deep water.

## SECOND AND SUCCESSFUL ATTEMPT, 1866.

The results of the last expedition, disastrous as they were from a financial point of view, in no wise abated the courage of the promoters of the enterprise, but only served to increase their energy by demonstrating the probability, if not the certainty, of an early and complete success. During the heaviest weather the "Great Eastern" had shewn exceptional "stiffness," whilst her great size and her manœuvring power (afforded by the screw and paddles combined) seemed to shew her to be the very type of vessel for this kind of work. The picking-up gear, it was true, had proved insufficient,* but with the paying-out machinery no serious fault was to be found. The feasibility of grappling in mid-Atlantic had been demonstrated, and they had gone far towards proving the possibility of recovering the cable from similar depths.† Last, but by no means least, it was proved that the low temperature and enormous pressure at ocean depths greatly increases the insulating power of gutta-percha, whilst the conductivity of the copper was substantially increased by the low temperature, but apparently unaffected by the pressure.

To overcome financial difficulties, the Atlantic Telegraph Company was, practically speaking, amalgamated with a new concern, the ANGLO-AMERICAN TELEGRAPH COMPANY, which was formed, mainly by those interested in the old business, with the object of raising fresh capital for the new and double ventures of 1866. The ultimate capital of this company amounted (as before) to £600,000.‡

---

* This was specially pointed to by the representatives of the "Atlantic" Company when considering the carrying out of the agreement by the contractors.

† It may be mentioned in passing also that Professor W. Thomson had in the interval delivered an address before the Royal Society of Edinburgh on "The Forces concerned in the Laying and Lifting of Deep Sea Cables" (see *Proc. Roy. Soc.*, December 1865).

He had previously contributed an article to *The Engineer* relative to the catenary formed by a submarine cable between the ship and the bottom, during submergence—under the influence of gravity, fluid friction, and pressure. In this communication Professor Thomson pointed out that the curve becomes a straight line in the case of no tension at the bottom—the normal condition, in fact, when paying out.

‡ In raising this, Mr Field first secured the support of Mr (afterwards Sir Daniel) Gooch, C.E., M.P., then chairman, and previously locomotive superintendent, of the Great Western Railway Company, who, after what he had seen on the previous expedition, promised, if necessary, to subscribe as much as £20,000. On the same conditions, Mr Brassey expressed his willingness to bear one-tenth of the total cost of the undertaking. Ultimately, the Telegraph Construction Company led off with £100,000, this amount being followed by the signatures of ten directors (as guarantors) at £10,000 a-piece, viz. :—Henry Ford Barclay, Henry Bewley, Thomas Brassey, A. H. Campbell, George Elliot, Cyrus W. Field, Richard Atwood Glass, Daniel Gooch, John Pender, and John Smith. Then there were four subscriptions of £5,000, and some of £2,500 to £1,000, principally from firms participating in one shape or another in the sub-contracts. These

It was now proposed not only to lay a new cable between Ireland and Newfoundland, but also to repair and complete the one lying at the bottom of the sea. A length of 1,600 miles of cable was ordered from the Telegraph Construction and Maintenance Company. Thus, with the unexpended cable from the last expedition, the total length available when the expedition started would be 2,730 miles, of which 1,960 miles were allotted to the new cable, and 697 to complete the old one, leaving 113 miles as a reserve. The slack, or difference between the total length of cable paid out and the distance overground covered by the ship, having been only 8 per cent. on the previous expedition, the length now provided appeared to be sufficient.

The core of the new cable was precisely similar to that of the old one of the year before, but the testing of the core under pressure was abandoned, finding that this practice, in several instances, instead of revealing the weak places, had a tendency sometimes to conceal them. The sheathing wires

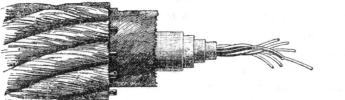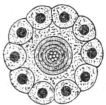

FIG. 40.--Atlantic Cable (1866).

in this cable were galvanised (besides being of softer iron to start with), and again separately covered with five strands of Manilla yarn, which in this case were left untarred.* The cable (Fig. 40) weighed 1 ton 11 cwt. per N.M. in air,† and 14¾ cwt. in water. Its total diameter was

---

sums were all subscribed before even the prospectus was issued, or the books opened to the public. The remaining capital then quickly followed.

The Telegraph Construction Company, in undertaking the entire work, were to receive £500,000 for the new cable in any case, and if it succeeded an extra £100,000. If both cables came into successful operation the total amount payable to them was to be £737,140.

* The tar compound which had been used for the Manilla yarn outside the wires had given much trouble in the handling of the previous cable during coiling and laying operations. It was, in fact, the occasion of several exciting scenes, from the layers of cable in the tanks sticking together and becoming entangled, thus causing "foul flakes," as they are termed. It would have been better, probably, if the compound had been changed instead of being abolished altogether, both on account of its tendency to make the finished cable less "lively," and owing to its preservative character. Some considered, however, that a tar compound tended to have an injurious effect on Manilla and all hemp yarns, and to screen faults of insulation.

† That is to say, nearly 5 cwt. less than the previous line, which was considered an advantage, its specific gravity being correspondingly lower.

1.1 inch, and its breaking strain 8 tons 2 cwt., being thus a little higher than that of the previous cable, due, no doubt, to the general improvements in manufacture as regards the iron wire and the class of hemp used. The iron wire, besides being stronger, was also less hard and brittle—in fact more pliable.

The shore-end cable (Fig. 41) determined on in this case was of an entirely different description.* It had only one sheathing, consisting of twelve contiguous iron wires of great individual surface and weight; and outside all a covering of tarred hemp and compound. The part of this cable which was intended for shallow depths was made in three different types. Starting from the coast of Ireland, 8 miles of the heaviest was to be laid, then 8 miles of the intermediate, and lastly 14 miles of the lightest type, making 30 miles of shoal-water cable on the Irish side. Five miles of shallow-water cable, of the different types named, were considered sufficient on the Newfoundland coast.

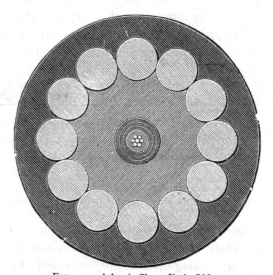

FIG. 41.—Atlantic Shore End, 1866.

The previous paying-out machinery on board the "Great Eastern" was altered to some extent by Messrs Penn to the instructions of Messrs Canning and Clifford. It again consisted of six vertical sheaves placed in a line one after the other, each supporting a jockey pulley with bevelled edges, narrow enough to fit between the flanges of the sheaves below. The bearings in which the spindles of the jockey pulleys revolved were mounted on frame levers pivoted at one end, so that the pressure of the jockeys riding on the V-sheaves, and on the cable passing between, could be regulated by weights attached to the opposite extremities of the lever arms. A small strap brake, worked by hand, was fitted round the shaft of each V-wheel. The cable on leaving the tank (Fig. 42) led in a straight line through the free space between the V-wheels G, and their riding jockeys, and then over a roller to a large drum P, six feet in diameter, round

* The Electric Telegraph Company had found over the Anglo-Dutch cable that a stranded sheathing was less durable than the previously used solid wires as above.

which it took four turns,* thence, through guiding rollers, under the dynamometer pulley D, and finally over the stern sheave A, into the sea. Two
Appold brakes, working in troughs filled with water, were fitted on the
shafts of both drums, similar to those controlling the V-sheaves ; and, to
avoid heating, water was kept pouring on the cable whilst passing under
the jockey pulleys and round the drums. The brakes were adjusted by a
chain wound on the axle of a spoked wheel under the control of an
assistant stationed at the dynamometer. In fact, as stated before, with
regard to the expedition of the previous year, the apparatus was very
similar in its ruling principles to that of 1858. †

Though different in details, the main improvement over the 1865 gear
consisted in the fact that a 70 horse-power steam-engine was fitted to drive
the two large drums in such a way that the paying-out machinery could be
used to pick up cable during the laying if necessary,‡ and thus avoid the
risk incurred by changing the cable from the stern to the bows. This addition

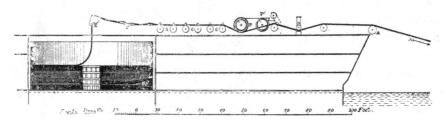

FIG. 42.—General Arrangement of Paying-out Machinery at Stern of "Great Eastern" (1866).

of Penn "trunk" engines—as well as the general strengthening of the entire
machinery—was made in accordance with the designs of Mr Henry Clifford.

The picking-up machinery forward (Figs. 43, 44), after the previous
expedition, was considerably strengthened and improved with spur

---

* The duplicate drum and brakes P¹ were merely an auxiliary for emergencies—*i.e.*, in
the event of the others being disabled, or for hauling back. Upon the overhanging ends of
the drum shafts driving pulleys were fitted, which could be connected by a leather belt
for the purpose of bringing into use the duplicate brakes, if the working brakes should get
out of order.

† The general similarity between the paying-out gear aboard the "Great Eastern"
and that fitted to H.M.S. "Agamemnon" and U.S.N.S. "Niagara," is referred to in the
complete account of the 1866 machinery given by Mr Elliot (afterwards Sir George
Elliot, Bart., M.P.) in the course of a paper on the subject read before the Institution of
Mechanical Engineers the following year. See *Proceedings Inst. M.E.* This gentleman
was a partner in the firm of Messrs Glass, Elliot, and Co., and afterwards director
of the Telegraph Construction Company, ultimately becoming chairman and remaining
so up to the time of his death in 1893.

‡ It is a little remarkable that no steam gear had been applied to the paying-out
apparatus of 1865 for this purpose, seeing that the corresponding machinery of 1858 was
so furnished.

[PLATE VII.

FIG. 43.—Plan and Elevation of Picking-up Machine aboard S.S. "Great Eastern," as used for recovering the 1865 Atlantic Cable.

FIG. 44.—General View of Picking-up Machine aboard "Great Eastern," 1866.

[*To face p.* 94.

wheels and pinion gearing. It had two drums 5 feet 7 inches in diameter, worked by a similar pair of 70 horse-power trunk engines. This formed an exceedingly powerful machine, and much credit is due to those who devised and constructed it.

Similar gear was fitted up on board the two vessels—S.S. "Medway" and S.S. "Albany"—chartered to assist the "Great Eastern."

For the purpose of grappling the 1865 cable, 20 miles of rope were manufactured, which was constituted by forty nine iron wires of No. 14 gauge, separately covered with Manilla hemp. Six wires so served were

FIG. 45.—Buoys, etc., used on Atlantic Expeditions of 1865-66.

laid up strand-wise round a seventh, which formed the heart or core of the rope. The breaking strain of this rope was about 30 tons.*

In addition, five miles of buoy rope were provided, besides buoys of different shapes and sizes, the largest of which (see Fig. 45) would support 20 tons. As on the previous expedition, several kinds of grapnels were put on board, some of the ordinary sort, and some with springs

---

* The comparatively short length (six or seven miles) of grappling rope provided for the expedition of the previous year was said to possess a corresponding breaking strain of only $8\frac{1}{2}$ tons—being, in reality, buoy rope, though used also for grappling.

to prevent the cable surging, and thus escaping whilst the grapnel was still dragging on the bottom (Fig. 46) ; others, again, were fashioned like pincers, to hold (or jamb) the cable when raised to a required height,* or else to cut it only, and so take off a large proportion of the strain previous to picking up.†

FIG. 46.—Type of Grapnel mainly used for Recovery of the 1865 Atlantic Cable during Expedition of 1866: the Cable is shewn hooked on one prong.

The propelling machinery of the "Great Eastern" had similarly received alteration and improvement in the intervals of the two expeditions. The paddle wheels were reduced in diameter (six knots being the maximum speed required‡), and made to work independently, to enable the ship to turn rapidly without headway. The screw propeller was surrounded with an iron cage, to keep the cable and ropes from fouling it, as had been provided by Sir Charles Bright for the "Agamemnon" and "Niagara" in 1857.

The testing arrangements were perfected by Mr Willoughby Smith, in such a way that insulation readings could be continuously taken, even whilst measuring the copper resistance, or while exchanging signals with Valentia.§ Thus there was no longer any danger of a fault being paid overboard without instant detection.

On the 30th June 1866, the "Great Eastern," from the Thames, followed by the "Medway" and "Albany," arrived at Valentia, where H.M.SS. "Terrible" and "Racoon" were found, under orders to accompany the expedition. The "Medway"

---

* Most of this apparatus was furnished by Messrs Brown, Lenox, and Co., the famous chain, cable, anchor, and buoy engineers, several of the grapnels being to their design, as well as the "connections.

† This was a grapnel with a cutting edge in the bed between the flukes, or prongs, and the shank, which tended to cut through the cable under a given raising strain rather sooner than would be the case by ordinary methods of breaking. This, however, was not a cutting grapnel, as at present understood, for cutting the cable at the bottom.

· Mr Latimer Clark had just devised an alternative holding or cutting grapnel (the two operations being also combined, if desired), but this was not adopted for the expedition. It was the first attempt in the above direction, and was of a somewhat elaborate character, modified subsequently by the late Mr Frank Lambert.

‡ This limitation of the ship's speed was made a special point of on this occasion, partly to provide for the ship being brought to a standstill quickly in the case of mishap during paying-out operations, and also to ensure the cable being laid in such a way that it properly accommodated itself to the irregularities of the bottom. The ordinary speed ultimately adhered to was five knots for the ship and six knots for the cable.

§ Professor Thomson's reflecting apparatus, for testing or signalling, had been considerably improved since the previous Atlantic cable of 1858. Again, this was the first cable expedition on which the above, now famous, ship and shore system of combined

had on board 45 miles of deep-sea cable in addition to the American shore end.

The principal members of the staff acting on behalf of the contractors in this expedition were the same as in that of the previous year, Mr Canning being again in charge, with Mr Clifford and Mr Temple as his chief assistants. In the electrical department, however, the Telegraph Construction Company had since secured the services of Mr Willoughby Smith as their chief electrician, whilst he still acted in that capacity for the Gutta-percha Company. Mr Smith, therefore, accompanied the expedition as chief electrician to the contractors, his chief assistant being Mr J. May. Captain James Anderson and Staff-Commander H. A. Moriarty, R.N., were once more to be seen on board the great ship, the former as her captain, and the latter as navigating officer. Professor Thomson was on board as consulting electrical adviser to the Atlantic Telegraph Company, whilst Mr C. F. Varley was ashore at Valentia as their electrician. Mr Latimer Clark was also at Valentia, representing Messrs Bright and Clark as consulting engineers to the Anglo-American Telegraph Company,* Mr J. C. Laws and Mr Richard Collett† being respectively aboard and ashore at the Newfoundland end representing the same firm. Mr Glass was ashore at Valentia for the purpose of giving any instructions to his (the contractor's) staff on board, whilst Mr Gooch and Mr Field were on board the "Great Eastern" as onlookers, and as watchers of their individual interests. ‡

On the 7th July, the "William Cory" landed the shore end in Foilhommerum Bay, and afterwards laid out 27 miles of the intermediate

---

testing and signalling was put into practice. Both Mr Cromwell Varley and Professor Thomson are said to have devised similar methods to that of Mr Willoughby Smith.

On this occasion, also, condensers were applied by Mr W. Smith to the receiving end of the cable, having the effect of very materially increasing—indeed, sometimes almost doubling—the working speed. Though Mr Smith was the first to do this in a practical way, it transpired afterwards that Mr Varley had taken out a patent in 1862 embodying such a principle for working long cables. Mr Varley was a man of many patents.

* Sir Charles Bright, M.P., was engaged elsewhere at the time of the expedition.

† At a later period, after both the 1865 and 1866 cables were in working order, Mr Collett actually sent a message from Newfoundland to Valentia with a battery composed of a copper percussion cap, and a small strip of zinc, which were excited by a drop of acidulated water—the bulk of a tear only! This was during some experiments carried out by Dr Gould on behalf of the Astronomer-Royal (in concert with the Magnetic Telegraph Company), between Greenwich and Newfoundland, via the Atlantic cables, for the verification of longitudes in the United States.

‡ There was also on board Mr J. C. Deane, the secretary of the Anglo-American Telegraph Company, whose diary proved of much use to *The Times* and other newspapers in the absence of Dr Russell, who had so vividly and thrillingly described the events of the previous expedition.

cable. On the 13th, the "Great Eastern" took the end on board, and, having spliced on to the remaining three miles of similar cable coiled on the top of the deep-sea section, started paying out. The track followed was parallel to that followed the year before, but about 27 miles further north. There were two instances of fouls in the tank, due to broken wires catching neighbouring turns and flakes, and thus drawing up a whole bundle of cable quite close (forming an apparently inextricable mass of kinks and twists) to the brake drum. In each case the ship was promptly got to a standstill, and all hands set to unravelling the tangle. With a certain amount of luck, therefore, coupled with much care, neither accidents ended fatally; and, after straightening out the wire as far as possible, paying out was resumed. Fourteen days after starting, the "Great Eastern" arrived (see below) off Heart's Content,* Trinity Bay, where the "Medway" joined on and landed the shore end partly by boats, thus bringing to a successful conclusion this part of the expedition. The total length of cable laid was 1,852 N.M., average depth 1,400 fathoms. After much rejoicings† during the coaling of the "Great Eastern,"‡ the Telegraph Fleet once more set out to sea.

---

* This is situated on the opposite side of Trinity Bay to Bull Arm, where the 1858 cable had been landed, and not so far up. It was supposed to be even more protected than Bull Arm, from which it is some 18 miles.

† These were at first somewhat dampened by the fact that the cable between Newfoundland and Cape Breton (Nova Scotia) still remained interrupted, and that consequently the entire telegraphic system was not even now complete. However, in the course of a few days this line was repaired, and New York and the rest of the United States and Canada put into telegraphic communication with Europe.

‡ Six steamers, laden with coal, had set out from Cardiff some weeks in advance to feed the "Great Eastern" on her arrival on the other side of the Atlantic.

[*From the Painting by Henry Clifford.*

S.S. "Great Eastern" approaching Heart's Content, Trinity Bay, in Completing the Laying of the 1866 Atlantic Cable.

### RECOVERY AND COMPLETION OF 1865 CABLE.

It now remained to find the end of the cable lost on the 2nd August 1865, situated about 604 miles from Newfoundland, to pick it up, splice on to the cable remaining on board, and finish the work so unfortunately interrupted the year before. The difficulties to be overcome can be readily imagined, the cable lying 2,000 fathoms deep without mark of any kind to indicate its position. The buoys put down after the accident had long since disappeared, either their moorings having dragged during various gales of wind, or the wire ropes which held them having chafed through, owing to incessant rise and fall at the bottom. The position of the lost end had to be determined by astronomical observations, which necessitate clear weather, and can then only give approximate results, unless frequently repeated, on account of the variable ocean currents, which sometimes flow at the rate of three knots—*i.e.*, three nautical miles per hour. Moreover, for grappling and raising the cable to the bows, the sea must be tolerably smooth, and in that part where the work lay a succession of fine days is rare, even in the month of August. However, they still had on board Captain Moriarty, one of the ablest navigators in the world. Added to this, the greater portion of the cable in deep water had been paid out with about 15 per cent. slack.

The chiefs of the expedition, fully confident of success, hastened their preparations, and on the 9th of August 1866 the "Great Eastern" again put to sea, accompanied by S.S. "Medway." On the 12th, the vessels arrived on the scene of action, and joined company with H.M.S. "Terrible" (Captain Commerell\*) and S.S. "Albany," these vessels having left Heart's Content Bay a week in advance to buoy the line of the 1865 cable and commence grappling.

Mr Canning's plan was to drag for the cable near the end with all three ships at once. The cable, when raised to a certain height, was to be cut by the "Medway," stationed to the westward of the "Great Eastern," so as to enable the latter vessel to lift the Valentia end on board.†

When the "Great Eastern" arrived on the grappling ground, the "Albany" (with Mr Temple in engineering charge) had already hooked

---

\* Now Admiral of the Fleet Sir J. E. Commerell, V.C., G.C.B.

† This being, of course, before the days of cutting and holding grapnels as we now have them. These render it possible for a single ship to effect repairs, even where it is out of the question to recover the cable in one bight. Mr Claude Johnson, Mr F. R. Lucas, Mr W. F. King, F.R.S.E., Mr H. Benest, and Professor Andrew Jamieson, F.R.S.E., have—amongst others—devised special grapnels of this character.

and buoyed the cable; but the buoy chain having been carried away, they not only lost the cable, but 2,000 fathoms of wire rope besides.

On the 13th of August, the "Great Eastern" made her first drag, about 15 miles from the end; and, after several vain attempts, the cable was finally hooked and lifted about 1,300 fathoms. During the operation of buoying the grappling rope, a mistake occurred which resulted in the rope slipping overboard and going to the bottom.

The "Great Eastern" now proceeded six miles to the eastward, and commenced a new drag, for raking the ocean bed with 2,400 fathoms of wire rope. About eleven o'clock at night the grapnel came to the surface with the cable caught on two of the prongs. The cable thus successfully brought up was parti-coloured like a snake—half a muddy white with the ooze of microscopic shells on which it had rested, and half as black and shiny as when it left the factory—affording positive proof that hereabouts, at all events, the composition of the bottom was such that the cable had not sunk entirely into it. Boats were quickly in position alongside the grapnel. Shortly afterwards they were endeavouring to secure the cable to the strong wire rope, by means of a nipper, when the grapnel canted, allowing it to slip away from the prongs—like a great eel—and disappear into the sea.

On the 19th, the cable was once more hooked, and raised about a mile from the bottom, but the sea was too rough for buoying it. During the following week all three vessels dragged for the cable at different points, according to the plan previously arranged, but the weather was unfavourable, and the cable was not hooked, or, if hooked, had managed to slip away from the grapnels. The ship's company about this time became discouraged —in fact, more and more convinced of the futility of their efforts.

On the 27th, the "Albany" signalled* that they had got the cable on board with a strain of only three tons, and had buoyed the end; but it was soon discovered that her buoy was 13 miles from the track of the cable, and that she had recovered a length of three miles which had been purposely paid overboard a few days before.

Shifting ground to the eastward about 15 miles, the vessels were now working in a depth of 2,500 fathoms.

As the store of grappling rope was diminishing day by day, and the fine season rapidly coming to an end, Mr Canning decided to proceed at

---

* During this expedition much use was made of the invaluable system of day and night visual telegraphy introduced to the Navy by Captain Colomb, R.N., and already referred to with regard to the Persian Gulf cable, during the laying of which it was first employed in connection with cable expeditions.

INK-PHOTO. SPRAGUE & C? LONDON.

Published by Crosby Lockwood & Son, London.

RECOVERY OF 1865 ATLANTIC CABLE, BY S.S. "GREAT EASTERN,"

FROM A DEPTH OF TWO MILES,

At 0·50 a.m., on Sunday, September 2nd, 1866.

(After the Painting by Henry Clifford.)

[To face p. 100.

once 80 miles further east, where the depth was not expected to exceed 1,900 fathoms, and there try a last chance.

After repeated failures, the cable was hooked on the 31st of August by the "Great Eastern" (when the grapnel had been lowered for the thirtieth time), and picking up commenced in very calm weather.* When the bight of cable was about 900 fathoms from the surface, the grappling rope was buoyed. The big ship then proceeded to grapple three miles west of the buoy (Fig. 47), and the "Medway" (with Mr London on board) another two miles, or so, west of her again. The cable was soon once more hooked by both ships, and when the "Medway" had raised her bight to within 300 fathoms of the surface she was ordered to break it. The "Great Eastern," having stopped picking up when the bight was 800 fathoms from the surface, proceeded to resume the operation as soon as the intentional rupture of the cable had eased the strain, which, with a loose end of about

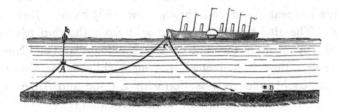

FIG. 47.—Diagram illustrative of the final method adopted for
picking up 1865 Atlantic Cable.

A, Point where cable was buoyed by " Great Eastern."
B, Point where cable was broken by " Medway."
C, Bight of cable ultimately brought to surface by " Great Eastern."

two N.M., at once fell from 10 or 11 tons to 5 tons. Slowly but surely, and amid breathless silence, the long-lost cable made its appearance at last (see plate opposite), for the third time, above water a little before one o'clock (early morn) of 2nd September. Two hours afterwards the precious end was on board, and signals were immediately exchanged with Valentia.

This was at once led into the testing-room, where Mr Willoughby Smith, in the presence of all the leaders on board, applied the tests which were to determine the important question regarding the condition of the cable, and whether it was entirely continuous to each end.

In a few minutes all suspense was relieved, the tests shewing the cable to be healthy and complete; and immediately afterwards (in response to

---

* The monster vessel did her work admirably. To quote the words of an eye-witness : " So delicately did she answer her helm, and coil in the film of thread-like cable, that she put one in mind of an elephant taking up a straw in its proboscis."

the ship's call) the answering signals were received from the Valentia end, which were received with loud cheers that echoed and re-echoed throughout the great ship.

Let us now look at those patiently watching day after day, night after night, in the wooden telegraph cabin on shore, the experiences of whom may be taken as a fair sample of those of the electrician ashore during repairing operations in the present day.

Such a length of time had elapsed since the expedition left Newfoundland that the staff at Foilhommerum (under the superintendence of Mr James Graves) felt they were almost hoping against hope. Suddenly, on a Sunday morning at a quarter to six, while the tiny ray of light from the reflecting instrument was being watched, the operator observed it moving to and fro upon the scale. A few minutes later the unsteady flickering was changed to coherency; the long speechless cable began to talk, and the welcome assurance arrive: "Canning to Glass, Valentia. I have much pleasure in speaking to you through the 1865 cable. Just going to make splice." The glad tidings were also sent from the ship *via* Valentia to London, and by means of the 1866 cable to Newfoundland and New York. Thus for the first time it happened that men, whilst being tossed about in a stormy sea, could hold conversation with Europe and America at one and the same time, without being able to discern either!

The recovered end was spliced on without delay to the cable on board, and the same morning at seven o'clock the "Great Eastern" started paying out about 680 N.M. of cable towards Newfoundland.

On the 8th September, when only 13 miles from the Bay of Heart's Content, just after receiving a summary of the news in *The Times* of that morning, the tests showed a fault in the cable. The mischief was soon found to be on board the ship, and caused by the end of a broken wire which, bending at right angles under the weight of the men employed in the tanks, had been forced into the core. This occurrence explained the probable cause of the faults (of same character) which had shewn themselves during paying out the year before, tending to remove all suspicion of malicious intent. The faulty portion having been cut out, and the splice made without delay, paying out again proceeded, finishing the same day at eleven o'clock in the forenoon. The "Medway" immediately set to work laying the shore end, and that evening a second line of communication across the Atlantic was completed. The total length of this cable, commenced in 1865, was 1,896 miles; average depth, 1,900 fathoms.

In connection with the above expeditions, Mr R. A. Glass and Mr Samuel Canning received the honour of knighthood, Professor W. Thomson, F.R.S. (since Lord Kelvin), receiving a similar distinction. Captain James

Anderson was also knighted, and Staff-Commander H. A. Moriarty, R.N. (as a Government servant), received his C.B. on this occasion.* Mr Curtis M. Lampson, the deputy-chairman of the original Atlantic Company,† had conferred on him the dignity of a baronetcy, as had also Mr Daniel Gooch, M.P.

[*From the Painting by Henry Clifford.*

S.S. "Great Eastern" at Night bumping against Mark Buoy No. 1, on predicted line of 1865 Cable, just after hooking it.

The main feature and accomplishment in connection with the second and third Atlantic cables of 1865 and 1866 was the recovery of the former

---

* A striking instance of this officers navigating skill is as follows :—The observations taken during the partial laying of the 1865 cable having been made by him, when embarking on the recovery and completion of this line in 1866, he went out in the "Albany" to place a buoy, marking the line of cable at a suitable (two-mile) distance from the broken end. Others were afterwards put down along the supposed line of cable.

When the "Great Eastern" subsequently succeeded in hooking the cable, and when, at 11 P.M., she was in the act of buoying the grapnel rope, the first mark buoy previously put down in accordance with Captain Moriarty's observations was discovered to be bumping against the ship's side, near the port paddle, as seen in the illustration given above.

† The Right Hon. James Stuart-Wortley, Q.C., M.P., chairman of the company, refused the honour offered him in connection with this great undertaking. It need hardly be said that Mr Cyrus Field would have been similarly graced by Her Majesty, but for the fact of his being a United States subject, with a strong appreciation for his citizenship.

in deeper water than had ever been before effected,[*] and in the open ocean,[†] just as in the first 1858 line it was the demonstration of the fact that a cable could be successfully laid in such a depth and worked through electrically.

Never in the history of the application of science to the arts have more wonderful results been obtained, or more singular difficulties surmounted, than in the Atlantic oceanic cable work between 1857 and 1866.[‡] From this date, at the latest, the pioneer stage may certainly be said to have ceased. Although there have since been many prominent names connected with submarine telegraphy, they are those of men who have gained reputation, not by any startling improvements, but rather by patient elaboration and application of principles laid down by the early pioneers, under whom they have in some instances served in early days.

Owing to the uniformly low temperature (about 39° F.) at the great depths at which they were laid, as well as to the tremendous pressure of water above them,[§] both these cables shewed an enormous improvement in insulation and carrying capacity by submergence. A great part of the 1865 cable having been down at the bottom for over a year, this, when completed, worked better even than the 1866 line.

In the course of testing these cables previous to the owning company taking them over, the following (as an example) was found in proof of their high and greatly improved condition subsequent to submergence:— If either of them was disconnected from the earth and charged with electricity, it required more than an hour for half of the charge to escape through the insulating covering to the earth.

In illustration of the degree of sensibility and perfection attained at this period in the appliances for working the cable, the following experiment is

---

[*] As already shewn, cable repairs had been previously carried out in the Mediterranean in 1,400 fathoms.

[†] The recovery and repair of a cable from the depths of the open ocean are now matters of ordinary everyday occurrence, forming part and parcel of cable operations generally. Thus, in 1882, Mr Edward Riddle—who, on behalf of the Telegraph Construction Company, has been connected with work of this sort almost from the beginning —succeeded in picking up and renewing the Falmouth-Lisbon line in 2,600 fathoms in the Bay of Biscay. This is probably the greatest depth in which repairs have been accomplished, though it is true that one end was in comparatively shallow water.

[‡] In 1867 a highly interesting and instructive paper regarding these expeditions was read before the Royal United Service Institution (see *Journal*, vol. xi.) by Captain H. A. Moriarty, R.N., C.B., who had played so conspicuous a part on each of them.

[§] The effects of both conditions being that of consolidating the gutta-percha dielectric.

of somewhat striking interest :—Mr Latimer Clark had the conductor of the two lines joined together at the Newfoundland end, thus forming an unbroken length of 3,700 miles in circuit. He then placed some pure sulphuric acid in a silver thimble,* with a fragment of zinc weighing a grain or two. By this primitive agency he succeeded in conveying signals twice through the breadth of the Atlantic Ocean in little more than a second of time after making contact. The deflections were not of a dubious character, but full and strong, the spot of light traversing freely over a space of 12 inches or more, from which it was manifest that an even smaller battery would suffice to produce somewhat similar effects.

Notwithstanding the dimensions of the core, these cables were worked slowly at first in ordinary practice, and at a rate of about eight words per minute. This, however, soon improved as the staff became more accustomed to the apparatus, and steadily increased up to fifteen and even seventeen words per minute on each line, with the application of condensers at the receiving end.†

Unfortunately both these cables broke down a few months later, and one of them again during the following year. These faults were localised with great accuracy from Heart's Content by Mr F. Lambert, on behalf of Messrs Bright and Clark, the engineers to the Anglo-American Telegraph Company.‡

---

* Mr Clark borrowed the thimble—which was a very small one—from Miss Fitzgerald, the daughter of the Knight of Kerry, living at Valentia. This gentleman shewed great interest in, and offered every assistance in furtherance of, the Atlantic cable enterprise from the very beginning.

† Condensers were not used at the *sending* end until the introduction of the siphon recorder.

‡ In some instances the faults were found to be due to grounding icebergs at the entrance to Trinity Bay.

## SECTION 2.—FURTHER DEVELOPMENTS.

In the last section, the history of submarine telegraphy was brought down to the close of the ever-memorable double achievement of successfully laying the Atlantic cable of 1866, and recovering and completing the cable of 1865.

With the success attending these important operations, began a new era in submarine telegraphy. The period of first attempts was virtually over. It had been demonstrated that not only in narrow and shallow seas, but across the great oceans of the globe, telegraphic lines could be laid and maintained. Improvements in details might still be found necessary, but it was now believed that a suitable type of cable had been discovered for deep water,* whilst proper methods had been devised for recovering and repairing in great depths.

Again, improved instruments for transmitting signals with increased rapidity had been constructed, and more precise methods had been invented for testing cables, as well as for localising faults during manufacture and laying. Moreover, an accurate conception had been arrived at as regards the requirements for giving a certain signalling speed on a submarine cable, the result being decisive (at any rate so far as concerns the necessary dimensions of the core—*i.e.*, the section of the conductor and the thickness of the dielectric) for given lengths. Various enterprises which followed upon the successes of 1866 may now ·be referred to.

In 1868 the ANGLO-MEDITERRANEAN TELEGRAPH COMPANY was formed for the purpose of establishing fresh communications between Malta and Alexandria, by means of a direct deep-water cable of about 900 miles across the Mediterranean. This was found necessary owing to the constant failure of the old line between these points, which had been laid on a bad bottom in shallow water, touching at intermediate points along the north coast of Africa.†

This new cable was laid with complete success. The Telegraph Construction and Maintenance Company were the contractors, with Sir Samuel Canning and Mr Willoughby Smith as their chief engineer and electrician respectively. Sir Charles Bright acted in the double capacity of engineer and electrician to the Anglo-Mediterranean Company.

---

* The since-discovered inherently weak features in this type have already been dwelt on.

† The old cable was acquired at the same time by the newly-formed company, an additional undertaking of which was a land line through Italy.

The cable gave every satisfaction afterwards as regards its working. The core was composed of copper conductor = 150 lbs. per N.M., and gutta-percha dielectric = 230 lbs. per N.M.* The speed obtained was nineteen words per minute, with one Menotti cell direct to line—*i.e.*, without condensers.†

In 1869 France was put into direct telegraphic communication with America by means of a cable from Brest to the island of St Pierre, and another from St Pierre to Sydney, Nova Scotia.‡ The former length was manufactured by the Telegraph Construction and Maintenance Company, and the latter by Mr W. T. Henley. The Telegraph Construction and Maintenance Company were the contractors for laying the whole cable on behalf of the FRENCH ATLANTIC CABLE COMPANY (Société du Cable Transatlantique Francais).§

This work was successfully accomplished from the "Great Eastern" (Captain Robert Halpin), by the same staff as had laid the 1866 cable aboard the same ship. Owing to the route, this line was materially longer than the previous Atlantic cables, its length (from Brest to St Pierre) being as much as 2,685 N.M.‖

The working speed attained on the French Atlantic cable was $10\frac{1}{2}$ words per minute. The conductor of the Brest - St Pierre section was composed of seven copper wires stranded together, weighing 400 lbs. per N.M., covered with a gutta-percha insulator of the same weight. The core of the St Pierre-Sydney section was made up as follows :—Copper = 107 lbs. per N.M.; gutta-percha = 150 lbs. per N.M.

Unfortunately this cable also (as well as several others following after) was of the same type, mechanically, as the 1865 and 1866 Atlantic cables, each iron wire being enveloped in a serving of hemp-yarns. As previously stated, the result of practical experience shews this type to have been a

---

* The armour of a great part of this line was close-sheathed.

† This cable afterwards formed the European end of that vast world-wide system of electro-metallic nerves to the East and Far East, which is now owned by the "Eastern" and "Eastern-Extension" Telegraph Companies.

‡ This enterprise, although mainly on behalf of France and the rest of the European Continent, was principally advanced by financiers in this country : the working of the cable was also chiefly under British direction and management. The concessionaires were Baron Emile d'Erlanger, of Paris, and Julius Reuter, of London.

§ Afterwards, in 1873, merged, with its cable, into the Anglo-American Telegraph Company and its system.

‖ This forms the greatest length of cable which has ever been laid in one continuous length up to the present time. The length of the section is now, after various repairs, 2,717 N.M.

mistake; although at that time it was considered by many to be the best possible for deep water, not only with view to the process of laying it, but also to its durability and the facility of its recovery.*

Another long and important length of cable manufactured and laid by the Telegraph Construction Company, from S.S. "Great Eastern," in the same year (1869), was a direct line for the BRITISH-INDIAN SUBMARINE TELEGRAPH COMPANY† between Egypt (Suez) and India (Bombay). This passed, of course, down the Red Sea, touching at Aden, and then crossing the Arabian Sea.‡

The armour in the above cable was composed of the same iron and hemp combination which was adopted for the 1865 and 1866 Atlantic lines, as well as for that of 1869, etc. In this instance, however, a hempen whipping was tightly applied round the completed cable with a lay equivalent to about ¾ inch between each turn, so as to prevent any broken wires becoming disengaged.§

Following this came the extensions carried out by the same firm (Telegraph Construction and Maintenance Company) on behalf of the BRITISH-INDIAN EXTENSION and CHINA SUBMARINE TELEGRAPH COMPANIES respectively, in which Sir C. Bright acted as engineer for the companies; also, in 1872, a cable from India (Madras) to Australia, for the BRITISH-AUSTRALIAN TELEGRAPH COMPANY, *via* the Straits Settlements.

---

* This cable, which has been "down"—electrically speaking—for some years, has proved a costly line in repairs. One expedition alone is said to have run into as much as £95,000, but this (in 1894) was a record case.

† This was the outcome of the ANGLO-INDIAN TELEGRAPH COMPANY, formed for the purpose of establishing direct telegraphic communication to India by means of submarine cables, instead of relying upon land lines to the Persian Gulf, and a cable thence as heretofore. The Anglo-Indian Company, however (which had acquired the Egyptian landing rights previously granted to the Red Sea Company, and had secured as their engineers Sir Charles Bright, M.P., and Mr Latimer Clark), failed to raise the capital required for carrying out their enterprise. Mr H. C. Forde acted as engineer to the expedition ultimately undertaken by their successors, the British-Indian Company.

‡ Mr J. C. Parkinson has described this expedition in an interesting narrative, entitled "The Ocean Telegraph to India" (Wm. Blackwood and Sons, Edinburgh).

§ Considerable trouble of this nature had been experienced with the preceding cables of this type, as has already been shewn. Wires breaking at a weld, for instance, would get loose, and the broken ends become entangled with, or actually pierce, the next turn or flake in the tank under pressure. Thus, not only was paying out rendered most hazardous, with serious prospects of a "foul," but the core, also, was liable to become fatally damaged at any moment by the piercing of a broken wire. This actually occurred at least three times during the previous undertakings, with fatal results as regards insulation. The above plan being found to meet the difficulty to a great extent, it was again adopted for all subsequent cables of this description, though, of course, out of place for the ordinary close-sheathed cable of the present day.

It was about this time that the MARSEILLES, ALGIERS, AND MALTA TELEGRAPH COMPANY was promoted. This project—viz., the telegraphic connection of these important Mediterranean places by means of a cable touching the Algerian coast at Bona—was also successfully accomplished.

A few months later the FALMOUTH, GIBRALTAR, AND MALTA TELEGRAPH COMPANY was formed, to complete a direct submarine communication by telegraph between Great Britain and her Eastern possessions. As the result of pressing advances on the part of the Portuguese Government, this cable was ultimately taken into Carcavellos, Lisbon, on its way to Gibraltar. The starting-point chosen for it eventually was not Falmouth but Porthcurno, a quiet spot about ten miles from the Land's End,* the company leasing a land line between there and London.†

All the above-mentioned schemes were put into effect during that peculiarly busy telegraphic period characterising the end of the seventh and the beginning of the eighth decade of this century. The cables were, in each instance, laid by the Telegraph Construction Company,‡ although owing to the great pressure of business at that firm's works at this period, the manufacture of certain portions of them was undertaken on their behalf by Mr Henley. It was over this group of cables that Willoughby Smith's process of gutta-percha manufacture was first employed for the core.§

It is perhaps worthy of note that none of the companies who ordered the construction and submersion of these lines, and who worked them (or are still working them) to the great benefit of the world's commerce, were assisted by any Government monopolies, subsidies, or guarantees—in this respect differing from the original Red Sea cables. It may truly be maintained, therefore, that the Governments and mercantile communities of the

---

* Boasting of a large telegraphic population, and including a telegraphic training school, it now belongs to the " Eastern " and its allied telegraphic companies.

† The cable between Porthcurno and Lisbon passes through very deep water, and lies for some distance at a depth of nearly 2,700 fathoms—one of the greatest depths in which any cable has been laid. It has, moreover, been *repaired* in these waters.

‡ During most of this work Captain (afterwards Rear-Admiral) Sherard Osborn, C.B., F.R.S., acted as managing director to the Telegraph Construction and Maintenance Company in succession to Sir Richard Glass. Captain Osborn was succeeded later on by Rear-Admiral Sir G. H. Richards, K.C.B., F.R.S., and, still later, by Mr William Shuter, the present managing director. Sir George Richards became chairman of the company on the death of the late Sir George Elliot, Bart., M.P., in 1893. He, however, only lived to occupy this position for three years, the present chairman being Sir Robert Herbert, G.C.B., formerly Under-Secretary of State for the Colonies.

§ Almost all these cables have since been duplicated, several of them triplicated, and one quadrupled.

world owe the vast benefits they have reaped from direct submarine telegraphic communication during the past twenty-five years entirely to the enterprise of a few British merchants and the original shareholders in these companies. Without claiming any very lofty philanthropic motives (or, indeed, anything more romantic than shrewd pluck and enlightened self-interest) for these gentlemen, they certainly risked their money in undertakings which at the time appeared to many of their fellow-countrymen to be the maddest of speculations. It will be observed that all these cables were laid along the principal trade routes, where, under hostile conditions, the navy would be able to afford some protection, since these would naturally be the routes most carefully patrolled by our men-of-war for other purposes.

Another point deserving incidental notice, is that all these companies were being floated just after the Telegraph Purchase Bill of 1868 had been passed. The promoters rightly calculated that this afforded an opportunity of securing for new telegraphic ventures a good deal of the capital now let loose by the "winding up" of the "Magnetic," "Electric," "United Kingdom," "Reuter," and other telegraph companies which had, up to that time, shared amongst them the control of the land lines of Great Britain and Ireland. For this result of the acquisition of our land telegraphs by the State had the necessary further consequence of liberating something like £8,000,000 sterling for re-investment by those who looked favourably on electric telegraphs as a subject of safe and sure remuneration. Moreover, the publicity which the proceedings of the Parliamentary Commission, appointed in connection with the Government Purchase Scheme, gave to the lucrative nature of telegraphic enterprises generally, together with the recent success of the Atlantic cable in deep water, emboldened financiers and capitalists to create fresh investments of the same character by promoting and supporting new companies for the further extension of submarine telegraphic enterprise.*

Various schemes were promulgated from time to time for different forms of light cables, *i.e.*, cables without any of the ordinary iron wire sheathing. None of these, however, came to anything. It would seem that nobody cared at that period—any more than they do now—to risk so considerable a part of the capital of a submarine telegraph company as is employed (and, in the most literal sense, *sunk*) in the cable itself, by staking its success

---

* Former shareholders in the "Magnetic" Company were especially predisposed to take a pecuniary interest in submarine telegraphy, in virtue of their connection with the early Atlantic lines.

upon an experimental—*i.e.,* untried—change in its structure. Everybody preferred to wait for somebody else to make the experiment first. Iron-sheathed cables having been proved to be fairly satisfactory, the telegraph world generally thought it best to "leave well alone." It was uncertain, indeed, in the first place, whether an unsheathed cable could be laid at all; in the second, whether, if laid, it could ever subsequently be recovered. Thus, practically speaking, the original type of submarine cable has been adhered to throughout, with certain modifications and improvements to meet various practical requirements. Nevertheless, in several concerns which were started, light cables, to be laid between certain points, formed the salient feature of their programme, whether for the manufacture and laying, or for the owning and working of submarine telegraphs.

Quite a number of companies were "floated" about the same time for effecting telegraphic communication with the East, with America, and other parts of the world; but those mentioned here were the schemes actually carried out. It will be observed that these successful enterprises were almost all set on foot by the same financial group—men of shrewd business capacity, as has been clearly proved in the sequel.

In 1868 the INDO-EUROPEAN TELEGRAPH COMPANY was formed for establishing a more speedy and reliable line of communication between England and India than that hitherto afforded by the Turkish State land lines from Fāo to Constantinople *via* Baghdad. [The latter had been erected about the same time as the Persian Gulf cables were laid to India. Their communication with England was by the Continental European system of telegraphs, through Austria to Paris and hence on to Calais.*] The line † was constructed for this company by Messrs Siemens Brothers, who indeed had been the originators of the scheme, ‡ which was completed by them in January 1870. It passed through Germany and Lower

---

* The section from Constantinople to Baghdad had been open for traffic as early as 1861, but the extension to Fāo, for bringing Turkish Arabia (*via* the Persian Gulf cables and Indian land lines) into communication with Bombay and the rest of India, was not effected until eight months after the laying of the Gulf cables.

† In the course of this book, whenever an important land line of any length is in question, it should be understood that the "line" consists of at least two conductors erected along the same route and on the same poles, or not, as the case may be.

‡ More commonly than not in telegraph and cable matters, the contractors of an undertaking will also be found to have taken the first steps in its original promotion—their objects being, in the first place, to keep their plant and staff remuneratively employed, and in the second, to obtain a good profit from each such ancillary (or alimentary) enterprise.

Russia, a good traffic being picked up as far as Teheran in Persia, where it joins the system of the Indo-European Telegraph Department * of the Indian Government Telegraphs. This system extends, *via* Ispahan and Bushire (and including the Persian Gulf cables already alluded to), to Karachi (or *Kurrachee*, as it used to be spelt), where it meets the rest of the Indian Government telegraph systems. For British communication with the Continent of Europe, the Indo-European Company lease two Government cables, or rather two conductors, each in separate cables. The line across the Continent is entirely aerial, with the exception of a few miles under a part of the Black Sea.

In 1869 the GREAT NORTHERN TELEGRAPH COMPANY was established for the purpose of telegraphic communication between Europe, China, and Japan, by means of the Russian State lines through the Russian Empire, connecting up their cables in Europe with those in the Chino-Japanese waters. This company was the result of an amalgamation of the " Danish, Norwegian, and English," the " Danish-Russian," and the " Norwegian-English," Telegraph Companies. The first of these were owners of a cable between Denmark and Norway, laid by Messrs Newall in 1867, and of one between England and Denmark, laid by this firm in 1868, Messrs Hooper having supplied india-rubber core for the same. The second-named company were owners of a cable between Sweden and Russia, laid by Messrs Henley in 1869. The third company (namely, the " Norwegian-English ") had owned a cable between Norway and Scotland, which was laid in 1869. Messrs Henley were concerned in its construction and submergence, the core being Hooper's.

A few years later—namely, in 1872—the Great Northern China and Japan Extension Telegraph Company (formed in 1870 and owning cables and land lines round about China and Japan) was also amalgamated with the Great Northern Telegraph Company. Thus, at the present time, the Great Northern Company owns 3,518 N.M. of submarine cable in European

---

* This system was erected by Major (afterwards Sir) J. U. Bateman-Champain, R.E., and a staff of "Sappers." The object of the Indian Government in constructing it was partly to bring Teheran, the capital of Persia, into telegraphic connection with India and Europe, and partly to establish more certain communication between these two great sections of the world than the old Turkish route afforded, or at least, to create an additional line of communication in case of the latter's failure. This last-mentioned event was by no means rare, the Baghdad-Fāo section of the line having constantly been tampered with by the uncivilised Arabs of those parts. Another advantage anticipated from the new line was that it would be worked by educated and disciplined operative clerks instead of by the unreliable Turkish underlings of their Government lines. The political importance also of no longer trusting all our telegraphic eggs to one basket—and that a Turkish one—was sufficiently obvious.

waters, and 3,511 N.M. in those of the Far East—a very equal division. This company, partly on account of the tropical climate of the southern portion of the China Seas, have adopted india-rubber core, as made by Messrs Hooper, for some of their cables; in fact, their system is made up of about equal lengths of india-rubber and gutta-percha cores—the two cores meeting, in some cases, in the same length of cable. Their experience as to the relative merits of each under similar circumstances must, therefore, be most valuable. No signs of the teredo have, however, been met with in any of these cables, in the China Sea or elsewhere; and, as a matter of fact, some of the gutta-percha cores are not even protected by brass tape.

As regards the European system of this company—just as in the case of the "Eastern" Company's European system—a breakdown of one or two cables would hardly be felt, as the work can immediately be divided between other lines of communication, so ingeniously is the system distributed, besides being connected by Government land wires through Denmark, Norway, and Sweden.

The way in which the Great Northern Company's system came to extend itself in the Far East is as follows:—As early as 1854, in view of the (at that time) apparent impossibility of spanning the Atlantic by a submarine cable, a project was mooted for establishing telegraphic communication between Europe and America by means of land wires through Siberia on the one hand, and through North America terminating at Alaska (which was then Russian territory) on the other; these two systems were to be joined by what would be only a comparatively short length of cable—less than 100 miles—in the shallow water (maximum about 300 fathoms) of the Behring Strait.* This project, of course, had to receive the assent and support of the Russian Government before it could be put into practical effect, and it was not until 1865 that the Western Union Telegraph Company obtained a concession for carrying it out. The Russian Government undertook to build the line through Siberia to the Behring Strait, whilst the American Company were to construct one across the North American Continent to the nearest point on the opposite side, and to connect the two up by submarine cable. When both the land lines had been erected for some distance, the success of the Atlantic cables of 1865 and 1866 caused the Americans to abandon the above projected

---

* This route would have been quite impracticable owing to the constant snow and ice at the approaches. Further south a cable might have been laid across the Behring Sea *via* the numerous Aleutian Islands, the distance here being some 750 miles between the two continents, and the maximum depth—avoiding certain parts—about 800 fathoms.

connection with the Old World. The Russians then became anxious
turn their land line—already partly constructed—to some useful accoun
and thus they conceived the idea of utilising them for overland commur
cation with China and Japan, to be completed by submarine cabl
Eventually the concession for the latter was granted to the Great Northel
Company, who had, in fact, already worked in the interests of Russia ar
its Government by establishing communication between that country ar
Denmark, and thence on to England.

These cables between the Russian Empire, Japan, and China (Posiet
Bay, Nagasaki, Shanghai, and Hong Kong), were not actually laid unt
1871.* The Hooper Company contracted for their construction, Hooper
india-rubber core being used, whilst the sheathing was carried out by Mess
Siemens Brothers. They were laid by the staff of the Great Norther
Company, from their new telegraph steamer " H. C. Örsted,"† Mr C. I
Matheson being the engineer in charge, assisted by Captain (afterward
Colonel) V Hoskiær of the Royal Danish Engineers, who subsequentl
succeeded him. ‡

It was afterwards thought desirable to include Amoy in this far-easter
*réseau*. Instead of taking the cable into Amoy from one direction and ou
of it again in another, it was determined to 'tap" the line between Hon
Kong and Shanghai with a double-core branch cable to Amoy; by thi
means a considerable saving was effected in a heavy and expensive typ
of shallow-water cable. This T-piece method was adopted at the sugges
tion of Mr J. R. France, following the example set at Bushire in the cas

---

* It may be remarked that neither the Chinese nor the Japanese had expresse
any anxiety for this telegraphic connection with Europe. Permission, indeed, was nc
obtained from them until after the work had been executed, the shore ends being secretl
landed in drain-pipes. The Chinese, when the cable had been laid, displayed a complet
scepticism as to its uses, and refused to have anything to do with it. They were soo
convinced and converted, however, by various practical demonstrations.

† This was the first case of a cable-repairing ship built expressly for a cable-workin
company. The "H. C. Örsted" was christened after the famous Danish electrical *savat*
of that name. Other ships fitted out for cable work and employed for the same kind c
work were the "William Cory " and the "Chiltern," engaged for the "British-Indian
Company in 1870 (both of which had previously done Atlantic cable work in 1866 an
1867), and the "Hawk," engaged for the Falmouth-Gibraltar Company. The "Chiltern
still does satisfactory duty for the "Eastern" Company. In the early days of submarin
telegraphy very few such boats were fitted out at all, and still fewer specially built fc
cable-owning companies, as it was originally intended that the Telegraph Constructio
and Maintenance Company should maintain in repair all the cables which they laid—
hence the full title of that firm.

‡ Captain Hoskiær was the author of two useful little volumes relating to submarin
telegraphy, one from the engineer's point of view, and the other from that of th
electrician.

of the Persian Gulf line. The required bi-cored cable was made for the Great Northern Company by Messrs W. T. Henley, and successfully laid for them by Mr France in 1873, the length being about 40 N.M., with the T-piece in 30 fathoms of water. This latter device was a complete success, and has given no trouble whatever down to the present day.

The Eastern Extension Telegraph Company's system reaches as far north on the coast of China as Shanghai. There is thus a double telegraphic communication between this place and Hong Kong—namely, one by the Great Northern Company's cable, and the other by the Eastern Extension's; the latter touching, *en route*, at a point near Foochow, instead of at Amoy. These two lines are, however, worked together by both companies under a "pooling" arrangement for their common benefit, so that the whole telegraphic system of each is placed in communication with all these important towns and seaports.* Moreover, an additional working arrangement exists between these two companies, by which both systems may be said to receive individual extension.

All the Great Northern Company's other cables have subsequently been duplicated (A.D. 1883, 1884, and 1891). Unlike similar cable-owning companies, they have usually laid their own cables from their repairing ship, but in 1891 the Telegraph Construction Company laid (as well as made) their duplicate cables, together with some extensions. The "Great Northern" was the first instance of an entirely non-British company promoting submarine telegraphy purely on its own account, and the Danes deserve every credit for this their—far from inconsiderable—share in the work of building up the vast and magnificent system of ocean telegraphs which now encompass almost the whole world.† The same company has also been entrusted with the laying of an independent cable for the German and Danish Governments between those two countries, the manufacturers of which—at any rate of the sheathing—were the eminent German firm, Messrs Felten and Guilleaume. The Great Northern Company, it may be remarked, have always had the reputation of drawing up very detailed specifications for their cables, as well as for very carefully overlooking and indeed controlling the manufacture of them. ‡

---

* The Eastern Extension Company's original cables to China were laid the same year as those of the Great Northern Company's Eastern system—*i.e.*, 1871—and previously referred to.

† The only other foreign nation that has since taken any important part in the extension of submarine telegraphy is the French, who have recently made and laid a considerable length of cable.

‡ It should be noted, before leaving the subject of the Great Northern Company, that they were the first to succeed (in 1881) in establishing land lines in China, in the teeth of a very strong native opposition. Since then the Chinese have built lines for themselves—

The company's engineer-in-chief, Mr P. C. Dresing, originated a special form of cable suitable for shore ends on very rocky coasts. It is of the ordinary shore-end type, but the outer sheathing wires are laid up with a very short lay, so as to get a maximum weight in a given circumference, and thus obtain a more pliable cable than would be the case if the same weight were obtained by means of larger wires laid in the ordinary way. A shore end was made according to Dresing's specification by the Telegraph Construction and Maintenance Company, and laid on the Chinese coast (at Hong Kong) to a distant lighthouse station in 1892; its weight was 28 tons to the mile. This type having lately been adopted by the Anglo-American Company, for their Atlantic cable of 1894, was fully described in the course of an able article concerning that enterprise by Mr Arthur Dearlove, in *The Electrician* of 12th October of the same year.

In 1869 the WEST INDIA AND PANAMA TELEGRAPH COMPANY was formed, with the objects, first, of uniting telegraphically the West India Islands themselves, and the Spanish Main; and, second, of bringing them (as well as Brazil, Colombia, etc.), through the Western Union land lines at Florida,* into communication with the whole of North America and Europe. In connection with the same system, and under the same financial auspices, another similar enterprise, viz., the CUBA SUBMARINE TELEGRAPH COMPANY, was floated for the purpose of laying a cable along the coast of Cuba, thereby avoiding the necessity of using the Spanish Government land lines through that island. This completed† the submarine *réseau* from Florida (U.S.A.) and the Gulf of Mexico in the north, by the Greater and Lesser Antilles to Georgetown, Demerara, on the one hand, and to Colon, and thus on by an aerial line to Panama, on the other. From the

---

employing Danish engineers for the purpose—and have been working them with the assistance of Danish operators, who have gradually trained a large staff of Chinamen in all branches of the service. In 1894 the length of the lines in China consisted of 15,000 miles of wire to about 21,000 miles of country. There were upwards of two hundred stations, worked by a staff of eight hundred clerks.

* As a first part of this scheme, cables had been laid the year before between Florida and Key West (an island), and between Key West and Havana, Cuba. These cables were the property of the International Ocean Telegraph Company of New York, and are now worked for them by the Western Union Company as a part of the latter's system —by far the largest telegraphic system in the United States, indeed the largest land system in the world. The Western Union was founded in 1852 to work the "House" type-printing telegraphs, but in 1854 the competing Morse system amalgamated with it. Since then the Western Union Company have absorbed nearly all the other telegraphs throughout the United States, its principal rival being the Postal Telegraph Cable Company.

† With the help of a short wire across the narrowest part of Cuba.

latter it was already contemplated to run another cable to the West Coast of South America.*

In connection with this West Indian enterprise, some 4,100 N.M. of submarine cable were manufactured and laid, being the greatest length ever dealt with, so far, in any single concern of the kind. The system (see map below) comprised no less than twenty separate sections between the islands from Florida to the north coast of Central and South America. The cable was made by the India-rubber, Gutta-percha, and Telegraph Works Company, of Silvertown† (this being the first important cable contract which this now well-known and flourishing firm had undertaken), and was laid by Sir Charles Bright. No less than five steamers were

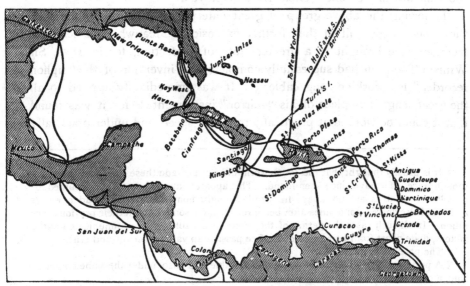

Map shewing the West Indian Cable System and its Connections (1897).

required for the work, including H.M.S. "Vestal" (Captain J. E. Hunter R.N.) as consort, besides three large sailing ships.

These cables had to be laid on what is undoubtedly the very worst bottom that any submarine cable has ever been deposited on, for, in order

---

* This was the scheme of the Panama and South Pacific Telegraph Company, which was formed to connect up the chief towns of South America with Central America and the rest of the world, by submarine cables joining the previously existing land lines through Peru (of the National Telegraph Company), and thus communicating, through other land lines, with every centre of civilisation in the South American Continent. This scheme, however, was never carried out by the above-named company, though, as will be seen, it formed a part of another undertaking later on.

† For brevity's sake, to be henceforth referred to in this work as "the Silvertown Company"—a title by which it has become familiarly and generally known in the telegraphic world.

to reach most of the islands, it has to pass over a chain of coral.* In the manufacture of this great length of cable, the core-compound employed was different from that known as Chatterton s mixture. This new departure did not, however, prove such a successful experiment as had been anticipated ; its failure, indeed, was sufficiently conclusive not to warrant repetition, and since these West Indian lines, Chatterton's compound has always been rigidly adhered to.† Mechanically, the above cables were all that could be desired, constituting, in fact, the first return to the ordinary close-sheathed type for deep water. ‡

Bright's Bells used to be employed as the signalling instrument on some of these lines ; and this is now the only system of cables in which the "mirror" instrument is exclusively employed.

Following the above group of great enterprises in oceanic telegraphy, there was a slight lull in their further extension. Meanwhile its scientific facilities were brought to a greater state of perfection, for in 1870, Sir William Thomson had successfully completed his invention of the "siphon recorder" for working long cables.§ It was immediately applied to all the great lengths in place of his "mirror" instrument ; ‖ for it was found that it could be worked at practically the same speed, and under practically

---

* It is usually thought that the *teredo* worm did not pervade these waters at that time though traces of it have since been found. This apparent anomaly may be accounted fo⁻ as follows :—Cables are nowadays frequently brought home from teredo-ridden waters and relaid elsewhere—sometimes after being re-sheathed, sometimes in their previous condition. In all such cases the germs of the teredo are liable to be communicated at the factory, quite apart from being afforded a free passage to various oceans and seas.

† The patent expired in 1873.

‡ A portion of the second Malta-Alexandria cable had also, under the same engineership, a close armour.

§ The first patent for an early form of this instrument had been taken out as early as 1867.

‖ The mirror is now entirely replaced by the siphon recorder, except on the cables of the "West India and Panama" and "Cuba Submarine" Companies, as above mentioned, and on some of the northern sections of the Western and Brazilian Company. The French company, while employing the siphon recorder on their long sections use the Morse with Siemens' permanent magnet recorder (similar to the Brown-Allan relay) for working some of their shorter lengths in the West Indies. The Great Northern Company, however, are in the habit of using for signalling purposes an instrument known as Lauritzen's "undulator," invented in 1876 by a former member of their staff. This is based upon electro-magnetic principles, and, as regards the recording arrangements, presents points of resemblance to the siphon recorder. This instrument (described fully in Part III.) is capable of being worked through very great lengths of cable and land line joined up together with translating apparatus between. This, moreover, at a high-working speed, such as 65 words per minute through 1,000 N.M., and in another case 90 words a minute *simplex*, and 60 *duplex*, through 550 N.M. Previous to this, the "Great Northern" Company had used Wheatstone's receiver in connection with his transmitter.

the same conditions of length and battery-power as the latter, but with the advantage of yielding an actual record of the signals as received, thus enabling the source of error to be traced—besides other important advantages.* Both these instruments will be found dealt with at some length in Part III. of this work. †

In 1871 the Silvertown Company made a cable for the French Government to be laid between Marseilles and Algiers. This cable was laid for the company by Sir Samuel Canning. In 1872 the DIRECT SPANISH TELEGRAPH COMPANY was formed, principally by those interested in the Silvertown firm. Its object was to connect England and Spain by a direct submarine line, and thus avoid the delays and errors to which traffic, passing over Government lines in foreign countries, is always liable. Accordingly, in the same year, a cable was laid for this company by the newly-formed Silvertown staff, headed by Mr F. C. Webb as engineer-in-chief. ‡ This line extended from a point near the Land's End, Cornwall, to the north coast of Spain, in the neighbourhood of Bilbao.§ Subsequently, in 1885, this company's system came under different direction. and it now forms a part of the "Eastern" group, or network of cable systems.

In 1872 the four companies owning the cables on the direct route to India were amalgamated into the now world-famous EASTERN TELEGRAPH COMPANY, which was registered in June of that year. These companies and their cables (already referred to) were the so-called "Falmouth, Gibraltar, and Malta"; the "Marseilles, Algiers, and Malta"; the "Anglo-Mediterranean," and the "British-Indian." Their amalgamator and successor, the "Eastern" Company, now possesses by far the largest and,

---

* Another strong point in favour of the recorder is that the eyesight is not so liable to be damaged by reading the slip as it is by the strain of following the movements of a "shivering" beam of light from one side to the other. Thus, a constant supply of good operative clerks is more readily obtainable for working the recorder than for the mirror.

† It is not unusual for a company to start with the mirror instrument and then to substitute for it the siphon recorder as soon as all the traffic arrangements are in thorough working order. This plan is especially convenient when the contractors work the cable for private ("service") work during the month's guarantee, in the course of which the proprietary company have a good opportunity of putting their system into working order, the recorder clerks getting the new instruments, etc., set up ready for regular public traffic. The mirror instrument, moreover, is always useful as a reserve at each station in case of anything going wrong with the recorder.

‡ Succeeded in that capacity in later years by Mr Robert Kaye Gray.

§ This cable was originally laid over what proved to be a very remarkable hole about 30 miles north of Bilbao. Here, within a distance of about one N.M., the depth varies from 1,000 to 2,000 fathoms. In the laying of the present cable by the Telegraph Construction and Maintenance Company in 1884 this hole was avoided.

from a national point of view, the most important telegraphic system in the world. It was promoted under the chairmanship of Mr Pender (afterwards Sir John Pender, G.C.M.G., M.P.), with Lord William Hay (now Marquis of Tweeddale) as vice-chairman. Sir James Anderson * became the general manager (afterwards managing director), and Mr George Draper the secretary.

This consolidation having been accomplished, in the following year (1873) the EASTERN EXTENSION, AUSTRALASIA AND CHINA TELEGRAPH COMPANY was formed for absorbing those companies which owned the extension lines to the further side of India, the Straits Settlements, China, and Australia, already referred to. The companies thus incorporated were the "British-Indian Extension," the "China Submarine," and the "British-Australian." The board of this amalgamating company was an equally strong combination to that of the "Eastern" Company, being, in fact, very similarly composed, and presided over by the same chairman, with the Right Hon. W. N. Massey, M.P., as vice-chairman.†

It may be mentioned here that an important extension of this company's system was carried out in 1876, by the submersion of a cable for them (acting in concert with Sir Julius Vogel, K.C.M.G., who represented the Government of New Zealand), by the Telegraph Construction Company, between Australia and New Zealand

Within recent years telegraphic communication with the East, by the Eastern Telegraph Company, has been further strengthened through the leasing of a line for their service between London and Marseilles, and thence by cable to Bona and Malta. The principal portion of the traffic between Great Britain and the East now passes over this line, in preference to the more oceanic and circuitous route *via* Lisbon, Gibraltar, and Malta.

As a part of the "Eastern" Company's system, in 1874 the BLACK SEA TELEGRAPH COMPANY was formed for the purpose of working a cable across the Black Sea from Constantinople to Odessa. At the latter point it was to meet the Russian State lines, communicating with the Great Northern Company's system at Moscow, and thus to connect Constantinople,

---

* Succeeded, since his death in 1893, by Mr John Denison Pender, the present managing director.

† It is perhaps hardly necessary to add that the directorate of these two great companies represents a similarity very nearly approaching to identity between the financial groups which control each of them, and which used to control the original seven companies from which they sprang. These enterprising capitalists—some of whom from the very beginnings of ocean telegraphy had staked their money on its success—may be justly described as the commercial pioneers of the above great undertakings for telegraphic communication to the East and Far East. In addition to those already mentioned, Mr Julius Beer was prominent in the financing of these important schemes.

etc., with St Petersburg and the Russian Empire generally, besides constituting an additional telegraphic route to the Far East. At Odessa, moreover, the cable also meets the Indo-European Company's system, forming an extra connection with Great Britain, Germany, etc. The requirements of the Russian Government necessitated the formation of a separate company for the purposes of this cable. The latter was laid the same year by the Telegraph Construction Company.

In 1873 Sir John Pender (then Mr Pender), with other promoters of the principal telegraph companies, formed the GLOBE TELEGRAPH AND TRUST COMPANY. This was intended to provide the public with a means of investing in submarine telegraphs in such a way as to avoid the risks due to fluctuation in the value of shares in any particular cable company, or companies, by spreading them. Each share in the Globe Company includes a proportion of shares in nearly all the various telegraph and cable companies. By this means the chance of an investor coming off badly is reduced to a minimum, since it could only occur in the highly improbable event of the shares of all, or most, of the companies concerned being heavily depreciated at the same time.*

During the first few years of working of the Eastern submarine telegraphs, several prolonged and complete interruptions of communication took place, lasting in some cases for as long as five months. The two allied companies soon recognised the serious effects which the continuation of this state of affairs must produce, both directly upon their own pockets, and indirectly upon the estimation in which they were held by the Governments and the telegraphic public. Accordingly they set to work duplicating, and in some cases triplicating, their lines.† The same fruitful and far-seeing policy has constantly been followed ever since by these and other cable-owning companies.

**Duplex Telegraphy.**—The time has now arrived for noticing another remarkable development of the art of signalling by electricity, namely, duplex telegraphy, the first practical application of which to a submarine cable was made in 1873. This was effected by the late Mr J. B. Stearns,

---

* THE SUBMARINE CABLES TRUST had been formed a short while previously (in 1871) with the same object and under similar financial auspices ; but dealt only, as it still does, with three or four of the largest telegraph companies' shares.

† Indeed, the Suez-Aden cable, down the Red Sea, has been even quadruplicated.

whose general *modus operandi* is described in the specification of a patent which he took out the year before.* The cable in question was one of the short sections (about 300 miles) of the "Anglo-Atlantic"—between Newfoundland and Cape Breton. The duplexing of the main section gave some trouble, and was not successful. No "Anglo-Atlantic" cable

---

* The great feature of this patent was that it contained the first suggestion of the use of condensers in the artificial compensating circuit of a duplex arrangement—to reproduce the effects of the static induction of such a line—thus rendering duplex telegraphy for the first time applicable to submarine or subterranean cables possessing considerable capacity. It is true that this was a mere *replica* of a patented arrangement of Mr Cromwell Varley's in 1862, with condensers or induction plates—as they were then called ; but Varley's apparatus was invented for a different object, and with no idea of duplex working. It was devised, in fact, to serve the purpose of a local test-circuit at each station. The artificial line (of which he was the originator) being of exactly the same electrical value as the cable, could be signalled through at will in place of the latter—as a test of the signals which would pass through it.

The first attempts in the way of duplex telegraphy, as applicable to land lines (without any appreciable capacity), were made by R. S. Newall (1854), C. W Siemens (1854), and W. H. Preece (1855). These differential two-circuit systems of Preece and Siemens —the latter being, in effect, very similar to Newall's—were turned to some account. Preece's method consisted of adjusting the local current to the line current without the use of resistance coils—viz., by varying the influence of the local current on the recording part. This was effected by varying the distance of the coil from the magnetic needle of the galvanometer through which the local current passes. Combinations of a somewhat analogous description to Stearns' (what was then known as a "double-speaking" system) had also been published by Farmer in America, by Gloesener in France, and by Zetsche and Frischen in Germany, as well as by Vaes in Holland. Siemens and Frischen took out a joint patent over this work. It was Frischen's method, in fact, that was adopted by Siemens in the first duplexing of land lines (in Germany) by artificial adjustable resistance. This, however, was not a complete success, and the nature of the balance very rough and variable, owing to the circumstance that even land lines always possess a certain capacity, especially in wet weather.

Patents for the duplexing of aerial lines also stand in the names of Dr Gintl (1854), who was probably the first to conceive the general idea, and who, in his early experiments, used a doubly-wound (differential) coil, with two separate batteries and a double-pointed key. The first to suggest duplexing by the Wheatstone bridge arrangement, in contra-distinction to the differential galvanometer plan (as first prescribed in Newall's patent of 1854), was M. Maron. The term duplex telegraphy, as we now use it, was first employed by Isham Baggs in 1856. The first *published* suggestion of a method for duplexing cables (with capacity) was included in a patent of Professor Thomson (now Lord Kelvin) which he took out in 1854.

Another early one—if not the second—was that comprised in the patent (No. 2103, A.D. 1855) of C. T. and E. B. Bright. This simple circuit method for the Bell instrument was applied extensively (and with complete success) to the underground cables of the Magnetic Telegraph Company, between London and Liverpool, up to the time that they began to fail in insulation. More particulars concerning some of the above systems of duplexing will be found in Part III.

Duplex telegraphy being a profitable subject for inventive enterprise, a great many attempts have been made and patented which it would be impossible to describe here ; and those who are specially interested in the subject must be referred, for full data, to the Abridgments of Specifications published by the Patent Office, or to the specifications themselves.

was entirely duplexed by Mr Stearns until 1879, when, with the assistance of Mr James Graves, he succeeded in duplexing the 1874 cable.

In 1875 duplex telegraphy was successfully applied to a part of the "Eastern" Company's system, viz., the Marseilles-Bona cable. This was effected by Messrs John and Alexander Muirhead and Mr Herbert Taylor, ably assisted by Mr John Munro.* The method adopted was based upon the patents of Messrs Muirhead and Taylor of 1874, for an artificial line *combining* capacity and resistance, instead of imitating these two properties of an electric cable *separately* as heretofore. In the following year this cable was "put through" with that going from Bona to Malta, and the entire length, running into some 850 N.M., successfully duplexed by the late Mr J. Muirhead and Mr H. A. Taylor. Later on, in 1878, the Madras-Penang cable, although subject to great "retardation," was similarly and successfully balanced by duplex apparatus on the Muirhead system.

It should be remarked, incidentally, that Mr C. V. de Sauty had previously, in 1873, applied duplex apparatus to the Lisbon-Gibraltar cable (365 N.M.), but this partook more of the nature of an experiment, and was not applicable to actual practical working on a large scale, the duplex balance being secured at one end only.

Following upon this, Messrs Muirhead and Taylor then duplexed the greater lengths of cable from Suez to Aden (1,400 N.M.), and from Aden to Bombay (1,900 N.M.), the latter being effected in 1877. Then, in 1878, the Direct United States cable—alluded to further on—was successfully duplexed (by Messrs A. Muirhead and H. A. Taylor), this being the first line across the Atlantic so treated, with complete satisfaction, moreover, a speed of sixteen words a minute each way being obtained over a length of 2,420 N.M.

The Messrs Muirhead and Taylor have now duplexed altogether some 75,000 N.M. of submarine cable, including all those in operation under the Atlantic, *i.e.*, about 50 per cent. of the total length laid up to date.

The whole of the Stearns and Muirhead and Taylor duplexing work is now superintended by Dr Alexander Muirhead.†

---

* Part author of that exceedingly successful pocket-book, Munro and Jamieson's "Electrical Rules and Tables."

† In 1879 an agreement had been entered into between Mr J. B. Stearns on the one part and Messrs Muirhead and Herbert Taylor on the other, with a view to avoiding litigation in the future with regard to the validity of the Muirhead and Taylor patents, it having been questioned whether the latter might not possibly be held to be an infringement of those of Mr Stearns. Under this agreement for their mutual benefit and protection, their interests as regards the working of their respective patents, existing and prospective, were to some extent "pooled."

The Bridge system, as perfected by Messrs Muirhead and Co., is now invariably adopted for *cable* duplexing, an accurate balance being thereby more effectively secured.*   Mr Stearns, however, employed the differential system in his early work on submarine cables, the coils of the signalling instrument (whether recorder or mirror) being wound differentially for the purpose.   On land-line circuits the differential method is always used, partly on account of its greater simplicity and convenience—especially in this country, where the Post Office still employs a form of the differential galvanometer for testing, etc.

Mr Benjamin Smith† and Mr (now Professor) Andrew Jamieson, F.R.S.E., M.Inst.C.E., should also be mentioned here as practical pioneers in the field of duplex telegraphy applied to submarine cables.   They both have their individual methods of working.   Among other systems differing in various details from those of Stearns and Muirhead are some of Mr Harwood (standing in his name, coupled with that of Sir James Anderson, to represent his company), and also the device of the late M. Ailhaud of the French telegraph service.

*Quadruplex* telegraphy, although largely applied to land-line systems, has never yet been found applicable, for practical purposes, to submarine cables in those cases (the majority) where it is necessary for the siphon recorder to be the instrument employed.   In 1878 Dr A. Muirhead and Mr G. K. Winter carefully studied the question of quadruplexing telegraph cables, but the result was not favourable.   The fact is, that to make "cable-quadruplex" feasible, an instrument has yet to be devised which will respond to variations in a different manner to that of the recorder. Possibly the solution of this problem may be found in some application of the principle of the electro-dynamometer.

**Further Undertakings.**—In January 1873 the BRAZILIAN SUBMARINE TELEGRAPH COMPANY was established, and in April of the same year

---

* By their methods, Messrs Muirhead and Co. are able to obtain at the present time an increase in the carrying capacity of a cable amounting to at least 90 per cent.—indeed, as a rule, it is very nearly doubled thereby, which means that in transmitting signals both ways at the same time by the above instrumentalities scarcely any loss in speed is incurred in either.

† Mr B. Smith's name has also, more recently, been associated with the application to cables of manual translation.   By the apparatus which he has devised the operator is now enabled to send on or "translate" direct to another line from the received signals as he reads them.   This avoids the necessity of the message being first written out, as taken from one line, and then handed over to another clerk for repetition on the other, thereby saving both time and occasions for error.

the WESTERN AND BRAZILIAN TELEGRAPH COMPANY.* The former had for its primary object the promotion of telegraphic communication with Brazil and the rest of the South American Continent, as well as the West Indies; while the latter undertook the interconnection by telegraph of the principal ports of Brazil itself. Thus, by means of these two systems working together, each Brazilian port would be put in communication with England. An agreement was, in fact, entered into between the two companies that they should "work in unison" as one system.

The Western and Brazilian Company's cables, consisting of five sections, touching at Pernambuco and other ports between Para and Rio de Janeiro, were laid in 1873. These cables, making up a total length of some 2,500 N.M., were manufactured by Hooper's Telegraph Works Company,† who were also entrusted with the laying of them. Mr J. R France acted as their engineer in charge of the expedition. ‡

Shortly afterwards the CENTRAL AMERICAN TELEGRAPH COMPANY was formed, to connect up the Western and Brazilian Company's system with that of the West India and Panama Company, thus placing the former company in a doubly-secure position as regards its communications with Europe. These cables were laid in 1874, forming an object of the

---

* The latter was the outcome of the Great Western Telegraph Company, promoted by the same capitalists, which had for its object the laying of a cable from England to Bermuda, and thence to the United States on the one side, and to the West Indies on the other, thus establishing a more direct communication than heretofore with the latter. Though a large portion of the cable destined for this purpose was manufactured, and a ship specially designed for the work, the original scheme was never carried out ; but the "Western and Brazilian" was financially substituted for it.

† This firm, indeed, had been largely interested in the promotion of the enterprise, as well as in the original Great Western scheme, referred to in the last note.

‡ Most of these cables were of the light type, each iron wire being enveloped in hemp, and originally intended for use in deep water in connection with the Great Western scheme. They subsequently proved to be peculiarly unsuitable for the Brazilian coast, partly on account of the attacks from fish which they were subjected to in the shallow water there prevailing. This "open-jawed" pattern is always liable to be victimised in that way. In this particular instance, however, it is only fair to remark that the close-sheathed type of cable, laid in the same fish-ridden waters at the mouths of rivers, appears to have experienced no more immunity from these attacks than the other. Indeed, in the close-sheathed cable more faults could be actually *traced* to that cause, owing to bits of fish having been broken off inside ; whereas, in the case of the open type, the fish appear to have succeeded in entirely extricating themselves from the wires, after effecting their mischievous purpose.

It has also been suggested that the core of the old open-sheathed type had an insufficient thickness of india-rubber. Although sword and saw fishes seemed to be disagreeably plentiful on the north coast of South America, yet, contrary to expectation, there were no definite signs of the *teredo* either there or in the West Indies. Notwithstanding the many faults that have been experienced and repaired, the *teredo* was never given as their cause.

same expedition—being, in fact, manufactured and laid by the same hands as the Western and Brazilian. The length was about 1,000 N.M., from Para, the most northern point of the "Western and Brazilian" system, to Demerara, the most southern point of that of the "West India and Panama." The main cable had a T-piece running into Cayenne about midway between Para and Demerara. This was the first T-piece of its kind, that is to say, where the branch is composed of a single conductor serving for both circuits in either direction. In this case the T-joint—about 50 miles from shore, in about 50 fathoms of water—was made in the conductor (and insulator) itself at the point where the branch from Cayenne met the main conductor stretched between Para and Demerara. This T-joint was protected by a corresponding T-splice.*

These "Central American" cables† subsequently, and after several repeated repairs at different points in their course, failed altogether. In 1876 they were abandoned, and the company wound up, thus leaving no connecting link between the Western and Brazilian and the West India and Panama systems.‡

---

* Here we have the way in which all T-pieces are effected nowadays, and there have been at least three made since—one at Bathurst in the West African Company's system, one at Panama (Central and South American Company's), and one at Mole S. Nicholas in the French Company's West Indian system, not to mention another at Cayenne in the same *réseau*. All of these have been deposited in quite shallow water —*i.e.*, under 100 fathoms.

By this plan, when it is worth while, the greatest economic advantage can be secured; for whenever it happens that the branch cable only is damaged, the break of communication is confined to that. By the original system, on the other hand (*e.g.*, of carrying the main line in and then out of the station in question), if the branch happened to get broken—being in this case really part of the main, the whole cable, between the two stations on each side, was rendered useless until the branch was repaired. The old method is, in fact, not truly a T-piece method at all, for there is no T-joint. The two conductors of two ordinary shore ends are simply enclosed in a single cable, the so-called T-splice being effected where each conductor is jointed to its corresponding end of the main cable.

The new plan has proved perfectly successful in all instances, and, with improvements in T-joints, and increased skill in jointing work, it is found to be just as reliable. It is true that it involves a little more trouble in test-calculations (as may easily be understood), so that it is liable to render the localising of faults somewhat irksome. This, however, is a comparatively small matter. Outside the entire T-piece the two halves of an iron box are now often placed round each side and fixed on firmly, thus giving the T-piece increased security. This addition, by taking the strain off the joint, renders a finished splice—a somewhat lengthy, troublesome, and, at the best, unsatisfactory operation at sea—quite unnecessary. Such a plan was first devised by Mr M. H. Gray for the Bathurst T-piece.

† Not to be confounded with the true Central American system of the Central and South American Company, hereinafter referred to.

‡ Telegraphic communication between the West Indies and Brazil was restored, however, in 1891 (*via* Brazilian land lines) by LA COMPAGNIE FRANCAISE DES TÉLÉ-GRAPHES SOUS-MARINS, the scope of whose system will be described later on.

In 1874 a series of cables were laid in southerly extension of the Brazilian Company's system down the coast of Brazil and Uruguay, from Rio de Janeiro to Montevideo, calling at three points *en route*. These cables—running into some 1,200 N.M.—were laid for the COMPANHIA TELEGRAFICA PLATINO-BRASILIERA,* and were constructed and laid by Messrs Siemens Brothers. Under an agreement, the Western and Brazilian Company were to work these cables for the local company; and subsequently (in 1879) they took them over altogether, under the title of the " LONDON-PLATINO-BRAZILIAN TELEGRAPH COMPANY." At Montevideo this system connects up with that of the RIVER PLATE TELEGRAPH COMPANY,† reaching to Buenos Ayres and a short way up the Rio de la Plata.

A few years before the Platino-Brazilian cables were laid, namely, in 1872-73, another company—the MONTEVIDEAN AND BRAZILIAN TELEGRAPH COMPANY—had been formed to establish a direct cable between Montevideo and Chuy. This cable was made and laid by Mr W. T. Henley. It was subsequently bought up by the Western and Brazilian Company (into which the Montevidean Company was "absorbed"), and now forms part of their system—a continuation, in fact, of the "Platino" cables—together with the other duplicate line laid by the Telegraph Construction and Maintenance Company in 1892 between Chuy and Santos.

In connection with the River Plate cables a land line runs across the South American Continent to Valparaiso, a distance of some 700 miles, thus putting the Western and Brazilian system in communication with the cables running along the West Coast of South America.

In 1895 the Western and Brazilian Company promoted an extension of their system up the Amazon River, and a company called the AMAZON TELEGRAPH COMPANY was formed with this object. The cables for this new company were laid last year by Messrs Siemens Brothers. The line extends from Para to Manáos, nearly half-way up the great river, in the heart of Brazil and the plains of the Amazon. There are sixteen stations on the line altogether, its total length being over 1,300 N.M. This cable is expected to further develop a large india-rubber, coffee, and sugar trade. All previous attempts at telegraphic communication by means of land lines

---

* Entirely promoted in Brazil, but sometimes known as the "River Plate and Brazil Telegraph Company."

† A company of completely local origin, with which the Western and Brazilian have a working arrangement. The construction and laying of this line was carried out by Mr W. T. Henley as early as 1866, the length being 30 N.M. only.

had proved unsuccessful, owing to the rapidity and density of forest growths in this part of the world.*

The cables of the Brazilian Submarine Telegraph Company, although this concern was launched a few months previous to the Western and Brazilian, were not laid until the following year, 1874, by the Telegraph Construction Company. These cables extend from Carcavellos (Lisbon) to Madeira,† thence to St Vincent (Cape Verde Islands), and thence to Pernambuco, in Brazil;‡ the system being placed in communication with Great Britain by means of the "Eastern" Company's line from Porthcurno to Lisbon. The latter forms part of their Anglo-Spanish-Portuguese system. The "Brazilian Submarine" Company, it may here be remarked, is (financially and in its *personnel*) closely allied with the "Eastern."

In 1873 the DIRECT UNITED STATES CABLE COMPANY was formed,

---

* Indeed, in the case of the cable, the new experience of laying one up an immense river whose bed is subject to very considerable and frequent changes, has in itself turned out a formidable business ever since the line was completed. Notwithstanding that Messrs Siemens Brothers and Co., the contractors, and Messrs Clark, Forde, and Taylor, the consulting engineers, took precautions in the way of preliminary survey, soundings, and selection of route, several of the sections have in turn been interrupted—more or less continuously—subsequent to their submergence. The greater part of the trouble has been in the upper waters, where numerous subsidiary rivers flow into the Amazon, and where, in the rainy season, enormous masses of *débris*, in the shape of trees, etc., are carried down by the extremely strong prevailing currents, which latter have also the effect of scouring out the bottom, thus leaving the cable unburied in places. Elsewhere the cable has sunk so deeply into this unusually soft mud that much difficulty is met with in recovery. Rocks which had not been discovered in the survey have since been found to be an additional source of breakdown.

This line was laid in the middle of the stream with the idea that here there would be the least prospect of any vessel anchoring and destroying it. With the knowledge since obtained, it would seem that the cable might have had a better chance if laid more on the bank of the river, where it would soon become deeply buried in the soft bottom, and be less liable to disturbance at the hands of moving matter. Recovery-work here must, in any case, be a subject of immense labour. For further particulars regarding this line, see lecture recently delivered by Mr A. Siemens, M.Inst.C.E., at the Royal Institution.

† A very curious circumstance occurred in regard to this Lisbon-Madeira section in 1891. The cable was found to have broken in 1,500 fathoms, not far from Madeira, on the very day that an immense tidal wave was observed. It was concluded accordingly that the latter must have had something to do with the rupture of the cable, although beyond the strange coincidence there was no proof of any connection between the two occurrences. On the other hand, during the laying of the above, the T.S. "Seine" discovered a bank in latitude 33° 47′ N., and longitude 14° 1′ W., the depth being about 100 fathoms, with 2,400 fathoms in its immediate neighbourhood.

In those days soundings were not taken as closely as has since been found to be necessary, the result being probably that the cable became suspended in a festoon when submerged.

‡ This last section lasted nine years before any sort of repairs were necessary—or even before any fault proclaimed itself—a record case, in fact.

being the first competitor with the "Anglo-American" Company,* so far as concerns direct submarine communication between Great Britain and the United States.

This cable was originally intended to have been taken direct to the shores of the United States, without any intermediate land line over Newfoundland or any other part of British North American territory—hence the term "Direct" in the title of the company. Ultimately, however, it was decided that if the cable were laid thus, in a single length, the working speed attainable (with the core provided) would be too low to compete with the different lines of the existing company. Accordingly, attempts were made during the expedition to land at Newfoundland,† but landing rights being withheld there, the end was eventually taken ashore at Halifax, Nova Scotia (where several other Atlantic lines have since been landed), and from there another was laid to the United States.

Messrs Siemens Brothers, who had taken an active part in the promotion of the company, were the contractors, both for manufacture and for submersion.‡ It was, indeed, the first really important length with which this firm had been concerned as manufacturers. Mr Carl Siemens was placed in charge of the expedition for laying it, which was attended with complete success, and the line was opened to the public in 1875.

Later on, in 1877, the direction of the Direct United States Company passed into other hands — indeed, the company was reconstructed — and their system entered into the "pool" or "joint purse," established shortly after the 1869 Atlantic had been laid.§

---

* This company had just had two fresh cables laid for them (1873 and 1874) by the Telegraph Construction Company with some of their usual staff. The laying of the 1874 Atlantic line was the last piece of telegraph-work performed by the "Great Eastern." A sectional view of this leviathan vessel, as disposed for cable operations on this occasion, is given in Fig. 48 overleaf. She has since been broken up, after being employed (amongst other things) as a sort of floating "variety show"!

New cables were first rendered necessary—according to the joint-purse agreement previously referred to—by the final breakdown (after several repairs) of the 1866 cable in 1872. Later on (in 1877) the 1865 also succumbed, and another "Anglo" cable was laid by the same contractors in 1880. In this latter cable the shore ends of the old 1866 were turned to good account, but all the deep-sea type was new. With the exception, therefore, of the shore ends, all the old cable remains abandoned, and it is most inaccurate to describe the two cables (1866 and 1880) as one, as some have done. The Telegraph Construction and Maintenance Company laid this 1880 cable without any hitch or stoppage within the surprisingly short space of twelve days—the "record" up to date in Atlantic cable-laying.

† A suggestion of this may be gleaned from the chart by the circuitous route which the cable makes.

‡ This cable was of the same hemp and iron combination then still in vogue with some authorities for extreme depths. Here the hemp was again compounded before application to each wire. A whipping with about ½ inch or so between the turns was applied finally.

§ This plan originated as follows :—After working in opposition to the "Anglo-

In the same year (1874) the Silvertown Company laid a cable for the Direct Spanish Telegraph Company between Marseilles and Barcelona, thus extending their system to France, and thereby constituting a kind of Anglo-Spanish connection with the "Eastern" systems, of which now, since 1885, it forms a supplementary offshoot. The telegraphic communication of Spain with the East, as well as with the rest of the European Continent, was now doubly secured.

In 1876 the original WEST COAST OF AMERICA TELEGRAPH COMPANY was incorporated, whose objects were to purchase and work cables on the coasts of Chili and Peru, between their capitals Valparaiso and Lima, as well as to put the latter into communication with the Western and Brazilian Company's system, and thus with Europe, by means of the land line of the TRANSANDINE TELEGRAPH COMPANY,* and the "River Plate" and "Platino-Brazilian" cables.

This West Coast of America Company's cables had been made by the Silvertown Company, and were laid by the same firm in seven sections, during 1875 and 1876. The greater part was submerged previous to the actual formation of the "West Coast" Company.† The company itself was originally constituted almost entirely of, and by shareholders in, the Silvertown (or India-rubber, Gutta-percha, and Telegraph Works) Company.

---

American" for about a couple of months, the French Atlantic Company accepted their rival's invitation to form a pool or "joint purse," made up from the nett earnings of both their cable systems, the revenue accruing to each company therefrom to be fixed in proportion to its respective contributions to the said pool. Subsequently all the Atlantic cables, with two exceptions, have come into this arrangement, and now constitute—for practical purposes—one great financial combination.

A somewhat similar working arrangement now exists also—so far as regards the Indo-European traffic receipts—between the Eastern Telegraph Company, the Indo-European Telegraph Company, and the Indo-European Department of the Indian Government Telegraphs. This originated at the instance of the Indian Government, who (by their initial agreement with the British-Indian Submarine Telegraph Company) granted landing rights to the latter for the Suez-Aden cable in 1869. Another instance of the "joint-purse" arrangement is that which exists between the "Eastern Extension" and "Great Northern" Companies regarding the traffic by their respective cables between Hong Kong and Shanghai.

* This local (Chilian) company—COMPANIA DEL TÉLÉGRAFO TRANSANDINO—had obtained authorisation from their Government in 1870 to work a land line between Valparaiso and Villa Maria, Chili. This line was subsequently carried across the Argentine frontier, and right on to Buenos Ayres, with intermediate stations. The whole system was ultimately purchased (in 1891) by the Central and South American Telegraph Company, whose lines are hereafter referred to. The origin of the title "Transandine" is sufficiently obvious; but it is interesting to remark that in certain very high Andine altitudes the line has not been carried, strictly speaking, *over* the mountains, but rather *under* them— at any rate under *the soil*, subterranean cable having been perforce employed for 70 miles, where no aerial line could be expected to withstand the prevailing snow and wind.

† Probably the only case of its kind.

[PLATE VIII.

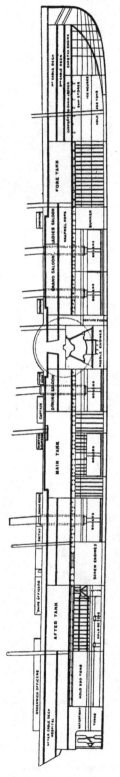

FIG. 48.— The S.S. "Great Eastern" when last employed as a Telegraph Ship.

DIMENSIONS OF CABLE TANKS.

Fore = 51' 8" × 20' 6"
Main = 75'  × 16' 6" *
After = 58'  × 20' 6"

\* The Main Tank was originally of the same dimensions as the After Tank, but was subsequently enlarged.

But subsequently, in 1877, it underwent a reconstruction, and has become closely identified with the " Eastern " and its allied companies.*

Another period of calm or inactivity in the further extension of marine telegraphy now supervened, lasting until 1879. In that year the EASTERN AND SOUTH AFRICAN TELEGRAPH COMPANY was launched under the financial auspices of the " Eastern " group. Its purpose was to establish and maintain telegraphic communication between Aden (where it would connect up with the " Eastern " Company's system) and the Cape of Good Hope, by means of a series of cables down the East Coast of Africa, joining, at Port Nàtal, a Government land line to Cape Town, which had been completed in 1878.† These cables were all manufactured by the Telegraph Construction Company, and laid by them in 1879. Their combined length was as much as 3,900 N.M., one section alone—that from Aden to Zanzibar—measuring 1,915 N.M.

The same (Eastern and South African) company have since placed one of the Seychelles Islands, as well as the important colony and naval station of Mauritius in telegraphic communication with the rest of the world. This extension, joining their main system at Zanzibar, was carried out for them (by the Telegraph Construction and Maintenance Company again) in 1893. Another extension of theirs down the lower part of the West Coast of Africa to the Cape—being, in other respects, a continuation of the West African Company's system—will be found referred to in its proper place.

In 1879 another French company was formed, to establish independent communication between France and the rest of the European Continent on the one hand, and the United States of America on the other. The —to English ears and lips—somewhat cumbersome title of this concern was LA COMPAGNIE FRANCAISE DU TÉLÉGRAPHE DE PARIS À NEW YORK, but it soon became styled in England as the " Paris and New York Telegraph Company," or, more commonly still, as the " Pouyer-Quertier Company," M. Pouyer-Quertier being its presiding genius. The cable, generally known as the " P.Q.," was made and laid in the same year by Messrs Siemens Brothers, Mr Ludwig Loeffler being in charge of the

---

* The West Coast of America Company's system has since been extended by the establishment of land-line communications with fresh points in the interior, the most important town so connected up being Santiago in 1889.

† The sea-bottom round the Cape is not at all favourable for a cable. Moreover, the land lines, completed and harmonised in one "system" the previous year, as worked by the Cape and Natal Government Telegraph Departments, had proved thoroughly efficient and reliable.

expedition.* This cable runs almost between the same points as the original "French Atlantic" cable of 1869 (subsequently absorbed by the "Anglo" Company), although the positions of its terminal stations on the American side are not quite similar. It was laid, however, along a different route.†

The "P.Q." Company afterwards joined the "Pool." Quite recently, however (in 1894), it was amalgamated with LA SOCIÉTÉ FRANÇAISE DES TÉLÉGRAPHES SOUS-MARINS, under the title of LA COMPAGNIE FRANÇAISE DES CÂBLES TÉLÉGRAPHIQUES. Since this amalgamation it has withdrawn from the Joint-Purse.

In 1881 an American company was formed, under the guidance of the late Mr Jay Gould, entitled the AMERICAN TELEGRAPH AND CABLE COMPANY, with a view to partaking in the profits of trans-Atlantic telegraphy by establishing another line of communication between the United States and Great Britain, and thence to the rest of Europe. This cable was also constructed and laid (in the course of that year) by Messrs Siemens Brothers,‡ who were part promoters of the enterprise, as well as another cable for the same system in the following year, 1882.§

This company's cables work in connection with—are, in fact, leased by —the WESTERN UNION TELEGRAPH COMPANY, which was practically Jay Gould's property, and remained so up to close on the time of his death a few years ago. In 1883 the above system entered the "Pool"—the happy destination for which, may be, it was originally launched into existence.

---

* This scheme had taken three years to "mature" before it reached contract point. It was in the first instance promoted and financed by several large Parisian banking-houses and French merchants, and further aided by Messrs Siemens.

† Both this cable and that of the "Anglo" Company's 1869 French Atlantic are connected with Great Britain by means of cables between Brest and Porthcurno.

‡ The reason why such an enterprising people as the Americans come to England for their cables is that no factory for the purpose exists in the United States. The same explanation applies to the two French Atlantics (of 1869 and 1879), both of which were made in England. Since then, however, submarine cable works have been established at Calais. A cable factory was also established at Milan in 1886 by Messrs Pirelli and Co., an Italian india-rubber and gutta-percha firm, and there are several factories for making core of various descriptions in France, the United States, and elsewhere, besides those in England. The absence of any submarine cable factory in the United States is probably due to the fact that they possess no colonies over sea to which to lay cables.

§ Notwithstanding the distance between the parties here concerned, the preliminary negotiations appear to have been remarkably quickly transacted. Report has it that an estimate was asked for and supplied by "wire"—exceeding a million sterling—on which the following message was sent with a view to concluding the business: "Make and lay two cables. Fifty thousand pounds at your bankers." In connection with this story, it must be remembered that Mr Von Chauvin—then associated with Messrs Siemens Brothers— was in New York at the time. Apart from this, however, the rapidity with which a conclusion was arrived at is a great compliment to the high position enjoyed by the house of Siemens, and to the renown attached to that name.

In 1880 the MEXICAN TELEGRAPH COMPANY was formed by a body of American capitalists, and in the following year the CENTRAL AND SOUTH AMERICAN TELEGRAPH COMPANY was floated by similar hands. The object of these two schemes was to bring Mexico and the West Coast of South and Central America into direct communication with the United States (*i.e.*, independent of the West Indian lines), thus also diverting through the States part of the aforesaid traffic with Europe.

The cables across the Gulf of Mexico were laid between 1880 and 1882, and those for the Central and South American Company, along the West Coast, in 1882. The latter started from Peru, at the spot where the system of the West Coast of America Company ended ; touching at Panama and various points of communication with other systems, they extended to the Gulf of Tehuantepec. By means of a land line across the isthmus of the same name, and a short cable on the other side, they there met with the Mexican Company's cables at Vera Cruz. The Silvertown Company were the contractors for the manufacture and submersion of all these cables,* those on the West Coast comprising eight sections, and having a total length of 3,200 N.M.

The Central and South American Company have within recent years (in 1891) extended their system to Valparaiso, and thence by land line to Santiago, thus entering into competition with the West Coast of America Company. They purchased at the same time the Transandine Telegraph Company's land line from Valparaiso and Santiago to Buenos Ayres (with its four intermediate stations), thus securing an independent connection of their own with the great Argentine and Uruguayan capitals, Buenos Ayres and Montevideo. By other land lines through Brazil they also reach Rio. This (Central and South American) company's connection with Europe is, however, as previously mentioned, exclusively by way of the Mexican Company's system, the Western Union land lines, and the Atlantic cables to England and France.

In response to this move on the part of the American group, the West Coast of America, the Western and Brazilian, and the Brazilian Submarine Companies joined forces to promote another company in the following year for erecting a new land line across the Continent in their own interests. It is entitled the PACIFIC AND EUROPEAN TELEGRAPH COMPANY. The construction of the required line followed soon after, along a route rather to the north of the Central and South American Company's, but connecting the same points with one intermediate station. This line was open for traffic in March 1894.

---

* Since this Messrs Siemens Brothers have also laid a cable across the Gulf of Mexico for the Mexican Telegraph Company.

In 1883 the SPANISH NATIONAL TELEGRAPH COMPANY came into existence. This company (with subsidies from the Spanish Government) was founded to create and maintain telegraphic communication between Spain and her possessions, the Canary Islands, as well as to effect—by the help of another subsidy from the French Government—an extension from the Canaries, thus connecting the European Continent with the French colony of Senegal, West Coast of Africa. The first part of the enterprise was carried out in the same year, and the second in 1884. The Silvertown Company, who had practically floated the "Spanish National," were the contractors for the manufacture and laying of these cables.*

The first part of this company's system, connecting Spain with the various Canary Islands, was taken over in 1893, after ten years' working (according to the original agreement), by the Spanish Government. The system in its entirety now forms a connecting link with continental Europe and England for the South American Cable Company's system—to be hereafter referred to—which it joins at St Louis, Senegal.

Two more African cable companies were registered in the latter part of 1885, namely, the WEST AFRICAN TELEGRAPH COMPANY in September, and the AFRICAN DIRECT TELEGRAPH COMPANY in December. The former was promoted by the Silvertown Company, and was to work in connection with the Spanish National Company's system; the latter was an offspring of the "Eastern" Company group. These two companies had each obtained subsidies or guarantees from the various European Governments—French, Portuguese, and British—to whom the different colonies belonged,† at which it was proposed the cables should touch, going along the West Coast and including St Jago and St Vincent.

---

* Before starting cable-laying, the T.SS. "Dacia" and "International," employed for the work, sounded from Cadiz towards the Canaries on two separate zig-zag routes—the virtue of which will be obvious. No less than 552 soundings were taken in this way at distances of 10 miles apart. In the course of this survey the "Dacia" discovered a bank in latitude 31° 9′ 30″ N., and longitude 13° 34′ 30″ W., the depth being 58 fathoms in the immediate vicinity of 2,000 fathoms — analogous to the relative heights of St Paul's Cathedral and some of the tallest Swiss mountains. These soundings were in what might otherwise have been taken as the line for the Canaries cable. Later on another vessel took over 1,000 soundings down the West Coast of Africa as a preparation for the cables to be laid there. These surveys have been described in detail in the course of two interesting papers by Mr J. Y. Buchanan, F.R.SS. (Lond. and Edin.), before the Royal Society of Edinburgh and the Royal Geographical Society respectively, with full details regarding the extremely stong currents from the rivers on the "West Coast." Mr Buchanan was a member of the famous "Challenger" Expedition, and took a special part in the above highly scientific submarine researches for which the Silvertown Company are now famous as a matter of ordinary routine and precaution previous to laying a cable.

† The West African Company had secured agreements with most of the foreign possessions and protectorates, while the "African Direct" had arranged with the British, and, in the case of the cable up the Cameroon River, with the Germans.

The West African Company's cables were made and laid for them by the Silvertown people in the course of two expeditions, during 1885 and 1886.* Starting from the southern terminus of the Spanish National at St Louis, Senegal, this system extends to St Paul de Loanda, serving eleven stations on the way, mostly situated in the Bight of Benin, where cables behave better than men.

The African Direct system was made and laid, about the same time, by the Telegraph Construction and Maintenance Company. Through the instrumentality of the two sections between the West Coast of Africa (at Bathurst) and St Jago and St Vincent, this company secured its own communications with Europe *via* the Brazilian Submarine Company's lines to Lisbon, and thence by the "Eastern" Company's direct cable to England.

Two years after this, the Eastern and South African Telegraph Company determined to extend their East Coast of Africa system by a cable from Cape Town, *via* Mossamedes and Benguela, to Loanda, the southern terminus of the West African Company's system. Going south, the first of the two sections—that between Loanda and Benguela—was made and laid by the Silvertown Company, and the other by the Telegraph Construction Company. Communication was thus completed, in 1889, along the whole West Coast of Africa down to Cape Town. These three systems comprise nearly 8,000 miles of cable.

It was now arranged that the West African Company and its cables should be taken over by the "Eastern" promoters, so as to constitute, with the African Direct and the Eastern and Southern African—and for their mutual benefit—one working system of alternative communications with the Cape. Each of the stations served by these three systems have an opportunity of transmitting, on the combined system, any messages they may have at stated times during the day.

In the field of Atlantic telegraphy, a fresh competitor arrived in 1884, in the person of the COMMERCIAL CABLE COMPANY. Two cables were laid across the Atlantic for this company in the same year, its promoters wisely foreseeing that, in view of the continual chance of a breakdown, this was the only way in which they could safely attempt to compete with their more firmly established rivals. The "Commercial" Company was mainly promoted by two American millionaires, Mr J. W. Mackay, the celebrated New York financier, and Mr Gordon Bennett, proprietor of the *New York*

---

* In the course of laying some 3,000 miles in a hot climate with many exposed ends and splices, it is satisfactory to note that in this instance not a single fault was revealed.

*Herald,** with whom were associated the Messrs Siemens, who became afterwards the contractors for the enterprise.†

These cables, like the Jay Gould lines, stretch from the extreme south-west point of Ireland (which is connected by special cable with England) to Nova Scotia, and thence to the United States—one of them direct to New York. The system is directly connected with that of the Canadian Pacific Railroad Company, thus affording ready communication with the Dominion.

Neither the Commercial Company's system nor that of the COMPAGNIE FRANÇAISE DES CÂBLES TÉLÉGRAPHIQUES is at present in the "Atlantic Pool"; they, in fact, constitute the two exceptions previously referred to.

In 1886 the Italian Government entered into a contract with SOCIETÀ PIRELLI, of Milan,‡ for the construction and maintenance of a number of cables—about 700 N.M. in all—to connect various islands in the Adriatic and Mediterranean Seas with the mainland, chiefly for military and naval purposes. For the execution of this contract, the Società Pirelli built a cable-sheathing factory at Spezia, and a telegraph ship—the "Città di Milano"§—in England, Messrs Johnson and Phillips supplying the machinery for both. The core was made at the Milan works, and sheathed at those of Spezia. M. Pirelli had also engaged the services of Mr W. S. Seaton, who superintended the construction and laying of these cables on behalf of the "Società."

This Italian firm has since carried out other similar work, on a small scale, for its own Government, besides constructing and laying a fresh series of cables between the Balearic Islands (in 1890) for the Spanish Government.|| In electric lighting it was still earlier in the field, that department of electrical industry having been quite practically developed in Northern Italy for some time past.

---

* Whence the concern is still often spoken of as "the Mackay-Bennett cables."

† The construction and separate laying of the two cables (in the course of different expeditions) within the same year, forms probably the smartest performance in cable work up to date, representing, as it does, over 6,000 N.M. of cable. At the factory some 50 N.M. were turned out each twenty-four hours from ten machines.

‡ A firm of india-rubber and gutta-percha manufacturers (previously alluded to) who had been making core during the last ten years.

§ In future, wherever a properly equipped telegraph steamship for permanent cable work is referred to, the letters "T.S.," as a prefix to the ship's name, must be taken to stand for "*telegraph steamer.*" This implies her having tanks, paying-out and picking-up gear, also deep-sea sounding apparatus, and that she is *still* afloat as a telegraph ship.

|| For further information concerning the Italian Government cables, and the work executed for them by the Pirelli firm, the reader who is familiar with the language may be referred to an excellent little Italian treatise recently produced by one of Mons. Pirelli's associates, Signor E. Jona, which is entitled "Cavi Telegraphici Sottomarini."

Down to nine or ten years ago the French had not troubled themselves much about cables to their colonies or foreign parts, and the few they had felt necessary they got made and laid for them by English hands. Latterly, however, they have concluded to render themselves more independent of British lines of telegraphic communication, and with this object have commenced in earnest the work of making and laying cables on their own account

Thus in 1887 the French Minister of Posts and Telegraphs signed a contract for the subvention of an extensive system of cables to connect their colonies in the West Indies with French Guiana, communicating at Cuba with the United States and Atlantic systems, and at Viseu, in Brazil, with the Brazilian land lines to Rio and Buenos Ayres. A change of government led to the rejection of this contract in the Chamber of Deputies. Subsequently, however (between 1887 and 1888), a part of this scheme was carried out—as far as La Guayra, Venezuela. The cables were constructed by W. T. Henley's Telegraph Works Company,* who were also contractors for the laying of them. Mr R. E. Peake (of Messrs Clark, Forde, and Taylor) executed the latter part of the contract for "Henley's," while Mr W. S. Seaton acted as engineer on behalf of the proprietors, LA PARTICIPATION DES CÂBLES DES ANTILLES.

In 1888 LA SOCIÉTÉ FRANÇAISE DES TÉLÉGRAPHES SOUS-MARINS was formed, to take over the above cables and extend them as soon as the Chamber of Deputies should ratify their contracts made with various Ministers of Posts and Telegraphs. In 1889 the necessary ratification was obtained. Accordingly the society proceeded to enter into contracts with the SOCIÉTÉ GENERALE DES TÉLÉPHONES for the construction in France of the said extension cables. The latter firm, which had previously made core only, for telephone and other purposes,† now (in 1891) set up cable-sheathing works at Calais. The cables were manufactured with complete success—partly by the Société Française des Télégraphes Sous-Marins and partly by the Henley Company—and were laid for the "Société Française" by Mr W. S. Seaton between 1890 and 1891.

---

* Half the required core, however, was made at the Bezons Works of La Société Générale des Téléphones. This was the first order that firm had obtained for submarine cable core, and it proved exceedingly satisfactory.

† This well-known company—originally MM. Rattier et Cie.—had been india-rubber and gutta-percha manufacturers, and had made and laid telephone lines ever since the commencement of practical telephony throughout Paris and other French towns. M. Menier had also had rival works of the same description near Paris, but joined in with the newly-formed Telephone Company (La Société Industrielle des Téléphones), when that was established three years ago, his works being taken over by them. The "Société Industrielle" constitutes the construction business of the former Société Générale, since the Government took over the exploitation part of the latter's concern.

It may here be noted that those last-mentioned cables went over some of the same ground as the old "Central American" Company's lines already referred to.   It was not intended, however, that the French system should work in connection with the "Western and Brazilian" or the "Brazilian Submarine."   By agreement its cable joins the Brazilian Government land lines for local traffic purposes, the European traffic going, as above mentioned, *via* the Western Union land lines and Atlantic cables.   Moreover, the French company may be said to work in opposition to the West India and Panama ; for its cables not only touch at the French islands, but at several of the others (both colonial and independent), for purposes of West Indian traffic generally.

As stated before in connection with the first West Indian cables, the sea-bottom here is undoubtedly the worst over which any cables have yet had to be laid.   In order to cope with this difficulty, the plan was adopted, in one case, of laying down quite a heavy type of cable—"shore-end," in fact—in the middle of some of the sections where coral reefs were known to exist ; and, in order to ensure of the heavy type being deposited in the right place, this part of the cable was laid first and then buoyed.   The plan was naturally both a lengthy and a costly one, but it will be probably found to pay in the long-run.

The main types of these cables were sheathed with rather stouter, and therefore more rigid wire than is usual.   The object of this was to defeat the attacks of saw and sword fishes, which, as previously shewn, tend to make such havoc of cables in these shallow seas, as well as to anticipate the intrusive attentions of any possible *teredo*.

The HALIFAX AND BERMUDAS CABLE COMPANY was formed in 1889, to give effect to a contract with H.M. Government for laying a cable and maintaining telegraphic communication between the two important naval stations designated in its title.   This contract has been, and is, carried out under a Government subsidy of twenty years' duration.\*

The manufacture and submersion of this cable were given to Henley's Company, on whose behalf Captain A. W. Stiffe (formerly of the late Indian Navy, and prominently connected with the first Persian Gulf cables) was engineer in charge of the expedition.   It passes through water as deep as 2,824 fathoms, one of the greatest depths in which any cable is known to lie.

---

\* The concession had been in the first instance obtained by a company formed just before, and called the International Cable Company, who assigned it to its present holders.   The International Company had attempted to float a more ambitious scheme— the same, in fact, as that of the old "Great Western" Company—and had actually laid a shore end at Lisbon to save their concession.

The object of this telegraphic connection was Imperial rather than commercial. It forms part of a scheme recommended by a Royal Commission in 1884 to provide a means of communication between the British West Indies and the Mother Country, passing entirely through British territory.* As the line connects up with several Anglo-Atlantic cables at Halifax, the direct communication demanded by the Commission has been effectually established.

In 1891 the French Government invited tenders for two cables to connect Marseilles with Oran and Tunis respectively. For the first time in the history of submarine telegraphy British contractors were excluded altogether. The Oran cable was allotted to LA SOCIÉTÉ GENERALE DES TÉLÉPHONES, and the Tunis one to Mons. Grammont.†

The Oran cable (the core of which was turned out at Bezons, while the sheathing was applied at Calais) was laid by Mr W. S. Seaton on behalf of the French contractors, under the inspection of Mons. E. Wünschendorff, the Government engineer, and author of the first complete work concerning submarine telegraphy ("Traité de Télegraphie Sous-Marine," Baudry et Cie, Paris), on a part of which this book is, to a great extent, based.

M. Grammont retained the services of a civil engineer and compatriot, Mons. Peltier, for laying the Tunis cable. He chartered the Telegraph Construction Company's T.S. "Calabria" for this purpose, and availed himself of the assistance of some of the latter company's staff. For the manufacture of this cable, M. Grammont erected a factory (core and sheathing) at St Tropez, near Toulon.

In 1891 the SOUTH AMERICAN CABLE COMPANY was promoted by those interested in the Silvertown business (India-rubber, Gutta-percha, and Telegraph Works Company), the objects of which were to form an extension of the Spanish National Company's system on the one hand, and, on the other, an alternative competing line for the Brazilo-European traffic. This was to be effected by means of a cable between Senegal, West Coast of Africa, and Brazil.

The cable ‡ was constructed and laid the following year by the Silvertown Company. It extends from St Louis, Senegal, to Pernambuco,

---

* The completion of this scheme—by a newly-formed company associated with the above—is now about to be carried out by the line being extended from Bermuda to Jamaica, *via* Turk's Island.

† M. Grammont, it appears, had not done any work of this description before. His regular business was that of a large contractor to the French Army.

‡ Probably one of the most perfect modern types of cable for deep water ever designed, combining, as it did, the contractor's requirements for easy handling and laying, with the still more important owner's requirements for durability and maintenance.

Brazil, and touches at the island of Fernando de Noronha *en route*. The last section (about 350 N.M. in length) was furnished with india-rubber core instead of gutta-percha, the locality being considered a suitable one for giving india-rubber a fresh trial. The main part of this line passes through one of the greatest depths in which any cable actually rests, *i.e.*, 2,830 fathoms, lat. 9° 53′ N., and long. 21° 24′ W.

At Pernambuco this company's Africo-American system meets that of the "Western and Brazilian," and also that of the Brazilian Government land lines, thus gaining access (by agreement) to all the principal Brazilian towns as well as to the rest of the South American Continent.

In 1893 the EUROPE AND AZORES TELEGRAPH COMPANY was established to effect telegraphic communication between Lisbon and the Azores group of islands. This company is worked on a subsidy from the Portuguese Government; indeed, the cable is almost exclusively a Portuguese national affair, the company simply acting as Government agents in the matter. It was laid for them by the Telegraph Construction Company in the same year, and connected two of the islands, Fayal and San Miguel, with one another and with Lisbon.

La Société Française des Télégraphes Sous-Marins* promoted in 1893 a scheme for a cable between Queensland and the French colony of New Caledonia—distant some 800 miles odd—their primary object being to bring this island into telegraphic communication with France *via* the "Eastern" Company's system. Ultimately they intended that it should become a first part of the then proposed Franco-German Pacific Cable scheme.

The contractors for the manufacture and submersion of this cable were LA SOCIÉTÉ GENERALE DES TÉLÉPHONES. This firm had recently gathered together a staff of French engineers and electricians, and this was the first cable really laid by French hands. Mons. Rouillard was the engineer in charge of the expedition, Mr Edward Stallibrass being also on board on behalf of Mr W. S. Seaton, who held a kind of "watching brief" as adviser to the contractors.

In the early part of 1895 the French Government ordered a cable to be made and laid between Madagascar and Mozambique, from which point it would communicate with France and the rest of Europe *via* the Eastern and South African Company's system. This cable is entirely a State concern, being worked exclusively by, and on behalf of, the Government which ordered it. Half its length was made at the works of M. Grammont, and

---

* Now LA COMPAGNIE FRANÇAISE DES CÂBLES TÉLÉGRAPHIQUES.

half at the Calais works of the Telephone Company. It was laid by T.S. "François Arago" of the latter firm, by M. Peltier, engineer to M. Grammont.

In 1894 yet two more additions were made to the list of Atlantic cables —one on behalf of the Commercial Cable Company, and the other for the Anglo-American Company. The new "Commercial" line was constructed and laid by Messrs Siemens Brothers, and the "Anglo" cable by the Telegraph Construction Company.* Fig. 49 shews the type adopted for the deepest water of the latter. Full particulars concerning all the types of this cable will be found in Part II., with further reference to the core in Part III. Special arrangements were made in the design of both these cables to meet the requirements of increased speed. Since the successful application to submarine cables of various modifications of Wheatstone's automatic transmitter (intended only for land telegraphy) † the limit to the speed attainable only depends, practically speaking, upon the type of cable employed. In a general way, therefore, it will be readily understood that, if funds are available at the time for the construction of one cable

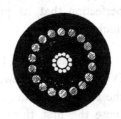

FIG. 49.—Anglo-American Atlantic Cable (1894), Deep - Sea Type.

which will do the work of two, a notable economy is the result. ‡ On these principles, the core of the new Commercial cable was composed of a copper conductor weighing 500 lbs. per N.M., covered with a gutta-percha insulating sheath weighing 320 lbs. per N.M., while the new "Anglo" (see Fig. 49) has a core with conductor weighing 650 lbs. per N.M., and gutta-percha insulator

---

* A detailed description of this cable was given in the course of an interesting article in *The Electrician* for 12th October 1894, and the full specification of it appears in Part II. of this work. It forms the fifth *working* cable of the "Anglo" Company. Their, three first cables—of 1858, 1865, and 1866—have been "dead" for some time, but the 1869 French cable has not yet been abandoned so far as simplex working is concerned, though not in operation at the moment of writing.

† This instrument (following others of Bain and Siemens) was invented in its original form in 1859. It came into very general use on H.M. Post Office telegraphs. In 1879 MM. Belz and Brahic adapted the same principle to an instrument suited to the exigencies of submarine cable work ; this was followed by other modifications devised by Herbert Taylor (1888) and T. J. Wilmot (1890), besides more recent ones. All these varieties of the Wheatstone automatic transmitter will be found described and illustrated in Part III., where the general principle of machine transmission and its advantages—alike in attainable speed and, from a "receiving" point of view, in the avoidance of the variable human element of manual transmission—are also discussed.

‡ A submarine cable system should, however, if it is to compete at all successfully, possess at least one spare "string to its bow" to provide against the contingency of a breakdown.

400 lbs. per N.M.,* involving a completed cable (main type) nearly double the weight of previous corresponding lines.

The actual speed obtained by automatic transmission with the latter cable is as high as forty-seven (or even up to fifty) five-letter words per minute.† On the previous lighter Atlantic cores twenty-five to twenty-eight words per minute was the usual maximum speed attainable; the former, say, by average manual transmission and average receiving, and the latter by automatic transmission, other circumstances corresponding. With a cable in good condition, practically the same maximum speed can now be obtained working duplex as in working simplex. Duplex telegraphy, as applied to cables, has, in fact, been developed to such a pitch of perfection that, as previously stated, their carrying capacity is practically doubled by it.

**Atlantic Cable Systems.**—As a part of the union between the Old World and the New, there are altogether fifteen cables at the bottom of the North Atlantic, which are usually termed "Atlantic cables." Out of these the first three are absolutely dead—as well as buried; nine are in perfect condition for duplex working; and three are, at the time of writing, rendering a fairly satisfactory account of themselves—with the help of occasional repairs—in simplex working.‡

In some cases the Atlantic companies have special cables of their own from the landing-place at the extreme south-west point of Ireland to points on the Continental coasts—those of France and Germany more particularly. These cables, as well as the European ends of the main sections of the various Atlantic systems, may be seen in the map on the opposite page (Plate IX.).§

---

* The largest core hitherto made for submarine cables was composed of 400 lbs. copper conductor, and 400 lbs. gutta-percha insulator. These were the core dimensions of the Malta-Alexandria cable of 1861, originally intended to connect up Falmouth and Gibraltar for the Government. This same type of core was also employed in the 1873 and 1874 "Anglo" Atlantics. Most of the Atlantic cores, however, had been smaller than this.

† The corresponding maximum speed on the new "Commercial" cable is said to be two hundred letters, or about forty-two words, per minute, by automatic-machine transmission.

‡ It may be mentioned here that if duplex be dispensed with (*i.e.*, in simplex working), a cable behaves rather *better* with a small fault in it than when perfect. No cable, however, can be satisfactorily worked in *duplex* unless it is in perfect order: at least it is then only that anything like the full advantage of duplex working can be secured.

§ Both this and the following map are brought up to date from the last issued by the International Telegraph Bureau at Bern. They are purposely reproduced in French, that being the language adopted by the Bureau and in the Telegraph Convention, besides still being the usual tongue in international matters.

[PLATE IX.

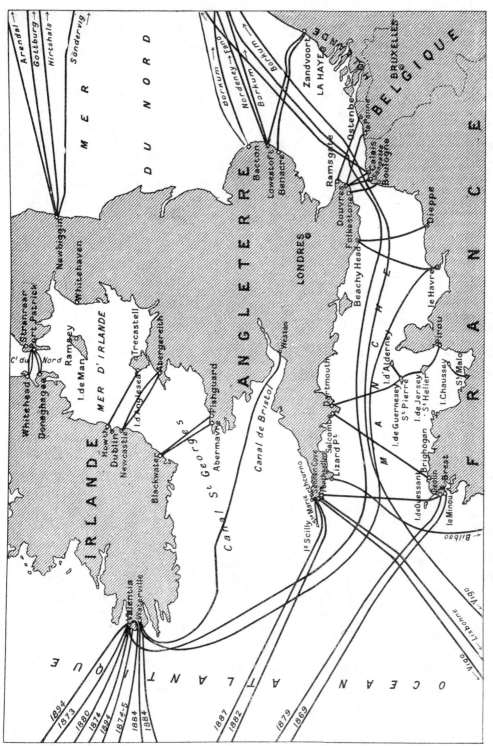

General European Connections of Atlantic Cables (from the Map published by the International Telegraph Bureau, Bern, 1897).

Plate X. (see overleaf) suggests a somewhat complicated state of affairs at the other, or American, end of these lines, through their ever-increasing numbers. Some of these cables, at each end of the corresponding main section, contain more than one insulated conductor.*

In the early pioneer days of ocean telegraphy, the Atlantic Telegraph Company † started with a *minimum* tariff of £20 for twenty words, and £1 for each additional word. This was first reduced to £10 for twenty words, and was further altered later on to £5 for ten words. After this it stood for a long time at a minimum of 30s. for ten words of five letters each. Subsequently, in 1867, the Anglo-American Company tried a word-rate of

---

* One great danger to which cables in the North Atlantic are continually subject is the grounding of ice-floes, besides the wear and tear of rocks, near the shore. For this reason special methods are now adopted for the protection of the shore ends. In the last (1894) "Anglo" cable, the sheathing wires of the Irish shore end, besides being of great circumference, are applied with a very short lay (see Fig. 50), the object of this being to increase the weight of iron (within a given surface), thus reducing the chances of its being shifted, and avoiding abrasion. This plan obviates the necessity of an exceedingly large, and therefore rigid, type of wire being used for sheathing. The idea originated, it is believed, with Mr P. C. Dresing, the engineer to the "Great Northern" Company, who in 1892 designed such a cable for communication with a lighthouse off the Chinese coast, near Hong Kong.

The section of a piece of cable, cutting through the wires thus laid up, gives them the appearance (as may be seen in the figure) of being elliptic. For further particulars, see Part II.

Lock-armoured cables have been devised by Messrs Felten and Guilleaume, as well as by others, with a similar object—*i.e.*, that of introducing the greatest possible weight into a given limited area. Cables of this class are, however, more especially intended for rivers subject to strong currents. They are also largely used for underground purposes, as being an excellent means of contending with interruption from pick-axes or animal life.

In the case of the new "Commercial," the shore ends are protected against the same danger by an armour consisting of some form of linked chain which is wound round them. At the Newfoundland end this is an especially desirable precaution, on account of the prevalence of ice.

A few years ago Mr H. Kingsford suggested an ingenious system of alarm wires for the shore ends laid on doubtful bottoms. These insulated wires, embedded in the serving between the inner and outer sheath, are in circuit with an electric bell, which, being to earth, only comes into operation in the event of one of the wires becoming chafed. This would naturally occur some time before the main core was affected inside the inner sheathing. Moreover, the said alarm wires furnish the means of a loop test being taken to localise the precise position of the fault. Again, with such timely warning a local repair may be made without involving the engagement of any repairing ship for the purpose. It also obviates the necessity of periodically underrunning the shore end— a practice open to several objections, besides that of expense. A plan like this may be the means of avoiding several weeks' interruption, and consequent loss of traffic. A full description and illustration of Mr Kingsford's device will be found in the *Journal of the Institution of Electrical Engineers*, vol. xix., p. 656.

† The tariff universally in vogue up to that time was, in fact, based on the message rate common to land telegraphs—*i.e.*, a charge was made on a message up to ten words, and so much for each additional word, depending upon its destination.

£1 for the 1865 and 1866 Atlantic cables; but it was not until 1872 that Mr Henry Weaver, their able manager, first instituted a regular word-rate system (without any minimum) of 4s. per word.* At the present time (1897), thanks to competition, to technical improvements in the plant, and to increased traffic, bringing in its train those economies in the working which are always possible in a larger scale of operations, the rate stands at 1s. a word with all the Atlantic companies.†

The twelve Atlantic cables now in use represent a total capital of something like £17,000,000 sterling. A knowledge of the profits derived from each system is not readily to be arrived at; but from a comparison of the traffic receipts or "money returns" of the oldest existing Atlantic company at different periods,‡ we are bound to conclude that the "takings" are, roughly speaking, very much the same now as they were twenty-five years ago. This is explainable by the fact that, although the number of messages now passing is much greater, the reduction of the rate

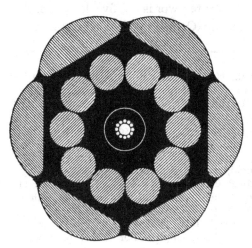

FIG. 50.—Irish Shore End of 1894 " Anglo " Atlantic.

(with the ever-increasing competition of rival lines) just about cancels the advantage, so far as receipts are concerned.

---

* Great advantage was taken of this by the public, and the word-rate system was almost immediately taken up by the other cable systems—to wit, by the " Eastern" in 1872, on the opening of the British-Australasian line. Though the number of messages conveyed at once enormously increased, it took some time for this increase to balance the effect of so large a proportion consisting of one or two words (chargeable only as such), as compared with the minimum ten-word tariff. In due course, however, as the use of the telegraph became more widely extended and appreciated, the increased number more than balanced the decreased average length of messages; and, ultimately, the introduction of the word-rate had the indirect effect of very materially adding to the earnings of not only the Atlantic, but of all cable systems.

† The shilling rate was first *permanently* adopted (down to the present, at any rate) by all the companies in 1888. Before that year competition had certainly had the effect from time to time of bringing the rates down to 1s., and even to 6d., for a few weeks or months at a stretch; but until 1888 they had invariably returned to their former figures, or something similar.

‡ The Atlantic and Anglo-American Telegraph Companies were not actually *amalgamated* until 1873. When the " Anglo-American" Company was promoted, in 1866, it was for the purpose of securing fresh capital to complete the work undertaken and started by the Atlantic Company. From that date until 1873 the two companies worked together as

[PLATE X.

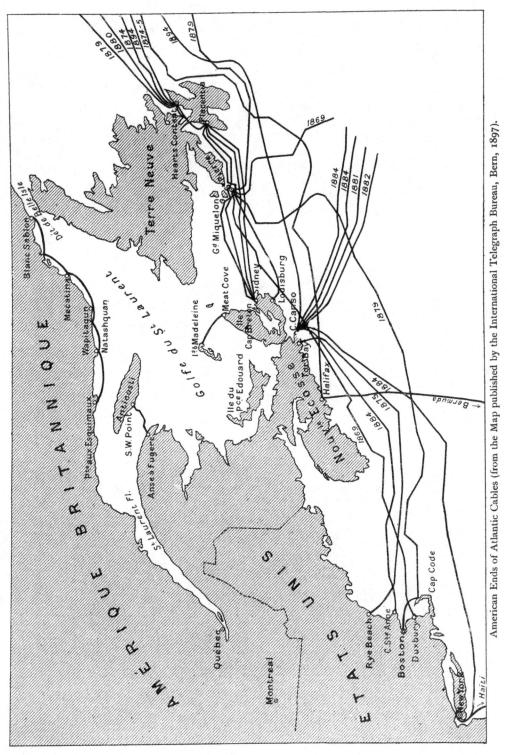

American Ends of Atlantic Cables (from the Map published by the International Telegraph Bureau, Bern, 1897).

[*To face p.* 144.

**General Retrospect.**—The reader has now been given an opportunity of surveying the whole, or at least the most important parts, of the vast network of submarine telegraphic systems with which the world is at present endowed. It is, of course, the original cables only, which first opened up telegraphic communication with the various countries, that have been most fully noticed, subsequent duplications being sometimes referred to, but not dwelt on in detail. As already mentioned, the important trunk-lines to the East and Far East have been more than duplicated, besides many of the branch and independent cables.

Two distinct lines, one of which is duplicated, now unite Europe (directly) with the South American continent and all its branch lines, thus indirectly giving additional lines of communication with the West Indies and North America. They also communicate with the West Coast of Africa, and *via* this branch with the Cape.

The folding map at the end of this Part gives an up-to-date idea of the main general telegraphic systems of the world—the network of its "electric nerves." This map indicates not only the original cables, but all their duplications and multiplications.* It is arranged in the way it is in order to shew more clearly the principal gap that is still open for repletion.†

A noteworthy void is observable in the North Atlantic, between the Azores and Bermuda, which, if filled up, would constitute another Atlantic cable, and thus an additional telegraphic highway between Europe, on the one hand, and the United States and Canada on the other. A still greater blank occurs between Mauritius and the West Coast of Australia, which might profitably be made good, so as to create an extra line to Australia. Another useful extension would be that of the Central and South American cables up to San Francisco.

---

one, sharing profits as soon as there were any to share. At the same time as this arrangement was effected, the "Anglo" Company entirely took over the New York, Newfoundland, and London Telegraph Company, with which the "Anglo" cables had previously been working in connection.

* Though, at the moment of writing, the "Halifax and Bermudas" Company's extension between Bermuda and Jamaica is not actually laid, from a statement of Mr Joseph Rippon (the general manager), there can be no doubt that it will be within a very short time of the publication of this volume.

† This is a specially prepared map, and the *lingua franca* has not here been adhered to, on the grounds that by far the greater part of the world's cables are English —made, laid, and worked by English hands—stretching in all directions to British territories and dependencies.

# CHAPTER IV.

## MISCELLANEOUS AND COMMERCIAL RÉSUMÉ.

### SECTION 1.—MAIN COMMUNICATIONS : PRESENT AND FUTURE.

**Pacific Cable Projects.**—How soon the greatest gap of all, namely, that across the Pacific Ocean, will be spanned by a telegraph cable,* it would be out of place to predict.† If, however, the present long peace

---

* When spoken of in the light of a direct line across the Pacific, from San Francisco to Japan and China, this is sometimes described as the "missing link" of the earth's girdle.

† Since the above lines were written, the Secretary of State for the Colonies (the Right Hon. Joseph Chamberlain, M.P.) has called together a conference at the Colonial Office to go into the question of an All-British Pacific Cable from every aspect. Over

between the great Powers of the world is not broken, there is no reason in the nature of things why this big enterprise should be much longer delayed.    It is difficult to foresee any insurmountable *engineering* diffi- culties in its way, albeit the ground has been only partially sounded over at present.    Along one or two of the proposed routes it is true that depths have been found* which exceed, by some 700 fathoms,† those in which any cable has hitherto been laid.‡    This, of course, would necessitate the construction of a well-adapted type of cable, as well as of suitable machinery for paying out and—in view of what must be called the usual eventualities—for picking up the same.§    But such matters will certainly not daunt any telegraph engineer or contractor worth his salt.‖

On the electrical side, again, it has been asserted that if the Pacific line

---

this the Under-Secretary of State (Lord Selborne) presided, and he was assisted in his investigation by various important colonial officials, as well as by the engineer-in-chief to H.M. Post Office and consulting engineer to the Colonies (Mr W. H. Preece, C.B., F.R.S.).    A number of experts were called to give evidence, and a report on the whole in favour of the project was duly arrived at.    This affair occupied from June 1896 to January 1897, Mr W. Hepworth Mercer, of the Colonial Office, acting as Secretary.

* Associated, moreover, with deposits of which no practical experience has been gained.

† Though the Atlantic runs the Pacific pretty close, the deepest sounding yet recorded was taken last year (1896) by H.M.S. "Penguin" about 550 miles north-west of New Zealand.    This gave 5,155 fathoms—or a depth of nearly six miles—but the proposed line would not, under any circumstances, require to go immediately into this region.

In these great depths the bottom is so soft that great difficulty is experienced in determining when the lead actually touches the bed of the ocean.    Thus it sometimes happens that the depth *registered* is far in excess of what it should be, owing to the wire continuing to run out and coil itself down after reaching bottom.    These remarks refer especially to depths over 2,500 fathoms, where, on account of the great length of line, the jump is less easily observed.

The Halifax and Bermudas cable, in approaching Bermuda, runs into over 2,800 fathoms, as does also the South American Company's line, and one of the lines spanning the North Atlantic; moreover, the Brazilian Submarine cable was laid in over 2,700 fathoms.

‡ Again, on what is the favourite route from a national point of view, the longest section would run into some 3,500 N.M., as against 2,717 for the longest existing section —across the North Atlantic between Brest and St Pierre.    This, however, would introduce no very serious difficulties.

§ It has been urged by those who are opposed to the scheme that recovery would be an extremely lengthy—if not impossible—task.    Regard must, however, be had for the fact that the Direct Spanish Company's cable has been picked up and repaired in some- thing like 2,000 fathoms within the course of two or three days, not to mention other repairs in much greater depths (upwards of 2,700 fathoms), though taking longer.

‖ Forty years ago the feasibility of Atlantic Telegraphy was the subject of similar incredulity at the hands of many of the greatest authorities until the cable was actually laid and worked in 1858.    In that instance there was practically no data to go upon, whereas the proposed Pacific line may be regarded as but a further extension of what has already been done, though certainly involving special arrangements and precautions.

were laid it would not work satisfactorily. This suggestion is not, however, in agreement with facts; though it is true that the maximum speed
obtainable on the long section of the All-British route (say 3,500 N.M.),
with quite a large core, would be low as compared with that attained on
the Atlantic cables, which are, of necessity, kept heavily burdened with
traffic more or less continuously. Still, with a core of the same proportions as that adopted in the "Anglo" Company's last cable, a speed
could be obtained well up to the minimum required by the Canadian
Government in 1894. The foregoing would be by ordinary manual transmission, whereas all the latest improvements in machine transmission with
curbing arrangements (as well as condensers) would naturally be applied
with something like 30 per cent. increase in the speed—quite apart from
the application of the duplex system. *

The more doubtful question, however, is whether the line could be made
a commercial success; whether, in fact—even within the next quarter of a
century, much more within the next ten years—such an expensive enterprise
could be made *to pay*. Although it is upon these commercial, as well as
what may be called quasi-political (Imperial, Anglo-Australo-American,
"Pan-Britannic") considerations, that its fate depends, rather than upon
any advances in the art of submarine telegraph engineering, some account
of what has been said and done in the matter may not be amiss here.

The scheme has been considered and discussed by many able authorities
for a long time. Thus it is now twenty-seven years since (in 1870)
a devious sort of trans-Pacific cable was first proposed by the late Mr
Cyrus Field and other American capitalists, who endeavoured to negotiate
financial arrangements for the purpose. Their plan was to connect Cali-
fornia with China *via* Alaska and Japan, but it had to be abandoned.†

---

* As a first venture the above cable would probably be sufficient to meet the ends in
view. It is conceivable that a larger core would be out of the question on the ground of
cost. Moreover, increase in the dimensions beyond this would have the effect of further
augmenting the mechanical difficulties as regards a suitable type of cable to enclose the
required minimum thickness of jute packing.

Experiments have already been made on two Atlantic cables looped together to test
this point of the possibility of working through a considerably greater length than that to
be dealt with in the longest section here, the results being perfectly satisfactory.

† By one of the schemes the total length of submarine cable required would only be
about 750 miles. This was by a line across the Behring Sea touching at various islands
of the Aleutian group (about a hundred in all) on the way. Most of the sections would
thus be in quite shallow water—say 30 fathoms—but some might have had to go into
very great depths in certain places, near some of the islands. The Behring Straits would
have still further reduced the length of cable involved to but little over 50 miles, indeed.
This route, however, was debarred partly on account of the presence of ice and snow
together with the absence of soundings by sea, but also owing to it being an impracticable route on each side for the erection or maintenance of any land lines.

Since that time the Pacific idea has been further developed in this country, and from a national rather than a private commercial point of view. The various routes which have found most favour in the English mind are those which provide in the first place for communication between British Columbia and Australia, with resting-places at one or another of the groups of islands mid-ocean. H.M. Government and the Colonial Governments most concerned have been urged, from time to time, to consider the matter in its naval and strategic aspects. Two Colonial Conferences (in 1887 and 1894) were largely occupied with this subject, as may be gathered from perusal of the bulky Bluebooks which record their proceedings. The Dominion Government took the matter up quite strenuously in 1893-94, and invited the various contracting firms to send in estimates for construction and laying, under condition of forming a company for working and maintaining the cable. Subsequently Lord Jersey, in the course of a long report on the Ottawa Conference,* strongly recommended the Home Government to take further steps, and to order a series of soundings to be taken on the various suggested routes.

The proposed tariff rate from Europe to Australia *via* the projected Pacific cables was 3s. a word. In view, however, of the losses said to have resulted from the recent reduction of tariff on the existing system, some authorities have expressed doubts whether such a comparatively low tariff could be rendered profitable.† This consideration raises the whole question whether the existing service *via* India ‡ does not meet all the practical re-

---

* Report on "The Colonial Conference at Ottawa," 1894, to Right Hon. the Marquis of Ripon, P.C., K.G., Secretary of State for the Colonies, by the Right Hon. the Earl of Jersey, P.C., G.C.M.G., Chairman of the Colonial Conference at Ottawa, appointed Delegate to represent the Imperial Government.

† Cheap tariff experiments have not been encouraging to the existing companies. The rate between Europe and Australia was reduced in 1891, under a Government arrangement between the Colonies and the Eastern Extension Telegraph Company, from 9s. 4d. to 4s. per word, which is said to have resulted in a heavy loss to the company. The Colonies then arranged for an increase to 4s. 9d., and this, with abnormally favourable circumstances, has considerably reduced the loss. Similarly, the reduction of the New Zealand cable tariff in 1893 from 8s. 6d. (and previously 10s. 6d.) to 2s. per word, resulted in a loss of nearly 60 per cent. of revenue. The rate at present stands at 5s. 2d. In any original agreement between a cable company and a Government, it is now always stipulated that the Government shall take upon its shoulders a certain proportion of the loss resulting from any abatements of tariff which may be insisted upon in any future international telegraph conventions to which the said Government may be a party. The highest existing cable-rate from England is that to Colombia (South America), which is still about 27s. per word, and the next highest that to Peru, at about 22s.

‡ By this service, provided by the "Eastern" and its allied companies, Australia and New Zealand, as well as India, are placed in communication with North America, not only *via* Europe, but (if necessary) *via* the Cape, West Africa, and South America. The further extensions of the great Eastern trunk-line also connect all these countries tele-

quirements of the case at present. It should be remembered that by this route messages from Australia to America only take a few minutes longer in transmission than those which stop short at Great Britain. This service, passing, as it does, over cables entirely duplicated, and in part triplicated, cannot be considered liable to interruption by any of the ordinary "accidents of cable life."

There is, however, one contingency—and a very serious one—in which this extremely undesirable eventuality might occur. In the event of a great war* between this country and another naval Power, or Powers,† it is quite conceivable that more than one of our Mediterranean cables (if no others) might be ruthlessly cut by the enemy, ‡ and our communications with Egypt, India, and other Eastern and Australasian stations entirely broken off.§

Not only can the cable be cut in shallow water near the coast by any small steamer with purchase gear that will raise an anchor, but lengths be removed in a manner that would tax the resources of a repairing vessel to

---

graphically with China and Japan. In fact, it will be seen, by referring to the Telegraph Map, that China, Japan, and our East Indian and Australasian colonies and dependencies, possess already—apart from duplications —at least *three* alternative lines of communication with the United States and Canada. These are—(1) The Eastern, Eastern Extension, etc., cables to India and Europe (to England *via* Continental land lines, or *via* Gibraltar, etc.), and the Atlantics thence to America ; (2) the Great Northern cables and land lines from China to Europe, *via* Siberia and Russia, in connection with the Atlantic cables as above; (3) the above-mentioned Cape route from India, *via* West Africa and South America.

* Mr C. Scott Snell has lately endeavoured to meet this by what he terms a cable defence scheme—*i.e.*, a plan for paying out cables rapidly from a man-of-war to any desired spot to take the place temporarily of an interrupted cable, or to establish communication for the moment between any two spots, or to an outlying station, hitherto unconnected. This scheme is referred to further elsewhere.

† The present lines to India and Australia are as follows :—(*a*) Lisbon, Gibraltar, Malta, Egypt, and the Red Sea ; (*b*) France, Italy, Greece, Egypt, and Red Sea ; (*c*) Germany, Austria, Turkey, Russia, and the Pacific Coast ; (*d*) Lisbon, and the West and East Coast of Africa. All these routes pass through foreign countries, and could at once be interrupted in the case of war.

‡ It is scarcely necessary here to go into the question as to how such interruptions can be effected. It suffices to say that, given sufficient occasion, means have already been found in times of war to effect this end—not only on, and close to, the shore, but also out at sea. A cable deposited in depths under 300 fathoms is peculiarly prone to interruptions of such a character, quite apart from the rough bottom so frequently to be met with in these depths.

§ A Russian journal, the *Novoe Vremya*, recently said: "In case of an armed conflict between this country and England, our first task would be to block England's communication with India and Australia."

It is, indeed, reported that when war with Russia was imminent some years ago, the Russian authorities had a ship at once equipped with the necessary cable-hooking apparatus.

replace—even if on the spot at the moment.* There are also several ways in which a cable can be readily (and, at first sight, innocently) interrupted at and near the water's edge—to wit, by bonfires on the beach, by fishermen, etc., supposed to be employed on their ordinary avocations. Several cables have already been interrupted by bonfires.

The International Telegraph Convention of 1884 recognises, it is true, the principle of protection by Government, but the effect of this recognition of responsibility is seriously qualified—perhaps practically annulled—by the following clause:—"It is understood that the stipulations of the present Convention do not in any way restrict the freedom of belligerents."

If this view of the value of the said Convention be correct, the argument that cables are as safe in times of war as in times of peace falls to the ground. An additional and entirely independent line of communication between North America, Australasia, and "the East"—such as the Pacific scheme would afford—becomes, therefore, a matter of greater moment than the ordinary commercial mind is able to appreciate. It would not only act as a spare string to our bow,† but would, for obvious reasons, be less at the mercy of the enemy than the others.‡ Thus, if the ordinary line of communication between England and Alexandria were broken, through the Mediterranean cables being cut,§ we should still be able to speak to the East, provided the Atlantico-Pacifico-Indian route round the globe remained intact. And, except in the deplorable and (let us devoutly hope) not very likely event of war with our American cousins, the chances of that route being also interrupted by an enemy's hand are not particularly serious.

The standpoint, therefore, from which any early necessity of laying a trans-Pacific cable is likely to be determined is the national, inter-colonial, and Anglo-American one, rather than the purely commercial. For their mutual benefit, the great English-speaking and English-governed countries —that Pan-Britannic "Oceana," living here under the "Union Jack," there under the "Stars and Stripes"—may decide sooner than either the tele-

---

* The only possible answer to this is that a man-of-war of our own nationality would always be at the right spot to keep guard or intervene at the right instant.

† Any scheme for further, and independently, reducing the chances of a total breakdown of telegraphic communication with our colonies should not fail, in the interests of the Empire, to commend itself to all British subjects—no matter what the efficiency of the existing system may be, which, however, as a pioneer system must receive every consideration.

‡ Though it is true that the cables to the East are laid on the trade routes protected by British war-ships, there would be a greater sense of security attached to a cable laid in the open, broad, ocean (away from other European Powers), at a three-mile depth, whose course need not be known, and which would have but few ends, all of which could be kept strictly under Government surveillance.

§ The Cape route, round Africa and *via* the Spanish National and Brazilian Submarine lines, would also be open to similar interruptions at the European end.

graphic or the financial world could hope for or anticipate, that they must and will have their Pacific cable.*

But although the *motives* which will probably have most weight in determining the rapidity or slowness of further ripening and the final realisation of this great scheme are politico-commercial and Imperial, rather than purely commercial ones, it is by no means so certain that it will be brought about through Government initiative.† In the case of the first Atlantic cable, there was a vast amount of talk about its possibilities, in high political circles and elsewhere, long before it was actually attempted. Its final translation into the realm of " concrete facts " was entirely due, as has already been shewn, to the enterprise of a few private capitalists.

The chief stumbling-block of the scheme ‡ is the want of a resting-place for the cable (along the most favoured route) between Vancouver and the Fanning Isles, where a section of 3,500 N.M. is involved. The most serious objection to a section of this length is the type of core entailed to afford the required working and earning capacity.§

---

* Some of our American friends have from time to time projected schemes for a Pacific line on their own account, with notions of concessions from the Government of Hawaii, with a view to landing the cable at that island. None of these have, however, so far come to anything. Hence here we ventilate the possibility of combining forces between all English-speaking countries—more especially as the main weakness in the "All-British" scheme is the land lines through Canada. This, moreover, may be regarded as a possible solution of any difficulties (or strained relations) with the United States, for would it not have the effect, so much to be desired, of once more firmly binding together the two countries—aye, and all English-speaking countries?

† So far Sir Sandford Fleming, K.C.M.G., M.Inst.C.E., has done far and away the most to promote the realisation of an Imperial Pacific cable. This gentleman, it will be remembered, was the main instigator of the Canadian Pacific Railway, besides being the engineer of that great project.

Again, this scheme has received the weighty support of the Hon. Sir Charles Tupper, Bart., G.C.M.G., C.B., and also of the Hon. Sir Mackenzie Bowell, K.C.M.G., formerly Prime Minister of the Dominion of Canada. Not so very long ago *The Times* opened its columns to an interesting and protracted correspondence between Sir C. Tupper and the late Sir J. Pender on the subject.

‡ Mr Herbert Laws Webb (the son of that famous cable engineer, Mr F. C. Webb, M.Inst.C.E., and of much experience in telegraph matters himself) contributed an excellent article on the Pacific Cable project to the *Engineering Magazine* for March 1894. Our prospective telegraphic isolation in times of war has also been discussed in an able manner in the course of a recent article in the *Contemporary Review*, recommending this scheme and an extended system of Colonial cables as a method of meeting the difficulty. Again, *Blackwood* has quite lately furnished an article on the subject.

§ The latest information, since the above was written, points to the probability of the All-British line being brought about in an entirely different way—albeit a distinctly satisfactory one, from a national and political point of view, at any rate. This is shewn in the map facing page 208 by a dotted line from Porthcurno to the Cape and hence to Perth, Australia, touching at various islands *en route*. It thus involves shorter continuous lengths than in the previous proposals. On the other hand, it does not include in its scope any connection with Canada.

## SECTION 2.—MILEAGE AND FINANCIAL STATISTICS.

The total length of telegraph cable laid at the bottom of the sea, down to the present time, is nearly 170,000 N.M.,* representing about fifty millions sterling (£50,000,000); as against some 662,000 miles of land lines (mostly consisting of several conductors), estimated to have cost sixty-two millions sterling (£62,000,000).† Adding the two together, we obtain a combined capital outlay of 112 millions sterling (£112,000,000) for the telegraphs of the world.

As regards submarine cables, the figure given is, of course, the whole actual capital, originally invested in the various cable companies for the purposes of the cables which they were formed to lay and work. The present total market value of their united capital greatly exceeds that figure. To give an idea of this, the shares of the Eastern Telegraph Company (which owns far the largest cable system in the world‡) are to-day worth 50 per cent. more than their original value. An idea of this company's system at the present time may be gathered from the map on the next page. The shares of some of the other companies have also risen enormously since allotment, although not so much as the "Eastern."§

---

* Made up by something like 1,500 separate lines, or sections of lines, varying in length from a quarter of a mile to over 2,700 N.M. A small proportion of these contain more than one conductor, thus constituting a multiple-core cable. Besides them, there are some 100,000 N.M. of electric cable of different descriptions belonging to various Governments. Only a small percentage of the latter is laid permanently at the bottom of the sea, the greater part being stored in tanks, etc., ready for use at a moment's notice. This total length includes the submarine-mine cables, field-telegraph cables, position-finding and electric-light cables, of all nations. The united length of its submarine cables would girdle the earth eight times, or reach nearly two-thirds of the way to the moon. The land wires joined end to end would reach four times to the moon and back again.

† Out of this total, 655,000 miles are aerial wire, and the remainder subterranean cables. The latter contain, in most cases, several insulated conductors (bunched together), thus bringing up the total length of *conductors* in land lines to 2,300,000 miles. In estimating all kinds of "telegraphic" communications, the 553,000 miles of *pneumatic tubes* must be added to the land and submarine electric conductors. The average cost of an aerial wire may be taken to be, roughly, one-seventh that of an average cable. The proportion is greater, of course, in the case of heavy types (or large cores) of the latter— that is to say, the cost then comes to *more* than seven times that of an aerial wire of the same length.

‡ This and the companies immediately allied with it (as a part of its system) own between them over 50,000 miles of cable, or nearly one-third of the total mileage of telegraph cables in existence, representing a capital of over £10,000,000. If other companies associated with the "Eastern," but not quite so closely, be included, it will be found that the capital involved in the combined systems of this alliance forms about a third of the entire capital engaged in submarine telegraphy.

§ As to the present control and management of all these valuable concerns, and the capitalists chiefly interested in their prosperity, the reader may be referred to the list of cable companies and their Boards given in the "Electrical Trades Directory," produced·

Of the total mileage of submarine cables as given above (over 165,000 N.M.), nearly 90 per cent.* has been provided by private enterprise, and the remainder by different Governments.    The actual length of submarine cable belonging to Government administrations is about 18,000 N.M.

Something like 120,000 N.M. of the whole have been manufactured and laid by a single firm of contractors, the Telegraph Construction and Maintenance Company, previously Messrs Glass, Elliot, and Co.    This includes nine cables across the Atlantic to North and South America,† as well as the great network of cables to the East, Far East, and South Africa.

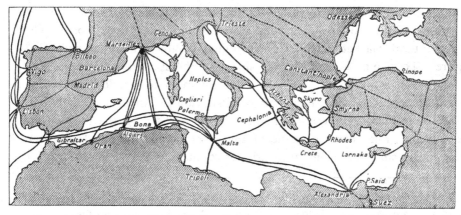

Cable System of the Eastern Telegraph Company and its Connections (1897).

### Section 3.—Engineers and Contractors.

It will not, perhaps, be out of place here to make a few more remarks with reference to the various contractors and engineers that have been principally concerned in carrying out submarine telegraphic enterprises, and for considering their mutual relations and positions at different times.

---

annually by *The Electrician* Printing and Publishing Company.    Those who care to ascertain the principal investors who risked their money on submarine cables in the early pioneer days—and who may, therefore, be distinguished as financial pioneers of this great industry—must study the original prospectuses of the companies then formed.    The above Directory will also be found most useful for giving the lengths of cable owned by each system from time to time, besides giving the date of laying of each section and between what points.    From the latter it will be seen that most of the various Government cables are of quite short length, and that the long sections are almost exclusively owned by the various English companies formed for working cables.

* Of which at least 75 per cent. is probably English capital.

† Messrs Siemens Brothers have constructed and laid seven North Atlantic cables. The first Atlantic cable was partly constructed by Messrs Glass, Elliot, and Co., and partly by Messrs R. S. Newall and Co., and was laid by the engineer to the Atlantic Telegraph Company, as already described.

In the earliest days of this industry, as already shewn, firms of cable manufacturers only contracted for the construction of the cables, engineers (independent or otherwise) being specially engaged—sometimes directly by the owners, sometimes by the contractors—for the purpose of laying them.

**Engineers.**—In certain instances the engineers had practically the entire control of the enterprise, and employed contractors to make the cable. The first Atlantic was carried out much in this way by Sir Charles Bright, on behalf of the Atlantic Telegraph Company. The relationship between engineers and contractors was afterwards reversed, when it was found that the latter had placed themselves (by engaging a permanent staff of engineers and electricians) in a position to carry out the entire work ; for, naturally, the cable companies could thus get served more cheaply. Then the contractors became the employers, and the engineers their servants. The turning-point in this matter seems to have come when two of the chief contracting firms (within a month of each other) increased their capital and improved their organisations by turning themselves into limited liability companies.

Just as some of the electrical engineers and electricians accepted service under the contractors, others of them—more especially those who had made for themselves a name as pioneers in cable work—felt that they would have a better field as independent advisers, or " consulting " engineers and electricians, to the various companies that were constantly being floated for carrying out new submarine telegraphic schemes, as well as to any Governments that might require-similar services. Thus, in 1860, Sir Charles Bright and Mr Latimer Clark entered into a partnership together, and acted as consulting engineers to a large number of the cable enterprises carried out or started between that year and 1871. Similarly Mr Lionel Gisborne and Mr H. C. Forde,* both being gentlemen who had made their reputation in connection with submarine telegraphy, now established themselves as a firm of consulting engineers for this class of undertaking, and to good effect. Shortly after Mr Gisborne's death, Mr Forde allied himself with Mr Fleeming Jenkin and later on with Messrs Bright and Clark, Mr Jenkin being also associated with this firm in certain matters. Again, Sir William Thomson and Mr Cromwell Varley, already combined in the matter of patents, became additionally united as consulting electricians to several important cable enterprises. Subsequently Mr Fleeming Jenkin joined them, and later on the firm of Thomson and Jenkin was established. About the same time Sir Samuel Canning—having previously

---

* Whose lamented death has so recently occurred.

left the service of the Telegraph Construction Company*—entered into partnership with Mr Robert Sabine for consulting work, in which they were successful in turning their engineering and electrical experience to good account. After the dissolution of the partnership between Messrs Bright, Clark, and Forde (the former joining his brother, Mr Edward Bright, in a consulting and general engineering business), the firm resolved itself into Messrs Clark and Forde. Subsequently both the late Mr Charles Hockin, M.A. (a Cambridge Senior Wrangler), and Mr Herbert Taylor (formerly of the Royal Engineers), became partners in this firm, which has now, in fact, been known for many years as Messrs Clark, Forde, and Taylor. They have acted as consulting engineers to the greater part of the recent cable enterprises.

Nowadays, however, through the greater experience and efficiency of the contractors' staffs, and to the gradual extension of the work generally, the scope for consulting engineers has tended to become more and more limited.

**Contractors and their Work.**—As to the contracting firms themselves, although there has already been occasion for referring to all or most of them, *apropos* of the various undertakings which they have carried out, some short *résumé* of the history of the chief firms may be of interest here.

In 1864 Messrs Glass, Elliot, and Co., of Greenwich,† who had shared with Messrs Newall the contracts for supplying almost all the telegraph cables so far constructed, amalgamated with the Gutta-percha Company of Wharf Road (City Road, London, N.)‡—with whom they had already had a regular business agreement for supplying the core for their cables—under the new title of the TELEGRAPH CONSTRUCTION AND MAINTENANCE COMPANY LIMITED. As already stated, this powerful corporation has been engaged in the construction and submergence of more than two-thirds of all the telegraph cables now at the bottom of the sea. Particulars regarding the founding of this firm, with the names of its founders, have already been given as matters of history with reference to the pioneering of Atlantic telegraphy.

---

* In which he was succeeded as engineer-in-chief by Mr Henry Clifford (already referred to as one of the first Atlantic veterans), who held this post for over thirty years. Mr Clifford has recently retired to a consulting position in the company's counsels, and has been succeeded in the engineership by Mr F. R. Lucas.

† This firm was founded in 1854, when it took over the business of Messrs Küper and Co., which had been established since 1847, originally as ordinary wire-rope makers.

‡ Following after Siemens, who was the first actually to make gutta-percha insulated wire, the Gutta-percha Company had for a long time been the only manufacturers of gutta-percha in this country.

Towards the end of 1863 the Berlin firm of Siemens and Halske,* subsequently so celebrated all over the world, established a branch factory at Charlton for the manufacture of cables and electrical instruments.  In 1865, Mr Halske having given up his connection with this part of the business, the London firm of SIEMENS BROTHERS was established.  In 1880 this was converted into a limited liability company, Dr Siemens (afterwards Sir William Siemens, F.R.S.) acting as its chairman.†  This distinguished firm‡ have not only carried out  many large cable contracts, as already shewn, but are universally esteemed for the excellence of their work in other departments of electrical industry, such as electric light, power, traction, etc.  More recently they have started an india-rubber core department, for electric light and other cables.

In 1864 the INDIA-RUBBER, GUTTA-PERCHA, AND TELEGRAPH WORKS COMPANY LIMITED was registered for the purpose of taking over the large works, for constructing india-rubber insulated conductors, and india-rubber goods generally, of Messrs Silver and Co. ;§ as well as for extending the sphere of operations.  This firm, in the telegraph world (and in this book) usually spoken of as the Silvertown Company, is still referred to on the Stock Exchange, and in City matters generally, as the India-rubber Company.  Previously Messrs Silver had only made comparatively short lengths of cable and india-rubber core, but since their conversion into a limited liability company, and the consequent extension of their sphere of operations, quite a number of contracts for the manufacture of cables of all sorts, as well as for their submergence and repair in various parts of the world, have been undertaken and successfully carried out from Silvertown. ||  Some of these undertakings have been dealt with in this work.  The idea of combining the solid but less romantic industries of

---

* In Germany, Messrs Siemens and Halske had constructed great lengths of gutta-percha covered conductors in 1848 (and earlier), when they made and laid a number of wires for the Prussian underground land lines.  They had also erected extensive aerial conductors ; indeed, in 1853-54, they supplanted the former by overhead wires suspended on poles.  Latterly, military considerations and others have prevailed with all the Continental authorities in favour of the subterranean plan, to which, consequently, there has been a return.

† Since Sir William's death, he has been succeeded in the chair by Mr Carl Siemens ; Mr Alexander Siemens, M.Inst.C.E., acting as director in this country. Besides the works at Berlin and Charlton, this firm also possesses large branch works at Paris, Vienna, and St Petersburg, all established many years ago.  At the Charlton works, Mr J. R. Brittle, F.R.S.E., is chief engineer of the Submarine Cable Department, and Mr Frank Jacob the chief electrician.

‡ Now Siemens Brothers and Company Limited.

§ The latter firm was established by the late Mr S. W. Silver soon after 1844, when the process of vulcanising was introduced.  They did not start covering electrical conductors till later, and then only on a small scale.

|| The firm has also large works at Persan-Beaumont, not far from Paris.

india-rubber and gutta-percha goods of all sorts—from india-rubber bands to waterproof coats—with the construction and laying of telegraph and other cables,* has turned out a very profitable one in this instance.

Mr W. T. Henley, an engineer and electrical inventor of old standing in connection with land telegraphs, started cable manufacturing works at North Woolwich in 1855. A fair length of the early telegraph cables were made here, some under direct contract with Mr Henley himself, some through the intermediary of the Telegraph Construction and Maintenance Company, who often had occasion to employ him as their sub-contractor for sheathing—especially between 1868 and 1873—when they had their hands too full to manage it all at the Greenwich works. In 1880 this business was "converted" into W. T. HENLEY'S TELEGRAPH WORKS COMPANY LIMITED, which, in addition to the firm's former work, now undertook the manufacture of gutta-percha core and the laying and repairing of cables. Since then they have established an india-rubber core department. Mr Henley died in 1882—a man of indomitable energy, who will never be forgotten in connection with the early history of telegraphy, land and submarine.

Although some short lengths of conductor, insulated with india-rubber, had been made previously, the first person to apply the vulcanising process to india-rubber core with any measure of success was Mr William Hooper, who thus first rendered it a competitor to gutta-percha for the insulation of different kinds of cables, etc.† Mr Hooper took out patents in connection with this invention, and improvements thereon, in 1859, 1863, 1868, and 1873. The time and money which he must have expended over the elaboration of his process, and the great perseverance which he must have exercised—meeting, as he did, with many rebuffs and disappointments before his efforts were crowned with success—deserve grateful recognition at the hands of the present generation of telegraphists.‡ The most im-

---

* In 1866 the late Sir Charles Bright, who had furnished plans for a cable department, strongly urged on Mr Silver the desirability of establishing a branch for insulating wires with gutta-percha.

† The general process known as vulcanising (i.e., the admixture of sulphur with rubber, and the subsequent heating of the mixture to a certain temperature) was first invented by John Macintosh some years previously. He applied it chiefly to such india-rubber goods as waterproof cloaks, to render them impervious to climatic changes of temperature—articles, the universal demand for which now advertises his name to countless multitudes. But it was Hooper who first succeeded in applying it to india-rubber covered cables.

‡ The manufacture of vulcanised india-rubber core is, of course, a more complex matter than that of gutta-percha core ; in the course of which latter no mixtures have to be made, and time and temperature have not to be so carefully regulated. The application of ordinary vulcanisation (i.e., without Hooper's special processes) is impossible in the case of electrical conductors, owing to the injurious action of the sulphur upon copper—even when "tinned."

portant details regarding the construction of this core, as well as a complete review of the relative merits of gutta-percha and india-rubber for the insulation and protection of electrical conductors by land and by sea, will be found in Part II. of this work.

The vulcanised rubber core, as made by Hooper's method, was soon recognised to be peculiarly well suited (as compared with gutta-percha*) to all cases where exposure to the atmosphere and great changes of temperature were unavoidable; also to cases where (as in military field-telegraphs) the cable was liable to a great deal of rough handling. For submarine purposes, however, it was some time before it was at all seriously taken up. In 1863 the Indian Government experimented with a short length, and soon afterwards adopted it for river-crossing cables. In 1864 Messrs Bright and Clark drew up a full report on Mr Hooper's core. This report, as well as another, about the same time, from Sir William Thomson,† had the effect of drawing attention to the (nowadays undisputed) excellent qualities of this type of core as a possible competitor to gutta-percha, even for submarine work.‡    At one time the late Sir George Elliot contemplated its adoption by the Telegraph Construction Company as a "second string to their bow" in case of the supply of gutta-percha failing. It was eventually employed for several of the cables now forming the European system of the Great Northern Telegraph Company, as well as for the India-Ceylon cable (1867) of the Indian Government Telegraphs, and later (in 1869) for a portion of their Persian Gulf cables. During this time Hooper's core behaved exceedingly well. It was found to lend itself admirably to the re-sheathing of cable picked up for repairs, whose armour had seriously deteriorated, as well as for general military purposes.§

Thus, in 1870, a company was formed to take over Mr Hooper's india-

---

* Which, if employed under the same circumstances, requires, by keeping it in water or otherwise, to be preserved from contact with the air.

† Other favourable reports were also obtained, mainly emanating from distinguished consulting chemists.

‡ It was found that it possessed a materially lower electro-static inductive capacity than ordinary gutta-percha, and a slightly lower one than the gutta-percha prepared by Mr Willoughby Smith's method; thus ensuring a corresponding increase in the working speed of cables insulated with it—a consideration which might be important in the case of long continuous lengths.

§ So well did this core appear to adapt itself to what were most unfavourable conditions for gutta-percha, that at one time it was hoped it might stand alternations of wet and dry conditions, besides dry storage and shipment without tanks, especially as it behaved well at high temperatures. However, although short lengths in some cases came out of the ordeal without damage, the core can by no means be considered absolutely proof against such treatment, which, indeed, in the case of gutta-percha, would nowadays be deemed madness.

rubber core works at Mitcham,* and at the same time add thereto the business of converting the core into cable form. For this latter purpose, sheathing works were established at Millwall. The title of the firm, registered under the Company Acts of 1862 and 1867, then became Hooper's Telegraph Works Company Limited. It manufactured the core for part of the Great Northern Company's Eastern system—that is, for the cables laid in 1870-71 between China and Japan. In 1874 and 1875 their first completely sheathed cables were manufactured respectively for the Western Brazilian and the Cuba Submarine Companies. The troubles which befell the former at the hands (or rather at the *snouts*, let us say†) of the saw and sword fish have already been referred to. The Cuba Submarine line—also laid as well as constructed by Hooper's—being in deep water, and of the close-sheathed type, did not suffer from the same depredations.

In 1877 Hooper's Telegraph Works went into liquidation, and in 1888 Mr William Hooper died. From the end of 1877 the business was for many years carried on as a private concern. In 1894, however, it was turned into a limited company, privately subscribed, under the title of HOOPER'S TELEGRAPH AND INDIA-RUBBER WORKS LIMITED, the form and title under which it still trades.‡

**Machinery and Implements.**—In 1875 Mr Claude Johnson, previously a member of the Telegraph Construction Company's engineering staff, associated himself with Mr S. E. Phillips, formerly chief electrician to Mr W. T. Henley's works, to found an independent business as constructors and manufacturers of all kinds of apparatus and machinery connected with telegraph work. The now well-known firm of Johnson and Phillips have earned a great part of their reputation in connection with the construction of gear employed for the manufacture, laying, and repairing of submarine cables, as well as for the necessary deep-sea sounding work. Within recent years Messrs JOHNSON AND PHILLIPS have established a department for the construction of india-rubber insulated cables. Mr Claude Johnson is himself

---

\* The Hooper Company's core works have since been re-erected at Millwall.

† These attacks are sometimes spoken of as fish-*bites*, which—though possibly correct in the rare case of a *shark* attacking the cable—is certainly a misnomer where the saw or sword fishes are concerned. In neither of these latter instances is it the *jaw* of the fish which does the damage, but the beak or snout. With this organ he attacks the cable on one side only, as may be seen from an examination of the faults in question. The sword-fish, indeed, as already mentioned, occasionally leaves a *pièce de conviction* behind him, in the form of a stray tooth, which gets lodged in the sheathing while he is extricating himself therefrom after his doubtless somewhat disappointing and innutritious attempt at a meal off the Western and Brazilian Company's property.

‡ The Hooper Company's work has been dealt with here at somewhat disproportionate length for the reason that, being—so far as submarine telegraphy is concerned—mainly historical, it is scarcely referred to in the subsequent pages of this book.

responsible for the design of a great deal of the various apparatus in use (as was Mr Phillips also in his lifetime), and some of the results of this firm's handiwork will be found referred to in Part II.

**Instruments.**—Though not pretending to deal with electrical instrument makers as a whole, it has been thought a suitable occasion to make light allusion to the now almost classic firm of Elliott Brothers, who constructed so many of the instruments used in the testing and working of the early submarine telegraphs. This house was started as Elliott and Sons in 1800 by William Elliott, for the manufacture of mathematical drawing and surveying instruments, later (in 1856) to be joined by Mr Charles Becker, under whose auspices most of the instruments used in the early cable expeditions were made,* as well as those for determining the unit of resistance. In 1873 the late Mr Willoughby Smith became a member of the firm. With his vast knowledge and experience of cable-testing apparatus, the foremost position always held by Elliott Brothers in these matters is scarcely to be wondered at. In subsequent years his son, Mr William O. Smith, has become the manager of this important concern.

### SECTION 4.—SHIPS.

For the purposes of laying and maintaining the world's vast system of submarine telegraph cables, a fleet of as many as forty-two steamers is at present afloat,† and find distributed employment here, there, and everywhere. Seven of these belong to Government administrations, and the remainder to the various contracting and cable-owning companies, the "Eastern" and its allied companies heading the list with ten. Out of the ordinary cable-laying (contractors') vessels, as distinguished from repairing ships, only ten are owned by the three largest English contracting firms, one by a French firm, and one by an Italian. Of these the three largest are the India-rubber Company's steamer "Silvertown" (4,935 tons displacement), Messrs Siemens' "Faraday" (4,917 tons), and the Telegraph Construction Company's "Scotia" (4,667 tons).

The T.S. "Silvertown" (launched in 1873) was originally designed by Mr J. R. France, at that time engineer-in-charge to Messrs Hooper, for the Great Western Telegraph Company's scheme, referred to previously. She

---

* Some of the most interesting of these may still be seen at Messrs Elliott Brothers' premises in St Martin's Lane—to wit, the enormous induction coils used by the late Dr Whitehouse for working through the first Atlantic cable of 1858.

† This Telegraph Marine represents a gross tonnage of over 60,000 tons. A full list will be found in "The Electrician's Directory" (aforementioned), under title "The World's Cable Fleet," and also in Munro and Jamieson's Pocket Book, alluded to elsewhere.

M

was, on this account, originally intended to be christened the "Great Western," and was at that time the second largest vessel to the "Great Eastern," possessing a 34 feet depth of hold, beam 56 feet, and length 338 feet.* She was intended to carry a large proportion of the entire length—some 5,000 N.M.—of the line, which was to be a light cable of the "open-jawed" type. A considerable portion of this was actually made, as already mentioned, but the scheme was abandoned and replaced by the "Western and Brazilian," which utilised the cable. The ship was then called the "Hooper," after its owner at that time; but being bought by the India-rubber Company in 1881, she was, in the natural course of events, re-christened the "Silvertown." She is still † the largest telegraph ship afloat,‡ and has a total carrying capacity of 8,000 tons. When fully loaded with the ordinary close-sheathed cable of the present day (carrying about 2,500 N.M. of it), her three tanks § are, on the average, not more than about half full.‖ She steams 10 knots (10 N.M. per hour). Her coal consumption is 30 tons per day. Thus she can remain at sea (under steam) for four months without requiring further coal.

The T.S. "Faraday" was built a few months later than the "Silvertown," and launched in 1874. She was the subject of Dr (afterwards Sir William) Siemens' special design. Her length is 360 feet, beam 52, and depth 35. The characteristic feature about her is that she has bows at each end, after the manner of the Steam Navigation Company's "penny steamboats"

---

* H.M.S. "Devastation," built about the same time, has very similar dimensions. The "Silvertown"—or, as she was originally named, the "Hooper"—was the first ship designed, at the outset, for cable work. She was built by Messrs Mitchell and Co. (now Sir William Armstrong, Whitworth and Co.), and in what remains absolutely the "record" time for this class of vessel, viz., only six months.

† At the moment of going to press it is understood that the Telegraph Construction Company have yet a larger vessel on order.

‡ It is said to be owing to the extra amount of cable she can hold, that this vessel has been recently chartered to lay a new Atlantic cable for the French.

§ One of the features about this vessel is that, on account of the enormous size of her tanks—each of which is capable of holding a fair-sized villa—she can pay out cable when going at her full speed, if necessary, without incurring any extra risk to speak of.

‖ It is said that when Mr France designed her, it was partly with an eye to the future that he gave her the enormous space which she has for cable. At that time the question of "light" cables—i.e., hempen cables without any iron armour—was seriously engaging the attention of experts, and it was thought that they might come into general use at no very distant period. In this case, one ship, with sufficiently large tanks, might be able to carry even the whole of a very long cable. This length would not be so great as might appear at first sight, owing to the fact that such a type though specifically light is one of great bulk. Moreover, being readily flattened under pressure, a limit is placed on the number of "flakes" permissible. It is possible, however, that by means of refrigerating machinery on board, the core could be preserved from any serious results.

On the other hand, in the absence of iron, spontaneous combustion could not well occur: watertight tanks would not, therefore, on this count form an essential feature.

on the Thames ; the object of this being to render her particularly adapted for the manœuvring necessary in cable operations, as well as for sounding work, *i.e.*, for readily going astern as well as ahead.   For similar reasons she was provided with twin screws.*   She is probably the only ship with funnels abreast of one another, instead of fore and aft, her great beam permitting of this irregularity.   The first job upon which the "Faraday" was engaged was the Direct United States cable, with which she was loaded up as soon as launched.

The T.S. "Scotia" was originally constructed (in 1863) as a passenger ship for the Cunard Steamship Company.   She was bought by the Telegraph Construction Company, and converted into a telegraph ship in 1879,† the first piece of cable work she did being the laying of the cable between Malacca and Singapore in 1879.   She has since laid a considerably greater length of cable than any other single telegraph ship, amounting to upwards of 25,000 N.M.   Like her companions, the "Seine" and the "Britannia," she is a very handsome vessel to look at—ornamental as well as useful.   As the "Silvertown" has the greatest beam (and mainly for this reason the greatest carrying capacity), so the "Scotia" has the greatest length, her dimensions being—length, 379 feet ; beam, 48 feet ; and depth of hold, 38 feet.

### SECTION 5.—MISCELLANEOUS FIGURES AND ESTIMATES.

**Cost of Construction and Laying.**—The cost of a submarine telegraph cable to its purchaser (taking the average proportion of the different types usually employed) may nowadays be roughly estimated at £150 per N.M.   The greater the proportion of core, and the larger this core is, the heavier, of course, will be the expense of the entire cable. ‡   The cost of its submersion—presuming average luck as to prevailing conditions and contingencies—may generally be roughly estimated at about half as much again as that of its construction.§

**Life of a Cable.**—Unfortunately, however, even though perfectly free

---

* The "Faraday" was, in fact, one of the first twin-screw ships built.

† The "Scotia" was originally a paddle-boat, but for deep-water cable purposes it was necessary to convert her into a screw ship—a performance at that time regarded as a rather remarkable piece of marine engineering skill.

‡ Cable for submarine telegraph purposes has been supplied at as low a figure as £50 a mile.

§ A cable costs between five and seven times as much as a land line, and the total cost of its construction and submersion comes to between seven and eight times that of the erection (including original cost of wires, posts, etc.) of the aerial wire.

from electrical faults when laid, no submarine cable will last for ever. No very precise limit can at present be assigned to the longevity of a cable of suitable type, carefully manufactured, and laid on the bottom and in the waters for which it was designed.* The history of submarine telegraphy is scarcely long enough to arrive at any very definite conclusions upon this point. All that can be said is that in cases where a cable has not been subjected to *casual* sources of detriment (such as ships' anchors, rocks, sharks, saw and sword† fish, teredoes, and other "common objects"‡ of the deep sea in different latitudes §), the records of cable existence have, of late years, been distinctly satisfactory. ‖

Several portions of the early lines laid in shallow water—between England and its neighbouring countries—from 1851 to 1855, are still working. In some of these instances the core seems to be as good as new, notwithstanding the fact that they have been resheathed—or at any rate relaid in a fresh place—and consequently subjected to a certain amount of "pulling about." The cable between Beachy Head and Havre, laid in 1870, broke for the first time in 1892.

---

* The life of a submarine cable can, of course, be almost anything, according to how it is laid. If laid on an absolutely harmless and smooth bed into which it should sink, its life might be interminable. Speaking as a heavy investor in submarine cable stock, the late Sir John Pender once put it down at as much as forty years ; whilst Mr C. W. Earle, when chairman of the "West India and Panama" Company, estimated it at thirty years.

However, the form of cable at present in vogue—provided it is of suitable type for its depth and bed, besides being properly laid—if not absolutely indestructible, may, at any rate, be fairly expected to do useful work for, at the very least, fifteen years. Indeed, under favourable circumstances, the deep-sea portion should last for that time without any repairs.

It may be added that an absolutely harmless bottom would be constituted by an entire absence of rock or other destructive matter, such as vegetation, plant life, or obnoxious minerals. These conditions could never be relied on where sub-surface currents prevail.

† Marine monsters of this order seem to regard the cable as their larder ; and, while raking off the marine life, either rupture the cable or penetrate the core with one of their formidable teeth. Again, sometimes a shark will bite the line savagely, leaving a few teeth in the interstices of the sheathing as a memento of the encounter.

‡ Shore deposits and marine growths are often very injurious to the sheathing wires of a cable owing to the iodine contained in seaweed and decaying vegetable matter, which is known to corrode iron rapidly.

§ Again, Professor John Milne, F.R.S., has recently—both at the Royal Institution and the "Geographical" Society—drawn our attention to the close connection between the interruption of telegraph cables and the occurrence of submarine earthquakes or volcanic eruptions. Moreover, Mr H. Benest, in a paper before the Institution of Electrical Engineers, has, only this year, shewn the effect which a submarine river, or spring, may have on a cable in its immediate vicinity.

‖ Sir Henry Mance, M.Inst.C.E., states that he has picked up a cable after resting nearly twenty years, and that he has found it in as good condition as the day it had left the factory.

Similar and still more favourable experiences can be related in the case of deep-sea cables.*

**Value as an Investment.**—Although at first sight a somewhat risky property, and in early days spoken of as gambling speculations, deep-sea telegraphs have generally proved a very remunerative source of revenue to their owners.†

Depreciation, of course, forms an important factor in the accounts of a cable-owning company, submarine cables having still to be reckoned from the investor's point of view as a comparatively quick-wearing plant.‡ A large renewal fund is therefore essential, if only to provide for necessary repairs from time to time.§ But the first cable laid between two given points is never regarded in these days as sufficient for its owners to rely upon solely

---

* In great depths, bottom specimens on being subjected to microscopical examination usually reveal in perfect form the shells, or coverings, of marine animal life. These have been preserved intact during the life of the world in the soft ooze—the product of their accumulation which has served as an undisturbed resting-place for countless generations. This accumulation of the remains of minute organisms increases with increasing depth. For all practical purposes, the telegraph engineer may regard the oozy bottom of the sea in mid-ocean as in a state of eternal rest ; and two inches of good substantial mud, or ooze, is found to be the best possible preservative for a submarine cable. Indeed, it may be taken as an axiom, the statement made by Sir Henry Mance, C.I.E., M.Inst.C.E., in his recent Inaugural Presidential Address to the Institution of Electrical Engineers (1897), that "the deeper the water, the better the bottom, the safer the cable."

† Regarding the value of cable shares generally as investments, see pp. 165-167.

‡ Probably the record case of freedom from repairs is the "Brazilian Submarine" Company's St Vincent-Pernambuco cable. This line lasted nine years after being laid before any sort of fault shewed itself—even in the shallow-water approaches.

§ There is a great element of luck in cable maintenance. When breaks *do* occur in great depths the cost of re-establishing communication is likely to be very considerable—to wit, the £95,000 entailed in repairing the 1869 "Atlantic." The cost of maintenance and repairs is, however, often put down at something between £6 and £8 per N.M. per annum.

Sir Henry Mance finds (see his Inst. E.E. Presidential Address) that during the first five years of the life of a shallow-water cable it is well to allow for $\frac{1}{2}$ per cent. per annum of its total length to provide for repairs. His further figures are :—

|  |  |  |
|---|---|---|
| From the 6th to 10th year | say $\frac{3}{4}$ per cent. | |
| „ 11th to 16th „ | „ 1 | „ |
| „ 17th year | „ $1\frac{1}{4}$ | „ |
| „ 18th „ | „ $1\frac{1}{2}$ | „ |
| „ 19th „ | „ $2\frac{1}{4}$ | „ |
| „ 20th „ | „ $2\frac{1}{4}$ | „ |
| „ 21st „ | „ 3 | „ |

He goes on to say :—"Three per cent. on the total length of the cable per annum after twenty-one years' sojourn at the bottom of the sea, does not, at first sight, appear excessive for a shallow-water cable ; but there comes a time when in consequence of the great public inconvenience, loss of revenue, and cost of repairs, it is cheaper to lay a new cable."

It will be seen, therefore, that Sir H. Mance—a telegraph engineer and cable company director of much experience—puts the average life of a shallow-water cable down at about twenty-one years, after which period its further recovery and repair is not worthy of the cost entailed, and the line had best be replaced by a new one—so far as working is concerned.

and permanently. They are bound to duplicate or even triplicate it in course of time, at least for their own security if they have competitors, and, if they have not, then also for the security of the public whom they serve, and whom they have induced to rely upon them as the sole caterers for their telegraphic correspondence between the points or stations in question.

What with duplications and repairs, submarine telegraphy is certainly a constant expense to those who have to find "the sinews of war." Nevertheless, as already stated, it generally brings them a good return for their money in the long-run. In some cases, just as railways in America and other new countries have often been used as pioneers of civilisation and prosperity, themselves creating a demand for transport rather than supplying one which already existed, so have cables been laid with a view to developing rather than merely facilitating telegraphic intercourse, and, in fact, laboriously building up the traffic which is to feed them in the future.

On the other hand, examples are not wanting of very quick and profitable returns from these enterprises. Thus the original Malta-Alexandria cable earned as much as £3,000 in a single week soon after it was laid, or at the rate of £117 per N.M. per annum. In one year the average earnings during the time it was open were at the rate of £90,000 (or £68 per N.M.) per annum ; allowing for interruptions, the maximum earnings in one year were £64,000, or £48 per N.M. Again, between 1864 and 1869 the Persian Gulf cable was earning at the rate of £100,000 per annum—this, moreover, under the disadvantage of a bad land-line connection through Turkey. At this time it had the monopoly of telegraphic communication with India, and it made the best of it. These halcyon days came to an end when the "British-Indian" arrived on the scene.

The great corporations in which the earlier Oriental lines were merged, or which grew out of them—namely, the "Eastern" and "Eastern Extension," with their allied companies—have already received in this book some of the attention which their importance demands. For the commercial and financial world generally, no less than for the telegraphic, their prosperous career will always have a fascination. The twenty-fifth anniversary of the laying of their first cables to the East and Far East was celebrated on 20th July 1894, by a banquet and reception at the Imperial Institute, presided over by the late Sir John Pender, K.C.M.G., M.P., at which the Prince of Wales and other illustrious guests took part.*

---

* On the occasion of this silver wedding of the East and West—as enthusiastic telegraphists may surely be permitted to call it—the Prince sent his greetings by cable to the representatives of the Crown in India and the leading Colonies (as well as to other political personages and officials), receiving their replies within the course of a few minutes. This was no doubt a proud moment for the "Cable King," who organised the function as host-in-chief.

It is worthy of note that none of these companies were assisted in their earlier days by any Government monopolies, subsidies, or guarantees* (as in the case of the original Red Sea cables); so that it may truly be said that the Governments and mercantile communities of the world owe the benefits arising from direct telegraphic communication during the past quarter of a century entirely to British enterprise.

The "Eastern" and its immediately allied companies (the "Black Sea," the "Eastern Extension," and the "Eastern and South African") own over 50,000 miles of submarine cable, or about one-third of the total cable mileage of the world. This is represented by a joint nominal capital of over ten millions sterling (£10,000,000), but which, at the present market quotations, represents some fifteen millions. These companies now carry about two million messages per annum. At the outset only about 400,000 were carried in the year, and there were but twenty-five stations and 900 miles of cable, the nominal capital being then only £260,000. In the case of the North Atlantic cables, a very similar increase of traffic will be found to have taken place in the same period—since 1870. When the "Eastern" Companies first entered upon their associated careers at the end of the sixties, the system of code-message, which has since come into such universal use, had not been inaugurated; consequently the estimate of receipts was based upon all telegrams being sent in open language. If the scale of this estimate had been maintained, the revenue of the companies would have been much larger than it has become. The existing cables would not, in the absence of the code system, have been capable of carrying the traffic, nor would their existing large staff have sufficed for it.†

In land telegraphy, the number of messages conveyed through the British Postal Telegraphs is at the rate of $1\frac{3}{4}$ messages per head of the population per annum. Nothing like this rate can be found in the case of "cablegrams,"‡ but during 1894 something like two million were conveyed over the "Eastern" and its directly allied systems. Roughly speaking, about six million messages pass over the entire network of the world's cables in the course of one (recent) year; which is equivalent to about 15,000 for each day of twenty-four hours.

---

* In this connection it must be observed, however, that exclusive rights were obtained from the Khedive for working cables to Egypt for an indefinite period—a privilege the value of which can scarcely be over-estimated.

† The "Eastern" group of companies employ a staff at home and abroad of some 1,800, exclusive of the 650 who are attached to their fleet of ten cable-repairing ships.

‡ The increase of our inland traffic during the last twenty-five years—ever since the acquisition of the telegraphs by the State—has gone on out of all proportion to the extension of the wires that carry it. Similarly in submarine telegraphy the rate of increase of the number of messages has been vastly greater than that of the mileage of cable provided for their transmission.

The average time occupied in the transit of cablegrams has been enormously reduced for public traffic. Where formerly it took an average of five to ten hours to transmit a message between certain given points, from thirty to sixty minutes now suffice.* This, of course, refers to extreme points. Between many stations the time is as low as from three to five minutes. A few instances of the average time now occupied in transit from England to some of the principal countries, compared with that occupied formerly, are given below:—

|          | At Opening of Line. | At Present Day. |
|----------|---------------------|-----------------|
| Portugal | 5 to 6 *hours*,     | 30 *minutes*.   |
| Spain    | 9 to 10 „           | 15 „            |
| Egypt    | 3 to 4 „            | 20 „            |
| India    | 5 „                 | 35 „            |
| China    | 8 „                 | 80 „            |
| Australia| 10 „                | 100 „           |
| Brazil   | 8 „                 | 25 „            |
| Argentina| 10 „                | 60 „            |
| Chili    | 10 „                | 70 „            |
| Peru     | 10 „                | 80 „            |

The record case of fast working up to date is probably that effected on the occasion of Li Hung Chang's visit to the works of the Telegraph Construction Company at East Greenwich, on 14th August 1896. On this occasion our distinguished guest had pre-arranged a series of lengthy cipher-code messages to Shanghai and elsewhere—in one instance addressed to *Shan Shun Yan Shea Nan*, almost a message in itself! These telegraphic despatches were no light reading, but Li was in a position to coo with pride and pleasure when he realised that he had sent a message 12,608 miles,† and had received an answer within seven minutes!‡

---

* A quarter of a century ago it was considered wonderful to receive a message from India in a few hours. Now the same message would only take a few minutes. The tariff, moreover, at that time stood for Bombay at £4 10s. for a minimum of twenty words, and proportionately for every extra word. Now it is 4s. per word without any minimum stipulation. Similar modifications have taken place elsewhere.

† It may be mentioned further that the message in question consisted of sixty-eight words, and that it was despatched in two and a half minutes! We must, however, remember that, both in this instance and in that of the Silver Jubilee Celebration of the Eastern Telegraph Company in 1894, every possible arrangement had been made beforehand to facilitate operations by Mr T. A. Bullock, Superintendent of the "Eastern" Company's London Stations.

‡ Another striking case of up-to-date telegraphy—subaqueous if not submarine— is the manner in which the evening newspapers in London are able to report the various stages of the University Boat Race whilst it is proceeding. This is brought about by the Press Steamboat paying out an insulated wire (connected to a receiving instrument ashore) as she follows the boats, having an operator on board who manipulates the sending key whilst observing the various incidents of the race. Thus the result can be filled in at the different Fleet Street offices within a second or two after—or *before*—the winner has passed the post! Yet another example of the kind is the recent Parliamentary Chess Match between the House of Commons and the United States House of Representatives. This

## SECTION 6.—EFFECT OF SUBMARINE TELEGRAPHY ON THE WORLD'S PROGRESS.

**Social and Political Influences.**—The great revolution which submarine telegraphy has effected in the world's progress (its rate and its nature) may be regarded from two main standpoints, the political and the commercial. Let us commence with the former.*

In the first place, then, it has accelerated—even more perhaps than the improvements in locomotion by land and sea—what may be called the practical *shrinkage* of the globe. The nations and peoples of the world, being in continual contact with each other through the telegraph and its powerful ally the Press,† know one another, and understand one another's actions, thoughts, and national aspirations, infinitely better than they did thirty or forty years ago. The effect of this better knowledge and insight upon their mutual relations may not always, in the first instance, be a happy one: there is certainly a seamy side to it, so far as the commercial ascendency of this country is concerned—*teste* the manner in which the Germans have been stealing our industrial thunder (and sometimes improving upon it) during the last two decades. The rapid rise of Japanese competition in the East is another case in point.‡ But if the whole world gains, as it undoubtedly *does*, by closer contact and the lessons which one nation is thereby induced to learn from another, we need not take very seriously to heart any *relative*—and may be, after all, quite temporary—decrease of ascendency in two or three departments of our national activities. Such "ups and downs" are the necessary incidents of social and industrial progress all the world over; we have had plenty of them in this country in the past, so must make up our minds to bear patiently with them in the

---

match was played over the cables and land systems of the "Anglo" Company and the "Western Union" Company. The smartest piece of transmission done during the play was London to Washington and back—a distance of 8,360 miles—in 13¼ seconds.

* Notwithstanding that a large proportion of the line is submarine, Dublin can now, by machine transmission, communicate with London at the rate of about 500 words per minute—a fact which somewhat discounts one at least of the stock arguments which used to be brought forward in favour of Home Rule for Ireland.

† Instances of the indebtedness of modern journalistic enterprise, for some of the strongest and most piquant food that it thrives upon, to the telegraph cable are given further on in connection with the commercial results of submarine telegraphy. It may safely be averred that without it the foundations of certain great newspaper fortunes could never have been laid.

‡ Can it be a mere coincidence that the same quarter of a century which has witnessed the phenomenal rise of Japan from the position of an insignificant Oriental nation to that of one of the (comparatively great) Powers of the modern world is also the period which has marked the completion and consolidations of the cable systems uniting the Far East with India and Europe?

present and to profit from them in the future. We may even yet have to pass through the fire of much greater tribulations and humiliations before we achieve our national destiny, but we shall not have the telegraph or any other modern instrument of progress to blame for that.

Meanwhile there is at least one political result of this great development of the world's system of electric nerves, which Englishmen may safely regard with unmixed satisfaction and pleasure. This is the much closer relations which have thereby been rendered possible—nay, are on their way towards being fully established—between the mother-countries of the United Kingdom and the daughter-nations, English-speaking, English-modelled (as to their institutions), and, in the main, of British and Irish stock, which have sprung up in the most distant quarters of the world.

The "Little England" idea, so fondly cherished by the old Manchester school of economists and politicians (who would gladly have seen all our young and vigorous Anglo-Celtic brood chased, as young birds from the parent nest, almost before they could fly), is practically as dead as a door nail. In its place, we hear on all sides of Imperial Federation and Inter-Colonial Federation schemes, of a Pan-Britannic Zollverein or Customs Union between the United Kingdom, its self-governing colonies, and India, and—grander, if less practicable, than all these—we now hear of negotiations for the establishment of a permanent arbitration tribunal for settling peacefully all future differences between the two main divisions of the English-speaking world. These movements may end in some form of British Imperial Federation, accompanied with a permanent *modus vivendi* with the United States. They may even lead, beyond this, to the constitution of a new nation, on a grander scale than any which the world has yet seen—a true Pan-Anglican Federation—embracing all the "free" communities in different parts of the world which, albeit of diverse races and even colours, are naturally united by the common bonds of the English language as their official and most prevalent tongue, and of religious and political institutions of European and mainly British origin. In a work like this, partly written for the rising generation of telegraphists in all these countries, from the United Kingdom, and its great "emancipated daughter" the United States, down to the smallest African and West Indian communities speaking and reading our modern *lingua franca*, it does not seem out of place to refer to such possibilities—especially as the extension of submarine telegraphy is doing more, perhaps, than any other single movement in the world to render their eventual realisation possible. To discuss them at greater length would, of course, be quite outside our province.

**Influence on Diplomacy.**—Another department in which submarine cables have produced a notable political effect, is the diplomatic. If the peoples have been brought more in touch with each other, so also have their rulers and statesmen. An entirely new and muchly-improved method of conducting the diplomatic relations between one country and another has come into use with the telegraph wire and cable. The facility and rapidity with which one Government is now enabled to know the "mind"—or, at any rate, the *professed* mind—of another, has often been the means of averting diplomatic ruptures and consequent wars during the last few decades. At first sight, the contrary result might have been anticipated ; * but, on the whole, experience distinctly pronounces in favour of the pacific effects of telegraphy. The most obvious risk, perhaps, in submarine telegraphy to a nation like Great Britain,† whose colonies and possessions—with the naval and military forces for their protection—are distributed over every quarter of the globe, is that her Government may, at one moment or another, *lean too much* upon this valuable means of rapid communication. In the event of a surprise war—carefully pre-arranged—declared against us by another naval Power, or of a well-prepared revolution in one of our colonies or dependencies, the inconvenience of suddenly-cut cables might conceivably take the Home authorities unawares.‡  On the subject of cable-cutting and of the value of International Telegraph Conventions more is said further on (see pp. 150, 151, also p. 179).

**Influence on Commerce.**—Let us now turn to the commercial results of these great developments of submarine telegraphy. These have been partly anticipated in describing certain improvements in tariffs and speed of trans-mission, and altogether the subject is so vast, so complicated, and so far-reaching, that to attempt a detailed, or systematic, account of it within the compass of a work like the present, would be but presumptuous. The fact is, the methods of conducting business between merchants and financiers in different countries have been completely *revolutionised* by the telegraph cable, which places the business man in touch with the money markets of the world. This is so patent and obvious to the older generation of business men now living, that it is the younger only that need reminding of it. Thus, fifty years ago it took a London or Liverpool merchant six months to get an answer to a letter addressed to a correspondent at Calcutta, and

---

* Indeed, it cannot be denied that there *are* occasions when rapidity in the inter-change of diplomatic communications may have had—aye, and may still have—the effect of producing ruptures which "a little more time to think" would have avoided.

† No other country can be said to possess such long, or acute, ears as John Bull.

‡ As might also indeed the temporary stoppage of telegraphic communication between two or more important parts of our empire by an enemy effecting a successful *coup de main* upon a station (or cable-end) in the main line of communication.

complete a piece of business: nowadays, by means of the telegraph, the same transaction can be effected within six hours.  Another result of the change of conditions brought about by the wire and cable is the partial elimination of the middle-man in some departments of international commerce.  This, again, is an item in the general revolution which can only be referred to, the discussion of it in detail being quite beyond our present scope.  It should be noted, however, that the phenomena of what may be called the telegrapho-commercial revolution are by no means the same between different trade-centres at equal telegraphic distances from each other.  *Longitude* is, of necessity, a powerful factor in the matter : * so are the political, religious, educational, linguistic, and other circumstances, which in one case may facilitate, in another impede, the rapid development of trade under the new conditions.  Thus, the previously existing commercial ties between Buenos Ayres, Cape Town—nay, even Yokohama—and London might be drawn (and, as a matter of fact, have been drawn) closer by means of the telegraph than those between Constantinople and London.

The total imports of the world amount to over 1,900 millions sterling.  As the late Sir John Pender has put it,† "considering that commercial operations are now begun and ended by means of the telegraph, it will be seen that these vast figures have an important bearing upon submarine telegraphy."  The converse is, of course, equally true and important.

Upon the progress of submarine telegraphy in the future as in the past, a great deal of the world's commercial and industrial progress must depend.  And not only the progress of the whole world collectively, but the relative importance and mutual relationship of certain parts of it.  The constant extension and duplication of the cables and land lines themselves, the fact that these communications have been effected chiefly by Englishmen, and retained for the most part in British and American hands, the great reductions made in tariffs, the improvements in speed and volume of telegraphic traffic (due to the successful application of the duplex system to cables, as well as various improvements in instruments, etc., already

---

* Owing to the difference in time between certain important centres, it is impossible for merchants in the one to obtain a reply on the same day to their messages despatched in the morning to the other.  To use a familiar bit of slang, their business hours don't "gee."  Even in the case of London and New York, notwithstanding the important negotiations constantly proceeding between them, there is only about one hour's simultaneous session of their two Stock Exchanges.  This means smart work for the operators, as an enormous traffic (in this and similar cases) has to be concentrated into a few minutes of time.  Obviously where longitude is *not* a barrier (*e.g.*, between London and Paris or Lyons) the *possibilities* created by the telegraph are relatively greater.

† See his speech at the Imperial Institute, while presiding over the banquet above referred to, held in celebration of the twenty-fifth anniversary of the completion of the "Eastern Company's" first cables to India.

referred to), and finally the new system of code and cipher signalling—all these developments may be continued, nay supplemented, by others in the immediate future. But how and by whom this is to be done; more especially whether chiefly by men of our race or institutions,* and in the interests of—or, *at least*, not to the detriment of—the continued growth of harmony and union in the English-speaking world; all this depends upon the private enterprise, as well as the political wisdom and decision of purpose, of the present generation.

**Proposed Reforms.**—Mr J. Henniker Heaton, M.P., has of late years been almost as indefatigable in his efforts to bring about a uniform reduction of rates for cablegrams to our Colonies and the United States, as in his earlier scheme for a universal penny postage. So far these efforts have not culminated in anything practical, but his day may yet come.† It must be remembered that the present (much varying) word-rate system itself did not come into universal and permanent use until after the St Petersburg Convention (of the International Telegraphic Conference), hereafter referred to, which took place in 1875. The adoption of a word-rate was then definitely agreed to by all the Government telegraph departments and all the companies represented there. ‡

**Dissemination of News by Submarine Cables.**—Before leaving the subject of the world-wide effects of submarine telegraphy during the present generation, one important class of them should be especially referred to, which is both political and commercial. This is the phenomenally rapid

---

* Let it not be supposed that it is intended in this book to deprecate the growth of French, Italian, Spanish, Japanese, or any other foreign enterprise in submarine telegraphy. Quite the contrary. Whether we call ourselves Englishmen, Americans, Australians, or what not, we shall prosper none the less, or the slower, because our neighbours—because the whole world in fact—are "moving." Therefore, by all means let every nation that wishes and is able to develop its own cable systems, and train up its own army of telegraph engineers and electricians, do so with our very best wishes. All that it is our business to see to is, that *we* at least don't lag behind. And if any ring-fence (of preferential rates or other privileges) is to be established, let us make sure that those admitted within it are also those who by kinship, community of language, or historical association, can be expected to get on well and harmoniously together both with ourselves and with each other.

† Mr Heaton has also recently urged the desirability of our cable system coming under State control, mainly on the grounds that if the Government takes over the working of the system, the rates could be reduced to half what they are at present, whilst the telegraph would, in his opinion, still remain a paying concern.

The time may, of course, occur when there will be one rate on all telegraphs—whether land or submarine—throughout the world.

‡ Regarding the earlier history of the evolution of the word-rate, see pp. 143, 144. Its introduction was no doubt largely brought about by the "packing" of messages by Reuter's Agency, their custom being to pack several cablegrams into one, thus enabling the public to economically send any less number than that corresponding to the company's minimum tariff.

dissemination in all quarters of the world of war and other sensational news. In old times it sometimes happened that battles were fought in ignorance of the fact that a treaty of peace had already been formally signed between the contending parties—sometimes long after it. Now, thanks to the telegraph, such dreadful mistakes would be impossible. The influence of this early news upon the policies of nations and the financial and commercial operations of individuals, upon the fortunes—indeed, the very existence—of a great portion of the daily press of modern times, is incalculable. Thus, during the Afghan campaigns of 1878, 1879, and 1880, the Indian authorities and our own made large use of the telegraph cable and wires, thereby incidentally enabling the public at home to read full details of every action almost as soon as it took place. On the other hand, the disaster of Isandula, South Africa, in January 1879, was not known in this country until some weeks after, owing to the absence of telegraph communication. This probably did more than any other single event to advance the negotiations for establishing a submarine cable to the Cape. In 1881 all the negotiations with the Transvaal were conducted through the telegraph. By this means the British Government was in hourly communication with the Boer leaders, and the unfortunate dispute was settled—as it certainly would not have been otherwise—without further recourse to arms. A still more signal instance of the value and capabilities of the telegraph in war occurred at the bombardment of Alexandria. During this operation the Alexandria end of one of the cables was taken on board the Eastern Telegraph Company's S.S. "Chiltern," and the progress of the destruction of the forts wired to London from minute to minute. The military operations which followed in Egypt, and in the subsequent campaign in the Soudan, were announced at home with the same wonderful despatch, telegrams being sent off direct from the battlefield at every stage of the proceedings. More recent events have afforded, and are at the time of writing affording, fresh illustrations of the great boon conferred upon us by the telegraph under the able administration of the "Eastern" and other services.*

---

* 'In connection with the quite recent (1896) trouble in the Transvaal, there can be little doubt that but for the telegraph, South African affairs would be in a worse way than they are, for then Mr Chamberlain would have been unable at the right moment to explain that the famous but lamentable "Jameson Raid" on Johannesburg had been arranged entirely without the authority, or knowledge, of Her Majesty's Government. About this time the two cable systems down the East and West Coast of Africa were each in turn subject to interruption. But for the happy chance that their breakdown was not simultaneous, our understandings with President Krüger might have been less satisfactory. All this points to the desirability of having direct communication with South Africa—independent of the existing lines touching at a number of foreign colonies *en route*. It is understood that, with commendable foresight, the Eastern and South African Company, recognising this necessity, are about to institute such a system.

## SECTION 7.—BUSINESS SYSTEMS AND ADMINISTRATION.

**Code and Cipher Messages.**—As has already been mentioned, one important change which has contributed very much to the increased use of submarine cables during recent years, is the development of a system of private codes. Secret language always took, as it does now, two forms, code and cipher. Code, or pre-arranged language, is composed of dictionary words, the context of which has no meaning, but each word of which represents a phrase or a sentence. Any two persons may arrange a code for private use. Several such codes have been published, some of which are adapted specially to a particular business, others to the affairs of daily life. The cipher system, on the other hand, is constituted by a number of letters or figures arranged in a manner quite unintelligible to the ordinary reader, or even of letters and figures combined. This system is mostly employed, in the shape of figures, in diplomatic or other Government communications. It is the most expensive form of secret language, only three letters or figures being allowed to the " word " for tariff purposes.

Various methods of building up a private code have been introduced from time to time with explanatory books of reference.* Probably the first was that of Reuter, followed some time after—in 1866—by that of the late Colonel (afterwards Sir Francis) Bolton, R.E.† The telegraph companies at that time could but accept code on the same terms as ordinary messages. At the Rome International Telegraph Conference of 1870, however, certain regulations were laid down regarding the use of code words; and again at the St Petersburg Conference of 1875. At the latter it was decided that code words should not contain more than ten characters.‡ Words of greater length in code messages are liable to be refused. Some telegraph companies, however, accept them at cipher rates, *i.e.*, three or five characters to a word, according to *régime*. Subsequently the Bureau of this International Congress was authorised to compile a complete vocabulary of the words to be recognised and admitted

---

* Almost from the very beginning of submarine telegraphy, temporarily improvised forms of codes were used both by Governments and by merchants. On the English land lines code messages were in vogue among the great mercantile firms as early as 1853, if not earlier.

† The telegraph codes of the present day are built on somewhat the same principle as the above. They are improvements mainly in the sense of being perfectly simple instead of extremely complicated—and yet they are equally, if not more, trustworthy, from a secrecy standpoint.

‡ A "character" consists of one letter only, except in the case of the combination *c h*. Almost invariably this is telegraphically expressed by four dashes (— — — —), and treated as a single character.

for code purposes. This vocabulary was duly printed and issued. Fresh editions of it are brought out now and again, and three years after date of issue it becomes obligatory upon all parties to the St Petersburg Convention to abide by it.

The transmission of submarine code messages is liable to be partially, or entirely, suppressed at any moment by the Government of the country which granted the concession for the cable in question. Moreover, Government messages at all times take precedence (immediately on handing in) before all others. These conditions, under which all such concessions are granted, are very obvious and natural precautions, if only in view of war; indeed, whether expressed as a stipulation or not, it is certain that any Government would be acting within its rights in suppressing code messages at such a time, and would almost certainly exercise this privilege.

From the point of view of the general public, the *economy* effected by the use of code is often even a more important consideration than its secrecy. A single code word, charged for only at a slightly higher rate than one ordinary word, may be made to convey the sense of a good many.* The telegraph cable thus becomes available for business and other purposes by many people who could not otherwise afford it, and the number of messages which pass over it daily have enormously increased in consequence. And with this increase in the number of them, there has not been the corresponding decrease in their length which might have been anticipated. The public has simply become educated to the more liberal use of the telegraph, and has availed itself of its facilities in the measure and in the spirit in which they have been granted to it. The increase of the total volume of traffic, and of business leading to still greater traffic in the future, has more than compensated the companies for the economies effected by its code-using customers.

The fact is, but for the code system, the existing number of cables

---

* The following examples, taken from a certain mercantile code, may be of interest here :—

| Code Words. | | Plain English Equivalents. |
|---|---|---|
| ELGIN, | = | Every article is of good quality that we have shipped to you. |
| STANDISH, | = | Unable to obtain any advances on bills of lading. |
| PENISTONE, | = | Cannot make an offer ; name lowest price you can sell at. |
| COALVILLE, | = | Give immediate attention to my letter. |
| GRANTHAM, | = | What time shall we get the Queen's Speech ? |
| GLOUCESTER, | = | Parliamentary news this evening of importance. |
| FORFAR, | = | At the moment of going to press we received the following. |

A striking example of the unlimited application of the code principle is the word "unholy," which was used to express *one hundred and sixty words*. Another English word, which we cannot recall, was made to stand for no less than *two hundred !* This is economy with a vengeance.

would, in many cases, be quite inadequate for the demands of the present traffic.*   This remark applies most conspicuously to the case of the North Atlantic, and will be readily understood when it is stated that, whereas prior to the univérsal recognition and adoption of code transmission, the average length of telegrams used to be thirty-five words, it is now only eleven.   In other words, but for code, the companies *might*, by now, be asked to transmit more than three times as many words as they are trans-mitting within the same time.†   More probably the proportion would no: be so great in practice, for reasons already given.   But even an addition of only half as many again would be embarrassing to the operators—and, indeed, to all concerned, excepting telegraph engineers and contractors, who would, in consequence, have extra cables to lay.

**Statistics.**—For a further insight into certain commercial aspects of telegraphy, the reader will find it worth while to refer to the extremely inter-esting paper on "Statistics of Telegraphy," which the late Sir James Anderson read before the Statistical Society in 1872.   This paper, in fact, gives a good idea of telegraph business in general down to that date, and forms an excellent introduction to the study of its later developments.‡   The latter may be conveniently followed—to a great extent, at any rate—in *The Statist*, which, from time to time, publishes interesting statistics and comparisons.

**International Administration.**—The International Bureau of Tele-graph Administrations, already referred to in these pages, was founded in Paris in 1865, under its French title of BUREAU INTERNATIONALE DES ADMINISTRATIONS TÉLÉGRAPHIQUES.   It was originally set afoot by the leading telegraph authorities of the various European Govern-

---

* Unless, indeed, as it is likely enough would have happened, the *absence* of code—or its suppression by unwise restrictions on the part of the companies—had starved and stunted the natural development of trade itself.   All commercial traffic, practically, is nowadays "coded."   Seeing that this custom began to grow up with the establishment of trans-Atlantic telegraphy, it is difficult now to estimate where we should be without it.

† This is to a great extent assuming that the total traffic, gauged by the number of messages, would still have gone on increasing at the same rate.

‡ Sir James Anderson was the originator of more than one apparently practical proposal, the adoption of which has been delayed, but which, it is to be hoped, may still be carried out.   Such was the abolition, by international consent, of Custom House formalities, and the suspension of quarantine regulations, in the case of ships engaged in cable enterprises.   A grander, but perhaps hardly so *feasible* a scheme of his, was one for the joint working, under international control, of all the cables of the world !   A some-what similar proposal, for Cosmos cables, is, by the way, attributed to Dr Cornelius Herz.   Such a telegraphic millennium still looks very remote ; the contemplation of it from a distance may, nevertheless, be a useful exercise for telegraph delegates assembled in international conclave.

ments.* Their main object in view at that time was the promotion of a uniform system of traffic exchange—in fact, it was formed for the purpose of arranging a universal tariff and zone system in such a manner that the sender of a message secures a large rebate on each country (after a certain distance) that the message passes through, in virtue of each country taking less than they would do by their individual tariff. Thus, the tariff per word being 1 franc for the first zone and 2 francs for the second, at the third it was to be $2\frac{1}{2}$ francs, and at the fourth zone $2\frac{3}{4}$ francs, and so on. It was only natural that Paris—as a central point of Europe, and as the headquarters of the international tongue—should have been selected as the common meeting-ground at first; but subsequently Bern was decided on as the permanent headquarters, it being thought that Switzerland was perhaps the most neutral of all the European countries, and Bern a fairly convenient centre for all the European telegraphic authorities—who, of course, formed, as they do still, the vast majority of those concerned. The conferences are held about once in every five years, at the different capitals in turn, of the various countries represented. Not only is each country represented by one or more Government delegates, according to its importance, but each of the submarine telegraph companies are now also invited to contribute delegates—although it is only the national delegates who are allowed to vote.

One of the most important matters now subject to the jurisdiction of this International Administration is that of landing-rights for cables.† In nearly all countries, not only does the " foreshore " or beach belong to the State, but the sea itself, up to three miles outwards from the shore—and, where there are promontories, the whole water-space enclosed between

---

* When the above International Telegraph Union was first promoted, none of the great overland routes had been as yet completed. The telegraphs of each country were isolated, doing very well for internal traffic, but very badly for external. A message which was sent across several boundaries was, in those days, subjected to an infinite number of annoyances and delays, and its cost was exorbitant. The original convention, to which the Governments and private companies then assented, required that each party shall devote a certain number of direct lines to international telegraphy, and that everybody shall have the right to use them. It guarantees the privacy of correspondence; permits that it be sent in secret language if the sender desires; and arranges that messages shall be transmitted, under certain circumstances, in the order of their importance. It aims at securing unity of rates each way between every two points, dictates certain monetary standards for international tariffs, and makes all regulations which will ensure quick transmission and delivery. At the successive conferences, held every five years, all changes in, and additions to, the original convention which are found necessary are made.

† One of the conditions under which a cable concession is granted, is usually that the cable company shall be adherents to the International Telegraph Convention, the tariff thus also coming under control.

them—is considered within the domain of the same nation.· In most countries landing-rights have to be paid for in some form or other.* The United States† forms an exception. The Government there refuse to grant *sole* rights, any one being allowed to land a cable on its shores, provided that the privilege is reciprocated in favour of American citizens.‡ The Bern Bureau also publishes about every three years official maps both of the world generally and also of Europe, shewing the telegraphic systems in operation at the time. The maps given on Plates IX. and X. (see pp. 142 and 144) are taken from the latter, and are, therefore, independent of any particular cable company.

Having regard to the not improbable contingency of a big European war, it has been doubted by many whether the publication of maps shewing the route of entire cables is politic. It certainly seems as though the entire route of cables should not be exhibited on any charts—such as Admiralty charts—on a large scale. Here it would serve the purpose equally well if only the *shore ends* were marked, to denote where ships must avoid anchoring. § Unfortunately, however, even at the very outset when laying a long length of cable, the position of the vessel is telegraphed home from time to time, by way of report to headquarters, together with the usual statement of the length of cable paid out, and this in itself (if repeated) could provide much material to the inquisitive foreigner for arriving at the route of the cable.

One thing is quite certain, however, and that is, that the safety of our cables at the bottom of the sea could no longer be relied upon in the event of an outbreak of war. This is pretty clearly implied by the International

---

* Landing-rights in France cost the Commercial Cable Company £8,000. Similar facilities were secured in England for a nominal £1, in response to a letter to the Board of Trade.

† The United States of America are not adherents to the International Telegraph Convention ; neither are the Atlantic companies, strictly speaking. Though some of the latter are represented by delegates, it is with no active or binding intent, the conventions not practically affecting them, partly for the reasons just stated.

‡ This, to a great extent, explains the large number of lines across the Atlantic, in comparison with what there are elsewhere.

§ There can be no objection to the *shore ends* of cables being made known on charts, for if an enemy were searching for the cable (in times of war) at all near land, they would certainly trace it from the cable-hut or beach. Indemnity can only be claimed against a vessel for fouling a cable—by her anchor or in some other way. Thus it behoves the telegraph companies to send to the Admiralty the positions of their shore ends for marking on charts. It seems as though there should be a regulation, subject to international legislation, whereby any vessel fouling a cable should report the same, with whatever nautical observations and bearings possible at the time. Thus thè spot for repairing a break would be readily indiċated. Moreover, a penalty should be inflicted when fouling of the above description was not reported, if it could be established that damage resulted

Convention itself, in the clause which we have already had occasion to quote (see p. 151). It is this sense of insecurity as regards our telegraphic system with the Colonies which appears to provide, perhaps, the strongest arguments in favour of an additional cable, or cables, communicating with Australia (and thus, also, with the Far East and East) by an entirely different route, and at the depths of the ocean far away from other European Powers.

### SECTION 8.—INSTITUTIONS, PAPERS, AND PRESS ORGANS.

In 1871* the SOCIETY OF TELEGRAPH ENGINEERS† was founded, mainly as a result of the efforts of Dr (afterwards Sir C. W.) Siemens, F.R.S.; General C. E. Webber, R.E.; Colonel (afterwards Sir Francis) Bolton, R.E.; Mr C. V. Walker, F.R.S.; Mr Latimer Clark, M.Inst.C.E.; and Mr W. H. Preece, M.Inst.C.E.‡ This was during the absence of several eminent telegraph engineers and electricians engaged abroad at the time, who would, no doubt, have co-operated in promoting and organising

---

therefrom. A complete discussion on this subject is one worthy of the International Telegraph Congress, especially considering the large number of so-called accidental injuries inflicted by the anchors of fishing craft which swarm in busy channels—such as the

FIG. 51.—Specimen of Cable torn by an Anchor.

English Channel—and so frequently cause interruption to our Government's and other cables. An idea of the injuries so caused may be gleaned from the above illustration.

* Thus, as was pointed out by Sir Henry Mance, C.I.E., in his inaugural address as President for the current year (1897), the Institution has just completed its quarter of a century. Sir Henry also took occasion to indicate the useful work done by the Society during its existence. Throughout nearly the whole of this period Mr F. H. Webb—who is about to retire in favour of a well-earned rest—has acted as Secretary, and to his unwearying efforts its success, in holding together many diverse interests, may be largely attributed.

It has been thought, therefore, that—if for these reasons alone—a few lines with reference to the history of the Institution from the telegraph engineer's point of view might not be out of place.

† Ten years later named the "Society of Telegraph Engineers and Electricians." In 1883 it attained sufficient importance for incorporation under the Companies Acts.

‡ Shortly after this occurred the death of that distinguished electrical *savant* and experimentalist, Sir Francis Ronalds, who bequeathed to the Society his almost perfect collection of books on electrical subjects, covering the very commencement of telegraphy. This was, indeed, a rich heirloom for so young a society to come in for.

the society had they been in England.   Its main objects were to hold meetings of the telegraph engineers and electricians established in or within reach of London ; to read and discuss at these meetings papers dealing with telegraphic and cognate subjects ; and to print and circulate among its members (in all parts of the world) a journal reporting the proceedings of such meetings.   It will be seen, by reference to the early numbers of this journal, what a large proportion of its contents relate to submarine telegraphy.   For some time past, indeed, it had been felt that this was a branch of civil engineering which, by its rapid growth in importance and in the number of those engaged in it, required a special society of its own, such as mechanical engineers already had.   This was recognised by the Institution of Civil Engineers, who rendered the same practical assistance to their new electrical offshoot as they had done to the Institution of Mechanical Engineers (and do now in several other instances) by allowing the meetings to be held on their premises.*

Later on, as the general evolution of telegraphy in all its departments

---

* It may here be remarked that the Institution of Civil Engineers had itself produced some most valuable papers connected with cable work—from the historical point of view perhaps the most valuable extant.   The following are their titles, with the volumes of the Minutes of Proceedings of the Institution in which they are to be found :—

"Submarine Electric Telegraphs." By F. R. Window, A.Inst.C.E.  Vol. xvi. (1857).

"Submerging Telegraph Cables." By T. A. Longridge, M.Inst.C.E., and C. H. Brooks.  Vol. xvii. (1858).

"The Practical Operations connected with Paying-out and Repairing Submarine Telegraph Cables." By F. C. Webb, Assoc. Inst. C.E.  Vol. xvii. (1858).

"Electrical Qualifications requisite in Long Submarine Telegraph Cables." By S. A. Varley.  Vol. xvii. (1858).

"The Maintenance and Durability of Submarine Cables in Shallow Waters." By W. H. Preece, Assoc. Inst. C.E.  Vol. xx. (1860).

"The Malta and Alexandria Submarine Telegraph Cable." By H. C. Forde, M.Inst. C.E.  Vol. xxi. (1862).

"The Electrical Tests employed during the construction of the Malta and Alexandria Telegraph, and on Insulating and Protecting Submarine Cables." By C. W. Siemens, M.Inst.C.E.  Vol. xxi. (1861).

"The Telegraph to India, and its Extension to Australia and China." By Sir Charles Tilston Bright, M.P., M.Inst.C.E.  Vol. xxv. (1865).

Thus, the Institution of Civil Engineers may be regarded somewhat as the nurse of submarine telegraphy.   When the Society of Telegraph Engineers was formed, the cardinal points had been settled ; indeed, submarine cable work was, to a great extent, an accomplished fact.

The Institution of Mechanical Engineers had also had some valuable papers upon machinery employed in connection with submarine telegraphy, viz. :—

"Description of a Machine for Covering Telegraph Wires with India-rubber." By Dr C. W. Siemens, F.R.S. (1860).

"On the Construction of Submarine Telegraph Cables." By Fleeming Jenkin, F.R.S. (1862).

"Description of the Paying-out and Picking-up Machinery employed in Laying the Atlantic Telegraph Cable." By George Elliot (1867).

had passed through the youthful and more rapid stage, and attained
something like maturity, a lull occurred in its further progress, during
which other branches of electrical work attracted more public attention—
especially electric lighting.    As the Society embraced all engineers and
*savants* who occupied themselves with electricity and its various applications, the papers read at the Great George Street meetings now began to
deal more and more with electric lighting, telephony, and the distribution
of electric power, until—about the time of the Paris Electrical Exhibition
of 1881—telegraphy began to take, comparatively speaking, a back seat.
Accordingly, in 1889, the Society was re-christened the INSTITUTION OF
ELECTRICAL ENGINEERS—a title more suggestive of its widened scope.

To students of submarine telegraphy, however, its proceedings in earlier
days under the former title had perhaps a greater interest—at any rate
until becoming adepts and past masters in their profession.   The following
may be considered to be the more important papers relating generally to
cable work which have been read before the Society of Telegraph Engineers from time to time :—

"Contributions to the Theory of Submerging and Testing Submarine
Telegraphs."   By Dr Werner Siemens.   Vol. v. (1876).

"The Working of Long Submarine Cables."   By Willoughby Smith.
Vol. viii. (1879).

"Cable Grappling and Lifting."  By A. Jamieson, F.R.S.E.  Vol.vii. (1878).

"Submarine Telegraph Cables : their Decay and Renewal."   By
Samuel Trott and Frederick Adam Hamilton.   Vol. xii. (1883).

"Deep-Sea Sounding in Connection with Submarine Telegraphy."   By
Edward Stallibrass, A.M.Inst.C.E.   Vol. xvi. (1887).

Most of the above papers, as well as several others, have already been
alluded to in this book in connection with those parts of our subject to
which they more particularly refer.   Some of the other contributions to the
journal of the Society, concerning the various methods of electrical testing,
the instruments to be used for this purpose, and for working cables, have
also been referred to, or are in Parts II. and III.   The papers by Mr J. J.
Fahie, Sir Henry Mance, M.Inst.C.E., and Mr A. E. Kennelly (in 1874,
1884, and 1887 respectively), on their methods of fault-localisation and the
discussions upon them, will be found especially interesting and instructive,
though somewhat beyond the scope of the present volume.*   The discussions on such papers will invariably be found well worthy of study—often

---

* It may, however, be remarked in passing that largely owing to these tests, coupled
with the "Fall of Potential Test" of Mr Latimer Clark, this art has been brought to such
a pitch of perfection that in the present day an electrician can often localise a fault in a
submarine cable closer than the captain can navigate his vessel.

eliciting more practical information than the papers themselves. Again, many of the annual inaugural addresses presented by the various Presidents of this Institution naturally contain much of interest concerning matters telegraphic—to wit, those of Dr C. W. Siemens, Sir William Thomson, Mr Latimer Clark, Mr C. V. Walker, Professor Abel, Lieut.-Col. J. U. Bateman Champain, Mr W. H. Preece, Lieut.-Col. Webber, Mr Willoughby Smith, Mr C. E. Spagnoletti, Mr Edward Graves, and Sir Charles Bright. The last-named, however, and that we have recently had from Sir Henry Mance, are two which deal more especially with submarine telegraphy, historically and otherwise.

For some years before the establishment of the Society of Telegraph Engineers, a certain means of intellectual intercourse for members of that profession, as well as for electricians and electrical engineers generally, existed in the Press. *The Electrician* (original series, weekly) was started as early as 1861,* and the *Telegraphic Journal* (original series, monthly and then fortnightly) in 1872 ; † the main title of the latter having since been changed to the *Electrical Review* as a weekly publication due to Messrs Alabaster, Gatehouse, and Co. Previous to the above *The Engineer* was practically the only technical organ available to telegraph engineers, its rival, *Engineering*, not having been established till 1866. Some numbers of *The Engineer*, about the time of the first Atlantic cable project, contain a good many articles of interest to submarine telegraphists. The early numbers both of *The Electrician* and of the *Telegraphic Journal* abound in valuable contributions—such as, naturally, we never see now— from the ablest authorities on ocean telegraphy, and should be consulted by all students of its evolution and early history. The above and the *Electrical Engineer* (at one time monthly, but now weekly) are the only English journals to which special attention need to be called to on the historical side of this department of electrical work. But telegraph engineers and electricians who wish to keep thoroughly up to date in their knowledge of the most modern improvements in plant, testing-room instruments and methods, may also read with advantage the *Electrical Engineer*, the *Electrical*

---

* The original *Electrician*—which ceased to exist in 1864—is said not to have been in any way connected with the present journal (appearing first in 1878), though got up in the same style.

† A weekly publication bearing the same title had previously made its appearance in 1864. Again, as far back as 1845, the late Mr C. V. Walker, F.R.S., had edited, for a short time, a highly interesting journal called the *Electrical Magazine*. This latter— somewhat resembling a "quarterly," reporting meetings and papers, besides reviewing books—was probably the first instance of periodic electrical literature, apart from actual books.

*Review*, and the *Electrical World* of New York, also the principal German electrical journal, *Electrotechnische Zeitschrift*, as well as the back numbers of the now defunct *La Lumière Electrique.*\* The foregoing, as well as many other useful periodical publications† dealing with the subject in English and other languages, may be seen not only at the libraries of the great professional institutions already referred to, but at that of the Patent Office, Southampton Buildings, Chancery Lane, London—which is open to the general public. The current numbers of most of them, we believe, may now be also found in the reading-rooms attached to all the chief public libraries of the great provincial towns of the United Kingdom.

## SECTION 9.—RETROSPECT.

With the exception, perhaps, of more careful preliminary surveys before the laying of a cable, and the ground being more completely sounded over previous to the selection—and in some cases the actual marking-out—of a route for the line,‡ the operations of constructing,§

---

\* Some years ago *La Lumière Èlectrique* published a series of admirable articles on submarine telegraphy from the pen of Mons. E. Wünschendorff, the French Government engineer. These were much appreciated at the time, hence his book (compiled from these), on a portion of which the present work is partly founded.

† The *Quarterly* and *Edinburgh* (formerly the *North British*) have both from time to time published admirable articles regarding matters connected with submarine telegraphy, some emanating from that able writer, the late Professor Fleeming Jenkin, F.R.S. Other magazines and journals have done likewise. Amongst those of comparatively recent date may be cited excellent articles, of the popular sort, by Mr Herbert Laws Webb (author of a useful little book on "Electrical Testing") in *Scribner's Magazine*, and some by Mr A. P. Crouch in *Cornhill*, as well as the *Nineteenth Century*. There are also various papers germane to the subject read before the British Association and Royal Institution from time to time, besides a capital series of Society of Arts Cantor Lectures delivered by Professor (then Mr) Fleeming Jenkin, F.R.S., in 1866.

‡ No doubt, partly owing to the light brought to bear upon the bottom of the ocean by the "Challenger" Expedition of 1873-76, in which so many distinguished men of science participated. The work of this expedition was fully recorded in a series of large volumes (edited by Sir Wyville Thompson, F.R.S., and by Dr John Murray, F.R.S.), which are entitled "Reports of the Scientific Research Exploring Expedition of H.M.S. 'Challenger, 1873-1876." The subject of sounding for telegraphic purposes has been very fully treated by Mr Edward Stallibrass, A.M.Inst.C.E., in a paper—referred to above—read by him in 1887 before the Institution of Electrical Engineers, on "Deep-Sea Sounding in connection with Submarine Telegraphy." Mr Stallibrass gives a complete sketch of the history of sounding work in all its aspects, and a detailed description of the apparatus used at various times by different parties.

§ Various experimenters have from time to time suggested cheaper classes of insulation other than that of gutta-percha or india-rubber. These suggestions, however, come from people who appear to be ignorant, or oblivious, of the fact that the above costly materials are only employed because others, though equally efficient electrically under normal conditions, altogether fail to carry out the mechanical requirements for submarine cable purposes, quite apart from their physical failures, lack of durability, and subservience to

laying, testing, and repairing telegraph cables are carried out in much the same way now, and with much the same appliances, as they were some twenty-five years ago. That is to say, the improvements which have been introduced during the last quarter of a century, are improvements in details rather than in general principles. More radical changes may, perhaps, be anticipated in the not very remote future. There are some, at any rate, who look forward to the time when, through the further development and practical application of the theories of such learned investigators as Mr Oliver Heaviside, F.R.S., Professor Silvanus Thompson, F.R.S.,* Professor Oliver Lodge, F.R.S.,† not to mention that *doyen* of electrical science, Lord Kelvin,‡ submarine telegraphy may be effected in an entirely different, and at the same time a simpler and less expensive, fashion than at present.§ The Cooke and Wheatstone of a new "root-invention" may even now be on the point of success. When so many able minds are found working about the same time upon similar lines—in electrical matters, *teste* the comparatively recent cases of Gramme and Siemens, Swan-Edison and Lane Fox, and, again, Hughes, Graham Bell, and Edison—there generally has been some important practical outcome of their contemporaneous efforts.

Meanwhile, one of the next important improvements in cable-working seems likely to take the shape of an alteration in the form of the cable itself, rather than in the instruments used with it. The latter have now

surrounding influences. These matters were recently dwelt on by the author in the course of an article in *The Electrician* ("Problems of Ocean Telegraphy," by Charles Bright, F.R.S.E., vol. xxxix., p. 6).

* The advent of ocean *telephony*, as well as of high-speed ocean telegraphy, have been prophetically discussed by Dr Thompson, and particulars of his suggested systems will be found in many of his scientific papers and patent specifications, as well as in Part III. of this book. He regards the primary cause of retardation of current impulses to be "capacity"; moreover, he points out that this is a *distributed*, and not a *local*, grievance. Hence, whatever means are applied to a cable to counteract "capacity" must, in his opinion, be *distributed* also. He further believes that such a distributed remedy is to be found in bobbins possessed of electro-magnetic induction. His proposal is to construct cables having such bobbins judiciously placed at intervals throughout the length of the conductor.

† Author of that admirable treatise, "Modern Views of Electricity," one of the *Nature* Series, which is responsible for teaching us more about electricity (from an enlightened standpoint) than is to be gathered elsewhere, and a publication that does the above journal and its publishers great credit.

Another most helpful work, which marked the introduction of an entirely new order of text-book, is "Practical Electricity," by Professor W. E. Ayrton, F.R.S. (Cassell and Co.). The latter sets aside the oft-repeated experiments with pieces of sealing-wax, a cat's back, etc., in favour of experiments having a more useful bearing.

‡ Besides Mr Nikola Testa, and, still more, the late Dr Heinrich Hertz.

§ The suggestions of Mr Heaviside, and those more recently made by Professor Silvanus Thompson, for high-speed telegraphy, are referred to further in Part III.

probably been brought to their highest attainable degree of efficiency—
quite beyond that required or justified by the cable itself under present
conditions.*

The human element scarcely enters into our calculations when machine-
transmission is in question; for, although the eye can only read at a
certain rate, the recorded message can be divided up in such a manner
that different parts of it are being taken down by different clerks at the
same time.

**Inductive Telegraphy**—Compared with the improvements last mentioned, INDUCTIVE TELEGRAPHY† is truly "a big order." Yet even this
achievement may not be very long delayed, if we can judge by recent indications. Within the last ten years or so various experiments have been made
in this direction, across the broad rivers of India, and across certain channels
of this country, as well as between boats and the shore, by Mr Melhuish of
the Indian Telegraph Government Department, by the late Mr Willoughby
Smith,‡ by Mr Charles Stevenson, F.R.S.E., and latterly, on a more extensive
and practical scale, by the engineer-in-chief to our Government telegraphs,
Mr W. H. Preece, C.B., F.R.S., assisted by Mr J. Gavey and Mr H. R.
Kempe, A.M.Inst.C.E.§  Mr Preece has made this one of his special
subjects, and attacked it with the business-like tenacity for which he is
well known. He has, from time to time, given us full descriptions of his
experiments in this direction. ‖  He has, in fact, kept the world well posted

---

* A certain economic limit being placed on the dimensions of the core.  Modifications
affecting the form of the conductor within the core have, however, been proposed recently,
and have constituted the subjects of patents by Mr O. Heaviside (1880), Mr W. H.
Preece (1892), and others. These consist in multiple-wired cores, the arrangement of
the conductors being designed to reduce their electro-static capacity as well as their
resistance, and thus, according to the KR law (see p. 203), to materially increase the
working speed.  These devices were principally intended to meet the special requirement of long-distance telephony—viz., low capacity—but they are, of course, applicable to
electric signalling generally over long distances.

† The first suggestion in this direction appears to have emanated from Mr J. W.
Wilkins, a telegraph engineer of the earliest day.  In 1849 Mr Wilkins proposed inductive
telegraphy as a means of establishing electric communication between England and
France—though never as yet put into practice on so large a scale.  In those days this
method—similar in principle to later suggested systems—was spoken of as "telegraphy
by means of earth conduction."

‡ Assisted by Mr W. P. Granville.  Their experiments were later on extended by Mr
Willoughby Smith's son, Mr W. S. Smith, the present manager of the Gutta-percha
Works at Wharf Road, attached to the Telegraph Construction Company.

§ Author of the famous "Handbook of Electrical Testing" (E. and F. N. Spon), which
has now run through several editions; also of "The Electrical Engineer's Pocket-Book"
(Crosby Lockwood and Son).

‖ In the course of papers read before the British Association's Meeting of 1894, and
the Society of Arts in the same year, the titles of which were "Signalling Through
Space" and "Telegraphy without Wires."

up, in this matter—as in others of an electrical nature—with reference to the stage of progress so far attained.

This method of telegraphy between places at short distances has frequently been spoken of as "telegraphy without wires"; but, though an attractive title to those who look forward to the cheapening of practical telegraphic methods, it is scarcely an accurate description (as a rule) of its present application.   This necessarily involves the employment of lengths of wire running parallel, or nearly so, at right angles to the direction of the communications to be effected, and of at least the same length as the distance between the points.*    The principles upon which it is based, and the manner in which it is carried out, are treated in greater detail elsewhere in this book.

At its present stage of development, inductive telegraphy does not seem likely to prove applicable to submarine working on anything like an extensive scale †—that is to say, between points at any considerable distance from each other.    Moreover, if it were, there would not appear to be any economy realisable by its adoption ; rather the reverse—as already shewn—since two long cables (one along each shore) would be necessary instead of one.    For this reason, then, even if the able and experienced scientists now engaged upon the subject see their way to making the inductive system (on its present lines, *i.e.*, the long parallel insulated conductors) *applicable* in all other respects to long-distance signalling, it is hardly probable that it will be adopted for practical work on an extensive scale.

If inductive telegraphy, as at present understood, is ever rendered thoroughly practicable for transmarine signalling over any considerable

---

* Indeed, seeing that the conductor has to repeat the same length on each side as the distance across, the total length required by this method should be double that involved by the ordinary direct submarine telegraphy, though allowing of a considerable cost reduction in obviating the necessity of an expensive armour.

† Its most promising sphere of utility for the present seems to be that of signalling between lightships, lighthouses, and the shore, a problem which it bids fair to solve successfully before long.   The ordinary method of communication by a submarine cable, connecting up any two or more given points, is apt to break down in these cases through the constant chafing of the cable against the moorings of the light vessel under certain tidal conditions, etc.   Mr H. Benest, A.M.Inst.C.E., read a very complete and interesting paper on this subject before the Balloon Society in 1892, under the title of "Coast Telegraph Communication."

In establishing telephonic or telegraphic communication along our coast-line—which has never yet been completely carried out—all conducting wires should be subterranean. The War Office should recognise the force of this, and make it their business to see it properly and completely carried out, for in time of war all telegraph lines above ground would be seriously imperilled.   Just as the Meteorological Department are in a position (by means of telegraphic communication with certain outlying stations) to issue warnings of the coming storm, so should the Military and Marine Departments be in a position to obtain immediate notice of what is taking place within sight of any part of our coast.

distance, the main way *in* which, and the general purpose *for* which, it seems most likely to be applied, is in the hands of the State, and (primarily at all events) with a view to naval and other national contingencies. In times of war, two coasting lines in the hands of the same naval Power, or of two allied Powers, would possess obvious advantages, as a secure means of communication, compared with a single, or even a duplicated or tripli-cated line across the ocean. As has been already pointed out, International Cable Conventions may prove rather broken reeds for a great commercial nation like Great Britain to lean upon, in the event of a war with other naval Powers. It is quite conceivable—nay, more than conceivable—that, without having by any means lost command of the seas immediately sur-rounding our coasts, we might be unable to prevent our trans-Atlantic cables from being cut by the enemy. Now this would be the very time when— not to speak of the ordinary exigencies of naval communication in warfare —it might be of most vital importance for us (*e.g.*, for avoiding the cutting off of our corn supplies through any unwarrantable panic among shippers on the American side) to keep in constant touch, telegraphically, with the seaports, if possible, even of Canada and the United States. Again, strategically speaking, it would also be important for us to possess alter-native methods of communication with Ireland, the Hebrides, and the Channel Islands; perhaps, also, with certain neighbouring countries of the European mainland which happened to be allied with us, or "friendly neutrals." Under such circumstances, we should certainly bless the wise prescience of our Government—possibly in collaboration with those of the United States, the Canadian Dominion, and the others concerned—if it established and maintained a practical system of inductive coast cables. Granting that greater national security is at all likely to be attained thereby, expense (in the way of experiments, etc.) should scarcely be "an object" in the calculation. Afterwards, the only substantial item, treated as a question of national expenditure, would be the initial one of construction and submersion. This, of course, could then be readily estimated. Even in the case of the trans-Atlantic communications, it may safely be averred that the combined cost of the requisite British and American coast cables would be small compared with many of our little Admiralty or War-Office "experiments."

This question, then, in the humble opinion of the present writer, will very soon become ripe for practical consideration by "the powers that be." As to the still greater question of inductive telegraphy for general use, who shall say that some entirely new system, such as may be more truly entitled to be called "telegraphy without wires," will not be evolved before long by which it may be successfully solved? Is it possible that one of the inventions ascribed about a year ago to Dr Cornelius Herz can relate to inductive telegraphy? The indications which he has at present given regarding one of them certainly suggest some entirely new departure in telegraphy. As it is stated they are about to be patented, the world will probably not be long kept in ignorance on this point. The past work of Dr Herz certainly entitles us to hope for further excellent and wide-reaching achievements from him in the domain of electricity.*

Quite recently a young Italian electrician, Guglielmo Marconi, has invented an electro-static method of telegraphy which would be independent of any continuous length of wire, and has, therefore, been spoken of by Mr John Munro (in the course of an able article) as *Ethereal* Telegraphy. Such inventions as those respectively of Herz and Marconi might, of course, in the end, deal a veritable death-blow to submarine telegraphy as at present known, and as described in this book.† On the other hand (if it ever came to that) they would bring about such an enormous extension of telegraphic work all over the world—constructive, administrative, and operative —that they would not necessarily be unwelcome to those who are professionally or industrially engaged on it.

**Past and Future.**—Submarine telegraphy is just one of these achievements of human science and perseverance which will never be forgotten, even if it comes to be superseded in any form. When many of the passing crazes of the present time, many of its sensational inventions, have been consigned to limbo, our great cable enterprises, and the men who carried them out, will be noted by the historians of posterity as some of the most characteristic features and personalities in the civilisation of the latter half of this century.‡ Still more sensational developments of material progress in the way of communicating with our fellow-men may be in store for humanity. We may learn to fly through the air on wings, or Mr Maxim may construct for us a new aerial leviathan, while Mr Edison or somebody else teaches us, with some new fish-like craft, to cleave swiftly and noiselessly through the still depths of the ocean, instead of pitching and tossing in a painful manner on its summit. But looking ahead, say, as far as the

---

* In the course of an interesting article in the *Fortnightly* for January 1897, Sir E. J. Reed, K.C.B., F.R.S., made special allusion to these inventions of Dr Herz.

† Signor Marconi's invention was dilated on at some length by Mr Preece in a lecture at Toynbee Hall, at the end of last year, on "Telegraphy without Wires," and again, quite recently, at the Royal Institution in the course of one entitled "Signalling through Space." From these it would seem that, so far, nine miles is the greatest distance at which this form of new telegraphy has been successfully accomplished. This was a material advance on what had been done by the previous inductive methods already referred to.

‡ The Committee of the International Submarine Telegraph Memorial have recently decided to establish a Jubilee Commemoration of Submarine Telegraphy in the year 1901.

rising generation and its immediate successors, it is extremely doubtful if any advance in applied science of such comparative importance to material civilisation will be seen for many a year as was marked by the successful laying of the first great submarine telegraph lines.

In the words of Rudyard Kipling :—

"The wrecks dissolve above us ; their dust drops down from afar—
Down to the dark, to the utter dark, where blind white sea-snakes are.
There is no sound, no echo of sound, in the deserts of the deep,
On the great grey level plains of ooze, where the shell-burred cables creep.
Here in the womb of the world—here on the tie-ribs of earth—
Words, and the words of men, flicker and flutter and beat."

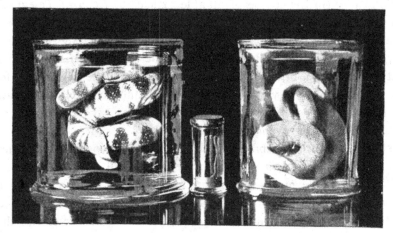

Some Enemies of the Cable.

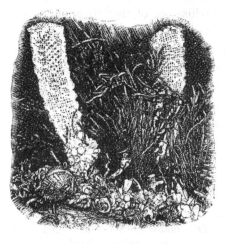

A piece of Deep-Sea Cable covered with
Shells, etc.

Barnacles on a Cable.

"THE WONDERS OF THE DEEP."

# APPENDICES.

# APPENDIX I.

___

## SUBMARINE TELEGRAPHS UNDER H.M. POST OFFICE.

WHEN the Government took over the land telegraphs in 1870 they at the same time became masters of all those submarine cables (belonging to the "Electric" and "Magnetic" Companies respectively) which went to form a part of the system—*i.e.*, certain cables between England and Ireland and between Scotland and Ireland, which united the land systems of each country.* The Electric Company's cable ship, the old "Monarch," was also transferred, along with the cables.† A few years later the late Submarine Telegraph Company's cable ship, "Lady Carmichael," was also acquired. Built in 1869, she is one of the first vessels in the telegraph service kept *permanently* for cable work. This small paddle steamer has since been re-christened by the Post Office under the patronymic H.M.T.S. "Alert."

There are in all, at the present time, fourteen cables (as shewn in the map on the following page) running from England to the various Continental countries around her, being the joint property of the different

___

* For several years past it had been felt that the welfare of the nation as a whole would be best provided for by giving the administration of all its telegraphs to the Post Office authorities—in other words, *nationalising* the telegraph service. At last, in 1870, this was effected by Act of Parliament. The companies were bought out, and their valuable property acquired by the State for a sum total of £4,182,362, including compensation to certain of the railway companies who had enjoyed the privilege of transmitting public messages. The total mileage of land lines transferred by this transaction was 48,378, exclusive of railway wires; the total mileage of cable was 1,622; and the total number of stations was 2,488.

† The Post Office at the same time took over those members of the staff of the principal companies who were available. Thus, Mr R. S. Culley became the engineer-in-chief, and on his retirement later, he was followed in this capacity by Mr Edward Graves, with Mr W. H. Preece as chief electrician. Both these gentlemen had served with the "Electric" Company. Since Mr Graves' death, Mr Preece has assumed the double responsibilities of engineer and electrician.

Similarly, in 1889, when the Submarine Company's property was absorbed by the Government, some of their then staff were simultaneously given places in H.M. Postal Telegraph Department.

administrations affected.    These are irrespective of the Great Northern
Company's lines to Norway, Sweden, and Denmark, of the Continental
connecting links for the trans-Atlantic telegraph service, as well as of the
Eastern and Direct Spanish Companies' cables.

Out of these, six run between England and France, three being maintained by the English Government, and three, now, by the French, all being
equally owned by the two countries.    Included in this total, there are two
cables reaching to Belgium.    These are maintained by us (half the cost being
charged to the Belgians), though the joint property of the English and

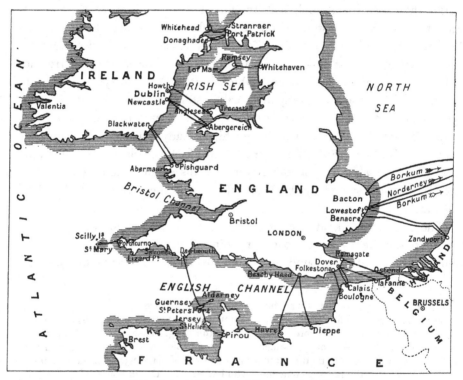

English and Anglo-Continental Government Cables.

Belgian Governments.    The two Anglo-Dutch cables are, however, in the
sole possession of our Government.

Then come, finally, the three Anglo-German cables, regarding which it
has already been stated that each Government owns one cable exclusively,
while the remaining one is held jointly by the two.    Each Government
looks after its own cable, and the joint cable is attended to by the English
Government, half the repair expenses being chargeable to the Germans.

The various cables owned, worked, and maintained by H.M. Government,
are now laid, and attended to generally, by H.M. telegraph ship " Monarch,"

drawings of which are here appended, and also by H.M. telegraph ship " Alert " referred to above.   The *original* "Monarch," after which the present ship of that title was named, only served Her Majesty for a very short period.   In the very same year that she was taken over by the Postal Telegraph Department, and on the very first piece of work she was sent out upon under the new *régime*, her engines broke down so hopelessly* that she had to be altogether abandoned as a ship.   She is now doing duty as a coal-hulk.

Though the then engineer of the Post Office (Mr R. S. Culley) had from the very first strenuously urged on the Government the advisability of replacing the old "Monarch" by a new Government telegraph ship, it was not till a good many years afterwards that this recommendation was put into effect.   Thus for some time all the few cables that were then under Government control were attended to as occasion required by various telegraph ships specially chartered for the purpose from the various cable contractors—indeed, in some cases (where proper cable ships were not available) ordinary steamers were hired and temporarily fitted with the necessary gear.   It will easily be understood that this, after a time (as the number of cables gradually increased), became very expensive.   This state of things would, indeed, have been practically impossible now that the Submarine Company's lines form a part of the system requiring Government attention—thereby bringing the total number of cables at the present time under the Post Office supervision well over a hundred.†   In 1883, however, the Postmaster-General, at the instance of Mr Edward Graves (the then Post Office engineer-in-chief), succeeded in persuading the Treasury to consent to the necessary expenditure, and the present "Monarch" was built and equipped at a total cost of about £50,000.   She is the result of carefully thought-out designs, largely based on the extensive practical experience of the marine superintendent, Mr David Lumsden,‡ and of the assistant marine superintendent, Mr W. R. Culley,§ the services of Mr J. H. Ritchie,

---

* This was under stress of weather, as graphically shewn by the *Illustrated London News* at the time, and also in Mr Wilkinson's " Submarine Cable Laying and Repairing," published at *The Electrician* office.

† The total number of conductors, including those in cables actually *owned* by H.M. Government, being above four hundred.   This includes all short lines to the various groups of small islands in the neighbourhood of Great Britain, as well as those across lochs, rivers, canals, etc.

‡ Formerly of the Electric Telegraph Company.

§ Son of Mr R. S. Culley, late engineer-in-chief.   Mr Culley has since become marine superintendent at the Dover Station of the Postal Telegraphs, in succession to the late Mr J. Bordeaux, Mr Lumsden still remaining chief superintendent, with headquarters at Woolwich.

the eminent naval architect, being also enlisted. H.M. telegraph ship "Monarch" was built by Messrs Dunlop and Co., of Port-Glasgow, under the supervision of Mr Ritchie. Illustrations of the "Monarch" are given in Figs. 52 to 56. Fig. 52 is a general elevation; Fig. 53 a section; whilst Figs. 54, 55, and 56 shew plans of the various decks. Her principal dimensions are:—Length between perpendiculars, 240 feet; breadth, moulded, 33 feet; depth, from keel to deck, amidship, 20 feet; tonnage, builder's measurement, 1,348$\frac{45}{94}$. Her load draught, with 600 tons of cable and 120 tons of fuel on board, would be 15 feet 6 inches, but as she rarely has reason to carry more than 100 tons of cable, this is considerably in excess of her usual draught. Her ordinary speed is about 10 knots, but she is capable of attaining 12 at a push. She has a raised quarter-deck of the height of the main rail about 65 feet long, a monkey forecastle about 20 feet, and a hurricane deck running from the front of the quarter-deck about 100 feet forward, below which are the deck-houses.

The "Monarch" is rigged as a two-masted schooner with pole masts. Unlike many telegraph ships afloat, which have a great sheave bolted on to the bows for effecting cable operations, the "Monarch's" bows are themselves so constructed that the bearings for the reels are part of the ship itself, and great strength and convenience are obtained, besides the fact that the ship is not made to look ugly and top-heavy at the bows by the use of the conventional bow-sheaves.* The plates of the vessel are gradually formed into two powerful box girders for the support of the bow-sheaves, and the tops of these girders are decked so that the officer conducting cable operations is enabled to walk out even beyond the sheaves for the purpose of effecting all necessary observations as to the lead of the cable, etc. When under steam in the ordinary way, the masthead stays are fixed to the forecastle head in the usual manner, but means are provided for moving them to the port or starboard bows when cable work is going on, so as to be clear of the cable leads, according to requirements.

Messrs Johnson and Phillips are responsible for the design and construction of all the cable gear.

The picking-up machine (as shewn in Fig. 57) is the largest of the kind ever made, weighing close on 64 tons. It is fitted with two overhung outside drums 6 feet in diameter and 2 feet 4 inches on the face.

These drums, which are used for picking up or paying out the cable,

---

* The ships of the "Eastern" and allied companies are now similarly constructed with the bow-sheaves built into the hull. Indeed, it is believed, that this plan originated with the patent specification No. 18,841 of 1889, taken out by Mr Percy Isaac, M.I.N.A., of the Eastern Telegraph Company.

[PLATE XI.

FIG. 52.—Her Majesty's Telegraph Ship "Monarch."

[To face p. 196.

[PLATE XII.

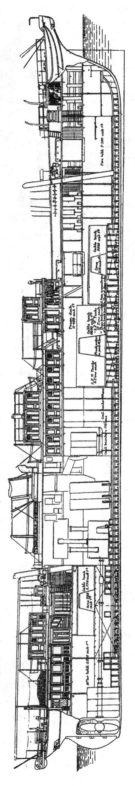

FIG. 53.—Section of H.M.T.S. "Monarch": General Elevation.

[PLATE XIII.

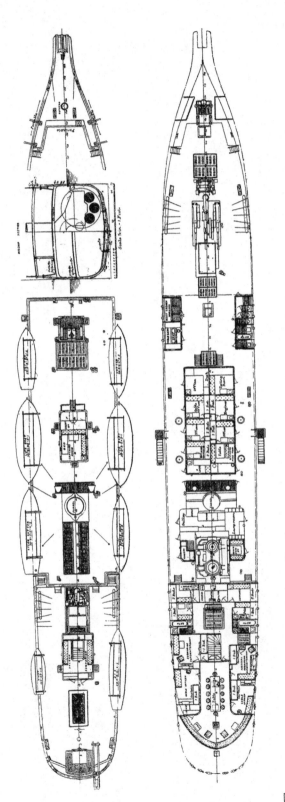
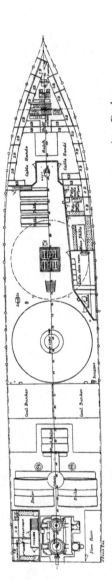

FIGS. 54, 55, 56.—Her Majesty's Telegraph Ship "Monarch," shewing her various Decks.

can be worked either quite distinctly or together at will. The indicated brake horse-power of this gear is 150, with a steam pressure of 80 lbs. That such a machine has enormously heavy work to do (and, therefore, requires to be exceedingly powerful) will be readily understood when it is remembered that some of the cables round our islands weigh nearly 30 tons per N.M., while the very lightest Post Office type is as much as 7 tons per mile.*

Again—as another reason why the "Monarch's" gear is, of necessity, exceptionally strong—when grappling in such shallow water as is met with

FIG. 57.—Cable Machine, H.M. Telegraph Ship "Monarch."

where the Post Office cables are laid, foreign bodies attach themselves to the cable in great quantities.†

---

\* Cables in shallow waters must necessarily be much heavier than those laid on the ocean bed, which should be, indeed, out of all harm's way—provided they are once properly laid down—and not likely to be disturbed. Thus, the main (deep-sea) types of our Atlantic cables do not weigh over $1\frac{1}{2}$ tons per N.M. as a rule—or even less sometimes ; whereas in the line from Holyhead to Howth, near Dublin — the heaviest submarine telegraph cable—the shore-end type weighs as much as 28 tons per N.M.

† A case is actually known where as many as thirteen anchors in all were picked up on a four-mile length in the Firth of Forth. Once, indeed, an actual ship was brought up with a cable, during repairs, in the North Sea—a small one, no doubt, but still a ship. It was raised sufficiently near to the surface to enable those on board the repairing vessel to recognise it as a schooner, before it fell off the cable and sank again to its bed below.

The length of cable to be raised on the bight in shallow-water repairs is small as compared with that dealt with in ocean repairs. The specific weight and accessories involved are, however, invariably much greater than in the latter case. Thus, the cable machinery of such a ship working in shallow water requires to be every bit as strong as the gear for ships working on ocean cables—and, as before stated, the "Monarch's" is actually the heaviest cable machinery in existence.

Great advantages are claimed for the comparatively modern overhanging drums, such as this gear provides.* These drums can be geared for different speeds. There are two speeds on the port drum—$\frac{1}{2}$ and $1\frac{1}{2}$ knots (N.M. per hour), the slow speed being applicable to lifting heavy cables, and the quick for light ones. The starboard drum has only one speed—*i.e.*, 1 mile per hour—being intended for intermediate types of cable. These drums, through the medium of a series of spur wheels, are driven by a double-cylinder engine which runs at from 150 to 160 revolutions per minute.

Fixed to the quarter-deck, on the port side of the rudder-head, is a large cable sheave, but it is only screwed down to the deck, being but seldom employed. Indeed, it is very usually unshipped, along with the appertaining leads, from the cable machine, and kept amongst the stores in the ship's hold until required.† Ships which are only intended for paying out comparatively short lengths of cable—such as "repairing ships"—are seldom provided with any cable gear at the stern, as it is considered that the greater convenience and safety of paying out from the stern for a short time only, is, in such instances *more* than balanced by the risk always involved in passing the cable along from bows to stern on the completion of picking-up operations, previous to starting paying out.‡

The holding-back gear of Sir Charles Bright forms a part of this vessel's equipment for paying out cable: a full description of it will be found in Mr Wilkinson's "Submarine Cable Laying and Repairing."§

Under the hurricane deck are situated (1) a large office for the engineering staff on the starboard side, and (2) the spacious electrical testing-room on the port side.

On account of the special character of the work on which the "Monarch,"

---

* The usual practice, hitherto, had been to place the drums (which carry the cable) inside the framework.

† This sheave is mainly used for passing the cable round with a view to slack, and in order to avoid it being subjected to jerks at the bows by the pitching of the vessel.

‡ Any advantage in paying out from the stern consists mainly in greater convenience. The vessel, however, always steers better when laying a cable from the bows.

§ *The Electrician* Printing and Publishing Company, London.

like other telegraph ships, is employed, where all manner of speeds and all kinds of steering are required from time to time, she has a very complete system of telegraphs between different parts of the ship, so that the engineer in charge may be enabled to telegraph instructions from the monkey forecastle to the steersman, as well as to the officer in charge of the engine-room.  It is especially on account of the great noise made by the cable gear * on the fore-deck that some kind of telegraphic communication between the forecastle and the bridge is so indispensable.  A semaphore worked by hand is, however, very often found to answer admirably, with the advantage that there is practically no mechanism to get out of order and which cannot be readily repaired as in the case of elaborate mechanical or electrical telegraph systems.

There are also many points in favour of ordinary speaking tubes over any electrical or mechanical methods of communication.

The particulars of the " Monarch's " cable tanks are as follows :—There are three of them, all as large as the breadth of the ship in each part permits of.  The after tank, which is seldom used, has a capacity of 2,730 cubic feet.  The cone in the middle of it, with a capacity of 200 cubic feet, is used as a store for keeping meat, &c., in good condition, which purpose it answers excellently, owing to the usual supply of cold water kept in the surrounding tank.  The main tank, situated a little forward amidships, has a capacity of 6,600 cubic feet, and is about 28 feet in diameter by 10 feet high.  Its cone is used as a fresh-water tank, being able to carry some 10 tons.  The fore tank is just forward of the main, and has a capacity of about 3,890 cubic feet.  It is the same diameter as the " main," but little more than half as high.

All the cable tanks stand on a deep double bottom of cellular construction.  These double bottoms are divided into five compartments, so that any part may be filled with water to help to maintain the trim of the vessel. This arrangement is particularly valuable on a telegraph ship, in compensating for the weight of cable as it is paid outboard from any particular part of the ship.

Forward of these tanks is the hold for storing large telegraph buoys, buoy chains, grapnels, mushrooms, etc.  The smaller buoys are fixed in the rigging except when under great stress of weather.  On the top of the forward cable tank is the " 'tween decks."  Between the main cable tank and the men's quarters in this 'tween decks is a large space which is used as a workshop.  At the after end of the main tank are the coal bunkers,

---

* More recent machines are, however, very quiet in their action.  Thus it is that no deck-signalling apparatus is employed on the " Eastern " Company's repairing vessel.

which are capable of carrying about 260 tons of coal should so much be required. Between the coal bunkers and the after cable tank is the engine and boiler space. A bulkhead is built just behind the main tank, and the corner spaces between it and the tank—as well as the spaces between the two forward tanks—are used for storing fresh water.

The cable tanks all drain into the ballast tanks below, and when they contain cable it is necessary to keep the cable covered with water. The side plates of the tanks are ⅜ inch thick. On construction they were subjected to a pressure of 10 lbs. per square inch, as a test of their water-tightness.

Like the jubilee of the Queen's accession to the throne, the jubilee of the telegraph in this country was commemorated in 1887, in virtue of Cooke and Wheatstone having established the first practical telegraph—based on an application of their combined patents—in 1837. This telegraph jubilee—mainly a Post Office affair—was celebrated by a banquet at which the then Postmaster-General (the late Right Hon. H. C. Raikes) presided. He was supported by all the leading telegraph engineers and electricians of the day, several of whom had taken part in the pioneer work of the early stages of electro-telegraphy.*

---

* There are two or three men still living who have been connected with electro-telegraphy ever since its establishment. Thus, a short time ago the jubilee in the telegraph service of Mr W. T. Ansell was commemorated by a banquet given in his honour. This gentleman, till recently the traffic superintendent of the Eastern Telegraph Company, may be said to be the *doyen* of telegraphy. He was one of the earliest members of the "Electric" Company's staff.

# APPENDIX II.

### SUBMARINE TELEPHONY.

INASMUCH as in 1880 it was determined by a court of law that a telephone is a telegraph,* it has been thought that a brief summary concerning the application of the telephone to submarine cables in its present stage might be suitable here, without actually entering upon the principle or construction of the telephone instrument itself.

The advent of the telephone began a new epoch in the progress of electrical communication.

In 1876 Professor Graham Bell exhibited his original "speaking telegraph" at the Philadelphia Exhibition, where it was seen and heard by Sir William Thomson, who introduced it to the notice of Europe at a meeting of the British Association held that year in Glasgow. The instrument by itself was not well adapted for speaking through long lines. The transmitter and receiver were both alike, and the sounds were feeble. It partook, in fact, more of the nature of a scientific toy than anything else. It has since become a useful adjunct to the telegraph, of which it may be regarded as the flower and culmination. In 1877, however, Professor D. E. Hughes, F.R.S., discovered the microphone, which has, up to the present, proved the best transmitter to employ in conjunction with the Bell receiver. The carbon transmitter, invented by Edison in the same year, similarly rendered the telephone a practical success. Various companies were soon promoted in the United Kingdom for establishing telephony throughout towns. In 1880 the United Telephone Company† was incorporated in

---

* Besides Mr Preece, the expert witnesses on behalf of the Crown were Sir Charles Bright, Mr W. H. Barlow, Mr Latimer Clark, Professor D. E. Hughes, and Mr Warren de la Rue. On the side of the Company there were Lord Rayleigh, Sir William Thomson, Sir G. G. Stokes, Dr John Hopkinson, Dr J A. Fleming, and Professor Tyndall.

† Now, by amalgamation, the National Telephone Company, which eventually absorbed all the other telephone working concerns in this country. The "National" Company works under a Government license for working telephones between people in the same city or town.

This was the outcome of the aforesaid legal action by which the Post Office estab-

England, for purposes of telephone exploitation, since which most of our large cities have been connected by trunk telephone lines. Central exchanges for intercommunication by word of mouth have been established in all the larger towns, and the telephone is now in constant use in almost every business house. A merchant in London can talk with one in Birmingham; a citizen of Edinburgh is now in earshot with a friend in Dundee. Corresponding, and still greater, progress has been made abroad—indeed, England was comparatively tardy in taking up the telephone properly, and even now the service can hardly be said to be satisfactory.

Paris has for some time been able to "speak through" to Brussels and Marseilles; likewise Chicago to New York, by a line over 1,000 miles in length.* Honolulu has a telephone exchange; and recently this instrument made its appearance on the Congo. Some years ago the late Mr R. A. Proctor predicted that ere long a whisper would be conveyed beneath the ocean, which none of the million waves that tossed above it would be able to drown. We are still some way off this imaginary consummation. It must be remembered, however, that after the opening of the first telegraph line in England, and after the first telegram was sent, twelve years elapsed before the first submarine cable was laid. Since then, practically the whole of our earth has been strung with electric wires, and time and space have both been annihilated.

Articulate human speech has been transmitted over wires many hundred miles in length—or (to speak with greater scientific accuracy) the vibrations of the air caused by the voice at one end of the wire, have, by the agency of the electric current, been faithfully reproduced at the other.

The subject of long-distance telephony and submarine telephony has been very closely studied by several eminent electricians for a considerable period, and various papers dealing with the difficulties and requirements have been read from time to time by Professor Silvanus Thompson, F.R.S., Mr W. H. Preece, C.B., F.R.S., and others, before the Royal Society, the British Association, the Institution of Electrical Engineers, and elsewhere.†

---

lished their rights over the telephone as a form of telegraph within the meaning of the Telegraph Purchase Act of 1869.

It is only a question of time when the telephone system generally will come under the direct management of the Post Office. The sooner this takes place the better, for the telegraph and telephone are obviously intended to work together.

* In America the telephonic systems are owned by private companies, but on the European Continent chiefly by the Governments in question.

† One of Dr Thompson's schemes for meeting the difficulties of long-distance telephony (submarine or subterranean) is to have a series of leaks and induction coils along the line, with a view to freeing the cable rapidly of some of its superabundant charge. This is referred to elsewhere in Parts I. and III. amongst the recent novelties not yet fully developed in practice.

The whole trouble in telephony is induction, which is a much more serious matter in this case than in ordinary telegraphy, on account of the delicate nature of the instrument and the delicate currents suitable for working it. We have here to deal with ripples corresponding to the infinitesimal vibrations of the voice, rather than *waves* of electricity. The former are naturally more liable to get blended together by the attraction and holding back effect of induction, in proportion to the electro-static inductive capacity, than in the case of stronger currents producing waves.*

To put the matter in a popular light, the electric current has the curious, and often inconvenient, habit of setting up other and different currents on surrounding conductors. In the instance of a cable, for example, the charge of electricity as it rushes along the wire induces an opposite charge on the wet serving or sheathing, or on the sea-water itself. These two charges attract each other—stay to exchange views, as it were—so that the original current —or rather, what proceeds of it—is very much delayed. The transmission of the delicate currents produced by the vibrations of the human voice is under these circumstances liable, in extreme cases, to be practically impossible.

As we know, the speed at which signals can be made to follow on one another in such a way as to be read separately at the other end (instead of getting thus blended or blurred together) is dependent on the amount of what is known as the retardation of the cable. This retardation is made up by the combination of the total resistance offered by the conductor $(=R)$, and the total electro-static capacity (for induction) of the cable $(=K)$. It is, in fact, the *product* of the two $(=KR)$. The maximum speed practically attainable (for separate distinguishable signals) is, indeed, inversely proportional to this factor as announced by Lord Kelvin as early as 1856.

Mr W. H. Preece has shewn that this same law (usually spoken of as the KR law) is the sole factor which governs the practicability of being able to hear through a telephone at the receiving end of a telephonic circuit —*i.e.*, in such a way that the separate voice ripples shall be distinguishable and not "blurred" or blended together on arrival at the further (receiving) end. Thus, in order that speech through a conductor may be practically possible at all, the retardation, or product KR, must be very much smaller than for fair working speed on an ordinary telegraphic circuit.

In calling attention to this fact, Mr Preece gave, as the result of

---

* The limits are also much more confined with the telephone, owing to the fact that so far the application of any form of relay has not been found possible as in a telegraph circuit.

experiment, the outside figures for "KR" at which speech becomes practicable or impracticable, as the case may be. The results shewn were obtained on some of the Post Office cables, the trials being made with a view to testing the practicability of a telephone circuit between London and Paris for the proper interchange of speech between those capitals. The figures given by him were as follows :—

When KR = 15,000 (or over) speech becomes impossible.
   „    = 12,500   „    „    „    possible.
   „    = 10,000   „    „    „    good.
   „    =  7,500   „    „    „    very good.
   „    =  5,000   „    „    „  . excellent.
   „    =  2,500 (or under)  „    „    perfect.

The point of difference in the conditions for telephony between aerial and subterranean or submarine lines is, of course, that the latter involve induction to a much greater extent; indeed, overhead conductors, under favourable conditions (owing to the distance of the suspended wire from the earth), introduce hardly any at all.

As, moreover, the induction requires to be much less on a telephone circuit than on an ordinary telegraph circuit on account of the difference in the nature of the instrument and current employed—and, since the retardation (as expressed by the symbol KR) varies directly with the length of the line—it is obvious that the length of underground or submarine cable (still more than the length of overhead wire) is of a strictly limited character—that is to say, with the present known devices. Again, as a metallic circuit for telephonic purposes is absolutely essential in long circuits on account of the extra induction introduced when using the earth as a "return," the distance between which telephonic working is possible is still further limited.*

The going and returning wires of a metallic loop must also be alike exposed to this influence from without. The practice is, therefore, to cross the wires so that they exchange places at intervals along the route. The electric echoes, sputtering noises, or "hypnotic suggestions" from other lines are thus annihilated, and overhearing, theoretically speaking, should be impossible. This looping, however, has the effect of something like doubling the KR by, roughly speaking, doubling the length of wire in

---

* The extra inductive phenomenon here referred to is the "cross-talk"—as it is popularly termed—from neighbouring circuits using the same (earth) return. Moreover, the earth is objectionable for the return in long circuits owing to the effect of earth currents—however weak—on so delicate an instrument worked only by the feeblest of currents. Again, the employment of the earth is often ruled out of court on all telephone circuits subjected to neighbouring wires conveying stronger currents—also on account of induction.

the circuit. Initial cost has also to be taken into account. The smaller (and shorter) the wire, the less it costs, and the lower the inductive capacity, but on the other hand the higher the conductor resistance. To complicate the matter further, we know that the resistance increases in a higher ratio than the capacity for induction diminishes by the above. Thus the resistance and capacity have to be so adjusted (by the type of wire or cable selected) as to give the most effective combination, compatible with a certain limit of initial cost.

## LONDON-PARIS TELEPHONE LINE.

From what has preceded it will be seen that the question as to whether a telephone circuit could be satisfactorily established between London and Paris was not by any means one which could be settled off-hand. Thus, no little credit is due both to the French and English Government officials— and especially to Mr Preece—who carefully went into the matter before it was a *fait accompli*, by the greatest city in the world being put into telephonic communication with the gayest.

The idea of connecting London with Paris by telephone first emanated from the French people. M. Coulon, who was the Minister of the French *Administration des Postes et Télégraphes*, thought telephonic connection both feasible and desirable, and communicated with the authorities at St Martin's-le-Grand upon the subject. Having been proved experimentally by them to be scientifically possible, the Treasury made no difficulty in coming forward with the necessary funds.

This undertaking was actually carried out in 1891—fifteen years after the invention of the telephone—and was the first instance of submarine— or rather *partially* submarine—telephony being established.* Though, of course, only on a small scale, it has proved entirely satisfactory, and up to, or even beyond, all expectations.

Inasmuch as the London-Paris telephone circuit would be partly composed of aerial wire (at each end) and partly (in the middle) of submarine

---

* In 1889 the telephonic system of France came into the hands of the *Administration des Postes et Télégraphes*. The English Post Office and the above French Administration in that year purchased, as a joint property, the Channel cables of the Submarine Telegraph Company, and their next step was, as stated above, to project a telephone line and cable from London to Paris.

Mr Preece had just designed a *river* cable to cross the estuary of the La Plata and connect the cities of Buenos Ayres and Montevideo by telephone. This cable was 28 miles long, but the bronze overhead wires brought the length of the entire line to 186 miles and the KR to 10,400. The speaking attained on this line is, however, very good.

cable, a number of experiments were tried with various types of overhead wires applied to artificial copies of the electrical components of different types of submarine cable in order to arrive at what would give the best result.

In Mr Preece's experiments he found that speech would be good on an overhead copper wire if the KR did not exceed 10,000, on an overhead iron wire if the same did not surpass 5,000, and on a submarine cable or underground wire if it be within 8,000. This last figure, in fact, determines the design of a telephone cable—*i.e.*, the type of conductor and the thickness of its insulation—just as the previous figures determine that of an aerial telephone line according to the metal used. Iron is clearly unsuited for long, or trunk, telephone lines, because its specific KR is so large, and, in fact, it is never employed except on short, or exchange, circuits. Copper, having a low electrical resistance, is used for long trunk lines and cables.

The total distance from London to Paris by the railway routes on each side is 271 miles, comprising 70 miles from London to Dover, 21 miles from Dover to Calais, and 180 miles from Calais to Paris. The loop circuit being essential, the length of wire needed is, of course, twice 271, or 542 miles, including 42 miles of submarine core. It was decided on each side to make two separate land lines with a common cable of four conductors, the joint property of both Governments. The English land lines are of copper wire, weighing 400 lbs. to the mile; the French are of the same material, but of a gauge represented by 600 lbs. per mile. In both cases the wires are elevated on poles 30 feet above the ground.

The connecting cable, as designed by Mr Preece and his assistant, Mr H. R. Kempe, contains four separate insulated wires, or cores, each consisting of a strand of seven copper wires (each wire being 35 mils. in diameter), weighing 160 lbs. per N.M. The above stranded conductors are individually coated with three layers of gutta-percha alternated by Chatterton's compound. The dimensions of this dielectric are represented by a weight of 300 lbs. per N.M., bringing the total weight of each core (conductor and dielectric) up to 460 lbs. per N.M., and the diameter to 390 mils.

The copper used for the conductor is of the very purest obtainable for yielding the highest specific conductivity possible.

The resistance of each conductor in the cable is about 7.5 to 7.6 ohms per N.M. at a temperature of 75° F. The capacity for induction of each N.M. of core does not exceed 0.305 microfarads, its insulation being not less than 500 megohms at 75° F.

The four cores, as above described, are laid up together, wormed by a serving of tanned hemp; and the diagonal pairs are used together in the same metallic circuit. The cable sheathing consists of sixteen galvanised iron wires, of 280 mils. diameter, their individual strength being represented by a breaking strain of 3,500 lbs.—or over 1½ tons.

An illustration of the cable—which was constructed by Messrs Siemens Brothers—is given here (Fig. 58). It was laid by H.M.T.S. "Monarch" and the Post Office officials in March 1891, when the heavy snowstorm and other inclement weather then in force greatly interfered with the operations. The length of the cable is 20 N.M.: it extends from St Margaret's Bay, near Dover, to Sangatte, near Calais. The land lines on each side had already been completed by the French and English Governments respectively.

The total length of the line is 31·1 English miles.

The copper resistance of each metallic circuit is 1,380 ohms; and its capacity (calculated) 5.3 microfarads. The total KR is, therefore, 7,314, which brings the circuit well within the limit of good speaking as defined by Mr Preece.

The instruments used in London are the usual Post Office telephones, each being fitted with two double-pole Bell receivers. In Paris, the Ader, the D'Arsonval, and other telephones are used, subject to the approval of the French Administration.

The first circuit was opened to the public on 1st April 1891, the traffic being considerable from the outset— in fact, more than could be dealt with on a single circuit. In February 1892,

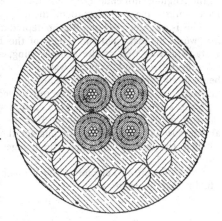

FIG. 58.—The London-Paris Telephone Cable.

the second circuit was thrown open for traffic, and both circuits are now always fully occupied.

The present charge for a conversation of three minutes' duration is 8s.* The line is largely used by members of the London Stock Exchange and Paris Bourse,† as well as for Press work and general commercial purposes,

---

* Two consecutive periods of three minutes may be arranged for ; but in no case can more than six consecutive minutes be allowed, except during slack portions of the day ; and then only by arrangement at the time of application for a conversation, and on the understanding that if the line is required at any time after the expiry of six minutes it will be taken. The line cannot be pre-engaged. It is customary for users to either write or telegraph to their correspondents to be in attendance at a certain time, and then wait their turn.

† Thus during the middle portion of the day the traffic is heavy, and users of the line occasionally have to wait a short time for their opportunity.

both circuits being always open day and night. The average number of messages per day (on the two circuits) is about 250.

This telephone line has since been followed by another* in 1893 between Scotland and Ireland (Glasgow to Belfast), and there is a further prospect of our being put into telephonic communication with Brussels and Berlin.

In spite of the electrical difficulties of the problem, inventors and experimenters are constantly engaged in endeavouring to better these beginnings, and even hold out some hope of some day talking to America by telephone.

For full information regarding the whole subject of telephony, the reader is referred to Messrs Preece and Stubbs' excellent treatise thereon.†

---

* The submarine cable in circuit here is of precisely the same pattern in every respect as the Anglo-Continental cable just described, except that the served and wormed cores have a brass tape sheathing round them.

This is in order to cope with the depredations of the *Limnoria terebrans*—a mollusc-like worm distinct from the teredo of the Eastern seas. It has a predilection for the gutta-percha as well as the hemp serving, and has been found within recent years to infest even these (non-tropical) waters. This is possibly owing to cables from the tropics being brought home in the course of repairs, and the germs being communicated at cable factories; indeed, the core is often resheathed for second use.

Thus, since 1893 it has been the custom with the Post Office to apply metal taping to the core of all Post Office cables. It is also partly on this account that the Department always specify hemp instead of jute for the inner serving; though it may be doubted whether the former withstands the ravages of submarine borers any better than the latter.

† "A Manual of Telephony," by William Henry Preece, C.B., F.R.S., and Arthur J. Stubbs (Whittaker and Co.).

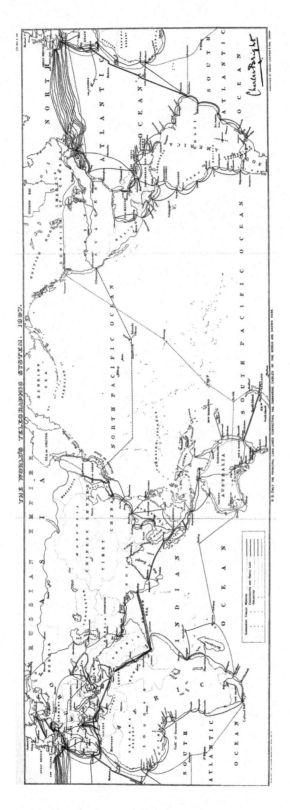

The material originally positioned here is too large for reproduction in this reissue. A PDF can be downloaded from the web address given on page iv of this book, by clicking on 'Resources Available'.

# PART II

THE CONSTRUCTION OF SUBMARINE TELEGRAPHS

# CONTENTS OF PART II.

---

## CHAPTER V.—COMPLETED CABLE.

---

## APPENDICES.

Eastern and Brazilian-Submarine Companies' Telegraph Station at Carcavellos, Lisbon.

# PART II.—CONSTRUCTION.

## COMPOSITION AND MANUFACTURE OF SUBMARINE TELEGRAPH CABLES.

NOTWITHSTANDING the modifications which have been proposed from time to time, all submarine cables laid up to the present* have been of the type first adopted in 1851 for the Dover-Calais line, comprising :—

1. A central conductor for conveying the electric current.

2. A covering of insulating material round the central conductor to prevent leakage.

3. An outer metal sheathing to protect the insulation from possible injury during and after submersion ; and to give sufficient tensile strength to permit of the cable being subsequently raised, if necessary.

The insulated conductor is generally termed "core." Where this expression is used here, it is in that sense.

---

* The exceptions are the Black Sea cable, which only lasted a very short time, the uncompleted cable from Oran to Cartagena, and some other less important lines of earlier date which never did public duty.

# CHAPTER I.

## THE CONDUCTOR.

---

## SECTION I.—COPPER.

**Electrical Qualities.**—Every material may be said to be a conductor of electricity; it is only a matter of degree. The question first arises, What constitutes a good electrical conductor, and why are the metals found to be the best conductors? Probably the fact of metals being highly homogeneous and very dense* in their texture, as compared with other substances, has something to do with it. Mixtures of good conducting materials will not conduct as well as each material when pure. This is, no doubt, owing to the variation in texture and a certain chemical action between the molecules setting up a complete change in the resultant material. For instance, 100 yards of pure silver 1 inch thick melted down and mixed with a similar 100 yards of pure copper of the same thickness, each having the same conductivity when alone, would not conduct twice as well as either wire by itself, though it would otherwise (having double the thickness for the same length) but for some specific change having taken place. Similarly, the conductivity of the alloy of the two metals would be distinctly less than the wire of the same dimensions composed of either metal in the pure state.

Copper, the electrical conductivity of which is higher than that of any other metal commercially available,† has been invariably made use of for the

---

* Physically speaking, electrical conductivity may be said, as a rule, to go with opacity, and electrical resistance with transparency.

† The only metal which when pure has, generally speaking, slightly higher electrical conductivity is silver, and this is naturally ruled "out of court" for conducting purposes on any large scale if only on account of cost—£3 5s. per lb. as against 7d. to 8d. per lb. for copper.

central conductor of submarine cables. The following table gives the conductivity (or conductance, as it is now sometimes termed) of the principal metals and alloys relatively to that of pure silver, at 0° Centigrade :—

| | |
|---|---:|
| Silver (annealed) | 100 |
| „ (hard) | 94 |
| Copper (pure) | 100 |
| „ (annealed) | 96 |
| „ (hard) | 94 |
| Gold | 74 |
| Aluminium (annealed) | 53 |
| Platinum | 17 |
| Iron (pure and soft) | 16 |
| „ (telegraph wire) | 10 |
| Lead | 8.2 |
| German silver | 7.5 |
| Mercury (liquid) | 1.6 |
| Selenium (annealed) | 1 |

**Mechanical Qualities.**—Besides its electrical suitability, copper has excellent mechanical qualities. To begin with, it has great powers of extension. It will, in fact, stretch 10 to 15 per cent. of its length before breaking; so that, if a cable is subjected to heavy strain, the conductor, permanently stretching* much more than the rest, will remain intact until after all the sheathing wires have parted, communication being thus maintained to the last.† The copper used for conductors of submarine cables is selected, however, rather on account of its electrical than for the sake of its mechanical features. Copper wire does not, in fact, usually bear a greater strain than the equivalent of 17 tons to the square inch‡ when annealed—the soft form in which it is used for purposes of electrical conduction; the hard-drawn, "camperneal," copper wires (unannealed) standing a strain tantamount to some 30 tons per square inch. The

---

* Amongst the curiosities in early inventions was a special cable, devised in 1858 by Captain Drayson, R.A., and Captain Binney, R.E., the object of which was to provide for the contingency of a longitudinal strain. It was called an "Elongating Tunnel Submarine Telegraph," and consisted of a single wire conductor in the form of a helix surrounded by india-rubber in which it was capable of moving freely. No further protection or strength appears to have been provided. Various other spiral conductors have been designed of tubular shape, by way of securing elongation and elasticity.

† It is unfortunate that the insulating material outside the conductor (gutta-percha or india-rubber) does not also confine itself to stretching instead of being correspondingly elastic, and therefore having a tendency to leave bare places on the copper wire with a certain amount of strain.

‡ Thus the tensile strength of an ordinary average strand is represented by a breaking strain of about 200 lbs. This is, however, practically never turned to account owing to its accompanying elongating qualities being far beyond that of the rest of the cable.

density, or specific gravity, of pure annealed copper at 60° F. is generally taken at 8.9,* and its melting point at 2000° F.

**Conductivity of Different Species of Copper.**—In the early days of submarine telegraphy the purity of copper was supposed to be sufficiently well indicated by the degree of facility with which it passed through a die. This, of course, would be governed by different degrees of softness, but the causes of the widely varying conductivity values, as determined by physicists at that time, remained entirely unexplained.

The table below serves to give an idea of the tremendous variation obtained by different early experimentalists for the conductivity of copper, as compared with that of pure silver, taken at 100 :—

| Becquerel - | - | 95.3 | Snow Harris | - | - | 200 |
| Riess - | - | 67.2 | Buff - | - | - | 95.4 |
| Leuz - | - | 73.4 | Pouillet - | - | - | 73 |
| Davy - | - | 91.2 | Arndsten - | - | - | 98.7 |
| Christie - | - | 66.0 | | | | |

As a matter of fact, in the above, the only results of observations made at a uniform temperature (0° C.) are those of Becquerel, Leuz, and Arndsten. It is believed that Mr Latimer Clark was one of the first to bring to light the wide variation existing between the conductivity of different species of copper.

In 1857, whilst the first Atlantic cable was being manufactured, Professor William Thomson (now Lord Kelvin) made the discovery † that the presence of foreign substances, even in minute quantity, reduced the conductivity of copper to a very marked degree, and he pointed out the effect this would have on the speed of signalling.‡

In 1860 Dr A. Matthiessen made a complete investigation of the subject, having been requested to do so by the Committee of Inquiry appointed by the Board of Trade in conjunction with the Atlantic Telegraph Company.§ He first experimented with pure copper carefully prepared by the method of electrolysis, and then observed the effect pro-

---

* One cubic foot weighs about 556 lbs.

† This discovery was made owing to a suspicion that the conductor was not electrically uniform. The suspicion was followed by an electrical test on various lengths, and has led to the present practice of a test for conductivity being applied to short lengths of a certain proportion of the hanks of copper on arrival at the cable factory.

Since the above, a branch of copper manufacture has grown up for producing what is known in the trade as "Conductivity Copper," to meet the requirements of all specifications subsequent to this date.

‡ *Proceedings of the Royal Society*, June 1857.

§ Joint-Committee to inquire into the Construction of Submarine Telegraph Cables, 1859. Bluebook, with Report and Evidence, published 1861.

duced on the conductivity by the separate addition of the principal substances in natural combination with copper ore; he subsequently analysed various samples of commercial copper, measuring their different conductivities. Dr Matthiessen found (what is now generally recognised) that the merest trace of arsenic is sufficient to reduce the conductivity 40 per cent.—*i.e.*, from 100 to 60*—and when the proportion of arsenic is increased to 5 per cent., the conductivity falls as low as 6.5. It is reduced as much as from 100 to 78 by 0.05 per cent. of carbon, though the latter is itself a very fair conductor; to 70 by 0.13 per cent. of phosphorus, and to 92 by 0.18 per cent. of sulphur, which also renders the copper very brittle, forming sulphuret of copper in some degree.

The presence of either bismuth or silicon have since been found to have damaging effects on copper—both electrically and mechanically speaking—almost equal to that produced by any of the above metalloids.

The action of oxygen is still more cogent: by melting pure copper with a very short exposure to air, traces of sub-oxide of copper formed, such as were too minute to measure by the then known methods of chemical analysis, had the effect of reducing the conductivity to as low a figure as 76: it rose again to 96 after hydrogen had been forced into the molten metal for three hours, thereby reducing the sub-oxide.

The influence of metals, though generally less apparent than that of the metalloids, tends in the same direction. The conductivity of pure copper alloyed with 0.48 per cent. of iron is only 35; with 1.33 per cent. of tin it is 50; and with 3.2 per cent. of zinc, falls to 59.

Strangely enough, the addition of 1.22 per cent. of silver — the best conductor, so far as we know at present—reduces the conductivity of pure copper to 90. It is thus proved that the conductivity of pure copper cannot be augmented by artificial means, but that, on the other hand, the presence of any foreign substance, to however small an extent—no matter what its own specific conductivity may be — always has the effect of increasing the electrical resistance.

It seems at first sight a curious anomaly that the resistance of a metal is actually increased by incorporating with it another metal possessing a lower specific resistance, but such is the case, quite apart from how low the resistance of the latter material may be. Indeed, it can be said that alloys have a particular resistance of their own, in no way governed by the specific resistance value of each of their component metals. This suggests

---

* Copper containing any considerable quantities of arsenic is, moreover, rendered too hard and brittle to be drawn into wire at all. Arseniferous ores are, indeed, quite out of the question for electrical conductor purposes.

that a distinct chemical and consequent physical change occurs, due to the mixing of different metals.

The alloys of copper, besides having such a damaging effect to electrical conductivity, appear also to lower its mechanical properties, inasmuch as impure is not as extensible as pure copper—breaking sooner, in fact, under a strain.*

Copper is met with in the largest quantities in Chili, Japan, Australia, Cuba, and Canada, and still in small quantities in Cornwall, Devonshire, the island of Anglesea, and in Ireland. Most of the copper used for electrical conductors, however, comes from either Chili, Australia (Burra-Burra), or Canada (Lake Superior), particularly from the latter, as it is found there in nature in the metallic state. For this reason, Lake Superior copper is generally spoken of as "native metal," the preparation in smelting and refining being comparatively small. Copper—like other metals —is, however, usually found as an ore; and the most common ore of copper is copper *pyrites*, a yellowish ore, being a combination of sulphur with iron. Other copper ores are, of course, met with; such as those in which tin and nickel are amongst the alloys, and in which oxygen is present in large quantities. Copper is so found in Australia (Burra-Burra) and in Chili.

In the processes of copper refining, good results, in the way of softness and its accompanying ductility, are, however, obtained by the addition of a small quantity of lead or tin, and this is explained by the following experiment of Dr Matthiessen's. A sample of copper, when melted and exposed to the air, gave a conductivity of 87; on adding 1 per cent. of lead or tin, and remelting in a current of carbonic acid gas, the conductivity rose to 93; but at the same time the residual quantity of lead or tin was found so slight as to be scarcely traceable. It seems that the lead or tin, as the case may be, reduces or drives off the sub-oxide of copper, leaving the metal relatively purer than at first.

---

* Theoretically, at any rate, the copper conductor should have a stretching capacity as high as possible, in order to approach as near as may be that of its insulating sheath. When a strain comes on the core, the insulating material stretches to an almost unlimited degree (especially if composed of india-rubber), and when once the conductor ceases to stretch also, it has to take the full force of the strain. Again, whereas gutta-percha or india-rubber return to their original form (especially the latter) on account of their elastic qualities, the copper conductor maintains its elongated length, and is, therefore, liable to knuckle through the insulation at any spot. Owing to its lessened rigidity, as compared with a solid wire of same bulk, the strand form of conductor avoids this defect to a certain degree. The above were, however, some of the original causes of failure in the earliest cables, from which experience has been gained since the Board of Trade Inquiry (alluded to above) concerning the construction of cables.

Eight samples of commercial copper obtained from various sources, which were experimented on by Matthiessen, gave measures of conductivity varying as much as between 98.78 and 14.24; the chemical analysis of which revealed, in the different instances, traces of silver, lead, iron, nickel, antimony, arsenic, and sub-oxide of copper, as the case may be, and in very different degrees. The sample at the head of the list in order of conductivity contained traces of silver only; the last but one, whose conductivity was 59.34, came from the Demidorff Mines in Russia, and the last one from Rio Tinto in Spain; * both the latter contained only arsenic, the Demidorff sample having a mere trace, and that from Rio Tinto 2 per cent.

It was found, again, that, apart from their nature, the greater the number of alloys in mixture with copper the greater the decrease in conductivity; and, what is also natural from the foregoing, the greater the proportion of the alloys with the pure metal, the less will be the resultant conductivity, as has, indeed, already been demonstrated.

In concluding his report, Dr Matthiessen pointed out the desirability of securing, for submarine cable conductors, the very purest copper obtainable in the market. He further shewed that, in order to secure this, the surest plan would be to specify in contracts that the copper should yield a certain conductivity. Thus, it would be to the manufacturer's interest to supply the purest copper, for otherwise he would be obliged to provide a greater bulk of copper to make up the required conducting power, which would probably not be to his advantage. Hitherto the only requirements that it was customary to specify were that the conductor should be up to a given gauge, trusting to the contractor for the purity of the copper.

As has been shewn, the electrical conductivity of copper, or of any metal, is governed both by the degree of purity of the metal (as well as by the nature of its composition) and also by its sectional area. As a matter of fact, both of these factors require to be specified for—the one as much as the other. If the specified conductivity is secured by a larger-sized wire, it will be accompanied by an increased electro-static inductive capacity, unless the dielectric be correspondingly increased.

Thus, by securing the purest copper, the bulk, and therefore the electro-static capacity, is reduced to a minimum—and in the most economical fashion—for a given conducting power, and for a given thickness of insulating covering. The result is that nowadays the gauge, as well as the

---

* The conductivity of this ore is, indeed, no higher than that of iron. The same remark applies to some other coppers which, like this, contain large quantities of other metals and metalloids, such as arsenic, lead, iron, and nickel.

Q

electrical conductivity of the wire, is tested on arrival at the cable factory as a check on the wire-drawers. Moreover, as a further and more accurate check on the material, and for additionally calculating the diameter, the weight of each hank of wire is also measured.

The great, though gradual, improvements made in the conductibility of copper within recent years—since the establishment of submarine telegraphy—is largely due to the advances made in smelting and in refining to a complete state of reduction or purification.

The conductivity of copper metal after smelting is, however, a great deal below what it is afterwards made to attain by the wire-drawers. The principal means of raising the conductivity is by the after-effect of heat in annealing the wire ;* but there are other means taken which are sacred to the wire-drawer. This particularly applies in the " best selected " (or " extra smelted ") copper, as prepared at the smelting works of Messrs Vivian and Sons at Swansea, where very large quantities of the Lake Superior copper are dealt with—*i.e.*, electrolytically refined—this being probably more largely employed in the crude form for electrical conductors than any other ore. This copper has only 87 per cent. conductivity in its original state, but is worked up to what it is at the present time by the wire-drawer and metallurgist. As a matter of fact, almost absolutely pure copper is now produced on a large commercial scale, though *ordinary* commercial copper still has a conducting power as low as 30, compared with pure copper at 100.

What is known as *annealed* wire is that which is " annealed," or made soft,† after " drawing." The last operation (in this country mostly performed at Birmingham) actually strengthens the copper somewhat—laterally, at any rate—but the process of annealing reduces it again to a point far below what it was originally.

With *copper*,‡ annealing consists of heating first and then cooling *suddenly* —by plunging into cold water, or by some other means. For electrical conductors the operation is necessarily applied after the copper has been drawn into wire ; partly because if done before, the copper might become too much weakened to permit of successful drawing into wire, as well as for the reason that the subsequent hardening effect by drawing would, in

---

* The beneficial effect of annealing copper—so far as its electrical conductivity is concerned—was first observed by Dr Matthiessen, during his exhaustive experiments already alluded to. He then found that after being annealed the conductivity of the same sample wire had increased by $2\frac{1}{2}$ per cent. over what it was when in the hard state.

† Copper when annealed will allow of a dent being made on it when using only moderate force.

‡ In this respect unlike iron and other metals.

this instance, tend to reduce the conductivity again to something like its original figure.

The process of annealing has a weakening effect on any metal, reducing its tensile strength to nearly half what it was before. Great longitudinal strength is not, however, aimed at for the conductor of a submarine cable, where the sheathing wires are intended to take all the strain.

The purest copper is obviously that obtained by electrolysis—*i.e.*, depositing copper (obtained by an electrolytic system) in a trough from, say, a number of dynamo machines. Such a method has the disadvantage, however, of slowness and costliness, and is not, therefore, much favoured as yet for practical work on any large scale.*

Nowadays commercial copper is readily obtained, which gives a conductivity equal to 97 per cent. (and over) of that of the pure metal.† American (Lake Superior) copper is remarkably free from objectionable impurities, and yields one of the highest conductivities; so does Australian (Burra-Burra) copper; and perhaps still more the copper from Chili, as well as that from Japan. In Burra-Burra copper, however, one of the impurities is bismuth, which is difficult to separate from the ingots, and is almost as objectionable as arsenic. It frequently happens in the present day, with the various samples obtainable, that by the test calculations the wire under test gives higher conductivity than that afforded by the standard, owing mainly to the latter not really being so pure as what we are now able to obtain commercially.‡ In fact, electro-deposited copper wire made according to the Elmore process has on the average a conductivity quite 2, or even 3, per cent. above that of Matthiessen's standard.

The ultimate conducting power (and, therefore, the resistance) of an ordinary copper wire depends very much on the drawing of the copper into the form of wire. Compression naturally has a tendency to harden the

---

* This method is, on the other hand, exclusively adopted for making the small quantities of absolutely pure copper required for electrical standards, and experimental laboratory work generally; moreover, it was the means by which Matthiessen obtained his standard of pure copper.

† On the first Dover-Calais cable of 1850 the conductivity of the copper conductor was as low as 30. On that of the 1857 Atlantic it was exactly half of what is quite ordinarily obtained at the present time for electrical conductor purposes. Thus, as the signalling speed attainable on a cable is directly proportional to the conductivity, a cable of to-day of the same dimensions would be capable of carrying just double the number of messages in a given time. This would imply double the earning power—a consideration which naturally appeals to the investor.

‡ This, of course, points to the desirability of a fresh standard of pure copper being established.

copper, and this again has the effect of reducing its conductivity.   This effect may be accounted for by the increased rigidity of molecules thereby produced, just as the process of annealing renders the molecules more free to move in response to any electrical strain.*

**Tests and Calculations.** — Electrical measuring instruments are usually arranged (for greater convenience) to give the value of a quantity which is the reciprocal of conductivity.†   This quantity is termed resistance."

The relative conductivity C of a wire, referred to that of pure copper which is represented by the number 100, ‡ may be obtained by multiplying its calculated resistance $\rho$ (on the basis of purity) by 100, and dividing the product by its actual (observed) resistance $r$.   Thus :—

$$C = \frac{100\rho}{r}$$

The specific resistance of pure copper, or the resistance of a cubic centimetre at 0° C. = .00000164 ohm ; and that of a cubic inch at 0° C. = .00000063 ohm.

The resistance of 1 foot of pure (soft) copper wire, weighing 1 grain at a

---

* In 1883 Mr W. H. Preece, C.B., F.R.S., read a very complete paper on the subject of "Electrical Conductors" before the Institution of Civil Engineers.   This paper, with the discussion thereon, gives very full information regarding the different phases of copper and other metals for conducting purposes.   More recently, Mr Bucknall Smith's book on "Wire" (published by *Engineering*) has furnished a great deal of up-to-date matter concerning copper wire, as well as an exhaustive description of the process of wire-drawing, which, it may be remarked, is somewhat out of our scope, even if the necessary space were available for treating with the subject.

† However, some years ago, Lord Kelvin had boxes constructed for joining up conductors in parallel in different ways, so as to give direct measurements of conductivity in different substances.   He gave the name of "mho" (ohm spelt backwards) to the unit of conductivity in the British Association or B.A. system of standards and units.

‡ In practice—for commercial purposes—the conducting value of copper wires is always compared with that of a pure wire of the same metal as the standard, the latter being, therefore, taken at 100.   By this means—instead of silver (of the highest conductivity) being the standard—an idea of the purity is also implied, though degrees of purity and degrees of conductivity cannot by any means be said to go hand in hand, in consideration of the fact that the merest trace of arsenic reduces the conductivity of a pure (100 per cent. conductivity) copper wire to 60 per cent.   This is the most striking example, but in *no* instance does the degree of effect in conductivity accord at all closely with the degree of alloying.   A copper wire may have a 90 per cent. purity and yet (owing to the 10 per cent. impurities being, say, partly composed of arsenic) its percentage of conductivity might not be more than about 30.   Thus the percentage purity of copper, as determined by chemical analysis, cannot be relied upon as strictly indicative of the percentage conductivity.   Moreover, the electrical test for conductivity often gives a more accurate insight into the degree of purity than can be attained chemically.

temperature of 0° C., being .2064 of the legal ohm,* that of a wire having a length $l$ expressed in feet and weighing $w$ grains, will be—

$$w = \frac{.2064 \times l^2 \dagger}{w}.$$

The conductivity, therefore, of a wire at 0° C. $l$ feet long and weighing $w$ grains, with a measured resistance of $r$ ohms, will be—

$$C = \frac{.2064 \times l^2}{w \times r}.$$

The resistance of a naut (N.M.)‡ of pure copper wire, or strand, weighing 1 lb. at a temperature of 75° F. is taken as 1196.7.§ From this the resistance per N.M. ($r$) of any other pure copper wire or strand weighing $w$ lbs. becomes

$$r = \frac{1196.7}{w} \text{ at } 75° \text{ F.}$$

where resistance per N.M. multiplied by the weight equals resistance per naut pound.

The resistance per naut of a cable conductor at 75° F. is in fact, roughly speaking, equal to 120,000 divided by the product of the percentage conductivity of the copper and its weight per naut, in pounds. It is also equal to the resistance of a naut-pound corresponding to the particular percentage conductivity divided by the weight in pounds per naut. Thus, the resistance of a wire of given material varies directly with the bulk or area—*i.e.*, decreases with the *square* of the diameter. For instance, if the diameter wire is doubled, the resistance will be a *quarter* what it was, and the conductivity quadrupled. Moreover, with a given *specific* resistance—*i.e.*, with a

---

* The legal ohm—or *ohm*, as it may be termed—is equal to 10⁹ C.G.S. units (centimetre-gramme-second electro-magnetic fundamental unit system) of resistance. It has now been agreed to take, as the practical unit of resistance, the resistance of a specified column of mercury (B.A. Committee on Electrical Standards, 1892; Report of Electrical Standards Committee of the Board of Trade, 27th October 1892).

To obtain the relation between resistances measured in B.A. units and resistances measured in ohms we have :—

1 B.A. unit = .9866 ohm.

1 ohm    = 1.0358 B.A. units.

Thus to reduce B.A. units to ohms, we have to multiply by .9866—*i.e.*, deduct 1.34 per cent.

† The explanation of the variation by length being *as the square* in this formula, rather than merely directly as the length, based on unit proportions, is as follows :—If a unit foot-grain of copper be drawn out to double its length, its section will consequently be halved over the process. The result is that the resistance will be not only doubled, on the score of the length being doubled, but will be again doubled owing to halving the section of the wire.

‡ Throughout this book either the term "naut" or the letters "N.M." are used in abbreviation for "nautical mile" in place of the oft misapplied term "knot," which is in reality a rate—representing a nautical mile per hour.

§ The equivalent at 32° F. (=0° C.) is 1091.22.

given quality of material and conductivity—the resistance of a wire varies inversely with the weight, and consequently the conductivity directly with its weight.

The percentage conductivity of any wire may be found by multiplying the resistance calculated for a pure wire of the same length and weight at the same temperature by 100, and dividing the product by the resistance of the wire as measured. Matthiessen's standard was a pure annealed copper wire 100 inches in length, possessing a resistance of 0.516 ohms at 60° F., and the conductivity of this wire was taken as 100. Thus, the conductivity of any other wire, in relation to this standard, may also be determined by taking a standard having a resistance (0.1516 $w$) equal to 100 inches of the above pure copper wire, weighing 100 grains at 60° F. The conductivity of any other wire again but of similar resistance will then be as the square of its length in inches, divided by its weight in grains.

**Influence of Temperature on Electrical Resistance.**—Becquerel shewed that the electrical resistance of metals increases with a rise of temperature. These variations were measured in 1862 by Matthiessen and Van Bosc, who, after numerous experiments, were able to state the following formula, the conductivity of a metal at 0° C. being represented by 100, and C. being its conductivity at any temperature $t$,

$$\frac{C}{t} = 100 \times at \pm \beta t^2,$$

$a$ and $\beta$ being positive or negative constants for different metals, according to circumstances. For copper—

$$\frac{C}{t} = 100 - 0.38701t + 0.0009009t^2.$$

The resistance offered to a current of electricity through any conducting wire varies, in fact, in proportion to the temperature, more or less according to a logarithmic curve, $i.e.$, by compound interest.

This variation is different for different metals—though in the same direction—and is largely dependent on the purity of the metal. The conducting properties of alloys are very much less influenced by temperature than in the case of pure metals. Moreover, the former are not effected to any great extent by exposure to the atmosphere.

The resistance of copper and all the metals increasing with a rise in temperature and *vice versâ* is possibly due to the changes of density thereby produced, thus possibly affording a more perfect conduction at a low than at a high temperature; though it must be admitted that in the former condition the molecules are more rigid—$i.e.$, less free to move—and can therefore only act *inductively* on one another.

The tests which submarine cable cores undergo during manufacture being made, as a rule, at a temperature of 24° C. or 75° F.,* the factor $t^2$ may be neglected in practice, and the resistance $Rt$ of a wire at any temperature $t$ calculated by the simplified expression

$$Rt = Ro(1 + at)$$

where $Ro$ is the known resistance at 0° C. The number 0.00388 obtained from the experiments of Van Bosc and Matthiessen is generally accepted by electricians as the value of $a$, being the resistance co-efficient for 1° C. change of temperature. The corresponding co-efficient for 1° F. is about 0.0021, equivalent to 0.21 per cent. resistance per degree F. It should be remarked, however, that these values are, strictly speaking, only applicable to pure copper wire. Alloys are invariably affected rather less by temperature than pure metals, tending to further prove that a physical as well as chemical, change is set up by the mere act of alloying. Similarly the best conducting metals are generally the most influenced in this way by changes of temperature.

With the following table, based on the above value for $a$, the resistance of a conducting wire composed of pure copper at different temperatures F. is easily found :—

TABLE FOR CALCULATING APPROXIMATELY THE RESISTANCE OF COPPER AT DIFFERENT TEMPERATURES F.

| To Increase from Lower Temperature to Higher, multiply the Resistance by the Number in Column 2. | | | | To Reduce from Higher Temperature to Lower, multiply the Resistance by the Number in Column 4. | | | |
|---|---|---|---|---|---|---|---|
| No. of Degrees. | Column 2. | No. of Degrees. | Column 2. | No. of Degrees. | Column 4. | No. of Degrees. | Column 4. |
| 0 | 1. | 16 | 1.0341 | 0 | 1. | 16 | 0.9670 |
| 1 | 1.0021 | 17 | 1.0363 | 1 | 0.9979 | 17 | 0.9650 |
| 2 | 1.0042 | 18 | 1.0385 | 2 | 0.9958 | 18 | 0.9629 |
| 3 | 1.0063 | 19 | 1.0407 | 3 | 0.9937 | 19 | 0.9609 |
| 4 | 1.0084 | 20 | 1.0428 | 4 | 0.9916 | 20 | 0.9589 |
| 5 | 1.0105 | 21 | 1.0450 | 5 | 0.9896 | 21 | 0.9569 |
| 6 | 1.0127 | 22 | 1.0472 | 6 | 0.9875 | 22 | 0.9549 |
| 7 | 1.0148 | 23 | 1.0494 | 7 | 0.9854 | 23 | 0.9529 |
| 8 | 1.0169 | 24 | 1.0516 | 8 | 0.9834 | 24 | 0.9509 |
| 9 | 1.0191 | 25 | 1.0538 | 9 | 0.9813 | 25 | 0.9489 |
| 10 | 1.0212 | 26 | 1.0561 | 10 | 0.9792 | 26 | 0.9469 |
| 11 | 1.0233 | 27 | 1.0583 | 11 | 0.9772 | 27 | 0.9449 |
| 12 | 1.0255 | 28 | 1.0605 | 12 | 0.9751 | 28 | 0.9429 |
| 13 | 1.0276 | 29 | 1.0627 | 13 | 0.9731 | 29 | 0.9409 |
| 14 | 1.0298 | 30 | 1.0650 | 14 | 0.9711 | 30 | 0.9390 |
| 15 | 1.0320 | | | 15 | 0.9690 | | |

* The exact comparison between the two scales is 23.88—*i.e.*, 75° F. = 23.88 C.

**Suggested Substitutes for Copper.**—More or less recently it has been proposed to replace copper for telegraphic and telephonic purposes by silicious bronze, which has about the same conductivity as commercial copper, and much greater tensile strength. This proposal—like that of aluminium and aluminium bronze—mainly applies to aerial land lines, where it has already been put into force in place of iron as well as copper. It has, however, also been suggested for submarine cable purposes. In this latter application the idea is that the iron sheathing might then be reduced in weight—*i.e.*, smaller wires, cheaper type, and less of them—which would have the effect of lessening the cost of the cable, besides rendering it easier to lay and repair.*

---

\* M. Vivarez, on the basis of the weights of copper and gutta-percha forming the core of the French Atlantic cable, has drawn up the following specification for a deep-sea cable with a silicious bronze conductor :—

|  | Pounds. |  |
|---|---|---|
| Silicious bronze | 485 | |
| Gutta-percha | 397 | |
| Jute | 176 | Per nautical |
| 28 iron wires of No. 18 B.W.G. | 1,102 | mile. |
| Hemp and compound | 550 | |
| Total weight in air | 2,710 = 1.2 tons. | |

Such a cable, about 1 inch in diameter, would weigh 0.3 ton in water per N.M., and would resist a tensile strain of 2.8 tons, half of which would be taken up by the conductor. It would, therefore, be strong enough to support nine miles of its own length hanging vertically in water. (*Zeitschrift für Elektrotechnik*, 1885.)

Eastern Telegraph Company : Gibraltar Station.

## SECTION 2.--THE CONDUCTING WIRE.

**Comparative Features of a Solid and Stranded Wire: Ultimate Reasons for the Latter Form.**—The conductors of the earliest submarine cables were composed of a single solid round wire, usually No. 16 B.W.G. =.065 inch in diameter. This, the simplest form, is also at first sight the most suitable and economical, presenting, for given volume, the smallest possible surface area. In the case of a single solid wire, the weight of insulating material required for covering the conductor with a given thickness, and the electro-static capacity of the cable, are both kept at their lowest for equal weights of copper.* Thus, in obtaining a certain result, the cost of manufacture is less than can be the case with any other form of conductor ; or, to put it in another way, as the speed of signalling attainable is in inverse ratio to the electro-static capacity, the nett returns of the cable are greater with a solid wire conductor than with any other form, owing to the speed being greater with the same weight of copper.

These electrical, and consequent commercial, advantages are, however, more than counterbalanced by several important mechanical drawbacks ; and ever since the year 1856, the solid conductor has given place to a strand composed of several finer wires laid up together, the centre wire being usually surrounded by six other such wires.† In point of fact, the metal from which the wire is drawn can never be absolutely homogeneous, consequently the line is liable to be of unequal strength and electrical resistance at various places. It may also be absolutely defective, and contain flaws at certain points, due to brittleness, or may contain foreign matters, such as both increase the resistance, and tend to bring about chemical action, and, in consequence, gradual corrosion. Moreover, any soldered joint may also give way at a point. It will be readily seen that in any of the above events, the resultant rupture of a single solid conductor would mean a total stoppage of communication.

* The electro-static capacity of a stranded conductor is, in fact, usually considered to be equivalent to that of a solid 'circular wire, whose diameter is that of a circle enclosing, or circumscribing, the strand, diminished by 5 per cent. This is so on account of the interstices involved in the strand—*i.e.*, of the greater compactness of the solid wire. As regards the conductor, the capacity is only dependent on its superficial area. Conduction, on the other hand, takes place through the whole mass of the wire. Thus, for a given total weight of copper the conductor resistance should be the same ; but for a given limit of diameter, the resistance of the strand will be greater, owing to there being less copper in a given area.

† The strand form of conductor was first adopted by Messrs Glass, Elliot, and Co., for the New York and Newfoundland Telegraph Company's cable laid across the Gulf of St Lawrence, between Cape Breton and Newfoundland, in 1856, and which ultimately acted as the connecting link to the first Atlantic cable.

In the instance of a stranded conductor, on the other hand, there is but little likelihood of a mechanical or electrical defect occurring in each wire at the same spot; thus, the chances of a complete break of continuity in all the wires is very considerably reduced. Another advantage of the strand form is that it introduces less rigidity and greater pliability than in the case of a solid wire of the same total diameter. The result is that the strand conductor is less liable to buckle up, and force its way into—or, indeed, through—the insulating envelope. It, in fact, yields to bending more easily, and is, therefore, less liable to break.* On this score (of increased pliability), and on the decreased chance of a mechanical or electrical defect being—or becoming—serious, several wires, laid side by side, are obviously superior to a single wire of the same total area. Theoretically, in this way, it would be unnecessary for them to be made up together in strand form, but for the fact that they could not otherwise (*i.e.*, if only placed side by side) be relied on holding together sufficiently compactly for electrical and mechanical purposes. Moreover, by stranding them together with a definite lay, a certain margin for drawing out is provided for, thus affording extra stretching facilities in the case of a strain on the cable.

However, a disadvantage of the stranded conductor is, that if one of the small wires happens to break, or separate itself from the rest, the fine points are able to seriously pierce the insulation (and even penetrate to the sheathing) much more readily than is the case with a solid wire of large diameter.† Moreover, the wires not being in continuous contact with the

---

* Moreover, wires of small section are always less likely to be brittle than large wires.

† It is for this reason, mainly, that the ordinary strand conductor, in submarine cables, is not, as a rule, made up out of wires of less than No. 22 gauge. In the ordinary seven-wire strand this makes the weight of conductor 107 lbs. per N.M. with compound.

There is also the objection of small gauge wires being so much more liable to break during stranding up, thereby causing delays, as well as subsequently when under strain.

It would seem, however, as though a strand composed of No. 23 S.W.G. were feasible. Seven wires of this would make up a conductor equivalent to 75 lbs. per N.M., which would be ample for many of the short lengths of cable to which 107 lbs. per N.M. for the copper conductor has usually been assigned.

This point is rendered more particularly worthy of consideration in view of the decreased quantity of insulating material involved to cover the conductor with the same required thickness, *i.e.*, about 100 lbs. per N.M. instead of 140 lbs. As a matter of fact, 700 N.M. of $\frac{107}{140}$ core would be capable of yielding a signalling speed as high as eighty-nine (five-letter) words per minute, thus exhibiting a manifest waste of material under the ordinary conditions of manual transmission.

As a further argument in favour of lighter cores for, say, coast cables of some 500 N.M., it should be remembered—if the line connects up unimportant places for local traffic only—that not even the maximum hand-working speed is really necessary where there is only about an hour's traffic per day at the outside. In such a case a speed of fifteen or

central heart-wire, water or moisture, having once found access, is more liable to creep along the whole length of the core as in a tube, should it chance to penetrate at a loose end, or through the insulation at any weak spot ; repairing a single fault* might thus entail picking up and relaying a long length of cable. The latter difficulty is, however, dealt with by coating the central wire, and also the outside of the completed strand, with a resinous composition whilst the strand is being formed, so that the outer wires mould themselves into the compound as described hereafter. All the excess of compound is squeezed out through the interstices of the wires, and afterwards adheres so firmly to the first coating of insulation, that even under the pressure of the sea (over 2 tons per square inch surface in the greatest depths) water never penetrates for more than a few inches along the conductor of cables manufactured in this way.† This becomes specially important in the case of a broken or buoyed end to which no seal has been applied, for otherwise it is liable to act as a tube.

Again, another disadvantage in the strand form is that it involves more copper being used—especially if the lay of the wires be short—i.e., a greater length of each surrounding wire‡ for a strand of given section, for a given length of cable. It is also most costly on account of the extra time and workmanship involved by the operation of laying up the several wires composing it.

**Segmental Conductor.**—In the design of the first Persian Gulf cable, Messrs Bright and Clark, as we have seen,§ endeavoured to—at any rate partially — combine the electrical advantages of the solid conductor with the mechanical advantages of the strand, by the conductor being com-

---

seventeen words per minute should be ample, especially where inexperienced clerks are engaged who cannot work above that speed.

Of course, it is another matter where the whole line (of several sections) is periodically "put through" for a long length in all, or where a Brown-Allan relay (in connection with a Morse system) is in force. Then a light core would not do, if heavy traffic be the order of the day. Moreover, the operations preparatory to submerging a cable, render less than a certain thickness of dielectric material risky in anything like a hot climate. For this reason also, the conductor should be limited in weight under such conditions. Again, a limit must be put on its size (even when composed of several wires stranded together) on the score of rigidity and buckling through the gutta-percha envelope.

* The effect of sea-water on copper is complicated and uncertain. Sub-chloride of copper, is, however, probably formed. Any sort of molecular disintegration reduces the electrical conductivity in some degree.

† Unfortunately, however, this compound itself tends to very gradually absorb salt-water in small quantities, where opportunity occurs.

‡ As a rule, the length of each wire surrounding the central wire is, indeed, some-where about 12 to 15 per cent. longer on account of the lay.

§ See Part I.

posed of four longitudinal quadrants drawn down through a copper tube :
thus embraced, they fitted closely together so as to form a stout composite
wire of circular section.*   This, however, proved initially to be a very
costly type of conductor on account of the labour, time, and machinery
involved in construction, though an almost perfect combination of the
electrical and mechanical requirements.

The strand form of conductor has of necessity (for mechanical reasons)
now become universal in the construction of submarine cables.

The strand usually consists of seven wires of the same gauge, one
central wire, and the remaining six laid up round it.   Other combinations
of strand are possible, but the above makes the most compact type for
electrical reasons, and is the best mechanically, as a rule.†

**Solid-Strand Wire.**—The conductor for the Direct United States
cable, manufactured by Messrs Siemens Brothers in 1874, consisted of a
stout central wire 0.091 inch in diameter, surrounded by eleven smaller
wires of .035 inch.   If we study the section of this conductor (Fig. 1), we

see that a given quantity of copper is contained in a
smaller circle than would be the case with a strand formed

Fig. 1. — Siemens' of wires all of equal diameter, the electro-static capacity
Solid-Strand Con- for a given thickness of insulation being, as a result,
ductor.
correspondingly less than in the latter case ; or with equal
areas, the quantity of copper in the conductor is greater, and the conductivity thereby increased.   According to the late Sir William Siemens,
the gain in conductivity, within a given area of copper, amounts to 10
per cent. ; and the corresponding gain in speed is distinctly sensible.
Messrs Siemens Brothers have since adopted this form of conductor for
all the subsequent Atlantic cables in the design and construction of
which they have been concerned.   Moreover, the conductor of the last
" Anglo" Atlantic cable was made by the Telegraph Construction Company after this pattern.   This modification of the ordinary "strand" may
be regarded as the nearest approach to the solid wire conductor that is, for
mechanical reasons, advisable in practice.   The segmental conductor, above
alluded to, is only ruled out of court (in most instances) on the score of
cost of construction, though otherwise a more perfect imitation of the
ideal form, electrically speaking.

* To meet the same ends, Sir Charles Bright also devised a conductor consisting of a
number of wires wormed into the interstices of larger ones, the whole being drawn down :
but the above plan was preferred at that time.

† This form of equal-wire strand was eventually favoured as the outcome of a series of
exhaustive experiments, both from the combined electrical and mechanical points of view.

**Other Suggested Types.**—The electro-static capacity involved in the case of an ordinary equal-wire strand may also be reduced either by drawing the finished strand through a die, and thus compressing each wire against one another,* so as to fill up the interstices, and thus reduce the outside circumference of the entire strand ; or else by filling up the interstices by worming them with intermediate fine wires in the process of stranding, as already shewn.

Where the insulation of a submerged cable is imperfect at some point or other, the positive current, passing from the copper to the sea, decomposes the water and the salts in solution ; soluble chloride of copper is then formed, which is carried off by the water as fast as produced. A negative current sent through the conductor forms a deposit on the copper in contact with the water, of hydrogen, sodium, etc., which are, electrically, positive components of water ; any solid substance in contact with the conductor is reduced and free hydrogen given off, the bubbles of which enlarge the hole or fissure through which the gas escapes. In either case aggravation of the fault ensues. The rapid succession of signals through long submarine lines unfortunately involves the use of alternating currents, the combined effects of which tend to bring about the destruction of the cable all the more quickly.†

As a remedy for this, the late Mr C. F. Varley proposed placing a fine platinum wire in the copper strand to maintain continuity in the event of the copper conductor being entirely eaten away. This device has not, however, in any instance been turned to account.

**Formulæ and Data.**—The weight per naut of any round copper wire is about $\dfrac{d^2}{56}$ lbs. subject to its degree of purity ; and for a strand of the same $\dfrac{d^2}{73.3}$ lbs., where $d$ is its diameter in mils. ;‡ in other words, the weight of a naut of copper strand wire varies as its sectional area or bulk, § divided by the numerical expression 73.3.

---

* Though the compression of the wires together will naturally have the effect of hardening the copper, thereby seriously reducing its conductivity, this may be subsequently corrected by the entire strand going through the process of annealing—*i.e.*, softening by heat and then, in the case of copper, cooling very quickly.

† In practice, however, this has never been found to be a serious matter.

‡ 1 mil. = $\frac{1}{1000}$th of an inch.

§ This, however, can only be regarded as an approximate method of arriving at the weight of a wire, depending—as it does—so much on the density (or specific gravity) of the metal, as well as on the lay and power of " gauging " a strand.

‖ Where possible, however, the weight should be ascertained from actual test measurements rather than by calculation from the diameter, except as an extra check. This measurement may then form a suitable basis for subsequent calculations -amongst

The diameter of a copper wire weighing $w$ lbs. per naut is about

$$7.1 \ \sqrt{w} \ . \ . \ . \ \text{mils.,}$$

and for an ordinary, equal wire, strand,

$$8.1 \ \sqrt{w} \ . \ . \ . \ \text{mils.*}$$

The first of these varies (as in the weight formula) with the degree of purity; the second also additionally with the formation of the strand—*i.e.*, its lay†—and the amount of compound incorporated therein.

It may be again remarked in this connection that more reliable and accurate data will be secured by calculating the diameter from the weight—where the latter is accessible—than by arriving at the weight from the diameter, owing to the variation of the diameter at different parts and in different lengths of wire, and also on account of the fact that the weight of a wire can be more accurately measured than the diameter, as a rule.

**Leading Principles Involved in Design.**—The speed of signalling practically attainable through a cable depends inversely on the total resistance of the conductor, and also inversely on the total electro-static capacity of the cable. These, again, vary directly with the length. Thus, in the case of long and busy cables it becomes especially important that the above equal factors shall be reduced to the lowest possible figure.

The conductor resistance per unit length is governed, as has already been shewn :—

(1.) By the material of which the conductor is composed, and its degree of purity.

(2.) By its section or diameter.‡

In increasing the diameter to obtain a low conductor resistance, we at

---

others that of the diameter, which, as a rule, may be more accurately arrived at thuswise than by actual measurement with a gauge.

* This is equivalent, in fact, to the diameter being 12 per cent. greater in a strand than in a *solid* wire for the same weight of copper, with a proportionate increase in the inductive capacity, and, consequently, also a corresponding decrease in the signalling speed.

With the Siemens combination solid-strand conductor (as a mean between the two) we get a $5\frac{1}{2}$ per cent. decrease on the diameter and capacity, resulting in the same increase of speed—an advantage which should represent something material in a long cable, where the electrical constants become more and more serious items with every slight increase in length.

† The shorter the lay the greater becomes the area, and consequently also the greater the inductive capacity.

‡ When the diameter is doubled, the area $(d^2)$ is four times what it was. Thus the power of conduction is also quadrupled, its resistance being a fourth. In fact, the conductivity varies directly with the sectional area, and the resistance correspondingly varies inversely with the area.

the same time increase the capacity, unless the thickness of the dielectric be sufficiently increased also.*

For this reason, also, it is essential, to start with, that the reduction of the conductor resistance to a minimum be effected, as far as possible, specifically—*i.e.*, by the very purest of the highest conducting metal being employed.

However, apart from this rule (which is a *sine quâ non*, and means economy in the end), increasing the size of the conductor is, in practice, more beneficially effective as regards the conductor resistance than it is the reverse as regards capacity.

It is, indeed, more economical to attain a high working speed by a low conductor resistance than by a low capacity. In the former, the conductor alone is involved ; but in the latter, in order to obtain the same degree of effect, the thickness of the dielectric has also to be increased, and it must be remembered that, whereas copper costs about 7d. a lb., gutta-percha (the usual substance used for insulating submarine cables) costs somewhere between 3s. and 6s. a lb., according to the state of the market at the time.

Thus, the usual recognised policy is to adopt a dielectric of just sufficient thickness to ensure mechanical safety, according to the type of conductor used ; and the speed is got up to the required degree through a given length by the conductor being made of sufficient section accordingly.†

Quite recently, in answer to the ever-increasing demand for trans-Atlantic telegraphy, it has become necessary not only for the competing companies to lay extra cables to meet the contingency of break-downs, but also that these new cables shall be rendered capable of working at a higher speed—effected by mechanical transmission. To meet this, the core of the last two Atlantic cables (of 1894) are, in each case, of a much heavier type than anything hitherto adopted for submarine telegraph purposes.

In the case of the new "Commercial" cable core it is, in fact, composed of a type represented by 500 lbs. copper per N.M. to 320 lbs. gutta-percha ; and the core of the last "Anglo" cable is constituted by as much as 650 lbs. for the conductor to 400 lbs. for the gutta-percha insulating envelope.

Inasmuch as there is practically no limit to the speed afforded by the

---

* By doubling the diameter of the conductor, the capacity is doubled where the thickness of dielectric is built up to exactly what it was previously. Thus, mainly on the score of the heavy cost of the dielectric material, it is essential that the required low conductor resistance shall be secured as far as possible specifically—*i.e.*, by copper of the highest conductivity—rather than by a large area of wire, as well as for mechanical reasons.

† The only factor which places a limit on the extent to which this principle can be pushed is that of fault liability.

various forms of Wheatstone automatic transmitters—up to, say, 1,000 words per minute—the dimensions of the core (where warranted by the traffic) might be still further increased so as to give the cable a still higher earning capacity, but for the already enormous initial cost of a long cable, and the great increase of the same thereby entailed.*

**Initial Tests previous to Manufacture at Cable Factory.** — The copper wire † arrives at the cable factory in hanks or bundles of varying size, weighing from some 15 to 70 lbs, the length being from $\frac{1}{4}$ to $2\frac{1}{2}$ miles or more, according to the type of wire. With a view to having as few joints as possible, the wire is required in the longest lengths the wire-drawer can produce, with a proper regard to convenience of carriage.‡ The weight and gauge of the wire on each of these hanks is measured on arrival. A sample length is then usually cut off 10 per cent. of them, which is carefully tested for weight, diameter, and specific conductivity. At certain factories the practice is for the samples to be 30 feet in length ; at others 12 feet only is cut off. In some ways, within limits the greater the length the better.

These different measurements, besides acting as a check on the wire-drawer (to see that the wire supplied comes up to specification in each particular), also act as a check on each other. It is, in fact, essential that the conductor should be true to measurement in every respect. Thus, if the conductivity were up to specification, but the diameter of the wire were larger than what was specified, when made up into strand the electro-static capacity would be too high on this account ; moreover, with the weight specified for the dielectric, its thickness (to cover the increased conductor area) would be less, which would be the cause of a still further increase in the capacity, besides prejudicially affecting the mechanical properties of the core. Again, if the diameter were correct, but the weight were not,

---

* If, however, the dimensions of the core be further increased so as to yield a still higher working speed—or to obtain the similar speed on a greater length—it is certain that the weight of the cable will be so much augmented by the increased number of sheathing wires involved, that it will become necessary to modify the existing form of paying-out machinery so as to maintain sufficient control over the cable during paying-out operations. Moreover, designing a suitable type for laying—and still more for recovering —in deep water would, if this principle be pushed much further, become no easy matter.

† If for an india-rubber core, the surface of the copper wire requires to be efficiently tinned, to prevent any sulphur which may percolate through the inside covering of pure rubber acting on the copper, as explained elsewhere.

‡ With the comparatively small wires used for stranding in telegraphic conductors, the length in which they can be supplied by the wire-drawers is, as a rule, in no way restricted by the limit of length which a single bar will produce, but only by considerations of portability, and. more especially, by the weight admissible for each bobbin of the stranding machine, of a universal type for all submarine conductors.

this would in itself imply that the metal was not to specification as regards conductivity owing to a discrepancy in the quality—*i.e.*, purity—of the copper used. For this reason, it would very likely be also wanting in tenacity.* Thus we see that all of these points require specifying,† and that tests for each should be made—partly as a check on one another. The conductivity must, of course, be specified and measured, for it is on the basis of the purest copper available being used that, with a given sized conductor (limited with a view to a low electro-static capacity) and a given circuit length, the required speed is attained.‡

The diameter is measured with an ordinary wire gauge ; the weight of the sample, of a certain chosen specific length, by a grain scale ; and the conductivity by the metre bridge, an arrangement on the Wheatstone balance principle suited for testing the resistance of short lengths of low resistance. This apparatus, and the procedure of the test, will be found fully described elsewhere, amongst other electrical instruments and tests. The general principle is that of comparing the resistance with that of a sample (standard) of pure copper wire whose resistance is known.

The measurements of diameter can never be very reliable—and certainly not as a clue to the weight—owing to its variation in different parts. As a rule, it is more accurately calculated from the weight, the purity of the material being either constant or known.

**Joints in the Single Wire.**—After the above tests have been applied,§

---

* The weight would, in any case, require to be specified and measured, if only because all articles of commerce are paid for by weight.

† A margin of about 2½ per cent. is usually given each way for the weight. Low conductivity copper—which is cheaper, of course—usually weighs more than pure copper. This would involve an increased diameter for securing the required conducting value, or else a lower conducting value, as compared with that obtainable from high conductivity copper. The former defect would be objectionable both mechanically as well as electrically, and with a given weight of dielectric, the latter electrically so.

‡ It is very usually specified that the conductivity shall be within 95, 96, or even 97 per cent. of that afforded by Matthiessen's standard. Sometimes the term "purity" is used instead of conductivity ; but, as has already been shewn, though the purity might be nearly up to the mark, its conductivity may fall far short of what is wanted. In present practice there is never any difficulty in meeting the specification in the matter of conductivity ; indeed, usually the wire tests materially above the requirements, and, generally speaking, it actually *averages* almost as high as the standard itself.

§ The copper wire is not generally tested at all for its mechanical qualities, though very often—and always in the case of torpedo cables—it is specified that it should be tough, and able to bear many turns without breaking. This seems to be a distinctly useful stipulation, as the copper might happen to be excellent electrically, and yet more or less brittle. Occasionally a clause is inserted in the specification requiring a certain breaking strain for the core. In view of the high degree of elasticity of gutta-percha (and still more of india-rubber), this would be practically represented by the breaking strain of the copper conductor alone, as shewn by tests on it when made up into strand.

the hanks are arranged in order, ready for joining up previous to being applied to the stranding machine. The system adopted is to join up a specifically heavy coil to a light one, and so on, in order that the average weight may be as near as possible to specification at different parts of the conductor. The wire is then drawn off from the hanks on to bobbins afterwards to be applied to the stranding machine. These bobbins hold from 20 to 30 lbs. of copper wire—about $1\frac{1}{4}$ to 2 N.M. of No. 22 S.W.G., the wire which makes up a "seven-strand" of 107 lbs. per N.M. As soon as all the wire is run off from any one bobbin of the stranding machine, that bobbin requires to be replaced by a fresh one all ready full of wire, whose first end has to be jointed to the end of the wire from the previous bobbin.[*] This metallic joint involves some skill, and care has to be taken to see that it is properly effected. It partakes of the nature of a braze.[†] The ends are slightly "scarfed," and then joined together by very strong hard solder (brass,[‡] or now more usually silver [§]) in a state of fusion, no binding wire being applied, thereby avoiding increased bulk.

## STRAND MANUFACTURE.

**Stranding the Wires.**—The machinery for stranding wires together is very similar to an ordinary small vertical rope-making machine. It is shewn in plan and elevation by Figs. 2, 3, and 4.[||]

In laying up the wires in the form of a strand, the wire which is to constitute the centre, or heart, of the strand, unwinding from a bobbin A (Fig. 2) situated beneath the framework of the machine, passes under a guide pulley and upwards through a receptacle B containing a molten com-

---

[*] Matters are so arranged that the wire from each bobbin runs out at materially different periods. Thus the joints in each wire composing the strand occur at some distance from each other.

[†] These joints have been made by welds effected electrically in some of the factories of late years. This is the only method by which such very small wires could be *welded*.

[‡] Brass has been, practically speaking, entirely superseded for this purpose, largely on account of its low conductivity as compared with silver and lower fusibility.

[§] Silver solder is very fusible and non-corrosive. It is also much employed in fine work connected with laboratory experimental apparatus.

[||] These machines are sometimes of horizontal form—as in those for heavy electric light conductors—but set vertically they take up rather less space, and when the weight is not a serious item this form is preferable. Moreover, where a vertical machine is practicable, it is more easily attended to and more readily charged with fresh bobbins of wire.

[Plate XIV.

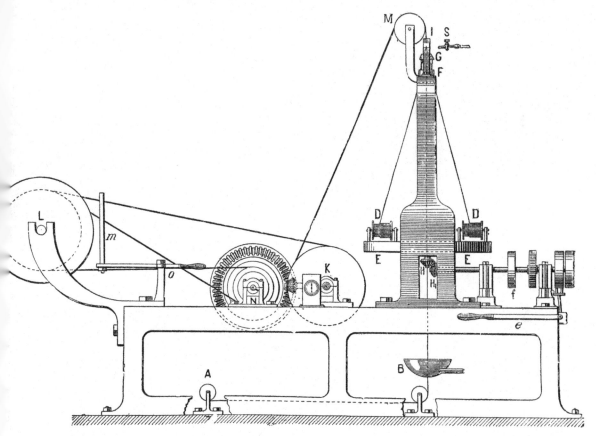

Fig. 2.—Stranding Machine (General Elevation).

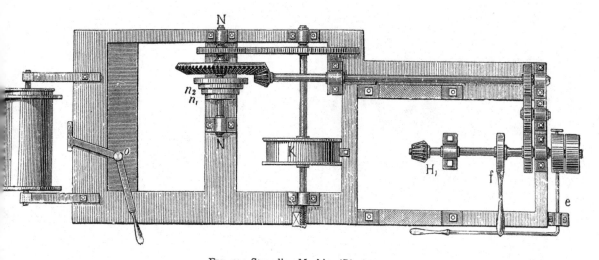

Fig. 3.—Stranding Machine (Plan).

[To face p. 236.

position commonly known as Chatterton's Compound,\* which is heated by steam. This compound is constituted by the following : †—

| | | |
|---|---|---|
| Stockholm Tar ‡ | - | 1 part by weight. |
| Resin § - - - | - | 1 ,,    ,, |
| Gutta-percha - | - | 3 ,,    ,, |

Receiving a coating‖ of the above,¶ the central wire passes vertically

---

\* This compound may be regarded as a sort of mastic. It is previously brought to a refined state (after the above proportions have been mixed and melted in a steam-heated vessel) by being forced through a wire gauze strainer, after the manner of the hydraulic press, described fully under the gutta-percha chapter.

The specific gravity of the above is about the same as ordinary gutta-percha. Its insulating capacity is, however, very much less, whilst its inductive capacity is somewhat greater. Moreover, it unfortunately tends to absorb water slightly, and for this reason amongst others some object to its use for this purpose.

† It was actually the invention of the late Mr Willoughby Smith—Specification No. 1,811 of 1858—but Mr John Chatterton, of the Gutta-percha Works, with whom Mr Smith was in collaboration, took out a patent a few months later for a method of applying it. Ten years previously Mr Charles Hancock had devised something similar for a like purpose.

‡ This is a vegetable tar from pitch pine, more especially growing near Stockholm. Unlike mineral (coal) tar it is not a solvent, and therefore does not tend to dissolve the gutta-percha. It was, in fact, especially selected on the above account. Moreover, it is thinner, finer, and clearer than coal-tar, though containing no naphtha or other spirit. Unfortunately, however, it slowly, very slowly, dissolves in sea-water where sufficient opportunity arises.

§ Resin (or rosin as it used to be originally spelt) is a kind of pitch or vegetable bitumen, being the residue of oil.

‖ This is of the thinnest possible character, more like a film or varnish than anything else, and may be taken at about 1 lb. per N.M. for a 107-lb. copper strand, composed of seven wires of No. 22.

¶ Besides avoiding the percolation of water between the interstices of the stranded wires, by filling them up, as already referred to, this application of compound round the centre wire and outside completed strand gets over the perhaps still more important difficulty of air spaces which, under the pressure of the ocean, are liable to lead to actual faults of insulation—either owing to original manufacture, or to subsequent joint-making at sea.

Moreover, the application of the compound here also acts as a suitable adhesive in such a way as to still further ensure a compact strand not readily separated even in cases where the lay of the wires is distinctly long.

The longer the lay, the larger the quantity of compound required at any spot to adhere the wires properly : on the other hand, with a short lay the number of turns being greater in a given length, an increased total weight of compound would tend to be involved. These two factors about counterbalance one another, the result being that a uniform weight of compound is usually employed in practice, independent of the lay, in a stranded conductor.

In 1867 Mr Matthew Gray took out a patent (No. 1,772 of that year) for coating each of the surrounding wires (as well as the central wire) with a compound of his own device, with a view to more thoroughly filling up the air spaces as above. This, however, is not adopted in practice—possibly on account of the ingredients suggested.

through the hollow axis shaft C (Fig. 4), and becomes encased by the outer wires on emerging through the die-plate G at the upper end of the shaft C.

The outer wires are wound on bobbins D (Figs. 2 and 4), supported on a horizontal turn-table E, which revolves with the axis shaft. From the bobbins they pass directly through the two dies F and G (at the upper end of the axis shaft), which have the form of flattened cylinders or prisms, pierced with as many holes as there are bobbins. The number of bobbins applied naturally depends on the number of outer wires. Various combinations are possible, but a seven-wire strand—*i.e.*, six round one —is found to make the most compact conductor. It is therefore the best electrically and mechanically, and, for ordinary strands, is universally adopted in submarine cable conductors.

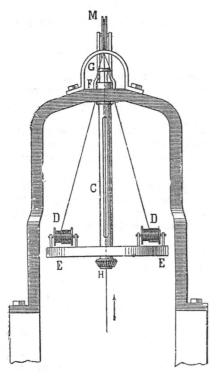

FIG. 4.—Stranding Machine : elevation of turn-table, frame, etc.

The holes in the dies curve slightly in a spiral direction to conform the wires to the required curve as they pass through. The shaft and attached turn-table are rotated by bevelled cogs at H, and the outer wires, which pass through both dies, are laid up round the central wire in more or less elongated spirals, the length of lay being governed by the rate at which the centre wire, or the strand as a whole, is drawn away. A gas jet S (Fig. 2) warms the wires at the point of junction and softens the compound again, so that the outer wires can mould themselves into it. The application of heat here also has the salutary effect of drying up any prevailing moisture.

The now finished strand passes through a hole I (Fig. 2) and over the pulley M to the drum K, round which it takes several turns, and is finally wound on to the reel, or metal carrying drum L, through an upright guide-fork *m* pivoted at *o*. As the flakes of strand are successively wound on, the rotary speed of the reel is diminished in the required proportion by gearing the driving belt to pulleys of gradually decreasing diameter, which are keyed to the shaft N. The machine can be quickly brought to rest by means of the brake $f$, after the driving-belt has been thrown out of gear by the clutch $e$.

Figs. 5 and 6 shew in detail the manner in which the bobbins D are fitted. A spring P R, one end of which is secured to the support P, presses against the cheek of the bobbin at R, and checks the rotation so as to keep the wires under slight tension.

The supports for the bobbins are movable in radial slots cut in the turn-table, being secured with a nut underneath, so that bobbins of different length can be used when necessary.

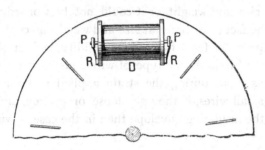

FIG. 5.—Stranding Machine : arrangement of bobbins on turn-table (plan).

One end of the shaft which carries the drum K is screw-threaded (Figs. 2 and 3), and gears into the first wheel of a counter which registers the length of strand constructed.

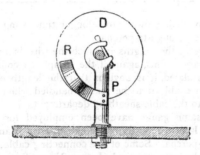

FIG. 6.—Stranding Machine : details of bobbin (elevation).

**Length of each piece**.—The stranded wire intended for submarine cable core is usually made in two or three mile lengths, according to the specific weight of the conductor and of the dielectric.* In very heavy conductors,

---

* In the first place, the length in which the copper is stranded is governed to a great extent by the length which can be covered in a continuous length with a single coating of the insulating material during a working day, thus avoiding any joint between new and hard gutta-percha. If the coatings are to be thick, or the area to be covered be great, the operation has to be conducted at a slower rate, and therefore a shorter length

however, such as that of the last " Anglo" Atlantic, it is made in lengths
of 1 N.M. only.

**Gauge of Wire adopted.**—The type of wire employed to form an
ordinary seven-wire strand in submarine conductors varies according to the
conditions from .028 inch (No. 22 S.W.G.),* as used to make up an ordinary
107 lbs. per N.M. strand,† up to something like .056 inch (No. 17 S.W.G.)
for the heaviest conductor (in an equal-wire strand) so far employed,‡ *i.e.*,
400 lbs. per N.M.,§ with an equal weight of gutta-percha.   There is a limit
each way, however, for practical mechanical reasons.   The wire must not
be over a certain size and weight, or it could not be worked into a strand
properly, besides the fact that the total size and weight as completed strand
must not be too great for fear of excessive rigidity, lack of pliability, and
tendency to buckle.   Similarly the type of wire must not be so small that
continual breakages occur during the strain applied in stranding up, and
also because very small wires, if they get loose or broken, are more liable
to pierce through the insulating envelope than in the case of wires of bigger
section.

**Speed of Laying up.**—The rate at which a stranded conductor can be
manufactured depends, of course, on the lay adopted, and to some extent
on the size and weight of the wire and machine.   Speaking generally, however, with an average lay, such a machine as we have described is capable

---

only can be covered in a day, mainly owing to a longer time being required for cooling
the greater surface, or thickness, of gutta-percha.

Another point which governs the lengths in which the wire is made up into strand is
the question of ready porterage.   This, again, if the completed conductor be of a heavy
type, and if it be heavily insulated, it is obvious that the length of each coil must be
limited accordingly, so as to enable it to be readily handled when coiled on a wooden
drum, and taken as desired to the cable-sheathing department.

* Wires of about the same gauge have been employed for a *three*-wire strand
conductor for a short cable laid in 1890 from Ireland to Tory Island, making up 47 lbs.
per N.M. with 59 lbs. gutta-percha.   Some of the connecting cables of the Commercial
Cable Company contain conductors composed of seven No. 23 S.W.G. (the next gauge to
that used in the ordinary 107-lb. seven-wire, strand), constituting a weight of 75 lbs. per
N.M., with 75 lbs. for the insulator.

† The weight of the stranded conductor as specified by the engineers is invariably
exclusive of any compound the contractors may apply to the central wire.   This usually
varies from 1 lb. to 4 lbs. for an even thickness, according to gauge of wire : consequently
the weight of a length of stranded conductor when made up exceeds the specified weight
for the wire alone by this amount.   1 lb. is about the weight for an ordinary 107-lb. strand.

‡ The central wire of the Siemens form of conductor used in the last (No. 11 S.W.G.)
"Anglo" cable was .122 inch in diameter, being larger than the single solid-wire conductor
of the first Dover-Calais line.   This was surrounded by 12 wires each .041 inch in diameter,
making up the largest submarine conductor so far made and laid.

§ Here the number of wires composing the strand is seven.

of turning out about 50 miles of medium gauge stranded conductor during a working day, and there are somewhere about a dozen such machines in any of the larger cable factories.

**Length of Lay.**—When a specification is drawn up, the weight of the conductor per unit length is invariably specified. This is partly for reasons already gone into, but it also acts as a check on the lay, otherwise an excessively long or short lay might be employed. It is usual to give a margin in this respect of $2\frac{1}{2}$ or 5 per cent., within which limits the average weight of the entire length must agree.* Thus, on the completion of each length of completed strand, the weight is measured before being sent (if correct) to the covering shop ready for receiving its insulating sheath.

**Formula for Lay.**—The percentage extra length necessary for each surrounding wire for enveloping a given wire with others of the same section with a given lay may be arrived at as follows :—

Here, let the base = length of lay.

perpendicular = circumference of centres of enveloping wires.

hypotenuse (Hp) = length of wire necessary to envelop centre wire.

Then $\text{Hp} = \sqrt{\text{base}^2 + \text{perpendicular}^2}$ = percentage extra length of wire required.

There are many conflicting points in detail connected with the question of the best length of lay for a conductor strand. Eventually this is almost entirely settled on purely mechanical considerations. That lay is fixed on which provides the required secureness for the completed strand and yet at the same time a given amount of pliability. The electrical points on each side are, however, of some interest. It is first of all evident that one item is the question of contact between wires composing the strand. It will be

---

* The above is not unusually expressed by saying that the total weight of conductor in the total length of the cable shall average not less than so much per N.M., or within, say, 5 per cent. This precaution is mainly taken for mechanical reasons as a check on dimensions and lay, the electrical qualities being tested in samples by the galvanometer. Moreover, if this were not done, the total length of conductor might be seriously at fault electrically, owing to the margin allowed on each coil—usually $2\frac{1}{2}$ per cent.—being taken full advantage of throughout. Similarly, as an extra check, it is frequently stated that the combined weights of the conductor and dielectric in each coil (or of the entire length) must be up to a certain percentage of the separate weights of each in every coil—or in the entire length—added together. This as an assurance that each do not fall short by the margin allowed.

obvious that some of the current tends to pass laterally from wire to wire so as to pervade the entire section, whilst some is conveyed only along each wire individually, and that the proportions of each depends upon the length of lay and the closeness of the wires. Theoretically, the best result should be obtained in this respect by the current all passing equally along the wires as though they were one single solid wire, thereby evading the extra resistance due to increased length by the lay of each surrounding wire. This condition is approached more nearly the fewer the interstices laterally in a given length of strand, and by their being as small as possible.

A long lay attains the first effect more closely, whilst the short lay produces the second, besides necessitating less compound round the central wire sometimes: and therefore, for this additional reason, very often yields a higher electrical efficiency. However, a short lay, by involving usually a greater area, tends to increase the electro-static capacity of the core. This increase in section (S) may be ascertained by the formula—

$$S = \sqrt{\text{length of lay}^2 + \text{diameter}^2} \ldots \text{mils.}$$

Again also, the shorter the lay the greater the quantity of wire required to make up a given length of stranded conductor.

**Principles which Govern the Lay.**—However, ultimately, as already stated, the lay is determined by mechanical considerations, and these are to some extent provided for in specifications by the stipulation that a given length of the stranded conductor is to have a certain weight, which condition can, with a given material, only be met by a certain lay—that which has been found to be the best suited for the particular type of wires.

**Usual Lay adopted in practice.**—As a matter of fact, this question of the most suitable lay is one that has been pre-determined for all classes of strand many years ago by those daily engaged in the operation of laying up wires. It varies from about $1\frac{3}{4}$ inch for an ordinary seven-wire strand of No. 22 S.W.G. (107 lbs. per N.M. copper) to about 3 inches for an equal strand of seven wires for building up a copper conductor represented by a weight of 500 lbs. per N.M.

Speaking generally, the length of lay is adjusted as required by the speed of drawing off the central wire: as that of the turn-table with revolving bobbins is a more or less constant quantity—being limited to some extent, according to circumstances, by questions of centrifugal force.

# CHAPTER II.

## THE INSULATING ENVELOPE.

Section 1.—Early Methods of Insulation in the First Underground Lines : Glass Tubes, Cotton, Pitch, Tar, and Resin : India-rubber and Gutta-percha.

Section 2.—Gutta-percha : Where and How Obtained : Historical Data regarding its Discovery, General Uses, and Electrical Application—Collection and Preparation : Classification : Importation.

Section 3.—Chemical, Physical, Mechanical, and Electrical Properties of Gutta-percha : Favourable and Unfavourable Conditions—General Electrical Formulæ—Effect of Pressure : Effect of Temperature : Ageing Effect—Diameter and Weight.

Section 4.—Method of Purifying Gutta—First Stages on Collection and on Arrival at Cable Factory : Mastication : Straining : Calendering—Rate of Work.

Section 5.—Willoughby Smith's Gutta-percha—Nature and Effect of the Process—Data and Formulæ.

Section 6.—Manufacture of Core : Covering Conductor with Gutta-percha—Preliminary Re-Mastication — Principle of Gutta-percha Covering Machine : Gray's Machine —General Operation of Covering Wires with Gutta-percha : Cooling and Hardening Process : Examination of the Insulated Wires—Relative Advantages of Single and Multiple Coats : Siemens' Multiple-Die Machine — Weight of Insulation : Outside Diameter—Rate of Covering — Mechanical Tests : Electrical Tests — Standard Values : Records—Specification Requirements and Limits—Siemens' Pressure Test —Localisation of Faults—Heavy Proportional Cost of Insulator.

Section 7.—India-rubber—Where and How Obtained—Method of Purification—Physical Characteristics—History of Electrical Application—Vulcanised Rubber—Historical Memoranda—Hooper's Core—Modern Practice of Vulcanised India-rubber Core Manufacture—Physical, Mechanical, and Electrical Data—Deterioration.

Section 8.—Relative Merits of Gutta-percha and India-rubber.

Section 9.—Other Suggested Insulating Materials — Particulars of Alternatives : Advantages Claimed — Gutta-percha and India-rubber Combined—Failures—Main Requirements.

------

## Section 1.—Early Methods of Insulation for Subterranean and Subaqueous Lines.

There has never been quite the same unanimity of opinion regarding the material to be used for purposes of insulating the conductor that there has been with reference to the conductor itself.

It may be fairly said that the first problem which had to be considered for underground telegraph systems when originally contemplated, was how the conductor should be covered to avoid dissipation of the current direct to earth.

Inasmuch as to some extent what applies to underground systems of insulation also applies to under-water systems, a short *résumé* of what had been done in the way of insulating underground lines previous to the commencement of submarine telegraphy may not be out of place.

In connection with the systems of the old telegraph companies of this country and of the Government telegraphs of other countries, the question of insulating a conductor so that it could rest on the ground arose soon after the establishment of electro-telegraphy generally.

An early attempt was made on a small scale by the Electric Telegraph Company, based on Sir Francis Ronald's plan (1816) of enclosing the conducting wire in glass tubes coated with pitch.* For obvious mechanical reasons, this method never came into anything like extended use for underground purposes, and was naturally never available for submarine lines, owing to the want of ductility of the glass, though possessing exceedingly high insulating qualities when dry.

In 1837 underground lines were laid down through the Blackwall Tunnel, on the London and Blackwall Railway, between Euston and Camden Town stations, followed by others laid down by the "Electric" Company in connection with Messrs Cooke and Wheatstone's patents. In these cases, the conductor was enveloped in cotton or silk, applied spirally, after being previously steeped in pitch, tar, and resin. Several such conductors were let into the side of a thick wedge of wood in wooden troughs—sometimes in wrought-iron pipes or lead tubing, where the line required to pass under water, across canals and rivers, or in a damp atmosphere.

Cotton itself, when perfectly dry, has perhaps a higher insulation resistance than anything, but unfortunately it is also highly hygroscopic. The resinous mixture (with which the cotton was previously saturated as well as after application) was intended to check this absorptive tendency, besides acting as a preservative — principally from damp — and further increasing the insulation. A line consisting of several conductors, insulated as above, let into a lead tubing taken through the Primrose Hill Tunnel, was in use for many years. However, most of these lines

---

* At another time (1842) Jacobi is said to have laid down an extensive system of underground lines in St Petersburg, insulated with india-rubber and run through glass tubes at intervals.

very soon failed, mainly owing to the absorption of moisture from the ground.*

Actual penetration of water occurred along the troughs, split tubes or pipes added to the deterioration of the resinous substance used (owing to the sun's heat and to other causes), thereby involving gradual decomposition generally, as well as a general fall in the insulation to a very low ebb. Again, great difficulty was experienced in the construction of these lines, as well as in laying them down. They were, therefore, soon replaced by an aerial system of bare wires,† after various other materials had been tried, including bamboo and ratan cane, sometimes steeped in pitch, which was found to answer the purpose very well for temporary— and usually military—purposes.‡

Covered wires were still pronounced indispensable for leading into stations, under bridges, and especially under tunnels, where the constant humidity of the atmosphere rendered an open-air system useless. It was in overcoming this difficulty in long tunnels that india-rubber was first successfully employed for insulation purposes instead of cotton and resinous compounds.

In the same year (1837) that Cooke and Wheatstone first realised the practicability of electro-telegraphy on *terra firma*, Professor Wheatstone (afterwards Sir Charles Wheatstone, F.R.S.) was actually contemplating the practicability of submarine telegraphy. It occurred to him that the same wires insulated for underground purposes might be let down under the sea to carry signals to neighbouring countries. This was not much doubted for short lengths or for quite shallow water, as subterranean lines had already been sometimes inadvertently taken across swamps, dykes, and indeed rivers; but whether or no any considerable length of such wire could even be successfully submerged, and whether it could stand the pressure, in any material depth of water, was considered very doubtful.

However, between 1837 and 1840, Wheatstone was engaged in devising

---

* Subsequent experience has, moreover, taught us to look upon tar and pitch as somewhat unreliable insulating materials (though they may do their work well at first under advantageous circumstances), and only as temporary preservatives. Moreover, they are frequently found to act injuriously on anything with which they are brought into immediate contact.

† Some of these aerial lines were again, later, replaced by gutta-percha insulated lines laid underground by Reid for the "Electric" Company.

‡ More or less detailed descriptions of these other methods may be found in the Report of the Committee (together with the evidence adduced) on the Construction of Submarine Cables, already alluded to; as well as in two eminently practical papers by Mr C. T. Fleetwood, on "Underground Telegraphs," read before the Society of Telegraph Engineers in 1875 and 1887 respectively.

a plan for suitably insulating a wire for submergence across the Channel His plan was for tarred rope to act as the insulating medium ;* in fact, the conducting wire was to form the core of a rope line well saturated with boiled tar, which was to be. lapped round the wire. Wheatstone also tried worsted and marine glue, and encased the whole in a lead tube.†

In India, Dr O'Shaughnessy (afterwards Sir William Brooke, F.R.S.), conducted a great many telegraphic experiments about this time, and as Superintendent-General of the Indian Government Telegraphs, put his ideas into practice. In 1839 he read a paper on "Subterranean Telegraphy" before the Royal Asiatic Society, and in the Proceedings of the Society these experiments were recounted, dealing with the means of carrying on telegraphic communication across rivers. In this paper Dr O'Shaughnessy says : "Insulation, according to my experiments, is best accomplished by enclosing the wire (previously 'pitched') in a split ratan, and then paying the ratan round with tarred yarn. Or the wire may, as in some experiments made by Colonel Pasley at Chatham, be surrounded by strands of tarred rope and by 'pitched' yarn. An insulated rope of this kind may be spread along a wet field, even led through a river, and will still conduct without any appreciable loss of the electrical signals."

These practical experiments, and those previously of Colonel Pasley, R.E., are both further described and fully illustrated in Part I. of this book. Sufficient has been said here, however, for it to be seen that Colonel Pasley and Dr O'Shaughnessy were respectively the first and second to definitely experimentalise in subaqueous telegraphy, and to actually lay short lengths of insulated wires across swamps, rivers, etc., in a crude and temporary form, previous to the introduction of india-rubber and gutta-percha for the purpose.

In 1846 the Electric Telegraph Company, under the auspices of Mr C. F. Varley, F.R.S., attempted to lay a submarine wire between Gosport and Portsmouth of over one mile of copper, covered thickly with cotton, and further insulated from wet with a mixture of pitch and resin. The failure of this experiment may, however, be said to be due in the first instance to

---

* A full description, with illustrations, of Wheatstone's project has already been given in Part I.

† Again in 1845 (only two years after its introduction into this country) Wheatstone appears to have experimented with and determined on gutta-percha as his insulating medium in place of the above. If this be so, he discovered the electric insulating qualities of gutta-percha two years previously to Faraday and Siemens pointing it out. These inventive schemes of Wheatstone were not, however, known of publicly till after his death ; or, at any rate, till after the first submarine cable had already been laid.

the specific gravity of the completed wire being insufficient to effect its sinkage.

It was not till india-rubber and gutta-percha were introduced into this country, and their excellent insulating and mechanical qualities were pointed out, that any complete system of underground telegraphy was embarked on —or, indeed, meditated. Even then, though it first made its appearance and was applied for insulating purposes a year or so sooner, but little was done with india-rubber prior to the introduction of gutta-percha—or, in point of fact, at any time, till Hooper successfully effected its vulcanisation for cable purposes. On these counts, and owing to its more general use for submarine cable insulation, gutta-percha will be dealt with first.

The Eastern Telegraph Company's Station at Platris (Mount Troödos), Cyprus, 3,200 feet above sea-level, connecting up the Commissariat and Ordnance Depots with the whole of the telegraphic world.

## Section 2.—Gutta-percha: How and Where Obtained.

Gutta-percha* is the natural product as a gum or milky juice, which oozes from certain sapotaceous, wild-growing trees, when an incision is made in the bark.

**Introduction to this Country.**—The first specimens of gutta brought to Europe from the Malay Peninsula and Malaysia, where it grows, were presented to the Royal Asiatic Society by José d'Almeida (a Portuguese engineer) in 1843. They included horse whips, knives, hats, basins, piping, and other forms as made up by the natives, besides a few pieces of the substance in its natural state. A few months later, Dr W. Montgomerie, a surgeon in the service of the East India Company, had also remarked on the peculiar character of the tool handles used by the natives of the Sunda Islands, and had been much struck with the ready manner in which they were fashioned, the substance softening when plunged into hot water, becoming sufficiently plastic to mould into any shape, and hardening again when cool. He (Dr Montgomerie) brought back a few specimens, which were exhibited at the Society of Arts towards the close of the same year. This was the first that was known of the substance in England, and Montgomerie pointed out its probable utility for various commercial purposes, and especially for surgical splints. It was immediately imported, and turned to practical account in numerous ways by Messrs Keene and Nickels, by the Brothers Walter and Charles Hancock, and later by the Gutta-percha Company on taking over the business of the first-named firm.

The Hancocks were india-rubber manufacturers of stoppers, corks, &c., and in dealing with gutta-percha they at first adopted the purifying processes applicable to rubber.

**Electrical Application.**—It was not, however till 1848 that the electrical (insulating) properties of gutta-percha were first publicly pointed to in this country by Professor Faraday.†

Many chemists and engineers then experimented on its insulating qualities, and immediately turned their attention to making use of it for insulating wires. The problem, in fact, soon became how to lay it on a

---

* The name borne by this elastic gum has its origin in the Malay word *guetta* (a gum) and *percha* (cloth).

*N.B.*—*Percha* has also been said to be merely the name of the particular tree.

† According to Dr Werner Siemens, he suggested the use of gutta-percha as an insulating medium to the Prussian Government two years previously to Faraday.

wire on an extensive scale in a thoroughly permanent, workmanlike, fashion, in such a way that it could be relied on to stick to the wire. It was thought that whoever attained this with either gutta-percha (which received the most attention) or with india-rubber, would make large quantities in answer to a probably extensive demand.

The first attempts, however, to insulate wires with gutta-percha were not, by any means, attended with immediate success either in this country or in the German Empire. This was, no doubt, principally owing to the defective form of joints adopted—that is to say, the gutta-percha was rolled round the wires in spiral, or longitudinal, strips, thus necessitating a continuous joint between each strip, which probably had insufficient adhesion, and would, after a short time, give out.

The first patent in this country in which gutta-percha was proposed to be used for insulating purposes was that taken out in 1848 by Messrs Barlow and Foster, in which several wires insulated from one another were drawn between two sheets of a compound of gutta-percha, New Zealand gum, and sulphur, which were pressed together between two rollers.

The sheets were, however, never properly united; moreover, the gutta-percha was not sufficiently purified of the wood contained in the raw material as it came from the tree.* Mr Charles Hancock, in the same year, patented a gutta-percha mixture for covering wires, consisting of gutta-percha mixed with muriate of lime, passed between heated cylinders and sprinkled over with rosin. This patent also covered a mixture of gutta-percha, shellac, and borax as an insulating compound. Neither of these branches of the above patent were, however, turned to any practical account.

Messrs Siemens and Halske certainly appear to have been the first to use gutta-percha for insulating conducting wires on any extensive scale, and to have eventually brought the practice to a successful issue.†

**First Die-covering Machine.**—According to the late Dr Werner Siemens, he, in 1847, first devised and introduced a machine for covering

---

* Barlow and Forster's patent was followed in the same year by one somewhat similar taken out in the name of John Lewis Ricardo, as chairman of the Electric Telegraph Company. This was adopted to some extent for subterranean lines. Moreover, a submarine line, insulated according to this system, was laid in Portsmouth Harbour by the "Electric" Company: this did not last long, however, owing no doubt to the continuous joint between the seams.

† One of the principal features discovered in connection with gutta-percha was that it is capable of being laid on a wire to a required thickness; and, when heated, that it becomes plastic. It can then be easily worked into any shape, and will remain firmly in that shape when cool.

wire with gutta-percha (rendered plastic by heat), pressed round the wire through a cylinder and die without a seam, thus obtaining a homogeneous covering, thereby avoiding the continuous joint existing throughout the length of previously made core composed of gutta-percha strips. Siemens described it as being similar to a macaroni machine, and it (together with Bewley's tube-making, or lead-pipe drawing, apparatus of 1845*) was undoubtedly the germ of the gutta-percha covering machines as we now have them in a perfected form.

Owing to Siemens being at the time a Government officer (lieutenant in the Prussian Artillery), no patent was taken out for this machine in England till 1850, when also a slight improvement was effected.

**First Gutta-percha Underground Telegraph Lines.** — In 1847 Werner Siemens laid the first gutta-percha covered underground telegraph line between Berlin and Grossberen. This was immediately after Mr C. W. Siemens (subsequently Sir William Siemens) had seen Dr Montgomerie's specimen of gutta-percha at the Society of Arts, and had suggested to his brother its possible utility for purposes of insulation in place of india-rubber, with which the line was to have been covered. Everything tends to show that Messrs Siemens are at any rate fully justified in claiming priority in covering wires with gutta-percha in such a manner that no seam was involved, with a machine specially devised by them for the purpose.

In 1848 and 1849 Messrs Siemens and Halske obtained contracts from the Prussian Government for making and burying many hundred miles of insulated conducting wires throughout the kingdom of Prussia, thus demonstrating for the first time in that country the practicability of sub-terranean telegraphy on an extensive scale. The lines did not, however, last long. This was, in some cases, owing to the admixture of sulphur with the gutta-percha insulating envelope—the sulphuretted gutta-percha being supplied by the Gutta-percha Company—and also to the lack of experience and perfection in the joints and general manufacture.†

---

* Henry Bewley was originally a lead-pipe drawer, but joined the Gutta-percha Company, for whom he applied the above machine to making gutta-percha tubes, bottles, etc. This was not, however, used for covering conductors with gutta-percha till 1847, and again in 1849 over the first Channel submarine line, laid the following year. Previously this company had, for over a year, applied the gutta-percha to connecting wires in strip form only. It was more than ever necessary that a single, complete, seamless covering should be employed for any line deposited at a certain depth under water.

† These gutta-percha covered lines were protected by a leaden pipe, and were laid under the curbstone and gutters. The lead was gradually eaten away, and the gutta-percha desiccated. Ultimately, in 1853 and 1854, these lines were replaced by bare overhead wires, on earthenware insulators, attached to poles.

**First Subaqueous Gutta-percha Insulated Line.**—In 1848 Messrs Siemens laid under water one of these seamless gutta-percha covered wires in the Port of Kiel, forming a part of a submarine mine system—sometimes less aptly termed a "torpedo system." Another short cable (quarter mile) in which gutta-percha was employed for insulation, is said to have been laid by the above firm the following year across the Rhine from Dentz to Cologne.

About the same time the Brothers Hancock and the Gutta-percha Company were doing their best in this country to design some machine that would lay the material on the wire in a more homogeneous manner than that of strips or spirals, as hitherto.

No pronounced or lasting success in this direction appears to have been recorded here till some time after Siemens' German lines had been laid; though a considerable quantity of wire was covered for railway tunnels, and later, for the lines laid in iron pipes under the streets of towns.*

**Experiment in the English Channel.**—In 1849 Mr C. V. Walker (electrician to the South-Eastern Railway Company) laid an experimental telegraph line in the English Channel. Full particulars of this venture have already been given in Part I. It suffices here to say that it consisted of a two-mile length of a single No. 16 B.W.G. copper wire coated with gutta-percha. The experiment was quite successful, and proved, at any rate, the possibility of signalling through a gutta-percha covered copper wire under a certain depth of water.

**First Anglo-French Line.**—In 1849-50 the Gutta-percha Company manufactured the first practical Channel line,† consisting of a copper conductor thickly covered with gutta-percha by means of a cylinder and

---

* After the gutta-percha underground lines laid by Werner Siemens (for Siemens and Halske) on behalf of the Prussian Government, the next complete system of subterranean telegraphs was that of the "Magnetic" Company laid down in 1851 between Manchester and Liverpool along the Manchester and Yorkshire Railway, under the supervision of their engineer, Mr (afterwards Sir C.) Bright. These lines were laid in the six-foot way between the rails, where they would be least liable to be disturbed. The following year similar communication was established by the same company, under the direction of Mr Charles Bright, between Manchester and London; but in this case the lines were laid along the public roadway. As a part of the "Magnetic" system, other subterranean telegraphs were laid to Glasgow and also through a great part of Ireland.

In 1853 the Electric Company laid their first long length of underground line, between London and Manchester, along the London and North-Western Railway.

All the above underground conductors were insulated with gutta-percha, as in Siemens' lines. They were made at Wharf Road by the then newly-formed Gutta-percha Company, and worked well for a number of years.

† See Part I.

die machine, thus avoiding, for the first time in this country, any continuous seam in the gutta-percha covering. The failure of this cable, the
day after submergence, was in no way due to any fault in the construction
of the gutta-percha insulating covering of the main (deep-water) part of the
cable. However, as regards that portion of the line intended for the first
half-mile, the copper wire was covered with cotton saturated in a solution
of india-rubber, and enclosed in a leaden tube. This form of insulating
covering at the bottom of the sea—even at that small depth—would be
quite sufficient to account for the short life of the line, apart from other
misfortunes.

**Dover-Calais Cable, 1850.**—The following year, the first actual complete submarine cable, as at present constructed, was laid between Dover
and Calais.* In the construction of this, at the works of the Gutta-
percha Company, the six conductors were each covered with two coats of
gutta-percha up to a ¼-inch diameter, under the supervision of Mr Samuel
Statham.†

Hitherto the custom had been to apply the gutta-percha envelope in a
single coating; but it was thought that a greater chance of detecting any
mechanical "flaws" or incipient electrical faults would result from applying
the required total thickness in separate coats. This plan is now very
generally adopted; indeed, three, or even four, coats (according to the
required total thickness) is the usual thing.

The general character of the die-covering machine, adopted by the Gutta-
percha Company for the construction of these lines, was not unlike that of
Siemens.

### Collection and Preparation.

We have now traced the history of gutta-percha, from an electrical point
of view, up to the period at which submarine cables were first constructed—
as regards the method of insulation—on the same general principles as at
the present day; moreover, we have reached the stage at which the gutta-
percha envelope was applied in the same general way as it is now. It is
not, therefore, necessary to pursue the historic details of each subsequent
cable core, and we will now proceed to consider the collection and manufacture of gutta-percha. Before doing so, however, it may be remarked

---

* See Part I.

† This gentleman may be said to have led the way in this country in applying himself
to the problem of the manufacture of gutta-percha for submarine cables. In connection
with the manufacture of this cable, Mr Statham enormously improved the machinery for
the manipulation of the gutta-percha.

that for some ten years after the introduction of india-rubber and gutta-percha into this country, and subsequent to the starting of submarine cables as articles of commerce, various patents were taken out for mixing other vegetable substances with the pure masticated india-rubber and gutta-percha, with a view to beneficially effecting or preserving the material, and also from a point of economy ; * but certainly as regards gutta-percha, they were, without exception, a distinct failure. As a matter of fact, the purer the gutta-percha, the better article it is physically and mechanically—moreover, the higher and—what is more to the point—the more reliable the insulation resistance, besides the electro-static capacity being lower very often.

**Different Species.**—There are several different species of gutta as a resultant gum from different percha (or gutta trees) met with in different Malay Islands and Peninsula. These various trees may, in fact, be met with between latitudes 4° N. and 3° S. of the Equator, and between 100° and 120° E. longitude.

**Isonandra Gutta.**—The principal tree from which gutta is at the present time obtained—at any rate, for insulating purposes—is known as *Isonandra gutta*,† and is illustrated by Fig. 7 (opposite next page).

It yields large quantities of the milky juice or sap which, when indurated, forms the crude gutta-percha of commerce. The *Isonandra gutta* grows principally in a rocky subsoil at the foot of hills and jungles. The wood is soft, fibrous and spongy, pale in colour, and traversed by longitudinal receptacles or reservoirs filled with the gum, forming ebony black lines.

---

* It should be stated, however, that at the time of the first Atlantic cable (1857) gutta-percha cost less than half what the corresponding material does now, owing to its greater scarcity.

It is to be regretted that no cheaper insulator than gutta-percha can be found to suit the purpose sufficiently well. The cost of gutta-percha in its raw state varies on the average as much as from 4s. 6d. to 7s. a lb. It, indeed, makes up a large proportion of the total cost of material in an electric cable—quite apart from the expenses in preparation and manufacture, which may be taken roughly to increase the cost by about half as much again. So great is the variation of gutta-percha that as raw material it has been known to fetch 10s. a lb.—and even materially more on special occasions—whilst at other times it is only worth 3s. 6d. Many years ago it was once as low as 9d. a lb.

On the other hand—though, of course, liable to be disturbed by economic and social considerations—the cost of manufacture is, on the whole, a *comparatively* constant quantity. The difference between the price of raw gutta and the manufactured article depends upon the loss in cleaning, *plus* labour: the loss in cleaning varies with the quality from anything between 20 to 80 per cent. One penny per lb. may be taken as representing the cost of washing on the raw weight.

† An Englishman named Burbridge, travelling in the Island of Singapore, was the first to discover this particular gutta-percha tree in 1848. He sent branches of it, with leaves and flowers, to Sir William Hooker, the famous botanist, who classified the plant under this name.

It is a hard tree to grow, and, being delicate, requires very careful attention, more especially now that it is becoming so scarce.

Being a slow-growing tree, it is sometimes twenty-five years before it yields any gum whatever.*　However, in order to expedite its growth, it is usually taken from the jungle up to the hills.　Gutta-percha was first of all found from this tree, in Singapore, but the resources of the island soon became exhausted, and gutta-producing trees had to be sought for in its neighbours, the Malay and Sunda Archipelago.　Thus, at Sumatra, and in the southern part of Borneo, as well as in the Malay Peninsula, trees were soon found, which, when " tapped," exuded a milky sap very similar to the Singapore gutta.　Owing to the great variety of species, to deceptive similarity of appearance, and to diversity of nomenclature among the natives, guttas of very variable quality, from the electrical point of view, appeared in the markets under the same designation.　On the other hand, trees belonging to the same species were known in the various localities, and sometimes even in the same forest, by different names.　Again, different qualities of gutta were frequently mixed together by the natives, so causing still greater confusion; and it was not uncommon to come across two guttas of the same name, and from the same locality, widely different in quality.　Neither has there ever been a precise botanical classification, owing to the difficulty of obtaining complete portions of the plant; and until quite recently, no naturalist in Europe possessed a perfect botanical specimen of a gutta-percha bearing tree, or, as it is sometimes termed, *guttifer*.　The trees flower but once a year at the oftenest, and do not bear fruit and flowers at the same time.　The fruit is very small, and extremely difficult to perceive from below, as the trees grow to a considerable height.　About thirty years are required for a gutta tree of this species to arrive at full growth, and not until then does it flower or bear fruit.†　The felling of the trees has been carried out in such a ruthless manner that an adult specimen is now rarely seen, and in Sumatra, the gutta-seekers themselves cannot recognise either the flowers or the seeds by sight.

**Classification.**—Thanks to the labours of Dr Beauvisage and of Dr W. Burck, as well as to the results of an exploring expedition undertaken by the latter to the table-lands of Padang, in Sumatra, we are now enabled to distinguish with some sense of accuracy ten or twelve species of trees, several of which yield gutta of excellent quality.　Arranging the several

---

* It only yields seed after thirty years.
† The height to which gutta-percha trees grow varies from 8 feet to 70 feet, according to the species.

[PLATE XV.

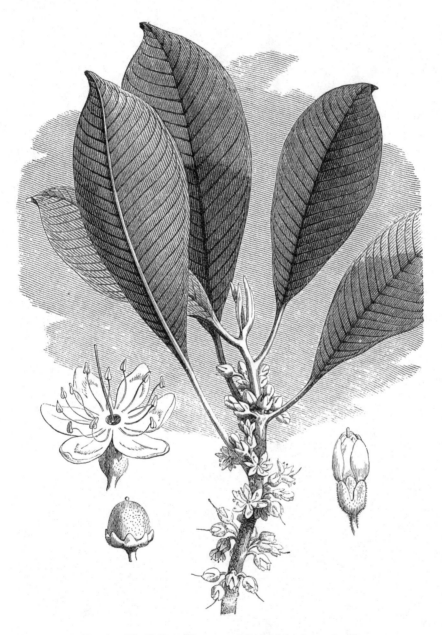

FIG. 7.—The Foliage, Flower, and Fruit of the *Isonandra Gutta*.

[*To face p.* 254.

kinds in order of merit, first comes the *Palaquium oblongifolium*, known to the natives under the following titles :—

Njatoeh balam tembaga,

Njatoeh balam sirah,

Njatoeh balam merah,

Njatoeh balam doerian, etc.

The words "njatoeh balam," or simply "njatoeh," usually imply a tree which exudes a milky sap, without reference to any particular kind; and similarly the solidified sap is known as "getah balam," or merely as "getah." The "njatoeh balam tembaga" is a native of Sumatra, Borneo, and Malacca, growing on the slopes where the soil of the damp virgin forest is rich in humus, the vegetable or organic constituent of the soil. It attains a height of 65 to 70 feet, being known by its leaves of regular oblong shape terminating in long sharp points, the largest of which measure about 8 inches in length and 2 inches across.

This tree gives gutta of the very finest quality, which is known commercially as "gutta merah," "gutta taban," "gutta taban merah," and "gutta doerian." The sap, like that of all other guttas, is milky white when it exudes from the tree, and afterwards turns brown. This is due to particles of bark and fibre, which stain the gutta with colouring matter during the process of cleansing.* "Gutta taban" or "gutta doerian," when thoroughly purified, is very elastic, bending easily without breaking: when plunged into hot water it softens, takes any shape given to it without becoming sticky, and hardens again perfectly when cold.

Second in order is the *Payena leerii* or "njatoeh balam bringin," also called "njatoeh balam soendai," "njatoeh balam bipit," "koelan," or "balam tandoek." This is found in Sumatra, Banca, Malacca, and sometimes round about Rio. The leaves are alternately oval and pointed at the base. The flowers are white, but there is a substance in the tissues which blackens when touched with an alkali. It grows to a height of 60 feet, arriving at maturity in less time than the last species described. The gutta, known as "gutta balam bringin," or "gueutta seundek," or "balam pipit," or "koelan," is the second best in quality. The milky juice is extremely liquescent, and can be collected free from bark; consequently this gutta, when purified and solid, is more white than the "gutta taban." It is close-grained, plastic when softened, and becomes as solid as before when cold. It is, however, less homogeneous than the "taban," easily becoming fibrous and thready.

Thirdly, the *Palaquium gutta*, once called *Isonandra gutta Hookerii*,

---

* There is, however, a grade of gutta-percha which is almost white in the hard, raw, state. This is owing to the bark of this species discolouring the gum so much less.

which came from Singapore, and formerly ranked as the gutta *par excellence*, is now no longer to be found; ruthless destruction by the natives for many years having, some time since, caused its entire disappearance from the forests of the island.  However, about thirty years ago, two specimens from Singapore were planted in the Botanical Gardens at Buitenzorg (Java), where they have recently flowered and yielded seed.

Fourthly, the *Palaquium borneense*, from Pontianak, the products of which closely resemble those of the *Palaquium oblongifolium* and *Palaquium gutta*.

Fifthly, the *Palaquium treubii* from Banca, which has much in common with the *Payena leerii*.

Of the remaining guttifers, those which are tolerably well known and classified, such as the "njatoeh balam tembaga de soepayang" and the "njatoeh balam doerian de soepayang," produce gutta of only inferior quality, sticky when warm, and soft enough when cold to be indented with a finger nail.

The natives of Borneo (where *Isonandra gutta* is now mainly met with) mingle the sap from some of the neighbouring trees with that obtained from this tree, either as a supplement or as an adulterant.  Sometimes saps from as many as five different species are thus mixed together to make up the ordinary commercial "gutta taban"* as we get it for most purposes, amongst others, for insulating electrical conductors.†

**Seat of Growth, and Appearance.**—Gutta trees are nowadays only found in the densest parts of the virgin forests, where the gutta-seekers discover them with marvellous skill.  They are guided, as to species, by the colour of the trunk, the thickness of the bark, or by the hardness of the wood.  Occasionally they make a cut in the bark and test the quality of the sap which trickles out by working it between their fingers.  When the verdict is satisfactory, the tree is felled with an axe, the top also being sometimes cut off to prevent the sap diffusing through the leaves and branches.

**Collection.**—The sap is collected in various ways in the different parts

---

* "Gutta taban" is, however, the term often applied exclusively to the product obtained from *Dichopsis (Isonandra) gutta*.

† The "gutta baton" described by Mr Seligmann-Lui, after his voyage to the East Indian Archipelago in 1882, appears to be obtained from the *Palaquium calophyllum Pierie* (Borneo), a variety of the *Palaquium oblongifolium* species of tree.  In quality this kind is second only to the "gutta taban," but it is of a lighter and redder colour, somewhat stiffer and of coarser tissue.

[PLATE XVI.

FIG. 8.—Collection of the Juice of the *Isonandra Gutta*.

[*To face p.* 256.

where it is to be found. Fig. 8 gives an idea of the collection of the gutta as it should be performed, with the tree undisturbed from its root. The usual plan is, however, to cut the tree down, and after laying it full length on the ground, to cut rings right round the bark along the entire length, at distances of a foot or so.* The sap thus readily exudes, and is collected before it sets hard in calabashes, gourds, or cocoa-nut shells, which are sometimes fastened to the trunk immediately under each wound; or else banana leaves are laid on the ground to receive the milky exudation as it trickles down.† At each visit of the gutta-seeker to a tree which has been felled, the sap (should it not have set naturally), is stirred with a stick, and kneaded up by hand into a small solid lump.

**Cleansing.**—With a view to roughly cleansing the gutta, all the sap collected during the day—whether from trees of the same species or not— is thrown into convenient pans filled with boiling water. Subsequently, it is sometimes actually boiled in steam, in order to get rid of as much water as may be at the earliest stage possible, for water naturally tends to lower its insulating qualities. About 20 per cent. of scraped bark (which is found to have an improving effect) is, at the same time, very usually added, as well as cocoa-nut oil and lime juice—the latter for the purpose of coagulating the gutta-percha in the course of exposure to the sun and air— immediately on ebullition.‡ The softened gutta is then spread out on a hardwood plank, well worked up to ensure thorough mixture of the different qualities, and reduced to as thin a sheet as may be: at this stage the woody particles scattered over the surface are removed either by hand, or by cold-water washing. Gutta which has been twice cleansed in this way acquires

---

* There is often no means available of turning the tree over, if it be a big one; and when this is the case all the lower half next the ground remains inaccessible, and its sap is therefore wasted.

† So careless are the natives in some of the gutta-producing districts that they often do not trouble to place any receiver in readiness until the work of bark-scoring is finished, all the sap which may trickle out in the meantime being lost. The exudation generally continues for several days, during which time the gutta juice is liable to become contaminated by earth, dry leaves, insects, and litter of every kind falling into it.

‡ The process of coagulation may be described as that of transforming the gutta-percha (or india-rubber) from the gum state into that of a clot as follows:—A certain quantity of lime is placed in the receptacle in which the gutta-percha is collected, which tends to solidify the milky juice more quickly by its oxidising heat than the air itself could possibly do. By this means all the solid matter is pretty well separated from most of the watery matter, and is then disengaged from it. Alum is also sometimes employed with the same object, though acting in a different way.

It is most important that as much of the water in the sap as possible should be got rid of at this early stage in order to decrease its weight for transport, as well as for the reason that gutta-percha with much water in it oxidises very rapidly, and in addition to the fact of its comparatively low electrical insulation.

an increased market value, becoming what is called "Gutta No. 1." Whilst still plastic it is cut into pieces of different size and shape, and so finds its way to the market.

**Importation.**—The gutta-percha sap, thus mixed and coagulated into a solid cake, when dry, of about 1 foot by 4 inches, arrives in this country in forms varying roughly between that of a ball and a large block, and weighing about 30 lbs. At this period of its existence, it frequently contains as much as 25 or 30 per cent. of impurities.

**Superiority of Borneo Gutta.**—The best gutta-percha now comes from Borneo, as a rule. Borneo gutta-percha is superior to others, mainly because its collection and coagulation is carried out on more systematic principles than elsewhere; also, for the reason that it is more carefully freed, and less tampered with by the negroes, thus not requiring so much careful washing on arrival. Consequently, as a rule, its cost is higher than that of gutta from elsewhere.

**Average Produce.**—The quantity of gutta which may be obtained depends on the nature and age of the tree. A fully grown specimen, twenty-five to thirty years old, and $3\frac{1}{2}$ to 4 feet in girth at the base, may be expected to yield one "katti"[*] of good clean gutta; and from a tree of the same species, two feet in girth a short distance from the ground, two-fifths of a "katti" may be obtained. About 21,000 peculs[†] ( = 1,248 tons) of gutta having been exported from Borneo in 1881, Dr W. Burck estimates that upwards of five million trees must have been felled on this island. Bearing in mind the destruction caused by felling trees indiscriminately in forests where the vegetation is very dense, the total number of guttifers annually destroyed in Borneo cannot be less than twenty-six million. In the year 1881, more than 60,000 peculs (about 3,600 tons) were exported from Singapore, which is the principal depôt for gutta; we can therefore easily appreciate how serious have been the effects of a few years' reckless felling in the very narrow belt of forest lands where the gutta trees thrive.

**Serious Scarcity of Gutta.**—The gradual scarcity, and threatened extinction of the guttifers, has for some years been the cause of much anxiety. The Dutch Government, desirous of preserving for its colonies this valuable source of revenue, has tried planting gutta trees of the best

---

[*] This Malay measure = $1\frac{1}{3}$ lb.

[†] A pecul, or picul (Malay measure) = $133\frac{1}{3}$ lbs.

species in the Botanical Gardens at Buitenzorg, to induce the colonists to undertake their general cultivation. Important results have already been obtained, 4,000 plants being under cultivation in the open fields as early as 1886. Besides this, experiments have been made with a view to ascertain what quantity of gutta can be obtained by making yearly incisions in standing trees, as compared with the method of felling hitherto in vogue. The results go to prove that the yield may be doubled by the new system, if it is properly carried out.

The cultivation of guttifers has also been, within the last few years, introduced into French Cochin China, under the direction of the French Colonial Office.

It is to be sincerely hoped that these efforts will be attended with success, and that electrical industries may not suffer from scarcity of this almost indispensable substance.

**Other similar Gums.**—Some years ago, Mr Edward Heckel discovered a milky sap which exudes from the *Butyrospermum Parkii kotschy* when incisions are made in its trunk and branches, and this substance, though not chemically identical with gutta-percha, appears to possess most of its physical qualities. This tree is very common in the forests bordering the Niger, and the equatorial portions of the Nile, and may replace, to a certain extent, the *Palaquium* and the *Dichopsis* (*Isonandra*) species. Some attempts have already been made to make use of *Bassia Parkii* gutta for industrial purposes. The nature of this substance must, however, be better understood, and the difficulty and cost of collecting it must be less, before it can be used in the manufacture of submarine cables.

**Balata.**—Again, in 1887, in the course of a Colonial and Indian Exhibition Report,* Mr Thomas Bolas drew special attention to the merits of a species of gutta coming from British Guiana. This frequently goes by the name of Balata, and was thought by some to be a "cross" between gutta-percha and india-rubber,† but actually it is a form of gutta coming from the Bullet (or "Bully") tree—hence the name Balata—of that country. Mr Bolas was of opinion that the method of collection of bullet-tree gutta in British Guiana was actually superior to that elsewhere, besides the gum itself being of rich quality. In view of the reduced supply of gutta-percha, this

---

* Report on Gums, Resins, and Analogous Substances, by Thomas Bolas, F.C.S.

† This may have been due to the different milks being mixed—always an undesirable process in this instance.

gentleman has recommended the careful preservation and transport from all parts of the world of large samples of any species of the gum whilst yet in the milky state (immediately after collection) for examination and experiment. He considers that in this way a fresh source of supply might be discovered.

**Collection from Leaves.**—M. Sérullas, who for many years has made a close study of gutta-percha, suggested a short time ago the more careful and systematic cultivation of gutta-percha, in such a manner as to render it possible to extract the juice from the leaves and twigs of the tree on a large scale, thus obviating the necessity of disturbing the trunk and of endangering its further growth. The leaves, after being dried, are first ground to a fine powder and mixed with one-tenth of their weight of caustic soda, dissolved in water and heated to boiling point. So far, however, it does not appear to have been found practicable to turn this excellent idea to account at all extensively for commercial purposes, or, indeed, to place it beyond the pale of laboratory interest and experimental investigation.*

**Chemical Analysis of Specimens.**—Again, M. Lagarde has pointed out the advantage of examination and chemical analysis of large specimens of the raw gum direct from the seat of growth, in addition to electrical tests on the manufactured article when about to be used as an insulator.† He has drawn attention to the fact that though two materials might test much the same electrically, they may really be very different in their nature, one perhaps containing more water or resin—both a sure source of subsequent deterioration, if in excessive proportion. It would thus seem in every way

---

* Too little of the juice could be obtained from each leaf by incisions to make the process worth going through. It could, therefore, only be collected by chemical means, introducing applications that would probably act injuriously on the gum itself. Again, this would involve machinery or materials being taken out to the scene of action, unless the whole plant were uprooted and taken home to be operated on—an extermination which would prove expensive. The latter appears to be the only way in which the leaves could be got at conveniently, but even then it would be a lengthy process, such as would increase the scope for early decay due to oxidation, partly owing to the substance being brought into a somewhat laminated condition. We must remember, moreover, in this connection, that in certain out-of-the-way places there are only a few trees, and that those which are present often yield but little sap—as is testified by the comparatively small lumps in which it is often sent home.

In any case it is probable that any juice collected from the leaves of the tree would be more open (during extraction) to oxidation, and consequent resinification, than by the ordinary means of collection.

† As a matter of fact, it is the custom at all the large cable works to submit samples of gutta to chemical analysis previous to selection.

desirable—especially from a durability point of view—that specifications for the dielectric of a cable should be based not only on electrical, but also on chemical, tests of the best materials available at the time.

**Reference for Further Particulars.**—Finally, we may add that in the *Electrical Review* of December 1890 and January 1891, will be found a series of interesting and instructive articles* concerning gutta-percha, to which the reader may be referred for further information.

Moreover, probably before this book is published, Dr Eugene Obach, F.C.S., will have delivered, at the Society of Arts, a course of Cantor Lectures on the same subject. These will, without doubt, prove both interesting and instructive.

---

* Prepared by Mr E. March Webb, Electrician-in-chief at the Silvertown Works.

Eastern Telegraph Company : Station at Suakin.

SECTION 3.—CHEMICAL, PHYSICAL, MECHANICAL, AND ELECTRICAL
PROPERTIES OF GUTTA-PERCHA.

**Native Gutta-percha.**—In the pure state, gutta-percha is a porous
substance of a milky-white colour, which however resists the encroach of
water, and is indestructible by chemical agents. It is, moreover, insoluble
in water, alcohol, alkalies, and dilute acids, but extremely soluble, especially
when hot, in ether, benzine, chloroform, turpentine, creosote, naphtha,
and generally in all resinous liquids. Strong sulphuric acid attacks it,
especially when moderately heated, and leaves nothing but carbon dust;
strong nitric acid changes it into a yellowish resin. As a hydro-carbon,
having the same chemical composition as oil of turpentine ($C_{20}H_{30}$), it
contains:—

$$\left.\begin{array}{llllll} \text{Carbon} & - & - & - & - & 88.92 \\ \text{Hydrogen} & - & - & - & - & 11.08 \end{array}\right\} \text{parts in 100.00.}$$

When subjected to the oxygen of the air, it turns brown and becomes
brittle, changing to a resinous substance soluble in alcohol. This oxidation takes place more readily in the presence of light, heat, and especially
under alternating dry and wet conditions, or great and sudden changes of
temperature.* It is still further accelerated under a combination of all,
or any, of these influences.

Absolutely pure gutta-percha can only be obtained by having recourse
to chemical methods of purification—far beyond that of mastication. As
already stated, the material is then absolutely white. This would, however,
be much too costly a plan for practical work on anything like an extensive
scale, and can, indeed, only be adopted for laboratory experimental work.

A small quantity of a certain very superior class of Borneo gutta-
percha (almost pure) is obtained which is quite pink in colour. This
costly species is principally used for the base of false teeth.† The Telegraph Construction Company, in the early days of submarine telegraphy,
imported some of this material from Singapore, the general Malay market
for gutta-percha.

---

* Thus, when applied to open-air conductors exposed to sunlight, it rapidly deteriorates,
and even cracks off the wire, or gradually drops away as so much powder; but when
enclosed in iron pipes, or submerged under the sea, it is practically indestructible. The
degree of effect produced by light is in proportion to its actinicism and intensity. It
is independent of the source. Thus the electric arc is sometimes found to be as
destructive to gutta-percha as daylight.

† Most of the gutta-percha employed as mouth mouldings is, however, rendered pink
artificially. India-rubber is still more commonly used nowadays as above.

The supply of gutta-percha generally being much smaller—and the cost accordingly much greater—than in former times, the average gutta-percha now commonly met with is usually of a more mixed, and, therefore, often of an inferior character to that which was used for the same purpose at that period. The mere fact of several different guttas of an inferior quality being mixed with a given high quality, but scarce and expensive, gutta, though sometimes to a certain extent beneficial, is often sufficient to account for this inferiority; not to mention the by no means uncommon practice of adding resinous matter in order to produce specifically certain electrical results, with a view to meeting the difficulty of the scarceness and increased cost of gutta-percha.

The result is that the material was in those days much more durable.* It is true that the total insulation on a line of given dimensions was not nearly as high in those days as it is at present†—notwithstanding the above facts. This was mainly due to the lack of experience in joint-making, but partly to the now more general presence of resin.

**Commercial and Manufactured Gutta-percha.**—Gutta-percha of commerce is not a true vegetable substance, but a compound, in varying proportion, of pure gutta, resin, water, and woody particles. According to Dr Miller, the composition of good commercial gutta is as follows:—

| | |
|---|---|
| Pure gutta | 79.70 |
| Resin | 15.10 |
| Woody matter | 2.18 |
| Water | 2.50 |
| Ash | 0.52 |
| | 100.00 |

---

* It is, in fact, found that the purer the material the more durable it is; for when homogeneous and unmixed, there is practically no scope for chemical action. Thus it is the purest gutta-percha that is always sought for—on account of its durability rather than on account of its specific resistance, which is, indeed, exceeded by resin being added to it, besides the inductive capacity being thereby reduced as a rule.

Apart from durability, it may fairly be said that the purer the material the better it is mechanically and electrically. Often, however, mixtures of various guttas have to be made, either owing to the scarcity of any particular species, or in order to give strength to what is perhaps the best species electrically but weak mechanically.

† Where a core yielded a dielectric resistance of 10 megohms per naut at that time, it would now give something up to 1,500 megohms. Again, the inductive capacity has been very much reduced, partly on the above account and partly owing to the greater amount of mastication now effected.

There is a strong belief with many authorities that the high mastication to which gutta-percha is often subjected to obtain low inductive capacity lessens the durability of the material. It should be secured by increased thickness of dielectric, in preference to specific means of this description.

Speaking generally, its average density is between 0.969 and 0.982, being, therefore, just below that of water.* The best qualities are yellowish and fibrous, the inferior kinds are whiter or of a reddish tinge. Solid at ordinary temperatures, gutta resists a stress of about 1,000 lbs. per square inch before elongation occurs. It does not break till a strain of some 3,300 lbs per square inch has been applied to it, with an elongation of 50 or 60 per cent.—having in fact a breaking strain of about 1¼ tons, when stretched.†

Gutta-percha is also to a certain extent elastic. This may be seen when a submerged cable is picked up. On examination of the end the conductor will be found to protrude beyond the gutta-percha envelope, showing that the conductor only has remained permanently elongated under the strain. Gutta is not, however, nearly as elastic as india-rubber, though almost equally extensible. Gutta-percha may be bent, knotted, or drawn through a die without difficulty, but, on the other hand, is easily injured with a pointed, or cutting, instrument.‡ It begins to soften at 37° C.: indeed, a gutta-percha insulated cable should never be unavoidably exposed, after manufacture, to temperatures much above 32° C. (90° F.).§ Between 50° and 60° C. (or about 125° F.) gutta-percha is absolutely plastic; it, in fact, possesses the uncommon feature of becoming quite soft when plunged into boiling water. In this state gutta-percha can be moulded on to any object, taking the exact details of form, however delicate, and retaining them when cool. It is thus rendered peculiarly well adapted to being laid on a wire, in a permanent manner, to any desired thickness. Manufactured gutta-percha begins to melt at a temperature of 100° C. (212° F.), when it is also especially liable to become resinous if exposed to the air; and, absorbing a quarter of its own weight of oxygen, to become brittle, shrink, and crack.

---

* Its density (or specific gravity) naturally varies with the temperature, both as regards the water with which it is measured and the material itself.

† It should be remarked, however, that this breaking strain of the gutta-percha insulator is of no practical value in a submarine cable. Owing to the gutta-percha stretching so much more than the conductor, as well as the other component parts of the cable, its breaking strain does not come into force at the right moment—*i.e.*, with the conductor, much less with the rest of the cable.

‡ Thus also it is readily injured by the boring tools of various marine creatures. A special armour (described in Part II.) is now applied to meet this source of damage.

§ Great care is therefore necessary with a gutta-percha line to keep it always as much as possible from the direct rays of the sun—especially in tropical climates—or from hot places of any sort, though the outside covering of hemp (or canvas tape), as now applied outside the iron wires, tends to protect it from direct heat more than the bare iron-wire cables of early days.

**Favourable and Unfavourable Conditions.**—The most durable gutta for electrical insulation purposes is that which contains the least resin. The qualities which rank highest are said to be those which emanate from Java and Macassar; the Sumatra species coming next, and lastly those from Borneo. Certain samples of Borneo gutta, viscous and of a whity-brown colour, contain a white juice which ferments, and gives off butyric acid and butyrate of amyl; thus, in course of time there remains nothing but resinous powder.* On the other hand, the methods of collection are better in Borneo than elsewhere.

Ozone decays gutta-percha very rapidly; and if—as has been suggested by some authorities—there is more ozone in the sea-air than elsewhere, it would appear that exposure here, if prolonged, would be an unfavourable condition for this material.

On the other hand, up to the present, the only known way to prevent oxidation of gutta is to keep it under water. It is then practically indestructible;† indeed, there has never been a case known of the gutta suffering through oxidation in the immersed portions of submarine cables. It is on this account that some companies whose cables are landed where the climate is hot, enclose the subterranean lines (joining up to the cables) in pipes which are kept filled with water under pressure from reservoirs on the higher portions of the ground. When tubes without water are used, or where the cable is buried in dry sand, if the wire be insulated with gutta-percha,‡ the gutta becomes brittle and perishes in course of time, leaving the conductor absolutely bare in places. Thus, land lines, or even beach cables, should be either kept in water (if gutta-percha insulated) or

---

* Resin is a most dangerous insulating substance. Besides rapidly evaporating, and tending to gradually render gutta-percha brittle—especially in a dry atmosphere or soil— it also acts chemically on gutta-percha, and tends to destroy its natural gumlike qualities. By rendering it less homogeneous, the gutta is more subject to oxidation.

† It has sometimes been thought that salt-water had in itself an absolute preservative influence on gutta-percha. This is not so. Gutta-percha remains constant at the bottom of the sea only on account of the absence of oxygen, both as regards air and light. It is probable that light tends to have an even greater deteriorating effect (by oxidation) than air, though it has never been actually proved. The sun's rays, however, lose their penetration at some 100 fathoms below the surface of the sea. Again, a core which is in any way armoured is not likely to be much troubled with the presence of light. In a cable at the bottom of the sea, gutta-percha is, moreover, at its best, mechanically and electrically, on account of the pressure and low temperature to which it is subjected.

‡ In this country, when gutta-percha covered wires have been laid underground, certain precautions were found necessary in quite early days in the way of protecting them from sun and air. The gutta-percha dielectric was taped externally, which taping was then tarred and sanded. The tar used at first was *coal* tar, but this was found to act injuriously on the percha. Subsequently *Stockholm* tar (applied to yarns) was adopted, and this was found satisfactory.

else covered with india-rubber.* In 1852 Mr Edwin Clark found that gutta, during the necessary purifying process, takes up mechanically a certain amount of water, which again evaporates partially during mastication, leaving in its place a resinous substance more or less porous. A fair sample of gutta taken from a newly manufactured cable, and analysed by Dr Miller in 1860, contained 15 per cent. of resin, and 2.5 per cent. of water.

Although the methods of purifying gutta have since been much improved, it also appeared from the 1876 experiments of Professor Abel† that as regards the products of oxidation formed at the expense of the gutta and of the water in·suspension, much the same results were obtained.

The analysis of some high-class guttas, which had been exposed for several years to light and air, shewed that oxidation takes place but slowly under the influence of these two agents in cases where the gutta has derived a certain degree of homogeneity and solidity, through prolonged mastication. The proportion of water in suspension thus indicates—at any rate, approximately—the degree of perfection of the gutta in this latter respect, upon which the insulation resistance so largely depends.‡

The proportional quantities of salt and fresh water absorbed by gutta-percha under the same conditions, at ordinary temperatures (according to the late Sir William Fairbairn's 1860 experiments §), are as 3 to 5.‖ This absorption takes place somewhat rapidly at first; after some time it proceeds very slowly but still goes on—indeed, it is doubtful whether even in the course of a year a submerged core has absorbed all the water it will ever do—depending upon the nature of the material and the prevailing circum-

---

* "Engineering and Electrical Report on the Yof Bay-Dakar Repairs," by Charles Bright, A.M.Inst.C.E., M.I.E.E., 1894.

† Now Sir Frederick Abel, Bart., K.C.B., F.R.S.

‡ The denser and more homogeneous a vegetable fibrous material is, the less porous it is, and consequently the less it tends to absorb moisture or actual water, with the result that the electrical resistance becomes higher. Viewed physically and electrically, the gutta-percha dielectric has been likened to a sponge (and also to sand) in which the conductor is enveloped.

§ These experiments were conducted in connection with the Board of Trade Inquiry (of that year) into the Construction of Submarine Cables.

‖ The quantity of salt-water absorbed by a given bulk of gutta-percha being thus less than the quantity of fresh-water which it will similarly absorb, is naturally accounted for by the greater density of salt-water. This was arrived at by the material being plunged into fresh and salt water each in turn, free from pressure. In the case of fresh-water, after the gutta-percha had absorbed all it could do, it was found that the weight of the gutta-percha was 5 per cent. greater than previously ; whereas when the same experiment was made in salt-water, the weight was increased by 3 per cent. only.

The author assumes that these experiments were made with raw gutta-percha, believing that manufactured gutta-percha would not absorb anything like this amount.

stances. The rate of absorption is said to be somewhat higher under a vacuum than under the sea. The absorption of sea-water doubles when the temperature rises from 4° C. to 49° C.—no doubt owing to expansion : for fresh-water the increase is rather more rapid. Pressure makes no sensible difference in this connection.* The water absorbed appears only able to penetrate laterally to a very slight distance,† and when this limit is reached, the quantity of water does not afterwards increase, no matter how great the thickness of gutta.‡ Water mechanically suspended in the pores of gutta-percha has, apparently, no effect on its electrical qualities until the weight of water exceeds 2 or 3 per cent. of the weight of gutta, consequently in ordinary practice this question of absorption by a gutta-percha cable core § need never be considered electrically, except in the case of a flaw.||

The main question which decides whether or no any of the materials such as are mechanically adapted for the purpose are also good insulators electrically is their degree of denseness and homogeneity. This is largely on the grounds of durability, as already shewn ; but it is also based on actual electrical values.

Viewing all substances as conductors of electricity, there are certain facts which it is well to bear in mind regarding good conductors and bad conductors, or insulators. Thus, the purer and more homogeneous the good conductor is, the better it conducts ; and, on the other hand, the purer and more homogeneous the bad conductor, or insulator, is, the worse

---

* The explanation of this rests, probably, in the fact that when material pressure is applied, though there would be a greater forcing in tendency as regards the water, yet, on the other hand, the material would become correspondingly denser in its texture, and therefore less porous and absorptive. Thus it is doubtful whether at the bottom of the ocean the absorption by gutta-percha is at all greater than in a bucket of salt-water.

If the experiment could be conveniently carried out, the degree of absorption might with advantage be noted at various depths, so as to determine at what point the structural influence outweighs the tendency towards increased absorption.

A proof of this increased homogeneity of gutta-percha under a given column of water exists in the reduced bulk, which may be observed in a cable core when picked up from the bottom of the ocean after a certain period of submersion.

† It is not thought that even the smallest particle of water ever reaches the conductor through the insulation of a submerged core (though under great pressure) except at an actual fault.

‡ Fairbairn found that the absorption by thin sheets was actually greater than in thick sheets, pointing to capillary attraction on the surface being the main cause. In any case, this is an extra reason for a certain minimum thickness of insulation for a submerged core, quite apart from questions of specific resistance.

§ Manufactured gutta-percha invariably contains a certain small percentage of moisture —indeed, from a durability point of view, this is a desirable feature.

|| Many years ago it was shewn by Professor Fleeming Jenkin, F.R.S., that the insulation resistance of a sound gutta-percha covered wire, enclosed in an iron sheathing (with the ordinary intermediate packing), is little, if at all, affected by submersion.

it conducts, as a rule, *i.e.*, the better it insulates. This may, to a great extent, be accounted for by the fact that the former is always a metal, and the latter invariably a vegetable fibrous substance with a certain degree of porosity.

It must, however, be admitted at the outset that there are many features regarding insulating materials,* which with our present knowledge we cannot attempt to explain—as, for instance, the many contradictory facts regarding gutta-percha and india-rubber when brought into comparison. It can only be said generally that what appears to constitute a good conductor in a metal, at the same time constitutes a bad conductor (or good insulator) in a non-metal, or vegetable substance. When a fibrous material is made up from several different substances it cannot possibly be so homogeneous, and therefore, in that case it must be more porous. Again, actual chemical action occurs between its components; and, in fact, the nature of the material is liable to be entirely altered, being governed by the composition and the proportions of the same, electrically as well as physically. A high insulation resistance is, however, only a secondary consideration for the dielectric of a submarine cable as compared with durability, and a *permanence* of its insulation at a fixed minimum.† If low resistance mixtures could only be relied on to remain constant they would answer the purpose admirably as dielectrics; but, unfortunately, owing to chemical action between the constituent parts, or to lack of homogeneity, the material tends to gradually alter—indeed, to deteriorate—physically and electrically, until a point is reached of complete breakdown in its insulating powers.

Speaking generally, it is found by experience that permanence of electrical, as well as mechanical, qualities can only be safely looked for from the pure (homogeneous) materials,‡ which also happen to have a

---

* As a contribution to, and illustration of, the molecular theory of conduction, Mr Rollo Appleyard recently read an interesting and instructive paper on "Dielectrics" at the Physical Society, during which he shewed experiments with a gutta-percha sheet having, in turn, various proportions of brass filings incorporated in its texture. In the course of the discussion the present writer suggested that the phenomena of the resistance being decreased so suddenly when more than a certain proportion of the filings were worked into the gutta sheet, was accounted for by the circumstance that the conduction as the result of induction became possible on their being sufficiently close to one another to permit of electric influence (*Proc. Physical Soc.*).

+ As a matter of fact, a cable with a comparatively low resistance dielectric should signal through somewhat better, within certain limits, than one with a very high resistance, other conditions being equal—though practically this seldom, if ever, comes in.

‡ As a rule, too, the greater the density of the texture, the higher the resistance on account of decrease of moisture suspended in the pores. Speaking generally, there is a

fairly high specific resistance at the outset, and higher than most of the compounds.* Thus, though different classes of gutta may be sometimes advantageously, or necessarily, mixed together, other materials, of an entirely different chemical composition and different texture, cannot be advantageously allied with gutta-percha for permanent electrical and mechanical purposes.† A certain minimum is, therefore, usually stated for the resistance of the dielectric, and this is best fixed from a knowledge of the corresponding specific values of the most suitable guttas to be obtained in the market at the time.

A fairly high insulation resistance has also the advantage of shewing up the presence of any incipient faults more readily during manufacture, and also of permitting a lower battery power for subsequent "working" purposes, than if a material were employed which, at the very outset, had but a comparatively low resistance. Inasmuch as an abnormally high resistance is nowadays usually suggestive of either a comparatively large proportion of resin or else an over-masticated‡ gutta (for purposes of low capacity), it is very customary also to specify a certain *limit* for the dielectric resistance.§

---

connection between density and homogeneity ; and, in the case of a material like gutta-percha, homogeneity usually implies a *maximum* electrical resistance, as well as a *regular* value in this respect. Where, however, the pores are filled with resin oil, the resistance then would usually be greater than with a more dense and homogeneous structure of the same material.

Incidentally it may be mentioned that attempts have been made to re-juvenate old gutta-percha (from land lines) with many cracks by forcing resin into its pores ; but, for chemical, mechanical, and other reasons, success has not been met with in this respect, except by thoroughly mixing the two materials together, after reheating the old gutta to a plastic condition, by being taken off the wire and practically re-manufactured.

Resin oil (unlike many oils) has a higher resistance and, as a rule, a lower capacity than gutta-percha. Thus, *electrically* speaking, its admixture with gutta-percha should be distinctly favourable. Unfortunately, however, it is of a non-durable character, as it tends to evaporate. Gutta-percha with a large proportion of it is, therefore, not serviceable for any length of time.

* Steadiness of insulation is what is mainly looked for, however, as a sign of mechanical and electrical permanence, rather than any abnormally high specific value.

† Broadly speaking, density and homogeneity—besides opacity, as a suggestion of purity for a given thickness and colour—may be taken as being collateral with a more or less high electrical resistance ; and the latter effect is, therefore, likely to be promoted by low temperature, pressure, and the mechanical "setting" due to age.

‡ In throwing off moisture, mastication, as a rule, has the effect of increasing the resistance and decreasing the capacity of gutta-percha up to a certain point, at which its physical and chemical character is altered. To use a trade expression, the gutta-percha when in this last condition is said to have been "killed."

§ A high rate of electrification also often implies the presence of a large proportion of resin. It is sometimes, therefore, specified that the electrification between the first and second minute should not be above some 5 or 6 per cent.

### GENERAL FORMULÆ.

**Dielectric Resistance.**—The insulating power of gutta-percha, or the resistance which it offers to the passage of an electric current, that of pure copper being taken as unity, is, for equal volumes, at 24° C., approximately,

$$60,000,000,000,000,000,000 \text{ or } 6 \times 10^{19}.$$

We may give an idea of this vast difference by remarking that light, with a velocity of 111,834 miles per second, would take more than six thousand years to traverse the distance in yards which these figures express.

The various guttas in their natural state having very different properties, manufacturers often find it advantageous, if not actually necessary, to mix the best fibrous qualities which are the most durable, with inferior kinds which are better insulators, and have less electro-static capacity. Accordingly the insulation resistance and electro-static capacity of gutta, per unit of volume, vary within certain limits and have to be determined for each particular case. Commercial guttas being themselves mixtures of different natural kinds, manufacturers desirous of ascertaining the precise value of the samples they buy, not unusually make about 500 yards of core from each sample, and then compare the electrical qualities of the different lengths.

The insulation resistance R of a hollow cylinder composed of any dielectric is given by the formula

$$R = \frac{A \log \frac{D}{d}}{L}$$

where A is a constant, D and $d$ the exterior and interior diameters of the cylinder, and L its length. Thus, for a cable with a gutta-percha core, the approximate resistance per naut will be

$$R = 750 \log \frac{D}{d} \text{ megohms} *$$

after one minute's electrification, and twenty-four hours in water at 24° C. When the cable is new, the value for A may be only two-thirds of that here given. In any case, this constant varies a good deal with the par-

---

* A plate of ordinary gutta-percha, 1 square foot surface and 1 millimetre thick, has a resistance = 1,066 megohms.

The resistance of a cube foot = $12.8 \times 10^6$ megohms.

The resistance of a cube naut (usually taken in practice as the specific value of the dielectric of a cable core) = 2,100 megohms.

ticular gutta-percha in question, and is therefore only approximate for gutta-percha generally.

Even with the imperfections in core-making of quite early days, it was noticed that the insulation of a fairly sound cable tended to improve after submersion. In view of water being a good conductor—and especially salt-water*—and bearing in mind the absorptive tendencies of the gutta-percha dielectric (particularly in any weak places), this appeared strange, and set various experts thinking. It was soon, however, discovered that both low temperature and pressure (especially the former), such as are experienced by a cable at the bottom of the ocean, increase the resistance of gutta-percha enormously. The separate effect of these two factors must now be dealt with.

**Pressure.**—The late Sir William Siemens first demonstrated the effect of pressure on insulating fibrous vegetable substances in the course of a paper on the subject read before the British Association in 1863, based on experiments made with the gutta-percha core of the first Malta-Alexandria cable (of 1861), as well as on certain india-rubber cores. Herein he pointed out how the resistance of gutta-percha was materially increased by pressure, and in direct proportion to the degree,† which was applied hydraulically and varied as desired, with a known constant temperature in Reid's pressure tank, described in Part I. of this book.‡

The formula given by Siemens is

$$R_p = R\,(1 + 0.00023\,p)\S$$

where R = the original resistance at ordinary atmospheric pressure, $p$ = the

---

* At one time it was thought that the salt had an improving effect, but it was soon found that salt-water at an ordinary temperature, in the absence of pressure, had no such influence, indeed rather the reverse. There is no difference in the effects of fresh and salt water under otherwise common conditions.

† According to another authority, however ("The Electric Telegraph," by Robert Sabine, 1867), the increase of resistance becomes slightly more marked at higher pressures with the same amount of pressure increment. This seems highly probable, if only on the grounds that the material would be rendered so much less porous; indeed, when the pressure gets over a certain figure, it may tend to actually squeeze moisture out of the gutta as it would a sponge. In this case a point might somewhere be reached at which permanent deterioration would occur.

‡ At first sight it would appear that the natural method of determining the effect of pressure on the resistance of a gutta-percha core when submerged would have been to lower it gradually in the sea: but here, again, the temperature would decrease as well as the pressure increasing; consequently the temperature would be taking part as well as the pressure in altering the resistance—more so, indeed.

§ In order to be entirely independent of the opposite effects of absorption, and to arrive at a strictly true coefficient in this formula, the pressure should, really speaking, partake of a solid form. This, however, would be impossible in practice.

pressure in lbs. per square inch, and $Rp =$ the increased resistance resulting from subjection to such a pressure.

Thus, 0.00023 was the amount of increase in resistance found to occur on unit resistance (1 megohm) by unit (lb.) pressure.*

Messrs Bright and Clark tested the first Persian Gulf cable under pressure, finding the gutta-percha to be increased in resistance 2.6 per cent. for every 100 lbs. per square inch pressure ; whereas Siemens' experiments shewed the dielectric resistance of the Malta-Alexandria core to be increased only 2.3 per cent. thereby.

The Malta-Alexandria core was composed of 400 lbs. gutta-percha to 400 lbs. copper per N.M., whereas the Persian Gulf core was a corresponding 275 gutta percha to 225 copper. It is possible, therefore, that pressure cannot influence the resistance of a thick covering of gutta-percha as much as a thin one, as is the case, indeed, with regard to water absorption.†

This question of increase of resistance by pressure has never been dealt with since, and it is possible that, with modifications in material and manufacture, the above constant may not strictly apply to all gutta-percha cores of the present time. ‡

---

* It has, however, been suggested by the present writer that the formula is probably not applicable for quite shallow depths of water, on the ground that here the effect of absorption—partly due to capillary attraction—may come in as much as, or more than, the effect of pressure, owing to the latter not being sufficient in such a case to alter the texture of the material by consolidation. In great depths, on the other hand, a submerged core is more influenced electrically by the weight of water rendering the gutta-percha more homogeneous by bringing its molecules into closer contiguity ; and thus this influence entirely overcomes that of absorption, besides actually rendering the material less porous.

It may, at any rate, be fairly said that the absorption of water by a gutta-percha core becomes less rather than greater under a great pressure of water ; by reason that the water pressure has a greater effect in making the gutta-percha denser in its texture (by closing up its pores) than in forcing the water into the gutta-percha (see *Trans. Roy. Soc. Edin.*).

† Bright and Clark's tests were made under a maximum pressure of 600 lbs. only— equivalent to a column of about 224 fathoms of salt-water, the greatest depth the Persian Gulf cable was likely to experience. In Siemens' experiment on the Malta-Alexandria cable, a pressure equivalent to 1,700 fathoms was applied—that cable being originally intended for much deeper water than it was eventually laid in.

‡ As a rule, low resistance gums and mixtures of gutta-percha improve more than very high resistance gums under the influence of pressure. This may be explained by the fact that there is more scope for increasing the homogeneous character of a low resistance mixture. The suggestion here is, therefore, that an inferior insulating mixture might be employed whose resistance could be increased by pressure. It would very likely, however, prove too risky an experiment—even by artificial means, on a small scale, beforehand.

The *thickness* of the covering is not supposed to make any difference as regards the influence of pressure on its electrical resistance.

The most useful way, in practice, of expressing generally the increase of the resistance of gutta-percha by pressure is to say that it is increased 6.15 per cent. for every 100 fathoms of sea-water, *i.e.*, roughly about 60 per cent. per N.M. pressure—equivalent to something like half as much again when submerged under 1,000 fathoms as that which it was when under atmospheric pressure only.

As a matter of fact, in the practical application in which we are concerned, when a cable is submerged and rests on the bottom, it is a case of pressure due to the weight of a column of water of a certain height over, say, a square inch of the cable, rather than uniform pressure all round.

The pressure of sea-water is usually taken at 2.676 lbs. per square inch per fathom = 2,715 lbs. per N.M. of depth. This is equivalent to 1 ton per square inch for every 800 fathoms, or nearly 3 tons in the case of an Atlantic cable.

Thus, when it is remembered that the pressure in 2,000 fathoms is equivalent to about 2¼ tons per square inch,* it will be seen that, apart from the decreased temperature, the improvement in prevailing conditions of a gutta-percha insulated cable, by increase of dielectric resistance, is something considerable—amounting, in fact, to 123 per cent. on the original resistance owing to the pressure alone. Thus, a dielectric resistance of 600 megohms per N.M. (at 75° F.) would, under these conditions, become converted into 1,339 megohms; and a resistance of 6,737 megohms at 36° F. (the approximate bottom temperature) would, due to pressure, be 8,287 megohms.†

For the purpose of comparison, therefore, whilst reducing the results obtained from a test on a laid cable to the standard temperature (75° F.), they should also be reduced to the standard (atmospheric‡) pressure by applying the correction for the pressure experienced, based on the above formula.

---

* The effect of pressure on a cable picked up from the bottom of the ocean is sufficient to give actual evidence of the core having been reduced in bulk, by the compression of the insulating envelope.

† The greater the depth to which a cable with a gutta-percha core is submerged, the greater the reduction in the difference between the share taken by temperature and pressure respectively in increasing the insulation resistance. This is owing to the fact that whereas the temperature decreases rapidly at depths near the surface, it falls only very gradually at greater depths, until often below a certain depth—usually about 2,000 fathoms—it falls no further. The pressure, however, on a submerged cable increases in direct proportion to the depth.

‡ In experimental investigations the pressure employed is still often expressed in atmospheres, so as to be readily comparable with the original unit of pressure—1 atmosphere ( = 15 lbs. per square inch).

Messrs Siemens Brothers are still always in the habit of applying the pressure test to their core immediately after manufacture, and the manner in which this test is carried out, in modern practice, by them will be dealt with later.

**Temperature.**—The electrical resistance of gutta-percha is, in practice, still more affected by a change of temperature.

The law of variation is briefly stated by the formula

$$\frac{R}{r} = A^t$$

in which R and $r$ stand for the respective resistances at the lowest and highest given temperature, the difference between which is $t$ degrees, A being a constant obtained by experiment.

Inasmuch as the resistance of copper as well as, in some degree, that of other metals was known to be influenced by temperature, it was thought that probably the resistance of gutta-percha would also not be entirely independent in this way.

Thus, in 1863, during the construction of the first Persian Gulf cable, Messrs Bright and Clark determined to make a series of tests on the gutta-percha core at various temperatures. The core was, therefore, placed in a tank of water. The tests were made on the core at every 2° C. between freezing point (secured by the admixture of ice) and 38° C. (100° F.).

This iron tank was "felted" outside, and facilities were provided for heating the water uniformly by steam pipes. Every precaution was taken for accurately securing the desired temperature, the water being kept in constant agitation throughout the test. The tests were made on four separate one-naut coils (free from pressure), and occupied thirty-three days, during which nineteen series of observations were taken starting from freezing point.

From the mean results* of these experiments, the following formula was obtained

$$R \times .8944^t = r$$

where R is the resistance at any given temperature ; $t$, the increase in degrees Centigrade ; .8944, a constant deduced from the experiments ; and $r$, the resistance at the higher temperature.

It was found, in fact, that the law of variation coincided—within the

---

* "The Telegraph to India, and its Extension to Australia and China," by Sir Charles Tilston Bright, M.P.

limits of observation—with a logarithmic, or compound interest, law ;* as in the case of the metals, only at a much greater rate.†

With different gums, qualities and mixtures, this coefficient of variation for gutta-percha is substantially different ;‡ moreover, it varies very much according to the age of the material, *i.e.*, according to the length of time it has been manufactured into core. Thus, though the law has never been questioned since first laid down as above, the coefficient is only roughly applicable to gutta-percha generally. Strictly speaking, the coefficient requires to be determined for each quality of gutta. Indeed, if we are to obtain absolute accuracy, it should be freshly arrived at by experiment on every new set of coils where a change is made in the gum, or gums, used. Even then it is only applicable to newly made core, and not strictly accurate when applied afterwards to a submerged cable some time after manufacture. An allowance should then, in fact, be made for the increase in resistance by ageing or maturation—supposed to be due to a natural "setting" of the texture.§ Such exactitude is not, however, really necessary in practice, and scarcely comes within the limits of the tests. The variation is a substantial one in any case, and should, therefore, always be allowed for with wide ranges of temperature, such as are experienced by the dielectric of a cable between what it is when laid, and what it was just after manufacture. ‖

For ordinary gutta-percha as now commonly met with, the resistance variation per cent. per degree Fahrenheit ¶ is usually somewhere between 7 and 8, *i.e.*, 100 megohms becomes about 107.5. Thus the coefficient to be applied to the resistance of gutta-percha at 1° below the standard tempera-

---

* A logarithmic curve was drawn, and another curve marked off based on the experimental results obtained at various temperatures. On connecting up the points, the experimental curve—except for irregularities due to slight errors involved in the experiment —practically coincided with the true logarithmic curve.

† The resistance of gutta-percha at 36° F. is about ten times as much as it is at 75° F., other conditions being the same.

When a gutta-percha core is submerged under a depth of 1,000 fathoms, the resistance of the dielectric is increased about 300 per cent. owing to the decreased temperature, and about 60 per cent. owing to the pressure of the water above it.

‡ As a rule, the less pure the material, and the lower its electrical resistance, the greater the variation by temperature.

§ This molecular drawing up, or contraction, is probably brought about by a species of crystallisation resulting from the previous absorption of moisture.

‖ Near the surface the sea is often quite warm, but at the bottom of a great ocean depth the temperature is very usually, owing partly to cold under-currents, as low as 36° F., which becomes a valuable item as regards the dielectric.

¶ In submarine cable work all practical temperature measurements and observations are made according to the Fahrenheit thermometric scale.

ture (75° F.)* will be 0.075; that is to say, an original resistance of 1 megohm under a decreased temperature of 1° F., would become 1+0.075 = 1.075.†
Then for a further decrease of temperature this coefficient is naturally applicable (by compound interest laws) to the last existing resistance—*i.e.*, in this case to 1.075—and not merely to 075. The same principle naturally applies when the resistance is some multiple of unity.‡  Thus, based on the coefficient for variation obtained by experiment, a complete set of tables may be drawn up, giving the correct coefficients applicable to each number of degrees of difference of temperature from the standard temperature—in fact, the coefficient for each degree Fahrenheit, already calculated. Thus also with the coefficients for temperatures on the other side of the standard temperature. Such tables (applicable only to the particular material in use) are to be found in every cable factory all ready drawn up, as well as at every testing station.§

The author has attempted an explanation of the phenomenon of the resistance of gutta-percha being increased by a fall in temperature and decreased by a rise thereof.‖  On the principle that all such electrical changes are due to physical changes, it comes back, as in the case of pressure, to varying degrees of density and homogeneity. The nature of the physical change produced by a change of temperature on vegetable fibrous substances of the gutta-percha (and india-rubber) order is that of cold

---

* This is the standard temperature at which tests are made during manufacture, and to which subsequent tests are reduced to, for purposes of proper comparison. It has been for many years universally agreed on as being a convenient average temperature capable of being readily provided during manufacture.

† Where 0.075 is the coefficient for 1° change of temperature on 1 megohm, then for, say, 20° difference the coefficient becomes

$$(0.075)^{20}$$

and in the case of the 20° being that amount *below* the original temperature (at which the resistance was 1 megohm), we get

$$(1+0.075)^{20} = (1.075)^{20}$$

Or, expressed in a general way for the compilation of a complete table

$$= (1.075)^n$$

where $n$ varies according to (*i.e.*, represents) the number of degrees difference.

‡ If the original resistance, instead of being 1 megohm, be, say, 600 megohms, then with a decreased temperature of 20° we get

$$(1.075)^{20} \times 600 \text{ or } 600 \times (1.075)^{20}$$

If, however, the temperature be *increased* by that amount, then the coefficient, on an original resistance of 1 megohm, becomes

$$(1-0.075)^{20} = (0.925)^{20}$$

and with an original resistance of 600 megohms this becomes

$$(0.925)^{20} \times 600 \text{ or } 600 \times (0.925)^{20}$$

§ In practice they are usually made out, in a simplified form, for uniform subtraction (or addition) of logarithms for either side of the standard temperature.

‖ *Journal Soc. Tel. Engrs.*, vol. xvii., p. 679; *Electrical Review*, 16th November 1888.

contracting the material, and heat expanding it. Viewing all substances, in some degree, as conductors, by contraction the pores of the material tend to be closed by its molecules being brought closer together. Consequently it is more compact, and less sponge-like or "leaky" in its texture: its absorptive tendency is proportionately decreased, and, indeed, any moisture previously contained by the pores is, more or less, squeezed out. The effect of heat, on the other hand, is that of expansion separating the pores, and thus increasing its natural absorptive tendency as well as the disposition towards leakiness, *i.e.*, conduction.

The whole effect may be, in fact, traced to mechanical causes as in the case of pressure, though the result is much more marked.* The fact that temperature exerts an opposite influence on the conductivity of vegetable fibrous materials to what it does on metals† may be to some extent explained, as the author has previously suggested,‡ by the fact that fibrous materials such as gutta-percha contain moisture, a good conductor in itself, and naturally tend to take up all the moisture they can. By pressure or low temperature this moisture is expelled ; by heat, more is absorbed, for reasons already shewn.§

The fact that the resistance of gutta-percha (or india rubber) varies by change of temperature according to a logarithmic curve, rather than by direct proportion—*i.e.*, is increased at a rate so much higher at low than at high temperatures—can be physically accounted for by the circumstance that it is only when low temperatures are reached that the expulsion of moisture begins to take place.‖

---

* When a cable is laid at the bottom of the sea, the cold water by convection takes away any heat established by pressure.

† This is sometimes less accurately expressed by saying that electrical *conductors* and *insulators* are oppositely affected (as regards their resistance) by temperature. Carbon, however, is a good conductor, but being a porous vegetable substance, it is affected by temperature accordingly, *i.e.*, its electrical conducting power is increased by a rise in temperature, and *vice versâ*.

‡ *Electrician*, vol. xxiv.

§ But for this, no doubt temperature would influence such gums the opposite way— *i.e.*, the same as it does metals—for, viewing everything as a conductor in some degree, increased homogeneity should imply more efficient molecular action, and, therefore, better electrical conduction—inasmuch as it is not unusually recognised that the high conducting power of the metals is based on their denseness and homogeneity.

The author recognises that this assumption is not in keeping with the vibratory theory, but the latter does not appear to be borne out in all its aspects as regards the actual facts concerning the effect of temperature on metals, though it does in the case of temperature and pressure in connection with fibrous substances.

‖ The conductivity of copper is affected by temperature according to the same rule. This is probably by reason that metals are not materially condensed until low tempera-

We have already stated that, as in metals, the low resistance gums and mixtures are more affected electrically by temperature than those of a high specific resistance.* The explanation of this fact, as regards fibrous materials like gutta-percha (or still more, india-rubber), is presumably that the lower resistance gums, being less pure and homogeneous, very usually contain more moisture, and are more absorptive. Thus pressure or low temperature increases their resistance more, by driving out moisture and preventing absorption.

The view that the difference in the effect of temperature on the resistance of fibrous materials to that on the metals may be accounted for by the presence of moisture and absorptive tendency of the latter is further supported by the fact that change of temperature takes much longer to affect the gutta-percha insulator of a core than it does the copper conductor.† It might, in some ways, naturally be supposed that, the copper conductor being enveloped inside the gutta-percha insulator, the gutta-percha would —especially in the case of a laid cable—the soonest adapt itself to any alteration in the prevailing temperature. This does not, however, appear to be the case ; in fact, the gutta-percha, as a rule, lags uniformly behind the copper in this respect by about forty-eight hours. That is to say, in the instance of a fall (or rise) of temperature of, say, 30° F., the gutta-percha would not give its true resistance value (due to change of temperature) till about forty-eight hours after the copper has done so ‡ In the copper conductor it is only a question of the degree of density and homogeneity of the metal being altered by changes of temperature ; but in gutta-percha the effect of temperature change is (by increasing, or decreasing, its density) that of gradually driving out, or absorbing, moisture, and thus gradually increasing or decreasing its resistance.

Inasmuch as it takes gutta-percha so much longer to adapt itself electrically to a change of temperature—or to become " temporised," as it has been expressed—there must be a time when, in reducing the observed dielectric resistance to 75° F., it is inaccurate to correct the resistance by the temperature calculated from the conductor resistance. Under these

---

tures are reached. No degree of pressure expels any large amount of moisture from gutta-percha or india-rubber, and the conductivity of copper is not affected at all by pressure—so far as we know, at any rate.

* As regards metals, no ready explanation offers itself—indeed, the opposite might with more reason have been looked for.

† "Cable Testing, Copper and Dielectric Resistance," by Charles Bright, M.I.E.E., *The Electrician*, 7th December 1888.

‡ This statement is based on actual experience on a cable expedition, during the course of which the temperature varied from about 32° F. to about 90° F.

circumstances it will be found sometimes useful and instructive—if only as a check—to calculate the temperature of the dielectric from its own resistance,* and also, as an *additional* check (though only a rough one), from the observed temperature where the core, or cable, rests.†

As it happens, the conditions at the bottom of the ocean are, ordinarily speaking, in every way favourable to a gutta-percha insulated conductor, as regards the conductor itself as well as on account of the insulator. Though temperature affects the copper and the gutta-percha the reverse ways electrically, inasmuch as one is intended to conduct and the other to insulate, this is in every way favourable. Thus, the low temperature at the bottom of the ocean increases the conducting power of the conductor, and still more increases the insulation resistance of the dielectric.‡ However, at very low temperatures gutta-percha is liable to crack. Thus, in some parts of the Pacific Ocean (where the bottom temperature is said to be below freezing point), it is possible that gutta-percha might lose its insulating qualities altogether instead of the dielectric resistance being further increased, as it would otherwise naturally tend to be. India-rubber has not this tendency to crack with low temperatures, and might, therefore, recommend itself under certain circumstances of this description, as much as in instances of abnormally high temperatures.

**Age.**—Another feature which governs the electrical resistance of prepared gutta-percha is its age, *i.e.*, the length of time for which it has been in the manufactured form.§ This influence is due to the maturing of the material causing a natural tendency to mechanically "set," which operation increases the resistance appreciably, though not to anything like the same extent that moderate decreases of temperature or average pressures do This mechanical "setting" goes on quicker at low than at high tempera-

---

* In this case the effect of any prevailing pressure on the resistance of the gutta-percha at that depth would, of course, require to be included in the calculation.

† It may be mentioned, in this connection, that the late Sir William Siemens introduced in 1860 what he called a "resistance thermometer." This was a plan for obtaining a more accurate idea of the temperature of the place where the cable rested by measuring the resistance of a coil placed alongside the cable, any given resistance being known to correspond to a certain temperature. This may prove a useful clue sometimes when a cable is obliged to remain dry in a tank, say; and where, therefore, the gutta-percha dielectric may be in jeopardy.

‡ The pressure of the bulk of water above the cable—though it is not supposed to perceptibly affect the conductor—also increases the insulation resistance, as has already been shewn.

§ When *raw*, gutta-percha tends, of course, to oxidise and deteriorate if exposed to air or light.

tures, and quicker out of water than in water,* where no evaporation of moisture can occur.

It mostly takes place immediately after manufacture,† owing partly to the moisture taken up during mastication. Thus, the insulation resistance of any particular coil can be easily brought up to specification by keeping it long enough (preferably at a moderate temperature out of water) before final testing, if it shews signs of being slightly wanting in this respect.

This factor of variation of resistance by age naturally tends to become merged with the other changing influences—temperature and pressure. As in the case of temperature and pressure, it varies with different gums and mixtures—and much in the same way.

At present the coefficients applied for temperature are used for gutta-percha generally, irrespective of age. This cannot really be correct, for, as we have already seen, the coefficient is different for gums of different specific resistances, and the effect of age is to increase its resistance— indeed, practically to render it a different texture of material by "setting." In the first place, in order to separate these two influences, the final tests at 75° F should be taken at the latest possible moment after manufacture. (2.) Experiments for determining the temperature coefficient should be made on the same manufactured gutta-percha as soon after construction as may be, so that the subsequent resistance change in the core shall be as far as possible independent of age. (3.) A resistance curve and coefficient should be drawn up based on experiments on gutta-percha at various periods after manufacture, the temperature and pressure having been main-

---

* Thus, again, the age effect of gutta (such as causes it to increase in resistance) is that of becoming more homogeneous. Moreover, the moisture is gradually forced out under certain conditions by the material drawing itself up. It will be understood, therefore, that a low temperature naturally favours these conditions; whereas a high temperature, by expansion, would be unfavourable to such a change.

Again, exposure to air and light tend to bring about oxidation as fast as the moisture leaves the gutta-percha, which also tends to increase the specific resistance of the material until cracks begin to appear, and then a rapid fall in resistance occurs, by the material gradually losing all its essential oil by evaporation.

A physical picture of what probably occurs would be supplied by covering a wire with cold cream, butter, or any material containing a certain proportion of water, especially if the experiment be carried out under ordinary atmospheric conditions and at a low temperature.

The fact that the absorption of oxygen (of high specific resistance) has something to do with the increase of resistance of gutta-percha by time in the presence of air and light is somewhat borne out by the circumstance that the material gradually becomes darker by exposure. Indeed, pure, white, gutta-percha eventually becomes as black as any other gutta-percha, and the sap when left out in the sun to coagulate soon darkens at the very first—this latter being due to oxidation.

† *I.e.*, almost entirely within the first 100 hours—usually assuming a permanent state, at any rate, whilst still at the factory.

tained constant. This would then enable the necessary allowance for variation of resistance due to age alone to be applied. (4.) On the above grounds a standard age should really be fixed on, for the purposes of speci-fication, and of after comparison and reduction, in the same way that a standard temperature and a standard (atmospheric) pressure is adopted, *i.e.*, all standard tests at factory should be made a certain stipulated time after manufacture of core. Sometimes, but not always, provision is made for this in specifications by a clause to the effect that the core tests are to be made a certain number of days—usually fourteen—after manufacture.*

**Inductive Capacity.**—The specific electro-static capacity of gutta-percha, relatively to that of air taken as unity, is about 4.2. The electro-static capacity K of a hollow cylinder, D and $d$ being the exterior and interior diameters, is given by the formula

$$K = A \frac{L}{\log \frac{D}{d}} \text{ microfarads}$$

where L represents the length of the cylinder, and A is a constant ; or, more definitely expressed,

$$K = 1.38 \, \kappa \frac{L}{\log \frac{D}{d}} \text{ microfarads}$$

where $\kappa$ = the capacity of a plate 1 square foot surface × 1 mil thick = .1356 ; and 1.38 a constant which varies materially with the particular gutta used.

The electro-static capacity of a cable core depends upon the surface of the conductor and the thickness of dielectric. It, in fact, varies inversely, and by compound interest, in respect to the relationship existing between D and $d$, which is expressed in a formula by

$$K \, L \frac{1}{\log \frac{D}{d}}$$

This law was first enunciated by Professor Thomson (now Lord Kelvin) in connection with the construction of early Atlantic cables.

It will be evident that the capacity varies also with the quality of the dielectric material, *i.e.*, with its *specific* inductive capacity. This is usually represented by the capacity of a cube naut of the core reduced from the

---

* A minimum dielectric resistance is often specified for the cable when laid. It is not unusually stated that when reduced to 75° F. the values shall be equal to that as core at the factory. It ought, however, to be materially greater on account of age as well as pressure.

measured capacity of a cable composed of the same dielectric material. It varies considerably with the quality of the gutta.

Where A is a constant—the specific capacity multiplied by some number—deduced from the above

$$K = \frac{A}{\log \frac{D}{d}}$$

Thus, in practice, the electro-static capacity of a cable per naut may be calculated from the specific electro-static capacity of the dielectric material and the dimensions of the conductor and insulator in this manner :—

$$K = \frac{0.177}{\log \frac{D}{d}} \text{ microfarads}$$

the constant 0.177 being an empirical figure.

Neither temperature, pressure, nor age * are supposed to influence the electro-static capacity of gutta-percha, or of india-rubber, so far as we know at present.

Inasmuch as both pressure and low temperature have the effect of reducing the thickness of the dielectric†—the former by compression, the latter by contraction—the electro-static capacity should naturally be increased on this count owing to the two plates of the Leyden jar being thereby brought closer together, just as an increase of temperature should correspondingly have the opposite effect.

The probability is, however, that any such results accruing from the above influences are counteracted by a specific change in the texture of the dielectric material being also thereby established—in fact, by an increased density, or the reverse, affecting its power of permitting induction in the same way that its resistance is effected thereby.

In any case, whichever influence predominates, the effect is so slight, that it may certainly be neglected in practice ; indeed, it is imperceptible usually, and its direction variable.‡

As in the case of insulation resistance, the specific electro-static capacity of the dielectric—i.e., the degree to which it permits induction between the

---

* This is assuming that climatic conditions are not favourable to any extreme structural change ; and that the degree of moisture is not excessive, or has not an opportunity of escaping very freely.

† This may be distinctly observed on any core after being picked up from a great depth.

‡ *Proceedings of the Institution of Civil Engineers*, original communication by Charles Bright, F.R.S.E., A.M.Inst.C.E.

plates on each side of it*—varies very much with the material ; and, if it be gutta-percha, with the gum or gums used in its composition, and the condition of its texture.

As a general rule, those materials with the highest insulation resistance offer also the highest resistance to induction—*i.e.*, have the lowest electro-static capacity—owing, also, no doubt, to their greater purity and density. This does not always, however, apply in the case of inductive capacity, for many inferior insulating gums and mixtures of gums have quite a high inductive resistance—more so, indeed, than some of the pure and highly insulating gums. This being so, in the author's opinion, a thin but sufficient coat of high quality insulating gutta supplemented by a sufficiency (for mechanical purposes) of lower quality gutta outside it, might be found a serviceable form of dielectric. The capacity might then be further decreased —and the working speed correspondingly increased—by some more low quality gutta being incorporated with the jute serving, thus converting the latter into a part of the dielectric instead of forming, when wet, the inside surface of the outer (Leyden jar) plate. Thus, by increasing the thickness of the dielectric, the capacity would be materially reduced, and the speed correspondingly increased. The initial cost of the cable might also be appreciably lessened in this way.†

Moreover, the inductive capacity being thus decreased, the size of the conductor (for low resistance) might be reduced for a given speed. Should the electro-static capacity of a core be too high, it can usually be got down within the required specification limits by the gutta being subjected to further mastication. Owing, however, to the circumstance of high insulation resistance and low inductive capacity being liable to be more or less inseparable, it is very often found that when the capacity has been sufficiently reduced, the insulation resistance has gone up to a point above the specification requirements, in cases where a limit is placed in this direction with the object already named.

**Weight and Diameter.**—The weight of gutta-percha required for a core of D outside diameter, the conductor having $d$ diameter (both D and $d$ expressed in mils.), is approximately

$$\frac{D^2 - d^2}{490} \ldots \text{lbs.}$$

---

* It might appear as though the reciprocal to this—*i.e.*, the inductive *resistance* of the dielectric, or its *resistance* to induction—would be the more natural feature to consider in this connection ; however, the inductive *capacity* is what is always taken and stated.

† Such a plan would provide for the localisation of faults in a manner that would scarcely be practicable with some of the suggested methods for obviating the static charge and so increasing the working speed.

When the ratio between the weights $w$ of copper and W of percha—or relative diameter $d$ of conductor and D outside core—are given, the capacity and resistance of the core are the same (with given specific values), no matter what the absolute values are. Thus it is only necessary to ascertain the quotient $\dfrac{W}{w}$ or $\dfrac{D}{d}$, and the capacity or resistance is arrived at from the specific values of each.

Eastern Telegraph Company : Aden Station and Quarters.

## SECTION 4.—METHOD OF PURIFYING GUTTA-PERCHA.

**Exportation and Importation.**—Most of the raw gutta-percha imported into this country and Europe generally comes from Sarawak, from British North Borneo, and from Java. Large quantities are sent from these places to Singapore as well as to England for preparation for commercial purposes.

From the Malayan Archipelago it arrives at the merchants, dealers, and agents in England, Germany, India, and elsewhere, in forms varying roughly between that of a ball of about 1 lb. weight and a large block of $\frac{1}{2}$ cwt., the average weight of each lump being about 30 lbs. These irregular lumps are packed and sent home in baskets (in which it arrives at the cable factory) about $3\frac{1}{2}$ feet high by $1\frac{1}{2}$ feet wide, and holding about $1\frac{1}{2}$ cwt. of gutta-percha.

**Storage.**—As already stated, gutta-percha rapidly oxidises by exposure to air or light, and in a less degree by exposure to either alone, as well as by variation of temperature. For these reasons, then, it is not as a rule kept at the cable factory until it is actually required, when it is ordered from the dealers. It is then stored in dark rooms or cellars, maintained at a fairly uniform temperature, as free from air as practicable,* and is passed through manufacture as quickly as possible.

**Impurities and First Stages of Manufacture.**—Being sold by weight, the noble but enterprising savage is prone to incorporate with it all sorts of impurities, such as bark, clay, sand, and stones, or any other substance more plentiful and ready than gutta-percha itself—indeed, even pieces of iron have been added as "make-weights." The raw gutta often contains as much as 25, or even 30, per cent. of impurities.† Thus, to render it fit for any of the uses to which it is turned, it has to pass through a series of purifying or cleansing processes. In order to render the gutta-percha "workable" for these said processes it requires to be given a certain temperature—just above the boiling point of water. It is highly desirable,

---

* It may be remarked, however, that the tendency towards oxidation, when in this form, is but small as compared with its oxidising tendency during, and after, manufacture, when the surface exposed is so much greater, and the tendency is further encouraged by the presence of heat.

† It is the first care of the manufacturer to eliminate these foreign substances ; and should this process not be effectually carried out, the germs of decay may be left within the manufactured article, which would speedily lead to its destruction. The value of raw gutta-percha depends almost entirely on the care with which it has been collected and the honesty of the collectors—or, in other words, on the quantity of foreign matter mixed with it.

however, that this temperature be not exceeded—*i.e.*, it must be just sufficient to soften it and to drive off the moisture, but not to *melt* the gutta—especially when intended for insulating purposes, for fear of affecting its electrical qualities.

The first process of manufacture*—or rather, of purification—is that of boiling the lumps of raw material in a large covered tank (about 8 feet long by 4 feet broad by 3 feet deep) full of water maintained at boiling temperature by the injection of steam, or sometimes heated to a moderate temperature by a steam-worm. About 12 cwt. of gutta-percha can be put in each of these tanks at a time. This operation occupies some two, three, or even four hours, according to the proportion and nature of impurities, after which the hard lumps of raw material are converted by heat into fairly soft, spongy, porous, and plastic lumps, capable of further "working." By this process, moreover, a great deal of the foreign matter has been loosened (thus facilitating its extraction afterwards), and a certain proportion of the roughest impurities actually got rid of, inasmuch as any large lumps of earth or stone or other heavy foreign matter will have dropped to the bottom of the tank, whilst the gutta-percha (with a specific gravity of about 0.97) will float on the surface of the water.†

**Mastication.**—The lumps of plastic gutta then go through the process known as mastication by being put into a sort of "devilling" machine, technically termed a masticator, and originally the invention of Mr Thomas Hancock.‡

---

* Sometimes lumps of raw gutta are first chopped into smaller and more conveniently uniform pieces, either manually or by a machine, and this operation is, under some conditions, and at some factories, followed by one in which the gutta-percha is operated on by what is known as a "scarifier" or "ticker," by which it is torn up into still smaller fragments. This latter process has, however, more usually given way to the above, and is seldom employed in addition, being considered unnecessarily tedious nowadays.

† During the first cleansing process the gutta-percha has absorbed a considerable amount of water—as, indeed, it tends to during all subsequent washing, or steam-heating, processes—partly owing to expansion. This moisture requires, therefore, to be got rid of again before the gutta-percha is applied to the conductor. Otherwise, if only excluded under pressure when laid, it will leave holes in the dielectric sufficient to establish a serious fault. Moreover, moisture prevents adhesion—*i.e.*, if the gutta-percha contains more than a certain amount it will not adhere to the wire. This moisture is, therefore, thrown off by means of the drying masticator, the air which it has also absorbed being got rid of at the same time.

‡ It may be mentioned here that most of the machines we now have—mainly masticators in one form or another—for the purification of rubber and gutta-percha are but modifications of machines due to either of the Brothers Hancock, who did more than any one, perhaps, towards turning gutta-percha (as well as india-rubber) to a practical use generally, immediately after its introduction into this country by Dr Montgomerie. The first system of purifying gutta-percha on a large commercial scale was mainly due to Charles and Walter Hancock.

This (Fig. 9) consists of a cylindrical cast-iron casing B in which revolves a grooved or fluted cylinder A.

The material to be masticated being placed between the rotating cylinder and the outer casing, is carried round under such a strain as subjects it to a very powerful kneading action, so that it is ultimately reduced to a compact mass in the form of a long solid lump, somewhat similar to a sausage,* thus at the same time mixing and grinding together the molecules of the texture, as well as wringing the water out. The principle of the "masticator" is illustrated by Fig. 10. It represents, in

FIG. 9.—Ordinary First Masticator.

fact, a sectional end view of the machine, G being the lump of gutta which has been inserted at O, and is being worked, under considerable pressure and heat, between the revolving roller and outside frame, in order to expel the occluded moisture.

As will be seen hereafter, there are several different forms of masticating machines applicable to different stages of manufacture. That which we are at present dealing with is a type of "drying" masticator, its principal object being to wring out the moisture absorbed by the gutta during the

---

* The blown-out appearance of the gutta when in the masticator is due to the expansion of moisture in the gutta-percha, under heat, to the form of steam, which naturally inflates the material.

previous operation. This process, besides getting rid of further impurities,
also renders it still more plastic by working it into a sticky mass. By thus
thoroughly mixing the various qualities employed, a compact and fairly
homogeneous lump is secured.

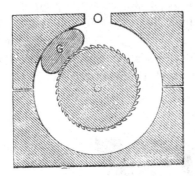

FIG. 10.—Sectional Diagram Illustrative of the Action of Masticating Machine.

Though the drying operation can be carried out in time by the different
portions of the gutta being, each in turn, exposed to the air, this machine
is sometimes steam jacketed to further effect this object, and so hasten the
operation. In such a case the pipe C (Fig. 9) serves to conduct the steam
to the inside of the hollow cylinder B for this purpose. The length of time
required for this operation depends upon circumstances, and is determined
by the foreman according to the appearance of the substance.

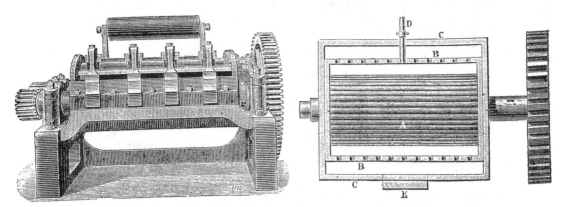

FIG. 11.—Washing Masticator.

The size of these masticating machines is represented by a weight of
about 2 cwt. Each of them is capable of working about 1 ton of gutta
per day.

The gutta-percha, in its inflated condition, is then put into the washing masticator, having water in it. This machine is shewn in Fig. 11, and is about 3 feet 4 inches long, by 2 feet 4 inches broad, by 3 feet 2 inches deep. It is, in fact, larger than the drying masticator to the extent of the extra space required for the water (say 100 gallons), so as to hold about the same quantity of gutta (roughly 50 lbs.)—indeed, each piece of gutta-percha is "worked up" in the different processes systematically and independently throughout. In construction, the washing machine is very similar to the drying masticator. It is not, as a rule, steam jacketed, but means are provided for the injection of steam into the body of the water in order to maintain it at the required (boiling) temperature, though the friction which the gutta experiences in the process goes a long way towards giving it the necessary heat for softening purposes. This machine (Fig. 11) consists of a solid cylinder A, grooved on the surface,* revolving inside a hollow cylinder B, which is pierced with holes and is itself enclosed in a second hollow cylinder C. The gutta is placed in the space between the grooved cylinder and the cylinder B, the space between the two outer cylinders being full of water heated by a steam jet D. The rotary motion of the grooved cylinder forces the gutta against B, bringing every portion of its surface, in turn, into contact with the water which carries away the impurities. These fall to the bottom of the cylinder C, which can be emptied through the door E.

The first boiling process in the tank has only the effect of separating the distinctly heavy impurities, such as stones, etc., that might be present. This last washing operation frees the gutta of a quantity of less hard and heavy impurities, such as particles of earth and sand, by thoroughly kneading them out, during the course of one to two, or more, hours, as the case may be.

An outside view of a similar machine,† for the same purpose, is given in Fig. 12. This shews the hinged cover, or lid, associated with every form of masticator, having two large holes R R. The entire arrangement is, in the above apparatus, enclosed in a large iron tank M N O P, which is itself filled with water. The impurities which are washed out fall to the bottom of the tank, and are removed at the close of the operation.

---

* The roller of the washing machine very often has much larger flutes than in the preceding (drying) masticator. Hence, though the entire machine is bigger, there is room for rather less gutta-percha at a time. These flutes are made larger with a view to more thoroughly tearing up the gutta-percha, so as to thoroughly free it of all impurities.

† Very usually known as the "Truman," after its designer, Mr Edwin Truman, a surgeon-dentist, who exercised himself much in such matters, and coated experimental lengths of wire with gutta-percha prepared by his process.

The gutta-percha is next placed in a drying masticator again,* of the same description as that previously set forth in reference to the second process, but of a smaller size. The outer casing of this machine is steam jacketed (instead of being subjected to the injection of steam) according to

FIG. 12.—Truman Masticator.

requirements, as in the previously described masticators. There are two reasons for this. One is so as to prevent any fresh moisture settling on the surface of the gutta, this being the last operation for its exclusion. The other is in order to more thoroughly and uniformly heat the whole

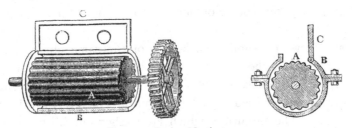

FIG. 13.—Drying Masticator.

material throughout—the degree of heat being much greater—with a view to "blowing off" nearly all the remaining moisture and absorbed water

---

* It may be said that the process of gutta-percha manufacture or preparation consists of alternate washing and drying operations. The latter is a necessary accompaniment to the former, in order to throw off the moisture previously taken up before serious oxidation takes place, while it is in a condition in which it can be readily got rid of without doing other damage.

from its pores. This masticator is usually about 1 foot 10 inches long, by 1 foot 4 inches broad, by 1 foot 4 inches deep.

In the last process, on account of the greater heat confined in a smaller space, and on account of the machine being lighter, the gutta squeezed between the revolving cylinder A (Fig. 13) and the outer casing B, forces open the entire lid C at each revolution, thus exposing every time fresh surfaces to the outside air, and in so doing giving off some of the water contained by the gutta-percha.

Moreover, at each turn, the sausage-like lump of gutta is cut up by hand, so as to more thoroughly effect moisture exclusion, as well as to prevent the gutta sticking at any part of the machine by the reduced amount of moisture at this stage of manufacture. Besides tending to entirely throw off all water taken up by the previous (washing) process, this operation also helps to knead out any remaining impurities.

We have now arrived at the stage at which gutta-percha is sufficiently prepared for ordinary commercial purposes, when it is then rolled into sheets in a manner described later, thus rendering it ready for storage and for sending away for further special manufacture as required.

**Straining.**—However, when great freedom from moisture is essential as well as a very high degree of purity—as in the case of the electrical insulation purposes, with which we are dealing—the gutta passes through a special (fifth) process of purification.

This process is, as will be seen, a purely mechanical one, it being found that the small residue of organic matter still remaining is better removed by such means—rather than by any further prolonged washing, favouring at the same time, as it does, the absorption of a further quantity of water, which becomes more and more difficult to extract later on.

The arrangement employed to effect this operation consists of a strainer or filter, in conjunction with an hydraulic press (Figs. 14 and 15).

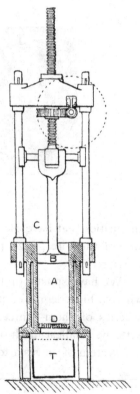

FIG. 14.—Hydraulic Gutta-percha Strainer.

Here A is a thick cast-iron cylinder (about 1 foot 3 inches diameter by 1 foot 6 inches deep), open at one end, and capable of holding 100 lbs. of gutta-percha. The gutta is placed in this immediately after the last (drying-mastication) process, whilst still in a thoroughly plastic state. In order to maintain the required plasticity, and to prevent the gutta-percha getting cool and hard—which is, of course, particularly important in this process—the walls of the cylinder are hollow, and steam caused to circulate in the intermediate space. Fitted to the bottom of this cylinder is a fine iron wire gauze plate, or sieve, D. The web of this gauze, or netting, is so fine that none of the gutta-percha finds a way through its meshes until the enormous pressure employed is brought to bear from above. The iron lid, or piston B, of the cylinder—connected to the piston rod C—is then fitted in its place, and the hydraulic ram system brought into force. By a slow, but powerful downward motion, at a pressure of nearly 1 ton to the square inch, the exceedingly fine, practically pure—or almost pure—gutta-percha is little by little forced down through the meshes of the sieve into the tank (or pan) T, leaving at the top—retained by the mesh—any

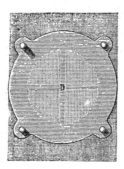
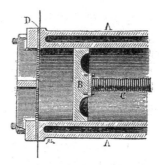

FIG. 15.—Horizontal Strainer.

impurities that have escaped the previous cleansing processes, such as the web of the wire gauze plate will not allow to pass, even under hydraulic pressure.

We have spoken of the steel wire gauze sieve, but, as a matter of fact, where a high degree of purity is desired, as in this case, there are usually a series of them—three, at any rate—placed one above the other,* that with the smallest mesh often having as many as some 10,000 holes, and on this system of sieves a total pressure of about 118 tons is brought to bear.

---

* The strainer, composed of three steel wire sieves, is usually composed as follows :—One with a mesh having twenty holes to the inch ; another, above that, with sixty-eight holes to the inch ; and then on the top of that one which has forty holes to the inch. The object of the top sieve plate is to take off the first heavy strain of the gutta from the finer wires of the sieve below it ; that at the bottom is for the purpose of holding up the middle sieve better. These sieve plates are 15 inches in diameter.

By this means, then, those lighter impurities are separated from the gutta, which the previous operations have failed to extract.

The receptacle T for the gutta freed from its impurities is of such a size as will contain a conveniently portable amount. Thus, when one pan is full, another is put in its place to receive the gutta-percha, until it is all strained through—or rather, all that will come through. The enormous hydraulic pressure must then at once be taken off, in order to avoid putting a strain on the wire netting. One cylinder-full of gutta-percha (100 lbs.) takes about a quarter of an hour to be strained off by this hydraulic machine. About a ton of gutta can, therefore, be easily strained during a working day.

This machine could do much more, but where there are very usually about half-a-dozen such machines at a large gutta-percha factory to some thirty "washing" and twenty "drying" masticators, one ton per day is the most that is ever required from the strainer to keep in time with the much slower rate of mastication, the strainer holding about three times as much as any of the masticators.

There is a certain convenience in this straining machine taking the horizontal form where space permits, as facilitating arrangements for working. On the other hand, a less even pressure is liable to occur than with a vertical machine, owing to the gutta tending to settle on the lower wall of the cylinder.

Where the gutta has been passed through this last refining process, it is then usually also passed through another steam-drying masticator process, though neither of these would be necessary, or desirable, where the gutta is only for ordinary commercial purposes.

The preparation of the gutta may now be said to be really complete, even for the finest purposes of refinement.

**Calendering into Sheets.**—With a view to the lumps being transformed into sheets,* for the sake of general convenience and portability, the gutta is then taken direct to what is commonly known as a calendering machine (Fig. 16). This is constituted, in the first place, by a pair of smooth or "chilled" faced horizontal iron rolls one above the other, which revolve in opposite directions. The lump of gutta, being placed between the faces of these rolls, is drawn in between them, and thus pressed into coarse sheet

---

* In some ways it appears as though this process might suitably be dispensed with under conditions where the gutta is at once required for covering wires, especially in view of the fact that it is when in sheet form that gutta tends to decay most, the degree of oxidation being directly proportionate to the surface exposed to the air.

form, the thickness of which is according to the distance they are set apart,* which varies with the requirements.†

As it emerges from the rolls in the form of sheet, the gutta is carried forward along an endless cloth-web band (of the same width as the rolls) actuated by "live" rollers. As it is drawn from the rolls the sheet is cut into convenient lengths for storage. The width of the sheet is usually about 3 feet (the length of the masticator cylinder and calender rolls), and it is cut across, on drawing off the cloth of the "calender," at distances of about 5 or 6 feet.‡

Quite apart from storage, the gutta, even when it is going to be used at once—as is often the case in a cable order—requires (by present arrangements) to be thus rolled into sheets for the sake of convenient porterage to

Fig. 16.—Calender Machine.

the core shop. In sheet form several sheets can be conveyed at the same time by being stacked one above the other on a truck. If not required immediately the sheets are piled up on shelves under cover.

---

* These rolls are hollow, and steam is passed into them by suitable pipes entering at the axis. Thus the gutta is slightly reheated, and is maintained in a plastic condition.

† The distance between these rolls requires to be very finely adjusted. Gutta-percha sheet for subsequent core purposes is usually made about $\frac{3}{4}$ inch or $\frac{1}{2}$ inch thick.

‡ According to one method of carrying out this process there are two sets of rolls, or "calenders," used. The gutta is first passed through one set three or four times—which roughly rolls it down to about $\frac{1}{2}$ inch—and cut into convenient lengths during the operation. It is then taken to a pair of much heavier and stronger calenders (steam-heated), which roll it down by degrees to exactly the required thickness, with a fine smooth surface. This modification is usually, however, for experimental, or fine sheet, purposes.

**Purifying Processes; Rate of Work, etc.**—Thus end all the operations of refinement and preparation which the gutta-percha goes through in the masticating shop from its first appearance at the cable factory as raw material. In a day's work (of fourteen hours) from 2 to 3 tons of gutta can be passed through all the above-described operations in an average factory, the exact amount depending on the degree of impurity of material, and still more, of course, on the number of the machines.

Owing to its plastic and comparatively non-elastic nature—which allows it to readily assimilate foreign matter—the preliminary preparations in the way of washing, mastication, etc., are, speaking generally, much more extensive than in that of india-rubber. Subsequently, on the other hand, its manufacture in any particular form for a particular purpose is much simpler, there being no nicely-proportioned admixtures to make with it.* There are, however, various methods of treatment of gutta in the masticating shop at the different factories. We have only attempted here to describe one routine. By some methods less washing—or less drying—may be done, even when the gutta is for electrical purposes, and, in some instances, actually more, perhaps. Again, the order of the various processes described may be sometimes reversed, though alternate gradual washing and drying usually forms the general basis of operations in any case. And yet again, the time periods assigned for each operation, and the temperatures adopted, are all matters of opinion based on practice with different gums and mixtures under different circumstances.

As has been before stated, the hydraulic strainer, and the corresponding process of dry mastication following on it, are only necessary where great purity and great freedom from moisture are essential, as is the case when the gutta is required as an insulator.

**Driving off Moisture: Temperature Suitable.**—Moisture has a most damaging effect on a core, being a conductor, and as being indirectly the cause of air-holes very often—or at any rate of holes, whether air fills them or not—on account of rendering the gutta-percha spongy after mastication. It is for this reason that most of the moisture contained by the gutta in its early stages of manufacture is driven off in the form of steam by the

---

* Though in present practice no ingredients form a part of the typical gutta-percha manufacture—which should be preserved as a hydro-carbon—in early days sulphur has been freely mixed with the gutta-percha insulator, with the idea of increasing its insulating qualities. It was, however, found to act injuriously on the gutta as well as on the wire, even when "tinned," unless some protective "jacket" were used (as in india-rubber core), and nothing was found to work homogeneously with gutta-percha, or behave well, chemically, in its company.

application of heat during the various processes of mastication.　If the moisture is not so driven off before the gutta is applied to the wire, it is liable to be expelled subsequently when the core is submitted to any pressure under the water, thus leaving a hole in the insulating envelope.　In such a case it probably causes a serious electrical fault at a time when the remedy is, at the best, difficult and costly.

What are technically termed air-holes in gutta-percha core are due to the air forcing its way out of the core under pressure on submergence. This is a very usual cause of that which is commonly spoken of as a "factory fault," being generally brought about in the course of construction, owing to some air not being excluded from the gutta-percha during manufacture.　Not escaping till forced out under pressure, it then bursts through the gutta, and thus causes a hole, perhaps extending as far as the conductor, and so placing it in more or less direct electrical communication with the earth.

On the other hand, a certain amount of moisture must be retained in order that the gutta may preserve its physical qualities.*　The fact is that besides aiding plasticity and rendering a material less liable to physical decay, a film of moisture has a tendency to prevent a material oxidising readily by preventing air getting into its pores.

Again, gutta-percha, like everything else, has a greater affinity for oxygen when at high temperature ; indeed, if the heat be sufficient, the chemical action takes the form of actual burning.

Thus, though a certain temperature is desirable for driving off the greater portion of the moisture, and for rendering the material sufficiently plastic and workable, there is a limit to the degree of heat, and to the length of time, during the course of manufacture, for which the gutta may be subjected to such heat, on both of the above counts.　These are most important matters in the preparation (or manufacture) of gutta-percha.

---

* Unfortunately, in getting rid of the wood, sand, stones, etc., in raw gutta during the process of washing and masticating, much of its natural oil also leaves it.　Indeed, if masticated sufficiently, it would (by oxidation) become powder ; and in this state, though offering the highest resistance and usually the lowest electro-static capacity, is very dry, hard, brittle, and, consequently, non-durable.

## SECTION 5.—WILLOUGHBY SMITH'S GUTTA-PERCHA.

By far the greater length of cable at the bottom of the sea is insulated with gutta-percha prepared in the masticating shop according to a plan of the late Mr Willoughby Smith, introduced by him in 1869, and first adopted over the manufacture of the Suez-Aden-Bombay cables in that year. This method has been in daily use at the Gutta-percha Works of the Telegraph Construction and Maintenance Company, to the exclusion of any other, ever since its introduction.

**Nature of the Process.**—Mr Smith's method consists mainly in freeing the pores of more moisture than is done with so-called ordinary gutta for the same purpose. This is effected in the drying of the gutta-percha, after being treated in the washing masticator, by submitting it to a dry heat (above 100° F.) for a short time.

**Effect of the Process.**—The important practical result thereby obtained is that of reducing the electro-static capacity,* and thereby increasing the speed in the same proportion.†

**Data and Formulæ.**—The electro-static capacity per N.M. ($k$) of Smith's gutta-percha is, in fact, approximately

$$k = \frac{0.1516}{\log \frac{D}{d}} \text{ microfarads}$$

or about one-fifth that of ordinary gutta, though its specific capacity is about 100 as against 98 for vulcanised india-rubber.

In throwing off more moisture oxidation is also further promoted.‡ This, however, does not appear to affect the material at all seriously, for its physical qualities are lastingly quite equal to those of the "ordinary" gutta, and its actual mechanical (tensile) strength—on account of slightly greater textile density—is said to be quite 10 per cent. greater than that of "ordinary" gutta-percha.

---

* It has been stated that this decrease of capacity is secured by the admixture of tar with the gutta, but this assertion is incorrect, it is believed.

† The only way that this could be actually proved would be by laying two precisely similar cables side by side at the same time, the dielectric of the one being composed of "ordinary" gutta-percha and that of the other core, of equal dimensions, of Willoughby Smith's improved gutta. The relative speeds obtained on each under similar circumstances would then determine their true relations in this respect.

‡ This may be seen from its somewhat darker colour.

The dielectric resistance of this gutta is invariably a good deal below that of ordinary gutta.* Thus, where an ordinary gutta-percha core, of very usual dimensions, will offer a dielectric resistance represented sometimes by as much as 2,000 megohms per N.M. at 75° F., a gutta-percha core of the same dimensions prepared by Willoughby Smith's process would yield 600 megohms per N.M. insulation.

The dielectric resistance per N.M. ($r$) of Smith's gutta at the standard temperature (75° F.) after one minute's electrification is, in fact, approximately

$$r = 350 \log \frac{D}{d} \text{ megohms}$$

or scarcely half that of ordinarily prepared gutta.

The rate of variation of resistance by temperature is rather greater in Willoughby Smith's prepared gutta-percha than with gutta as ordinarily prepared, though by logarithmic law, of course. The coefficient for resistance variation by temperature of Smith's gutta-percha, as determined by the inventor of the process, used to be taken at 8.0 per cent. per degree Fahrenheit. It is supposed to be rather less under present prevailing conditions, or, at any rate, more nearly resembling that of ordinary manufactured gutta.

It might be thought that inasmuch as Willoughby Smith's method of manufacture gets rid of more moisture that it would be more capable of absorbing water on submersion afterwards, and that on this account the advantage previously gained as regards reduction of capacity would be lost again—or more than lost again—after being laid at the bottom of the ocean for some time.†

This does not, however, appear to be the case. Anyhow, the working speed is found to remain as constant with Willoughby Smith's as with an ordinary gutta-percha dielectric.

---

* This is not what would be expected from what has already been said regarding the effect of freeing gutta of moisture, and also on account of the additional oxygen. Moreover, the resistance being lower it is somewhat remarkable that the capacity should also be lower. Any increased amount of oxygen and reduction in moisture would account for this, but only consistently if the resistance be greater instead of less.

† There would, in any case, often be great difficulty in actually settling this point, owing to the difficulty of accurately arriving at the capacity of a great length of cable on account of retardation and electrical absorption. Here again, this could only be absolutely settled by laying two cables, one insulated with ordinary gutta-percha and the other with W. Smith's, of precisely similar dimensions and length, and compare their capacities (or speeds) with one another both after manufacture (before laying) and also at a certain period—*subsequent* to submersion.

## Section 6.—Manufacture of Core: Covering Conductor with Gutta-percha.

Following the gutta-percha from the masticating department, it comes to the core shop in sheets (as previously described), where they are laid up ready for use, soon becoming more or less hard.

**Boiling down from Sheet Form.**—As any sheet is required, it is placed in a tank of boiling water. The tank is similar to that in which the lumps of raw gutta-percha are first placed, but rather smaller. The gutta-percha sheet is just laid on the top of the water, and its surface washed. By the heat of the water it is softened, and once more rendered plastic, ready for being "worked" once more in a convenient form and suitable condition for its future use.

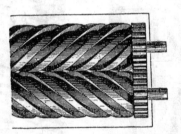

FIG. 17.—Final Drying Masticator (Plan of Rollers).

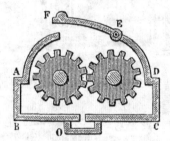

FIG. 18.—Final Drying Masticator (End View).

**Final Mastication.**—The gutta is then taken to a drying masticator again, similar to those in the masticating shop.* In this last drying machine, however, a slight modification has been made of late years. There are usually two cylinders instead of one, which are made to revolve in opposite directions, thus drawing the gutta between them. Moreover, these rollers—in recent forms—are, as a rule, spirally grooved (Fig. 17). These cylindrical rollers are enclosed in a rectangular iron tank with double walls (Fig. 18), between which steam circulates. It is steam jacketed, in fact. The tank is closed at the top by a half cylindrical cover A E D, which has a movable lid E F kept closed by strong iron bars. The gutta being drawn between the rollers is forced into the grooves, fresh surfaces

---

* When a big cable order is being effected, it is very usual to take the gutta straight from the hydraulic strainer to the masticator in this, core, shop (instead of converting it into sheet and storing it, etc.), to save time.

being in turn more thoroughly and continually exposed to the hot air in the tank.   The condensed water collects at the bottom, and can be drawn off at the outlet O, as required.

A more recent pattern of the double cylindrical rollers (as commonly used in some of the French gutta-percha factories) is shewn in Fig. 19. Here the spiral grooves on the cylinders are interrupted at intervals of a few inches, so as to renew the surfaces of exposed gutta still more rapidly.

Thus the gutta-percha is now rendered ready, worked up to a proper consistency for covering the conductor by insertion in the core, or covering machine.

The final process of mastication, besides rendering the gutta plastic

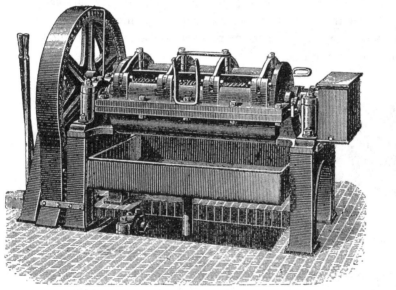

FIG. 19.—Special Drying Masticator.

and again ready for further application, also has the effect of thoroughly squeezing out any additional moisture taken up by the washing process. The latter is most important at this, the last, stage of preparation.

Fresh surfaces being continually exposed to the air in order to get rid of the moisture, a certain amount of oxygen is at the same time unavoidably absorbed in exchange.   The gutta then gradually turns brown, and the workman who examines it at intervals knows by the colour and consistency when it is ready to be taken out.*

---

* In point of fact, by stretching a fragment of gutta between the fingers until extremely thin it appears translucid, and the slightest trace of remaining impurity can be detected, even by unskilled eyes.

**Requirements for Insulating Gutta-percha.**—For the manufacture of insulated telegraph wire—as well as, in some degree, for many other applications of gutta-percha—it is necessary to have the gum absolutely free from impurities, and almost free from moisture. Thus, these processes of alternate washing and drying mastication require to be repeated several times, as has already been shewn. The last operation (in the core shop) is, however, final, and, at the same time, has the effect of converting the material into a perfectly homogeneous paste.

The efficiency of a submarine telegraph system is governed very much by the value of the insulator—*i.e.*, its degree of purity—the nature of the mixtures of gutta-percha employed, the various kinds and proportions of gum adopted, besides the manner in which the working and mixing is performed immediately before being placed in the coring machine. Indeed, the success ultimately attained largely depends on the care of those in charge of the successive stages of manufacture.

The gutta used for purposes of insulating the copper conductor is generally composed of various qualities, some selected for the sake of their physical, and others for their electrical, properties, in such proportion that will meet the conditions specified for the cable, and at the same time be a durable material.

**Principle of Gutta-percha Covering Machines.**—As already stated, Werner Siemens (for Siemens and Halske in Germany) was probably the first to apply gutta-percha under pressure whilst in a plastic state by means of a die in a *seamless*, tubular, form on a practical scale. His machine may, therefore, be looked upon as the prototype of those in use nowadays in their various modified and improved forms.* Siemens' tubular core machine was, of course, very similar to machines for making —(1) macaroni ; (2) lead-piping ;† (3) bricks from clay.

There are several different gutta-percha covering machines in the present day. An ordinary cylinder and piston is, perhaps, the oldest form amongst those still in vogue for feeding the conducting wire with a coating of gutta-percha. In this, the gutta is placed in the cylinder, and the piston

---

* Previously, both Bewley on behalf of the Gutta-percha Company, and Siemens and Halske had covered conductors with gutta-percha in strips, after the manner of india-rubber ; but much trouble was experienced with this type of gutta-percha covering owing to moisture getting in at the seams. Bewley had the first patent for covering wires, but Siemens' machine for seamless coats was the first practical success met with in this direction.

† The late Mr John Chatterton was originally a lead-pipe drawer in Wharf Road. Soon after joining the Gutta-percha Company he applied this form of machine to gutta-percha core manufacture.

—usually actuated by a screw motion—forces it out again into a die,
through the holes of which the wire is drawn; and, therefore, then draws
with it a certain thickness of gutta, according to arrangements. Such
machines have been set up both vertically and horizontally, the gutta being
forced through either the bottom or the side of the cylinder as the case
may be.*

The objection to a machine of this sort in its crudest form is that air
enters the cylinder with the gutta, and thus tends to make a way into its
pores—being hot and plastic, whilst inside the cylinder chest—when drawn
through the dies round the wire.† As has been already pointed out, air
getting into the body of gutta tends to force its way out under any outside
pressure, or under molecular expansion due to increased temperature, the
result being what is termed "blowing" of the gutta, leaving holes or
chambers which, if sufficiently deep, may be the cause of serious electrical
faults, and into which more air may find access, leading to further trouble
of the same description.

**Gray's Gutta-percha Covering Machine.**—However, in 1879, a patent
(No. 5,056) was taken out in the name of Mr Matthew Gray, for an
ingenious gutta-percha covering machine—being a new combination of
well-known mechanism—which very perfectly overcomes these difficulties,
and which, on account of its general perfection, is selected as a specimen
for detailed description here.‡

It is illustrated in end elevation by Fig. 20, and in side elevation by Fig.
21; a plan view is given in Fig. 22. The plastic gutta-percha, when it is
brought to this machine, is laid endwise, in cylindrical lumps of about 4 lbs.
at a time, between the pair of horizontal steel rolls D and D¹ (Figs. 20 and
22) heated with steam or hot water,§ whose axes are in the same horizontal

---

* These usually consist of two cylinders side by side, with a common channel
between them for the entrance to the die, so that whilst one has just been drawn from and
is being recharged, the other full one is supplying the die chest with more gutta for being
drawn off by the wire continuously.

† It was thought that any defects caused by the presence of air or moisture in one
coating could be remedied by a subsequent coating, but this has proved, in practice, to
be fallacious.

‡ Mr A. Le Neve Foster also appears to have been prominent with regard to the
application of machinery of the above type for the purpose in question.

§ The object of these rollers (with their bearings in the main frame F) being heated, is
partly to maintain the plasticity of the gutta by again heating up the thin skin, but also to
assist in working out the air and moisture from the gutta-percha in the drawing of the
gutta between their adjacent surfaces.

To further assist in keeping up the temperature of the percha, a shade is usually
affixed at a little distance above the rollers where the gutta is inserted.

plane, and which act as a feeding apparatus for supplying the material to the rest of the machine. These rolls are so actuated that they revolve towards one another (their rotary motion being obtained from the main shafting), and the gutta-percha laid on the top is thus drawn in between them. The distance between the rods is adjustable as desired, but $\frac{1}{4}$ inch is about the usual maximum distance separating them, and $\frac{1}{8}$ inch is, perhaps, more frequently adopted. The result is that but quite a thin sheet of gutta is drawn in at a time, the object being that any air or extra

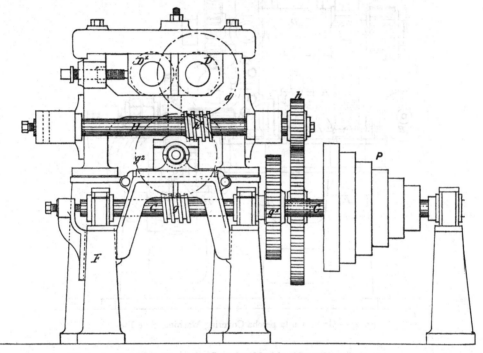

FIG. 20.—Gray's Gutta-percha Covering Machine (End Elevation).

moisture which, in the course of manipulation, has found its way into the body of the gutta may be effectually squeezed out, besides causing there to be no room for anything to enter besides actually the percha itself. Indeed, the advantage of this over the earlier forms of similar machines (in which the lump of gutta-percha was merely forced into a cylinder by an hydraulic ram) can scarcely be over-estimated.

**Practical Application of Gray's Machine.**—As the gutta is drawn down in a thin sheet by the rollers it is conveyed into end guides E E, which, tapering off at their lower extremities, convert it into the form of tubular cord,* in which form it is forced down the vertical pipe E¹, leading into a horizontal cylindrical receiver A (Fig. 21), carried by the frame F, and kept at a uniform temperature by steam, or hot water, jacketing. In this cylinder a worm, or endless Archimedean screw C, working on a shaft, is kept uniformly revolving in a certain direction and at a definite required speed.† The tube of gutta falls into, and fills up, one of the threads of

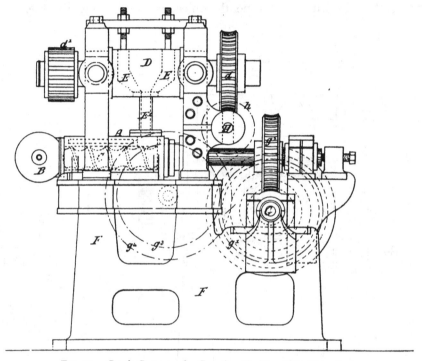

FIG. 21.—Gray's Gutta-percha Covering Machine (Side Elevation).

this screw, supplying it with a continuous length, under a definite pressure and at a regular speed. The object of this screw is that, as fast as the gutta-percha arrives in the cylinder, it shall force the gutta uniformly forward

---

* The object of placing a limit to the bulk of gutta-percha supplied to the cylinder is partly to avoid a strain on the worm or the cylinder.

† The speed at which this propelling screw is run is regulated by the cone-shaped pulley P (Figs. 20 and 22), which is on the main shafting. The strap on the main shafting is adjustable on to any section of the pulley, according to the speed required for the screw. This again—as will be seen later—depends on the speed of drawing off the wire, varying according to the thickness of insulation in question. Each section of the pulley usually corresponds to a particular speed of drawing off such as is adopted for a given type of core.

There are certain previously arranged and fixed rates for the screw according to

along the length of the cylinder, under a required pressure, through the entrance of the die chest or "nose piece," connected to it at its further end, through which the conducting wire is being drawn. Thus, in fact, the die box is continually fed with a regular supply of gutta at an unvarying uniform pressure, rate, and consistency.*

As the conducting wire under operation is drawn through a hole of a

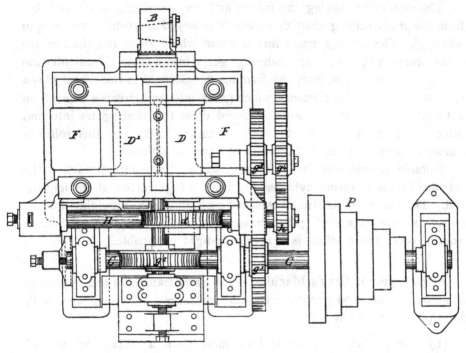

FIG. 22.—Gray's Gutta-percha Covering Machine (Plan).

certain size at the further end of the die chest, it thus draws with it (under a uniformly high pressure) a coating of a definite thickness—depending on

---

above requirements. In a similar way the speed at which the horizontal rollers are worked to feed the cylinder is also thus regulated—being, in fact, from the same shafting.

These rates, together with the speed of drawing off, may be adjusted to a nicety so as to give exact results—*i.e.*, so that the gutta-percha may get thoroughly cooled in passing through troughs; and yet that there shall be sufficient, but not too much, gutta uniformly forced on the wire.

* The only mechanical point which limits the speed at which the propelling screw can be run is that of safety. Thus, if it were run too fast, it would be liable to strain, or even burst, the die cylinder by forcing more gutta-percha in than there is room for. On the other hand, in order to give a certain required thickness of coating (as provided for by the die holes), the screw must be run up to a given speed in proportion to that at which the conducting wire is being drawn off.

the pre-arranged relationship between the diameter of the die hole to that of the wire—of the plastic gutta-percha, which, under pressure, adheres very firmly, especially as the conductor is first coated outside with compound. It is of the utmost importance that the gutta be kept thoroughly warm and plastic—in fact, in a "tacky" (*i.e.*, sticky) condition—right up to the time that it is applied to the wire, so as to ensure it laying on properly. For this reason, therefore, more or less the whole of the machine is steam jacketed.

The motion for driving the rollers and the propelling screw is taken from the main driving shaft G, to which is keyed a worm $g$ and a spur wheel $g^1$. The worm $g$ gears into a worm wheel $g^2$ on the shaft of the screw propeller C, and the spur wheel $g^1$ gears into a wheel $g^3$ mounted on a stud axle projecting from the framing F. Secured to this wheel $g^3$ is a spur wheel $g^4$ of larger diameter, which gears into and drives a wheel $h$ on a horizontal shaft H. A worm $h^1$ keyed upon this shaft gears into and drives a worm wheel $d$ on the axle of the roller D, and this roller is connected with the roller $D^1$ by means of a pair of pinions $d^1$.

Suitable provision is made for the escape of any air at starting the machine by the receiving cylinder, or chest, being pierced at its forward end. The die chest has also an overflow pipe and cock, providing for the escape of any excess of gutta under certain circumstances. When in working order, this machine is as nearly air-tight as possible.

**Advantages of Gray's Machine.**—The great features of this apparatus as an improvement on previously existing machines for covering wires with gutta-percha are briefly :—

(1.) The exclusion of air and of more than a certain amount of moisture,* by there being only space for a thin sheet of gutta to be drawn in between the feeding rollers.

(2.) An unvarying pressure of gutta being maintained by means of the screw being worked uniformly. The result of this is a uniform supply of gutta, both as regards quantity and consistency—in other words, a uniform homogeneity and thickness of coating throughout.

---

* The ill effects of any material quantity of moisture remaining in gutta-percha about to be applied to a conducting wire in the form of core is the same as that of air; namely, when afterwards it is driven off—either by heat, by the pressure at the bottom of the sea, or by being subjected to some abnormal temperature, say, in the ship's hold—spaces will be left in the core, and a little hole, possibly extending to the copper conductor, will be the result.

A certain proportion of moisture must, however, be retained in the gutta in order that it may preserve its physical qualities, and not decompose.

**General Operation of Covering Wires with Gutta-percha.**—We have now described the manner in which the gutta-percha is led on to the wire, taking the above machine (in use at the Silvertown Works) as a specimen for carrying this out. We will, therefore, now proceed to consider the manner in which the conducting wire is drawn through such a machine for receiving its gutta-percha insulating covering.

As each length* of copper strand is completed, it is taken on its carrying drum, or reel, to the core shop ready for being covered with gutta-percha.

At the back of the gutta-percha covering machine there is an inclined wooden frame, or stand, A (Fig. 23), for the reception of several such iron carrying drums, the spindles of which are mounted on this frame, one in

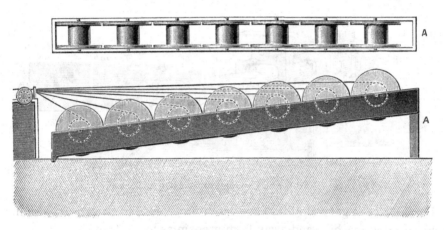

FIG. 23.—Copper Wire Drums mounted on Stand.

front of the other, the exact number depending on the number of wires the machine is designed to cover at a time. From an economical point of view it is very usual for these machines to be made capable of covering at least six wires at one operation—*i.e.*, by having holes in the die for six separate conductors to be led through. The wire is very slowly hauled off its drum through the die of the covering machine by gear connected with shafting.†

On its way it passes through a series of metal combs placed over jets of

---

* Varying from about 3 miles to about 1 mile, according to specific weight of stranded conductor.

† Brakes suitably check the motion of each bobbin, and thus ensure an equal tension on each wire.

gas, and then through a small tank of Chatterton's adhesive compound,* maintained, by steam jacketing, in a semi-liquid state—of the same description as was applied to the central wire of the strand previous to "laying up." The object of the gas flame is to warm the wire before receiving a thin coat of the compound, so as to cause it to stick better, and also with a view to driving off any moisture or foreign organic matter that may possibly have deposited itself on the surface of the wire. The object of the compound is to cause the gutta to adhere to the wire in a thoroughly compact and permanent manner, which, at any rate—in the case of a stranded conductor—cannot, on account of the interstices, be otherwise absolutely relied on. By means of the compound these interstices are effectually filled up, the compound acting as a complete and fairly permanent bond of union between the wire and the gutta-percha.†

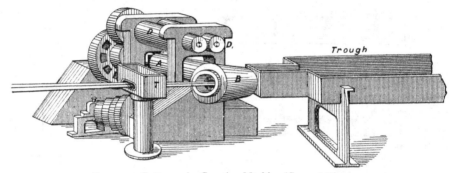

FIG. 24.—Gutta-percha Covering Machine (General View).

Fig. 24 gives a general idea of the wire's progress from its carrying drum to the covering machine and away from it again as "core."

---

* The above compound, as previously prepared—and referred to already—is supplied in a perfectly refined state from a small cylinder, which is filled from time to time, and is fixed just above the small tank. At the bottom of this cylinder there is a strainer, and at the top a tightly-fitting piston lid, which, by means of a screw, is gradually worked down the cylinder by hand, thus effectually pressing the pure compound through the strainer into the tank immediately below it.

† It may be mentioned that this compound is also sometimes used purely for purposes of adherence, as in the case of solid wire conductors about to be covered with gutta-percha.

Before Chatterton's compound came into use the copper conducting wire was liable—after the subjection of any strain on the core—to start out of the gutta-percha envelope. This would come about owing to the gutta having so much more elasticity than the copper. The percha tends, in fact, to return to something like its former length, whereas the copper remains pretty nearly at its elongated length due to the previous strain. However, by the application of Chatterton's compound to the wire, the gutta-percha was soon found to adapt itself to the copper wire in this respect.

Where the gutta-percha is applied in more than one coat it is highly desirable that the successive coats should be equally well united ; or the inner coat *alone* would conform with the wire as above, and a general disruption occur.

The wire is led into the compound tank T (Fig. 23) through a frame with holes sufficiently large for the passage of the wire, and out again by another frame with holes of a size just large enough to permit of a very thin coating, or film, of compound remaining on the wire.

It is essential that this compound shall be of exactly the right thickness, as if too liquid it would not adhere properly, and, if too thick, it is unworkable and liable not to produce satisfactory or uniform adhesion.*

The so-compounded wire, on its way to the gutta-percha covering machine, passes over another set of gas jets so as to warm up the compound immediately before-hand in order to effect a satisfactory union with the gutta from the die box of the covering machine.

We have already dealt with the feeding of the gutta-percha die chest by the covering machine. We will now, therefore, consider how the gutta is laid round the wire to the required thickness by means of the said, specially constructed, die mould.

The wire passes through the middle of the die box B (Fig. 24) by a pipe.

FIG. 25.—Die Box of Gutta-percha
Covering Machine.

FIG. 26.—Die and Cap of Gutta-
percha Covering Machine.

The gutta-percha, on arriving at the nose-piece, enters the die (Fig. 25) inside by the channels each side of it. It is not allowed to come into immediate contact with the copper wire, as by so doing it would, by its weight, be liable to bend the latter, and thus the gutta would not lie round it evenly.

The wire leaves the pipe by a nipple (shewn in *a* of Fig. 26), and then passes through a corresponding hole of a cover, or cap (shewn in *b*), there there being about $\frac{1}{16}$ inch space between the nipple and the hole, which is also slightly larger than the nipple.

The gutta-percha is, at the same time, forced through the channels on each side and through a series of very small holes about $\frac{1}{8}$ inch in diameter.

Having passed through these under great pressure, the gutta then meets the compounded copper wire in the allotted space between the die and its

---

* The quantity used for the required film is extremely small, and is, for instance, represented by somewhere about 1 lb. per N.M. for every successive coat—maintaining the same *thickness* for each—in an ordinary $\frac{107}{140}$ core. This quantity is always included in the total weight of the dielectric.

cap, which is just slightly in excess of the length of the nipple. The size of the cap holes are arranged according to the thickness of coat required. The wire, as it passes from the nipple to the cap hole, draws with it a coating of the hot, plastic, gutta-percha, which is pressed round it under considerable force, in the small space between the end of the nipple and the cap hole. A final pressure of the gutta-percha round the wire is effected in passing through the cap hole immediately in front of the trough.*

**Cooling and Hardening Process.**—The coat of gutta-percha having been applied round the wire in a hot and plastic state, the wire—which has now become what is termed "core"—is liable to get damaged as regards the insulating covering during its further movements, especially on being drawn round the collecting drum. Steps are, however, at once taken to cool, and so harden, the gutta-percha covering. Accordingly the covered wire, on leaving the die, is drawn very slowly through a long trough, some 200 feet in length (the near end of which is shewn in Fig. 24), filled with clean water, maintained at an exceedingly low and constant temperature.

The temperature aimed at is very usually 40° F. (4.4° C.). This should not at any rate exceed 55° F. Well-water answers the purpose; but large cable factories usually employ refrigerating machinery for feeding the various troughs (belonging to different machines) with water at the required temperature. This consists of ether machines worked by horizontal engines, as a rule.

According to a very generally adopted system, the core is drawn down the length of this trough (Fig. 24) to a drum at its further end, which, being driven very slowly from the main shafting of the shop, gives the necessary

---

* It is not improbable that means could be adopted for increasing the resistance of newly manufactured gutta-percha to its full limits at once—*i.e.*, to as high a figure as mechanical set, or drawing up, increases it in time.

This might be effected by each coat being, at this stage, drawn between two rollers at various consecutive points—say, along the trough.

The result would be the forcing out of moisture, especially at the first set of rollers just beyond the die.

Care would, of course, have to be taken not to drive out too much moisture, or the material would lose its gum-like qualities, and would become subject to serious decay by absorbing oxygen into its now loose pores.

The advantage gained would be that of effecting the mechanical set without keeping the core back after manufacture and before testing. A certain amount of drawing-up action is at all times desirable to render the material durable. It is possible, moreover, that by this means an inferior and cheaper insulating substance might be rendered suitable, mechanically and electrically—as regards capacity as well as resistance.

hauling-off motion. The core takes one turn round this drum, and is then drawn back again through the trough and round another drum just above, near the covering machine. It then repeats its journey down the trough to the drum at the bottom, and, after passing round this, it once more returns up the trough and half-way down again to a large carrying drum mounted on a high framework or scaffolding. This reel is revolved by belting from shafting under the water trough, its rate of motion being maintained in proportion to that of the other intermediate hauling-off reels worked from the same shafting. On this last reel the core is finally collected, by a process of automatic coiling, ready for subsequent operations.

To ensure even winding, the core is led to this collecting reel between upright guides, which are moved to and fro by a double-threaded screw parallel to the axis of the reel. The belting, which transmits the required drawing-off motion to all the various reels, is usually made of gutta, bands of this material being naturally at hand.

In passing three times up and down this trough, in the manner described, it is thus, in reality, passing through as much as some 1,200 feet of water at about 40° F., and as the drawing-off motion is very slow, the hardening effect is quite perfect, though the precaution is usually taken of testing the covering in this respect, by hand, from time to time.

**Examination of the Insulated Wire.**—The core now passes through a process of examination, in a special room, with a view to the detection of any possible flaws. This is carried out by the core being uncoiled from its reel on to another, and on the way being held between the fingers by an experienced workman, who, by practice, becomes very expert at detecting by touch any defect in the insulating covering. In the event of a hole, crack, or any sort of mechanical irregularity being discovered, the place is made good by the application of a hot iron to the surface, or by slightly warming the gutta over a wood naphtha lamp and working it up between the fingers. If necessary, fresh gutta-percha can, of course, be added to the weak spot. Should the defect consist in the presence of any foreign substance, this is, of course, extracted.

**Measurement.**—Where this completes the entire core—*i.e.*, where no further coating is required, or where the total thickness of gutta is laid on at one operation—it is then also measured whilst undergoing examination.

In the event of subsequent coats, however, being applied, this measurement is more usually only taken on the completion of the entire core, after it has received all the strain which comes into force in drawing off for each process.

Where subsequent coats are to be applied, the core is very commonly kept for one or two days before it is taken to the covering machine for its second coat, for purposes of maturing.

**Thickness of the Covering.**—As regards the thickness of the insulating covering, theoretically speaking—so far as electrical requirements go—a thin film, or "varnish," of gutta would be all that is necessary in many instances for the purposes of insulating the conductor sufficiently, as well as for acting as an efficient dielectric from an electro-static capacity, and signalling, point of view, especially if the jute packing be worked up as a dielectric substance in the manner already pointed out.

For efficient mechanical protection, however, something more than this is necessary, though the minimum thickness considered essential has been gradually reduced down to something much less than used to be—in fact, even for submarine purposes, cores with a thickness represented by 59 lbs. per N.M. gutta-percha, to 47 lbs. copper, have been constructed.* The ordinary core for lengths of submarine cables below about 1,200 N.M. is 140 lbs. gutta-percha to 107 lbs. Cu. It used to be thought that 160 lbs. gutta † was the least that could be prudently relied on, for mechanical reasons, with that sized conductor.‡

---

* A core of the above dimensions was made at the Silvertown Works in 1890 for a lighthouse cable laid between Tory Island and the mainland, off the northern coast of Ireland.

Again — besides the Brest-St Pierre "P.Q." cable — there is the case of the Commercial Company's Channel cables of 1884 and 1885—respectively; from Canso to Rochefort, from Waterville to Weston, and from Waterville to Havre.

Both of these have cores represented by 70 lbs. copper (solid wire) to 75 lbs. gutta-percha per N.M. The amount of insulating material in either of the above cables is, no doubt, quite sufficient for short shallow-water sections—indeed a like practice might be extended elsewhere with perfect safety.

† Other instances of a similar waste of material have occurred. For instance, the author has met with a case in which a thickness of gutta represented by 245 lbs. per N.M was applied to a 130-lb. conductor, where 170 lbs. gutta for the insulator would have been ample.

‡ In practice, as a matter of fact, this usually resolves itself into a question of the least number of coatings deemed safe. The gutta is very generally laid on the conductor in as thin coatings as possible, so as to decrease the chance, as far as may be, of weaknesses escaping notice. Three coatings, each of 30 mils. thickness (equivalent to about 140 lbs. gutta-percha), is considered by many to be the least number which can be used to safely insulate a conductor of 107 lbs. per N.M.

However, the insulation of certain cables laid in the Gulf of Mexico in 1881—though of 166 lbs. per N.M. (to 107 lbs. Cu)—was constituted by two coats only. There seems to be no reason why the two-coat dielectric should not be ample for many short sections, even up to 750 N.M. As a signification of mechanical fitness, it may be added that all the Post Office underground conductors have, for many years, been covered with only two coats of gutta-percha.

**Relative Advantages of Single and Multiple Coats.**—In the early days of insulated conductor for underground lines, and in the first submarine line laid from Dover to Calais in 1850, the insulating covering was applied at a single operation.* In the Dover-Calais cable of 1851, however, it was thought that by laying the required thickness of gutta-percha on the wire in thin coats, at separate operations, the chance of incipient faults escaping notice till after the subjection of pressure on submergence would be thereby considerably reduced. The gutta-percha insulating material was accordingly applied in two separate thicknesses; and this practice has been very generally adhered to ever since. Nowadays, the usual number of coatings for an average-sized core for submarine work is three, but with heavier cores, such as those employed for long Atlantic cables, the gutta is often applied in four, or even five, coats.†

By the application of the insulating envelope in thin coatings at separate operations, it can be examined at different stages of its manufacture on the completion of each covering.‡

Moreover, any holes, due to air bubbles—the principal cause of incipient "factory" faults—in one coating may possibly be efficiently filled up by the subsequent coat, thereby reducing the chance of such faults close to the conductor, which, on development, are liable to completely destroy the insulation at that point.

Another advantage in applying the gutta in parts at separate operations is that the chance of the conductor at any part becoming eccentric in its covering—thus, possibly, seriously weakening the insulation at a given point—is materially reduced. There is a tendency for this to occur where the core is being drawn round the drum at the foot of the trough, inasmuch as the inside edge of the core is liable to become flattened between the wire and the drum. In the case of a wire covered with its full thickness of gutta all at once, this is more than ever liable to take place, as such a mass is less likely to be cooled and hardened throughout its section to the

---

* In the latter case the thickness of the dielectric was .208 inch.

† Experiments have been made to see how thinly the gutta could be applied. With a corresponding thickness to the above, the gutta has been applied in as many as twenty separate coats, adherence between them being effected by means of coal-tar naphtha, a powerful solvent of gutta-percha. For practical purposes the results were not satisfactory, however.

‡ Again, the workman in running the core through his hand, by way of inspection, can much more easily detect a weak spot (due to foreign matter, air bubbles, or what not) in a thin coating. Thus, by covering the wire in separate layers, increased confidence of everything being satisfactory is obtained, though there is still the chance of the core getting damaged between the application of coats after examination of the previous layer.

same extent as a thinner coat ;* moreover, there is a greater surface exposed to the risk of flattening out on the drum, if not thoroughly hardened.†

Another point in favour of the multiple-coat system is that it gives an opportunity of varying the material of the dielectric at different depths. Thus, whereas the first concentric layer is composed of pure gutta, a cheaper but equally durable or even better protective material mechanically, may be employed for the subsequent layers—a mixture of various gums, for instance, or an actual compound, provided that a sufficiently high insulation and low capacity is still maintained to meet the electrical requirements.

On the other hand, the original system of applying the whole of the required thickness at a single operation has one advantage,‡ and that is, that it gets over the necessity of using Chatterton's compound in the body of the dielectric for adhesive purposes. This compound, though it has a high resistance, is considerably less than that of gutta-percha, and, what is more, it is a substance with a comparatively high capacity, thereby reducing the working speed—theoretically at any rate. Moreover, the Stockholm tar therein contained tends to slightly reduce the resistance of gutta when brought into contact with it, though not as much so as gas tar. These, however, are not in practice very serious matters, but there are other considerations which render it desirable to limit its use as far as possible.

When introduced by Mr Willoughby Smith in 1858, it was considered that it would give the necessary adherence between separate sheets of gutta (in place of spirits and harmful mixtures) to imbue the entire envelope with mechanical solidity without producing any ill effects. In this way it has done very useful work, coming as a godsend—not only for the above purpose, but also for the central wire of the strand conductor—at a time when so many cables were about to be made, and when the stranded form of conductors had just been adopted.

Subsequent experience has, however, led us to suppose that this compound is liable to have a deteriorating effect on the gutta-percha, though perhaps only to a small extent.

---

* It is true that this difficulty may be met to some extent by the adoption of correspondingly longer troughs; or by drawing the core through the cold water at a slower rate, or a greater number of times. There are, however, often found to be practical objections to all of these alternatives, as may be readily imagined.

† Again, when the gutta-percha is applied piecemeal, any eccentricity of the wire that may have occurred can be corrected in the subsequent coat by the wire being passed through the machine the other way up, and therefore the other side being drawn next the drum. It may be added, also, that any such irregularities can be more easily detected during and after manufacture in a thin coat than in a thick one.

‡ It also has the obvious advantage of taking less time. This difference, however, is not very marked, owing to the slower rate at which the operation can be performed with a regard for safety.

Stockholm tar—being a vegetable (wood) tar—is said to be harmless, but as all tars invariably contain some creosote or other similar solvent, this is not quite certain. Again, Stockholm tar slowly dissolves in sea-water. It also absorbs water slightly in excess of gutta-percha; besides tending to absorb the moisture—some of which is necessary for purposes of durability—from the gutta, leaving the latter more porous and also correspondingly more evaporative. Again, this compound is itself highly evaporative; and, absorbing oxygen freely, tends to go off into powder.

The most important point, however, in this relation is that it is now commonly considered that this adhesive mixture between the separate coats of gutta-percha in the ordinary multiple-process system tends to fill up temporarily any air holes in the previous coat. The result is that any such fault (which would otherwise shew itself) is, for the time being, effectually "masked" by the compound * until such a time as the core is subjected to pressure—usually at the critical period of laying, or perhaps not even till after submergence at the bottom of the ocean.†

These deterrents to the use of Chatterton's compound are certainly obviated by applying the entire thickness of gutta at one operation, besides the general economic advantage of the wire only requiring to be passed once through the covering machine. Owing to the arguments against Chatterton's compound, the one-coat system has of late years received consideration; and it has been even further suggested—where, for pur-poses of safety, it was thought desirable to apply the gutta in more than one layer—that all intermediate compound might be dispensed with, on the ground that if the gutta is made hot enough it will itself be suffi-ciently "tacky" for adhesive purposes. There is, however, a strong objection to overheating gutta, as great heat tends to drive off too much moisture, thus causing serious deterioration. Thus, another plan is to roughen and very slightly rewarm the surface of the gutta by way of adherence in preparation for a subsequent coat.

These notions and devices for returning to the single-coat system—or at any rate for doing away with the use of Chatterton's compound between

---

* It is the tar in the compound—though small in amount—which tends to effect this "masking" action.

† Where Chatterton's compound is used between the concentric layers, one way in which this difficulty could be overcome would be by electrically testing the core on each layer being applied, and that after at least a day's submersion in water. Again, another solution of the problem would be to apply the compound *under pressure—i.e.,* under a similar pressure to that which the core is to experience when submerged. Here, any defects in the gutta-percha might be made good—enough to stand the pressure of the ocean. It is a question, however, whether, when sufficiently heated, the compound would be solid enough for application under pressure: the experiment might prove too costly a one.

the layers—have not, however, as yet met with at all general acceptance as regards submarine cables.*

**Multiple-Die Machine obviating objection to Chatterton's Compound.**—An ingenious compromise which combines some of the advantages of the two systems was effected by Mr Ludwig Loeffler, of Messrs Siemens Brothers, in 1880.

With a view to avoiding the above disadvantages of the adhesive compound between the coats, Mr Loeffler proposed to lay the gutta-percha on the wire to the required thickness at one operation with separate streams of plastic gutta-percha, one immediately in a line with the other, by means of what he described as a multiple die (Provisional Protection No. 905 of that year).

This die—applicable to any covering machine—is furnished with a series of central holes, each one of successively larger diameter, through which the wire is drawn. To the space between each pair of these holes lateral passages or channels are provided, so that as the wire travels along it receives successive layers of covering limited only by the number of successive holes with which the die is provided. The lateral passages may all proceed from one vessel in which the covering material is subjected to pressure, or each set of these passages may be supplied from a separate vessel in order that a different material may be used for each layer.

This device and system is exclusively adopted by Messrs Siemens Brothers, who cover all their gutta-percha insulated conductors in this way. By this means the gutta presents the appearance of having been applied to the wire in three coats at one operation without any adhesive mixture being used. The success of such a plan depends on the gutta being sufficiently "tacky" to ensure the coats properly adhering of themselves: it therefore involves a slightly higher temperature.

In order that this method may be carried out with a good result it is essential that the multiple-die machine be one that will permit of no air entering between the streams of gutta, as such an occurrence would be more liable to be fatal in this than in any other system.

Practically speaking, the multiple-die system may be looked upon as being the same in effect as the single-coat system, except that it provides for a variation in the material used in the various layers applied immediately

---

* In the case of gutta-percha subterranean cables, where the risk is comparatively small, and the dielectric thickness less, the material (where it used to consist of at least two layers) is now often applied in a single coat in order to obviate the use of the compound.

after one another. Thus, it has the advantage of obviating the use of any compound between the coats,* but it also entails the disadvantage and lack of advantages which the single-application system involves, especially where a great thickness of insulation or a heavy type of core is in question.

**Separate Coat System, with Alternate Tests and Chatterton's Compound.**—What may be termed the "feel-your-way," separate-coats, plan recommends itself most to the author, using the usual compound, moreover, to ensure perfect adhesion of the layers. Here, on account of the "masking" action of the compound, it seems as though each coating should be tested separately in order to render this system really perfect.† This would render examination unnecessary, and therefore the extra time involved would not be so very great.

Where the required thickness of insulation is laid on in more than one coat, the subsequent concentric layers are applied in a precisely similar manner. The same operation is, in fact, gone through for each coat, with the dies changed and reset for larger ones every time according to requirements—*i.e.*, according to the additional thickness of covering aimed at.‡ The above is first of all preceded by the reheating of the previous coat over the gas jets at the head of the machine. This has the effect of rendering the surface of the previous covering fit for taking a similar thin coat of Chatterton's resinous compound, as was applied to the conductor previous to laying on the first coating.§

The proportions of gutta, resin, and Stockholm tar which go to form Chatterton's compound have already been stated elsewhere. As this was found to answer the purpose admirably, and to apparently have no ill effects on the gutta, it at once came into general use, though the royalty on the patent proved to be somewhat heavy on a long length of cable. Departures from it in the form of other resinous mixtures have been tried,

---

* To some extent balanced by the somewhat increased heat necessary for the gutta, to ensure each "stream" producing effective cementation.

† Preferably both on the completion of each coat and again immediately before the application of the compound for the succeeding layer.

‡ It is very usual to keep the core (in a dry place) for a day or so between the application of each coating, in order to drive off external moisture, which prevents adhesion.

§ The intermediate coats of Chatterton's compound—usually at least two in number—is a more important factor in contributing to the total weight of the core than might at first sight appear probable considering the comparative thinness of each. It is usually equivalent, in weight, to about 1 lb. per N.M.

but—in the case of one undertaking, at any rate—with disastrous results. In some of these costly experiments the Stockholm wood tar of Chatterton's mixture was substituted by coal tar. This tended to dissolve the gutta; besides which, holes were blown in the gutta-percha envelope when under pressure, on submergence. In other instances the naphthaline used as a solvent caused similar trouble by evaporation. In any case the insulation resistance of the gutta-percha covering was materially reduced by being in contact with a mixture containing coal tar. Unfortunately these bad features did not, as a rule, come to light till after the core had been submerged under the pressure of the ocean.

After passing through the tank of compound, the core is then led through the covering machine, and thus, by means of the compound, a fresh layer is firmly united to the previous one. The core is then again slowly drawn through the trough of cold water—three times, in fact— preparatory to being again automatically coiled on its collecting drum and once more examined for the fresh coat.

**Examination after Application of each Coat.**—On the completion of each successive layer, the core undergoes the same rigid examination throughout its entire length, and any flaws discovered are made good. Both of these processes have already been described.

**Measurement.**—Whilst this examination of the core is being per- formed—by being rapidly run through the hand during drawing off from one reel to another—it is at the same time passing through a process of measurement. This is effected in a doubly certain manner, by both the drum that it is being hauled off *from* and that which it is being hauled off *on to*, each having a revolution indicator in connection with their axles.*

Moreover, an additional measurement is also very often obtained during the process of covering, by an indicator being geared into some part of the hauling-off machinery—not uncommonly in connection with the drum at the foot of the cooling trough.

**Weight of Insulation.**—Besides the length measurement being taken, the weight of each completed coil is also tested. By deducting the pre-

---

* A comparison of the lengths obtained on the completed core (and on each individual coat) with that given by the conductor alone during the stranding process, shews the amount of stretch which has been involved, in the meanwhile, by "drawing off" during the operation of covering.

viously measured weight of the length of copper strand forming part of this coil from the total weight of the coil, a knowledge is obtained (by inference) of the weight of the insulator alone. This is not very accurate, as a rule, inasmuch as the machine used for weighing the entire coil is not by any means as sensitive as that for measuring the conducting strand by itself. It is, however, the only means of arriving at the weight of the dielectric.*

The weight of the dielectric is specified for (in addition to that of the conductor) as a further check on the material employed to meet the specified electrical requirements, besides ensuring the mechanical qualifications, as regards dimensions, being properly carried out.†

**Outside Diameter of Core.**—Similarly, the outside diameter of the completed core is usually specified for (in addition to that of the conductor), and is checked from time to time for this, and, still more, to ensure the mechanical proportions and individual requirements being secured.

There is an additional reason why the weights should be specified and tested—*i.e.*, because the cost of materials is always estimated and made out by weight.

Moreover, it is usual to measure the diameter of the completed core (the conductor having been previously gauged), or on a certain proportion of the coils, at any rate, to see that it is up to specification. It is highly desirable—as in the case of the weights—that both the diameter of the conductor and that of the completed core should be specified for, not only for mechanical reasons, but also as a check on the method of securing the required limits of capacity and resistance—*i.e.*, by a sufficiency of dielectric material, rather than specifically. The difference between the conductor diameter and the outside diameter gives the thickness of dielectric.

**Recording of Data concerning each Coil.**—Every coil, after being measured and weighed (and in some cases gauged), is given a rotation number from the commencement of any particular undertaking, and various information respecting it tabulated under different headings,

---

* A nautical mile is always taken as the standard length in this respect—as in other matters—partly on account of convenience, and partly because it is a fair mean length such as can be measured within a reasonable percentage of accuracy.

† It is very usually stated in specifications that the average weight of the entire length of the dielectric shall not be less than so much per mile—or within a certain percentage (generally 5) thereof.

besides being noted in a book for after-reference. These particulars usually consist of something like the following:—(*a*) The number of the coil; (*b*) its length in yards and nautical miles; (*c*) weight of conductor and weight of dielectric; (*d*) date of leaving covering machine; (*e*) date of immersion in water at 75° F.; (*f*) date of testing electrically; (*g*) date of leaving 75° F. tanks for serving shop.\*

**Rate of Covering.**—The particulars regarding length and weights are stamped on a gutta-percha (or leather) label which is then attached to the coil itself.

As regards the rate at which a conductor is covered with gutta-percha, the wire is drawn off through the "coring" machine on the *average* at a speed of 750 yards per hour.† This is, in fact, the rate of covering with one coating, for the very ordinary type of core constituted by a conductor of 107 lbs. copper per N.M. and a dielectric of 140 lbs. gutta, made up in three separate coats. This amounts to about 3 N.M.‡ per day of twelve hours.§

From this it will be seen that the ordinary machine with six dies—capable of covering six wires at a time with a single coat—can cover about 18 N.M. of wire with a single coat, in the course of a twelve-hour day's work, or 18 N.M. of three coats in three days.‖ Where the said covering is

---

\* It will readily be understood that some of this information has to be filled in afterwards.

† In the case of the Gray and Gibson covering machine previously described, the corresponding speed for the horizontal rollers forcing the gutta into the receiving cylinder and for the worm forcing the gutta-percha into the die chest is about fifteen to seventeen revolutions per minute.

‡ It may be remembered that it is on these grounds that the complete (stranded) conductor is usually made up in about this length. If the length of the conductor was shorter than that which could be covered in a continuous working day, scarf joints in the strand would be involved. These take a sensible time, during which the gutta at the end of the covered length would become more or less hard. If, on the other hand, the length of the strand were longer than could be properly dealt with by the covering machine during the time of continuous work, on resuming, a joint in the covering would be involved between the already hardened gutta (which would require abnormal heating) and the gutta in a hot plastic state.

§ It should be added that, as the wire is not usually supplied in any *exact* lengths, the length of each completed coil varies. It is sometimes the practice, however, to specify that each coil shall be of an exact length—2 or 3 N.M. as the case may be. This has the effect of simplifying calculations in reducing values to the standard unit, length 1 N.M. On the other-hand, however, this involves the conductor being cut at an exact point, leaving a number of odd ends which either introduce waste or increased time and expense in jointing them together.

‖ Were an attempt made to start on another coat in any one day, a further change of dies would be involved—apart from other objections.

constituted in its entirety by three separate coats, this is equivalent to about 6 miles of completed core in a day.

At a large cable factory there will be something like half-a-dozen such machines, the result being that somewhere about 50 N.M. of single-coated core can be turned out during a working day.

This, however, largely depends on the thickness of each coat, as already explained. In any case, the rate of core manufacture should be well ahead of the sheathing. There should, in fact, be no chance of the cable department having their machines standing still for want of core to feed them with.

**Mechanical Tests.**—It has not been the custom, as a rule, to apply any actual mechanical tests to the completed core any more than to the conductor alone, though on the last "Anglo" Atlantic cable of 1894 the core (Fig. 27) was tested for breaking strain. This core, composed of 650 lbs. copper and 400 lbs. gutta-percha per N.M., bore a strain of 1,000 lbs. before breaking.

FIG. 27.—Core of 1894 "Anglo" Atlantic Cable.

The fact is—as regards breaking strain, at any rate—the mechanical qualities of the core can scarcely be said to come into the question in practice. This is owing to the circumstance that the extensibility of the gutta-percha is such that its point of rupture is quite different to that of the rest of the cable ; and, therefore, it does not tend to increase the total strength of the cable, however high its breaking strain may be. Thus, any breaking strain obtained for the core is really that of the conductor alone ; and this would be quite insignificant as an addition to that of the rest of the cable, even if it came into action at the same moment.

**Maturing.** — The core so completed is usually kept — for maturing purposes — often as much as six, or even fifteen, days before anything further is done with it ; but sometimes only one or two, if required at once, or where further maturing might put it on the wrong side of the specification.*

**Electrical Tests.**—With a view to ascertaining whether the work of

---

* The specification should always state the length of time after manufacture at which the core is to be tested. This is desirable, in order that the results obtained may be absolutely independent of the influence of this important, and varying, factor.

covering the copper conductor has been successfully carried out,[*] the core is next taken on its large wooden carrying drums to the core testing department, where the coil is bodily submerged in cisterns, or tanks, filled with fresh-water,[†] uniformly maintained at 75° F., with the help of steam circulating through pipes, or else by a double-bottom steam-jacketing system.

The core is subsequently tested whilst in the tank of water, but it is first kept immersed there for at least twenty-four hours. This is in order to give the water a fair chance of percolating at any weak spot to the conductor, and also to ensure the core truly taking up the fixed temperature of the water. It is very usually stated in the specification the length of time after immersion in water or after manufacture that the core is to be tested—or, at any rate, the minimum length of immersion (usually twenty-four hours) is given. The length of time necessary to meet the above requirements varies, of course, with the type of core, a great thickness of gutta taking longer than a small one; indeed, thirty-six, or even forty-eight hours, may sometimes be found desirable;[‡] but with an average core, twenty-four hours' immersion should be sufficient, as a rule. The importance of this is only equalled by the importance of rigidly keeping the water at the temperature named. This temperature is taken as a standard, or basis, for subsequent comparisons of electrical tests: the factor of the variation of both conductor and dielectric resistance by temperature is thereby eliminated.[§]  75° F. was selected as a fairly convenient temperature for reproduction under any circumstances.

The core then, whilst still under water, passes through a very severe and exhaustive system of tests regarding the electrical condition of the core.[||]

---

[*] It would be useless to make this test in air, since even without any insulator at all the current does not pass readily into air.

[†] Theoretically, salt-water would answer the purpose better, being a superior conductor electrically and a more perfect imitation of what the cable has to experience on submergence; but there would be difficulties to effect this in practice.

[‡] Messrs Siemens Brothers are in the habit of keeping their core for two days (forty-eight hours) before testing it.

[§] Thus—a record being kept of the results of tests on each coil of core as made at the factory—the electrical value of any portion of the cable, at any other subsequent temperature, may be compared with these original values (as core) at the standard temperature, by applying to the observed value the coefficient for the reduction of the resistance to the standard temperature.

[||] This was carried out systematically for the first time by Messrs Bright and Clark over the manufacture of the first Persian Gulf cable of 1863 ("Notes on Telegraphic Communication between England and India," by Charles Bright, F.R.S.E., *Journal of the Society of Arts*, vol. xlii., Nos. 2,152 to 2,155).

This is made under a stress of some 500 volts,* and with the highly sensitive Thomson reflecting (astatic) galvanometer; so that the slightest imperfection in the insulation—which may have passed manual examination—would allow the current to pass through the defect to the water and tank, and so reveal itself in the form of a deflection on the galvanometer scale.

These tests will be found described in detail elsewhere. For our present purposes, one of the simplest and most easily explained tests for the insulation of the core may be described as follows:—The current is measured as it passes into the cable, and then the supply is cut off. The cable is left in this condition for, say, one minute, and then the current is measured as it passes out of the cable through the galvanometer. The difference between this measurement and the measurement of the current that was conveyed into the cable is the loss, and shews the condition of the cable; a bad or leaky cable exhibits a greater loss than a good, well-insulated cable.

In addition, there is also an ordinary insulation test kept on the cable —under the above-mentioned heavy electric pressure—for some fifteen minutes or more, according to the system. This tends very perfectly to bring to light any weaknesses.† A high degree of insulation is not so much what is sought for as a perfectly steady leakage—or rather, *apparent* decrease of leakage by time, due to gradual electrification, absorption, or

---

* It is sometimes stated that the insulation resistance of a cable will not be the same when measured with different potentials—*i.e.*, that it will be reduced by an increase of voltage—but it has been found that such is not the case. As long as the dielectric acts as a conductor only, there being no sparking action through it, no change in the insulation resistance can be detected.

† The same system is now adopted at all cable works, although (as already shewn) the period of immersion is in some cases different. At certain factories—notably at the Telegraph Construction Company's Gutta-percha Works—the core is also tested finally at a lower temperature of 40° or 50° F., with the object of more certainly detecting any small incipient faults. This is based on the principle that weakness in an insulating material reveals itself more readily in an electrical test when the material elsewhere offers a high electrical resistance, as would tend to be the case at a low temperature. Moreover, if a temperature more nearly approaching that of the sea bottom be taken as a standard, the error necessarily occurring in any subsequent reduction to the standard temperature— owing to a general coefficient not being applicable to any particular batch of gutta—is considerably lessened.

On the other hand, the higher the temperature, the greater the general strain put on the electrical resistance value of the core, besides being a severer test mechanically, though a minute fault may not be actually as easy to discover. It must also be remembered that in tropical climes, at shore ends—up to the hut—the cable often has to experience exceedingly high temperatures, when, for instance, unavoidably running under a long stretch of dry sand. Hence 75° F. is usually adhered to as the standard temperature.

polarisation of the dielectric. The electrification of the dielectric is more closely watched at this stage than at any other.

Besides the above test on the insulation, the first test made is that on the conductor for its resistance.

Perhaps the main object of interest in testing at this juncture is the electro-static capacity of the completed core. It is important, at the earliest possible stage, to know, for signalling purposes, that it is not above the specified limit in the above respect. This, like all the other electrical tests, will be found detailed elsewhere.*

**Improvements in Manufacture.**—The manufacture of cable core has so far improved within the last twenty years, that a core with an insulator weighing 150 lbs. per N.M., which then had a dielectric resistance of some 250 megohms per N.M. at 75° F., can now (principally owing to greater perfection in the joints) be obtained giving 2,000 megohms. This would not necessarily imply perfection or durability with regard to electrical and mechanical qualities, as already explained; though a moderately high insulation may usually be taken as a sign of excellence of material for the purpose.

**Re-making Core.**—Where a high capacity, due to insufficient mastication, is accompanied by a dielectric resistance some way above the limits, this can be remedied by the somewhat troublesome course of making the core up again after the gutta has been stripped off and re-masticated.†  Care must, however, be then taken in getting the required capacity, that the insulation resistance is kept within the required limits each way, where so specified.

**Standard Electrical Values at 75° F.**—These tests are taken as the standard electrical values of the cable for purposes of after-comparison, any subsequent tests made being reduced to the corresponding value at the same (standard) temperature, 75° F.

**Records.**—The results of these standard tests are recorded on special forms. Sometimes, indeed, curve sheets are drawn up for shewing the insulation resistance from week to week after original manufacture (to bring

---

* See Mr H. D. Wilkinson's up-to-date work, published at *The Electrician* office ; also Kempe's " Handbook," *ibid. ;* still more a forthcoming treatise by Mr J. E. Young on " Electrical Testing for Telegraph Engineers," of the same series as Mr Wilkinson's book.

† Old gutta may be rejuvenated in this way, by being mixed afresh with new gutta containing the essential oils, and the whole masticated and "worked up" together.

to light the improvement by age), as well as curves setting forth the rate of electrification* during a certain test.†

After any coil of core has been tested, if it is "passed," it is kept under water and screened from light as far as possible until it is required for use in the cable.

**Placing of Coils in a Section.**—Owing to the variations of gutta now in the market, and to the enormous difference in the electrical results accruing from different mixings and methods of manufacture, there is often a great discrepancy between the dielectric resistance of the various coils for a particular cable order. Some coils may have an insulation of 100

---

* Electrification—more closely defined as *absorption*—may be viewed as something between capacity and conduction, or leakage. It takes place very much quicker at low than at high temperatures.

From a wide range of experiments no broad or definite law can be arrived at regarding the relationship between the specific resistance and rate of electrification in any given insulating material : in other words, the absorption rate of a gum cannot be said to betoken any particular specific resistance. Speaking quite generally, however, rapid absorption is most usually an accompaniment of high resistance, and *vice versâ*—just as it is more liable to be with a given gum at a low than at a high temperature. Paraffin wax, india-rubber, and gutta-percha, when laid at the bottom of the ocean, are all examples of this in comparison with gutta-percha under normal, factory, conditions, as regards temperature, pressure, and age.

It should be explained, however, that a material possessing a high rate of electrification is synonymous with saying that it absorbs very little electricity, and that what little is absorbed soaks in within a very short time—*i.e.*, at a high rate. With such a material then, the *true* insulation resistance—denoted solely by a permanent leakage current—is arrived at comparatively quickly.

It seems probable that there is a closer connection between dielectric absorption and signalling speed than is at present known of; and that research in this direction might lead to important results. Theoretically, it would appear that from a *capacity* standpoint, a rapid absorption—*i.e.*, a small total absorption—should be favourable to a high rate of working. With prevailing practice, however, this does not appear to enter into the problem : moreover, the significance is not necessarily a good one with all gums ; there is another side to the question as is shewn further on. The great point to be aimed at, and ascertained, is that the electrification—*i.e.*, absorption—shall be perfectly *steady* in its gradual decrease subsequent to battery application.

† Air condensers have, of course, no absorption, and Mr R. T. Glazebrook, F.R.S. (B.A. Report on Electrical Standards, 1892), has compared mica condensers with them. He found the *apparent* capacity of the latter quite 1 per cent. smaller when the battery contact was very short than when the contact lasted for from five to twenty seconds. Clearly, then, a mica *cable* compared with an air condenser in the usual way at the factory would have a working speed—say 1 per cent.—better than that calculated, when the rate of signalling is high, and the contacts, consequently, short.

Though we have no mica cables in actual practice, there seems to be some interest —or even *importance*—attached to the question whether the effect is not in the same direction in the everyday comparison of gutta-percha cores with mica condensers ; and, if so, to what extent ?

megohms per naut where others have as much as 3,500. In such instances, the coils are arranged and joined up in order in the cable, in order that a coil with a comparatively low insulation may be balanced by one with a high resistance following it; so that, in the event of subsequent cutting, the average dielectric resistance will be about the same for any part of the entire line.

**Historical Records.**—In Appendix IV. at the end of this Part, Form B shews the sort of tests and general data that are made and tabulated in connection with a completed cable section when laid. A step-by-step statement of the factory " C. R." (conductor resistance), " D. R." (dielectric resistance), and " I. C." (inductive capacity), of each coil must also be drawn up for every *laid* section—just as it is done for every sheathed *factory* section; and this should be accompanied by a record of all bottom temperatures taken at intervals along the length of line as laid.

**Selection of Gums.**—To ensure the right electrical results being obtained, some systematic method of testing experimental proportions of gutta are essential. One plan is to make up several comparatively short coils not exceeding a quarter mile in length, each covered with a certain species of gutta. From the results obtained with each, the proportions of each gutta to be used to meet the specification of the core may be decided on.

Another plan is to test actual lumps of gutta between plates to which a connecting wire is attached. In any case, assigning proportions—such as produce the precise required result—is not so simple a matter as it might, at first sight, appear.

**Requirements.**—The exact insulation resistance of a material is only of importance in so far as it serves to signify the material being what is wanted, with a given thickness for mechanical requirements. The fact is that, apart from the necessities of electric insulation, a dielectric resistance, within certain limits, each way is significant of a certain, known, durable, gutta-percha material such as cannot be recognised by ocular external examination, or by any other test than an electrical one. The actual requirements, as regards electrical resistance, *for the successful working* of a submarine cable, do not require to be nearly so carefully looked after— that is to say, *any* gutta-percha would probably offer a sufficient specific resistance, and usually much more is obtained than is really necessary.*

---

* Indeed, as is pointed out elsewhere, a comparatively low resistance insulator would —if it could be relied upon—more closely realise the theoretical requirements of high speed signalling on a long cable.

As a sign of a durable form of gutta, it has already been pointed out that too high a specific resistance is every bit as bad as too much the other way. Indeed, extremes in either direction are a sign of the addition of other materials, or of widely different gums; * but a comparatively small divergence above the mark generally suggests the presence of a more or less large proportion of resin, which, when in good condition, usually has a higher resistance than gutta-percha. As has been shewn previously, resin readily evaporates into dust, and is, therefore, much to be avoided. It then not only has a low resistance, tending to absorb moisture, but also brings about general decay and mechanical destruction of the core, owing to the porous holes and cracks so set up.† Thus, it is quite as important to put a limit on the resistance of the gutta-percha as it is to specify a minimum; and this, nowadays, is almost invariably done by the engineers to the owners.

In driving off moisture, the degree of mastication adopted has a very considerable bearing on the resistance of gutta-percha, besides also governing its capacity.‡ It will be seen, then, that either of these electrical requirements may be readily met without signifying the employment of a durable material.§

**Specified Limits.**—Thus, properly speaking, a specification for the limits of dielectric resistance on each side should be based on experiments with all the gutta to be obtained in the market at the time. With the knowledge of what are the most suitable and durable guttas amongst the above,‖ samples of those selected should be electrically tested. After

---

* The admixture of a large number of widely different gums usually constitutes a non-homogeneous, and therefore non-durable, gum. This would be still more implied if other than gutta-percha form part of the mixture.

† Each of these in themselves constitute more or less of a fault in insulation if exposed to moisture.

‡ It does not always follow that the resistance of gutta-percha (or of india-rubber) is decreased by water absorption. In Willoughby Smith's process more moisture is driven off (by further mastication), but the dielectric resistance is lower—not higher—than "ordinary" gutta. As a broad rule, however, with what is termed *ordinary* manufactured gutta a low capacity is accompanied by a comparatively high resistance, and mastication usually sends the latter up, whilst bringing the former down.

§ Though almost any resistance—and, to some extent, almost any capacity—may be reproduced with many different qualities of gutta by the above methods of manufacture, it is not by any means always easy to combine both requirements in this way.

‖ A knowledge on this point requires to be based on examination, experience, and chemical analysis.

Mr Thomas Bolas, F.C.S., a great authority on gutta-percha, has pointed out that some sort of chemical test for determining the amount of oxidised products should certainly be applied to all gutta-percha proposed for use in insulation. He suggests the method of organic analysis. This gentleman considers that not only should any remaining resinous matter in gutta-percha sheet for the above be chemically removed by spirit, but

deciding on the proportion of each to be used, the electrical value of the whole combined should be calculated, and the specification drawn up from the results, to ensure getting much the same combination.    The latter should be selected mainly with an eye to durability—sufficient insulation being understood.    The above is assuming that for commercial, or mechanical reasons, a mixture of one or two various guttas is the order of the day, as is indeed usual.    Where, however, only one gum is determined on, it is merely a question of specifying, with a certain margin each way, the resistance obtained with that.

As regards capacity, it should be remembered that it is better, with a reasonable consideration of initial cost, to obtain a low capacity—as regards the dielectric—by extra thickness rather than by specific means.    This is on account of the fact that resin has a lower inductive capacity than gutta-percha, and that a low capacity obtained specifically usually implies the presence of a comparatively large proportion of resin in the gutta which is undesirable, as already explained.

Again, a low capacity is often obtained by an amount of mastication such as tends to draw out too much of the essential oils of the gutta as well as too much combined moisture, with the result that the material is rendered extremely liable to oxidise, and, therefore, gradually becomes brittle, and decays—owing to over-porosity and absorption.

Thus, as a check on the method of meeting the capacity limits, it is also very important that the outside diameter of the completed core, as well as that of the conductor alone, should be carefully specified.    It is, moreover, for this reason, essential that the outside core diameter and the difference between this and that of the conductor should be carefully calculated from the respective measured weights.    Failing this, direct measurements for the above must be made.

**Suggested Limitation to Specific Inductive Capacity.**—Here, again, it would appear—owing to the deteriorating effects of resin and to the doubtful durability of gutta with a high capacity *specifically*—that it would be well to carefully test all the gutta obtainable in the market at the time for capacity in the same way as for insulation resistance, and after selecting certain durable materials, to test them and draw up the specification as regards capacity founded on the results obtained from the materials determined on, and on their proportions in this respect.    There

that great pains should also be taken to ensure it not containing more than a small proportion of oxygen.  The electrical value of the material would then certainly be at its minimum, but a more reliable and durable substance would be secured.

are instances where, owing to the comparatively short length involved, the speed at which it is possible to manipulate the apparatus manually, is far below the theoretical speed with any reasonable capacity, and that, therefore, the question of a low capacity per unit length is of comparatively little value. At any rate, it should certainly not be sought after by *specific* means * at the expense of durability.

**Localisation of Faults at Factory.**—In the case of an actual fault, or weak spot, in any given coil which has escaped the hand examiner's notice but has been revealed by the above electrical tests, besides the "loop test" —of Varley or Murray—described elsewhere, there are various methods in which the coil is run on to another (insulated) drum, a good earth connection, with galvanometer in circuit, being pressed round the core as it passes. When the faulty place comes into contact with the earth connection—usually a wet rag at the end of an earth wire—the current, in suddenly having facilities for passing from the battery at the other end to earth, produces a more or less violent throw on the galvanometer. Various forms of this method of fault localisation in the case of moderate lengths of core (such as that of a single coil) are of daily application in the cable factory, and have proved eminently satisfactory in practice for shewing up and localising faults of even fairly high resistance.

**Pressure Test.**—With a view to bringing to light any incipient faults of insulation in the core before it is submerged at the bottom of the sea, and, in fact, to—in some measure—reproduce the conditions about to be experienced, the pressure test — briefly described in Part I. — is now invariably applied to the core by Messrs Siemens Brothers as soon as completed, *i.e.*, before it is electrically tested.†

Some of the faults which are liable to develop during the course of manufacture—for instance, holes filled by air—are very often not at the surface but in the first or second coat, say, such as become covered up by the subsequent layer. On pressure being subjected these will burst, and thus reveal the weak spot, or spots.

---

* As a check on this, it has been even urged that there should be a *limit* to the capacity in some specifications as well as a maximum, with given diameter measurements.

† The pressure test, as applied to core after manufacture, does not, of course, quite represent that which takes place when the core is afterwards submerged, as in the latter case it is ensheathed in an iron armour. Moreover, strictly speaking, uniform pressure round a core does not exactly represent what is afterwards experienced on submergence at the bottom of the sea, where it is a question of weight of a certain column of water resting on the cable, which is itself resting on the bottom.

If it were a matter of uniform pressure, the effect would be the same, no matter what the depth to which the cable is lowered below the surface of a given column of water.

The following constitutes Messrs Siemens' method of effecting the test. The core is wound on to a suitably constructed drum, which is then lowered into a steel cylinder containing fresh-water (always maintained at the standard temperature), with walls 18 inches thick. The open end of the cylinder is closed by a steel plug screwed into it, the threads being cut away at equal distances round the surface, so that the plug may be readily inserted and secured by a half-turn. This arrangement is, indeed, precisely similar to that employed in modern breech-loading guns. Stuffing boxes are provided in the breech piece through which the ends of the core are brought, to enable a connection to be made, so that the coil may be submitted to electrical tests whilst actually under pressure. The hydraulic pressure is applied by means of force-pumps, and its degree is adjusted according to the depth of the ocean in which the particular cable is about to be laid, the maximum available strength of pressure being equivalent to about 8 tons on the square inch*—well outside that of oceans' depths so far within our knowledge.

The above test is more particularly applicable to the system of laying all the gutta-percha on at one operation, whether by a multiple die or otherwise. This is so, owing to the increased chance of air bubbles and flaws generally, or of foreign matter otherwise escaping detection till laid at the bottom. It is also, in a less degree, applicable when the gutta is laid on in separate coats, unless an electrical test is made as each separate coat is applied.

The core may very suitably be thus subjected to hydraulic pressure whilst undergoing the already described standard tests in water at 75° F., or else the electrical test may be applied to the core soon after it has been subjected to the pre-arranged pressure. In any case it is well to keep the pressure on for some time before noting the effect on the dielectric resistance, inasmuch as the application of pressure has the effect of increasing the temperature for the time being; and consequently any alteration in resistance during that period would be partly due to the change of temperature. Similarly, the withdrawal of pressure tends to temporarily lower the temperature of the water, and this in itself would increase the resistance. Thus, sufficient time must be allowed for an accurate idea of the temperature to be obtained, so as to make the necessary correction.

This test is thought by certain authorities to be not only unnecessary in the present day—owing to the decreased chance of air bubbles, and other defects, with improved manufacture—but also actually objectionable, on the ground of tending to temporarily mask faults as well as to divulge

---

* Four tons per square inch represents, however, the pressure more ordinarily applied.

them. The late Mr Willoughby Smith shewed this to be the case in the course of some exhaustive experiments many years ago. What is perhaps of still more importance is the fact that instances are known where the insulation has been permanently damaged by the pressure test.* In any event, it is a somewhat expensive process, when the high pressures are reproduced as involved by the depth of the ocean. This, owing to the great strength necessary for the tank in order that it may withstand the heavy strain to which its walls are subjected by the pressure of the air, or water, until the required pressure is reached as indicated by the gauge. It is, however, a plan which has many points in its favour.

**High Proportionate Cost of Gutta-percha Envelope: Reason for Length of Chapter.**—If this chapter on gutta-percha and the gutta-percha covering appears to be unnecessarily lengthy, it must be remembered, on the other hand, that the cost of the gutta-percha dielectric in a cable forms a large proportion of the total cost. Moreover, there is the fact that this part of the subject—*i.e.*, the construction of the dielectric—has never been dealt with at all fully in any previous treatise.

---

* On the other hand, there are those who consider that pressure could be made to have a *permanently improving* effect on the resistance of gutta. This is supposed more especially to be the case with chemically impure materials of inferior quality. As some of these have quite a low electro-static capacity, it would seem that a field of research was opened up here for elucidating the problem of high speed, cheap, but durable, cables, and for meeting the requirements of long-distance submarine telephony. The difficulty in these matters is that the experiment can only be put to test in a practical way when the entire cable is made—aye, and laid at the bottom of the ocean.

## SECTION 7.—INDIA-RUBBER.

**Where and How Obtained.**—For submarine purposes, the only other material which has been in practical competition with gutta-percha for the dielectric, is india-rubber—or *caoutchouc*, as it was originally called.*

The importation of this gum to Europe dates back to the commencement of the eighteenth century.

La Condamine, during his mission to Peru, in the year 1735, was the first to make it known.

India-rubber is produced by the desiccation of a milky sap, which exudes from incisions made in trees of a certain species.

The most highly priced rubber comes from the central plains of South America, watered by the Amazon and its tributaries. Para, Clara, and (latterly) Manaos in the Brazils are the main sources of supply. This is always known in the trade as Para rubber, and is superior to rubber from other countries, especially for insulating purposes, largely on account of the more careful and reliable method of collection (Fig. 28). When transported from these countries, it is found to be much freer from foreign substances.

The lumps in which it is collected are usually loaf, or bottle, shaped, but occasionally—as after overflow from the collecting cups—in round balls called negro, or nigger, heads. These are not so clean, and more liable to decomposition than the fine kind, being subject to the heavy rainfalls, as well as to the admixture of bark, earth, etc., from the tree, and the ground on which it stands.

The trees which produce rubber gum are the *Siphonia* or *Hevea*, belonging to the family of *Euphorbiaceæ*; chief among these being the *Siphonia elastica* (Fig. 29) or *Hevea guayaulusis*.

Of other kinds of india-rubber trees, the following are the most widely distributed:—*Castilloa elastica*, found in Venezuela, New Granada, Ecuador, Peru, Panama, Costa Rica, Nicaragua, Honduras, and Mexico, yielding rubber of inferior quality, but which is frequently—and often necessarily (for mechanical, as well as economical, reasons)—mixed with the Para kind; *Ficus elasticus*,† met with in Java, Madagascar, Assam, and Australia;

---

* India-rubber—though of a far less oily nature than gutta-percha—contains, as a rule, a small quantity of oil known as "caoutchoucine."

† This is the species that is so commonly domesticated in our homes in this country in plant form. Being discovered in the Indies in the first instance, it is supposed that this tree gave the name of india-rubber, partly on the above account and partly because it was found when solidified to possess the property of rubbing out pencil marks.

[PLATE XVII.

FIG. 28.—Collection of the Juice of the India-rubber Tree in Para.

[To face p. 332.

[Plate XVIII.

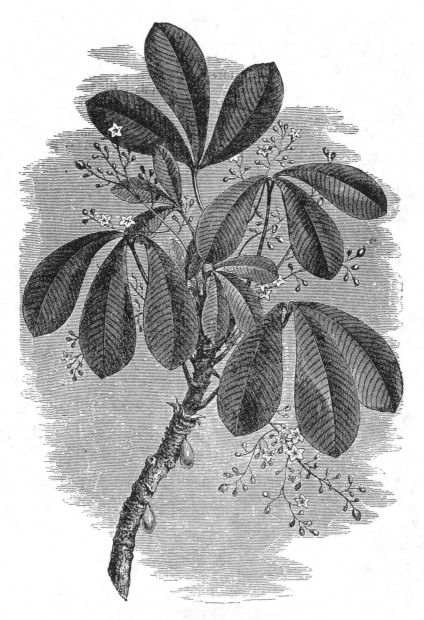

Fig. 29.—The *Siphonia Elastica* Rubber Tree.

*Urceola elastica*, confined to the Malay Archipelago, and principally found in Borneo,* Sumatra, and Penang. The latter variety is a sort of creeper, somewhat similar to the vine in appearance, having branches growing to a length of some 200 feet in the course of five years, though the trunk from which they spring rarely exceeds a few feet in height: it yields annually about 60 lbs. of rubber. Mozambique rubber is the least sought after: the natives wind it up on small pieces of wood, thus forming the spindle-shaped masses brought to Europe. Contrary to the practice in vogue for gutta-percha, so far as Para rubber is concerned, the trees are left standing, and merely notched with a hatchet, a trough, cup, or some such vessel being placed under each incision to collect the sap as it oozes out. About a quart of rubber is thus obtained from a tree in the course of a morning.†

The sap is then dried in the sun. With Para rubber the practice is to smoke it, instead of sun-drying. This smoking has much the same effect as smoking a ham or a haddock, the acetic acid acting as a preservative.

**Collection.**—The method of collection and its subsequent purification from clay, sand, grit, etc., is, in fact, in many respects, very similar to that of gutta-percha,‡ both being hydrocarbon gums.§ Though the mastication required (in the case of Para rubber, at any rate) is rather less expensive as a rule, the after-treatment is always much more elaborate and complicated; for, unlike gutta, a number of other substances—varying in character and proportions—require to be mixed with it (which require careful selection) in order to render it suitable for the particular purpose for which it is about to be used,|| of which there are many and various;

---

* The gutta coming from Borneo is usually considered second best to that from Para, and that which is brought in large quantities from Senegal (West Coast of Africa) perhaps third best. The latter is improving in quality.

† The African, East Indian, and Central American gums are, however, collected in a much rougher manner. In Central America the tree is sometimes felled and "ringed" at intervals, the milk being allowed to fall on the leaves placed under the trees. This is a brutal method of collection, as all the decomposed vegetable matter on the leaves and bark is mixed up with the rubber and tends to shorten its life. In the other countries named, though the method of collection is less crude, no trouble is taken over the purification.

‡ On the other hand, a certain difference of treatment is involved by the fact that, whereas gutta is naturally more or less plastic and not very elastic, india-rubber requires a distinctly increased temperature to become at all plastic, and is naturally highly elastic.

§ Moreover, as they are both employed for very similar purposes, they are often classed together as though more alike in character than is the case in reality.

|| In fact, unlike gutta-percha (which is used, more or less, in its raw state), india-rubber is employed as a manufactured article.

and according to the particular variety of gum or gums employed, of which there are at least twenty.*

Pure unvulcanised sheet india-rubber is prepared by cutting the slabs into very thin pieces, by means of a disc with cutting edges rapidly rotated. The fragments are washed in cold water and then broken up under rollers of different kinds. It is then dried in a room maintained at 100° F. ; after which it is masticated † to a thorough state of amalgamation, under a continuous stream of cold water. The result is a large lump of homogeneous dried, but pasty, substance. This is put into a straining machine with a plunger which is brought down by hydraulic power. After being left under this pressure a block is produced, from which thin sheets are cut, subsequently rolled out to the required thickness. The rubber sheet is then stretched on canvas and rolled up on itself so as to exclude the air and prevent the parts sticking together.

**Mastication, etc.**—Mastication is supposed to promote the absorption of oxygen, by increasing the porosity of the substance and the amount of air imprisoned in it. As an alternative, therefore, the slices of rubber are sometimes exposed to hot, dry, air, to evaporate the water in the pores. They are then compressed into a solid block by hydraulic power, the block being afterwards cut into sheets by means of a knife working rapidly to and fro with a slow downward motion at the same time.

**Physical Characteristics.**—Mastication has also considerable effect on the chemical properties of rubber, depriving it of strength and tenacity to a great extent.

Until the chemical and physical changes (partly due to oxidation) in the material has taken place, mastication—in throwing off moisture—tends, as a rule, to increase the electrical resistance of rubber, and to reduce its electro-static capacity.

India-rubber is a tenacious and extremely elastic substance,‡ grey in colour, and slightly less dense than water. Its specific gravity is about 0.92. When heated to 170° or 180° F. it becomes viscous and melts at 250° F., giving off a peculiar characteristic odour : at a still higher temperature

---

* Recently British Guiana has been looked to as a fountainhead for rubber, and Mr Thomas Bolas, F.C.S., has shewn (Colonial Exhibition Report, 1887) that the rubber trade here might be further developed. He also considers that there is scope for a considerable development in this direction throughout the Straits Settlements.

† By a machine consisting practically of a small cylinder revolving inside a small stationary cylinder.

‡ In both these respects it differs from gutta-percha, which is scarcely at all elastic or tough.

it burns with a very brilliant flame. Having been once melted, it never regains its former state, remaining soft and sticky when cold. On this account it cannot be moulded or laid on to wire (in a tubular form) when rendered plastic by heat, in the same way as is done with gutta-percha.* On the other hand, two freshly cut surfaces of rubber will unite again if firmly pressed together.

India-rubber absorbs a greater proportion of water than gutta-percha; thus, when slabs of india-rubber remain under water for three hundred days they increase in weight from 10 to 25 per cent. in fresh-water, and 3 per cent. in salt-water. Gutta-percha, under similar conditions, absorbs only 1.5 per cent. in fresh, and 1 per cent. in salt water. India-rubber is also to a slight extent soluble in water; as is, indeed, suggested by the stickiness at the surface acquired by slabs which have been some time under water. Absolutely pure rubber, like gutta-percha, is simply carburetted hydrogen; but the commercial kind contains resin, water, and ashes.

India-rubber, when exposed to light and air, takes up oxygen, changing into a white resinous substance, in the same way as gutta-percha; alternations of wet and dry conditions accelerate this action, which goes on more rapidly in the manufactured article than in the natural substance.

**Electrical Application as an Insulator: History.**—In 1842 Professor Jacobi, of St Petersburg, conducted a series of experiments with india-rubber, as regards its electrical qualities for covering underground conductors; but Mr Charles West appears to have been the most prominent in first drawing attention—in the same year—to its practical utility as an insulating material adapted for submarine cables.†

In that year Mr West covered a few fathoms of copper wire with india-rubber tape, and induced the "Electric" Company to submerge it from a small boat somewhere in Portsmouth Harbour. The experiment was a success, though the line did not work for long.

Notwithstanding that india-rubber was introduced for the purpose— almost exclusively on short underground lines—some time previous to gutta-percha, it never came into use on anything like a practical scale till

---

* The elasticity of rubber also renders this impossible, causing it to tend to resume its previous form and thus collect in lumps, leaving more or less bare patches of conductor elsewhere.

† Mr West seems to have thought that the insulating power of india-rubber being double that of gutta-percha, the signalling speed would be twice as great in a given conductor when covered with india-rubber and when coated with the same thickness of gutta-percha. This, of course, was incorrect, though, on the score of a materially lower capacity, a distinct advantage accrues to india-rubber, as is shewn elsewhere.

after gutta-percha had been similarly employed for many years. The first attempts—between 1848 and 1849—by Thomas Hancock,* Silver and Co., Siemens,† and others, did not, in fact, meet with much success, owing to the physical alterations above referred to (due to chemical changes) which take place in rubber by variations of temperature, especially by heat. It was found also that when in contact with copper it was liable to undergo a species of decomposition, and to become permanently softened.‡

Its hygroscopic qualities were also found to be a serious barrier to its use—for submarine work, at any rate.

## VULCANISED INDIA-RUBBER.

**Historical Memoranda.**—Nowadays, in all india-rubber insulated cables for submarine purposes, the rubber core has passed through the process of vulcanisation; indeed, in most instances when india-rubber is used as an insulator it is vulcanised rubber.

In 1843-44, Charles Goodyear, in the United States, and Hancock, in this country, independently introduced the system known as vulcanising for ordinary india-rubber goods of commerce.§

Sulphur combines with india-rubber, in various proportions, under the influence of heat. Roughly speaking, the operation of vulcanising—or curing, to use the "shop" term—consists of mixing from 3 to 8 per cent. of sulphur with the rubber, and heating the whole for about two hours up to

---

* This gentleman established the first India-rubber Manufactory, in Goswell Road. Later on he was joined by Mr John Macintosh, who had a patent for highly rectified coal tar naphtha as a harmless solvent of india-rubber.

† Messrs Siemens had introduced an ingenious machine for applying pure rubber by what was then an improved method. By this, two strips of rubber were placed longitudinally on each side of the conducting wire. The whole was then drawn through semi-circular grooves, which, in firmly pressing the strips round the conductor, caused them to unite at the edges (whilst clean and hot, immediately after they had been cut), and thus form a complete cylindrical casing for the conductor. The spare rubber from each strip was at the same time trimmed off at the sides by small circular cutters, one on each side, close to the rollers. This plan was a great improvement in some respects, as it overcame the objection of unevenness of covering due to an overlapping seam. The main principle of the machine may be seen repeated in many modern india-rubber covering machines. When more than one covering was applied, the joining lines of the successive layers were placed at right angles to one another—*i.e.*, the succeeding longitudinal layer was applied so that the seams of the two half tubes were at some distance from that of the preceding layer.

‡ This destructive action is checked and modified by tinning the copper conductor.

§ Brooman attempted to apply the same process to gutta-percha in 1845; but this, for various reasons, was a failure, as shewn elsewhere.

a constantly maintained temperature of somewhere about 250° F.,* but not, as a rule, exceeding 300° F. This has the effect of causing a chemical action, and, therefore, producing a complete physical change in the rubber, of such a character that it becomes almost entirely impervious to changes of temperature, thus overcoming the difficulties previously experienced.

In this condition—that in which it is said to be vulcanised—it, in fact, retains the same degree of flexibility at low temperatures, is better able to resist heat, does not oxidise in air, is more elastic, and absorbs less water; moreover, the ordinary solvents of rubber, when cold, have no effect on it, though it may be dissolved in boiling oil of turpentine.†

Vulcanising can be carried out on a small scale, by plunging the rubber into a bath of molten sulphur at about 250° F. If the temperature is gradually raised to 280° F., or thereabouts, a portion of the sulphur will combine with the rubber.

**Hooper's Core.**—In 1849 Mr William Hooper, a chemist of some distinction, who had turned his attention to india-rubber, determined that when vulcanised it would form a suitable dielectric medium for cables.‡

Having a materially lower specific electro-static capacity, and an appreciably higher specific resistance—especially, as a rule, when vulcanised —if it could be made a success mechanically, it would naturally be better suited than gutta-percha for dielectric purposes.

By vulcanising, Mr Hooper endeavoured to attain this result. The sulphur contained in vulcanised rubber was found, however, to have a

---

* The temperature adopted in curing varies considerably with the mixture to be vulcanised and its future use. It must always, however, be above the melting point of sulphur = 238½° F.

Similarly, the length of time for which this "curing" process should be applied depends only on the amount of sulphur in the mixture.

With the same proportion of sulphur, the higher the temperature the quicker the "cure"; but there are certain degrees of heat which are best for each rubber mixture.

† Vulcanising, in fact, restores to mixed (*un*-vulcanised) rubber the mechanical qualities of pure rubber, and renders them permanent.

‡ Thus the late Mr Hooper was the first to apply the principle of vulcanising rubber for insulating purposes. This he was successful in, and every credit is due to him as the originator of vulcanised india-rubber core; for it may certainly be said that until the introduction of what was then known as Hooper's core no success was met with in rubber as an insulator, owing to being unsuitable physically and mechanically, besides being seriously wanting as regards durability. Mr Hooper worked at the subject closely, and performed an extensive number of experiments with different mixtures—in which he was ably assisted by Mr T. T. P. Bruce Warren—before deciding on the constructive nature of his "separator."

The methods of the present day do not substantially differ from those evolved by Mr Hooper; for instance, the "separator" now in use is composed, similarly, of oxide of zinc.

deteriorating effect on the copper conductor, even when "tinned," injuring the conductivity besides rendering it brittle—which, again, also acted on the vulcanised rubber. He overcame this difficulty by introducing a coating of pure rubber next to the conductor, outside which was the "jacket" of vulcanised rubber, containing 6 per cent. of sulphur, and 10 per cent. of lead sulphide. To prevent the sulphur in the vulcanised rubber penetrating in anything like large quantities through the pure rubber coat, he applied what he termed a separator between the layers of pure and vulcanised rubber. This "separator" was made up of rubber mixed with about 25 per cent. of zinc oxide, mainly chosen as being one of the principal mixtures (or pigments) used in manufactured rubber. It was found to answer the purpose well, besides acting as a fairly adhesive union between the pure and unvulcanised rubber, which could not be made to properly stick together of themselves.*

Mr Hooper adopted somewhat similar methods to Siemens, Silver, and others, for applying his separate coverings to the conductor. For reasons already stated, it was found that india-rubber could not, like gutta-percha, be applied round the wire, by means of a die, in a tubular form. It had then to be laid round the wire, either in a longitudinal strip, or else in a spiral taping with overlap.† Each of these methods have their particular advantages (as will be shewn later), and a combination of the two was generally adopted by the succeeding layers being applied differently; a final binding of some sort being, as a rule, effected spirally to allow of greater mechanical pressure being brought to bear. The strip, of whatever form, was cut beforehand to the required breadth, ready for use.

In 1864 Hooper's core was very favourably reported on by Messrs

---

* Not only had difficulty been previously met with in adhering the successive layers of pure rubber, but also in making the first layer stick to the conductor, owing to the latter acting as a solvent on pure rubber and turning it viscid. Siemens had used Chatterton's compound for both purposes but with fatal results, a chemical action being set up which entailed viscosity in the rubber. An inner coating of gutta was also tried by Siemens as a protective union between the conductor and the rubber; but here, again, besides a chemical change occurring, a union of gutta-percha with india-rubber must at all times be bad, owing to the fact of heat affecting the two differently, thus producing, amongst other things, a tendency for the wire to decentralise.

Latterly heat alone was the agent—as now, to some extent. Heat, however, has often been found to fail to effect its object in a sufficiently strong and permanent manner, and great care requires to be exercised to apply the required pressure whilst the material is still sufficiently warm. If, on the other hand, too much heat is present, the insulation of the core is liable to be damaged.

† Adherence between the longitudinal half tubes, or the turns of the overlapping spiral, used to be effected by oil of naphtha; or else—more usually of late—by placing the core, on completion, in boiling water for half an hour.

Bright and Clark, as well as, in a less exhaustive manner, by other eminent engineers, electricians, and chemists.

Subsequently, this type of core was adopted in many miles of cable, being especially selected for tropical climates, where high temperature and the presence of submarine animal life were known to be contrary to the interests of gutta-percha, but found to be comparatively harmless in the case of india-rubber, so far as could be discovered. Hooper's core was, in fact, almost exclusively adopted—with complete success—in the China Seas when the Great Northern Company's Eastern System was being established in 1871. It, however, gave less satisfaction a few years after in connection with another important undertaking elsewhere.* Later on it fell out of favour for some time.

**Modern Practice.**—Nowadays, in the case of vulcanising for india-rubber core purposes, the practice at most factories on a large scale is somewhat as follows: — The rubber is first kneaded whilst warm in a masticator, and sprinkled now and again with flowers of sulphur till there is thorough mixture. This may be carried out by passing the rubber and the sulphur between two steam-heated cylinders (with jagged surfaces) revolving at unequal speeds. The rubber is thus torn as it is drawn between the points of contact, and simultaneously the sulphur is thoroughly and uniformly ground into it. French chalk is, at the same time, also largely mixed with india-rubber for insulating purposes, and sometimes other substances.† This being gone through, the mixture of vulcanising

---

* This has been to some extent explained by the fact that the type of cable adopted was originally designed for deep water, whereas the line was laid in comparatively *shallow* water, and subjected to various disturbances.

† For general commercial purposes many other substances—indeed, almost anything —may be added to render the india-rubber mechanically suitable according to the use to which it is to be turned. There are, in fact, over a hundred various mixtures, including Potter's earth, lime, plaster of Paris, sulphide of antimony, sulphide of zinc, sulphate (and various oxides) of lead, oxide of iron, magnesia, litharge, silica, chalk, Fuller's earth, linseed oil, lamp-black, etc. etc.—all in various proportions according to circumstances. Some of these may be only used as pigments to give a specified colour ; others also, or solely, to assign to the material the desired mechanical consistency ; whilst others again favourably influence vulcanisation, being in some cases solely incorporated on this account.

As a rule, these mixtures are only incorporated with rubber that is about to go through the process of vulcanisation ; but this is mainly due to the fact that unvulcanised rubber, other than pure, has nowadays an extremely small commercial use.

The mixtures vary also according to the gum (or gums) employed, in order to convert the whole into a working mass, and also with the purpose for which the material is to be turned. The relative proportions of these ingredients are varied according to the degree of flexibility, elasticity, and toughness desired. The manufacture of vulcanised india-rubber is, indeed, somewhat complicated ; and there is certainly plenty of scope for errors during the various stages, and especially in the mixing and curing. The details of these

rubber\* is taken to the calender (or rolling) machine. From there a uniform sheet is obtained which, after being cut into strips of the required breadth, is ready for applying to the wire. When this has been successfully effected, and the core finally surrounded by an even band of lapped calico (or cotton-felt) tape,† the completed core ‡ is placed in the "cure," where it is kept for the required time at a temperature of somewhere about 285° F. Besides completely vulcanising the two outer coverings, and partially the inside one, this process also has the effect of consolidating the whole mass.

The curing vessels (or "vulcanisers") have varied in form and shape very much of late years, at the different factories, and according to the material to be vulcanised, but the principle of all is the same. They are merely steam boxes—iron vessels, in fact—capable of being raised (quickly if necessary) to a certain temperature, and of maintaining it uniformly for a given time. The temperature of this oven is regulated by the pressure of steam indicated by steam gauges.

Perfect vulcanisation can only be secured by mechanically incorporating an excess of sulphur.§ The free sulphur, however, is liable to effloresce at the surface, turn acid, and cause deterioration. This is avoided by substituting for flowers of sulphur 15 to 20 per cent. of sulphide of antimony, obtained by boiling native sulphide in a solution of caustic soda and precipitating with an excess of hydrochloric acid. The sulphide should contain 20 to 25 per cent. of free sulphur, and, if necessary, further sulphur may then be added.

---

mixtures (as regards proportions for various purposes) on which success so much depends are naturally kept as trade secrets. In fact, the different mixtures and proportions are often given fancy names, the significance of which is only in the possession of those organising the work.

Some idea of the composition of a vulcanised rubber core may, however, be gleaned from its specific gravity, that of sulphur and other ingredients being about double that of pure rubber, which is about half as much again as water. To deal with the manufacture of rubber completely might well occupy an entire volume.

\* The term "vulcanising rubber" is very usually applied to mixed rubber capable of being vulcanised.

† In binding together the several layers of completed core, this outside tape acts somewhat as a metal mould. Moreover, it has usually been previously steeped in ozokerite, or in some preservative, insulating, anhydrous, and non-absorptive, composition to fill up its pores.

Ozokerite compound may be regarded as a partially vulcanised india-rubber, in which the colloid pores are filled up with the crystalloid particles of the hydrocarbon ozokerite, which is closely allied to solid paraffin wax—in itself a highly insulating substance.

‡ The curing is reserved for this stage, owing to the fact that if performed earlier the edges of the rubber coverings would not unite—being hard and possessing no "tackiness" when once vulcanised.

§ During this operation the coil is very usually covered up in powdered plaster, to ensure uniform heating and to prevent any turns adhering, should the rubber become sticky.

When the proportion of sulphur reaches 25 to 50 per cent., the rubber becomes remarkably plastic, and can be moulded into any shape.*

The degree of "cure" adopted is a most important factor in determining the physical, mechanical, and insulating properties of rubber. Over-curing is as fatal as insufficient "cure." †

**General Retrospect.**—No two india-rubber cables are exactly alike. Cores made and tested on exactly the same lines which do not behave alike have some difference in their manufacture—possibly over-curing, possibly over-mastication, or some error in the intermediate stages of the one or the other.

The degree of success met with largely depends on the care and regularity applied in every stage of manufacture.

When any great variation occurs in the results, it is sometimes put down to alteration in the raw material. It is usually, however, due to variations in one or more of the operations of manufacture.

For reasons already stated, a seamless tube of rubber being impracticable,‡ the first layer (of pure rubber) is applied either spirally or longitudinally—usually the former. The separator, in any case, is, as a rule, applied longitudinally; but with the seams of the half tubes at right angles to that of the previous covering of pure rubber, if the latter be applied longitudinally. The jacket is sometimes applied spirally (with a very good overlap), and more often longitudinally; but in any case there is, almost invariably, an outer cotton taping about 1 inch in width, usually steeped beforehand in

---

\* If the core is properly "cured" there is practically no unvulcanised portion, as the inner coat takes up a certain proportion of the sulphur from the two outer layers.

If, in addition, 3 per cent. of lamp-black and the same quantity of burnt magnesia is incorporated, and the mixture be vulcanised by steam at a temperature of 275° F. for periods of six, eight, or ten hours (according to the thickness of the rubber), a hard black substance called ebonite—formerly vulcanite—is the result. This substance interests us only as being (like ordinary india-rubber) an especially bad electrical conductor. It is, therefore, exclusively used for mounting electrical instruments.

† Indeed, in "curing" it is usually thought safer to err on the under-side than to overstep the mark. In the latter case it is difficult to tell, except by the action of time, whether the material is over-cured or not; whereas the substance being under-cured is brought to light at once.

‡ This statement more particularly applies to high-class rubbers. Owing to their stiffness and to the tearing action of a die, the material gets "killed" in the operation. The advantages of the tubular form have already been made evident, but the vulcanising of the present day so perfectly unites the seams that they are practically non-existent.

Owing to their greater plasticity—sometimes approaching that of gutta—low grade (cheap) rubber mixtures can be spewed on a wire through a die, as in the case of gutta-percha.

ozokerite or other compound, and applied spirally under pressure,* with a good overlap (and in an opposite direction to the previous covering, if spiral), in order to thoroughly bind the whole together whilst in the proper condition.†

There being a greater art in the successful application of india-rubber to a wire than in the case of gutta-percha, the manner in which this is carried out, in its successive layers, is usually kept, to some extent, a secret by those who have made a specialty of the subject.    This is more particularly the case owing to the fact that many of the machines used for the various operations do not form the subject of a patent.

Be this as it may, much valuable information regarding the practical details of rubber manufacture‡ may be gleaned by a perusal of Mr Stuart Russell's work on " Electric Light Cables," to which the reader is referred for further particulars of this description, if required.§

**Physical Qualities.**—India-rubber is a most curious material to deal with in many ways.    It alters in its physical properties so enormously under different conditions, that any description of it will only apply to that particular state of affairs.    When vulcanised, its physical properties are almost exactly the reverse of what they are when unvulcanised.    In the original raw state it is elastic ; in the manufactured (" mixed ") state it is inelastic ; but when " cured "—*i.e.*, baked—it becomes elastic again—indeed, more so than ever.

Thus, in speaking of india-rubber, it is very necessary to specify with some precision the exact form, or stage of manufacture, it is in.    Moreover, it is, in any case, impossible to speak of india-rubber collectively (with regard to its properties) owing to there being—as already shewn—over one

---

* As long ago as 1852, Mr H. V. Physick had a patented machine (specification No. 778 of that year) for applying "calico cloth or other fibrous material" to a wire in tape form.    This was by way of insulation, the tape being "previously saturated with gutta-percha, shellac, tar, pitch, or other insulating substance."    The machine was very ingenious, and actually applied the tape by a peculiar form of die.

† The various points regarding longitudinal and spiral applications of rubber are, roughly, as follows :—In the longitudinal half tubes the actual breaking strain of the core comes in more ; on the other hand, the continuous seam is of the weakest possible kind—being, in fact, a "butt" joint.    In the spiral tape the continuous joint thereby involved is much longer ; it is, however, of a stronger character, being of the overlap form.    More-over, more material is required here.    By the combination of the two methods alternately in each successive layer, the strongest and most reliable insulating envelope is obtained.

‡ It should be mentioned in passing that when once vulcanised, this material practi-cally never gets softened by heat ; thus it is unnecessary to draw a core so covered through cold water (as is invariably done with gutta-percha) for cooling and hardening purposes.

§ "Electric Light Cables and the Distribution of Electricity," by S. A. Russell, A M.Inst.C.E. (Whittaker and Co., London).

hundred different mixtures of it, with totally different ingredients and proportions, each mixture having its special peculiarity for the particular purpose for which the substance is intended.

Vulcanised rubber being so highly elastic, it cannot be successfully "worked" in this state. Thus when cured, less than ever can it be put into any shape by pressure whilst hot and plastic. If attempted, it only draws up and returns to its original form after a certain time. It is for this reason, then, that though the admixture of sulphur is effected immediately after purification, along with other "mixings," practically all the curing process—a given degree of heat for a specified time—is reserved till after the material has been made up into the form it is required. Thus, in the case in point, the rubber is not actually vulcanised to any extent till after it has all been applied to the wire in its various coverings ; the coil so made up is then placed in the curing chamber.

Several means have been suggested from time to time for testing the degree of vulcanisation in rubber. Most of these are in the nature of applying additional heat—*i.e.*, a sort of semi-cure—such as renders the rubber harder and less elastic. If there is an excess of sulphur in the specimen—or if it has been overheated in coating—the material at once shews signs of brittleness. *Insufficient* curing also easily shews itself.

Unfortunately there is no mechanical test for rubber that cannot be evaded in some direction or other. The reputation of an experienced contractor—together with electrical tests which are in any case essential—are best relied on for ensuring a good quality rubber, and one that is really *durable*, which is the main thing to be sought after here. It seems to be imagined by some that the purer the rubber the better for *all* purposes—partly, perhaps, because it is dearer ; inasmuch as pure rubber, when cleaned, costs about 4s. a pound, whereas some of the ingredients in mixed rubber cost about as many pence. However, where durability under unfavourable conditions is a consideration, well-made "compound"—*i.e.*, "vulcanising"—rubber will, as a rule, be found superior.

**Physical, Mechanical, and Electrical Data.**—Owing to the enormous variation in the nature of vulcanised india-rubber, as manufactured by various mixings and methods of treatment, it would be absurd to attempt to give any definite physical, mechanical, or electrical data, in the way of actual constants and specific values, with reference to vulcanised rubber generally. It can only be stated, that its electro-static capacity is generally taken as about 2.8 to 4.2 for ordinary gutta-percha*—where air = 1 ; and

---

* And Willoughby Smith's gutta-percha 3.1.

that its insulation resistance is usually at the very least double that of gutta-percha,* but often many times more.†

It is usually admitted that, if anything, pressure tends to actually decrease the insulation resistance of vulcanised india-rubber, though the effect is, in any case, exceedingly small.‡ The difference in this respect, as compared with gutta-percha, is probably owing to its greater absorptive power (even when vulcanised), due to the fact of being necessarily applied in such a way as to involve a continuous seam, instead of in homogeneous tubular form.§

Temperature influences the resistance of all classes of rubber and rubber mixtures enormously ‖ less than gutta-percha.¶

Again, vulcanised india-rubber is much less influenced by age than gutta-percha. If, however, it is left unvulcanised, rubber undergoes much the same molecular setting by age as in the case of gutta-percha, the result being a very similar increase in electrical resistance. On the other hand, with unvulcanised rubber this only goes on up to a certain point, after which it becomes "treacly '—by decomposition due to oxidation. When once "cured," rubber is no longer open to any material physical changes, and, therefore, its electrical resistance also remains fairly constant irrespective of time.

Yet, electrification takes place very much more rapidly with india-

---

* This is a curious circumstance on the face of it. Being a more or less pure, un-adulterated, and homogeneous gum, it might naturally be thought that gutta would have the higher electrical resistance. It appears, however, to be almost impossible to compare these materials owing to the ignorance still prevailing regarding the absolute physical conditions of each.

† Notwithstanding the comparatively low specific values of most of the mixing ingredients, the insulation resistance of many classes of vulcanised rubber is often considerably greater than that of pure rubber ; but this again varies tremendously with the nature of the mixture and cure. This can only be accounted for by the chemical and physical changes effected by the mixing and curing.

‡ When compared with gutta-percha, the above seems strange at first sight, as it might be supposed that the least homogeneous (most porous and absorptive) material would offer the greatest scope for improvement in this respect under pressure. ("The Physical and Electrical Effects of Pressure and Temperature on a Submarine Cable Core," by Charles Bright, F.R.S.E., *Journal Inst. E.E.*, vol. xvii.)

§ Still, in so far as pressure and low temperature tend to make such a material more *dense*, the electrical resistance of india-rubber tends to be increased on this count—though not to the same degree as gutta-percha.

‖ The proportion is, roughly speaking, about 1 to 4 ; but if the rubber be unvul-canised there is nothing like this difference. In fact, pure rubber is usually even more affected physically by temperature than gutta-percha. Hence, it has been found that when submerged, the absorption of a certain class of rubber was eight times greater at 120° F. than at 39° F., whereas with gutta-percha it was only doubled.

¶ For several reasons the opposite might reasonably be expected. All that can, however, be said is, that having a distinctly different textural character, they are affected quite differently ; and this explanation must be taken as equally applying to the effect of pressure.

rubber (whether vulcanised or not) than with gutta-percha. This again varies enormously with the mixture; but so great is it that it is not unusual for the apparent resistance after three minutes' battery application to be double that after one minute. With gutta-percha the electrification only caused a rise in the apparent resistance of about 25 per cent. during the first five minutes: for rubber it may well be over 100 per cent.*

The formulæ for obtaining all the above-mentioned data are naturally the same as with gutta-percha, differing only as regards the constants and coefficients.†

## DETERIORATION OF INDIA-RUBBER.

On account of its valuable electrical qualities, and from the fact that it is not supposed to be prejudicially attacked by marine organisms, india-rubber would seem to be quite the ideal form of dielectric for submarine cables, especially as, when vulcanised, it stands heat and exposure to air and light.

Unfortunately, however, it is subject to more or less serious deterioration, the exact nature of which is only very partially understood;‡ cables so insulated, supposed to be similar in every respect, giving sometimes altogether different results. Thus, armoured torpedo cables containing seven separate conductors insulated with india-rubber vulcanised throughout, have been found to seriously deteriorate after eighteen months' immersion. At

---

* Thus the apparent resistance after three minutes' battery application is, roughly speaking, twice that after one minute; and after five minutes it is threefold what it would be after one minute only.

† Though these can be given with some sense of accuracy in relation to gutta-percha, the mixtures of india-rubber and its methods of manufacture at different factories being so varied, it would be useless to attempt it with the latter. The best place for obtaining this information accurately at the time, for practical purposes, is at each individual factory for each material.

‡ Mr Bruce Warren—finding that bromine, iodine, and chlorine, instead of oxidising india-rubber in contact with water, produce an altogether different effect—once endeavoured to turn this to account in the manufacture of cable core. By his process, the conductor was covered with two coatings of india-rubber, which were first welded together in boiling water, and afterwards plunged into a solution of iodine in iodide of potassium, or of bromine in bromide of potassium, or else exposed to the action of chloral.

India-rubber treated in this manner is said to withstand a considerable amount of heat without deteriorating. It also resists the action of air, and that of its ordinary solvents, although no trace of iodine, bromine, or free chloral can be detected by chemical analysis. Containing no sulphur, it does not attack the copper, which, therefore, does not require to be tinned. It is said to retain a permanent elongation like copper when subjected to tension, so that the conductor should keep central when the strain on the core is released. It is stated that the electrical qualities of the india-rubber are improved by the above treatment.

It is a process, however, which, in common with other species of unvulcanised rubber mixtures, would necessarily be open to some measure of deterioration by time, and has never been adopted in actual practice.

the faulty parts, the dielectric had become sufficiently porous to permit of the water passing through to the conductor, with the result that some of the cores had much lower insulation than others in the same cable.

It would seem, then, that vulcanised india-rubber, as we now have it, cannot always be absolutely relied upon to provide any specific and un-varying insulation resistance, notwithstanding the improvements which experience have given us.

Vulcanised india-rubber cable, some protected with canvas for army field service and others with iron-wire sheathing intended for submarine-mine purposes, have been stored up, exposed to the air, for ten years or more without suffering in the least, though other similar cables stored in the same building—some under water and some in air—have behaved very differently. In one instance, a length of a few inches at the ends of the cables stored under water was left dry to meet the required conditions for testing, and here the inside coating had become viscid and oozed out at the ends. This deterioration ceased abruptly at the point where the cable entered the water, shewing that, in this case, the deterioration could not be due to the contact between the rubber and the copper. The same viscous exudation was noticed at the ends of the armoured cables which were kept exposed to the air, but here the mischief was found to extend over a much greater length. This deterioration was, no doubt, the result of serious oxidation, due either to too much heat or to sunlight—producing, in fact, precisely the same effect as over-vulcanising in manufacture. The direct rays of the sun have a distinct influence in this way.*

The gradual transformation which sometimes turns the inside covering of rubber into a semi-fluid, viscous, substance does not appear to have the effect of actually lowering the insulation of the cable—especially where next the conductor it remains quite pure. In course of time, however, it un-doubtedly has a solvent action on the outer coatings of vulcanised rubber, and must always be highly injurious eventually to the general mechanical constitution of the cable.

It may be stated generally that the deterioration of rubber is, as a rule, due to bad manufacture, or bad materials—usually the former, and some-times both.† Cables made and tested on exactly the same lines which do

---

* Indeed, in America the sun's rays are actually turned to account in a certain method of vulcanising, for what are called sun-cured goods—or fabrics spread with rubber and sulphur subjected to the direct rays of the sun. This evidently is the effect of a combination of heat and light which produces the same result as the action of heat alone at a higher temperature.

† Owing to the fact that rubber is a manufactured article, whereas gutta-percha is used almost in its raw state, there obviously exists a much greater scope for errors of manufacture in a core of rubber than in that of gutta.

not behave alike have some difference in their manufacture, probably over-curing, possibly over-mastication, or some error in the intermediate stages.

Since the original application of the principle of vulcanising to rubber as an insulator, the manufacture of vulcanised india-rubber core has been made a subject of close and careful study at the hands of the Silvertown Company and Messrs W. T. Henley and Co., as well as of Messrs Siemens Brothers and Messrs Johnson and Phillips, besides Messrs Hooper, the originators.* The result is that it is now, by experience, recognised as the most reliable —though initially, most costly—insulating material for electric lighting requirements, as well as for other underground conductors, especially where a comparatively high electric pressure is employed, and also generally where rough handling, frequent changes, and frequent storage may be looked for, as in the instance of torpedo cables.†

Vulcanised india-rubber has also been turned to good account as the dielectric for several more or less deep-sea telegraph cables, more especially in tropical waters and where the teredo and other enemies are known to abound.

---

* Latterly, also, the Telegraph Construction and Maintenance Company have enlarged their sphere of operations by manufacturing considerable lengths of india-rubber insulated cables for electric light and other purposes.

† Gutta-percha is, however, also employed as the dielectric for submarine-mine ("torpedo") cables. In either case, felt bands, or even tarred linen bands, are usually lapped outside the cores of torpedo cables as a preservative, and to give the core greater strength for pulling about. In this case of shallow water, the objections to the tar so close to gutta-percha even do not seem to seriously apply in practice, or are overcome by the advantages gained in the preservative tape.

Eastern and South African Telegraph Company : Mombasa Station.

## SECTION 8.—RELATIVE MERITS OF GUTTA-PERCHA AND INDIA-RUBBER.*

**Advantages of India-rubber.**—As previously stated, at first sight india-rubber would appear to be the best adapted material possible for insulating purposes—especially when vulcanised—even for *submarine* lines. This is on account of having a materially lower electro-static capacity than percha in any form, and usually more than double the insulation resistance.

Thus on the first count a materially greater signalling speed would be the result on a long length of continuous cable, where every little increase in the electrical constants beyond a certain point makes a serious difference.†

Secondly, vulcanised india-rubber is, as a rule, when in good condition, infinitely tougher than gutta-percha. Thus it stands rough treatment much better.

Thirdly, india-rubber, when vulcanised, undergoes heat better than gutta-percha, mechanically and electrically. The highest temperature which gutta-percha will stand before softening is about 115° F.; whereas no mixture of vulcanised india-rubber softens till about the boiling point of water is reached.‡ Thus, if the cable be anywhere exposed to the sun, the conductor is much less likely to fall eccentric, or the mechanical and electrical qualities of the dielectric to be otherwise detrimentally affected by an increase of temperature with a vulcanised india-rubber covering than in one composed of gutta-percha.§

Fourthly, india-rubber, when vulcanised, is almost unaffected by the atmosphere, besides being but little influenced by dry heat. Gutta-percha and nearly all resinous substances are composed in part of volatile ingredients, and after being exposed for a time to the atmosphere or the rays of the sun, these volatile substances evaporate, the compound becomes hard and brittle, and in consequence liable to crack. The cracks thus

---

* India-rubber and gutta-percha being both hydrocarbons, appear chemically similar, inasmuch as it is very hard to learn much about a hydrocarbon. In reality, however, they are perhaps as unlike in physical properties as is conceivable, for any two such materials, though both suitable insulating mediums.

† In the case of the proposed Pacific cable, for instance, an india-rubber dielectric would—on *electrical* grounds at any rate—be at a distinct advantage, with given dimensions.

‡ "Insulation Resistance," by Charles Bright, A.M.Inst.C.E., *Journal Inst. E.E.*, vol. xviii., p. 123.

§ Cables in shallow water often experience temperatures as high as 85° F.—or even nearer to 90° F. towards the beach—which, after a time, is more liable to affect a gutta-percha core prejudicially than one of india-rubber in the ordinary course of affairs.

formed admit and retain moisture, which has the effect of greatly impairing the insulation of the conductor, and finally of destroying it altogether.

When pure, india-rubber is no better than gutta-percha, if exposed to light and air, as regards durability; neither is it under alternating dry and wet conditions.*

When vulcanised, however, it is infinitely superior, especially on the first count; indeed, light and air are not supposed to materially affect certain forms of vulcanised india-rubber.†

**Advantages of Gutta-percha.**—On the other hand, high-class india-rubber cannot, owing to its great elasticity, be applied in a permanent manner round a wire in tubular form by means of a die apparatus, as is done with gutta-percha.‡ Being laid on in previously prepared longitudinal and spiral strips, a continuous seam occurs, which, though overlapped, must always tend to be a source of weakness, mechanically and electrically, as compared with the complete tube.§

Again, electrically, the insulation resistance of vulcanised india-rubber, when submerged, has been known to become reduced by time, pointing to a possible deterioration rather than improvement, as in the case of gutta-percha under the same conditions. This may be due to the somewhat extensive absorption of water by india-rubber of all forms; ‖ very likely partly owing to the continuous seam already alluded to.

---

* As regards gutta-percha, mechanical and electrical deterioration under these conditions is brought about by the evaporation of the natural oil under extreme dryness, causing cracks which are subsequently filled up by water under the opposite succeeding influence.

When new, india-rubber may also be said to be slightly volatile, though scarcely constituted by an oil in the general sense.

† These last three features apply where a cable is liable to experience much handling and changes of position, as in the instance of submarine-mine connections, sometimes at the bottom of the sea, sometimes stored ashore.

‡ Inferior mixtures of rubber, being more or less plastic, may be squirted on to a wire tubularly. Owing, however, to the tearing action of the die, this is a doubtful advantage with any class of this substance.

§ It may be remarked, however, that this disadvantage in rubber core has, of late years, been reduced down to rather a fine point by improved methods of manufacture, the seam being eventually so perfect as to be practically imperceptible when performed by certain machines. With definite materials and by a definite method of procedure the mixing and vulcanising renders the seam non-existent in reality.

‖ Henley's ozokeritted india-rubber core was intended to meet this objection, the pores of the india-rubber being filled up with ozokerite wax, which is, moreover, an excellent insulating medium itself.

The ozokerite may also be applied to tape as an outside covering to the core for the same purpose. This is very usually done, and with good effect.

The process of joint-making in vulcanised india-rubber must always be a more elaborate and troublesome affair than that in gutta-percha, besides the risk of defect being here correspondingly greater, where the precise time for vulcanising is so important a consideration.

On account of the higher specific resistance, faults of insulation in india-rubber should be easier to detect than in the case of gutta-percha. Owing, however, to the high degree of elasticity of the former material, any weak place is liable to be at times concealed—sealed up, in fact. Thus intermittent effects are produced, such as very often render anything like accurate localisation an extremely difficult, if not impossible matter, in comparison with gutta-percha under the same conditions.

In the event of gutta-percha becoming seriously scarce—as it has sometimes threatened to—it is at any rate satisfactory to know that there is a fair substitute in india-rubber, which, with further experience and improvements in manufacture, may still more closely realise the requirements even for submarine purposes. For *land* purposes it is all that could be desired, as has already been shewn. It is, moreover, well suited for exposed places—on the beach and elsewhere.

Eastern and South African Telegraph Company : Cable Hut, Mauritius.

### SECTION 9.—OTHER SUGGESTED INSULATING MATERIALS.

It may be fairly said that as regards the initial expenses attached to a submarine cable, the greatest apparent scope for improvement lies in the composition of the dielectric, owing to the heavy cost of gutta-percha (or india-rubber) as an insulating medium.

**Various Inventions.**—From time to time, subsequent to the employment of gutta-percha and india-rubber for this purpose, various inventors have been said to have discovered marvellously cheap substitutes for the above; but it is worthy of note that nothing is ever heard of these being adopted in practice, though it certainly does not follow from this that they are not possessed of great merit. However, on close inspection, they are not, as a rule, found to have quite the same qualities that experiment voluntarily suggested, or was made to suggest. On the other hand, where these "inventions" are found to be entirely satisfactory, it invariably turns out that they are in reality a form of rubber or gutta-percha—gutta-percha or india-rubber forming, in fact, the principal base—under a different, and more elaborate, name; or, at any rate, that this satisfactory result is brought about by the introduction of a more or less large proportion of rubber (or gutta) into their composition.

**Advantages Claimed over India-rubber and Gutta-percha.**—Some of these compounds have been said to be most capable of resisting oxidation, attacks of insects, and other objections; besides being less affected by changes of temperature. Many of them certainly appeared—at first blush—to possess valuable electrical qualities, but with none as yet have we found anything like the mechanical indestructibility and electrical fixedness of gutta-percha when kept under water, or the toughness and reliability of india-rubber on land when exposed to the sun. Up to date, indeed, nothing has been found which is likely to replace gutta-percha or india-rubber in their application to submarine telegraphy, as a dielectric material, for many years to come.

**Particulars of Alternative Materials.**—We will, however, briefly make reference to some of the principal substances which it has been proposed to substitute for these from time to time :—

Paper,* wool, cotton, silk, hair, canvas, flax, jute, hemp—in yarns or

---

* Comparatively recently Messrs Felten and Guilleaume have suggested the use of flat strips of prepared paper applied round a cross-shaped conductor with air spaces between, the whole cable being protected outside by a waterproof casing—lead tubing, or other

strips or braiding, as the case may be—all formed the subjects of early patents in various ways for insulating purposes, and were even suggested as a suitable means of insulation for *submarine* cables. It should be mentioned, however, in passing, that, in most instances, it was proposed to previously saturate the above with some preservative composition of a resinous nature that would tend to render the material anhydrous, and consequently non-absorptive.

Bitumen, or pitch—a residual product of coal tar—was the usual liquid material suggested for steeping any of the above fibrous materials in, or else some variety of the same. Vegetable and coal tar and their oils were also prominently put forward as a preservative anhydrous mixture, possessing a fair insulation resistance; as well as linseed oil, paraffin* and paraffin oil, *et hoc genus omne;* all possessing a very high degree of insulation when fresh.

Besides the plan of previously saturating the material with any of the above compositions, there was also that of pouring it into a metallic tube of some sort (usually lead) into which the conductor, lightly surrounded with the fibrous material, had been previously drawn.

Then there were the suggestions of forcing, or steeping, bee's-wax, paraffin, or other wax, rosin, shellac, and bitumen, as insulating substances, in a fluid or semi-fluid state, into the pores of wood or any of the previously mentioned fibrous materials. White lead, ozokerite (*i.e.,* paraffin in a natural state), spermaceti, tallow, cold cream, and butter, have all been made mention of in the same way, to render the fibrous material durable and anhydrous, as well as assigning extra insulation to the whole substance.

On the other hand, however, one or two inventors appear to have

---

protecting material of some description. This form of cable they suggested, on the grounds of low capacity combined with lightness and cheap insulation, more especially for long-distance submarine telephony. The joints might, however, present some difficulty, both as regards efficient carrying out and the attainment of non-absorption and subsequent water-tightness.

Again, the dielectric in the famous concentric cable of Mr S. Z. De Ferranti is composed of prepared paper. This cable is said to be doing useful work as a high-pressure main for electric lighting.

* Paraffin, a product of the distillation of certain kinds of coal and bitumen, is an excellent insulator and an admirable excluder of moisture; but being of a·brittle nature, it is generally acknowledged nowadays that it cannot be directly made use of in cable manufacture, even for land purposes.

It, however, serves to protect the bare ends of the leads used for testing purposes, and keeps them from contact with the air. To do this, the ends of the wires are dipped in paraffin previously melted in a small vessel; on this being repeated five or six times, the ends receive a white coating of solid paraffin, which prevents any loss of electricity along the surface.

abandoned the idea that the insulator, even for a submarine cable, need necessarily be waterproof, going in only for what they principally cherished, as being strong, fibrous, pliable, and durable materials, with a certain insulating power of their own, such as horse-hair, for instance!*

In Hearder's patent of 1858 (No. 444), he proposed layers of gutta-percha to be alternated by layers of some fibrous material, with a view to combining strength with insulation. He even stipulated that the fibrous material should be at the same time *porous*, on the grounds that porous materials have a lower electro-static capacity—especially if not possessing too high a specific resistance.† Mr Hearder further arranged that his fibrous and porous material was to be next the conductor, and was again to form the outside of the dielectric so as to be on the surface of each Leyden jar plate, with a view to reducing the static condensed charge effect based on the above notions. So far as increasing, at least possible cost, the thickness of the dielectric, and therefore the distance between the plates, his idea, from a *capacity* point of view, was a good one, and probably the first of its sort in this direction.

Even vitreous substances, such as glass tubes, have been proposed, and actually patented, from a very early date, not only for underground purposes, but as a method of submarine insulation. Sulphur was also suggested many years ago, mainly for vulcanising, and this was highly successful with india-rubber. On the other hand, in combination with gutta-percha, it proved a bad business—though tending at first to increase the dielectric resistance —besides the fact that gutta-percha would not stand the temperature necessary for vulcanisation.

Ground cork, lime, mortar, and various cements have all formed the subjects of early patents in connection with insulating methods.

The difficulties of effecting really efficient joints in any of the previously mentioned fibrous, resinous, and vitreous compositions and compounds, does not appear to have been fully taken into consideration by the designers and patentees.

Again, oxidised oils‡ have been tried in combination with rubber, it

---

* It should be remarked, however, that hair is impervious to moisture, and whilst being exceptionally light, possesses great strength, especially for its weight and toughness. Moreover, it was considered that it would not be attacked by the teredo, or other worm or insect.

† This is by no means usually the case, of course. Speaking generally, a high specific inductive resistance (low specific inductive capacity) more often goes with a high specific insulation resistance than otherwise.

‡ In the preparation of these, something containing oxygen is added in order to procure viscosity or the semi-solid state for ready adulteration with the rubber. The proportion, however, must not be overdone for fear of brittleness.

having been found that unless previously oxidised, all oils tend to attack the rubber, which is, in fact, a solvent in oil. The results, however, have scarcely proved completely successful, for (being oxidised) the oil tends (after drying) to become brittle and unreliable,* even though rendered more or less impervious to changes of temperature by vulcanisation.

Wray's composition—one of the most hopeful modifications—was tried on a small scale in early practice. This was in the main a mixture of india-rubber and gutta-percha, mostly the former. It had a high insulation, and behaved well at high temperatures, but, unfortunately, was found to deteriorate in the sea.

Okonite, kerite, and nigrite are all names for special preparations of vulcanised india-rubber.

In the first named it is difficult to say where any particular characteristic difference comes in.

Kerite is said to be a mixture of the products of the oxidation of drying oils, for instance that of linseed,† nut and cotton-seed, with vulcanised india-rubber, and certain other substances, such as ozokerite. Whilst having a considerably lower capacity than ordinary vulcanised rubber and a high insulation, it is also maintained for this material, that it offers extra resistance to the effects of air and heat, rendering it well suited for tropical land lines, especially as it also successfully withstands alternations of dry and wet conditions. It is, however, less reliable, mechanically and electrically, than ordinary vulcanised india-rubber.

By masticating together, at the lowest temperature that will give plasticity, india-rubber and "black-wax"—a residue product of partially distilled ozokerite—a substance is obtained, called nigrite, which is said to be even less affected by heat than india-rubber. This material has been successfully employed at times for submarine-mine cables.

As sulphur enters into the composition of most of these latter substances,‡ it is desirable, as a rule, that the conductor should be "tinned."

In order to meet the objection of india-rubber being so absorptive, except when so highly vulcanised as to have become ebonite, Mr E. T. Truman took out a patent, in 1878, for a cable with an india-rubber dielectric in this hard non-absorptive condition. Here he obtained the required flexibility, without the vulcanite cracking, by an outer lead tubing. Such a cable has not, however, ever been used in practice.

---

* The so-called "drying" of a paint is, in reality, nothing more nor less than an oxidising effect, in some degree or another.

† Linseed oil tends to absorb oxygen somewhat freely.

‡ Most of which are repetitions or modifications of what were experimented with many years ago.

**Gutta-percha and India-rubber Combined.**—There have also been a number of patents, from time to time, for covering a gutta-percha dielectric with a thin outside casing of india-rubber, with a view to protecting the gutta-percha against heat, air, light, atmospheric pollutions and climatic changes, as well as against the insidious attacks of the teredo for waters favoured by that enemy to cables. Some of the devices for attaining these objects took the form of alternate layers of gutta and india-rubber, the gutta-percha being next the wire, and the india-rubber outside the whole. However, as has already been mentioned, these materials will not unite well, if only owing to the different degrees to which each is affected by heat. The union of rubber with gutta must certainly also be bad on account of their having such different degrees of homogeneity, the former being used in an almost absolutely pure state.

**Failure of Alternatives.**—The various compounds that have been tried have almost invariably failed to retain their insulating qualities sufficiently, owing to one cause or another—though, initially, infinitely cheaper than either gutta-percha or india-rubber.

**Main Requirements.**—Finally, it may be remarked that—as regards *submarine* requirements, at any rate—whatever the material is, it must be—

(1.) A bad conductor of electricity.

(2.) A repellent of moisture.

(3.) Fairly tenacious and flexible at all temperatures, so as to allow of being bent and twisted without injury.

(4.) Not liable to seriously soften or volatilise under continued exposure to any temperature—as for instance, when subjected to the sun's rays.

Eastern and South African Telegraph Company : Station at Delagoa Bay.

# CHAPTER III.

## JOINTING.

---

## Section 1.—General Remarks and Implements Used.

THE various applications of this chapter will be apparent. It, in fact, refers equally to all the different occasions on which joints require to be made between lengths of the insulated conductor, *i.e.*—(1) the various unsheathed coils ; (2) the sheathed sections at factory ; (3) the same aboard ship in preparation for laying ; (4) during repairing operations.

The term "splicing" is sometimes applied to all the operations necessary for continuity between two cables ; the conductors, insulating envelope, hempen covering, and sheathing wires being each separately joined together. These operations are of paramount importance in submarine telegraphy, a single badly made joint often necessitating lengthy and costly repairs,* and so materially lessening the commercial value of the cable.

This kind of work at sea is of necessity carried out in a more "rough-and-ready" manner than on shore, owing to prevailing circumstances. The same remark still more applies with regard to the testing of joints effected aboard ship, owing to the impossibility of employing such instruments and methods as would give a satisfactory clue regarding the state of affairs. Sometimes, indeed, such tests are altogether delusive.

When two sections of cable already laid are spliced together on board, and the bight so formed quickly let go, it is almost impossible to obtain even the roughest indication as to the condition of the joint. There is

---

* Sometimes involving several months interruption to traffic, besides a direct cost of thousands of pounds !

thus no available means of verifying the workmanship. Consequently jointers—especially those employed at sea—should be skilled workmen of guaranteed integrity.

One of the essentials for making a good joint is extreme cleanliness. It is not every one who makes a good jointer, however clever he may be with his hands in other ways. In some persons there is always a greasy exudation from the pores of the skin. Should this be the case with the jointer, who frequently requires to touch the portions of the core he is at work upon, a greasy film forms on the copper or gutta-percha, preventing perfect electrical connection between the parts, and this in spite of every precaution to ensure cleanliness.

A jointer at work should always have an assistant, so that he may not have to touch anything unconnected with his work, after the joint has once been started.

## TOOLS.

The necessary tools for making an ordinary gutta-percha joint include*—

> Trestle bench.
> Plumber's stove, or firepot.
> Spirit lamp.
> Soldering iron.
> Tooling, or polishing, irons for working and kneading the gutta-percha.
> Flat pliers of different sizes.
> Knives, rags, emery paper, resin or stearine, etc.

The bench (Fig. 30) is furnished with a pair of small vices, or clamps, placed in a line facing one another. One of the vices is adjustable along a groove cut in the bench, to which it can be secured at any point by a nut underneath. These clamps are for firmly holding each conductor end together during the process of jointing.

The spirit lamp serves to warm the polishing, or tooling, irons, and to soften the gutta-percha, as well as to heat the soldering iron. Any kind of spirit lamp may be used, provided it gives sufficient heat, burns, when

---

* Many of the above become unnecessary where the joint in the conductor is effected by the process of electric welding. This, however, has not yet been adopted on a practical scale for the conductor of a submarine cable. It is said that the conductivity of the copper is affected thereby to the extent of about 5 per cent. For this reason it is sometimes debarred in specifications.

filled with naphtha,* for at least two hours, is portable, and finally, is convenient for handling.

The spirit lamps for this purpose are usually provided with a cap such as may, when required, be fitted on, so that the contents cannot escape during transport.

The spirit lamp may be placed under a hollow cylinder or hood of sheet

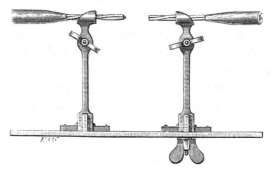

FIG. 30.—Jointer's Vice.

iron (Fig. 31). This hood protects the flame from any stormy gusts of wind such as would blow it out, and acts as a holder for the various irons, there being one or two windows cut into it, through which the irons to be heated are passed, so as to be just over the hottest part of the flame.

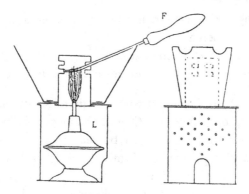

FIG. 31.—Spirit Lamp and Hood.

There are also a number of smaller holes pierced all round the hood, thus ensuring a sufficient draught of air through them and out at the top, with the result that the flame acts somewhat as a forge flame.

---

* Naphtha is more especially used in a jointer's lamp on account of being so invaluable for cleaning and drying the joint (or core generally) during the jointing process.

The tooling irons (Fig. 32) are oval in section, and made of polished steel. They are fixed in light wooden handles. A well-proportioned iron for convenient use should have the following dimensions :—

| | |
|---|---|
| Length of iron - - - - - | 6 Inches. |
| Breadth at the end - - - | $\frac{1}{2}$ ,, |
| Breadth near the handle | $\frac{5}{8}$ ,, |
| Thickness at the middle - - | $\frac{3}{16}$ ,, |

The iron must only be heated over a spirit lamp (of the description given) in a plumber's stove as described, and its use is confined to spread-

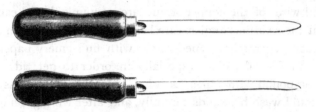

FIG. 32.—Tooling Iron.

ing out the Chatterton compound, besides polishing and filling up the seams in the gutta-percha. The proper temperature to which the iron should be heated is recognised by a gentle warmth perceived on holding the iron an inch or two away from the cheek—when also the end of the iron has acquired a pale blue colour.

The gutta-percha is cut with what is commonly known as a "trimming knife," the blades of which must be very sharp. Just before using, the tongue is passed along the blade to prevent the gutta-percha adhering to the metal. The waste edges of gutta-percha sheet used for the joint are cut off with a pair of bent scissors, previously moistened in a similar manner.

## Section 2.—Jointing the Conductor.

The gutta-percha insulation of the two pieces of core to be jointed is cut away (Fig. 33), so as to bare the copper conductor for about $1\frac{1}{2}$ to 2 inches from either end. Care has to be taken that the copper wires are not at any time nicked by the knife—even in the smallest degree * Should accidental injury thus occur, there should be no hesitation in snipping off the length of core already bared, and starting again on a fresh piece.

The several wires of the copper strand are now unlaid, and each pulled out straight at their various angles, care being taken to give them no twist. Each one is then separately cleaned either with fine emery paper or with the back of a knife-blade, more especially in order to get rid of all the Chatterton's compound with which each has become more or less covered. The jointer should wash his hands carefully, and after wiping, plunge them into naphtha, which is allowed to evaporate so as to leave the fingers

Fig. 33.—Conductor Ends Prepared for Jointing.

perfectly dry. The bared copper wires are also sometimes wiped with naphtha, so as to completely clear them of any grease or other foreign matter.†

The copper wires are now twisted together again into a true strand as before for about a third of their length, using a pair of small flat-nosed pliers, care being taken to twist them in their original direction, and preserve the proper length of lay.

This first part of the operation is common to all the methods of copper jointing. The two ends can be jointed together in any of the following

---

* After a little softening by heat and pressing round with the pliers to the required distance, the length of gutta-percha may often be neatly *drawn off* the conductor with the pliers, thus obviating the necessity of cutting with a knife.

† This is not, however, allowed as a rule, owing to the insulating properties of naphtha, which might thus prevent proper conduction between the various wires composing the strand. The cleaning of the wires is, in fact, usually alone carried out by emery paper, or by the back of a knife-blade as above.

ways, all of which give good results, so far as reliable continuity is concerned :—

1st. The seven wires* of each conductor are divided into two sets (Fig. 34), one containing four and the other three wires. The close portions of the two conductors are then brought together end to end, and the four sets of wires intercrossed. Each set of wires is then separately wound round the opposite conductor, covering the lay of its corresponding set, but

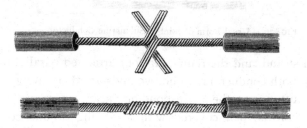

FIG. 34.—A Form of Conductor Joint.

without crossing each other. The ends are trimmed off with cutting pliers. The wires are wound in place, using small flat pincers held in one hand, whilst the centre of the joint is firmly held in the other hand with a larger pair of pincers.

2nd. The two ends of core to be joined being prepared in the usual manner already described, the central, or heart, wire of either end is cut, and the remaining wires opened out, as in Fig. 35. They are then interwoven so that each wire of one end passes between two wires of the other end.

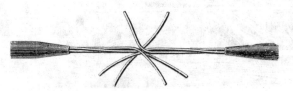

FIG. 35.—A Form of Conductor Joint.

The ends of the heart wires are then butted together, and the other wires of one end wound spirally round the opposite strand, and conversely. The splice is first made by hand, and afterwards finished off with the flat pincers.

The two above methods give excellent results with regard to continuity and mechanical strength, but increase the diameter of the conductor at the

---

* This is assuming the conductor to be composed of an ordinary seven-wire strand ; but these remarks would in this method equally apply for jointing any other form of conductor so far as the general principle is concerned.

joint, this, though of little consequence in land lines,* would become a serious drawback in submarine cables, where the core must be enclosed in an outer sheathing of fixed and uniform diameter. For this reason, one of the three following methods is to be preferred :—

(*a.*) The wires being first laid up again to form the original strand for about two-thirds of their length, three alternate wires of equal length are

FIG. 36.—Usual Form of Joint in Submarine Conductors.

cut out of each strand, and the remaining four arranged parallel to the axis of the strand in both conductors. The two ends are then brought together, and each wire butted into the corresponding vacant space in the opposite strand. The whole is then covered with a serving of fine copper wire, extending beyond on either side for a distance equal to the length of the joint.

(*b.*) The wires composing the strand are first carefully re-formed over

FIG. 37.—Jointer's Bench and Tools in Operation : Splicing the Conductor.

the entire length of each end of bared conductor, soldered together † and filed down to a long bevelled or tapered wedge (Fig. 36), so that when laid together they exactly fit, without increasing the size of the conductor

---

* Nowadays the above somewhat *rough-formed* joints are scarcely ever adopted even for land-line purposes, though perfectly efficient.

† This is best effected by dipping the bared end entirely into a lump of molten solder on the soldering iron, running it round the iron to ensure evenness, and finally shaking off any superfluous lumps whilst still in the fluid state.

at the joint.* The bevelled ends having been cleaned in the way to be presently shewn, a little further solder is applied to them ; † when the file is again used lightly to obtain smooth sharp surfaces. The conductor ends are now placed in the vice, ‡ the bevelled ends being fitted together (Fig. 37) and pressed against one another. These two wedge surfaces are then

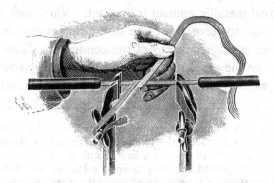

FIG. 38.—Applying Binding Wire to Metallic Joint.

soldered together. Over the joint and on either side, for a distance equal to its length, is wound (Figs. 38 and 39) fine copper wire of No. 30 S.W.G. (.012 inch), intended to bind the joint firmly together in a

FIG. 39.—Completed Metallic Joint.

permanent manner, and to maintain communication in case of separation of the bevelled surfaces by failure of the metallic joint. The whole of the

---

* This preparation for scarfing the two conductor ends together to form a scarf joint (the principle of this method) requires to be performed with great care to ensure the two scarves fitting properly and truly against one another, by their angles exactly correspond-ing as well as by a perfectly even surface being secured. To carry this out in practice it is very usual to have a niche cut somewhere (as a gauge) on the jointer's box (Fig. 37), into which the end of the wire is laid and filed away till flush with the block. The wedge or taper is usually about one-third in length, but longer for big conductors. It will be obvious that the longer the taper the better, with the limit that it must not be so long as to be disposed to break off, and this depends on the diameter.

† One of the main practical points in soldering is to remember always to clean the surface of the iron previous to using it, either by wiping or else, preferably, by spirits of salts and emery cloth. Another point is that the iron should be applied as hot as possible without risk of damage.

‡ In placing the ends in their respective clamps care must be taken not to allow the edge of the gutta-percha covering to come too close to the ironwork of the vice for fear of it getting seriously heated during the soldering of the joint.

joint is then finally soldered over.*   This latter is the method universally adopted for the conductor joint of submarine lines of the present day.

**Tin Soldering.†**—As already indicated, after a metallic joint is made good, it is tin-soldered.‡  This is obviously indispensable in the case of the last method described ; and by this means, in the instance of the other earlier methods, good metallic contact is ensured.  Variations of temperature have a tendency to force apart the interlacing wires, in unsoldered joints, and so increase the resistance of the conductor.  Great changes of resistance occurred on the old aerial lines, even during the same day, where the ends of the wires were simply twisted round each other without

---

* The spiral lapping of fine binding wire was originally intended in part to meet the conditions of the conductor being subjected to an undue strain, by allowing of a certain amount of drawing out—after the manner of a helical spring.  To turn this to account, therefore, it is very often considered best only to apply solder at each end of the spiral and not along its entire surface.  Or, again, very usually there are two wrappings applied reverse ways, so that the wires in each cross one another.  Here the first is soldered right along (the interstices completely filled with the solder), but the outside wrapping is only soldered at the ends.  Before applying the binding wire its surface is first cleaned usually by drawing the length intended for use through a piece of fine emery cloth.  It is applied (see Fig. 38) in the form of a flat band by being doubled into about four lengths.  This spiral lapping actually formed the subject of a patent in 1855 taken out by Messrs Statham and Willoughby Smith, as before stated, to meet the contingency of a temporary strain or actual rupture of the conductor at the joint where it is naturally less extensible and flexible, besides being more brittle, than elsewhere ; and of course a fracture of any single joint renders the cable useless till repaired.

† As Mr W. H. Preece has pointed out (*Electrical Review*, 29th October 1888), soldering is a term which should be applied solely to the process of uniting the surfaces of metals.

It is effected by the intervention of a more fusible metal than that which has to be united.  The soldering metal, called " solder " on account of the purpose to which it is turned, being melted upon each surface, serves—partly by chemical affinity, partly by the exertion of cohesive force—to bind them indissolubly together.  "Solder" is usually an alloy.  It must not only be more fusible than the metal or metals to be joined, but it must (as already explained) also have some chemical affinity for them.  Hence it is that different sorts of "solder" are employed for different purposes.  It is called either hard or soft, according as its fusing point is high or low.  The test of efficiency of "solder" is a peculiar creaking sound, called the "cry of the tin," which is heard when a stick of it is bent close to the ear.

‡ The soft solder used as the soldering agent for joining electrical conductors (which used to be known as telegraphic solder) is composed of equal parts by weight of ingot tin and pig lead melted together : expressed in *bulk*, this would come to about two of tin to one of lead.  Hard silver (or brass) solder is, however, invariably used for joining together the individual wires during laying up of conductor into strand form, being very fusible and non-corrosive.

It has been pointed out by T. P. Bruce-Warren (*Electrician*, 19th October 1878), that owing to the impurity and brittleness of many qualities of solder, it is advisable, so far as telegraph work is concerned, to mix the right proportions of the metals comprising the soldering fluid rather than to purchase it ready made.

soldering; and the difficulties in working these were notorious. There would be danger of the same sort of thing in the case of submarine cables, which are tested at 75° F., and afterwards laid on the sea floor, where the temperature may be as low as 35° F.

The conductor is cleaned just before soldering (as already described) with a little spirits of turpentine or naphtha, and then dipped into, or covered with, powdered rosin* or stearine, by way of a flux† for the solder.‡ A small quantity of solder is melted on the soldering iron and poured over the joint in a small continuous stream, until all the interstices of the wires are completely filled up.§ Immediately this is effected, two or three slight blows are given to the conductor as a test, and in order to detach any small portions of alloy which may not have properly adhered. The solder, when cold, is finally worked over, or smoothed down, lightly, with a fine file to remove any possible roughness or irregularities, and the jointing of the dielectric is then proceeded with.

---

* This is now sometimes termed resin, but the old name (rosin) is preferable for this, if only as a distinction from *resin* as used in a chemical sense.

† The flux action is that of keeping the surface of the conductor clean—and thus permitting clean metallic contact with the solder—by preventing the formation of a thin film of oxide (" scale "), which would naturally take place otherwise during the process of soldering owing to the heat introduced.

‡ Spirits of salts (*i.e.*, hydrochloric acid and zinc) have been used as the flux agent for this purpose, as it is more convenient and effective, besides cleaning the wire at the same time. It is, however, nowadays usually objected to on the ground of seriously eating into the copper conductor, by introducing a continuous chemical action between the former and the solder. Moreover, like sal-ammoniac (which is also sometimes used) it harbours dampness, owing to the fact that it does not dry up.

However, both of these latter are conveniently and safely used for the less vital matter of effecting a joint between the iron sheathing wires and an earth wire at the testing hut or cable station.

§ A very usual way of carrying this out in practice is to rapidly run the hot soldering iron underneath the bared conductor at the joint, and simultaneously run a stick of the solder along the top of the conductor so heated.

SECTION 3.—JOINTING THE INSULATION : GUTTA-PERCHA CORES.

The joint in the conductor—almost invariably a scarf joint—is after a little practice a fairly simple matter.

The joint in the dielectric however, introduces much greater scope for really serious mishaps due to improper workmanship, as will be explained later.

It suffices here to say, that there is considerably more practice required in the case of the dielectric joint, the difficulty being at first to keep all air bubbles excluded from the material while hot, and to finish with the jointed conductor perfectly central in its insulating cover.

First of all, the ends of the core are usually pared off to avoid joining to the gutta-percha those portions which have received heat from the soldering, and which may be in some degree injured.* The metallic joints and the gutta-percha ends are next cleaned by being wiped over with a clean bit of rag soaked in wood naphtha. The metal conductor is now gently warmed by waving the flame of the spirit lamp underneath, care being taken, however, not to burn, or injure, the gutta-percha at all.

Whilst this is being done, the assistant jointer has softened over another spirit lamp one end of a stick of Chatterton's compound.† The jointer now lightly dabs this stick along the bare conductor, afterwards spreading the adhering compound‡ evenly round the strand with a warm tooling iron.§

We now come to what is known as the first covering. This is con-stituted (Fig. 40) by the drawing of a portion of the gutta-percha from each end over the joint.

The ends of the gutta-percha to be joined together are warmed by the lamp moved to and fro underneath—always at some distance, however.‖

---

* Indeed, having become oxidised to a certain extent, these ends would not join properly with the new gutta-percha to be applied.

† In this operation care must be taken not to allow the compound stick to remain in the flame of the lamp long enough (at a stretch) to burn, or trouble will afterwards very likely ensue due to air-holes.

‡ As need, perhaps, scarcely be remarked, the Chatterton's compound is here employed for purposes of ensuring perfect adhesion between the metallic joint and the first covering about to be applied.

Some authorities object to it for joints partly on the same grounds as they do for the conductor generally—*i.e.*, that of injuriously acting on the gutta-percha, and also on the score of its application in joints tending to introduce air-holes afterwards.

§ After being heated over the lamp this iron should be thoroughly wiped with a clean rag before touching the " Chatterton " or gutta-percha with it.

‖ It is very usual to blow the end of the flame near the gutta-percha.

In this way the gutta-percha is softened on both sides of the joint for nearly 2 inches, and the two ends of gutta-percha are then gradually (with the fingers*) drawn towards one another till they are about ½ inch apart, the core being turned round with a twisting motion backwards and forwards throughout the operation. One end (*a* in Fig. 41) is then drawn down to a thin film, whilst the other (*b*) is gradually, by dexterous manipu-

FIG. 40.—First Stage of Gutta-percha Joint.

lation (Fig. 40), drawn over it, until the joint is enveloped in a covering materially thinner, but nearly as uniform—if performed by a skilled workman—as the rest of the conductor. The first covering is then completed. The thickness of insulation is about the same as that of the first coat of gutta-percha in the rest of the core, as is, indeed, shewn in Fig. 42, which

FIG. 41.—First Stage of Gutta-percha Joint.

represents the joint at this stage. The joint is then kneaded and smoothed down with the tooling iron.

After the whole of the "draw-down" has been thoroughly worked over with the thumb and forefinger, it is allowed to cool and "set" for a while, after which the surface of the draw-down is roughened with a knife. It is then slightly "lamped" over again, preparatory to Chatterton's compound

FIG. 42.—Completion of First Stage in Gutta-percha Joint.

being lightly applied (for adhesive purposes) as before, though in this case some consider it better to *roll* the hot end of the stick of "Chatterton" over the gutta-percha, instead of dabbing it on in lumps. In any case, this

---

* The fingers require usually to be slightly moistened with saliva to prevent sticking during manipulation of the semi-plastic gutta-percha.

process is followed by "lamping" and tooling over the compound until an even surface of compounded gutta-percha is obtained, especial care being taken not to overdo the heat for fear of damaging the gutta-percha, and causing subsequent air-holes : at the same time the heat must be sufficient, or the compound will be liable to remain in hard lumps.

One of the strips* of gutta-percha sheet roll kept purposely for jointing† is then carefully softened by the assistant jointer by warming over the spirit lamp. When sufficiently softened, a piece about 2½ inches wide is cut off the sheet with a pair of scissors, which is next stretched in order to reduce the thickness to the required limits ; the length of the sheet will now be just what is required for the joint.

The jointer then takes the strip into his fingers (which are moistened with saliva to avoid sticking) and applies it from underneath lengthways to one end of the core, and being firmly pressed is drawn along the length of

FIG. 43.—First Covering of Gutta-percha Joints.

the joint underneath ; the edges are raised (Fig. 43) and gradually by a uniform pressure brought together at the top so as to completely surround the core.‡ At this stage—and from time to time—it is usually found advisable to warm up the sheeting and the core by "lamping," to ensure better adherence ; as well as to facilitate good workmanship in the way of an even

---

* Before starting work the assistant jointer usually cuts several strips off from the roll of gutta-percha convenient for handling over the spirit lamp, of suitable dimensions respectively for the two coverings for which it is used—the thickness and number of each bearing some kind of corresponding relation very often to that of the separate coverings of the rest of the core. Usually, however, two such strips are applied to each joint, the last (outer one) overlapping that previously by an inch or more at each end.

† This gutta-percha sheet forms part of the "stock-in-trade" of a jointer's box. It is generally kept in an air-tight metal case, usually filled with water, to preserve its characteristic qualities. It is most important that it should be thoroughly clean as well as quite free from moisture at the time of application to the joint—as much as that the kneading tools be so kept.

‡ The reason for this procedure is to make sure there are no air-bubbles imprisoned by commencing at one end and gradually squeezing the strip against the core at every part till the joint is entirely covered flush with the rest of the core where any spare ends are cut off. Should there be any sign of an air-bubble or moisture, the hot tooling iron should be applied to the spot at once so as to draw it out, the place being made good afterwards.

and uniform covering. Indeed, it is best to turn the joint upside down* after heating on one side with a view to warming it on the other, thus helping to drive off any moisture from the hands that may have found its way amongst the core and sheeting. The jointer now pinches the edges tightly together between finger and thumb to obtain firm adherence at the seam formed between the new strip and the old gutta-percha inside. He then trims off the excess with a pair of bent scissors.† The heated tooling iron should then be passed over the seam so as to open it again. After this it is once more pinched up, the result being that any air is forced out that may have penetrated. In the last pinching up process, it is best to make one edge overlap the other very slightly, so that the warm tool finally run over the seam may more perfectly seal it up; or, if the edges are merely butted, great care must be taken that they properly meet. The joint is now again "lamped" over and kneaded with moistened thumb and forefinger, care being taken to preserve its shape and to knead evenly all round. The iron, heated to the required temperature, is next used to smooth off any inequalities in the gutta-percha, in order to make a better join between the two ends of the new layer and the adjoining portions of the old gutta-percha by working down smooth the ends of the strip so as to taper off gradually on to the core. The same operations—softening (by lamping), kneading, and smoothing over—are repeated a number of times until the outside surface is faultless, and there are no signs of air-bubbles escaping from the gutta-percha under the kneading influence. Finally, this second covering is perfected by the joint being rubbed into shape with the moistened palm of the hand.

Followed by the same warming, compounding, and tooling over, another gutta-percha sheet (forming the third covering) is then applied to the joint in precisely the same way,‡ equal pains being taken in the subsequent hand manipulation by way of kneading and tooling down irregularities, as well as in tapering off and smoothing over the joint at each end between the main core and the gutta-percha sheet.

In this covering the strip is longer by an inch or more at each end than

---

* In doing this, the jointer and his assistant should turn it over simultaneously, so as to avoid putting a twist in the joint. Similarly, when the operation is concluded, the joint should be turned back in the *opposite direction* in order to take the twist out again, otherwise a permanent twist will make its way into the core at each joint.

† In order to effect this, it is necessary to pass the blades of the scissors over the tongue, as already explained.

‡ It is, however, advisable to apply the outside strip, bit by bit—starting from the opposite end to what was done with the first strip—and thus working the strip on by hand in a *reverse* direction this time.

2 C

the previous one so as to ensure it completely overlapping the latter, thus bringing the joints with the rest of the core at either end to different points and consequently ensuring a more general tapering.

In applying this outside covering care must be taken, moreover, to see  that the centre line (or the full part) of each strip comes immediately over the joining line (or seam) of the strip preceding, so as oppose the lines of weakness in each layer (Fig. 44). It is necessary also to avoid softening the finished layers beneath, when warming the outside gutta-percha, in order that the kneading will only affect the layer actually under treatment.

FIG. 44.—Section of Finished Gutta-percha Joint.

When the kneading of the last layer is finished, the gutta-percha is once more and finally slightly softened. Then, with the palm of his hand well moistened, the jointer rubs down the joint. This rubbing must be done uniformly and equally all round; it tends to solidify the joint and inadvertently assigns to it that highly polished and finished appearance so characteristic of good work.

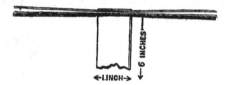

FIG. 45.—Alternative Gutta-percha Joint : Commencement of First Process.

We have here only endeavoured to describe one very usual method of gutta-percha jointing.

By another method, very largely adopted, the gutta-percha sheet instead of being applied as a longitudinal strip, after the core has been lamped and compounded, a strip, about 6 inches long and 1 inch wide is (after warming) wound round the core (Fig. 45) several times in the form of a knob (Fig. 46) over the centre of the usual draw-down, forming the first covering. The knob is then quickly worked for about $4\frac{1}{2}$ inches in both directions with the

FIG. 46.—Completion of First Stage.

finger and thumb, so as to make the new gutta-percha meet that of the rest of the core on about the same level, where the joins of the two gutta-perchas are then tooled down with the heated iron. This then, after being carefully worked and tooled to an even and gradual tapered surface, completes the joint so performed. This method has certain advantages. Probably, however, with equally skilled workmanship the two joints are equally efficient—in fact, as efficient as could possibly be.

A well-made joint is usually, when finished, somewhere between 6 and 8 inches in extreme length, and but very little larger than the adjacent core, tapering away gradually, as shewn in Fig. 47, on either side from the centre.*

The joint being completed, it is first allowed to cool by being placed in cold water—or having cold water poured over it—and afterwards hung up in the open air. It is important that steps should be taken at once in this direction, to prevent any damage or decentralisation of the conductor in its insulating envelope.

At sea—when time is of first importance—to accelerate the action, the core is usually placed in a trough of iced water ; or, still better, in a special

FIG. 47.—Finished Gutta-percha Joint.

cooling mixture composed of a solution of ammonia and saltpetre, the proportion being—

| | |
|---|---|
| Muriate of ammonia | 5 parts. |
| Saltpetre | 5 ,, |
| Water | 16 ,, |

This mixture must be kept well stirred until the joint has sufficiently cooled.†

The mixture is placed in a kind of tray or trough, and the joint held down in it by two rings at the bottom for somewhere about ten or twenty minutes, after which the joint is tested. The special construction of this tray (to meet the electrical requirements) will be found fully dealt with elsewhere.

---

* A section of the joint—see Fig. 44—even when examined under a lens, should show a perfectly smooth surface round the conductor, and there should be no suggestion of bubbles, or of moisture. When cut through longitudinally, the copper should be found exactly in the axis of the cylinder, and adhering firmly to the insulation. The several layers of gutta-percha should be traceable by very fine black lines of Chatterton's compound between them, and there should be every evidence of perfect adherence throughout.

† When, however, the above is not available, an excellent cooling mixture consists of 2 oz. sal-ammoniac, 2 oz. washing soda, 2 parts fresh-water. This has been found to cause the solution to drop in temperature as much as 21° F. (from 78° to 57°) in the course of two minutes.

At the factory, the joints in the core—which occupy about an hour*—are each rigorously tested, as soon as they are effected and sufficiently cooled, by means of an electrometer or galvanometer, and what is known as the accumulation method (due to Mr Latimer Clark), whereby the effect of the weakest current passing through the insulating sheath at the joint is intensified many fold in its effect—and, indeed, shews itself where it would not do otherwise. This test is very fully described in Mr Young's work on Testing, with all the different ways of carrying it out.†

On the other hand, as a rule, owing to the prevailing conditions, it is very often the custom not to attempt to test the joints made at sea.‡ Others, however, go through some sort of test (as a control on the jointers to a certain extent), as shewn elsewhere, though by the test which is not uncommonly applied, it would require to be an extremely weak joint to call any special attention.

By way of general review, it may be remarked that the jointer must be always careful to heat with the lamp any portions of the joint he may have touched with wet fingers, before laying on another layer of compound or gutta-percha, so as to drive off all trace of moisture. Should there be the slightest sign of perspiration on the jointer's hands owing to the heat of the weather, he should immediately wipe them with a clean cloth ; or, better still, dip them in naphtha, the rapid evaporation of which will cool the surface of the skin. This should be repeated during the continuation of the work as often as necessary.§

It is usually understood that all dirty work is to be done by the

---

* By practised hands they can sometimes be effected in forty to forty-five minutes, including the time taken (10 to 15 minutes) over the braze in the conductor. In the tropics—aboard ship or ashore—they require a quarter of an hour longer for cooling and hardening by immersion and bathing in iced water as aforesaid, unless one of the above mixtures be available. In the latter case, the joint is almost instantly rendered fit for covering up previous to the splice being started upon.

† "Electric Testing for Telegraph Engineers," by J. Elton Young (*The Electrician* Printing and Publishing Company, London).

‡ In present practice, jointers very rarely fail to make a perfect joint, having, as a rule, served their time in the factory, where joints, of necessity, occur at every three, or even two miles—according to the type of core. They are, moreover, specially picked out as reliable men for sea-work.

§ Some men are actually prevented altogether from making good joints on account of being so prone to perspiration.

A joint requires continual handling during manufacture, and it will be readily understood that perspiration is liable to affect the gutta-percha in such a way as to prevent a proper junction between the several coatings.

In further explanation it may be remarked that the perspiration, coming into contact with the hot gutta-percha, sometimes turns into steam, which, under pressure, bursts through the layers of gutta-percha.

assistant jointer, leaving the jointer free for the requirements of joint coverings, where cleanliness is so important an item.

In holding the core it should be taken firmly between the thumb and forefinger at such a distance from the joint as to be beyond the influence of its heat, for the gutta-percha here should always be kept hard. If the hands be too near the joint, they may press the material where it has been softened by the heat, and so cause serious damage.

It is of the greatest importance that the strips of gutta-percha used in making a joint should be, as far as possible, of the same quality as the gutta-percha employed for the rest of the core. With gutta-percha of widely different quality, it is almost impossible to produce a sound, homogeneous joint, and many faulty joints have been due solely to dissimilarity in the materials used, though each material may have been good of its kind.*

Faulty joints are generally caused by—

(*a.*) Unskilful cutting of the copper wires, causing nicks.

(*b.*) The wires being badly laid up, or not filed to a true scarf.

(*c.*) Faulty soldering.

(*d.*) The conductor being centrally displaced through unskilful kneading, or tooling.

(*e.*) Lack of efficient adherence between the conductor and the insulation.

(*f.*) Air-bubbles between the different layers of gutta-percha due to insufficient pressure when applying the strip.

(*g.*) Burns on the gutta-percha caused by careless use of the lamp, and overheating.

(*h.*) Imperfect adherence between the layers of gutta-percha or Chatterton, through moisture or want of cleanliness.

(*i.*) Accidental abrasions.

If any water, or even moisture, finds its way in any part of the core during the manufacture of the joint, it is sure to be the cause of trouble afterwards if the gutta-percha is applied over it.† This, owing to the water forcing its way out under the subsequent pressure (when laid at the bottom of the sea), and leaving a hole in the insulating envelope, which

---

* Thus anything like a permanent joint with an old cable, the gutta-percha of which has been exposed, is an almost hopeless matter.

† Hence the importance of thoroughly drying the joint under construction at its various stages by "lamping" it over continually with the naphtha spirit lamp.

may or may not fall a ready prey to more water reaching the conductor; but, in any case, leads to a more or less serious insulation fault. Another frequent cause of this "blowing" action of the gutta-percha is that of air finding its way into the core during manufacture. Should there be any such air-bubbles imprisoned in the joint, they are almost certain to lead to ruptures in the insulation sooner or later when under the great pressure at the bottom of the sea, which amounts to as much as $1\frac{1}{4}$ tons per square inch for every 1,000 fathoms depth. When the cable is, on submergence, subjected to this pressure, any places where air is imprisoned are, indeed, invariably burst open, thus causing serious faults. It is, therefore, as desirable that any chamber of air shall be exploded, and the air let out during manufacture of core, or joints in core, as that moisture should be.

Joint-making in tropical climates is a peculiarly troublesome and delicate operation on the above accounts; firstly, because the strong evaporation of a hot climate especially tends to draw out any moisture from the gutta-percha; secondly, owing to so large a quantity of moisture pervading the atmosphere; and, thirdly, on account of the difficulty sometimes experienced in getting, or keeping, the gutta-percha sufficiently cool and hardened.

However, the special conditions and provisions regarding joint-making at sea—and especially with reference to hot climates—are gone into more fully elsewhere, with reference to cable-laying operations.

Finally, it may be remarked that there is more in the actual carrying out of successful gutta-percha joint-making (as regards really *permanent* joints for submarine purposes) than might at first sight appear to be the case. There are, indeed, many practical details which make all the difference in the degree of success attained.

For this reason, jointing is not a thing which can be properly learnt from a book—perhaps even less than many other operations in engineering practice. To attain anything like-perfection, the student should put himself under instruction from a professional jointer.

These notes on the subject are only intended to give an idea of the main points involved, and some kind of notion of the principles in practice, as a guide for the amateur who may be suddenly—under stress of circumstances—required to act in the absence of professional comrades.* Practice alone will further help.

---

* The matter is gone into somewhat at length, on account of its vital importance in cable work.

It may be said generally that gutta-percha joints, as effected nowadays by first-rate jointers under favourable circumstances, are as a rule (for submarine cables), practically speaking, as perfectly homogeneous and absolute —mechanically and electrically—as the rest of the core.*

---

* Actually when compared electrically against a similar length of the same core, the joint almost invariably tests better. This may be to a great extent accounted for, however, by its slightly greater dielectric thickness.

Eastern and South African Telegraph Company : Quarters of one of the Staff at Delagoa Bay.

### SECTION 4.—JOINTING VULCANISED INDIA-RUBBER CORES.

The soldered scarf joint in the copper conductor having been made in the usual manner (as shewn in Figs. 48 and 49), and left with a smooth and clean surface, the external tapes and braidings should be stripped back from the rubber. The more or less bevelled surfaces at the edge of the old rubber, which have necessarily been exposed during the making of the

FIG. 48.—Scarf Joint in Conductor.

joints in the conductor, should be well scraped—or, indeed, pared away—and then wiped or painted over for an inch or so with pure benzole or naphtha to free it from dirt or grease by actually dissolving the outer surface.\*

A strip about ¾ inch broad—or more than one, as the case may be—of

FIG. 49.—Finished Conductor Joint.

pure rubber treated in the same way (so as to obtain the natural coherence of two similar fresh rubber surfaces) and having about the same thickness, as the original inner strip, is applied, under tension, either longitudinally or spirally round the conductor—usually the latter—so as not only to cover the entire metallic joint, but also (see Fig. 51) to overlap the inner edges of the rubber in the rest of the core. There is an advantage in the inside covering of the joint being applied spirally in that it permits of it being

---

\* After cleaning as above, it is very usual in present practice to first cover the metal joint and exposed conductor with one lap of ⅜ inch broad cotton tape (sometimes coated

FIG. 50.—Metallic Joint Enveloped in Tape.

with india-rubber) as shewn in Fig. 50. This prevents action between the different rubbers or between the rubber and the copper wire—in the event of any sulphur having percolated as far—and gives a fairly even bed for the rubber strip that follows.

laid on tightly under a fair pressure. In either case a joint is very often made with the inside layer of pure rubber, merely by pressure.

Where, as is usually the case, more than one layer of pure rubber is necessary in order to build up the joint to the same thickness as the inside coat of the rest of the core, the successive layers are each applied (under considerable tension) in alternate directions.

The ends of each of these pure rubber strips are very usually secured down by means of quite a small quantity of india-rubber solution.* This

FIG. 51.—First Rubber Covering.

solution may, indeed, be applied outside each tape to ensure their uniting together, care being taken to allow sufficient time for the spirit to evaporate before applying another tape.

Then over this and the bevelled surface of the old rubber are lapped layers of mixed and vulcanising rubber,† followed by an outside covering of waterproof tape, varnished over all, until the diameter of the completed joint (Fig. 52) is about equal to the entire thickness of the original core.‡

The rubber should be put on tightly and evenly, so as to leave no spaces filled with air—indeed, it is actually stretched slightly in the process, as a rule.

FIG. 52.—Completed India-rubber Joint.

The rubber joint so effected should first be covered with a prepared tape put on spirally, then with a piece of calico sheeting firmly rolled on

---

* India rubber solution consists of 20 parts of rubber which has undergone mastication in 100 parts of benzine or mineral naphtha oil.

† These are usually composed of two coverings of prepared tape, ⅜ inch broad, laid on in opposite directions, with strong shellac varnish between them.

‡ By the old method of joint-making in Hooper's core, the oxide of zinc separator was maintained in the joint by wrapping the pure india-rubber strip round with canvas strip impregnated with oxide of zinc. This was then covered with two strips of vulcanised rubber (*i.e.*, rubber sprinkled with flowers of sulphur), the last overlapping the first—a plan still often adhered to.

longitudinally. This again is tightly bound up with strong selvedge cotton tape applied spirally. The sheeting and selvedge tape are only applied to act as a mould and hold the joint in shape under pressure whilst it is being cured. They are removed again as soon as the vulcanisation of the joint is completed.

The process of curing—as has already been shewn in the chapter on india-rubber—involves apparatus for maintaining the article under operation at a high temperature for a given length of time. At the factory, this condition is readily met by the joint being kept in contact with steam.*

A convenient and fairly accurate test of the degree of vulcanisation is to try and indent the rubber, when cool, with the thumb nail. If properly cured, the rubber will yield to the pressure, but no mark will remain. If, however, the imprint of the nail is left, the rubber is not sufficiently cured; and if it is hard and unyielding, it is probably *over*-cured.

When it is known that the joint has been properly made with the right amount of vulcanisation, after being electrically tested in the usual way, it is finally protected externally by a lapping of strong tape, which should extend an inch or so over the braiding or serving. This latter very often finally receives a coat of varnish to further imbue it with the necessary waterproof qualities in correspondence with the rest of the core.

The length of time required for a vulcanised india-rubber joint is a varying quantity, according to the exact routine adopted and prevailing conditions. It usually occupies somewhere about three-quarters of an hour, but the curing process may run into as much as three hours, or it may only take an hour, depending on the type of core, etc. On the other hand, being practically independent of temperature, little or no time for cooling is necessary.

The principal causes of faults in india-rubber joints are the same as in gutta-percha — *i.e.*, air-holes and moisture. The same precautionary care is necessary to avoid them, including "lamping" over, from time to time.

Successful and reliable joint-making in india-rubber cables, especially as regards the vulcanising, is a much more uncertain matter than in the case of gutta-percha. It will, indeed, be obvious that the joint can never be as homogeneous an affair in the former instance as in the latter.

Thus, with rubber joints more than ever, instruction in their manu-

---

* For land-line, and aboard-ship, work, different forms of portable "cures" have to be used, but these are matters foreign to our immediate subject.

facture can only be satisfactorily obtained from a professional jointer and by much experience.*

These notes are only intended to give a general idea of the principles involved in practice: the student can best find out further details for himself by experience.

---

\* The main requirement in vulcanised rubber jointing consists in recognising the right degree of curing, besides care and experience in giving the right temperature, and time for the same. No great manipulating skill (as in gutta-percha jointing) is necessary ; for here the applying of spiral (or longitudinal) tapes is comparatively speaking a simple matter, as against the kneading with a hot iron and fingering required for gutta-percha joints.

Eastern and South African Telegraph Company: Superintendent's House, Capetown.

# CHAPTER IV.

A " Cable Shop."

## MECHANICAL PROTECTION AND STRENGTH.

## SECTION I.—METAL TAPING.

FOR those portions of any cable which are about to be deposited in waters where marine organisms exist, the core now very usually first receives a sheath of metal tape by way of protection.

**Submarine Borers.**—These enemies to cables consist mainly of what are known to naturalists as *Teredo navalis*,\* *xylophaga*, and *Limnoria tere-brans* or *Limnoria lignorum*.† They are all various species of marine ravaging insects, or worms, provided with boring tools, which they make use of to pierce into the cable. Beyond idle curiosity, these creatures appear to have a *penchant* for jute and hemp, judging from the extent to which they have been found to devour it. In satisfying this taste, and in order to burrow a passage, these borers have a habit of also "scoring" the gutta-percha in the form of an elongated groove, though the actual meal appears to be, usually, in the main, made off the jute or hemp.‡ Vulcanised rubber seems to be too hard for their proboscis, or else the sulphur disagrees with

---

\* This miserable little mollusc first made itself a reputation by eating up wooden ship hulks, until builders took to plating them. When the cable came, it took to it at once.

† Amongst the foes to submarine cables there are also, of course, the saw and sword fish, which have a disagreeable habit, whilst hunting for dinner, of attacking a cable —mistaking it, maybe, for some submarine monster bigger than themselves—with their beak or snout. Saw-fishes vary in size from 10 to 150 lbs., and sword-fishes (sometimes confused with the former) are usually of dimensions such as are represented by a weight of about 100 lbs. These are mostly met with on the Brazilian and other tropical coasts, and especially near the mouth of rivers.

The shark and the whale must also be included in this category—indeed, several curious cases of warfare between a whale and a cable have been experienced. However, the attacks of these monsters cannot be rendered at all less effectual by any such metal taping round the core ; though a sheathing of tough, close-sheathed iron wires is a natural precaution, in tropical shallow water, with this object.

Fish attacks and fish bites, as above, are dealt with in full elsewhere.

‡ Injuries to cables by marine organisms were first noticed in the Levant. The late Mr R. S. Newall found, in the case of a hemp-covered cable which had been down there only a few months, that the hemp had been destroyed by a species of " teredo."

Again, in 1859, Mr C. W. Siemens (afterwards Sir William Siemens) found on a similar cable, which had been laid rather less than a year, millions of small shell-fish or snails, accompanied by tiny worms, which had entirely demolished the unsheathed hemp covering. Moreover, small circular holes were found pierced here and there in the gutta-percha.

Since then similar havoc has been wrought on several cables, on a small scale, in all parts of the world ; notably in the Western basin of the Mediterranean, the English Channel, Irish Sea, Atlantic Ocean, and more especially on the coast of Brazil, in the Persian Gulf, East Indian Archipelago, etc., caused by the different species of ravaging insects above alluded to.

them, for it is only gutta-percha cores that are so affected: hence rubber cores never require metallic taping as an anti-boring sheath.

Though the core is, as a rule, only influenced quite superficially, cases have been known where these borers have penetrated down to the conductor, thus establishing more or less "dead earth," electrically speaking. In any case, a tendency towards weak insulation is liable to be set up at this point, especially as the serving becomes conspicuous by its absence, and particularly when a loose place for the water to more thoroughly act on the iron wires, and on the gutta-percha, is established—eventually culminating, perhaps, in the water reaching the conductor. Such faults are naturally likely to occur where the sheathing wires have — due to chemical, or mechanical, action—become to any extent corroded or worn, leaving the serving open for comparatively rapid consumption.

All these ravagers are sometimes spoken of by the general name "teredo." This is inaccurate.

The *Teredo navalis* is mainly to be met with in the Mediterranean Sea, and has probably been responsible for most of the experiences to which we allude. In higher latitudes, however, the "teredo" proper is scarcely known. For instance, it is the *Limnoria terebrans* which mainly affects our waters round about the English Channel and Irish Sea. This mollusc-like worm is found more particularly to confine its attentions to the jute, or hemp, serving. Thus, the cables under the control of the British Postal Telegraphs, since 1893, have been invariably brass-taped. The tape is applied outside the serving,* by way of evading these ravaging insects.†

Faults due to marine borers have been very usually found to occur at the mouth of a river, where, owing to various forms of food coming down the river, animal life is more prevalent. They also occur along more or less tropical coasts.

In any case the metallic riband is only applied to the core of that part of the cable likely to go into such insect-ridden waters, which more commonly means in depths under 250 fathoms,‡ though in warm climates the teredo has been known to exist at much greater depths—1,000 fathoms,

---

* It is partly for a similar reason, moreover, that hemp is always adopted for the inner (as well as the outer) serving of these cables, since the rest of them have been taken over by the Post Office from the Submarine Telegraph Company.

† This also has sometimes an economic advantage in the case of multiple-core cables (like most of the above), by saving the individual taping of each core.

‡ Thus it is not unusual in selecting a route for a coast cable to avoid a line of depth under about 500 fathoms, if possible. This is partly with a view to the cable being free from fish attacks, teredo bores, and vegetable and animal life generally; as well as to avoid strong currents and a rocky bottom.

or even more.* It is, indeed, almost entirely a question of bottom temperature which determines their presence, or otherwise ; and this again is largely governed by currents.†

The *Teredo navalis*—a sort of soft-bodied snail—is but little bigger than a pin's head.

Even when new, the sheathing wires of a closed cable are never likely to butt tightly together throughout the entire length owing to occasional irregularities. Thus, it can readily be understood how such a minute creature finds its way in ; and with any species of open-sheathed cable, it becomes a comparatively simple matter.

The other borers are, however, a good deal longer—about ¼ inch, as a rule. These are said to exist ashore, and to have been traced in underground lines.‡

Further particulars regarding marine animal and insect life—together with illustrations—will be found elsewhere : for full details, the reader is referred to an excellent early paper on the subject due to the late Mr George Preece.§

From what has transpired it will be readily understood that the metal taping here spoken of to ward off marine borers is only applied, as a rule, to the cores of the shore end, and intermediate types of a cable—and then only for certain waters in cases where such objectionable life is supposed to exist.


**History of Metal Tape.**—The first instance of a metal riband being suggested for a cable was that of the brothers Bright.‖ As early as 1852, a patent (No. 14,331) was taken out by E. B. and C. T. Bright, in the course of which was mentioned the covering of cables with hemp yarn, and a very thin metal spiral riband, or taping, between two fine flexible woven tapes.¶

---

* There is said to be no limit to the depth at which other harmless forms of animal life exist.

† Thus in the shoal waters on the West Coast of Africa, no doubt, the teredo would be found but for the existence of cold bottom currents.

‡ Indeed Mr Willoughby Smith once advanced the idea that they took a passage in the serving of the cable from the factory, and eventually turn into powder later, like the mite in a cheese.

§ "Cable Borers," by G. E. Preece, *Jour. Soc. Tel. Eng.*, vol. iv., p. 363.

‖ Previously the protection of underground lines from such ravages had been sometimes effected by embedding them in cement.

¶ Later on the bituminous silicated compound, which Sir Charles Bright patented in 1862 for the outer sheathing of cables, was partly intended to serve the same purpose. This was by the mixing of powdered silica with the pitch and tar ; thus breaking and

According to this device, the metal tape was applied outside the hemp serving. Theoretically speaking, this would be correct for warding off the ravages of insects, inasmuch as it is the yarn they have a special weakness for, and which they play the most havoc with.

However, in practice some difficulty is experienced in making the metal tape lay properly over such an uneven bed as that afforded by the serving of yarns; and, thus, if any boring insect should find its way anywhere, it would be liable to cause more trouble than ever.

It was, in fact, found to be essential that the metal tape should have an almost absolutely flat surface as a bed to lay on properly.* Moreover, an overlap at each turn is, in any case, necessary.

Mr Henry Clifford was the first to pay special attention to this particular point, and—whilst Engineer-in-chief to the Telegraph Construction Company—to introduce in 1878 a method of metal taping peculiarly suited for this purpose.

According to Mr Clifford's plan, the metal tape is applied outside the core instead of outside the serving.† It is this which is now in extensive use in the present day. Clifford's brass taping was first applied by the Telegraph Construction Company to a part of the Eastern Extension Company's Penang-Malacca cable laid in 1879. Since then it has been adopted for at any rate all the shallow-water sections of the Eastern, Eastern Extension, and Eastern and South African cables which are laid in at all tropical waters, where the teredo *et hoc genus omne* are liable to flourish, and where gutta-percha is the insulating medium.

In selecting a suitable metal for this purpose, it was necessary to combine lightness with as much toughness and strength as possible,‡ the

---

rendering useless the boring-tool of the teredo and such-like. Both systems have now been adopted for many years in the case of all cables exposed to these attacks—in depths under 100 fathoms, at any rate. These combined remedies have been found entirely effectual.

* In early days, when metal tapes were applied spirally outside the serving as above, a packing of flexible cotton tape used to be interpolated longitudinally between the hemp serving and the brass ribbon.

Messrs Siemens have used tapings of this description on some cables laid by them in the East.

† It was found necessary to give up all ideas of protecting the serving, owing to the difficulty of applying the tape to it in an efficient manner.

‡ This is so, largely owing to the necessity of keeping the bulk down as much as possible. Moreover, if the weight of the cable were materially increased by such a taping, the modulus of tension would be seriously reduced; though this would, as a rule, only apply to the case of fairly deep water.

question of economy being also considered of course. Thus, on these grounds (especially on the latter), as well as on that of strong chemical action, it was at once found that copper was unsuited for the purpose.

Certain forms of brass were soon found—partly on account of the hardness introduced—to be particularly suited to attain the object in view.*

Though this metal riband is often spoken of as brass taping, it is nowadays more generally composed of that variety of brass which goes by the name of Muntz metal, being a compound of copper, zinc, and tin, containing rather more tin and zinc in it than ordinary brass, and correspondingly less copper.

This metal mixture is found to afford a remarkable degree of hardness and toughness to a comparatively small weight, besides which it introduces little or no chemical action.†

Recently phosphor-bronze‡ has been occasionally used for this purpose instead of ordinary brass, or Muntz metal. This is with the idea of still further combining the maximum strength with the minimum weight beyond that effected by Muntz metal. Again phosphor-bronze is supposed to be more durable, though certainly more expensive at the outset.

**Present-day Application.**—According to Mr Clifford's plan, first of all a fine (wet) cotton tape (about ½ inch broad) is applied spirally with about half overlap by way of bedding and preservative; then (2) the metal riband of same breadth and 4 mils. thick is laid on with half overlap; and finally (3) a waterproof preservative and protective cotton tape, rendered so by being previously soaked in stearine (mineral wax), ozokerite, or other resinous compound.§

---

* Besides the metallic hardness defeating the borers' intentions, it is supposed that what chemical action occurs is offensive to the taste of the ravagers.

† No doubt the notion of the special suitability of this metal for the purpose first originated with its application to the bottoms of wooden ships. Here it was discovered that Muntz metal ensured such a permanently smooth surface that barnacles were unable to cling to it for any time, besides being so hard as to defeat all efforts on the part of borers. Moreover, a thin sheath of Muntz metal was found to be not only cheaper than copper as previously used, but, owing to the absence of chemical action, lasted as long as the ship itself, instead of having to be periodically renewed.

‡ Both phosphor and silicium bronze are in reality hard-drawn copper, with 3 per cent. of tin added for producing still further hardness. They are so styled on account of the materials which are employed, in some form, as a flux, in the process of manufacture.

§ The main object of compounding the tape thuswise is to check any tendency in the direction of a galvanic battery being set up, with consequent chemical action, between the

These three tapings are each applied* alternately†—at one operation to save time—with opposite lays (the length of which depends on the width of the tape adopted), with a view to more thoroughly covering, upon each side, every overlap of metal tape, for its better preservation and water-proofing, as a complete armour against the ravaging insects. Inasmuch as in a brass-taped cable the tape acts practically as the main "earth," it is undesirable that the inside tape should be compounded in any way, for fear of masking faults.‡

By the Clifford system all these tapes are applied at one operation in a manner which will be described later.

The Silvertown Company use wet cotton *threads*, or yarns, as a bedding for the metal tape in place of the calico tape. This, in the author's opinion, has certain advantages. It forms a softer bed, and avoids the ridges that are a very usual accompaniment with any tape. The difference in this respect mainly rests in the fact that the cotton threads, tending naturally to fill up the space on the surface of the core, do not require to be applied with any overlap. It also gets over the danger of ravelling (or "rucking") up, and of the riband pressing into the core at uneven places as above. Again, threads retain the moisture better than tape.§

---

iron wires and the brass—the wet serving forming the electrolyte by absorption of sea-water or due to its previous saturation. By compounding, the tape becomes more or less waterproof, and therefore should prevent any electrolytic action such as would constitute a battery.

Were the above action allowed to take place, iron being electro-positive to brass, the sheathing wires would be gradually eaten away, though less so than if the metal tape was formed of copper—more highly electro-negative than any other metal—where the greatest potential difference would exist. It may be added that Muntz metal is at a slight advantage over ordinary brass in *this* respect again. The above compounding of the outer tape appears to effect its object fairly well, as there has never been absolute evidence of chemical action between the metal riband and the sheathing wires in a cable with brass sheathing to its core—even after many years' trial. In the case of faulty insulation at any point—and especially if the salt-water has entered—the presence of brass tape would probably tend to accentuate it ; but, on the other hand, it would very likely render its localisation more easy.

* The tapings—especially the metal tape—have to be laid on very tightly, and in the process adopted to accomplish this the core is liable to become stretched beyond what it would do otherwise. This forms, in fact, one of the objections raised in early days to this method of protection, though not now considered serious in practice.

† By another arrangement, due to the same gentleman, the metallic and cotton tapes are stuck together before applying them to the core. This latter plan, though experimented with at first, has never been turned to practical account.

‡ In order to secure a still more regular bed, the Telegraph Construction Company now very often dispense with the inner tape bedding and apply the metal tape direct on to the core, followed by the outside tape as a preservative protection to the metallic riband.

§ Water is played on whilst being applied to the core.

Fig. 53 shews a recent form * of the ordinary three-head taping machine,†
capable of applying the three coverings in one operation after the manner
of either of the above descriptions.

As will be seen from the illustration, each head is provided with a
three-speed cone pulley, by which it is driven through the medium of
belting from the main shaft, which can be seen running longitudinally
along the base of the machine. The opposite lays of each taping are
arranged for by the tape head cone pulleys being run in opposite directions,
*i.e.*, by the method of reverse belting from the main shaft pulleys, in the
ordinary way. The taping heads are, as usual, each provided with a
counter-weight on the opposite side of the mandrel to the taping bobbin,
to balance the weight of the latter, and so do away with any vibration.

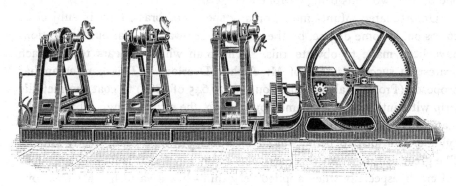

FIG. 53.—Taping Machine.

The machine has a powerful hauling-off drum (such as ensures perfectly
regular motion) driven by cog gearing round the inner circumference of its
flange. The pinion which drives it receives its motion from the main shaft
through bevel gearing, which is enclosed in the box seen to the right of the
illustration.

A set of spur change wheels is provided with the machine, and by
means of them the lay of the various tapes can be adjusted to the required
pitch, according to the size of core which is being taped ; thus also
allowing of considerable variation in the width of the tapes and the amount
of overlap. The machine also has an indicator, which records the length of
core covered. This, of course, varies materially with the width of tape,

* This machine has only recently been designed and set up for use by Messrs
Washington and Co., Engineers, of Sowerby Bridge, Yorkshire. A full detailed description will be found of it in *Electricity* for 21st February 1896.

† It need scarcely be remarked that where only two tapes are applied, a similar
machine would be used, but with two taping "heads" instead of three.

length of lay, etc., but three or four miles per working day may be taken as an average output.

The machine above described takes up very little floor space—only about 17 feet, indeed.

The various core tapings are sometimes applied at one operation, with the two layers of yarn constituting the inner serving. Where, however, the latter are applied at the same time as the iron wire sheathing and entire outer serving, the core tapes are usually laid on first as a separate process.

Each taping head holds about a mile of tape. The successive lengths of the metal riband (from fresh heads) are "brazed"—*i.e.*, united with brass solder. The cotton tape ends are secured by a dab of adhesive compound, and in the case of threads each successive length is attached to the previous one by the two ends being "wipped" together.

Unfortunately Muntz metal, or any species of brass, tape is subject to corrosion, in some degree, by the action of sea-water. Various suggestions have been made to obviate this. The plan which appears to approach nearest to success is that of Mr Arthur Dearlove. This gentleman has proposed (Provisional Specification No. 20,645 of 1889) to coat the metallic strip with anti-sulphuric enamel, by passing the strip through a bath of the material. It is then dried by being drawn through a steam-jacketed cylinder previous to being wound on a reel, or "tape-head," ready for use. The great feature about this particular enamel is that it is flexible and does not crack, especially when applied very thin—like a varnish Mr Dearlove has also suggested its application to the outer sheathing, with the same object in view. For neither purposes has it, however, been so far adopted in practice.

Provided it is properly applied, metal taping is supposed to be a fairly absolute protection against borers ; indeed, no cases of trouble have been known to be traceable to this cause with a cable, the core of which is so sheathed. Moreover, it is to some extent effective in preserving the dielectric from decay (by oxidation, evaporation, and the result of alternating wet and dry conditions) in the instance of gutta-percha insulated cables laid in dry soils ashore.*

However, in the recent Amazon River cable, Messrs Siemens Brothers,

---

* In multiple-conductor cables, metal taping outside each core has been adopted by the Engineer of the Post Office (Mr W. H. Preece, C.B., F.R.S.) in the 1896 Bacton-Borkum cable, with a view to reducing to a minimum the effect of static induction from wire to wire, and also in order to neutralise—by thus introducing *magnetic* induction—the electro-static capacity of each circuit. Previously the multiple cables of this department had only a single brass tape outside the serving of the wormed cores. Substantial benefit in working efficiency is said to have been derived from the above modification.

to meet these conditions, made what was a new departure as regards submarine telegraphy.* The core of the shore-end type, instead of being brass taped, was covered by a lead tubing,† thus rendering it absolutely teredo-tight, as well as air-tight. The core was followed by a cotton taping, over which the lead covering was applied‡ and externally compounded. Outside this were two servings of jute of opposite lays, followed by the usual sheathings, etc.

For the land connections in the form of trench cables the same plan was adopted, with a view to more perfectly obviating the decay which gutta-percha is subject to in dry soil, as already explained.

The trench cable consisted of core, cotton tape, leaden tube, compound,§ and two servings of tarred jute, the proper length of this being previously let into each length of shore end without any joint or splice.

**Alternative Suggestions.** — Some authorities have objected to the adoption of any form of metal covering to the core, not only on the ground of additional expense, but also of electro-chemical action with the galvanised iron sheathing.

Thus, from time to time, various other means have been suggested for defeating the ravages of insects. To preserve subterranean wires from the attacks of ants, etc., one inventor (as early as 1854) thought of incorporating in the insulating material "some resinous or bitter substance of a poisonous nature." An external coat of india-rubber and various mixtures with it have also formed the subject of a patent for the special purpose of protecting the gutta-percha core.‖

---

* Since the above lines were written, the author learns that Messrs Hooper adopted a lead tubing for protecting the india-rubber core of the " Cuba-Submarine " Company's Cienfuegos-Batabano cable in 1895.

† Probably lead could never be applied as a riband taping, though a provisional specification (No. 855 of 1877), drawn up in the name of Colonel T. G. Glover, speaks of a spiral of lead-foil (with the same object in view) covered with a hemp serving saturated in castor oil. In any case, a seamless tube is a much more complete, though more costly, protection. However—no matter how thinly it is capable of being applied—it can scarcely fail to be less flexible than the ordinary so-called brass tape.

For shore ends and in land cables the consideration of weight does not, of course, enter into the question practically; there is, therefore, no objection to lead on this account.

‡ There used at one time to be some difficulty and risk in applying a lead tubing to gutta-percha core, on account of the heat required for the molten lead, though Mr John Chatterton had a patent for effecting this many years ago. Since then, however, means have been devised for applying it cold and direct from a hydraulic press, and this is carried out on a large scale—more especially to india-rubber electric light cables.

§ The compound here would tend to act as a preservative against the decay to which lead is subject in certain soils.

‖ However, as has already been pointed out, any admixtures of rubber in close contact with gutta-percha are highly objectionable, even if they would properly adhere.

Again, in 1877 provisional protection (Specification No. 1,416) was obtained by Mr H. A. C. Saunders and Professor Andrew Jamieson, F.R.S.E., for a method of saturating the inner serving with andiroba oil. This substance having proved successful in overcoming the ravages of white ("termite") ants, it was thought that it would also defeat the teredo and other animalcula. Moreover, it was suggested that the iron wires should be so treated as a preservative against rust. Though apparently an excellent idea—it being the serving that forms the main seat of attack—nothing appears to have come of this plan in practice.*

Finally again, with the same object in view, the incorporation of powdered silica in the gutta-percha covering was suggested by Mr Willoughby Smith in 1875, and subsequently Mr Edward Stallibrass devised a method of blowing this material on to the outside of the core. Moreover, in 1878 the late Mr Frank Lambert secured a patent (No. 759) for applying a serving of silicated cotton—sometimes known as mineral, or "slag," wool—to the core (in place of the ordinary jute serving), mixed with a suitable binding of agglutinating material, such as pitch, tar, various oils, or resins. As an alternative, a cotton tape was to be impregnated with silicate compound, as above, to be applied spirally round the core, as in the metal tape.† It was suggested that the silicate cotton compound should also act as a protection from moisture (amongst other things) for the outside tapes of a cable. This latter claim would, however, have materially vitiated the rest of the patent, probably, if it had been challenged.

However, it is believed that none of these ideas have ever been put into practice on anything like an extensive scale; and the metal taping—the application of which we have described—is almost invariably the means now adopted to ward off boring insects.

---

* Possibly the above oil was found to act as a solvent of gutta-percha.

† This clause was on much the same lines as Provisional Specification No. 3,938 of 1868, standing in the name of Mr Henry Clifford.

## SECTION 2.—INNER SERVING.

**Historical.**—Theoretically speaking, from an electrical point of view, the conductor and its insulating envelope are all that is necessary.* For purposes of mechanical protection, however, it is absolutely essential that this "core" should be protected before being submerged at the bottom of the sea, both to enable it to be securely laid and maintained, and in the case of necessity, to be lifted for repairs.

The form of protection decided on from the very first in the second Anglo-Continental line (1851) was that of a complete armour of iron wires entirely encircling the core. It was at once seen, however, that these iron wires could not be applied directly to the gutta-percha insulating envelope for fear of damaging it. Some sort of packing, or bedding, was recognised as being necessary. Hemp was thought to best meet the requirements, being durable though fairly soft.

Quite apart from this, there is an additional reason for a certain thickness of packing between the dielectric and the iron sheath; and that is to increase the diameter of the cylinder, and thus form a support for that number of sheathing wires which will assign to the cable the required mechanical strength. Nowadays jute is more usually employed as the cushion for this purpose;† but practically the present type of cable is in

---

* At the birth of submarine telegraphy this was very naturally the only idea conceived, as is evidenced by the first line from Dover to Calais of 1850, besides that across the Black Sea which lasted rather longer—throughout the Crimean War, in fact. Subsequently, the late Mr H. C. Forde, M.Inst.C.E., and the late Professor Fleeming Jenkin, F.R.S., had a scheme for laying several such cores across the Atlantic, with the view of reducing the initial cost of a cable, by doing away with any strengthening material, besides obviating the necessity of holding back gear aboard the laying vessel. The work was to be shared by the different cores, no attempt being made to effect repairs when failure occurred.

It was, however, doubted whether such lines could even be safely submerged in a continuous length at the bottom of the ocean.

The objection to laying an unprotected core would be the length of time occupied in sinking to the bottom. In great depths this would be a question of days, probably. If, moreover, sub-surface currents prevailed, there is no knowing what might not happen. In any case, it would probably involve a very much increased length to withstand the strain due to it being taken considerably out of the direct line—even if escaping actual rupture under the heavy strain of such currents. In the comparatively shoal water of the Channel it was found necessary to attach weights to the historic gutta-percha covered conductor laid in 1850 between England and France, alluded to above.

† It may be remarked, however, that the Post Office still always specify hemp for the serving of their core. This, it is believed, is partly to combine strength with its other functions, and also as a greater security against the attacks of *Limnoria terebrans*—a mollusc-like worm which infests these waters. The yarns of the serving material must,

principle the same as that of the first submarine cable above alluded to. Modifications in the form of light cables without any iron sheathing have been brought to notice from time to time; but in practice the original pattern is pretty generally adhered to in its essential characteristics, though somewhat modified and improved on, as will be shewn hereafter.

**Constitution of Jute.**—Jute consists of the bark fibres of two plants, called the "chouch" and "isbund" (*Corchorus olitorus* and *Corchorus capsularis*), extensively cultivated in Bengal, and very largely prepared in factories at Dundee. The state in which it is used for the purpose in question is very nearly approaching the natural state.

As already hinted, jute tends to decay in water of any sort.

**Tarred Jute.**—By way of preserving the jute—or rather hemp in those days—and in order to prevent water reaching the dielectric, as well as with the idea of actually increasing the dielectric resistance, the custom used to be to saturate the serving with tar. The probability of injury to the cable by the application of even moderate heat being at that time altogether unsuspected, this was effected by running the served core through a vessel of molten pitch, the temperature of which could scarcely have been below 100° F. Now Stockholm tar always contains a certain proportion of creosote, and this substance being a solvent of gutta-percha, the combined effects of the high temperature and the pitch caused a loss of insulation often as much as 50 per cent. Again, this compound had the effect of "masking" the presence of small faults of insulation—or those near the surface at any rate—till after the cable was laid at the bottom of the sea. This came about by the tendency of the pitch to seal up minute holes, or cracks on the surface of the gutta-percha that might occur in manufacture until sufficient pressure burst them out again, or until the pitch was partially washed away, or removed mechanically. The result was that many minute "factory" faults which might have been discovered during the construction of the cable did not declare themselves till after they had become so inaccessible as to be practically irremediable.

**"Tanned" Jute.**—Mr Willoughby Smith was the first to point this out,* and to suggest another process in place of the above.

---

however, be applied with a fairly short lay in order to avoid the wires working through to the core. This, therefore, militates against any notions of combining strength with the initial motive of protection. Hemp, however, has the advantage of not breaking so readily during "laying up," thus involving fewer stoppages in the course of serving.

* "On the Means Employed to Develop Factory Faults," by Charles Bright, F.R.S.E., A.M.Inst.C.E., *Jour. Inst. E.E.*, vol. xvi., p. 457.

By this, the jute (or hemp) is previously steeped for a day or more in tanks containing brine or some conducting fluid, or spirit, which also has a preserving effect on the jute.

A modification of this plan was put into operation, at the instigation of Messrs Bright and Clark, over the construction of the first Persian Gulf cable in 1863.

According to more modern practice, the jute is soaked in a strong solution of "cutch" (*catechu*), which contains a large proportion of tannic acid (tannin),* the bitter element of oak bark,† and a strong preservative for any fibrous material.‡

Jute—and, still more, hemp—is very subject to shrinkage, especially in salt-water, and in early days trouble was sometimes experienced by the core being gradually strangled after immersion, on this account.§

By this method, however, the jute (or hemp) is thoroughly shrunk before being applied to the core.||

The jute serving,¶ thus "tanned," being applied to the core whilst still wet, and arrangements being made for a stream of water trickling over it immediately on application, it will be seen that the effect electrically of the serving to the core is precisely the opposite to that of former times when *tarred;* for in this case the result is that of shewing up any minute flaws at the earliest possible stage of armouring, and especially on the subsequent pressure of the wires at the lay plate. If the application of the wet serving has not already brought to light any flaw—such as may have been developed since the electrical tests were taken on the core—the compression of the iron wires should certainly do so.

---

* This process is, therefore, sometimes described briefly—in specifications, and so forth—by speaking of the serving being *tanned*.

† It is also obtainable from gall-nuts.

‡ Jute and hemp both contain a natural oil which tends to decompose, and in doing so, to rot the material. Tannic acid has the effect of killing this oil in jute or hemp. In thus tending to allay subsequent decay, it may be described as a preservative of jute.

§ This would be especially liable to occur in the case of a hempen cable without any iron armour, as here the hemp would on submersion be directly exposed to the sea.

|| A slight further shrinkage takes place after laying up. Evidence of this exists in the fact that the core may often be drawn out of a specimen length of a close-sheathed cable, owing to the inner serving fitting loosely in its armour after a time.

¶ The serving has been occasionally impregnated—*i.e.*, its pores filled up—with ozokerite by way of preservation, and with a view to increasing the insulation of the cable by some 10 to 12 per cent. The author has also suggested the possibility of working cheap inferior qualities of gutta percha in with the jute, mainly in order to decrease the capacity. Any such plans would, however, be contrary to the ideas of divulging minute faults at an early stage of manufacture.

**Data.**—The specific gravity of jute is slightly over unity*—1 cubic foot weighing about 67 lbs.

Being the softer material it absorbs, and holds, rather more moisture than hemp, a cubic foot of which, when dry, may weigh as little as 38 lbs.

The breaking strain of jute is, as a rule, distinctly low ; and some trouble is given by the constant snapping of the yarns when under a strain, whilst being laid up round the core. Even after being thoroughly soaked in a preservative mixture, it tends to seriously decay in water—in salt-water, moreover—and especially when combined with iron wires. Thus this part of a submarine cable is that which tends to perish soonest as a rule.

**Routine of Inner Serving Application.**—When the core comes from the testing (75° F.) tanks on its carrying drum, it is drawn off and coiled down into a tank at the back of the serving machine, ready for serving with jute or hemp.

As fast as one drumful has been coiled down, the end of another is jointed on to it, a complete joint being made in both the conductor and insulating envelope in a manner which has already been described in detail.

It is necessary to have about three miles of core ready to go through the machine, as joints cannot be tested satisfactorily until twenty-four hours after submersion in water subsequent to completion. Each successive joint is subjected to a searching electrical test by means of the highly sensitive Thomson quadrant electrometer, or, in the absence of this, by a Thomson galvanometer, the accumulation method due to Mr Latimer Clark, F.R.S., being invariably adopted. These instruments and all methods of joint testing will be found fully described elsewhere.

A rotation number is assigned to each joint as made, and the results of the tests are registered on forms.

During the entire process of manufacture a workman follows every joint in the core, marking its position on the outside of each successive covering. When the cable receives its final outside covering of canvas tape or hemp, a leather or gutta-percha "tally" is secured to it by spun yarn at each joint with the number of the joint marked on it.†

---

* It may be taken at 1.15.

† Mile marks of a similar character are fastened to the outside of the cable, as the completion of each nautical mile is indicated by the "counter" of the machine.

The position and distances of these marks in the several coverings are carefully noted at the time they are attached. Thus, in the case of a fault declaring itself, it is an easy matter to find any particular joint, or to cut the cable at any required distance by the mile marks.

The numerical order of the several reels of core in the cable* with their lengths, weight of conductor, and dielectric, as well as all electrical values (arrived at from the tests), are tabulated on forms and in books which follow up the construction of each " section " of cable separately—a separate book being very usually assigned to every machine for the construction of each separate section.

Each skein of jute yarn,† after being prepared as above described, is wound on to a wooden bobbin belonging to the serving machine.

These bobbins are taken down to the serving machine as required, and

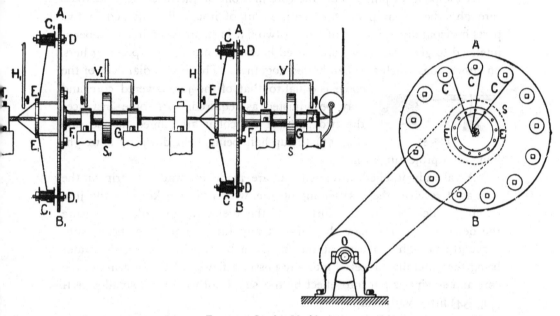

FIG. 54.—Serving Machine.

fitted to the " fliers " of the revolving disc, or cylinder frame, which carries them.

**Serving Machine.**—The machine for serving the core with jute yarns is, as a rule, similar to that illustrated in principle by Fig. 54.

---

* As already stated, some attention is paid to the sequence of the coils jointed together for cabling, so as to avoid a number of coils of a very low conductor, or dielectric, resistance, or of a high capacity, following after one another ; thus ensuring fairly average values at any given part of the section, in the event, say, of future cutting for repairs.

† In jute factory parlance, a thread, or fine yarn, consists of a number of fibres laid up together. In cable factory parlance, so many " ply " is several threads wound together for making up a yarn. A thread may, however, be constituted by a single ply.

This consists of a hollow shaft F G, which revolves on large iron disc\* A B (about 6 feet in diameter), pierced with holes disposed circumferentially near the outer edge. Into these holes fit the spindles of the bobbins C fitted with the jute yarns, each spindle being secured by a nut D at the back of the disc. A hollow ring E (pierced with as many holes as there are bobbins) is fixed to the disc by four iron rods, and therefore revolves with it. The jute yarns, as they are drawn away and unwind from their bobbins, are conveyed through the holes in the guide ring E, after which they meet with the core round which they are wound. The core itself is drawn forward from its resting-place through the hollow shaft F G by means of pulleys.

At the point of junction on the core in front of the disc, the yarns pass through a die, or lay plate T, which is a sort of iron collar divided in two parts horizontally. Each half having two flat lugs, strips of india-rubber— intended to give elasticity—are placed between the two corresponding lugs, on either side, which are then bolted together. The inside diameter of the

FIG. 55.

collar is equal to that of the jute-covered core, and may be adjusted as required. In some instances the two parts of the collar are hinged at one side (Fig. 55), the up half being held down by a weight hung on a projecting arm.

As already implied, arrangements are made for water to drip on the core as it receives the first serving of jute, with a view to keeping the jute damp, besides wetting the surface of the dielectric, in order to ensure the detection of any flaw therein — at any rate, on pressure being subsequently brought to bear at the lay plate by the iron wires, the water being then literally squeezed into any possible flaws. This watering of the core and serving is put into effect by the supply of water to a small pipe H (Fig. 54) fitted with a tap.

The rapid rotation of the hollow shaft and disc, combined with the slow longitudinal movement of the core, causes the jute yarns to be wound, or spun, on spirally.

---

\* Here it is a disc machine which is shewn ; but the same work may be equally well performed by a cylinder frame machine, as described further on in connection with the iron wire sheathing. Each have their advantage, and sometimes both principles are applied in the same machine. The disc is capable, generally speaking, of holding more bobbins—some being placed on each side of the disc—whilst also taking up less floor space. On the other hand, the cylinder frame—the centrifugal force being less serious by this arrangement—permits of a greater weight for each bobbin, and therefore a greater length of wire.

In order to get an increased number of yarns for increasing the size of the serving for heavy types, an extra disc loaded with bobbins may be conveniently applied at the back of a cylinder frame machine, thereby combining the two arrangements in one machine.

The bearings in which the shaft F G revolves are kept cool by water constantly dripping from the pipes V and $V_1$.

**Lay of Inner Serving.**—Fitted to the revolving part of the machine is a brake, consisting of a block of wood at the end of a lever arm, and this may be pressed against the edge of the disc carriage A, to stop it turning, as required—for instance, when it is necessary to feed the carriage with a fresh bobbin of jute, or to effect a join between the ends of a broken yarn.

The function of this serving round the core is, as before stated, almost solely that of forming a protection of a certain thickness to the core, from the iron wires, on the one hand; and, on the other, as a bed for the latter of such an outside circumference as will permit of the required iron (wire) arch being formed. This being so—any questions of strength being scarcely, if at all, relevant for the inner serving*—it is usual to apply the jute with a distinctly short lay (varying from 2 to $3\frac{1}{2}$ inches, according to circumstances), by way of making the firmest possible packing to the core, in order to prevent any iron wire finding its way between the yarns.

The lay is determined by the relation borne by the speed of drawing off to the revolutionary speed of the disc carriage.

The speed of the revolving disc is always maintained the same—a fairly high maximum speed,† with few and light bobbins and a small disc or cylinder—but the "draw-off" speed is altered by means of change wheels according to the speed required to give the desired lay.

If the lay is shorter than a certain angle, the outer diameter is liable to be too great for the wires to close round properly; and if too long, the packing is liable to be insufficiently dense, or secure. To correct either of these, the speed of the haul-off gear should be increased or decreased, as the case may be, or—under some circumstances—by the revolutionary speed of the bobbins being decreased or increased.

**Miscellaneous Particulars.**—The jute yarn used for this purpose is usually "single ply"—*i.e.*, one yarn—but sometimes two or even three ply is adopted when any appreciable strength is desired.

---

* In fact, the jute used for the inner serving, being employed only as a packing or bed, has generally a very low breaking strain : it is of so little value practically and is so unreliable that it is not, as a rule, considered.

† 130 revolutions per minute is very usual with an ordinary rate of draw off to give an average lay. This is equivalent to about four miles of core being served in the course of a working day.

The jute yarn is seldom, if ever, tested for breaking strain, but sometimes a limit to the number of twists in a given length of the yarn is specified.*

The various skeins of yarn are wound, by means of a swift, to fill up the bobbins. Each bobbin holds on the average somewhere about a mile length. Strength not being considered, joints are only limited as far as possible on the score of their involving very slight lumps, such as mean stoppages, though of extremely short duration.

These joints are usually effected by a whipping of single spun yarn (or twine) being bound round the join ends. Matters are so arranged in practice that the bobbins of yarn run out at different points so as to avoid the joints in the various yarns occurring coincidently.

The number of bobbins depends, of course, on the type of core to be covered and the type of yarns† used, as well as on the thickness of serving for the particular form of cable in question, the object aimed at being to have the closest possible packing, whilst maintaining perfect evenness. With an average core, about twenty-five separate yarns in all‡ (each mounted on a bobbin) is a very ordinary form of first serving.

Within recent years there has been a tendency in some quarters to reduce the quantity of inner serving for deep-water cables to as fine a point as possible in order to reduce the number of iron wires required, and, therefore, the weight and bulk of the entire cable. This may be safely pushed to a fine point, where a great thickness of gutta-percha surrounds the conductor, and where the cable is not likely to experience a high temperature. Thus, the deep-sea type of the last Commercial cable made by Messrs Siemens Brothers—an excellent type in every respect—had an inner serving of this description. However, such space as there was between the core and the iron wires was very thoroughly filled in§ by what may be termed a tight packing. It is usually an acknowledged axiom that, with a given

---

* For a "single," the Post Office authorities often specify a limit of ten to twelve twists in a 12-inch length.

† The type and number of yarns employed varies with the description of cable. Thus in intermediate or shore-end types, the yarn adopted is invariably thicker than that used for deep-sea cable.

‡ However, different factories have different practices in this respect, as in others. Some employ a larger (or smaller) number of yarns of less (or greater) size to produce the specified bulk of inner serving. A greater number of yarns introduces more stoppages and joins, but should secure a denser, more even, and surer bed. Generally speaking, as large yarns are used as is practicable for the space to be filled, whilst providing for a firm packing. The number of yarns may be, in fact, anything from ten to fifty, or more. If the former, it would, probably, be in one application only.

§ It must be remembered, moreover, that the wires are very small. Again, in any close-sheathed cable, the wires butting firmly against one another act as a complete tubular shield.

space between the core and cylinder of iron wires, it is better that there should be overmuch, rather than undermuch, packing; thus ensuring the jute fitting well into the interstices of the wires, whilst also securing good coilable qualities, though at the expense of material, bulk, and coiling space. Springiness in a cable is, in fact, obviated (1) by plenty of inner serving, and (2) by taping of each iron wire. The former must not, however, be pushed so far that the cable at all approaches an alternate iron and jute pattern.

The inner serving for heavy types must not only be sufficiently greater in quantity to secure (for safety) the same thickness as with a deep-water type, but must also have an increase in this respect to combat with the area of the large wires. It is not, however, necessary to have the same quantity in a given area—*i.e.*, the inner serving is not so tightly packed—as in a small type of cable with a limited area for the wires.

This inner serving is almost invariably applied in two separate applications, one above the other, laid up the opposite way, to ensure good packing.

Thus the core served as above, in being drawn from the preceding machine, is usually drawn through another precisely similar machine, just beyond it, as shewn in Fig. 54, which is made to spin in the reverse direction for applying an opposite lay. In other respects they are precisely the same with corresponding arrangements in each, and the two servings are exactly similar—indeed, they may be regarded as a single machine, each part of which effects a part of the same operation.

In the figure, the core is shewn coming from above; but, of course, it may be in the same shop. Moreover, the sheathing can be performed at the same operation from a part of the same machine as the serving—by a combined serving and sheathing machine.*

On the other hand, the serving is sometimes (where more convenient) effected in a separate shop, and the served core drawn from tanks as required —or direct from the serving machine—for sheathing. Anyway, the serving machine, in order to save ground-floor space, is very usually on an upper floor, and placed just above the sheathing machine it is intended to feed

---

* This is a very usual plan in modern practice, for it introduces a saving of time and hands—though perhaps less sure. By the other system, an opportunity presents itself of testing the core after being served, and this is here always taken advantage of.

Where the core is brass-taped, the operation is either performed separately or else the jute serving is applied at the same time. In the latter case, the sheathing and outer coverings form a separate process; and where the former is the system adopted, the jute is laid on contemporaneously with the sheathing wire and outer covering.

Thus by either plan it is a case of two operations in all.

with served core. Where the latter is not required at once for cabling —or where it is the practice to test it first—after the serving operation, the core is coiled down into tanks filled with water.

**Formulæ, etc., for Weight of Inner Serving.**—For calculating *the weight of inner serving in a cable*, in Fig. 56, let $D_1$ be the diameter of the centre line of the iron wires in inches; $D$ the diameter of the dielectric in inches; $d$ the diameter of a single wire also in inches; and $n$ the number of iron wires.

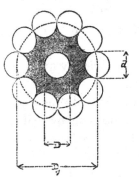

FIG. 56.—Inner Serving.

Then transverse sectional arc of the jute (or hemp) is

$$0.7854 \left(D_1{}^2 - D^2 - \frac{n}{2} d^2\right) \text{ square inches.}$$

And the weight per N.M. of jute, or hemp, serving required to efficiently fill up the space between a given core and cylinder of iron wires as above is *approximately* as follows :—

$$20 * \left(D_1{}^2 - D^2 - \frac{n}{2} d^2\right) \text{ cwt.}$$

It is also very usual, in the first instance, to make up various machine-made specimen lengths of a fathom or so, for purposes of examination and experiment. This trial is sometimes found useful in order to ascertain whether the amount of jute settled on for the inner serving could be properly got into the space between the core and the iron sheathing arranged for.

## MULTIPLE-CONDUCTOR CABLES.

In the case of a multiple-core cable, *i.e.*, a cable to contain several—two, three, four, five, six, or even seven—insulated conductors, the central heart which is drawn through the hollow shaft is usually a hemp or jute yarn (though sometimes an insulated conductor), and round this are laid up alternate insulated cores and hemp, or jute, yarns.[†] The number of bobbins will then depend on the number of conductors, with an intervening bobbin of hemp, or jute, yarn for protective, "worming," purposes.

---

* It should be pointed out, however, that the above constant varies very much according to individual views and practice. The formula is, indeed, only given as presenting an idea of the principles adopted in such calculations.

† These "wormed" cores are then served with jute in the usual way—*i.e.*, as in the case of a single core.

Here the cores to be laid up together are reeled on bobbins and placed in machines somewhat similar to those described in the next chapter for deep-sea sheathing *—*i.e.*, the same provision being made so that no torsion is given to the core. If one of the cores is to form a central heart round which the remainder are to be wound, it is passed singly through the hollow shaft. More usually, however, the central core is replaced by a hempen strand of requisite thickness, according to the number and diameter of the cores in use.

To completely fill up or "worm" the interstices remaining between the cores, a disc E, carrying small bobbins of hemp yarn, is fixed to the hollow shaft in front of the stranding apparatus (Fig. 57). The cores unwinding from the bobbins A pass through the openings B in the disc (Fig. 58), and enter the lay plate D side by side with the yarn, where all are stranded together. The so-formed multiple core next receives its serving of jute and

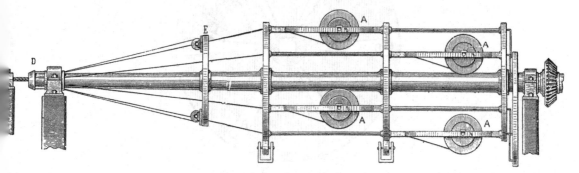

FIG. 57.—Worming Machine for a Multiple-Conductor Cable.

sheathing, according to specification, in the ordinary way, as described further on.

**Where Desirable : Advantages and Disadvantages.**—Such cables are almost entirely confined to the purposes of heavy traffic to neighbouring important islands and continents,† or across rivers, when the extra heavy initial outlay is sufficiently warranted : for nowadays it is not thought by the best authorities that a number of conductors laid together inside a necessarily heavy and expensive single armour sufficiently decreases the chance of total interruption to warrant the extra cost in insulated con-

---

* In special cases, however, each core is first surrounded with jute, and the so-served cores are then laid up together in a similar manner.

† Thus the old "Electric" Company used to own cables of the multiple-core type ; and the cables under the auspices of H.M. Postal Telegraphs are at the present time almost exclusively of this description.

ductors and armour. It is, indeed, usually considered better to guard against such a contingency by laying separate (single-conductor) cables between the points and on entirely different routes as far as possible; though multiple-cored cables are admissible in some cases where—owing to strong currents (as in certain rivers and in shallow and stormy seas), or to a shifting or very rough bottom—the cable in any case requires to be a very heavy one,* and where, therefore, the exigencies of traffic, if involving more than one conductor, may just as well be met by the several conductors being laid up together inside one such cable.†

In all other instances it is better to provide the required number of lines

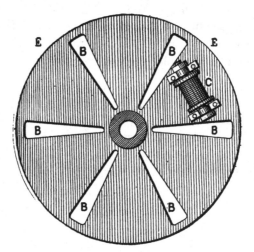

Fig. 58.—Disc of Worming Machine.

by laying separate single-conductor cables. The lighter cables are more flexible, easier to lay and repair; and in case of one breaking, communication is kept up by the others.

---

* It should, however, be remarked that, for shallow water where strong currents prevail, the cable should, as a rule, be heavy *specifically*, rather than by bulk, the latter involving increased surface subject to the prevailing elements.

† The relative merits of laying several separate ordinary cables, and of enclosing several insulated conductors in a single armour to do duty for the same purpose, was gone into at length at the Institution of Civil Engineers in 1860, in the course of a paper, entitled "On the Maintenance and Durability of Submarine Cables in Shallow Water," by Mr W. H. Preece. This pioneer paper, together with the discussion thereon, may be strongly recommended for a close study.

## SECTION 3.—SHEATHING.

**General Description of Ordinary Cable.**—As has been previously remarked, the type of submarine cable at present in vogue is practically the same (in its main characteristics) as that of the first ever laid *—*i.e.*, a close iron-sheathed cable, after the manner of an ordinary rope, of wires varying from .07† to about .400 inch in diameter,‡ and in number from ten to about thirty-six,§ thus forming a more or less flexible mail, according to depth and local conditions.‖

---

* This is probably the only instance of the sort in engineering practice.

† Wires of as small a diameter as .040 inch (No. 19 S.W.G.) have on occasions been used, in which case they are generally covered with Manilla, unless for "torpedo cable" purposes.

‡ No. 15 and 4/0 British legal standard wire (S.W.G.) respectively. In the Historical Sections (Part I.) of this book, the gauge referred to was the usual gauge of the time— *i.e.*, the Birmingham wire gauge. However, since those days the standard gauge adopted by the Board of Trade—as a compromise of the various gauges in use—has been legalised for general use to meet the difficulty. In the further text this gauge (see Appendix I. at end of this Part) has accordingly been assumed for up-to-date, non-historical, matter. In many instances the actual diameter of the wire in fractions of an inch has also been given.

§ This includes that form of cable whose outer serving consists of a number of three-wire strands, as used occasionally in heavy "shore ends." Twenty-four is a fair outside limit for an ordinary close-sheathed "deep-sea" type, and twelve about a minimum. The requirements here are sufficient flexibility, qualified by an assurance that the wires shall not be dangerously small in section, either from a piercing or corrosion point of view, though as regards tensile strength, the smaller they are the greater will be the breaking strain per square inch as a rule.

Actually, the number of wires adopted may be said to depend on the relation existing between the outside diameter of the served core and the gauge of wire employed.

In the "deep-sea" portion of the 1894 "Commercial" Atlantic cable there were twenty-four wires each of No. 15 S.W.G. (.072 inch)—the smallest type of wire used in any ordinary close-sheathed telegraph cable. It is doubtful whether wires of less diameter than this could be safely employed—not at any rate on a mud bottom, which (unlike sand) very often acts chemically on iron—on the score of leaving too small a margin for reduction of strength by gradual rusting. Small wires, however, have the advantage of making up into an *absolutely* close-sheathed complete arch cable without involving rigidity or springiness.

‖ The recent tendency towards heavy cores to give a high signalling speed (by machine transmission) introduces mechanical difficulties, inasmuch as it involves a greater number of iron wires in order to completely surround the serving, which means an increased weight. This necessitates stronger holding-back machinery for paying out. Though a cable of this description has its advantages as an investment, there is a limit to which this can be carried. Thus it appears that in the event of a Pacific cable being undertaken, it would not be advisable to attempt a high (machine transmission) speed, such as would involve an extra heavy type of cable, the difficulties of laying and of recovery being already sufficient in view of the great depth to be dealt with. This is quite apart from cost or prospective traffic considerations. If, however, extremely heavy cores are to be the order of the day, it is probable that in the case of extreme depths

In this the armour consists of iron wires laid helically on to the served core in a perfectly symmetrical form.*

The iron serves a double purpose: it affords tensile strength and also resistance to abrasion.

**Requirements for Different Conditions.**—In deep water †—for purposes of after-recovery—tensile strength, with a given limit of weight, is the main consideration; and here a comparatively fine wire of mild Bessemer steel, or homogeneous iron, is employed with a breaking strain represented by as much as 80 tons to the square inch—or more. ‡ In shallow water, however, a different set of conditions have to be considered. For instance, in the type of cable adopted for approaching the beach, the quantity of wire necessary to resist abrasion with rocks, and for giving the

---

some return to the alternate hemp and iron type of cable will have to be resorted to—or preferably, perhaps, the taping of each wire may be adopted on an extended scale.

* In a close-sheathed cable the diminution of diameter is prevented by the abutment of each iron wire against its neighbour, which, therefore, support each other laterally, after the manner of a complete arch or tube, so that the precious soft heart, or core, of the cable should undergo practically no compression on this count as regards solid pressure—though not, of course, water-tight against fluid pressure, such as that of the ocean.

Theoretically speaking, for purposes of tensile strength—so far as deep-water cables are concerned—the wires should be applied straight. By this plan, however, the wires would require careful binding, and the same result can be more effectually attained by a lay in the form of a helix, besides affording a greater facility for stretching in the event of a strain.

† Where a cable is, as a rule, unmolested by rocks.

‡ The Bessemer process consists in forcing air through molten iron, which takes off certain impurities and leaves only a given proportion of carbon. The difference between iron and steel is constituted by the difference in the proportion of carbon. Steel has a definite and very small proportion, such as best yields the strength sought for—less than that of wrought iron, though greater than that of cast iron.

Homogeneous iron (Bessemer process) is not exactly steel, but it is very nearly as strong—combining, in fact, some of the malleability of wrought iron with, practically speaking, the same strength as steel. It is this which is almost invariably specified for the sheathing wires of quite deep-water types, for it secures practically as high a breaking strain as that of steel, but at a much lower price than if the latter were actually asked for. Another advantage in the sheathing wires of an ordinary deep-sea cable being composed of "homo" (homogeneous) iron rather than of absolute steel, exists in the fact that the former are less springy, making up, therefore, a better cable for handling and all subsequent operations. The improvement in the manufacture of homogeneous iron wire within recent years is something enormous, especially as regards tensile strength. A stress of quite 90, or even 100, tons per square inch can now be applied to this class of wire before it breaks: twenty-five years ago there was difficulty in obtaining iron wire which would stand 50 tons per square inch. In "drawing down" to a small diameter the tensile strength is considerably increased, though, in being rendered harder, the wire becomes more brittle texturally speaking.

desired weight to avoid shifting under pressure from currents, etc., is such that an abundance of tensile strength is secured by wire testing at less than half, or even a third,* the above strength.† The same quality of iron is used for the sheathing wires of the "intermediate" type, as a rule. As it is usually a single-sheathed cable and there is altogether much less iron in it, the available strength is materially less than the shore end, and sometimes less than the deep-sea type; but this is of no importance for shallow

FIG. 59.—Marine Growths found on a Cable Fault.

water cables, as has already been explained. It is, indeed, in every way more advantageous to employ softer (annealed) wires of the commonest iron,‡ as above, partly on account of steel or "homogeneous" iron wires

---

* The breaking weight of the commonest iron rod is about 25 tons per square inch section. The breaking weight of drawn wires—especially if hard drawn—is very much greater than any rolled or annealed iron. This, however, varies so much with the particular quality that it would be absurd to attempt to give any rule or statement regarding its *actual* strength.

† Moreover, the paying-out or picking-up machinery will often not bear a stress much above 50 tons, so that any breaking strength beyond that would be wasted. It should, however, be stated that it is usually acknowledged as an axiom in the present day that all the cable machinery connected with a telegraph ship should be capable of standing a strain at least double the stress which the strongest cable in vogue will bear.

‡ Iron wire of a still commoner class is less annealed, and, therefore, rather harder. This is occasionally thought to be best suited for "shore-end" and "intermediate" types,

being too springy and rigid for handling and working a bulk of wires of this number and section, as well as for economical reasons. Such wire, whilst bearing a strain up to about 30 tons, has an elongation equivalent to as much as 18 per cent. of its length when stretched.*

It is usually known as extra-superior quality, and designated "best best" (B.B.). This wire is always carefully annealed, is very ductile, and, after galvanising, should stand several bends at the same point before breaking.

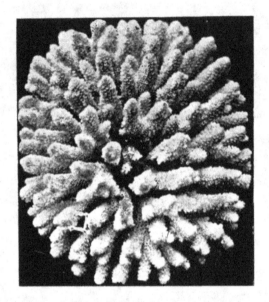

Fig. 60.—A Fine Specimen of Coral.

The specific gravity of ordinary bar iron is about 7.8. If suspended vertically in water, a wire composed of the above should not break under its own weight until it is made to support a length of about 3.5 N.M. In practice, however, it is not prudent to load iron much beyond a quarter of its actual breaking strain. Where the wires are all applied with equal tension, the tensile strength of a completed cable being approximately

---

under certain conditions, owing to its greater durability under water as regards abrasion or corrosive action, though more brittle.

Sea-weed and vegetable matter are known to corrode iron, more or less rapidly, owing to the iodine they contain. Thus, some shore deposits and marine growths (Figs. 59 and 60) are very injurious to the sheathing of a cable.

* It may be here remarked that the greater the elongation, the less will be the tensile strength, and *vice versâ*. In fact, in tempering iron, some of the tenacity is lost, which would naturally accrue to it when hard.

equal to the sum of the weights which each individual sheathing wire is capable of supporting,* it will be evident that cables, with the above ordinary iron, should not be laid in depths exceeding about 1.5 N.M.

It is for the above reason, therefore, that this class of wire is not employed for the deeper waters; in fact, in extreme cases, well-tempered steel is sometimes resorted to with a tensile strength equivalent to nearly 100 tons per square inch—or, at any rate, quite hard-drawn wire. The sheathing wires of all types of submarine cables are galvanised.

This was so from the very beginning as regards comparatively shallow-water cables, though not in the case of the first Atlantic cable. It was thought at that time to be unnecessary for deep-water cables, on the plea that there was no scope for chemical action at the bed of the deep ocean. Moreover, a strong idea prevailed that galvanising tended to reduce the strength of a wire.†

The object of galvanising is to preserve the iron wire from the oxidising influences of air and water—i.e., from rusting.

The process of galvanising an iron, or steel, wire consists in coating it with a film of molten zinc. ‡ Zinc readily changing to oxide of zinc, when exposed to the air or water, a wire so treated becomes covered with a thin coating of oxide of zinc.§ This, in absorbing carbonic acid from the air, passes into basic carbonate, which tends to protect the metal from further chemical change. Zinc is electro-positive to iron. Thus any result from these metals being in contact is at the expense of the zinc rather than of the iron.

Should any imperfection, however, exist in the galvanisation, the iron here tends to corrode more rapidly than if ungalvanised. It is, therefore, of the utmost importance that the very purest zinc should be employed for galvanising iron or steel wires for the purposes of a submarine telegraph cable.

As a general rule, the iron wire (of various descriptions) is ordered at the

---

* This approximation would not, however, be considered by many authorities to hold good where hemp *cords* are applied outside, it being contended that these add materially to the strength of a cable—to the extent of 1 ton very often—provided they are applied with a lay bearing such a proportion to that of the iron wires as accords with their relative elongation.

† As a matter of fact it is now usually considered to have slightly the opposite effect. In softening the wire, it certainly renders it less brittle.

‡ The wire is first cleaned with sulphuric acid, so as to free it from any grease, etc., which may be present. It is then passed through a bath of the molten zinc.

§ Oxide of zinc is insoluble in water, which fact would in itself render it a preservative to the wire.

cable factory ready galvanised.	Messrs Henley have, however, been in the habit of galvanising (as well as annealing) the wire themselves.*	The competition in the supply of ordinary ungalvanised wire being infinitely keener than in galvanised wire, they are thus enabled to obtain it at a lower initial cost.

In our submarine cable, as usually constructed in the present day, there are many and various forms which the armour may suitably take, according to the depth and nature of the bottom.

In a proposed submarine line between any given points particular types are invariably assigned to certain portions of the line.	These types vary as regards the number and class of iron wires used; and, again, as regards whether there is only a single sheath of iron or whether this (with an intermediate jute, or hemp, serving) is supplemented again by an outer armour of larger wires.	The latter type, in one form or another, is employed in shallow water, nearing shore.	The character of the cable varies still further according to distances from shore, nature of bottom, strength of any prevailing current, presence, or otherwise, of ground icebergs, ice-floes, etc.†

Just as for purposes of recovery, a cable for deep water should, first and foremost, be as light, and, at the same time, as strong as possible,‡ so also for purposes of *efficient* laying, the cable should, as a property or investment, have the greatest possible weight§ within a given area‖ in order to ensure sinkage into the irregularities of the bottom, at an angle not too obtuse with reference to the ship.	No doubt, from the contractor's point of

---

\* Indeed, in early days, Mr W. T. Henley was originally a wire-drawer in addition to a cable manufacturer.

† Specially powerful gear would, in extreme cases, be necessary for bearing the strain of picking up such a cable.

‡ That is to say, it should be capable of resisting a tensile strain somewhat greater than that involved by the weight of the greatest length of cable, which might reasonably be in suspension during the course of repairs.	Hence it is usually considered that the " modulus of tension " of a deep-sea cable intended for the deepest water ought to be such that it will support at the very least 6 N.M. of itself in water.

§ This is apart from the fact that the cable should also be able to support the complete length between the ship and bottom, during paying out, without fracture.	This naturally implies, on the other hand, that the lower the specific gravity the better.	However, such a cable as an investment can never be laid so *efficiently*—though more easily without accident.

‖ From a certain standpoint—that of skin friction in passing through water—the area should be limited as far as possible both from an efficient laying and a successful recovery point of view.	In the former case, friction tends to retard the rate of submersion and lengthen the angle, and in the latter to seriously increase the strain.

view, such a cable is more difficult to restrain in paying out,* and involves more precautionary measures to adjust the amount of slack within the required bounds.†

---

* As has been pointed out by Mr Thomas Gray, F.R.S.E., a cable is paid out too fast if it goes to the bottom quite slack.

When a cable is laid at a uniform speed, on a level bottom, quite straight but without tension, it forms an inclined straight line towards the position of the bottom that it ultimately occupies. This is precisely the movement of a battalion in line changing front.

When laying a cable in a three-mile depth, it is calculated that with the ship steaming eight knots the length from the stern of the vessel to the spot where it touches the ground is over twenty-five miles, and that it takes a particular point in the cable more than two hours and a half to reach the bottom from the time that it first enters the water.

† The difference between the speed of the ship and the rate of paying out gives the amount of slack. This varies in different cases between 3 and 12 per cent. But the mere paying out of sufficient slack is not a guarantee that the cable will always lie closely along the bottom, or be free from spans. Whilst it is being paid out, the portion between the surface of the water and the bottom of the sea lies along a straight line, the component of the weight at right angles to its length being supported by the frictional resistance to sinking in the water.

After a little consideration it will be evident that the angle of immersion depends solely on the speed of the ship : hence in laying a cable on an irregular bottom it is of great importance that this speed should be sufficiently low. This may be illustrated very simply as follows :—Suppose $a\ a$ (Fig. 61) to be the surface of the sea, $b\ c$ the bottom,

FIG. 61.

and $c\ c$ the straight line, then if a hill H, which is at any part steeper than the inclination of the cable, is passed over, the cable touches it at some point $t$ before it touches the part immediately below $t$; and if the friction between the cable and the ground is sufficient, the cable will either break or be left in a long span ready to break at some future time. It is important to observe that the risk is in no way obviated by increasing the slack paid out, except in so far as the amount of sliding which the strength of the cable is able to produce at the points of contact with the ground may be thereby increased. The speed of the ship must therefore be so regulated that the angle of immersion is as great as the inclination of the steepest slope passed over. Under ordinary circumstances the angle of immersion varies between six and nine degrees—*i.e.*, quite a small amount, but still an amount which under unfavourable circumstances may make all the difference in the safe deposition of the cable from an owner's point of view.

Thus, in practice, most telegraph engineers do not consider paying out an increased quantity of slack a good substitute for going slowly over the ground. Under *all* circumstances the laying vessel should approach banks and ledges at quite a moderate rate. If a large quantity of slack is paid out *after* it is discovered that a bank has been passed, the slack comes out at the wrong place, and only results in dangerous kinks (when afterwards picked up) by the cable collecting in coils at the bottom. There is also the disad-

The limiting feature to this weight is, of course, the absolute necessity of providing that it shall be sufficiently light in proportion to its strength to permit of successful raising from the depth at which it is to be deposited —not essentially on the basis of a single bight recovery ; indeed, attempts are seldom seriously made to pick up an entire cable in a greater depth than about 1,000 fathoms.

A cable laid in deep water, free from strong bottom current, should, unless under a strain, sink well into a soft bed of ooze, sand, or mud, which, in itself, acts as an actual preservative, unless some chemical agent is present which attacks the iron wires, and which indirectly decays the hemp or jute.

The nature and depth of the bed has been known to alter in medium depths owing to the action of sub-surface currents in causing shifting banks and so forth. These currents, which tend sometimes to remove all the soft microscopic shell deposit* from the surface, leaving bare hard rock, are not supposed to occur, even to a small extent, at greater depths than 1,000 fathoms.† Where no such deposit can be found on the bottom, a strong current and hard bottom may be fairly assumed to exist, and should be provided for with a cable offering greater metallic surface for withstanding abrasion.‡

Again, stouter—and therefore more rigid—sheathing wires are a wise precaution for that part of a cable which has to be deposited in waters that the saw-fish, the sword-fish, or the shark, are known to especially favour.

**Types.**—As a rule there are at least three types in any given length of cable. These are usually designated under the general terms " shore-end," " intermediate," and " deep - sea," or " main," cable, which latter,

---

vantage of waste by a superabundance of slack. On the other hand, if laid too tight, a cable remains suspended between the peaks of submarine mountains exposed to fracture by its own weight.

Where irregularities have to be contended with, the best plan is to make a very careful survey before setting out to lay a cable, and, after determining on the route, to carefully mark it off with buoys in the treacherous region, thereby obviating the necessity of going at a high speed in order to steer a particular course—i.e., thus ensuring the right route, currents and slow speed notwithstanding.

Occasionally, however, a high speed is necessary, in order to lay the entire cable before threatening weather becomes serious. In such instances extra attention must be paid to getting out the required slack. Unless the tanks are large, some risk is here run of a foul, or accident, during the operation.

\* This takes the form of what is commonly described as a sand, mud, or ooze bottom.

† This is the greatest depth at which they are at present known of.

‡ It should, however, be remarked that currents of great strength are mainly confined to quite moderate depths, where an " intermediate " type of cable, sheathed with wires of comparatively large diameter, would in any case be furnished.

generally speaking, forms by far the greater proportion of the total length, and varies again in character according to the depth.

To give a rough idea, the "shore end" is often used, say, for the first two miles (or less) from the beach in depths up to 35 fathoms, the "intermediate" into water up to something like 350 fathoms, and the "main" type for the remaining portion, changing in description according to the depths and conditions to be contended with. It should be remembered, however, that a cable may be built up from four, five, or even six types in all, where, in a great length, the varying conditions are of a very wide character. In this case, the different types are very usually split up by the expressions "heavy" and "light" "intermediate" and "heavy" and "light" "deep-sea," such further designations being applied in each instance by way of distinction: or, again, each type may be numbered A, B, C, D, etc., or 1, 2, 3, 4, etc., and such a plan is perhaps preferable, if only to obviate the apparent paradox, which sometimes occurs, of a "light deep-sea" cable being heavier in air (though lighter in water) than a "heavy deep-sea"—being materially larger, in fact, though having a lower specific gravity.

The "shore-end" type should have sufficient iron in it to successfully cope with the wearing action brought about by neighbouring rocks, anchors, fishing trawls, etc.; and, within a given limit of area, sufficiently heavy to prevent shifting. Owing to the incessant movement of the sea, due to wind and tidal influence setting up a continual sawing motion on the bottom, shore-end cables have been occasionally made to meet these requirements by a single sheathing of exceedingly stout wires. A type of this description can, however, scarcely be recommended, for it involves the wires being so stout that they become almost in the nature of rigid rods.

Shore-end cables are, therefore, more usually constituted by a double sheathing. This has many advantages. It is then generally arranged so as to be a case of merely re sheathing the deep-sea, or light intermediate, type (with a packing of compounded hemp between), the inner sheathing wires being maintained of the same gauge,* such as does not unduly compress the core, as is sometimes the case with the single sheathing of stouter, rod-like, wires.†

In very large shore ends the outer sheathing has sometimes been

---

* By the D.S. type running in this way right through, it can be conveniently used alone to form the beach cable to the hut in instances where no "intermediate" type is used.

† In a double-sheathed type, in fact, the inner sheathing takes off the squeezing tendency of the outer sheathing, assuming that the wires are "keyed up," and not of the "open-jawed" type.

composed of previously formed strands *—three wires laid up together—
with a view to offering greater surface to withstand attrition, and in order
to decrease the rigidity for so heavy a type. However, as a rule—even in
such extreme cases—single solid wires with a very short helical lay are
preferable, inasmuch as a greater weight is thereby secured in a smaller
area.†

---

* The above more particularly applies to multiple-core cables, where a certain bulk
is, of necessity, involved, and where, owing to the nature of the bottom, a great quantity
and weight of iron is entailed. This form of sheathing, in fact, combines a certain
amount of flexibility with the above conditions. The cable laid at the mouth of the
Seine in 1877, between Le Hoc and Penedepie—where great banks of sand shift bodily
during spring tides, and where, consequently, a light cable could never last—is an excellent
example of this type. Manufactured by M. Menier at Paris-Grenelle, it has lasted nearly
twenty years without requiring any sort of repairs. Apparently, owing to its great weight,
it has sunk deeply into the sand, so as to be below the layers which are periodically
washed away by the sea. This cable (Fig. 62) consists of five separate cores laid up

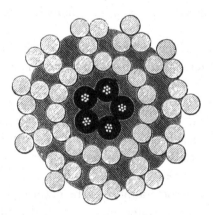

FIG. 62.—Multiple-Conductor Cable with Stranded Outer-Sheathing.

together. These are "wormed" with tanned jute, the whole being then externally served
with jute, round which are spirally wound fifteen wires 5 millimetres (= .192 inch) in
diameter (No. 6 S.W.G.). Outside this inner sheathing is a second covering of jute
(compounded), and over this again the outer sheathing, formed of eleven strands close
fitting, each strand having three wires of 5 millimetres (.192 inch) diameter, previously
laid up together, by an ordinary stranding machine. The methods of laying up such
cables and the machinery employed is, of course, precisely the same as that described for
others. The heart of "wormed" cores (laid up as already described) is merely drawn
through an ordinary sheathing machine—usually of the deep-sea type—for the inner
serving. After the next serving is applied, the whole is then drawn through a heavier
sheathing machine for laying on the previously stranded wires.

† Shore ends of this type will be found fully described and illustrated later on in connection with the specification of the last "Anglo" Atlantic cable (1894). Here the outer
sheathing wires are laid up so "short" that a photograph of the cable in section gives an
idea of their being oval rather than ordinary circular wires, as they are actually.

The Post Office Department have naturally had a great deal of experience in the direction of heavy shore-end types for multiple-conductor cables in instances of rocky, anchor-ridden, and boisterous approaches and landing-places round about the various islands of the United Kingdom; as well as across rivers, and in connection with coastguard lightships and lighthouse communication.

They have, indeed, made a special study of the question regarding the most suitable pattern: the result is that the shore-end type now adopted by this Department is invariably a very heavy, close, double-sheathed cable.*

The "intermediate" type usually consists of a single sheathing of more or less ordinary iron, the wires being rather stouter than that of the main cable—in degree governed by the existing requirements as regards depth and nature of bed. Very occasionally, in special cases of bad bottoms for some distance from shore, intermediate types have been composed by a deep-sea type with an outer sheathing of wires rather larger than, but not so large as, that of the shore-end outer sheathing.

## TESTING THE IRON WIRE.

When a stock of sheathing wire reaches the cable factory, each coil is first tested for weight and diameter, the latter being taken from the mean of two measurements, one in each direction. Should any coil be seriously at variance with what it ought to be in these respects—the former relative to the length of course, and the latter to the specified gauge required—it is rejected. A proportion of 10, or sometimes 20, per cent. are then tested (a sample of about 1 fathom length being cut off from each end of the coils) for tensile strength and elongation; or for torsion, as the case may be, according to the class of wire and type of cable it is intended for. Large, soft, iron wires are tested both for torsion and tensile strength; but the tensional test is alone applied to small "homo" or steel wires. About 6 inches of this sample length is also set aside for testing the galvanising only. In testing a wire for tensile strength—i.e., till it breaks—it is also, at the same time, tested for elongation; or rather, a note is taken of the amount of elongation which has occurred before the breaking of the wire

---

* For example, one of the P.O. types between England and Ireland is represented by a weight per N.M. of as much as 27 tons. This forms part of a seven-conductor system. Such a cable has a breaking strain of nearly 30 tons. There are several other very similar instances.

Correspondingly, the lightest form of close iron-sheathed cable weighs usually about 1 ton per N.M. wet in air, and bears a 6-ton stress—or more.

under the gradually increasing stress, the ultimate amount of the latter being additionally noted.

The small "homo" (or steel) wires of great tensile strength intended for deep-water cables are tested for their breaking strain and corresponding elongation, the requirements of the case involving both these factors, especially the former. An elongation of 4 per cent. on the original total length is usually specified for, about 5 per cent. or so being about the ordinary elongation which they will stand before breaking. A breaking strain, or modulus of fracture, represented by at least 80 tons per square inch, is very ordinarily required, and an equivalent of over 100 tons per square inch often secured.*   This class of wire is not tested for torsion, as a rule, inasmuch as it is known that when a number of such wires are laid up (even closely) into cable form, the result is sure to be sufficiently flexible—more or less, i.e., directly in proportion to the number, and inversely to the diameter, of the wires employed.

The larger, ordinary (or "best best") iron wires are, on the other hand, tested for torsion.   Representing a test of toughness and even quality, this is especially to the point here ; as the type of cable it is employed in— mainly "shore-end" and "intermediate"—is liable to experience a good deal of rough usuage generally,† in which brittleness might prove a fatal source of trouble.‡   It is also important on the score of the great bulk of wires, tending to make the whole cable very rigid.

The test is carried out in practice by noting the number of turns the wire will bear before breaking, the length of specimen being usually 6 inches.   The largest type of wires employed in cables generally stands about six turns or twists in the above-mentioned length, whilst those very ordinarily used to a "heavy intermediate" type bear some twelve twists before breaking, the smaller wires taking a much larger number.   The

---

* The true breaking strain of the wire should, strictly speaking, be obtained by testing the wire without its outer galvanising coat, which may be got rid of by dipping the sample in sulphuric or other acid ; or else, in working out the B.S. of the wire alone, an allowance of from .002 to .004 inch should be made for the thickness of the galvanising.

Again, the reduced area of the wire at the fracture is sometimes taken for calculating the B.S. per square inch, instead of the area of the full section elsewhere.   However—as it is uncertain that the former is ultimately any more accurate—the latter is usually adopted, or a mean between the two, as a basis for calculations.

† Besides being more open to local damages, as already shewn.

‡ Common malleable wrought iron—sometimes only rolled, not drawn—being so much less homogeneous, is more subject to brittleness than anything of the steel order.

Besides the objection to brittle wire as above, there is also a strong objection from a cable manufacturer's point of view, owing to the increased number of stoppages thereby involved during the operation of sheathing.

tension test is also applied to these soft wires,* their elongation being often noted at the same time. This latter may amount to a nearly 30 per cent. increase on the original length, 18 per cent. being very commonly specified for.†

We will now turn our attention to the machines used for carrying out these tests.

There are many different forms of machines employed as regards both

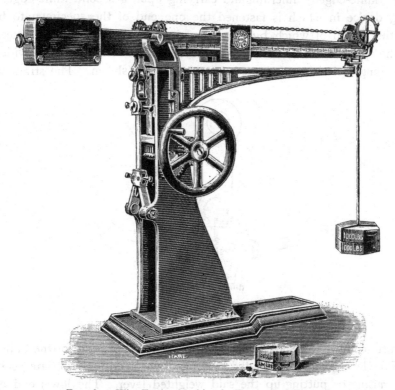

FIG. 63.—Wire-Testing Machine for Tension and Elongation.

the tensional and torsional tests, and recent years have brought to light several improved devices.

---

\* This test obviously also acts as a test of brittleness, and it is in this way that it is additionally applied to the above class of wire used for heavy cables. Indeed, till within recent years it was solely employed—before the torsion test came into vogue. A soft wire, if good, generally shews a breaking strain equivalent to some 25 or 30 tons per square inch.

† Soft iron naturally stretches more than hard under a given force or stress, just as it also naturally bends more readily in any direction—on account of greater ductility—though more likely with fatal results, owing to being less homogeneous.

Again, ordinary soft iron corrodes more quickly under water than hard iron.

**Tension and Elongation.**—Perhaps the best known and most efficient machines for testing a wire for tensile strength and elongation are those of Kitchen, Carrington, and Denison. The latter is probably the most accurate and beautifully finished machine for this purpose, and when prime cost is not of first importance, appears to recommend itself beyond all others. It is, therefore, selected here for description.

This machine (Fig. 63) consists in the employment of a lever working about a "knife-edged" fulcrum, and carrying upon a second knife-edge, a clipping device in which is fastened the one end of the specimen to be tested, the length of which, by convenience, is usually either 2 or 3 feet. Upon the opposite end of the lever an automatic counterpoise is provided for travelling upon the graduated "race," or scale, shewn. The straining

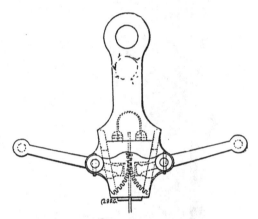

FIG. 64.—Wire-Clipping Mechanism.

mechanism comprises a slide provided with a suitable bearing for the knife-edge of the lever, which is actuated by a hand-wheel, and appropriate gearing for raising or putting up the said weighted lever. The lower end of the wire under test is secured in a clip bolted to the framing of the machine. The most important feature, however, in this machine is the automatic moving arrangement of the counterpoise, which is obtained by the action of the suspended weight which drives a chain attached to the travelling poise.

The chain further drives the governing mechanism, which consists of a revolving cataract of mercury, and a detent wheel which engages with a stop on the lever at a certain position. The extension, or elongation, as well as the point of rupture in the specimen is shewn on a scale fixed to the saddle piece, whilst a "vernier" divided into hundredths of an inch is attached to the body of the machine. The action of the apparatus is as follows :—The wire to be tested is inserted between the clipping contrivances described, and the slack taken up. It is then fixed and the pointer set to zero, the travelling poise being then released from its catch. Upon turning the hand-wheel a downward pull is exerted upon the specimen, so as to lift up the long arm of the scale lever; whereby, moreover, the detent wheel is released, and the poise set free to travel. This weight continues to move until the point of equilibrium is reached, when it is arrested by the re-

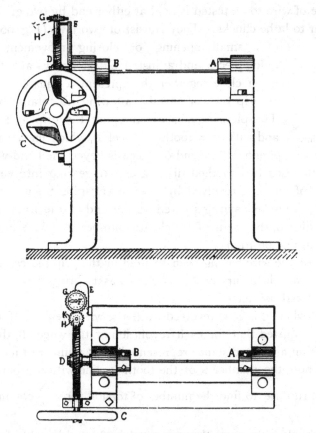

FIG. 65.—Wire-Testing Machines (Torsion).

engagement of the automatic detent device above described. The position of the vernier now indicates the exact amount of stress that has been applied to the wire in obtaining a certain elongation and final rupture.

Denison's patent wire-clipping mechanism, illustrated at Fig. 64, is an improvement upon the ordinary wedge box largely used. Here the wedges will be seen to be operated simultaneously by the intervention of levers and segmental teeth, the said wedge pieces being kept apart by means of a bow spring. These machines are fitted with two speeds of gearing, i.e., one for testing light wire, and another, fifteen times as fast, for actuating the more powerful mechanism.* The testing strains are applied by means of the hand-wheel which actuates worm and screw gearing so as to obtain a steady and uniform motion; and it is in this last respect that this machine is superior to others, though at the expense of extra elaboration.

**Torsion Test.**—Fig. 65 serves to illustrate a very ordinary machine for this purpose.

The piece of wire to be tested is held at either end by two grip pieces, A and B, similar to lathe chucks. They consist of two jaws (Fig. 66) capable

FIG. 66. — Jaws of Torsion Machine.

of a small opening or closing movement between guide pieces, and adjusted by set screws at either side; when close together they present a square section, and so grip the wire firmly, allowing no rotary movement. The piece at A is a fixture, but the grip at B revolves, and with it a toothed wheel D, which engages a worm pinion, to one end of which is keyed the hand-wheel C.

The counter has two toothed discs, K and G, gearing into each other, the lower one of which is revolved by a worm engaging the upper teeth of the wheel D. A hooked spring secured at one end to the frame acts as a pawl to the disc G, the teeth of which are numbered. At F is a pointer rigidly fixed in one position.

The ends of the wire having been secured in the grip pieces, the spring pawl is held clear whilst turning the disc G, so as to bring the zero tooth in line with the fixed pointer F.

The hand-wheel C is now revolved until the wire breaks. If the tooth gearing is so adjusted that for each revolution of the wheel D, the disc G turns through an angular distance represented by N number of its teeth, we have only to note the number N of the tooth in line with the pointer F, at the instant of rupture, to find the number of torsion turns $\frac{N}{n}$ communicated to the wire.

A more simple plan—and the one usually adopted in cable factories—is to trace a line with ink or tar along the wire between two marked points, before applying the test, and afterwards to count the number of turns shown by the line. The length of wire for the torsion test is generally 6 inches.

The heat developed in the wire during torsion is much less than that experienced during the elongation test; work being done, in the former instance, only on the outside fibres of the metallic cylinder. For the same

---

* This latter, however, would not usually come into use for testing such wires as are tested for tension in this connection.

reason the duration of the torsion test affects the result in a much less degree than in the case of elongation.

Speaking generally, the better qualities of mild steel wire usually stand torsion much better than tension ; and, moreover, better than other classes. However, the latter require—for the types for which they are used—to be tested more for torsion.

As the displacement of fibre in iron wire under torsion increases proportionally to the distance from the centre of the wire, the number of torsion turns which wires of similar quality but of different thickness can withstand should be in inverse proportion to their diameters.

Another practical test which affords a very precise measure of ductility, is to count the number of times a wire can be bent backwards and forwards at right angles, on a cylinder of constant diameter. Unfortunately the test affects only a very short length of wire, which may be of different quality at a little distance from the actual point where the experiment is made.

**Test of Galvanising.**—In order to test the galvanising, the wire is usually dipped four or five times successively, for a minute each time, in a saturated solution of sulphate of copper at 60° F. (with five times its own weight of water), the wire being wiped clean between each dip. Should the steel, or iron, be laid bare at any point by the coating of zinc being too thin—which is shewn by the copper depositing—the coil to which the specimen belongs is rejected on this count ; and if there be many such cases it may be found necessary to reject the entire batch.

Another test of the galvanising—and one very generally adopted at cable factories—is to coil the specimen round a cylinder about ten or twelve times the diameter of the wire, and note whether the zinc coating flakes off. Should this occur, unmistakable evidence is afforded of inefficient galvanising. It ought, moreover, to be seen that the galvanising is smooth and free from irregularities.

**References.**—Mr Bucknall Smith's very complete book on wire[*] will be found, amongst other things, to give exhaustive information regarding the testing (as well as the manufacture) of iron wire, and the reader is referred thereto for further details on this subject.

**Data.**—The Table on next page serves to give an idea of the present capabilities of some of the various classes of wire employed for the sheathing armour of submarine cables, besides indicating the tensional and torsional tests which it is now the custom to apply.

---

[*] "A Treatise upon Wire," by J. Bucknall Smith, C.E., published at the offices of *Engineering*.

RESULTS OF TESTS *on Specimens of Iron Wire used for Sheathing Cables, by Messrs Clark, Forde and Taylor, between 1892 and 1895.*

### No. 6 L.S.W.G. Wire.

| Number of Tests Averaged. | Diameter in Inches. | | | Breaking Strain in Lbs. | | | Elongation per cent. | | | Torsion Twists in 6-Inch Lengths. | | | Breaking Strain in Tons per Square Inch. | | |
|---|---|---|---|---|---|---|---|---|---|---|---|---|---|---|---|
| | Av. | Max. | Min. | Av. | Max. | Min. | Av. | Max. | Min. | Av. | Max. | Min. | Av. | Max. | Min. |
| 16 | .200 | .204 | .196 | 2434 | 2610 | 2220 | 24.8 | 27 | 22 | 14 | 15 | 13 | 34.6 | 35.6 | 32.8 |
| 14 | .198 | .201 | .194 | 2039 | 2160 | 1920 | 16.7 | 20 | 14 | 16.5 | 18 | 15 | 29.6 | 31.9 | 29.0 |
| 30 | .200 | .204 | .198 | 2435 | 2580 | 2220 | 20.2 | 24 | 15 | — | — | — | 36.0 | 39.0 | 32.8 |
| 30 | .199 | .203 | .197 | 2225 | 2460 | 2040 | 17 | 21 | 12 | 17 | 25 | 10 | 33.2 | 35.3 | 31.1 |
| 30 | .202 | .205 | .197 | 2324 | 2550 | 2100 | 23 | 29 | 17 | 11 | 20 | 10 | 33.7 | 36.9 | 31.7 |

### No. 9 L.S.W.G. Wire.

| Number of Tests Averaged. | Av. | Max. | Min. | Av. | Max. | Min. | Av. | Max. | Min. | Av. | Max. | Min. | Av. | Max. | Min. |
|---|---|---|---|---|---|---|---|---|---|---|---|---|---|---|---|
| 25 | .144 | .151 | .141 | 1143 | 1296 | 1065 | 17.9 | 21 | 16 | 23.3 | 26 | 20 | 31.2 | 34.3 | 29.4 |
| 40 | .145 | .147 | .142 | 1122 | 1298 | 1003 | 15.6 | 22 | 11 | 23.0 | 26 | 20 | 30.3 | 34.1 | 27.1 |
| 40 | .144 | .147 | .142 | 1102 | 1232 | 1007 | 14.7 | 20 | 10 | 23.3 | 26 | 21 | 30.1 | 32.8 | 27.4 |
| 15 | .144 | .147 | .142 | 1147 | 1272 | 1045 | 14.4 | 20 | 11.5 | 23.7 | 28 | 21 | 21.3 | 33.5 | 28.2 |
| 15 | .143 | .145 | .142 | 1122 | 1301 | 1004 | 14.6 | 17 | 11 | 24.1 | 25 | 22 | 31.0 | 35.7 | 27.1 |

### No. 13 L.S.W.G. Wire.

| Number of Tests Averaged. | Av. | Max. | Min. | Av. | Max. | Min. | Av. | Max. | Min. | Av. | Max. | Min. | Av. | Max. | Min. |
|---|---|---|---|---|---|---|---|---|---|---|---|---|---|---|---|
| 25 | .098 | .100 | .096 | 1454 | 1600 | 1360 | 6.8 | 8 | 6 | — | — | — | 93.5 | 105.1 | 89.4 |
| 30 | .098 | .101 | .095 | 1430 | 1600 | 1300 | 7.9 | 8 | 5 | — | — | — | 92.0 | 96.7 | 78.5 |
| 30 | .098 | .101 | .096 | 1404 | 16co | 1320 | 6.7 | 8 | 5 | — | — | — | 90.3 | 100.8 | 90.6 |
| 25 | .099 | .102 | .096 | 1506 | 1640 | 1360 | 6.5 | 8 | 4 | — | — | — | 94.8 | 107.8 | 91.3 |
| 30 | .099 | .101 | .094 | 1478 | 1610 | 1390 | 6.3 | 7 | 4 | — | — | — | 93.1 | 99.3 | 85.7 |

### No. 14 L.S.W.G. Wire.

| Number of Tests Averaged. | Av. | Max. | Min. | Av. | Max. | Min. | Av. | Max. | Min. | Av. | Max. | Min. | Av. | Max. | Min. |
|---|---|---|---|---|---|---|---|---|---|---|---|---|---|---|---|
| 30 | .083 | .085 | .081 | 1152 | 1280 | 1050 | 5.4 | 6 | 4 | — | — | — | 104.9 | 110.9 | 95.6 |
| 30 | .083 | .085 | .081 | 1119 | 1240 | 1020 | 6.1 | 7 | 5 | — | — | — | 101.9 | 107.4 | 90.6 |
| 30 | .083 | .085 | .082 | 1100 | 1300 | 1020 | 5.7 | 7 | 5 | — | — | — | 100.2 | 112.6 | 95.3 |
| 30 | .083 | .086 | .081 | 1078 | 1180 | 1000 | 4.9 | 6 | 3 | — | — | — | 98.2 | 99.7 | 95.9 |
| 30 | .083 | .087 | .081 | 1003 | 1060 | 950 | 4.8 | 6 | 3 | — | — | — | 91.3 | 99.0 | 86.5 |

### General Particulars regarding a Sheathed Cable.

**Routine.**—After weighing, gauging, and testing specimen pieces for tension, elongation, and torsion of a certain proportion of the coils of iron wire, the coils are then prepared for use in sheathing and made up in the proper form, ready for applying to the "closing" machine.

**Partial Failure of Galvanising.**—In quite the early days of submarine telegraphy it was discovered that the process of galvanising failed to act as a complete preservative against corrosion of the iron wires. After a more or less short period of submergence under the sea, they were found to have rusted considerably, and to be so weakened that recovery became a hazardous, if not impossible, operation.

Iron, even when galvanised, is acted upon by salt-water, especially when resting on ground which contains soluble sulphides. Here even quite stout wires are corroded away in a very short time, the ends being sharpened to needle points.

It is usually thought that the salt-water tends to decay the zinc coating in the galvanising, and having effected this, the iron wire rapidly oxidises. Thus, whilst the insulation remains intact and perfect, the cable is liable to break at the first attempt to raise it. Another sort of chemical action also causes deterioration in iron, giving a characteristic fibrous appearance to wires which have been under water for a long time.

**Failure of "Open-sheathed" Cables.**—It was partly to meet these drawbacks that the type of cable adopted for the 1865 Atlantic came into vogue, and was employed for a number of undertakings in deep water. One of the features of this type was that of preserving each of the iron wires by previously encasing them in tarred hemp. Here, again, a cable of low specific gravity was obtained, whose tensile strength as a whole exceeded the combined strength of the iron and the hemp taken separately.* However, as has already been explained, this type of cable did not prove durable, and was eventually abandoned. The hemp being in close contact with the once rusted iron, rotted away sooner than ever, even when of the best class †—not to mention its immediate contact with any possible

---

* This result is explained by remarking that the weak places in the wire and the surrounding hemp are probably seldom coincident, so that the strength of the covered wire is the sum of the average strengths of the wire and the hemp ; whilst each of these substances, when *separately* put to the test, will give way at its weakest point, where the strength is always *less* than the average.

† On exposing to sea-water a cable protected with wire so covered, the water, entering between the threads, attacks the metal ; and a rapid destruction of the hempen covering thereupon ensues.

mineral or organic matter on the ocean bed. Again, the iron wires not fitting together, any species of boring insect, or fish, can much more readily find its way through to the gutta-percha core* than is the case where the wires fit closely against one another. A cable so formed once losing its solidity—owing to the destruction of the hemp—a general decay rapidly sets in, on the entry of the water all round the iron wires, and its recovery for repairs soon becomes practically impossible.

**Bright and Clark's Cable Compound.**—The oxidising action of salt-water (and of air) on the zinc coating in the galvanised iron sheathing wires is nowadays to a great extent arrested by the application of what is known as Bright and Clark's bituminous compound, together with an outer binding, or bindings, of hemp, jute, or canvas tape, in a manner which will be gone into hereafter. The compound is incorporated with any of the above bindings, both outside the wires † and outside the entire cable, besides the materials being sometimes previously steeped therein. This system is based on patents of the late Sir Charles Bright (Nos. 466 and 538 of 1862), an additional object of which was that of firmly adhering and binding the various coverings together.‡

The above preservative compound (first used on the Persian Gulf cable of 1863) is a mixture of asphalte, or mineral pitch,§ silica, and tar.

---

* There are several insects which have a pronounced *penchant* for hemp and jute, as well as for gutta-percha.

† It is highly desirable that compound should be applied outside the sheathing—or preferably that each wire should be previously steeped in compound and also taped, if possible—as the outer compounding must go with the tape, yarns, or cords when they fail, which (except perhaps in the latter case) does not take long to occur.

‡ Previously, in 1858, Messrs Clark, Braithwaite, and Preece had taken out a patent—of which Mr Latimer Clark was the main author—for applying a covering of hemp and asphalte as a preservative to the iron wires of a sheathed cable.

The asphalte was to be applied whilst hot outside the finished cable, and was heated up by charcoal fires. This plan was tried on a short cable to the Isle of Man in 1859; but gave a good deal of trouble during manufacture, the insulation becoming seriously damaged by the process. No further use was ever made of this particular system of preservation.

§ Mineral pitch is the heavy residuum of coal tar, after various spirits have been drawn off. It may also be described as the solid variety, at a normal temperature, of artificial bitumen. *Natural* bitumen (or *vegetable* pitch), from Trinidad and elsewhere (formed as the heavy residuum from wood tar), is not usually turned to account in cable compounds. It has, however, been occasionally incorporated as the mollient in place of tar, being much clearer, finer, and softer than mineral pitch—having, in fact, more tar left in it and of the finer sort.

Pitch—more particularly *vegetable* pitch—is still occasionally termed by its original name, bitumen. It has, indeed, occasionally been spoken of as asphalte—or asphaltum, originally. The latter, however, is distinctly unsuitable in the present day, considering that the term is generally used to imply something very different.

In this concoction, it is the pitch which acts mainly as a more or less air-tight and waterproof preservative casing.   The tar is incorporated principally as a mollient—*i.e.*, to make the pitch workable, the latter being, by itself, too stiff; just as the tar alone would be too thin and liable to run, or get washed away afterwards by the incessant action of the sea.   Both, however, are similar in their characteristic preservative and waterproof qualities, the mixture being effected chiefly to obtain the required consistency.

The silica—made from calcined flints ground down to a powder, similar to sand—was added to the above composition especially with the object of evading the ravages of the teredo by damaging its boring tool, which object there is evidence of it having successfully accomplished.   It is now, however, very often left out (owing to the difficulty of keeping it properly mixed with the rest),* though particularly to the point in shallow-water cables infested by marine " borers."

It may be remarked that the actual proportions of this mixture are varied very much according to the requirements.   Special compound cements are made up for different and particular purposes, the result of experiment and experience ranging now over a number of years.   However, all the cable compounds in use at the various factories are based on the above original mixture,† with the same—besides sometimes additional —constituents therein.‡   Again, in all, mineral pitch forms the base, or main constituent; and the amount of tar added is comparatively small, as a rule, though in some instances it is larger.§

The specific gravity of Bright and Clark's compound varies, of course, according to the proportions.   However, one cubic foot of the original composition weighs about 100 lbs.

---

* This defect in practice is one which might suitably receive more attention in some of the cable factories.

† Largely owing to the method of application (to be described later), this patent proved perhaps the most lucrative of any connected with submarine telegraphy, yielding as much as £30,000 to the inventor (Sir C. Bright) and his partner, Mr Latimer Clark.

‡ Resin oil is now frequently incorporated in hot compounds.

§ As will be shewn hereafter, the tar employed is sometimes coal tar and sometimes Stockholm (wood) tar.   The former, which is cheaper, is invariably employed in the case of hot compounds for the outer serving.   But the latter, being of a lighter and finer consistency, is always adopted for the cold compound outside the sheathing wires, as it would not run at low temperatures with gas tar—in fact, here the amount of Stockholm tar used has to be about the same (by bulk) as that of the pitch.   This (cold) compound is troublesome stuff to deal with, and would be quite unsuitable for the outside serving, as it would not sink in readily, neither will it stick on like hot compound.   However, hot compound could not be applied outside the sheathing when once "closed," for fear of damaging the insulation.   The matter is usually obviated now by the more complete preservative system of dipping each coil bodily into hot compound before it is applied to the served core. This matter is dealt with in subsequent pages.

**Compounding of each Wire.**—Another device with a view to overcoming the decay of the sheathing armour was that described in a patent of the late Mr Willoughby Smith (No. 3,622 of 1878) for embedding the iron wires, in long open spirals, into an extra insulating coat outside the ordinary core, composed of a soft mixture of gutta-percha and shellac; either the wires or the outer surface of the core being heated immediately before, by way of ensuring efficient embedment. This plan was presumably found to be difficult of proper effection; for, it is believed, nothing was done with it in practice.

It, however, no doubt served to lead the way to present-day practices. Thus, in 1880, the " Anglo" Atlantic cable of that year (manufactured by the Telegraph Construction Company) was close-sheathed with iron wires each of which had been previously coated with a mixture of refuse gutta-percha and Chatterton's compound.

This method of preservation to each iron wire has been very commonly adopted ever since—indeed almost invariably except where each wire is surrounded with preservative tape (in a manner which will be explained hereafter) or even as a preliminary additional preparation. Nowadays— with the prices of cables so much cut down—the gutta-percha mixture is very commonly replaced by the ordinary bituminous preservative compound of Messrs Bright and Clark, just described.

In order to effect the complete coating of each iron wire with the compound, each coil, after being tested on arrival, is first heated for an instant in an oven to effectually drive off any moisture. It is then dipped bodily into a bath, or tank, of the hot compound.*

This plan very usually takes the place of the coating of compound outside the completed sheathing, thus rendering certain that the wire will be entirely covered with the mixture. When only applied externally, it probably never percolates to any extent through the interstices of the wires, and is confined, therefore, to the outside surface.

There are those, however, who assert that this method of compounding each wire separately (or that of enclosing it in a similarly compounded tape) has a tendency to effectually disguise the presence of faults brought about during the process of sheathing—due, say, to burning, or what not— by preventing water reaching until the cable is subjected to the pressure of the sea. The author has, however, never heard of any instance of trouble

---

* Possibly on account of the nature of the operation, the term " pickling" has been sometimes applied to the above " dipping in compound" process. This, however, is liable to lead to confusion with the process of plunging iron wire into acid, after being drawn, to cleanse it of all grease or other impurities on its surface previous to galvanising.

on this account. Again, it is asserted by certain electricians that the above plan—as well as that which is to follow—causes some inconvenience in reducing the efficiency of the "earth," the sheathing wires always being made use of in a submarine cable for this purpose, as offering a ready and excellent form of electrical connection.* It has been pointed out that if the sheathing wires are separately compounded in the manner described, they are no longer in direct metallic contact; and, inasmuch as the compound is not only of a waterproof but also of an insulating nature, the earth connection must, perforce, become considerably less reliable and effective. Extreme variability in this respect has certainly been experienced with cables whose wires are separately compounded or taped, besides "kicks," etc., during testing and signalling operations.

It would seem, however, as though such drawbacks must be either submitted to, or overcome in some other way; for the advantage of dipping each wire separately into compound—in the absence of taping each wire—is now so generally recognised as to be regarded as a *sine quâ non*.

**Taping each Wire.**—In 1870 a patent was taken out in the names of Messrs Matthew Gray and Frederick Hawkins, of Silvertown, for covering the sheathing wires of a cable separately with strips of woven cloth (made of cotton, flax, or other yarn) saturated with some form of bituminous (vegetable pitch) waterproof composition for purposes of protection and preservation. According to this patent the previously prepared tape was to be applied spirally to each wire before laying up, in "at least two coverings alternately in opposite directions."

Whilst the inner taping was by preference to be composed of cotton, the outer was to be of Hessian canvas.

In this original device the wires so lapped with tape, when laid up, completed the entire cable, and one of the incidental advantages claimed for it—in comparison with the previously prevailing open-sheathed type of cable—was the fact that a material diminution of the bulk was secured without reducing the breaking strain, whilst providing a more efficient and durable form of protection for the armour.

This type was first made, on an extensive scale, by the Silvertown Company over the Post Office Wexford cable, more especially in connection with the shore-end types.

Nowadays, this device, in a modified form, is turned to very general account with regard to deep-sea types.

---

* Especially in a long length of laid cable, the surface exposed to the sea being, in that case, enormous.

In present practice the outer covering of comparatively thick Hessian tape is replaced with a similar covering of cotton tape, laid on in the opposite direction ; or else—far more generally—it is applied singly, it being commonly considered that with more than one lap of tape there would be too great a distance between the wires, to meet the qualifications of a close-sheathed cable.

Again, there are always other coverings outside the sheath of taped wires, which, in this modification, butt pretty closely against one another. Indeed, in its present application the cable approaches more nearly to the close-sheathed cable of modern days, whereas the original device may be looked upon as a species of the " gridiron " type. Cables of this description have been made on an extensive scale—mainly, or characteristically, by the Silvertown Company, for the last twenty years or more.

A cable with taped wires may be regarded as some sort of compromise between the absolute close-sheathed and open-armoured cables—or, more properly, as an excellent modification of the absolute close-sheathed cable—combining in the main the great advantages of the former with certain conveniences and advantages in the latter. Taping each iron wire has the effect of reducing the resiliency of an ordinary close armour—making it more flexible, in fact, whilst rendering it less springy and rigid. It is, therefore, more pliable for handling during contractors' operations, and is less liable to kink—i.e., to throw itself into a loop when leaving the hold of the vessel in which it is coiled—or to any other species of accident during laying.

It is also claimed for the taped-wire principle, that a cable so constructed is more durable than with the ordinary close sheathing, on account of the better protection of each wire by the prepared tape closely enveloping it,* which latter is supposed to be more or less waterproof. This feature renders a taped-wire cable well adapted for corrosive bottoms.†

Where the sheathing wires are taped, the specific gravity of the cable is, moreover, appreciably reduced. This type is, therefore, peculiarly suited for extremely deep water—where the exigencies of recovery become a

---

* Thus—if only on this account—materially increasing the chances of successful recovery after years of submersion.

It may be suggested that the wires of grapnel and buoy ropes should also be treated in like manner, besides being compounded ; for the action of sea-water in tropical climates is most destructive to steel wire—as to iron chains.

† The last is a strong feature, inasmuch as it is often not long before the outside covering rots away—especially if only composed of tape or yarns.

serious consideration by rendering the cable capable of suspending a greater length of itself in water.*

On the other hand, it is also contended that the decrease in specific gravity and increase of skin friction here entailed, is not sufficient to seriously alter the angle in paying out in such a way as would prevent the cable adapting itself to the irregularities of the bottom.†

Where the sheathing wires of a cable are to be individually "taped," the coils are not generally dipped bodily into preservative compound, as previously described. ‡

However, in the operation of applying the silicated prepared tape to the wire, the latter in being drawn off from the iron bobbin on to which it has been coiled, is first led through a tank of hot compound, on its way to the taping apparatus, immediately previous to the application of the tape. This operation is performed by spinning round the wire a bobbin or disc,§ loaded with the prepared tape. || The speed of draw off in relation to that

---

* The fact that the coefficient of friction is slightly increased by the increased bulk due to taping does not usually apply here as much as the advantage of decreased specific gravity ; though it would if pressed further, unless the speed of picking up were proportionately reduced. However, where the wires of a cable are *untaped*, the complete arch is certainly better maintained when under heavy strain, longitudinal and lateral, during lifting operations over the bow sheaves of a vessel.

† Against this argument it must also be remembered that too much slack would be as bad as too little, if introducing lengths of cable coiled down at the bottom. The latter, when lifted, are liable to involve kinks such as break under any considerable strain.

‡ Nevertheless, owing to the very perfect preservation by the dipping in compound, preceded by heating, this process forms a most fitting preliminary in any case.

By this method, all moisture or moist matter being driven off, the compound, applied at an *exceedingly* high temperature, adheres almost as firmly as a varnish.

§ The disc here is mounted between two metal plates, which serve to keep the tape in its place.

|| The material is usually prepared by being led through a tank of hot preservative (cable) compound from its roll whilst in sheet form. In coming away from the compound tank it is drawn through a lathe where it is cut in strips to the required width, and then wound on to the discs here alluded to.

The above is the Silvertown plan, according to the Gray and Hawkins patent. Messrs Johnson and Phillips, however, patented a device in 1876 (No. 3,533) for cutting the tape direct whilst on the second (wooden) spindle, after passing through a compound bath. By this plan, the spindle itself was also cut right through into thin discs or spools, which are ready for fitting to the wire-taping machine or to the cable machine, as the case may be, with the tape of the required width all ready mounted on them. The operation of winding the tape from the entire roller on to separate bobbins is thus saved, which is certainly, on the one hand, an advantage. Messrs Johnson and Phillips have thus been in the habit of supplying canvas and cotton tapes, so cut and prepared, to the Telegraph Construction Company for all their cables in which it is used.

The outside covering of canvas tape has, however, of late years been again very much replaced by the original hemp yarns ; or by hemp *cords*, to be dealt with hereafter.

An advantage claimed for the ordinary Gray and Hawkins method of winding off the tape as it is cut from the roller on to bobbins, is that this winding puts the tape to a

of the revolving disc or bobbin is so adjusted as to give the required lay, with a very slight overlap—*i.e.*, about ⅛ inch. The width of the tape is, as a rule, about ½ inch, in which case the lay is rather less than ½ inch. The taping machine for this purpose usually takes a vertical form nowadays, as occupying much less ground space than the ordinary horizontal taping machine, which has already been described with reference to the metal tape sometimes applied to the core.

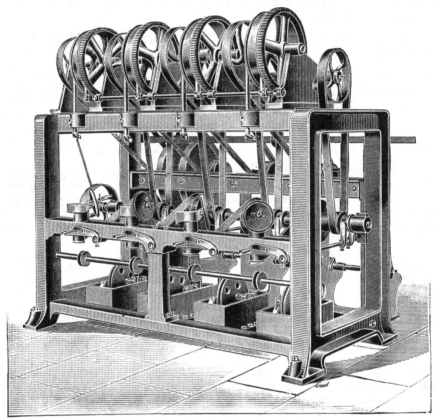

Fig. 67.—Johnson and Phillips' Wire-Taping Machine.

There are then a number of separate machines close together, all combined in one frame. Thus, the work being in a small compass, one

---

suitable tensional test before being applied to the machine, and thus avoids stoppages due to broken tapes during "cabling." By the Johnson and Phillips method each layer of tape is naturally more firmly stuck, and though this tends to induce breakages slightly, it has the advantage of putting a suitable check on the tape during application round any cylinder. Keeping it thus under tension prevents it getting slack, and avoids overrunning during application.

lad can attend to several "heads."*   One of Messrs Johnson and Phillips' compound frames, embracing several such taping machines, is shewn in Fig. 67.   As fast as the wire is taped it is led on to another bobbin, which is then ready for use in the sheathing, or "closing," machine.

A cable whose wires are individually taped, as above described, certainly involves a slightly increased initial cost; and this is one of the objections raised by the opponents to taped wires.   However, in the author's opinion —for deep-water † purposes at any rate—this increased initial cost is more than repaid on the various counts named.

Again, it is suggested that in time such taping wears off or decays, and consequently leaves an open space between the wires, thus repeating the objections, on a smaller scale, to an open-sheathed cable.

The inconvenience as regards defective or intermittent "earth" connection, and as regards "fault screening," in the case of each sheathing wire being separately compounded, applies also, of course, in the instance of a taped cable; but the nature of the inconvenience is not sufficient to warrant it being viewed as a positive objection (even if irremediable), provided that sufficient advantage in the system is established.

In the manufacture of the last (1894) Atlantic cable by the Telegraph Construction Company, besides the coils being first plunged into a hot compound of gutta-percha residue, each sheathing wire was further protected and preserved by a separate tape.   This cable may be taken as a specimen of an up-to-date deep-sea cable constructed on all the latest improved principles, mechanically and electrically.

As regards durability, the advantage of "taping" the iron wires may be put to the test by submerging a length of untaped and taped wire of the same quality in a vessel containing sea-water; or, still better, two specimens of completed cable, the one with taped and the other with untaped wires of the same construction in every other respect.   If possible the bottom of the containing vessel should be further lined with a sample of the bottom to be encountered.

---

* Roughly speaking, each "head" is capable of so "taping" two to four miles of wire per working day, according to length of lay adopted.

† The sheathing wires are individually taped in deep-sea types only.   With shallow-water types, the extra expense incurred by taping the larger wires would not be warranted. Here, decay and wear and tear are usually such prominent features that they cannot be successfully coped with in this way.

### SECTION 4.—APPLICATION OF THE SHEATHING WIRES TO THE SERVED CORE.

The iron wire reaches the cable shop on metal bobbins, on to which it has been wound from the original coils in which it came from the factory.*

As already stated, theoretically speaking, straight wires would be correct for purposes of strength—so far as a cable for deep water is concerned—but such an armour would be liable to get loose even with the most complete binding outside.† The wires are, therefore, wound, or laid up, around the jute or hemp serving, in much the same way as the core itself is embedded in its serving, by means of a wire-rope making machine of the eccentric type.

The object of the spiral "lay" is then (1) to ensure the wires holding closely together ; and (2) to afford greater facility for stretching or "drawing out," without injury, on the subjection of a strain.

In the case of deep-sea cable the lay is not, however, made any shorter than is absolutely necessary to meet these requirements, bearing the other necessary qualifications in mind.‡ In shallow-water types, the large wires used being on the whole stiffer (on account of their individual size), do not require such careful binding together ; and, for this reason, the "lay" may be longer. On the other hand, a long lay is not, as a rule, so necessary on the score of strain. Moreover, a short lay has the advantage of offering a great surface to withstand abrasion—the main consideration, generally speaking, which determines this question as regards cables in shallow water. Thus, though usually much longer than in a deep-sea type, in special cases the lay in the shore end has been sometimes actually shorter, for purposes of greater security against movement or abrasion, by introducing the maximum of weight in a given bulk.

**Cardinal Points Involved.**—To turn now in detail to the process of sheathing. The served core is drawn from the tank in which it is coiled,§

---

* In hauling from the coil (on a swift) to a bobbin, the wire—especially if of large type—is straightened out as much as possible, by a process of "killing," which is effected by being drawn through a die placed half-way between the coil and the bobbin to which it is being transferred.

† In the early days more than one patent was taken out for a cable constructed on these principles, having sheathing wires perfectly straight and parallel to the conductor, bound round with hemp or other serving to keep them in place.

‡ Moreover, a short lay—especially if of hard wires—has a great tendency to springiness ; and is therefore liable to kink, besides being inconvenient to handle.

§ As already stated in the article on "serving," the sheathing may be performed at the same time as the serving, or—where more convenient—at a later period, as here by means of pulleys, through a hollow shaft of the "closing" machine, round which is connected a skeleton cage frame—or sometimes a disc—which carries the bobbins of wire. The wires unite round the served core at a certain distance from the carriage, forming the sides, so to speak, of an elongated cone. The cable is drawn steadily forward in the direction of the axis, round which the carriage holding the bobbins of wire revolves. The wires are led from the bobbins through their respective holes in a plate arranged in a circle at uniform distances from one another. They are thus wound round the serving in a perfectly symmetrical form, after the manner of a long helix, the actual length of lay being governed by the relation existing between the velocities of the longitudinal (drawing off) and rotary motions of the machine for "closing up" the served core. The right tension is kept on the cable throughout by the proper adjustment of the hauling-off gear.

If, however, the machine worked in the simple way above described, each wire would be twisted through 360° at every revolution of the carriage, and would, therefore, very soon break.* In the case of the application of the jute serving, this torsion does actually take place, but here it gives rise to no serious inconvenience, owing to the pliability of the jute or hemp: the only precaution necessary in this instance is to see that the jute is twisted beforehand as little as possible. In the sheathing machines, on the other hand, special arrangements must be made to eliminate all torsion from the wire. This necessity was recognised in the manufacture of ordinary wire ropes many years previous to the days of telegraph cables, both by Mr W. Küper and Mr R. S. Newall.† Indeed, in 1840 the latter took out a patent (No. 8,594 of that year) in this connection to meet the difficulty in question, followed by another in 1843 (No. 9,656) to effect a similar object by an improved form of machine.

---

suggested. However, carrying out the two operations together should certainly have the effect of saving time and labour, and is, perhaps, more general. In instances where the core is first metal-taped, the inner serving is often applied at the same time as the tapes ; and then the sheathing invariably forms a separate operation. On the other hand, the tapings may be first performed separately ; in which case the serving and sheathing are usually carried out together.

* If the bobbins of wire are fixed rigidly to the plate or frame which carries them, it will be evident that when the machine revolves, they no longer keep the same horizontal plane. The wires, therefore, make a turn or twist, and in such a case may truly be said to be "twisted," rather than "laid up," round the central heart or core they are intended to envelop.

† Originally a miner in Germany, Küper appears to have been actually the first to conceive this idea with his device for an elastic hempen heart in the centre of iron wires laid up together; but Professor Gordon (Newall's partner) was the first to take the matter up and turn it to a tangible account for telegraph cables.

In these devices the bobbins—as they revolve round the axis through which the core passes—all turn on their own axes, and, remaining parallel to themselves, are always kept on the same horizontal plane, thus ensuring the absence of any twist in any individual wire.

This constitutes, in fact, a sun-and-planet motion ; and all sheathing machines,* whether of the disc or skeleton-frame pattern, are constructed (as from the first) on this principle†—*i.e.*, of the axis of each bobbin, during the revolutions of the carriage, remaining perpendicular to the plane of the earth and to one another, with the above object.‡

In early days the sheathing machines used to be—like most small (conductor) wire-stranding machines—of the vertical type, as was the

---

* Patents for somewhat similar devices were also obtained later by Mr W. T. Henley, Monsieur F. M. Baudoin, and others.

† The principle adopted in sheathing machines may also be likened to the "feather-edge" principle of paddles in a paddle-steamer.

‡ The method of effecting this is based on very simple geometrical considerations. Suppose a circumference (Fig. 68) turning on an axis passing through the centre A, and a second circumference of equal radius having its centre B on a line drawn vertically through A. Again, suppose the two circumferences to be connected by a series of rigid rods parallel to the line A B, and to one another. If the second circumference is capable of no other movement than that of rotation on its centre, it is evident that the rods will all retain their relative initial positions. The one extremity M of any rod M N, being brought to a point M by the rotation of the circumference A, the other extremity of the rod can only occupy one of two positions, $n$ and $n_1$—the points of intersection of an arc of a circle described about the centre $m$ with radius equal to M N, with the circumference B. Now $m\,n$ is parallel to M N ; for if we move the second circumference along the common diameter B $m$ so as to bring its centre over the point A, it will exactly coincide with the first circumference, and all its points will have moved through equal distances in parallel lines. Moreover, $m\,n$ is the only possible position for the rod M N to take up, for the points $n$ and $n_1$ being symmetrically placed with regard to the line B $m$, it would be impossible to pass suddenly from one position to the other ; unless, indeed, the rod M N happened to be in the position $m\,v$, and to be the only one. The simultaneous presence of several rods obliges them all to retain their original parallelism.

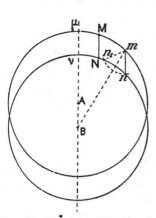

FIG. 68.—Principle of Machine for Laying up Wires without Twist : Geometric Demonstration.

In further explanation, it may be added that the second circumference is compelled to move with the same angular velocity as the first, by the rods, half of which are in tension and the other half in compression ; the turning effect being greatest at the sides and decreasing to nothing at the top and bottom. They are, in fact, "coupling-rods," which can have no other direction than one parallel to the line joining the centre of the two circles.

case with rope-making machines of that period.   These were, however, replaced after a time by horizontal machines, the first of this class being used over the manufacture of the Persian Gulf cable of 1863 at the works of Mr W. T. Henley.   This form has been adhered to ever since. Though the vertical machine naturally occupied less floor space on the same level, it involved either a materially greater diameter to carry all the bobbins or else a number of platforms to enable it to be attended to at various points ; thus, also, requiring, in consequence, a larger number of hands to look after it.  With the horizontal form, moreover, a higher working speed is possible, owing to being on one common foundation, with greater stability in consequence ; as well as—partly for reasons of centrifugal force—on account of the smaller diameter, by various sets of bobbins being placed one behind the other.   Again, in the early machines (of Newall, etc.) the revolving carriage to hold the bobbins took the form of large discs of great weight.

The speed at which it was safe to "run" such a machine was closely limited by principles of centrifugal force, varying, of course, with the outer circumference of the disc and the weight of loading.

These have, therefore, been gradually replaced at most of the cable factories by machines * whose carriage consists of a long and comparatively light framework—at any rate as regards those intended for the construction of cables with a light type of sheathing wire, where a high speed of sheathing may be safely and usefully adopted.

For laying up very heavy wires, the old disc carriage machine is still sometimes adhered to †  Here, notions of speed have in any case to give way to strength, without the chance of an uneven strain being placed on any part of the machine.

It may also be mentioned incidentally that in the class of wire here involved the breaks are of somewhat frequent occurrence, and, therefore, attempts at a high rate would be unsuitable—not to say wasted.

Moreover, the speed advisable with any machine is more closely limited in this case.  Again, there is not here the same call for a high speed of running, owing to the length required of heavy type cables being much less, as a rule.  With either class of machines the bobbins are always "geared" on the sun-and-planet principle already alluded to.

In the case of a double-sheathed cable—such as a shore-end, and some-

---

* Such as can be safely caused to revolve at a much higher rate under the different set of conditions generally prevailing.

† It is improbable, however, that any new machines, even for heavy wires, would be made on this principle.

times intermediate, type—the second sheathing is performed by the cable being run through a larger machine suited for loading with the heavy class of wires constituting the outer sheath.

## MACHINERY.

We will first proceed to describe the general principle and working of that form of sheathing machine which is in most common use in the modern practice of cable construction, especially—and characteristically—in connection with the manufacture of cables of a light type.

**The Skeleton-Cylinder Machine.**\*—In this machine—a general view of which is shewn in Fig. 69—the three frame wheels A, B, C (Fig. 70) are

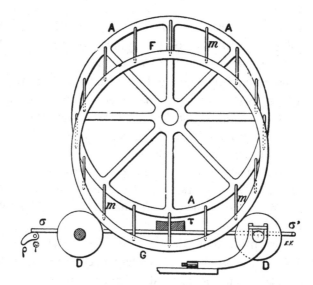

FIG. 71.—Skeleton-Cylinder Cable Machine : End View.

keyed on to the hollow shaft and turn with it. Round the circumference of the front wheel A are fitted sixteen cranks $m$ (Fig. 71) of equal length, which couple to a large ring F G, the diameter of which is equal to that of A; the centre of the ring is in the same vertical line as the centre of A, and the ring is kept in position by the two adjustable rollers D D. Sixteen rectangular horizontal iron frames H are placed, half between the wheels A and B, and half between B and C. These frames are pivoted longitudinally at

---

\* Known sometimes also as a skeleton-frame, or cylinder-cage, machine.

[PLATE XIX.

Fig. 69.—Deep-Sea Sheathing or " Closing " Machine.

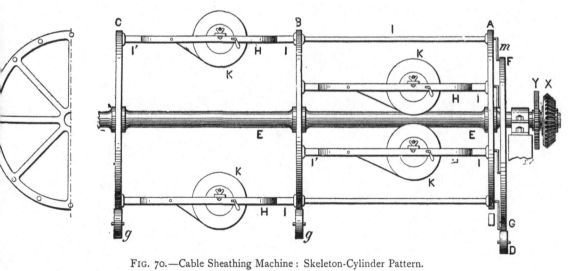

Fig. 70.—Cable Sheathing Machine : Skeleton-Cylinder Pattern.

[To face p. 434.

either end by horizontal rods, the front ends of which are secured to the crank heads in the front wheel A (Fig. 72). These frames carry the reels of wire K,* the spindles of which are horizontal, and revolve in grooves cut in the upper faces of the frames. The spindle ends J (Fig. 73) are kept in place by plates of iron M secured to the frame by two butterfly nuts, or by hinge and bolt.

**Routine.**—When a bobbin of wire is entirely wound off, the plate M is removed, the empty bobbin replaced by a full one, with the help of a crane, and the ends of the wire jointed together. With the large, soft, iron wires, this joint partakes of the nature of a weld,† no solder being used, but sal-ammoniac, or sand, as a flux, for cleansing purposes only.

The small "homo" and steel wires are, however, jointed together by "brazing," *i.e.*, a hard metal like brass being used as a reliable solder, where great heat is necessarily involved as in welding hard iron or steel.

Within recent years the joints in both classes of wire have been very successfully effected by the process of electric welding. Here, the weld should be absolutely pure, being free from sulphur, coal, dust, etc., and should, therefore, in this sense, be superior to—and more durable than—any ordinary forge weld, on account

FIG. 72.

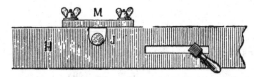
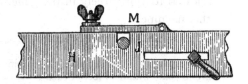

FIG. 73.

of the greater homogeneity at the joint thereby implied, with equally quick, and usually quicker, effection.‡

---

* The number of these vary very much according to the type of cable, the machine being fitted accordingly. However, the average number of wires for a single sheathing of deep-sea type is about fourteen—each from a separate bobbin.

† It is sometimes actually specified that all the joints in the iron wire are to be performed by welds. This is partly as a safeguard against pronounced scarfing, which entails skilled workmen and sharp edges such as might (in breaking) damage the core. However, it would scarcely be to the contractor's interest to attempt proper scarf-joints in the sheathing wires.

‡ Electric welds in iron wire can scarcely fail to be preferable to a *braze*, which latter must always tend to introduce chemical action when submerged in salt-water.

This system of welding electrically is now universally adopted by the
Telegraph Construction Company, as well as by Messrs Siemens Brothers,
in the operation of sheathing cables. There are those, however, who still
prefer the old forge method for various reasons.

These joints in the wires necessitate stopping the machine whenever a
bobbin has to be changed. The length of wire which each drum or bobbin
holds, varies considerably with the class of machine (*i.e.*, weight and size of
bobbin) and the type of wire. In any case, however, the stoppages are
fairly frequent; for, in the first place, arrangements are, as a matter of
course, made so that no wire terminates at the same point,* in order to
avoid the position of joints in the various wires coinciding.† A small
spring brake *d* (Fig. 74) presses against the frame or spindle of each bobbin,
thus restraining its motion and causing the wire to unwind with the requisite
amount of tension. The bobbins are made to always remain in the same

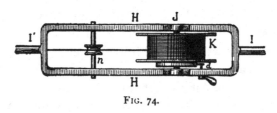

FIG. 74.

horizontal plane (so as to ensure, as already described, an absence of twist
in the wire during "laying up") by the carriage in which each rests being
allowed to work freely in the general framework, during the revolutions of
the machine.

---

* It is very often stipulated in specifications that no two joints in the different wires
are to come within 12 feet of each other.

† Moreover, a break sometimes occurs in a wire—at a joint, perhaps, after "laying
up." This more particularly applies to the large iron wires owing to their lower breaking
strain per square inch, and to their greater brittleness—so far as standing bending about
is concerned. In such a case, where a fresh joint is impracticable, the entire sheathing
has to be bound round here with binding wire (about No. 15 gauge) and spun yarn. In
some of the early cables of the "open-sheathed" type, a yarn "whipping" was applied
along the entire length in the form of a long spiral just sufficient to prevent the projection
of any broken wire such as would tend to catch into the next turn or lower flake, or into
the machinery, during paying out. This was found to meet the necessities of the case
very effectually.

In 1869 Mr F. C. Webb took out a patent (No. 3,489) for a metal tube, more especially
to obviate the piercing of the core by broken sheathing wires. This metal tube (steel for
deep-sea cables, to bear any material strain; iron or copper for shallow-water) was to be
applied as a riband outside the jute or hemp serving. After receiving an external covering
of tarred yarn, and the usual sheathing wires, the cable was completed. The principal
feature of interest in this patent was, however, the ingenious method of putting it into
effect.

On leaving the bobbin, the wire passes over a small guide pulley $n$, and on through the hollow pivoting rods to the "laying up" dies.

To avoid overloading the hollow shaft E E (Fig. 75), the weight of the two wheels B and C is taken by two pairs of rollers $g$.  As there is much wear between the wheels and rollers, the spindles of the latter are pivoted in a fork frame $f$ (Fig. 76) worked in or out by a screw $h$, and the rollers, by this means, are kept up to their work.  The roller on the side towards

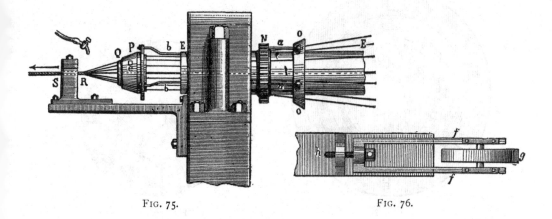

FIG. 75.                                    FIG. 76.

which the lower half of the wheel is moving usually wears away much more quickly than the other.

The actual "laying up" of the wires round the core naturally takes place at the further end of the machine, which we will now turn our attention to.   The end of the hollow shaft E E (Fig. 75) is supported by a plumber block L.    Just in front of the plumber block is keyed to the shaft a toothed wheel N, pierced with holes through which the several wires are led.  Connected to this wheel, by three rods, is a ring O (Fig. 77) which surrounds the wires and forces them into parallelism with the shaft.  On coming through the holes in the wheel N, the wires pass along longitudinal grooves cut in the body of the hollow shaft where it passes through the plumber block L.  A hollow cone piece P (Fig. 78) is fixed to the end of the hollow

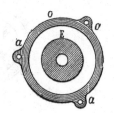

FIG. 77.

shaft E by three rods or brackets $b$ (Fig. 78) ; inside this fits tightly another solid cone Q (Fig. 79), secured by a set screw $e$, and having spiral grooves cut in its outer surface  The wires, in passing through these grooves, adapt themselves to the spiral form they have afterwards to assume, and unite round the core at the point R, where a jet of water keeps everything cool and moist.

The die block s, through which the cable passes immediately behind the point R, serves to force the wires into their places and keeps them close fitting. Sometimes it is simply an iron collar like the one in the serving machine of the same diameter as the cable, and in two halves bolted together with india-rubber cushions between. At other times, the die consists of four discs, or narrow rollers, placed crossways round the

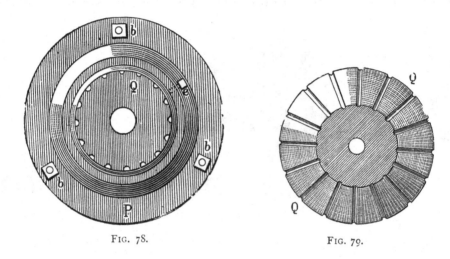

FIG. 78.                    FIG. 79.

cable (Fig. 80); the rollers can be adjusted to the diameter of the cable. Again, sometimes it is formed of two horizontal rollers, one above the other, both having semicircular grooves (Fig. 81): the spindle of the upper roller works between vertical slides, and is kept pressed down on the roller beneath, by a powerful spring.

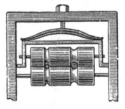

FIG. 80.                    FIG. 81.

On leaving the die, the cable takes three turns round a large drum T (Fig. 82), which obtains its rotary motion from the hollow shaft E, as we shall see further on, and which regulates the travel of the cable through the whole train of machinery—acting, in fact, as an intermediary "draw-off." It is on this drum, moreover, that the length of cable, as manufactured, is usually measured by means of a clockwork revolution indicator geared to, and controlled by, its shaft. A piece of hardened steel U—which, being edge-shaped, is called the "knife"—bears against one of the flanges of the drum on the inside, constantly forcing the turns outwards as they

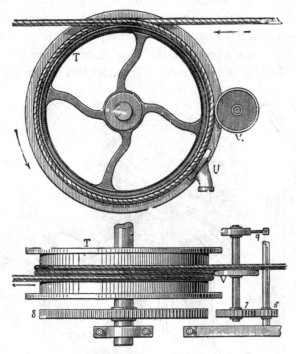

FIG. 82.—Intermediate Draw-off and Measuring Drum.

come on, and so making room for the next. The pulley V—also turned by the machinery—compresses the second turn against the drum with a force which can be regulated by slightly displacing one end of the spindle by means of the screw q.

The control of the various parts of the machine is effected by means of the wheel N (see Fig. 75), and also by two other toothed wheels X and Y on the hollow shaft E E, which are situated between two bearing brackets (Fig. 83) near the wheel F G (see Fig. 70). The bevelled cog wheel X (Fig. 84)

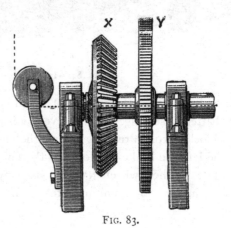

FIG. 83.

engages the cone W, the spindle of which, at right angles to the shaft E, carries two pulleys; the one marked α being keyed, and the other

$\beta$ being loose. The belt which drives the pulley $a$ passes between a fork in the form of two upright rods $\gamma \gamma'$ mounted on the upper of two horizontal battens $\delta \delta'$ working through the slides $\lambda \lambda'$ and bracketed together by the piece $\epsilon$. By forcing the battens over to the right, the driving belt is thrown off the fixed pulley on to the loose one; this stops the machine, which is set in motion again by moving the battens in the reverse direction. The use of the lower batten $\delta'$ is to work a brake on the wheel A, and so bring the machinery to rest directly the belt is thrown off. To effect this, the brake lever $\sigma \sigma$ is

FIG. 84.—Intermediate Driving Gear.

raised by a cam $\rho$, the spindle of which is rotated by a lever $\nu$, connected to the batten $\delta'$ (see Fig. 71); the lever $\sigma \sigma$ is pivoted at $\sigma'$, and the brake consists of a block of wood $\tau$ rubbing against the wheel A. When the battens are moved to the left, the brake is released—thereby allowing the machine to start.*

The toothed wheel N (Fig. 85)—by means of intermediate gearing shewn in Fig. 86—communicates rotary motion to the drum, and also to

---

* Sometimes, instead of acting on the outer circumference of the carriage (whether cage or disc pattern), the brake is placed so as to work direct on the driving shaft.

the pulley V (see Fig. 82). Finally, the cog wheel Y turns the two wheels numbered 10 and 11 (see Fig. 84). The spindle Ω of the latter is carried along the whole length of the machinery and transmits the motion, as

FIG. 85.—Intermediate Driving Gear.

we shall see later on, to the apparatus which lays on the outer covering, and also to the hauling sheave placed next the tank in which the cable is coiled.

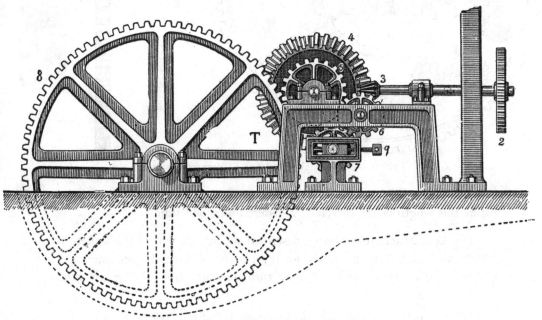

FIG. 86.—Hauling-off Drum.

In this way the onward motion of the cable is rendered exactly uniform with the rotation of the hollow shaft E, and the wheels A, B, and C. This uniformity of motion is necessary, on the one hand, to prevent the wires laying up round the core in bunches; and on the other, to keep the lay from becoming "long-jawed."

As already stated, a "counter"—to avoid complications, not shewn in the drawing—is fitted to the spindle of the drum T (see Fig. 82), and registers the length of cable manufactured. We may here remark, however, that the indications of this instrument are not very exact, for the drum very often revolves a little faster than the cable, which is held back to a certain extent by the wires. The indicator lengths are, therefore,

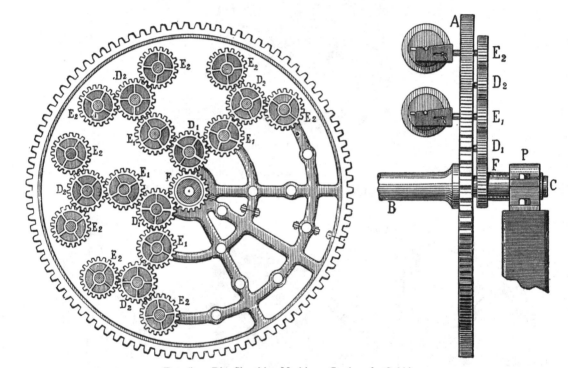

FIG. 87.—Disc Sheathing Machine: Carriage for Bobbins.

invariably too long, as the cable tends to occasionally slip back on the drum, this length being, therefore, registered a second time.*

**The Disc Machine.**—Having described the main points of the skeleton-frame machine—as used mostly for light-type cables and sometimes exclusively for all types—we will now turn our attention to the essential

---

* There is an additional objection to this "slipping" of the cable on the drum, i.e., that it tends to cause over-riding—or even kinking—of the wires at the die-block. Slipping may be, to a great extent, obviated by keeping the cable fairly tight on the drum with a sufficient strain in "drawing off," as effected by a high speed.

characteristics of the original disc machine as first devised in its original form for wire-rope manufacture by Küper and Newall. This machine is still largely used in the construction of heavy cables, for the reasons already stated.

The reels containing the wire are carried on a single cast-iron cogged wheel or disc A (Fig. 87) of large size, formed in sectors bolted together, and having a hollow shaft B C through which passes the served core. The framework of this wheel consists of radial and concentric bars which are pierced with holes in certain places, and on to which plates of sheet iron are secured. On the hinder face of the wheel are seen a large number of small cogged wheels, all of the same diameter, gearing into one another, and symmetrically arranged in concentric circles, as at $D_1 D_1 \ldots E_1 E_1 \ldots$ $D_2 D_2 \ldots E_2 E_2$. The spindles of these disc wheels pass through the holes in the framework of the wheel A, those of $D_1$ and $D_2$ being simply secured by a shoulder and nut; but the spindles of the wheels marked $E_1$ and $E_2$ are prolonged, and terminate in a fork to carry the reels of wire. The wheels $D_1$ also gear into the wheel F, which has the same diameter, and the same number of teeth as the others; this wheel is a fixture to the plumber block P, and has no rotary motion.

The sun-and-planet principle is perhaps peculiarly well illustrated in this particular form of machine, as regards its various parts being free to move during the revolutions of F.

Thus, it is easy to see that the wheels $E_1$ and $E_2$, when carried round by the rotation of the wheel A, will turn on their own centres through an angle equal and opposite to that which they describe in a given time about the centre of the hollow shaft B C; they will, in fact, retain their parallelism throughout the entire revolution of the parent wheel.*

---

* To prove this, let $m$ and $n$ (Fig. 88) be the points of contact at a given moment between the wheels F, D, E, of which F is fixed; assuming, for the sake of simplicity, the centres of the three wheels to be in the same straight line. Let $\alpha$ be the angle through which the wheel A has turned in a given time, D and E will then have arrived at the positions D' and E'. The new position at $m'$ of the former point of contact $m$ between the wheels D and F is found by making the angle F D' $m' = \alpha$: the point $n'$ at the opposite end of the diameter $m' n'$ is the new position on the wheel D of $n$, the former point of contact between the wheels D and E. To obtain $n'$ on the wheel E, we have only to make the angle F E'$n' = n'$ D' E' = F D' $m'$ $= \alpha =$ E' F E. The lines E' $n'$ and E $n$ are therefore parallel; consequently the wheel E has obtained its parallelism whilst being carried round from E to E'.

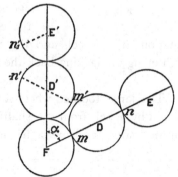

FIG. 88.—Geometric Demonstration of Sun-and-Planet Motion.

The same process of reasoning holds good when the centres of the three wheels are

From this it is evident if the spindles of the reels and the forks in which they revolve are horizontal at starting, that they will be retained in this position through an entire revolution of the wheel A; and the wires will, therefore, unwind without torsion—the result aimed at. This was, in fact, the manner in which Küper first overcame the tendency for the wire to twist during laying up.

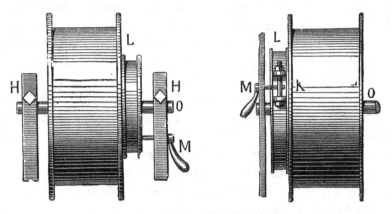

FIG. 89.—Bobbins of Disc Machine.

As in the previous machines, the reels are made of sheet iron (Fig. 89), having a steel spindle O. Here the ends of the bobbins revolve in notches on the sides of the fork; the spindle is kept in place by a screw bolt H which is turned by means of a key J (Fig. 90). A brake wheel is bolted to

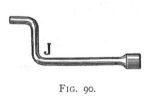

FIG. 90.

one cheek of each reel, and surrounded by a band of copper, the two ends of which are connected together by an adjusting screw K: this screw can be worked by passing a spanner through an oblong hole cut for the purpose in one side of the fork (Fig. 91). The friction of the copper band on the brake wheel checks the rotation of the reel, and the tension on the wire can be regulated by tightening or slacking back the adjusting screw K.

The large toothed wheel A is geared to a pulley N (Fig. 92) which is driven by a belt from the shafting. This pulley (Fig. 93) revolves loosely on the shaft Q, and has a friction cone on one side. Sliding along a short

---

not in the same straight line, and also when the intermediate wheel D is of different diameter to the other two wheels; in the latter case we must take into consideration the lengths through which the circumferences turn, instead of the angles they describe.

feather on the shaft is a friction cup which engages the cone, a frame U (Fig. 94) is pivoted on the rod X X, and receives the bolts V V. By forcing this in one direction, the cup and cone are geared together, causing the

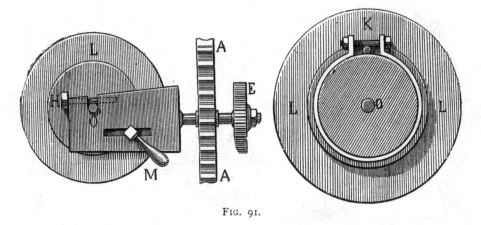

FIG. 91.

shaft Q to revolve with the belt pulley N ; by reversing the movement the belt pulley N is thrown out of gear, and the outer surface of the friction cup $s^1$ made to rub against wooden brake blocks Z Z, by which the momentum of the large wheel is soon absorbed.

The hauling-off drum (Fig. 95) in this machine differs only in one particular from that already described in connection with the cage machine. The wheel $a$, which presses the centre turn of the cable against the drum, is not driven, but simply rotates by friction with the cable. To prevent displacement of the wheel sideways, a second wheel $b$ is mounted on the same spindle, and grooved so as to engage one of the flanges of the drum. The bearing blocks of the spindle are enclosed in rectangular frames which admit of their being moved, to a small extent, to or from the drum by the screws $c$; the wheel is adjusted in this way according to the different sizes of cable in question. The drum

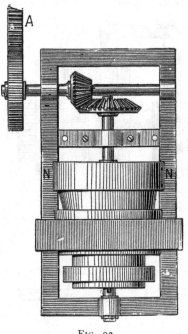

FIG. 92.

receives its motion by intermediate gearing (Figs. 96 and 97) from a toothed wheel on the hollow shaft B C in front of the "laying-up" gear.

When this type of machine is adopted for the outer sheathing of shore ends it is naturally, as a rule, of a heavier type than that used for "laying up" single-sheathed intermediate cable. Besides being larger and heavier in its

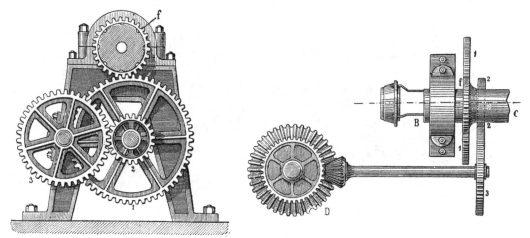

FIG. 96.—Driving Gear.  FIG. 97.

main parts, there are often other special arrangements for the construction of heavy (shore-end) cables, as regards the laying up of the outer sheathing.

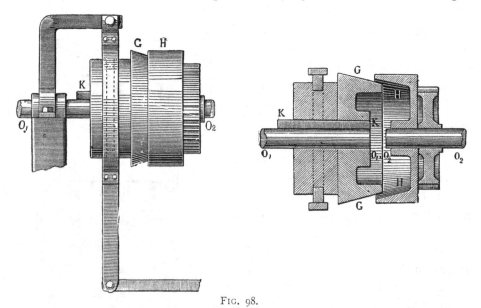

FIG. 98.

Thus the gearing and throwing out of gear of the machine is very usually accomplished by a cone and cup friction clutch G H (Fig. 98), somewhat similar to the one just described, mounted on a shaft parallel to

[PLATE XX.

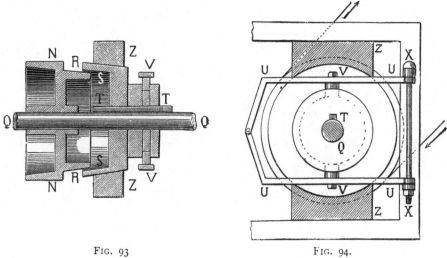

FIG. 93

FIG. 94.

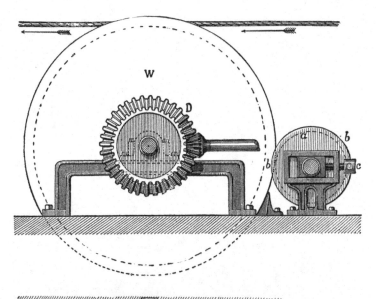

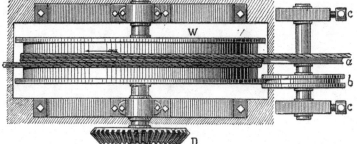

FIG. 95.—Hauling-off Drum.

the driving shaft. This shaft is in two pieces $O_1$ and $O_2$, in line one with the other; the friction cup H is keyed to the shaft $O_2$, and the cone portion K slides along a feather on the shaft $O_1$.

The brake is done away with, the resistance of the gearing wheels being

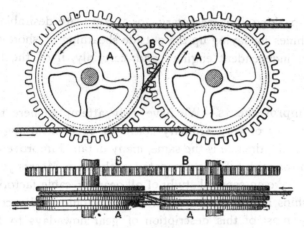

FIG. 99.—Hauling-off and Measuring Drums.

sufficient to rapidly overcome the momentum of the moving parts when connection with the driving shaft is broken.

As shore-end cable is too stiff to grip tightly round a small drum, it is

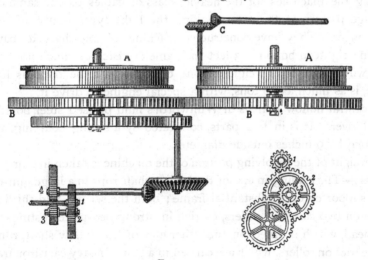

FIG. 100.

usually passed round two double-grooved drums A A, in letter S fashion (Fig. 99), the drums being placed close to one another in the same vertical plane, and turned opposite ways by two cog wheels gearing together. In Fig. 100 the method of communicating the motion of the shaft $O_2$ to the drums is shewn in detail. One of the cog wheels on the drum shafts gears into a pinion, which, by means of the bevelled cog C, transmits the motion to the machinery for the outside covering—usually of jute or hemp yarns.

Generally speaking, it is necessary—or at any rate desirable—to drive all heavy machines for laying up the outside sheathing of shore-end cables direct from an independent engine; and, certainly, if of the heavy disc pattern.

**Modern Improvements.**—We have only attempted here to describe the main principles of very ordinary forms of cable-sheathing machines. Though the main principle is the same, many detailed improvements have of late been introduced. These emanate largely from Messrs Johnson and Phillips—who make an especial study of all gear for cable factories, as well as for cable ships and general submarine telegraphic appliances. They, indeed, supply most of this description of gear nowadays to H.M. Post Office, as well as to other Governments and foreign cable contractors— indeed, everywhere outside the ordinary large cable contractors at home, who naturally supply their own wants, as occasion requires.

Within recent years Messrs Johnson and Phillips have been in the habit of making the machines for the heavier class of cables on the same cylindrical cage pattern as those for making the light types—only, of course, much stronger. They have constructed machines of this character capable of carrying eighteen bobbins (3 feet in diameter), each of which hold about a ton of wire. The total weight of one of these, with the bobbins loaded as above, is as much as 70 tons, with a special engine to drive it.

The central hollow shaft is of Whitworth's fluid pressed steel, bored and turned all over. It is in two parts, connected by a strong coupling, and is 28 feet long by 9 inches outside diameter.

The weight of the revolving portion of the machine is taken in four places, as follows :—The leading-in end of the hollow shaft runs in a large gun-metal bearing, supported on a substantial frame; both the second and third large discs run on two cast-iron rollers, carried in strong bed-plate castings; and the lay-head, which is fixed on the other end of the hollow shaft, runs on four large friction rollers, which are carried in a pair of heavy cast-iron frames. These bearing rollers are all adjustable, and they enable the machine to work very freely, and with very little loss of power by friction.

Turning again to the machines for light-type cables, it is now very usual for the central hollow shafts which carry the revolving part of the machine to be provided at each end with heads which run on adjustable anti-friction rollers, with the centre discs also on similar rollers. Thus, the machine requires comparatively little power to drive it at a high speed.

Again, the bobbins for all classes of machines are now invariably maintained stationary (as regards their plane) by means of eccentric gearing, together with spindle and cranks, in place of the old and objectionable spur-wheel gear. They are provided with tension bands or "rope frictions," and will usually each hold about 3 cwt. of wire.

Sometimes even the light machines for deep-sea types are—for security and convenience—run direct and independently from a separate engine, thus doing away with the ordinary counter-shafting system for several machines worked from a common source of power. This plan is, at any rate, always employed for driving the heavier machines.

To meet the requirements of special cables (more particularly for torpedo, telephone, and electric-lighting cables, such as require to be handled and moved about a good deal), machines have within recent years been devised to lay up an extra large number of wires beyond what could be managed by the ordinary machines.

In the latter, it has been already shewn that there is a limit to the number of bobbins which the cylinder frame, or disc, can safely carry (whilst revolving at a fair speed) both from a straining and centrifugal force point of view.

If, however, a large number of bobbins be divided amongst two or more frames, or discs, placed one behind the other in tandem-fashion, the requirements are met by the wire from the bobbins of the back frame, or disc, being drawn between two of the front carriages, so that wires from each are alternate with uniform tension and lays.

Messrs Johnson and Phillips have lately designed a triple tandem machine (with three skeleton carriages) on this principle to meet special cases.*

A novel feature in this machine is that the three sections are driven separately by chain gearing (as employed in bicycles) instead of the usual counter-shaft belt gearing, or spur-wheel gear.

**Lay of Wires.**—The wires are laid up vertically (as above described)

---

* This is mainly intended for laying up a number of cores, together with the ordinary intermediate jute, or hemp "worming," as well as for the exigencies of heavy electric-light conductors, composed of a large quantity of wires—up to as many as 127, for instance—such as could not be laid up by a vertical machine.

rather than straight (1) in order to hold them together ; and (2) for purposes of a certain amount of elongation to meet the case of a strain coming on the cable.*

For some years in the early days of submarine telegraphy (following in the steps of rope-making) the different contractors at various times laid up the wires opposite ways. This though advantageous, when carried out systematically for purposes of identifying two cables lying side by side (the lay of each being properly noted), was, on the whole, a distinctly undesirable state of affairs. It involved the chance of difficulties contingent on subsequent repairs, in requiring to effect and maintain splices between wires of opposite lays.

Fortunately, nowadays, submarine telegraph cables are invariably made with what is most commonly described as a left-hand lay as regards its main covering of iron sheathing wires†—i.e., the reverse of most hempen ropes.

How the method of nomenclature as regards right- and left-handed lays came to be first applied does not appear to be very clear. However, a left-hand lay is now usually taken to imply that the wires tend to go towards the left, from the person holding the cable.‡ Mr F. C. Webb once suggested (*Electrician*, 18th May 1880) a more scientific phraseology in the expressions " N.W." and " N.E." lays in place of " left-" and " right-handed " lays respectively.§ It has not, however, replaced the previous rule-of-thumb method of description.

There should certainly never be any question about the *direction* of the lay of a cable, for the reasons given ; and if necessary this should be clearly specified in a manner that is unmistakable.

To turn now to the *length* of the lay, this is always made as long as possible in the case of a submarine cable, the limit being that the wires must not be liable to open out, by not lying sufficiently close together. The lay is kept long for purposes of tensile strength, and to avoid the strangling

---

* As a matter of fact the sheathing wires of a completed cable stretch but little more under a given strain than an iron rod would do.

† The direction of the lay of a cable is always taken as referring to the lay of the *sheathing wires* in the usual iron-armoured cable.

‡ Another explanation of the nomenclature is to say that a left-hand lay is similar in its direction of lay to a screw with a left-hand pitch.

§ This is based on placing a piece of cable with its axis vertically, and imagining its axis to be a North and South line ; then the direction of the wires as seen will either run in a direction approaching North-East and South-West, or in the reverse direction—*i.e.*, North-West and South-East.

of the core, which is always a danger where a strain is put on a cable with a short, or "quick," lay.*

Though of distinct importance, engineers do not, as a rule, specify the length of lay of any of the separate coverings in a cable contract. Generally speaking, they leave such details to the contractors, who are usually—by close experience—the best judges of what is most advisable in these matters.

However, it may be remarked, as regards the sheathing wires, that the length of lay depends very much on the type of cable. It is usually longer for the larger wires of heavy-type cables than for the comparatively small wires of the lighter. Thus, the sheathing lay for deep-sea cables varies from about 9 to 11 inches; whereas with a shore-end type—composed of, say, 18 No. 1 S.W.G. wires—it may be as much as 18 inches.

Obviously there are two points which govern the length of lay obtained, viz. :—

(1.) The distance of the bobbins, on their universal carriage, from the lay, or closing, plate.

(2.) The relation existing between the speed at which the revolving frame, or disc, is run, and that at which the cable is hauled through the machine.

In practice the length of lay is adjusted as required solely by the speed of "draw-off." That of the revolving frame, or disc, is maintained throughout at the highest speed at which it is found safe to run, depending on the circumference and the loading weight. No attempt must be made to adjust the lay by the form of the machine alone. Thus, the size of the disc is pre-

---

* Incidentally, moreover, the shorter the lay, the greater the length of material used for a given cable. Mr F. C. Webb has pointed out (*Electrician*, 9th October 1880) a method for finding the length of a single convolution of wire, the diameter of the cable and wire and the "length of lay" being given.

If a line be drawn as a helix round a cylinder, so as to go exactly one turn round, the line will evidently be the hypotenuse of a right-angle triangle, of which the circumference of the cylinder is the base, and the distance between the two ends of the helix is the perpendicular. If, therefore, D is the diameter of the cylinder, and $l$ the distance between the two ends of the helix, the length of the helix will be expressed by—

$$L = \sqrt{(3.1416\,D)^2 + l^2}.$$

In the case of a telegraph cable, the length $l$ is the length of cable made by one turn of the machine, and is called "the length of lay." To get the length of the wire in one convolution, we must take for D in the formula the diameter on the centre line of the wires. Thus if $d$ is the diameter of one of the wires, and $D^1$ the outside diameter of the cable; then, to get D, we must subtract $d$ from $D^1$. Then the complete formula for obtaining the length of one of the wires in making a complete convolution becomes—

$$L = \sqrt{(3.1416(D^1 - d))^2 + l^2}.$$

determined by other considerations altogether, which latter have been already briefly alluded to. Similarly, the angle of the wire at the lay-plate must be a constant quantity. It is arranged to avoid all chance of what is known as "crippling" the wires by straining, ricking—or even of cutting into the wires due to too sharp an angle by the revolving disc and frame being too near to one another.

Indeed, as already stated, the lay must only be determined by the comparative speeds of the bobbin frame and "draw-off." Moreover, as the former is maintained constant—*i.e.*, the highest speed which is safe according to the circumference of the frame and the weight as loaded—the lay is practically adjusted by the speed of "draw-off" alone.

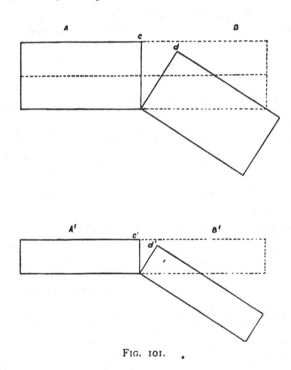

FIG. 101.

**Points in Construction.**—The revolving disc or cylinder must not have too small a circumference; for it is essential that it should carry all the bobbins at a certain distance apart to avoid their respective wires overlapping, or even touching, previous to arrival at the lay-head. This condition would, however, also be governed by the distance between the bobbins and the lay-head; besides being aggravated if the wires are allowed to "dip" at all.

In fact, the circumference of the carriage or disc is to some extent fixed at the largest which can be safely run at a certain maximum speed according to weight and number of bobbins. The distance of the bobbins from the lay-plate is arranged accordingly to meet, on both hands, the above previously mentioned requirements. Moreover, it depends, to a great extent, on the size of wire to be stranded, as well as on their number.

For large wires the angle at the lay-head must be longer (*i.e.*, a longer stretch between the bobbins and the lay-plate must be provided) than what can be permitted for smaller gauges, for the reason that large wires will bear less lateral strain than small wires. This is explained by the fact that wires with a large surface break more easily under bending. Fig. 101 will render the latter statement clearer.

Here A B is a piece of metal twice as thick as $A^1 B^1$. Therefore, $d$ is separated from $c$ to double the distance that $c^1$ is from $d^1$. Thus, the molecules being further separated—in fact, twice as much—the wire A B will break, whereas $A^1 B^1$ would bend to double the angle before breaking. It is also explained by the greater specific strength (or tenacity) of the material of which the smaller wires are composed. Indeed, the large wires, even though drawn from the same quality of iron, cannot possibly be made to give the same strength specifically. The relation borne in this respect is roughly one of inverse proportion to the gauge.*

Again, for big wires, the distance between the lay-plate and the revolving disc must be greater (*i.e.*, the machine longer) for another reason—viz., to permit of the revolving disc, or frame, being larger, without altering the angle of the wires, which should be greater rather than less, on account of the extra brittleness of the larger gauged wires. Such wires, moreover, require more straightening out in the machine; and this can only be attempted where there is a good length between the bobbins and the lay-plate, and where there is no chance of the angle being too sharp at the lay-plate. Again, the disc (or cylinder frame) requires to have a bigger circumference in the case of a heavy-type cable, so that the bobbins may be further apart. The latter is necessary in order to contend with the greater chance of overlapping before they reach the lay-plate, partly owing to their increased weight inducing an extra tendency to drop. This in itself must be corrected—if only for other reasons—by maintaining a certain longitudinal strain.

---

* This is due to the circumstance that any given process of annealing or tempering only affects a given wire to half the depth that it does one of half the section. Thus any material whose strength is equally distributed throughout, as here provided, being more homogeneous, has greater tenacity. It is on this account that suspension bridges are supported by stranded cables of many small wires in preference to the same quantity of iron in solid form. It may also be added there is always an increased chance of more than one flaw in the metal occurring at the same spot the bigger the gauge of wire.

**Formula for Weight of Iron Wire in a Cable.**—Clark and Sabine have given the weight of iron wire in a sheathed cable, allowing 3 per cent. for lay and waste, as approximately—

$$\frac{d^2 n}{6806}.$$

Where $d$ = diameter in mils. of the type of wire adopted; and $n$ = the number of such wires; and 6806, a constant attained empirically.

The diameter of any iron wire weighing $w$ lbs. per N.M.—

$$= 7.91 \sqrt{w} \ldots \text{mils.}$$

The diameter of the completed sheathing laid straight would be—

$$1 + \frac{\operatorname{cosec} 180°}{n} \times \text{diameter of wires.}$$

Where $n$ = number of wires adopted.*

When, however, the wires are laid up spirally, as in a cable, the cylinder of iron wires would be increased in bulk, for the section of each wire then becomes elliptical, and, therefore, the major axis takes the place of the diameter of the wire. This increase may be actually calculated from the angle of the lay; but in practice it is not unusual to allow, say, from 5 per cent. for twelve wires to $7\frac{1}{2}$ per cent. for twenty-four wires—*i.e.*, 2 per cent. for each additional wire.†

**Trial Specimens.**—It is also very customary to have machine-made specimen lengths of a fathom or so made up with differently composed sheathings previous to final determination of the number and type of wires, and of their lay. The same sample length will, of course, do for examining the inner and outer servings with a similar object.

---

* In designing the type according to prevailing conditions, the most suitable class and number of wires are considered together; though, to some extent, the type of wire is a predetermined quantity, very often.

† Deep-sea cables vary a great deal as regards the number of sheathing wires according to depth and build—in fact, as much as from twelve to twenty-four.

The intermediate type has usually from ten to twelve wires, their gauge differing according to whether it is "heavy" or "light" intermediate.

Shore-end types may be composed of from ten to twelve wires for the inner sheathing (of light intermediate or deep-sea types), and fourteen for the outer sheathing. The latter is, however, occasionally a three-strand wire, and there may be twelve such strands, making the total number of wires in outer sheathing $3 \times 12 = 36$.

## Section 5.—Compounding and Outer Serving.

As previously mentioned, where each wire has been individually compounded, or compounded and taped, the sheathing, when laid up, does not require any further compounding, but is at once covered with its outer serving.

**Bright and Clark's Cold Compound.**—When, however, neither of the above processes form part of the arrangements, the surface of the laid-up sheath is coated with Bright and Clark s compound previously described.

In the patent of Sir C. Bright (No. 338 of 1862), previously referred to, special arrangements were made for the application of the compound whilst hot so as to meet the difficulty which was found to exist of the hot compound damaging the insulation by the compound running over any one portion for a certain length of time during the various stoppages involved, or—still more—of the cable remaining stationary inside the vessel containing the molten compound. These arrangements consisted of various devices by which a fine stream of compound (such as would become sufficiently cooled in its passage through the air) was caused to pour over the cable in a perfectly regular manner, its flow over the cable being automatically shut off whenever the sheathing and draw-off machinery was stopped.

The cable was then passed through semicircular rollers—a stream of water being poured over them—by which the coating whilst in a plastic state was thoroughly pressed into all the interstices of the wires.

This compounding method—under the above patent of Sir C. Bright—has ever since been universally adopted, in one form or another, at all the cable factories.*

Here the coat of compound was applied at the same time as the sheathing and the subsequent outer serving, by a part of the same machinery; and the delay, cost, and damage entailed by the recoiling was avoided by making the compounding and laying on of the outer hemp, jute, or tape, serving one operation.

---

* An additional point which favoured the employment of this compound is the fact that the outer covering of hemp and pitch slowly perishes, and the wire rusts, whilst various marine growths weight the cable. Thus increasing weight and decreasing strength often end in its fracture.

For further particulars, see *Journal of the Society of Arts*, vol. xlii., No. 2152, Note on " Telegraphic Communication between England and India : its Present Condition and Future Development," by Charles Bright, F.R.S.E.

Fig. 102 represents a very commonly adopted form of the machine for effecting the first coat of compound outside the iron wires—being the pattern originally designed by Sir C. Bright.

In such an application, it is very usual nowadays, in order to avoid overheating of the core, for it to be made with compound in a cold state—*i.e.*, cooled down, after mixing, to an ordinary temperature of about 60° F., at which it has the consistency of treacle.

The main feature in the composition of the cold compound—as apart

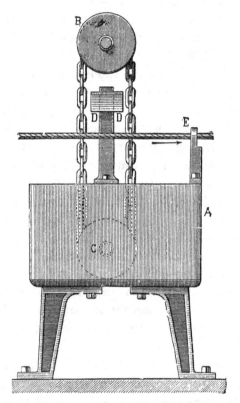

FIG. 102.—Apparatus for Applying Cold Compound.

from what it is, as a rule, when hot—is constituted by the tar used as a mollient being *Stockholm* tar instead of coal tar. Stockholm tar is employed here as being of a lighter consistency, which is necessary in order to make the mixture "run" sufficiently when cold.*

This compound, though awkward stuff to deal with, ensures an absence

---

* In fact, the ordinary (hot) compound with coal tar would not run at all, but for the application of heat.

of damage to the insulation by overheating. Again, being more sticky than a hot compound, it serves as a better adherent for the outer covering to the iron wires.

With further reference to Fig. 102, the bituminous mixture is contained in the iron tank A, to which is fixed an elevator arrangement, taking the form of a large-linked endless chain which works round two fixed pulleys B and C.* Some of the compound in being drawn up by the chain falls from the upper pulley into the inclined "shoot" D (so fixed to serve as a guide for the compound), down which it runs on to the cable. A die at E, shewn in detail by Fig. 103, regulates the thickness of the coating,† forcing a portion

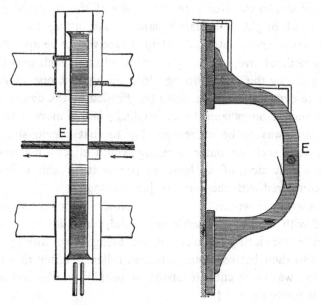

FIG. 103.—Die of Compounding Apparatus.

of the compound between the interstices of the wires, and causing all excess to fall back into the tank.‡

---

* This is found an excellent sort of elevator for the purpose, inasmuch as the chain, in continually passing through the compound, ensures it all remaining properly mixed, including the silica, if any be used. When the machine has stopped and the compounding shoot is thrown on one side of the cable, it is important that the chain and pulleys should still be kept continuously running—partly with the above object, but also to prevent the compound setting on the upper pulley and that part of the chain outside the tank, as this would clog up the apparatus.

† The proportion in bulk of compound to hemp or jute in one covering is, not uncommonly, nearly as much as 2 to 1.

‡ Sometimes a rubber clamp, fitting the cable fairly closely, is used—either instead of, or in addition to, this die.

By a modified plan (due to Messrs Johnson and Phillips) the elevator takes the form of a disc. This pattern is especially suited for *cold* and thin compounds, as above. The disc is made to revolve through the compound tank at a fairly high speed. The apparatus is closed in—except in front—in order to prevent the throwing off of the compound by centrifugal force. This is a very simple and compact machine, though not perhaps so well adapted for general use as the one previously described.

**Outer Canvas Taping.**—In 1872 Messrs Johnson and Phillips* introduced the plan of applying Hessian (canvas) tape as an outside preservative covering and binder to the iron wires,† in place of the hemp or jute yarns forming a part of Bright and Clark's patent. According to Johnson and Phillips' provisional specification, the strip of tape was to be applied spirally either with or without overlap, but preferably without, each turn forming a sort of butt joint by the edges fitting close up against one another, thus securing a more even surface than could be afforded by the overlap method. One or more such tapings were to be applied; and if more than one, the succeeding spiral was to be of reverse lay to that previously. It was claimed here that such an outer covering would effect a material saving in cost, with a reduction of at least 20 per cent. in the coiling space required, as compared with the hemp or jute serving.

The canvas was to be woven out of jute threads, or some such suitable fibre, and saturated with Bright and Clark's compound, which latter operation was to be effected by the cloth (of convenient size) being drawn through a steam-heated bath and then between steam-heated rollers. After this the cloth, so compounded, was to be cut into strips of suitable widths and wound on to the bobbins ready for use. ‡

The manner in which the canvas cloth was prepared in practice by

---

* Mr W. T. Henley is also said to have taped cables externally at an early date—possibly based on this device.

† Besides acting as a vehicle for the compound, by binding the wires firmly together with this prepared preservative tape, it was thought that insects would be as well if not better warded off than by the hempen casing. This outer serving was also regarded as an efficient safeguard against broken wires—at joints, for instance—getting loose during cable-laying operations.

‡ This plan was, in fact, somewhat similar to that set forth in the Gray and Hawkins patent, already alluded to, except that it was for applying tape to the outside of the laid-up cable, instead of to each individual wire ; and, moreover, that it was further mentioned as being applicable as a serving to insulated cores. For the latter purpose, however, it never came into practical use. The Gray and Hawkins patent stipulated "at least two tapings," whereas this speaks of "one or more."

Messrs Johnson and Phillips may be gathered by studying Fig. 104.   Here the cloth is drawn between the two hollow rollers A and B.  The compound,* contained in a double-bottomed trough D, is raised by the third cylinder C parallel to, and in contact with B.   Steam circulates inside the hollow rollers and the double bottom of the trough, to heat them and prevent the compound solidifying.

The compound is pressed into the fibres of the canvas whilst the latter is passing between the rollers, and all excess falls back into the trough. The sheet of canvas when cold is rolled up on itself and cut into strips of the required width—usually from 1½ to 2 inches—by a circular saw, or knife.

Subsequently (in 1876) Messrs Johnson and Phillips patented a plan for

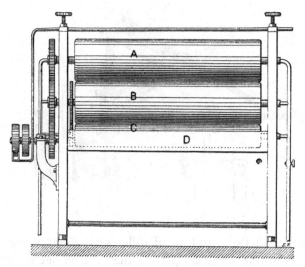

FIG. 104.—Machine for Preparing Canvas Tape.

saving the operation of winding from the spindle—on which the cloth has been cut—on to bobbins.   This consists in the roller, composed of wood, being cut right through with the tape into the form of discs convertible into a sort of reel (called a "tape-head") by having plates fitted to each side and a spindle run through them.   This plan has been already alluded to with reference to the individual taping of each iron wire.†

---

* Sometimes instead of Bright and Clark's compound, merely tar has been used—usually Stockholm (vegetable) tar.   Again, at other times quite different mixtures have been employed.   In the 1880 "Anglo" Atlantic, the outside covering of canvas strip was previously impregnated with "stearine," *i.e.*, ozokerite compound.

† The preparation of the cotton tape for each iron wire, and the canvas tape for outside the entire cable, are performed at most factories in precisely the same way, whatever the method be.

Messrs Johnson and Phillips have been in the habit of supplying to the Telegraph Construction Company (for use in their cables) large quantities of canvas tape, so prepared, on disc bobbins—ready for application to the cable-making machine.

The manner in which the canvas tape is applied round the sheathing immediately after the cold compound has been applied (if there be any) is shewn by Fig. 105.

The end of the compounded strip is drawn from its reel G into the hollow shaft or tube A B, supported by two standards P, and having a longitudinal opening $m$ at the opposite end, about a foot long and 2 or 3 inches in width. The shaft is rotated by driving a belt on the pulley S, to

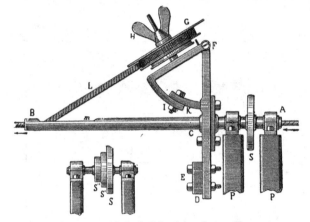

FIG. 105.—Method of Applying Canvas Tape.

which two other pulleys S' and S'' of different diameters are sometimes joined, so as to vary the speed by placing the belt on one or other of the pulleys. To the hollow shaft is fixed a collar C with two arms, one of which, D, has a counterpoise E, and the other, F, carries the "head" G on which the tape is wound. The tape-head is secured by a bolt and thumb screw H to a quadrant I, pivoted at F, and adjusted to the required angle by a bolt and nut at I. This bolt works in a slot in the bent bar K, which is itself fixed to the upright arm C F.

The strip of canvas L enters the hollow tube through the opening at $m$, and, carried round by the rotation of the shaft and reel, winds on the cable, which at the same time is being drawn steadily forward through the shaft. Under the combined effect of the two movements the strip is laid round the cable in spiral turns.*

---

* The tape-heads hold on the average about a mile at a time. When this is exhausted a freshly charged "head" is fitted to the machine, the end of the new tape being united to the end of the tape already on the cable by means of ordinary cable compound.

The requisite tension of the strip L is given by means of a small brake attached to the reel G.

To regulate the length of lay the angle of the lead at F is set as required.* This naturally varies with the width of tape adopted where there is no overlap. At different factories different widths of tape are used, varying from 1 to 2 inches. The length of lay of the tape varies correspondingly.

When, as is usually the case, a second taping follows this in the reverse direction, it is not the custom to apply the tape with any overlap, the two edges of each lap being made to fit against one another. By this arrangement, practically speaking, a perfectly even surface is procured, besides the coiling space involved being reduced to a minimum.

**Bright and Clark's Hot Compound.**—A second coating of Bright and Clark's compound is now applied hot over the compound or tarred canvas tape, after the latter has been wrapped round the cable and in the manner just described.

In this case it is a *hot* compound that is used, there being practically no risk—especially by this method of application—of the insulation being damaged where a bad conductor of heat, such as jute or hemp, intervenes. For purposes of firmly sticking the tapes or yarns together, cold compound would still always be better, but this is not of so much vital importance as in the instance of adhering the first layer to the iron sheathing. Moreover, cold compound being such troublesome material to deal with, its use is always avoided where possible—especially as when hot the compound is much cleaner.

The molten mixture is contained in the tank A (Fig. 106), the double bottom of which (not clearly shewn) is heated by steam from the pipe Q. As in the previously mentioned apparatus for applying cold compound, the method of application here given is that of a chain elevator. The fluid compound is raised by an endless chain of thick links moving round the pulleys C and D. Most of the compound so lifted falls from the pulley C on to the inclined shoot E, and so on to the cable.†

---

* The length of lay is also, very commonly, controlled by the revolutionary speed of the tape-head. The speed of "draw-off" cannot be adjusted for this purpose, as this has to be regulated according to the requirements of the lay of the iron wires, which again depends on their number.

† It will be seen that in this machine the chain is made to run in a diagonal direction instead of upright as in that for applying the *cold* compound. By this means, besides serving as a distinction, the less adhesive and heavier compound is less liable to inadvertently tumble off as it is drawn up from the tank.

When, for any reason, the cabling machine requires to be stopped, the workman who is attending to the compounding of the cable draws aside the inclined shoot by means of the lever L, so that the hot compound falls clear of the cable—and is, in fact, diverted into the tank again—thus avoiding heating it up at one spot during stoppage.* The revolving chain is, however, kept running continuously (throughout all stoppages) to prevent the compound forming solid blocks around the links and the upper pulley, such as would tend to choke the apparatus on re-starting. On resuming "cabling" operations the compound "shoot" is again drawn over the line of cable.†

By way of regulating the thickness of the coating of compound a man is commonly stationed just beyond the compound tanks with a pair of tongs

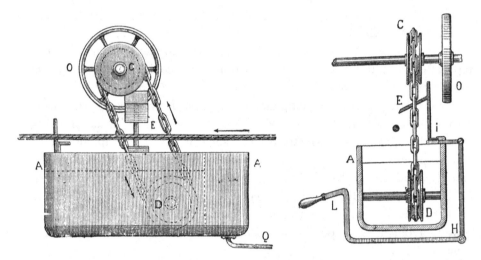

FIG. 106.—Hot Compounding Apparatus.

(Fig. 107), through which the cable is made to pass, and which therefore act as a die. These tongs require to be heated over a furnace from time to time in order to be effective in this way ; and it has been found that, on their application to the cable, they are occasionally liable to overheat it. To obviate this

---

* Moreover, if the stoppage is likely to last any length of time, it is very usual to cover up that part of the cable immediately over the compound tank—by means of a half-tubing of rubber, for instance, or by some other bad heat conductor. Failing this, it would be necessary to temporarily shut off the steam in the compound tank.

† The disc form of elevator, previously described, is also sometimes here employed. It is, however, less suited for hot compounds, which, owing to their greater consistency, require to be more continuously and thoroughly agitated. Moreover, the chain is usually found to draw the thick compound up better.

objection, and to more thoroughly and completely round off the surface of the compound,* at the Silvertown Works a superheated scraping die is put into gear with the rest of the compounding apparatus. This apparatus, introduced by Mr F. Hawkins some ten years ago, is shewn at P in Fig. 108. The die is formed in two halves, each half being steam-jacketed in connection with a flexible steam-pipe. The two parts may be drawn close

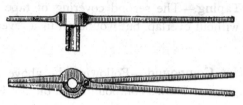

FIG. 107.—Hand Compound Die.

up to one another, or may be drawn apart from each other. To the back half is attached the trough or shoot, which catches the compound from the chain. The two halves are connected by levers as shewn, and in their normal position are kept apart by means of a weight on one of the levers. The levers are connected by a rope which is led overhead to the man whose

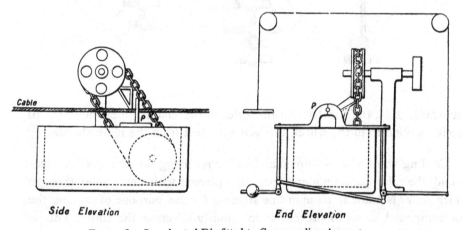

Side Elevation        End Elevation

FIG. 108.—Superheated Die fitted to Compounding Apparatus.

duty it is to attend to this part of the cable machine. When the machine is started, the man pulls on the rope, which action—by lifting the weight— brings the two halves of the die together and the compound flows on to the

---

* One of the features of Bright and Clark's cable compound is that of imbuing the completed cable with a smooth exterior, thereby reducing the coefficient of friction in picking-up operations. This is especially to the point in the case of a rough hemp-covered cable, and still more if it have an outer casing of hemp *cords*.

cable, any superfluity being readily taken off by the steam-heated die.　On the machine stopping, the man lets go the rope and the die separates.　The superfluous compound which has not passed the die is then pulled back by hand (previously dipped in water) whilst the compound is still hot.

This apparatus is applicable whenever the cable is covered with hot compound—*i.e.*, in every case except where the bare sheathing is uppermost.

**Second Outer Taping.**—The second covering of tape is applied in a similar manner and without overlap, but wound on in a reverse direction to the first taping.

**Final Coat of Hot Compound.**—Following the above again comes a third coating of Bright and Clark's compound laid on hot in precisely the same way as the last coat.

**Details of Compounding Apparatus.**—To obtain continuous rotation of the chains in the different compound tanks (for reasons already explained),

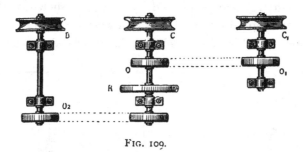

FIG. 109.

the wheels $O$, $O_1$, $O_2$ (Fig. 109) on the axles of the chain pulleys B, C, $C_1$, all receive motion from the wheel R driven by a belt from the main shafting.

**Cooling the Cable.**—Immediately after receiving the last coat of compound, the cable passes under a long pipe pierced with small holes, through which water is made to issue in fine streams for the purpose of cooling the hot compound, so as to prevent it injuriously affecting the core.　This is shewn in Fig. 110, which serves also to give a general idea of any "deep-sea"[*] sheathing and covering machine, in which the outside covering is

---

[*] Nowadays, neither shore-end nor heavy intermediate cables are taped outside, as a rule.　Instead of this, these types are usually enveloped in jute or hemp yarns, the yarns (when such heavy types are in question) being less readily damaged by cable machinery during paying out as well as by rocky bottoms and hard ground generally.　Otherwise, in principle—though not in detail—this illustration would serve, equally well, to give a good general idea of the heavier type of machinery used for laying up heavy (shallow-water) cables.

[PLATE XXI.

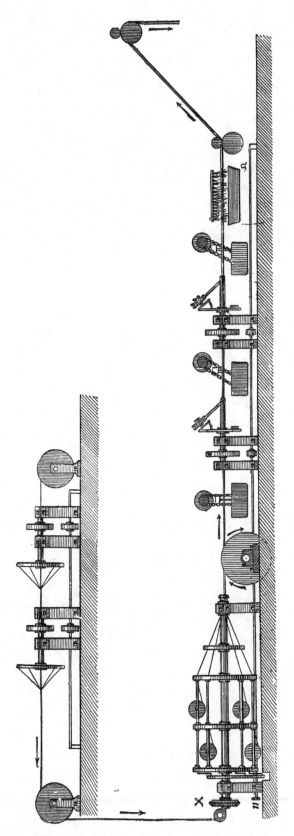

FIG. 110.—General View of Cable Manufacture.

constituted by two layers of canvas tape, besides the arrangements of the various parts. This illustration also shews the served core coming from an upper floor, all these various coverings, in this instance, being applied at one operation ; or rather, one behind the other during a single " draw-off." *

**Relative Merits of Outer Tape and Yarns.**—The main object of the outer covering of an ordinary submarine cable—of whatever it be composed—is that of holding the wires together, and of preserving them from decay.

The above system of canvas tapes applied as an outer covering to sheathed cables has been very largely used for a number of years ever since its first practical introduction by Messrs Johnson and Phillips for this particular purpose.

It has certain advantages over the previous hemp or jute yarn outer covering. Thus, the canvas (Hessian) tape is more durable, as a rule, than the loose yarns. Secondly, it·keeps broken wires better in place, partly owing to the fact that it can be applied with greater tension. Thirdly, it is perhaps rather better able to resist the attacks of living submarine organisms† for a longer time, at any rate.‡

On the other hand, tapes involve a more rigid and less pliable cable, which is an objection from a contractor's point of view—as regards general handling up to the time of actual submergence.

Again, yarns are found to adhere together better than tapes.§ They are, in fact, a superior vehicle for the compound, owing to it being better able to percolate between the threads.

For these reasons—and still more owing to the improvement in the durability of jute and hemp yarn—canvas tape is less used nowadays than it was, a return having been made to hemp yarns as well as to jute for this purpose, not to mention the introduction of hemp cords.

---

* Where the core is brass-taped, however, either this forms a separate process, or where the inner serving is applied at the same time, the sheathing and outer serving must together form distinct operations.

† Hemp yarns are somewhat readily demolished by the teredo, and others of its sort.

‡ Several of the cable-owning companies have found it a good plan to keep a stock of this Hessian tape at their stations, as well as on their repairing ships, all ready for renewing the outer servings of cables picked up for repairs. The gear for applying the tape to the cable (also fitted on the repairing vessel) is then rendered complete by the addition of draw-off gear, change wheels, and counter-shafting. Thus an old cable, when picked up for repairs, may be re-taped as necessity requires.

§ An outer covering of canvas tape tends to peel off somewhat readily after a comparatively short period of submergence.

**Outer Yarn Covering.**—A yarn is what in spinning parlance is known as a "single." Very usually for greater strength and compactness the jute or hemp used for this purpose takes the form of what is spoken of as "three-ply," which is constitituted by three "singles" laid up together. Sometimes, however, "two-ply" yarn is used—*i.e.*, two "singles."

The method of applying such yarns for the outer covering is naturally precisely the same as in the case of the inner serving. There are invariably two layers in reverse directions, alternated by coats of compound, as already described, the yarns being laid on with a somewhat longer lay than the inner serving, as a rule. This outer covering to the iron wires may consist of anything up to about sixty yarns, depending on type of cable, etc. The final layer involves more of course (out of this number) than that preceding it. The gear and general arrangements are the same as in the already described process of covering cables outside with canvas tape, except that the taping apparatus is replaced by carriages loaded with bobbins of hemp or jute yarn, as the case may be.

Moreover, the outer covering in this place may be—and invariably is—as in taping, effected simultaneously with the process of sheathing, as originally set forth in Sir C. Bright's patent, No. 558, of 1862. Thus, the carriages containing the hemp or jute bobbins are placed behind the sheathing frame; and the yarns—brought down from their respective bobbins, through guides on to the cable—are laid round it as the disc or frame revolves.

**Hemp Yarn applied Externally.**—We will now first deal with the question of an outer covering of yarns composed of hemp.

To begin with, hemp is a vegetable substance, coming from the bark of several plants in India, Manilla, Russia, Italy, etc. Hempen plants have been also cultivated in small quantities in Suffolk and Lincolnshire, as well as in Ireland.

The ordinary plant (*Cannabis sativa*) was a native of India in the first instance, but has since been largely cultivated in Europe as above. It is still, however, obtained in the largest quantities from India.

Russian hemp is of good quality, and is perhaps more used than any other for the purposes of an outer serving.

Italian hemp is of still better quality, and correspondingly more costly.

Manilla hemp (*Musa textilis*) is different, and, generally speaking, superior to all other classes. It is, in fact, usually spoken of as a separate material, and called "Manilla" simply. Manilla is, as a rule, the most expensive of all the hemps. It is invariably a good deal lighter than the others. It does not absorb water or tar readily, and, in fact, no attempt is ever made to tar it, hence it is often spoken of as white hemp.*

The tensile strength and durability of Italian hemp and Manilla are very often about the same.

As in the case of jute, hemp yarns are mostly prepared at factories in Dundee. The state in which they are used for submarine cable purposes is, like jute, very closely approaching the natural state.

At one time the hemp yarns employed for the outer covering used first to be impregnated with ordinary coal tar, or drawn through a bath of molten pitch, by way of preservation. It has been found, however, that the tensile strength of the yarns is thereby weakened—to the extent of about 25 per cent.—the fact being that the hempen fibres are partly destroyed by the tar.†

Ozokerite has been tried in the same way for purposes of preservation, and with satisfactory results in filling up all the interstices of the fibre. However, the general rule nowadays is not to treat the yarns (whether hemp or jute) for the outside covering with any preservative or anhydrous solution until the usual Bright and Clark's cable compound is applied after the yarns have been laid on the cable.

The specific gravity of either hemp or jute yarns is considerably below unity when dry, but on account of absorption, becoming readily "water-logged," they naturally sink of themselves. Again, Manilla and other hemp is rather heavier than jute, but the latter is much more porous and absorptive.

The specific gravity of ordinary hemp when dry is about 0.66, and 0.92 when moist,‡ some yarn being about 1.3.

In a cable as ordinarily applied for the outer serving, 1 cubic foot of Russian or Italian hemp yarn weighs about 39 lbs.; and 1 cubic foot of Manilla comes to about 41 lbs.

---

* However, when the term white hemp is used, it is sometimes only in contra-distinction to ordinary Russian or Italian hemp tarred.

† Grappling and buoy ropes are very usually slightly impregnated with *Stockholm* tar. Good results accruing from this plan, the "Great Northern" Company, in their cables, are in the habit of specifying Stockholm tar for preparing the yarns of the outer serving.

‡ Thus under all circumstances an outside hempen covering—especially if of some thickness—tends to reduce the specific gravity of a cable a good deal, though at the same time adding to the superficial area subject to friction during cable operations. Both these effects are almost as marked as in either of the open-sheathed types.

The tensile strength of ordinary hemp is represented by a stress of about 6,000 lbs. per square inch, but it varies very considerably with the class. When compounded the strength of the same material is sometimes so far weakened by the presence of tar (as above explained) as to have a breaking strain of only 3,300 lbs. per square inch.

The hemp yarns used for the outer covering are not, as a rule, tested for tensile strength or elongation, in the case of an iron-sheathed cable, as their values in this respect are not, practically speaking, considered to come into force. A certain proportion of them are, however, invariably tested for torsion. Being usually composed of two or three ply, it is naturally far less homogeneous than iron; and is, therefore, tested for torsion in much longer lengths.

As a test of subsequent durability on submergence at the bottom of the sea, Messrs Siemens Brothers have an excellent practice of keeping different kinds of rope—sometimes laid up into cable—in a tank of salt-water for a considerable period, and then examining and testing them in various ways.

As already stated, the main function of the outer serving is that of encasing the wires for purposes of protection and preservation, as well as for holding them together, more especially to meet the contingency of one of them breaking at any spot.

This being so, it will be seen that it is essential that the hemp yarns be applied with a comparatively short lay in order to effect an efficient binding.

This lay of the yarns is, in fact, very much shorter than that of the iron wires; but even if applied with a long lay, it is uncertain whether they would add any appreciable strength to the armour.

For these reasons, then, many consider the use of hemp yarns to be a needless expense, and actually prefer jute yarns for the outer (as well as the inner) serving of a submarine cable. It may, however, be pointed out that the use of hemp yarn involves much fewer stoppages in the manufacture of the cable than where jute yarns are employed, owing to the greater strength of the former and consequent less liability to fracture under the tension of laying up. Jute yarns are certainly a cause of frequent delay on this account.

**Jute Yarn applied Externally** —Jute has of late years been very largely employed for the outer covering. This is partly owing to improvements in its quality and preparation rendering it much more durable than it used to be, and materially more so than hemp as a rule. Jute yarn is also very appreciably less costly, though on the other hand much weaker. It has, to a great extent, replaced hemp yarn for this purpose as well as for the inner serving, though in the Post Office cables hemp yarns are always employed in both instances.

Jute is a variety of hemp, and is of a similar nature, though inferior in strength, coming from different plants growing in the same countries. Like hemp, it is mainly spun into yarns at factories in Dundee, and the condition in which we use it equally resembles the natural state.

Where jute yarns are employed, notions of strength for the outer covering are, of course, entirely abandoned—unless a very great number of yarns be adopted for the thread, making up a multiple "ply" instead of a single "ply" or yarn. And unless applied with a very long lay, in which case its main function as an efficient binder and protective preservative would, to a great extent, be gone.

Very usually jute is used for the shallow-water types of a given cable, and hemp for those which are intended for deep water, as regards the outer covering.*

**Outer Hemp Cords.**—To more effectually meet the above requirements, what are known as hemp cords came into use, some years ago, for the outer serving of deep-sea types, in place of the ordinary three-ply, 10 lbs. per N.M., hemp yarn.

This plan was first adopted by the Silvertown Company, who have always adhered to it where extra strength is aimed at, in deep-water types.†

These cords are made up of Russian hemp from several strong yarns— very often three‡—laid up together—i.e., "multiple-ply" Russian hemp. Thus, these superior yarns are about double as stout as an ordinary hempen yarn.

Theoretically speaking, however, the only way in which such cords (any more than any other outer serving) can take their part—to the smallest

---

* In any case—from a tensile strength point of view—hemp would be completely wasted on a shore-end or intermediate cable ; besides the fact that it is less durable than jute.

† Hemp cords were first employed in the St Vincent—St Jago cable, made and laid by the Silvertown Company in 1884.

‡ Hence, one hemp cord is frequently composed of three 3-ply of best dressed Russian hemp, weighing about 16 lbs. per N.M.

extent—in the total breaking strength of the cable, is by having such a lay as will ensure their breaking with the iron wires, notwithstanding the different elongating qualities of the two materials.

In order to meet these requirements there must, in fact, exist a certain relation between the lay of the cords and that of the iron wires, according to their relative rates of elongation.

Those who employ these cords contend—with some show of reason—that they actually carry out this principle in practice, though it may, at first sight, appear a difficult matter. Thus, hemp cords (or " strings " as they have been sometimes termed) possessing considerable strength in themselves are said to be applied with such a lay as has the effect of increasing the strength of the cable by the full amount of their individual strengths; thus adding, it is asserted, no less than 1 ton to the breaking strain of the completed cable, where each cord bears a stress of about 1 cwt. before giving way.*

The hemp cords are only applied in one serving as a rule, this often consisting of about twenty,† applied with the opposite lay to the previous serving, whether of hemp or canvas tape.‡

Such cords, possessing greater strength than the single, double, or treble, ply yarns, certainly have the additional advantage of acting as a more secure binding, and thus more effectually restraining the unlaying of the wires under strain. This especially applies when a ship is hanging on to the end of a cable in deep water for splicing operations.

On the other hand, it is contended by the opponents to hemp cords, that, placed outside a more or less complete metallic arch, they can never add to the strength of the sheathing, no matter what the lay be; though they might possibly if made to encircle each wire, after the manner of an open-sheathed cable. There are those who not only say that it is a mistake to suppose hemp cords as applied increase the strength of a cable, but who go further, and actually affirm that in giving a greater jar when they break than ordinary yarns, a more serious and sudden strain is thereby brought on the wires, causing them to break, where they might not otherwise.

The hemp cords certainly cannot be laid with anything but quite a short

---

* It will be seen, therefore, that the breaking strain of these cords is well worth testing before use for the outer covering.

† The number of cords depends, again, on the weight of each, as well as on the diameter of the cable. There are, however, invariably less than in the case of yarns, owing to the greater bulk and strength of the former.

‡ An outer covering peculiarly characteristic of Silvertown cables is a layer of canvas tape followed by a serving of hemp cords.

lay, if they are to perform the main function of the outer serving as an efficient binder and preservative. Thus, it is contended by some that if the latter requirements are to be properly met, the lay will be too short for the different actual elongations of the wires and of the cords, to permit of their breaking at all at the same moment; and that, therefore, the cords cannot add to the strength of the sheathing wires.

It may certainly be said—except where such hemp cords are applied with the proper lay—that the only actual strength that can be practically relied on in a cable is that afforded by the iron sheath. Moreover, that except where the above conditions are met with, the outer serving—no matter what it be composed of—cannot appreciably augment the total strength, or breaking strain, of the completed cable; its only use being, in fact, that of a preservative and efficient binder.

**External Canvas Tape and Hemp Cords.**—In the present day, the outer serving for a deep-sea cable not infrequently takes the form of a single covering of canvas tape, followed by a coat of Bright and Clark's compound, and then a single covering of the above hemp cords—some twenty in number, say—which are then finally compounded, thus combining the advantages of both the yarn and tape styles of covering in the best manner possible.*

Though the author has endeavoured here to deal with all the various forms of outer covering and methods of application, there may of course be other modifications and different routines with which he has not had an opportunity of becoming acquainted.

**Outer Covering: General Particulars.**—The quantity (by weight) of hemp or jute used for the outer serving varies, roughly, from 5 lbs. to 20 lbs. per yarn per N.M., according to the type of cable and thickness of covering desired. The number of these yarns varies also with the diameter. The two layers of yarn are invariably "compounded," the result being that the above weights are practically doubled. The weight of the complete outer covering is further increased by the weight of compound applied between each layer, which, indeed, may also even double it.

The only satisfactory way of putting the relative merits of each type of

---

* In several ways it would be better to apply the hemp yarns or cords first, to be followed by an external casing of canvas tape; but tape will not lay well over a bed of hemp or jute.

outer covering to the test would be, of course, to lay two cables at the same time side by side, externally covered with each respectively, whilst otherwise similar. To carry out the idea properly, the two samples should be laid in vessels containing the same water, and lined with a specimen of the bottom to be encountered.*

Alongside the whole length of the cabling machine at the floor level, a counter-shaft runs in connection with the main factory engine shafting; and from this the taping heads or yarn discs are driven by belts, and the draw gears by means of bevel and spur gearing.

A brake is usually fitted to the machine, by means of which it can very quickly be brought to a standstill throughout. As a rule, this brake

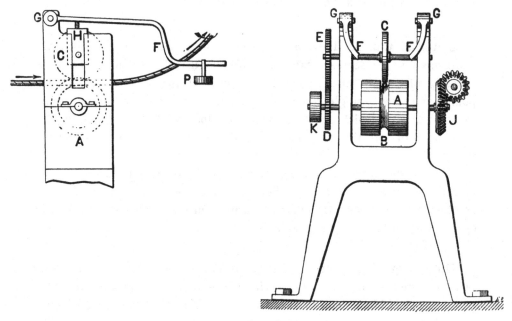

FIG. 111.—Hauling-off Gear.

almost entirely encircles a part of one of the discs or cylinder frames, and is actuated by means of a lever at the lay-plate.

All "cabling" machines, especially those for applying heavy wires—such as the outer sheathing machines—should be set up on a very strong bed-plate, continuous from end to end.

* Especially if containing mineral or organic matter.

**Hauling off Completed Cable.**—The gear which transmits the hauling motion to the cable through the "cabling machine" is shewn in Fig. 111. Here the cable is led over a deeply grooved sheave A, whilst a jockey-wheel C, also slightly grooved, rides on the cable and keeps it pressed down on the sheave. Both sheave and jockey-wheel are rotated by gearing-wheels D and E in opposite ways, so that the cable is gripped between them and forced onwards. To ensure sufficient friction between the jockey-wheel and the cable, in spite of variations of diameter during the different processes, the spindle of the jockey-pulley is made to work up or down in a vertical slot. A forked lever F, pivoted on either side of the framework at G, and weighted at P, presses down the spindle ends of the jockey,

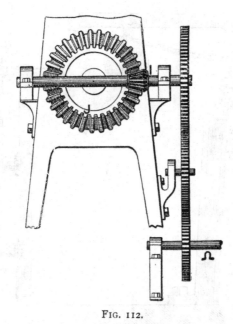

FIG. 112.

keeping the cable firmly grasped between the two wheels. The pressure of the jockey-wheel on the cable is regulated by moving the weight P in or out along the lever arm. The wheel D which drives E, receives its motion through intermediate gearing from the horizontal shaft previously designated by the letter $\Omega$ (Figs. 84 and 112) and so from the main shafting.

The preceding gives an idea of a very usual form of hauling gear for deep-sea cables. That for "intermediate" and shore-end types is the same in principle, but the various parts are constructed of stronger material, besides being of a heavier build, as may be seen from a glance at Fig. 113. Here, the hauling-wheel A is driven by the wheels D, E, F; and the jockey-pulley C by the wheels G and H. The pulley J is intended to take an endless rope, and by that means, drives similar draw-gear, placed, when necessary, immediately above the iron tank (in a direct line with the "cabling" machine) into which the cable is to be stored.

**Rate of Cable Manufacture.**—The iron sheathing wires having a considerably longer lay than the copper strand, it might appear that the operation of "cabling" would be performed at a faster rate than that at which the conductor is stranded up. As a matter of fact, however, it is

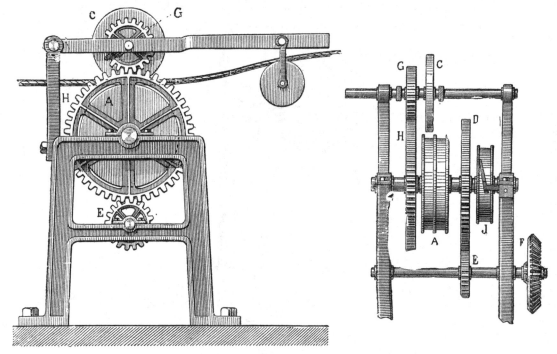

Fig. 113.

very much the reverse, the rate being about as 1 to 5 in favour of the stranding of the conductor.

With such a long lay as the larger iron wires are given, the point which limits the speed at which the wires can be laid up is the mechanical difficulty already alluded to, involved principally by centrifugal force— especially in the case of heavy wires and a large number. This, in turn, puts a limit to the rate of drawing off (with a given distance existing between lay-plate and bobbins) and thus to the speed of "cabling." The machine itself has, moreover, to be made heavier in proportion to the type and number of wires, and this in turn influences the speed of laying up, with whatever type of machine it be.

Again, mainly owing to the fact that the bobbins hold so much less wire, the stoppages for changing bobbins and for brazes are a great deal more frequent in a heavy type of cable with large wires than in a light type.* These stoppages † are, indeed, the principal cause of the great difference in speed of sheathing small and large wires. Thus, ¼ mile of a shore-end type of cable is sometimes all that can be sheathed (with outer sheathing) in a working day—though 1 to 1½ N.M. is more commonly done. The total time taken up for stoppages of one sort or another is, in fact, often as much as would be sufficient for sheathing a complete mile of cable; whereas in the case of a deep-sea type, 3 N.M. is a very ordinary, amount to be completely "cabled" in the same time, at a speed of about thirty revolutions per minute for the cylindrical cage frame carrying the bobbins.‡ The lightness of this carriage, as compared with the ponderous discs used sometimes for laying up heavy cables, in itself makes a great difference, of course, to the speed attainable with safety.

The larger cable factories have some ten to twelve cable machines, for all types. Each of these machines could, under pressure, turn out as much as 5 or 6 N.M. in a day. Under ordinary circumstances a large factory manufactures some 30 N.M. of finished cable in an ordinary working day of twelve or fourteen hours from an average of half-a-dozen machines, equivalent to about 180 N.M. in a week.§ In the case of the last (1894) Mackay-Bennett cable, the manufacture was proceeded with continuously day and night; thus some 50 to 55 N.M. were turned out at Messrs Siemens' works during the twenty-four hours. As a rule, however, night work is not allowed in a cable contract, owing partly to the light not being so good for the work; as well as to the fact that it involves a shift of "hands" of possibly less experience.

---

* In early days the slow rate of sheathing with heavy wires was further accentuated by their greater degree of brittleness, which was especially liable to cause breaks at a braze when passing round the hauling-off drum.

† Together with the obvious impossibility of running a heavy (disc) machine at as high a speed as a comparatively light (skeleton) machine.

‡ The *number* of wires and yarns has a great influence on the rate of manufacture as regards all types, by involving a greater or less number of stoppages for insertion of bobbins, joints, breakages, etc. Thus, the modern deep-sea types, with a large number of wires to enclose a heavy core, occupies a longer time in construction than former corresponding types.

§ The corresponding rate of manufacture over the first Atlantic cable was about 20 N.M. a day, or a little over 100 N.M. a week.

Cable Tanks at Factory.

## SECTION 6.—STOWAGE, TESTING, ETC.

**External Whitewashing.**—The cable is coiled into the factory storage tank by hand, ready for testing in water previous to, and ready for, shipment. Each "flake" (or layer) is plastered over, by means of a "swab," with whitewash (chalk and water) to prevent the turns of the same flake sticking together, as well as of adjoining flakes, either of which occurrences* may often be the cause of serious accidents—by a "foul" in the tank †— on the cable being drawn from the tank again. ‡

By this plan, however, it will be seen that the whitewash is liable not to penetrate entirely round the surface of each turn of cable.

---

* More especially liable to occur in warm surroundings.

† Such "fouls" are, however, almost entirely confined to laying operations, and seldom occur during shipment from the factory tanks, where the speed of coiling is usually less than half that adopted for paying out. For this reason no "crinoline" gear is fitted to the factory tanks as in the case of the ship tanks ; as well as owing to the circumstance that in the former the space between the top flake and hawse pipe is invariably sufficient for the turns to take themselves out.

‡ On the other hand, whitewashing *can* be overdone ; inasmuch as in paying out a cable a slight sticking action with the turn below serves as a suitable check to the rate at which the cable flies up the tank.

Thus, at the Silvertown cable works it has been for some years the custom to lead the finished cable on its way to the tank (in fact, just above it) through a wooden die-box loaded with whitewash, having a rubber ring at the entrance.  This plan (devised by Mr F. Hawkins) has, moreover, the advantage of saving labour in the tank, besides enabling coiling to be carried out a trifle faster.

**Coiling into Tank.**—The corresponding tanks (see illustration at head of this section) into which the cable is coiled from each machine are of enormous size, having a diameter of somewhere about 30 feet, and a depth of 15 feet or so.  Each tank is capable of holding something like 200 N.M. of deep-sea cable.*

In the centre of these factory tanks is a wooden cylindrical or conical framework—about the height of the tank, generally—round which the cable is coiled (right-handedly†) from the wall (or "outside") of the tank, turn by turn, towards the centre. ‡  On this wooden framework stands one of the coiling hands who guides the cable down to the tank, where it is taken in hand by the other "coilers," of which there are several, one of whom journeys round the tank with the bight of cable.  On the completion

FIG. 114.—Cable laid down in Tank, with "Feather-edge."

of each flake, the bight is taken out to the side of the tank for a new flake to be begun.§

The length which is laid across each flake (called the "lay-out") is protected from pressure by long wedge-shaped pieces of wood (suitably termed "feather-edge") being placed on each side of it, the thick side of the wedge being slightly thicker than the diameter of the cable itself.

This is shewn in Fig. 114, the circle representing the section of cable.

---

* Here the cable may be not inaptly likened to a sea-serpent, in gliding into the nuge tank with a wriggling, ceaseless motion.

† The reason that the iron wires of an ordinary submarine cable are made with what is known as a left-handed lay (going from right to left from the observer), is in order to ensure the turns put in during right-handed coiling being in the same direction as the lay, and thus obviating unlaying of the wires by the operation.

‡ If the framework in the centre be conical in form, it is advisable to leave about half a foot space all round it so as to avoid any chance of the cable getting hung up.

§ With a view to preserving evenness as far as possible, this "lay-out" is varied in position at each successive flake.

The feather-edge is composed of strips of deal wood about 6 feet in length —formed as described—and shewn so as to pack up the lay-out and form a *gradual* rise and fall for the turns of the upper flake where they cross it.

Any unevenness due to the lay-out, etc., is obviated by means of short flakes filling in the declines.

The rate of coiling is governed solely by the limit of speed at which the "cabling" is effected, and, in fact, naturally corresponds with it. A much higher speed for coiling is possible, as evidenced by the subsequent rate of coiling into the tanks of the laying vessel.

The ends of a length of cable that has been coiled down are taken up at the side of the tank to a test-box conveniently placed at the top.

**Testing during Manufacture.**—We have endeavoured to describe as far as possible some of the principal methods of cable construction. During the entire process of serving and sheathing a continuous electrical test is applied to the insulated conductor, very usually by connection to one of Lord Kelvin's quadrant electrometers, which accurately records the potential and its rate of fall from "full charge" in a manner described elsewhere. Regular observations (in the course of the above "control" test) are, in fact, made and noted from time to time as an index of the electrical condition of the cable by means of what is certainly a most sensitive test of the insulation. Thus, the slightest injury to the core during the course of conversion into cable may be at once observed, and made good at the earliest possible moment. Two complete galvanometer tests for dielectric resistance are also made, as a rule, before and after a day's work.

Moreover, at pre-arranged specified periods the cable is tested in a very complete manner (under every head) on non-working days, as well as on the completion of each factory "section." These latter tests are made after the cable (in the water-tight tanks) has been covered with water for a sufficient time to meet the requirements of a really crucial test.*

**Stretch of Core during Sheathing.**—During the process of "cabling" a certain tensional stress is necessarily put on the core in drawing off. This involves stretching to the extent of about 0.3 to 0.5 per cent. of its original

---

* When once a cable has been in water, it takes a long time for the serving to dry— taking place, indeed, only by evaporation. The custom is to keep the cable constantly covered with water as fast as it comes into the tank up to within 2 inches of the top flake, this limit being made for the convenience of coiling.

At the end of each day's work the water is turned on, and the cable is left to soak in its element. Indeed, it is never left long enough without water to become dry.

length in the case of deep-sea types.    Again, the length is increased—under the circumstances of a longer sheathing lay, and a greater weight round it—as much as 0.75 per cent. by the first sheathing only, in the instance of double-armoured (shore-end) types.

From the above it will be understood that the electrical results obtained on the cable will naturally be different to what they were when tested as core.    For purposes of true comparison (owing to the alteration in length) it is essential to arrive at a correct estimate of the new length.    Unfortunately, as has already been shewn, the length indicated by the hauling-off drum is not reliable as an index of the length of cable as sheathed, owing to the fact of the drum, or cable, slipping back occasionally, thereby resulting in a second registration of that amount of cable.*    The true length of the cable is, therefore—for the purposes of accurate electrical values, by reduction to standard values of unit length †—best determined by the addition of all the values obtained in terms of conductivities‡ and capacities for each coil composing the completed cable.    A certain percentage is then allowed (as already explained) for increased length by stretch.    Such a course is at any rate advisable, if only as a check on the cable indicator measurements.

---

* Messrs Johnson and Phillips have endeavoured to meet this objection by a separate measuring drum (with revolution indicator) which the cable merely passes over, there being a roller pressing on the top of the cable to prevent " slip."

† These tests being compared with the standard tests of the core at 75° F.—in order to detect at once any signs of deterioration, and for subsequent comparison—should be either made with the cable at the same (standard) temperature ; or, if this is impossible, a very correct idea of its actual temperature at the time of testing should be arrived at, in order to make the required allowance (or reduction) for this difference of conditions.

‡ In the case of the dielectric to be afterwards converted back into resistance, by use of reciprocal tables.

# CHAPTER V.

## COMPLETED CABLE.

### SECTION 1.—REQUIREMENTS AND TESTS.

**Matters of Importance: Data necessary.**—Besides the specific gravity of a cable and its weight in water—together with a knowledge of the " breaking strain "—being matters of importance both for purposes of laying and recovery,* it is also necessary to know, for loading requirements, what the weight will be of a given length when wet in air. This is in order to arrive at the dead weight of cargo involved by a given cable when coiled into the tanks of the laying vessel—inasmuch as it is wet at the time after lying in

---

* As already pointed out, a cable with high specific gravity is the best from a laying point of view. On the other hand, from a recovery standpoint a low specific gravity cable in deep water is preferable, provided that the coefficient of friction is not too great. Though a low specific gravity cable is also the easiest to pay out with the minimum of risk, it is less likely to be *efficiently* laid—*i.e.*, at such an angle as ensures it adapting itself properly to the irregularities of the bottom, so as to be of a durable description for the owner. These remarks only refer to deep-sea cables. With shallow-water cables such matters are of no special moment ; but their weight wet in air is usually taken for practical purposes, this being materially different from the corresponding weight when dry.

water in the factory tanks, besides being almost invariably maintained in this condition.*

A deep-sea cable weighs about 1 ton per N.M. when wet in air, or even less sometimes ; † the corresponding weight of a heavily armoured (double-sheathed) shore-end type being sometimes nearly 30 tons.

A knowledge is also necessary of the weight of the cable when dry in air. This is primarily required with reference to estimates—*i.e.*, in connection with the cost of materials according to the detailed weights of each component required for a given design. This latter information is, moreover, necessary for checking its component parts, as regards quality and quantity, when made up into completed cable. It is again sometimes required for loading purposes, as a mean between the weight wet and dry ‡ is often taken in estimating the total weight of cable as cargo.§ This latter applies where, as has been already pointed out, cables are sent out "dry" to their destination in spar-wood tanks, or drums, after being under water for testing purposes only.

The weight of a bulky cable with much serving may be as much as 10 per cent. greater when wet than when dry. Its weight in water, however, will be less than when dry in inverse proportion to its specific gravity.

The weight of a cable in salt-water is determined by the measurement of a given standard sample length (usually a foot, or a yard) in this

---

* Cables have been sent out in safety to their destination perfectly dry, but it is considered more prudent to keep them wet up to the time of submergence. When paying out a cable it is, at any rate, essential that it should be laid from a water-tank ; besides being covered with water a short time beforehand, with a view to the detection of any fault previous to submergence. The alternating conditions of wet and dry are the most dangerous possible for an iron-sheathed cable, by inducing chemical action and spontaneous combustion. Thus it has been pointed out that if there are not facilities for keeping a cable permanently wet, it had better go out to its destination absolutely dry, until about to be laid, after being tested (as core) in water at the factory. It may, however, be remarked that when a cable has once been covered with water, it takes some time to become dry.

† The weight of a cable in sea-water may be calculated approximately from its weight wet in air and from its diameter, as follows :—Find the cubic contents in cubic feet of, say, 1 fathom of the cable. Multiply this by weight per cubic foot of sea-water (64.5 lbs.). Subtract this result from weight of 1 fathom of cable dry in air, and the weight of 1 fathom in sea-water is obtained.

‡ The weight wet is usually from 5 to 10 per cent. more than when dry with an average close-sheathed cable of the present day.

§ When estimating the weight of a cargo of cable, the weight of water which is pumped into the tanks to cover the cable must be considered. The water in the tank naturally occupies all the space not taken up by the cable.

respect.* This is carried out in practice by hanging the sample in water from one arm of an ordinary balance.

To determine the specific gravity from the weight of the same sample dry in air and its weight in air, the formula is—

$$\text{Specific gravity} = \frac{W}{W - w.}$$

Where W = weight dry in air,
and       w = weight in water.

It has already been pointed out that the tensile strength of a submarine cable is of the utmost importance. In heavily sheathed cables an abundance is sure to be provided on account of the number of large wires involved by considerations of wear, having regard to the conditions and bottoms for which they are intended. Indeed, shore ends often weigh nearly 30 tons, or, at any rate, much more than can ever be actually made use of, especially considering the shallow water in which they are laid. Thus, no special note is made of the tensile strength of either shore-end or intermediate types as completed cable, though the wire of which they are composed is carefully tested for ductility in order to secure flexibility and good coilable qualities.

The stress which a deep-sea type of cable will bear must, however, be closely gone into—*i.e.*, calculated and even measured—preparatory to its ultimate adoption for the required purpose.

The breaking strain of an ordinary iron-sheathed cable may be calculated from the accredited breaking strain per square inch of the material composing the wires and from the sectional area of each wire. This latter is obtained from the formula

$$\pi \gamma^2$$

where $\gamma$ is the radius of a circle and $\pi = 3.1416$, the relation existing between its circumference and diameter.

Then the total breaking strain of the complete sheathing should be the result of the above multiplied by the number of wires composing the armour, assuming all to be of the same gauge and quality.

It may also be calculated, of course, from a knowledge—if such exist— of the actual breaking strain of each wire, which when added together will give the breaking strain of the entire sheathing: or, again, the measured

---

* This may, at the same time, also be turned to account as a test on the individual specific weights of the materials by weighing each separately.

breaking strain of a single wire may be taken as representing that of each of the other wires, and may be multiplied by the total number of wires to give the total breaking strain of entire sheathing.

Though, roughly speaking, the strength of an ordinary submarine cable is very often constituted solely by that of the iron armour, this is not always the case—at least not in the opinion of some authorities, who aver that, at any rate, hemp *cords*, when applied with the proper lay (as already referred to), add materially to the strength of the wires, and so contribute to the total strength of the completed cable.

This being so, it will be evident that absolutely accurate information regarding the breaking strength of an entire cable is only obtained by actual measurement—*i.e.*, by testing samples of it. Even this is not altogether satisfactory because of the necessarily short lengths experimented on.

In any case, the engineers on behalf of the owning company usually require a certain number of lengths of the deep-sea types of cable in a given contract to be tested to their satisfaction, as regards tensile strength and elongation. It need, perhaps, scarcely be remarked that no torsional tests are ever applied to any type of submarine cable when *completed*.

The ordinary breaking strain of a completed deep-sea cable intended for considerable depth should be somewhere about 6 or 7 tons at least, varying according to the type and depth it is intended for. Thus, the last "Anglo" Atlantic stood a strain up to 8.2 tons when tested, with an elongation of 4 per cent.

The breaking strain of the completed cable is invariably somewhat greater than what it is calculated at from the wires alone, the weakest point in a number of wires never being exactly at the same spot.*

From the breaking strain, absolute information is obtained as regards the depth of water to which such a type can be laid with some sense of sureness that it can be subsequently recovered if necessary.

---

* Moreover, this is so owing to the friction between the wires acting as a mutual support in a close-sheathed cable. In the case of a cable having hemp cords applied with the proper lay, this discrepancy is, again, further accounted for.

### SECTION 2.—MECHANICAL TESTING.

**Tensile Strength.**—Several kinds of apparatus have been devised for the purpose of applying a gradually increasing stress to a cable till it breaks, the value of this breaking stress in weight being then noted as representing the force, or strain, it will bear—or rather, the minimum at which it breaks.

The earliest and simplest form consisted of a tripod about 30 feet in height, supporting at the top a pulley round which the upper end of the piece of cable to be tested was secured. The lower end was fastened to a similar pulley, to the spindle of which was hung a platform, on which weights were placed one by one, until rupture occurred.

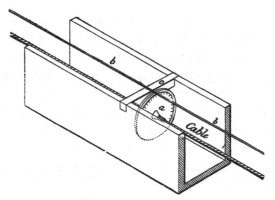

FIG. 115.—Early Tension and Torsion Apparatus.

In this machine the degree of elongation was ascertained by measuring just before rupture the distance between two small discs $a$ (Fig. 115) previously secured to the cable at about 3 yards interval, in line with two cross pieces $c$. These discs were graduated round the circumference, so by making the zero points coincide with a wire stretched parallel to the cable, before tension was applied, the amount of torsion was afterwards shewn by the angle through which the discs and cable had turned.

A more complete and reliable form of apparatus, originally designed by Messrs Gisborne and Forde (Fig. 116), has a solid beam A supported and firmly secured to masonry by two uprights. To one end is bolted an iron shoe, the upper end of which has a strong hook, or shackle, for securing one end of the cable; at the other end of the beam pivots a bent lever B, the long arm of which carries a hanging platform and weights, the short arm having a hook to hold the other end of the cable. The lengths of the

[PLATE XXII.

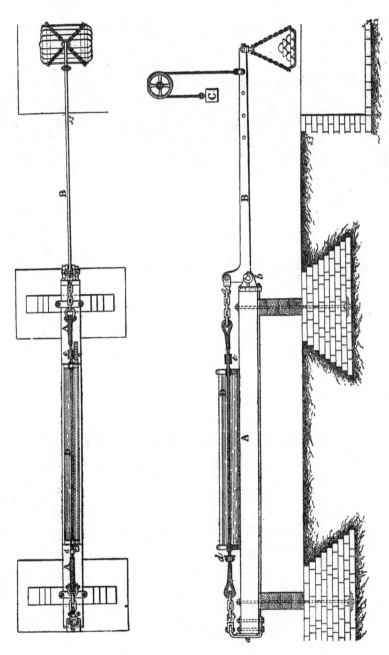

FIG. 116.—Gisborne and Forde's Tension and Elongation Machine for Testing Cables.

lever arms are as 1 to 10: the counterpoise c balances the weight of the lever arm and platform.

The cable passes through a trough D filled with water for observing the effects specially due to water—such as contraction in the case of hempen

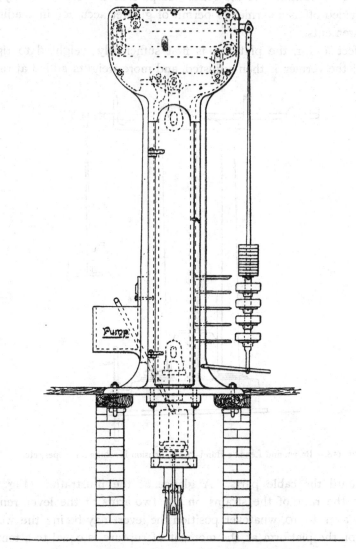

FIG. 117.—Brown and Lenox's Cable Testing Machine.

cables. The holes in the ends of the trough which the cable only partially fills up are rendered water-tight with tow.

The elongation is measured by means of a scale c which passes through the ends of the trough enclosed in a pipe placed close, and parallel, to the cable. To prevent the torsion of the cable, or stretching of the various parts of the apparatus affecting the position of the scale, the clip *d* on the cable is fitted with a small circular disc which engages at right angles in a notch in the scale, so causing the latter to move horizontally with the clip. At the other end, a small cylinder is fixed to the cable which, as the cable twists, turns in proximity to the graduated part of the scale : the cylinder itself is divided off as a vernier to permit of greater accuracy in reading off the measurements.

To effect a test, the platform is first sufficiently weighted to tighten the cable ; the vernier is then adjusted, and more weights added at regular

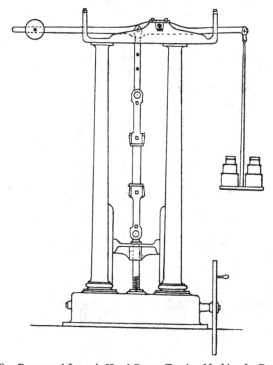

FIG. 118.--Brown and Lenox's Hand-Power Tension Machine for Ropes, etc.

intervals until the cable parts. A glance at the illustration (Fig. 116) shews that the ratio of the strains on the two arms of the lever remains constantly as 1 to 10, whatever position the lever may be in : the weight, therefore, on the platform at the moment of rupture is equal to a tenth of the strain on the cable.

The elongation is obtained by the displacement of the vernier cylinder *e* along the scale up to the instant of rupture.

Messrs Brown, Lenox and Co. are a firm of engineers who have always been famous for making a close study of the requirements involved in apparatus and work of this description—in fact, a large proportion of the telegraph cables at the bottom of the sea have previously undergone tests at their works.

Fig. 117 represents a recent type of hydraulic power machine of vertical form, as designed, constructed, and used by this firm. Here the power is applied by hydraulic cylinder and hand pump. This machine—designed for testing samples of chain, cables, iron, etc.—tests up to as high as 50 tons, though for the purpose in question such power would never be required.* Messrs Brown and Lenox have also another hydraulic machine —horizontal instead of vertical—capable of applying the same stress.

This firm has also a hand-power vertical machine (Fig. 118), with wheel-worm and screw-gear for applying the power. This machine tests up to 1 ton, and is especially designed for testing samples of canvas, rope, etc., and is not, therefore, turned to account in regard to submarine cables, though it might very suitably for testing the component parts.

A cable covered with good iron should bear at least 2 tons per lb. of iron wire per fathom ; and the equivalent elongation should be at least 3 per cent.†

The elongation of an ordinary iron-covered cable (the sheathing of which forms practically a metallic tube) when exposed to half of its supposed breaking weight varies from 0.5 to 1 per cent.—an amount, by the way, which could never injure the core. Whilst being laid, a cable generally untwists slightly, but the elongation from this cause is also insignificant. Moreover, it is not permanent, as a cable always reaches the bottom in exactly the same state as it was in when first coiled in the tank.

The modulus of tension (or figure of merit from a tension standpoint) of a cable is a most important matter, in that it represents the maximum length of itself which the cable will support in water:‡

In deciding on the minimum breaking strain allowable for a modern deep-sea cable, it is not unusual, where possible, to give the cable something like four times the tensile strength that is necessary for its own suspension in the greatest depth in which it is to be laid. Thus, for such a depth as 2,000 fathoms, a cable should have sufficient strength to support some eight,

---

* In connection with submarine telegraphy this machine is, however, frequently used for testing stranded wire and hemp grappling ropes, which are often made of a multiple type, bearing a stress of 25 tons and even over.

† The elongation of a completed cable is, however, almost invariably slightly greater than that of each individual wire composing it. This is partly owing to the cable's tendency to unlay and straighten out when subjected to a strain ; as well as to the mutual support of the wires—especially when fitting tightly together under strain.

‡ It is obvious that this may be calculated thus—

$$\text{Modulus of tension} = \frac{\text{Breaking strain}}{\text{Weight in sea-water}}$$

or more, miles of its own weight in water. This is highly desirable if only from a paying-out point of view—let alone picking-up requirements.*

Reverse lays, though they produce great compactness as an outer serving, and form an excellent binding to hold the wires together, cannot produce a breaking strain such as may be obtained where both coats are laid in the same direction. This is owing to the fact that, with lays of the same length in opposite directions, any elongating tendency in the one is exactly neutralised by that of the other, and thus under these circumstances the strength of the outer serving is at once put to the test—long before the sheathing wires, with which, therefore, it cannot possibly work. In some ways, it would seem, if the outside covering is intended to increase the strength of the cable, that the two applications of hemp cords (or yarns)—where more than one is adopted—should be laid with the *same*, rather than with reverse, lays; besides, of course, being applied with such a length of lay as bears the right relation to that of the iron wires according to the relative elongation of the materials. This latter, at any rate, is always made a close study of by those who seek to ensure the outer covering taking its share in the total strength of the cable by causing it to break *with* the iron wires, instead of some time before.

**Qualifications.**—For purposes of laying and repairing operations, the completed cable should be flexible, as well as strong, and not inclined to kink. † It should also, even when kinked, bear a considerable strain before parting.

An advantage in the sheathing wires of an ordinary deep-sea cable being composed of " homo " iron rather than of actual steel exists in the fact that the latter are exceedingly springy and " homo " wires are less so.

Finally, it may be remarked that in the present day under average circumstances—as regards conditions for laying—the success of a cable depends, perhaps, principally on its construction.

**Representative Deep-sea Cable.**—The actual available strength of the last (1894) " Anglo " cable is nearly 30 per cent. higher than that of the 1865-66 Atlantic (open-sheathed) type, though, of course, the specific gravity is not so low. On the opposite page (Fig. 119) will be found sections of

---

\* The above factor of safety is necessarily subject to modification in practice at *extreme* depths or where—owing to the introduction of high speed (machine) signalling—the weight of core becomes excessive without introducing practically any increase in the strength of the entire cable.

† Flexibility is of almost first importance in the instance of submarine-mine cables, which require constantly to be coiled and uncoiled.

[PLATE XXIII.

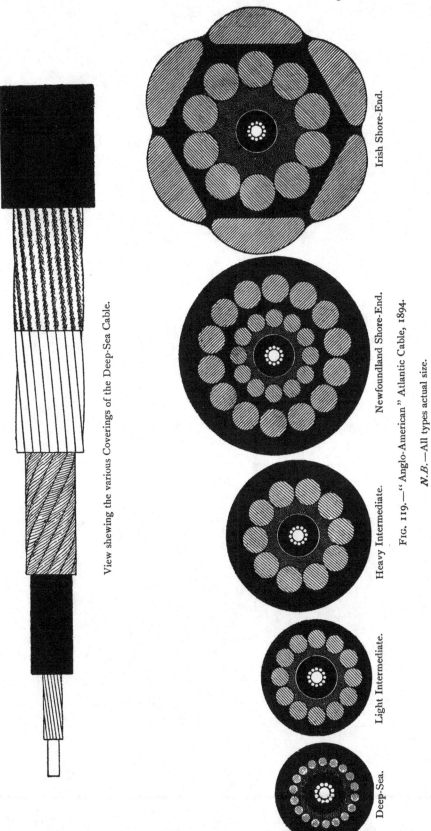

View shewing the various Coverings of the Deep-Sea Cable.

Irish Shore-End.

Newfoundland Shore-End.

Heavy Intermediate.

Light Intermediate.

Deep-Sea.

FIG. 119.—" Anglo-American " Atlantic Cable, 1894.

*N.B.*—All types actual size.

each type composing this latest* Atlantic cable †—a splendid specimen of up-to-date engineering in ocean telegraphic enterprise, and a vast improvement on this company's previous cables. The full specification of this typical and important specimen of deep-sea telegraphy is given in Appendix II. at the end of this Part. It is a good example of Messrs Clark, Forde and Taylor's careful specifications as engineers to the company in question.

The sectional elevation of the deep-sea, or main, type (D) shews well its various constituent parts, or coverings.

In the provision for the Irish shore end (type A in the specification)—where great surface and weight is necessary, owing to the prevalence of rocks—the iron wires, being laid up exceedingly short,‡ to meet the above requirements, do not appear in true section to be circular, though they are in reality.

The length of cable actually laid was 1,845 N.M., some of it running into 2,600 fathoms of water.

This cable was made and laid by the Telegraph Construction and Maintenance Company. For further data on the subject, the reader is referred to an able article by Mr Arthur Dearlove, which appeared in *The Electrician* at the time.§

**Representative Shallow-water Cable.**—As a representative of up-to-date shallow-water cables, the full specification will be found in Appendix III. of the Government Anglo-German four-conductor cable, made and laid by Messrs Siemens Brothers in 1891 between Bacton (Norfolk) and Borkum Island, over a sandy and fairly level bottom, with a maximum depth of only 20 fathoms. This cable is the joint property of the German and English Governments.

Another similar Anglo-German Government cable (with certain special

---

* At the moment of sending in proofs a new French Atlantic cable is about to be laid. This, having practically the same core as the above, is similar in character, except that in introducing an extra number of sheathing wires of the same gauge its weight is substantially— if not seriously—increased.

† The illustrations here given may be taken as a good example of the various types of cable suitable for different depths—up to about 3,000 fathoms—that is to say, where a heavy core of this description is employed for purposes of high-speed working on long and busy ocean cables. Where a shorter length is in question, or ordinary manual transmission suffices, the number of sheathing wires involved by the smaller core would be correspondingly less.

‡ The length of lay is actually about 4 inches, presenting an appearance of a spiral.

§ *The Electrician,* 12th October 1894.

modifications alluded to elsewhere) was constructed last year (1896) by the Silvertown Company.*

These specifications give fairly fully all that, as a rule, requires to be actually set forth—indeed, Post Office specifications are somewhat famous for their amplitude.†

**Specifications and Tests.**—The cable is invariably tested electrically after complete shipment in its various sections, either by the engineer to the owners, or in his presence. Indeed, there is sometimes a clause in the specification ensuring provision for this, whilst stating, moreover, that the cable must test steadily throughout, and yield uniform results with both currents.

It is also sometimes signified in the specification that every facility is to be given to the engineers for inspecting all the materials used in the manufacture of the cable, and for testing these materials if desired, besides a separate room and testing apparatus being provided at the contractor's works.

**Details of Construction.**—However, by those who, from experience, are best able to judge, it is not, as a rule, considered necessary, or even desirable, to go too closely into details of construction in the specification, but to leave them—in which there is, moreover, no finality—to responsible contractors who have a reputation to keep up,‡ and often a greater experience in details to work upon than any consulting engineer. Again, there is a certain prejudice to the publication of details of construction, such as might be involved by stipulations in a contract. Thus, a very ordinary specification clause with reference to the inner serving, runs somewhat as follows :—" The inner serving to consist of a good and sufficient serving of jute yarn steeped in cutch or other preservative solution ; " without any kind of reference to the lay of the yarns or any other similar detail in the method of carrying out the manufacture.

Once more, where an application of compound is to be applied imme-

---

* At the time of writing, this cable was still a subject of contract, and the particulars were not, therefore, available.

† Moreover, the construction of all Post Office cables is very carefully overlooked by officials of the Department at the contractor's works.

‡ Moreover, it must be remembered that as the contractors usually undertake the maintenance of the cable in good working order for a certain time, they are—if only for this reason—almost as interested in its complete success as the owners themselves.

diately outside the laid-up wires, it is most important that at the time of application it should be quite cool ; yet as often as not, no such stipulation is actually mentioned in the specification, it being known that the contractors are fully acquainted with the disastrous effect of the compound being applied hot straight on the iron wires, and that they are probably as anxious as the owners that the cable shall be a success.

**New Types.**—It need scarcely be remarked that manufacturers are always prepared to construct any sort of cable specified—within certain limits, at least. Sometimes, however, it happens that the contractors have opportunities of recommending a certain type of cable, such as they know by experience to suitably meet the requirements; or, may be, in which there is some manufacturing specialty of theirs—the result, perhaps, of a large range of practical experiment and experience.

**Cost of a Cable.**—In estimating the cost of a proposed cable, the first consideration is naturally the points to be put into communication ; secondly, the nature of the bottom between those points ; thirdly, the nature and length of various types suited for the different depths and the form of bottom to be encountered.

These points having been settled, it becomes necessary to calculate the cost of materials involved at the existing market prices.

The cost of the component parts of a core have already been touched on in a previous chapter. With regard to its serving, sheathing, and outer covering,* ordinary galvanised iron wire—standing 25 to 30 tons per square inch—as used for the heavy types, costs somewhere about £8 to £9 a ton ;† the "homo" wire, as used for deep-water cables, is sold at about £20 to £25 a ton ; and steel (of special quality), bearing a strain equivalent to 120 tons per square inch, sometimes runs into as much as £44 a ton.

Jute can now be got at £13 a ton ; but hemp varies in price from £30 to £40 a ton—the best species being even more.

The cost of a completed cable varies, of course, very much. Certain deep-sea types have of late years been known to be supplied at as low a

---

* It should, however, be remarked that owing to constant fluctuations in the market, all these prices are very variable within comparatively short periods.

† The extra charge for galvanising averages at about £1 a ton of shore-end wire.

figure as £75 a mile;* whereas shore-end types run into some £300 per N.M. and over.†

The cost of laying a cable may be often very roughly considered as somewhere about half as much again as the cost of construction; and may be estimated for (with a big margin for risk and contingencies) on that basis. Such estimate, of course, should include all the labour ‡ charges involved.§

The value of a cable when laid is the item of special consideration to shareholders and prospective shareholders; this including insurance. It is, however, sometimes a nice point as to what contingencies insurance would cover at a push.

**Data and Records.**—For some reason or other there is still a certain amount of difficulty in procuring full particulars respecting submarine telegraph lines that have already been laid for various companies. This is somewhat surprising in view of the fact that submarine telegraphy has many years ago emerged from the pioneer stage—if not actually from the sphere of science into that of routine. Moreover, the practice of cable engineering has now fallen into the hands of several big contracting firms, whereas for some time it was practically in the hands of one. However, that there *is* a difficulty, there can be no question.

In their excellent pocket-book—the up-to-date *vade mecum* of electric engineering practice—Messrs Munro and Jamieson‖ have given the results of electrical tests of various cables from the year 1872, together with such mechanical data as the individual weights of each component part.¶ Moreover, Mr H. R. Kempe's now famous, and very complete, work on electrical testing ** contains similar information regarding some of the more recent

---

* £250 to £300 per N.M. is a fair price for an Atlantic cable, with the heavy core involved, *without laying*. This would be materially more in the case of extra heavy cores providing for machine transmission.

† The average cost of a length of submarine cable may be approximately taken as nearly seven times that of a corresponding land wire.

‡ So long as Trades Unions hold sway—to the detriment of the country's trade, and, therefore, to its subjects generally—it would be well if a clause providing for the contingency of strikes were embodied in all contracts.

§ Atlantic cables have been made and laid for half a million of money. This figure compares favourably with the million sterling expended on several of our warships.

‖ "A Pocket-Book of Electrical Rules and Tables," by John Munro, C.E., and Andrew Jamieson, F.R.S.E., M.Inst.C.E. (Charles Griffin and Co.).

¶ The same sort of tables are also to be found in the "Handbook of Practical Telegraphy," by R. S. Culley (Longmans, Green and Co.).

** "A Handbook of Electrical Testing," by H. R. Kempe, A.M.Inst.C.E. (E. and F. N. Spon).

cables. The same class of particulars was presented by Clark and Sabine*
regarding cables laid between 1864 and 1870. In the author's opinion this
information—as regards the mechanical data—is of but little use by itself.
For the above combination of reasons, then, it has not been repeated here,
and owing to the difficulties already mentioned, no table whatever has been
given. However, in the Appendices will be found a form—already referred
to—which gives an idea of the class of information such as is of some value,
from an historical record point of view, for obtaining and filling in as
opportunity occurs.

On examination of the form in question, it will be evident, from what
has already transpired, that much of the mechanical data would only be
required as regards the deep-sea types ; and, on the other hand, some only
as regards shore-end and intermediate types. The individual length and
description of each type should be furnished, as well as other information
respecting each and all—such as gauge and number of wires in the
sheathing or sheathings, also the description of the outer covering.

Moreover, the results of the particular tests applied to the wires used in
every type should be set forth under that particular type.

---

* "Electrical Tables and Formulæ," by Latimer Clark and Robert Sabine (E. and
F. N. Spon, 1871).

Indo-European Telegraph Department : Staff Quarters at Fāo.

## SECTION 3.—LIGHT CABLES.

**Early Forms.** — There was a period in the history of submarine telegraphy when many devices in the form of what went by the name of "light cables" * (*i.e.*, without any outside armour whatever) were seriously thought of; and several undertakings based on their adoption were set afoot. These, however, were never turned to practical account, owing, presumably, to the heavy risk involved.

These devices were a very natural sequence to the various failures in the Mediterranean, and elsewhere, in the fifties and sixties—failures not only to recover a heavy ordinary iron-sheathed cable, but also actually to maintain control over it during the operation of paying out, for want of efficient holding-back gear on the laying vessel.†

Mr C. F. Varley, F.R.S., was always an enthusiastic advocate of light cables, and so was Professor Fleeming Jenkin, F.R.S.

According to most of the many devices in the way of "light" cables, the little strength considered necessary was put in the conductor itself,‡ the latter being iron very often. An early patent § was taken out by Monsieur F. M. Baudoin for a cable based on this principle, the conductor composed of six iron wires, each 079 inch in diameter (protected only by bitumen), encircling a heart of tarred hemp—a reversal somewhat of ordinary practice.

The light cable, however, which attracted the most attention was that of Mr Thomas Allan. This consisted in a solid copper wire surrounded by a

---

* Some of the above, however, scarcely partook of the dignity of a "cable," as the term is usually understood in the present day. Still, they represented the first principles of a light cable in its most essential features, and were, perhaps, on the whole the most sensible form.

† The number of different kinds of patented cables inspired in the minds of inventors about the time of the "First Atlantic" and between that and the "Second Atlantic" were simply legion. These strange devices were intended to overcome supposed impossibilities. Besides being of the "light-cable" description, cheap insulation formed a feature in most of these. In some the conductor—in one instance to be a spiral capable of being drawn out—was only to be enveloped in a serving of rope, composed of hemp yarn or other vegetable fibrous substance worked up with varnish, marine glue, pitch, or the like. This covering was to be applied with a very short lay. Such a cable would obviously be unsuited for the bottom of the sea—especially where at all irregular—even if capable of being laid.

‡ This was apparently, to a great extent, with a view to economy—as well as lightness—no outside sheathing wires being employed. In some of these schemes only just enough weight was furnished in the conductor to ensure it sinking when covered with its insulating material. Moreover, as a rule, no notions of after-recovery were indulged in, the point aimed at being to make a cheap line which could be successfully laid from point to point.

§ In 1858.

number of closely fitting smaller steel, or iron, wires,* applied spirally. The conductor so formed was covered with ordinary gutta-percha or india-rubber insulation—the latter at any rate outside — the core being encased in a thin covering of hemp, tarred string, or canvas impregnated with pitch, or marine glue.† All these were intended for insulating as well as preservative purposes.

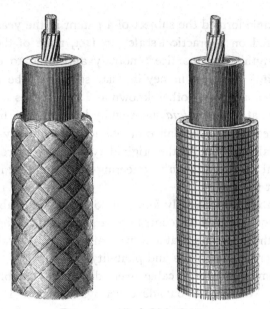

FIG. 120.—Allan's Light Cables.

It was claimed for this cable—two forms of which are shewn in Fig. 120—that it combined the utmost lightness (weighing only 353 lbs. per N.M. in sea-water) consistent with necessary tensile strength (2 tons).‡ It

---

* Mr Allan appears to have considered at that time that galvanising had a weakening effect on iron, and his wires were not to be galvanised. In point of fact, in the present day it is found to slightly increase the breaking strain.

† Marine glue is a mixture of one part of india-rubber dissolved and softened in mineral naphtha oil, and two parts of gum-lac.

‡ Thus it was said to be capable of vertically supporting in the sea the weight of 12 miles of its own length. Again, not being bulky, a ship of quite medium size could easily carry some 3,000 miles—more than sufficient to telegraphically connect Europe with America. Moreover, the laying could be effected with little risk and without elaborate "holding-back" machinery, the cost of construction added to that of laying being some 40 per cent. less than in the case of an ordinary iron-sheathed cable to meet the same requirements.

It was even claimed that, by such a form of line, the speed would be increased by as much as 50 per cent. The reverse might more reasonably have been expected.

However, in addition to the objections applicable to all light cables, it was remarked with regard to Allan's cable that the contact of the two different metals in the conductor would be likely to set up serious chemical action.

was enclosed, moreover, in as small an area as possible, in order to reduce the coefficient of friction to a minimum.

The electrical objections to a line of this description are, however, not inconsiderable. The combined metals would form a bad conductor to start with, quite apart from the chemical action which would be likely to occur. Again, the electro-static capacity with so large a conductor would be very great.

Mr Allan's cable formed the subject of a patent in the year 1853, but it was never adopted on a practical scale : in fact, none of these proposals ever came to anything in practice,* nobody appearing to care (when it came to the point) to invest money in what seemed to be at the best a risky affair as compared with other known and tried types of cable which had already been, at any rate, *laid* successfully.† With the light of experience it would seem that the balance of opinion in those days was correct. We have practically adhered to the original type of cable from the beginning—an unusual circumstance in engineering—shewing that it was not the cable that was at fault.

In some of the different early forms of so-called light cables, not only was the conductor intended to contribute largely towards the strength of the whole, but the dielectric material was also purposely chosen for its actual tensile strength, toughness and pliability.

About the same time several cables were designed—notably one by the late Mr Charles West—with outside coverings constituted by layers of

---

* There are great objections to absolutely light cables of any species if only on account of their tendency to sink so slowly under ordinary conditions ; and being seriously subject to surface currents, all control over them whilst paying out is lost. Thus, such a cable may be laid on an entirely different—and possibly less safe—route to that intended, and the length of cable wasted is liable to be, in consequence, of much greater value than the initial cost saved in construction.

Again, a cable of this type could never be laid as well at the bottom, and therefore would be of less value ultimately to its owners. In paying out, the ship would require to proceed abnormally slowly under any circumstances, though it must be admitted there would be less risk of putting an undue strain on such a cable (of low specific gravity) in the operation, and the "holding back," or brake, power required would be correspondingly less.

† In this connection it may be remarked that submarine telegraphy is somewhat at a disadvantage—perhaps more so than many other branches of "heavy engineering"—in that in investing money in any new venture it is being·*literally* sunk at the bottom of the sea ; and, if a failure, is not recoverable. Thus, unless a novel type of cable or original method of insulation bears success and advantage on the face of it, an inventor finds great difficulty in obtaining for it a fair trial in practice ; though contractors, of course, may often be in a position to try a novelty of their own. Furnished with an extensive practical experience, the latter are, however, certainly the best judges, and anything they attempt in the way of novelty is distinctly more likely to prove a success.

canvas, weaved yarn, etc., plaited up reverse ways, with a view to binding the whole more firmly together.

At a later date MM. Blondot and Bourdin, designed a type of light cable which was exhibited at the Paris Electrical Exhibition of 1881.

This was, in reality, merely an ordinary core; the conductor being a single solid copper wire .118 inch in diameter, and the dielectric constituted by three alternate layers of gutta-percha and Para rubber,* with a coating of Chatterton's compound between the conductor and the first dielectric layer. The gutta used for manufacturing the outer layer of the insulating envelope was mixed, whilst hot, with bichloride of mercury, in order to kill any submarine creature which might attack the insulation. The whole was covered with rubber-proofed cotton tape of great strength and durability.

The point of special interest in connection with this light cable was, however, the detailed manner in which, for the first time, the designers had endeavoured to cope with the difficulties of properly laying such a core.

They recognised that so far from any holding-back strain being admissible, they must actually force out their line, owing to its extreme lightness. Accordingly, their arrangements included a large reel or drum—mounted horizontally on the laying vessel—which was to be turned by an independent engine. The speed of this, however, was to be governed by that of the vessel's propelling machinery. The idea was that the rate of paying out would then vary in accordance with the ship's speed, and would only differ from it by a previously arranged proportional quantity, representing the slack—i.e., the percentage of excess cable over distance covered.† The cable passed over the ship's stern into the sea between two horizontal rollers geared together and covered with india-rubber. A small model illustrating the paying-out machinery for laying this light (core) cable was exhibited along with the specimen of the cable itself.

In order to obviate the necessity of joints of insulation in so unprotected a line, the whole length of copper conductor was to be prepared beforehand, in one piece, so as to enable the insulation covering to be applied in one continuous length in each coat. The joints of the conductor were

---

\* As before stated, the combination of layers of both gutta-percha and india-rubber in any dielectric could never be a success in practice on account of the different degrees to which each are affected by a change of temperature, and especially in the case of *pure*, unvulcanised, rubber.

† In this, however, the inventors appear to have overlooked two facts. First, that the ship's speed through the water bears no fixed reliable ratio to the speed of the propelling machinery; and secondly, that the amount of slack depends upon the difference between the length of cable paid out and the distance travelled *overground*. The latter can only be determined by reference to fixed outside points, such as land, or astral bodies.

soldered with silver, and protected with a gun-metal ferrule. Practical trials had proved that such joints possessed great strength and sufficient flexibility to allow of the so-formed continuous conductor being wound on a small cylinder.

The first form of light cable which was actually put on trial in the early days of submarine telegraphy was that designed by Mr Siemens (afterwards Sir C. W. Siemens, F.R.S.), which he endeavoured to lay between Oran and Carthagena in 1864.* This has already been described and illustrated in Part I.† Here the whole tensile strength was confined to the two layers of white (untarred) hemp which surrounded the core, the object of the metal riband being in this case that of holding the hemp firmly together, besides acting as a mechanical protection.

Although this cable was ultimately laid after several successive failures in depths of 1,300 and 1,400 fathoms, when another break occurred within sight of the Spanish coast, the undertaking was abandoned. ‡

In 1876 Mr F. R. Lucas—who has probably had a greater experience than any one at present engaged in cable-laying—took out a patent for a type such as is a modification of that of the 1865 and 1866 "Atlantics,"§ the object being similarly that of low specific gravity. In this device, instead of each iron wire (sometimes taped) being previously served with a strand of hemp, the iron wires were alternated with Manilla hemp or other yarns.‖ As in the Wright type above alluded to, one of the principles of this cable was that the breaking strain would be constituted by the

---

* It may be observed again here that the design was that of the contractor himself.

† This cable is mainly of interest on account of its ingenious, and beautifully carried out, construction.

The tape being of a material thickness, there was a continuous indentation at the edge of the metal tape for the overlap of each turn to fit into, thus producing a perfectly flat surface.

‡ The final breakdown was attributed to insufficient slack having been paid out on getting into shoal water. The cable must have been suspended over a great length, and being unable to withstand the strain brought on it at the points of suspension, parted before the laying vessel had reached the other shore.

The conclusion to be drawn here is that essentially light cables, by reason of their low tensile strength, must be laid with considerably more slack than those of ordinary type, so that they may experience the least possible strain, both during and after submersion. This being so, any looked for saving in cost in manufacture would probably not be realised in the end. Moreover, all other things being equal, the working speed attainable would be liable to be less for the same reasons—*i.e.*, by increased percentage of slack necessary for a light cable. Thus, the nett returns accruing to a light cable, if above a certain length, would tend to be below those of a cable as ordinarily constructed.

§ Based on the patented invention of the late Mr Edwin Payton Wright for a combined hemp and iron rope, and fully described in Part I.

‖ Such a cable is naturally laid up by each bobbin on the carriage of the sheathing machine being alternately of iron wire and hemp yarns.

sum of the breaking strains of the iron and the hemp—both being applied together with the same lay.* It has an advantage over the Wright cable in obtaining an equal strength in a smaller bulk, thus reducing the surface area. On the other hand, the hemp not being applied round each wire, but only between wires, the feature of protection to the iron wires—for however short a time—could not apply in this case. Moreover, after the hemp has decayed, the wires are more than ever free to buckle into the core.

Nevertheless cables of this description were adopted for several important lines—notably on one to Australia—and again between certain points on the north coast of South America. It was, too, a good deal less costly than the previous hemp and iron combination.

It may be remarked here that either of the above types in obtaining a low specific gravity by using a large proportion of hemp, are of necessity of great bulk, comparatively speaking.

Other combinations of hemp and iron for deep water—based on the principle of a low specific gravity for great depths—have, from time to time, been devised. This, with a view to meeting the difficulties of laying, and still more of recovery, experienced in deep water—both in respect to "holding back" during paying out, and to the weight in suspension whilst picking up.

**Requirements: Weak Features in Hemp and Iron Light Cables.** —For purposes of recovery, a cable should have as low an actual weight in water as possible, combined with a high modulus of tension,† but should not be of great bulk, as this introduces outside friction. ‡

The above low specific gravity combinations of iron with a large quantum of hemp are very light in air when dry. On the other hand, when wet—owing to the large amount of water absorbed by the hemp—they become materially heavier in air. Thus, in loading the tanks of a ship with such a cable, it not only requires a large amount of space (thereby involving extra cost) on account of its great bulk; but also, further, the limit of length which can be carried comes into serious consideration by reason of its weight when wet. In fact, a cable of this description must be regarded as a heavy type for transport purposes, owing to the dead-weight of water which it absorbs.

---

* Such a system of equal length of lays does not, apparently, take into consideration the different rates of elongation of the two materials. The breaking strain of the two together was, however, said to be the same as the sum of each of the two taken separately.

† Sometimes also expressed as the modulus of tenacity.

‡ In these hemp and iron cables, the strain due to friction may easily be greater than that due to the dead weight of the cable itself.

Such a cable has an actual breaking strain—of even greater importance for recovery work than a low specific gravity—far below that of a corresponding close-sheathed line.

Under strain, the wires of any open-sheathed cable are liable to break one by one—instead of all together as in the close-sheathed type—and to be pulled in on to the core.

In picking up a bulky, open-sheathed cable, the bight is likely to become flattened on the grapnel by the wires spreading out until each in turn parts under the unequal tension—or at any rate the core is liable to obtrude between the wires where they are flattened out. In fact, these open-sheathed—or "open-jawed" cables, as they are sometimes termed—are always more subject to faults. Thus, if a wire breaks, either end may pierce through to the core, which could not well happen with the close-fitting arch type of sheathing where each wire butts firmly against the next.*

Again, the "open-jawed" type lays itself open much more readily to the successful attacks of the teredo (and other similar enemies of cables), which cannot so well work their way into a close iron-sheathed line when in perfect condition.

It is true that the hemp may tend to preserve the wires, to some extent, for a certain length of time. But this soon decays ; especially as the oxide of iron—or *rust*, in common *parlance*—tends to destroy it. Thus, ultimately, the open-sheathed cable becomes, practically speaking, a cage composed of a bundle of loose iron wires, in which form they decay more rapidly than ever on account of their entire surface being exposed to the oxidising influence of sea-water.†

It was mainly to meet these difficulties that Bright and Clark's compound was introduced as an extra preservative for resisting oxidation beyond that of galvanising, the silica being added with a view to defeating teredo and other such-like attacks.

---

* The wires of a cylindrical metal tube, in forming a complete arch, retain their position even under a heavy strain, and so protect the core, by taking an equal pressure all round. The total resistance to stress is, in fact, in this type very much enhanced as compared with an open-sheathed cable, where the wires become deranged with fatal consequences sometimes, owing to the absence of any such metallic arch or cylinder.

† On the other hand, with the *close*-sheathed cable, when the outside covering becomes decayed, there is still a complete cable with many years of life remaining. Moreover, as the wires butt against one another, their inner side is protected from oxidation, even after the outside covering of yarn, or tape, has been removed.

Thus, the strength of a cable of this class is likely to be maintained for an appreciably longer time than is possible with the open-sheathed type. Indeed, from a durability or "wear-and-tear" point of view, the latter is probably unsuitable even for the deepest water—for which it was especially designed, on account of its low specific gravity.

An iron close-sheathed cable so dealt with now forms the usual type of cable adopted;* and this, therefore, is the type the construction of which has been followed up here in all its stages, subsequent to the serving of the core.

**Later Types.**—Some remarks are now due concerning the more modern notions of a light cable—after the pattern of a hempen sheath, pure and simple.

As early as 1864, the late Mr Willoughby Smith suggested to the Atlantic Telegraph Company a type of light cable, the core of which was to have two coverings of hemp wound on in opposite directions, the inner covering consisting of eight strands and the outer of seventeen. This cable was to weigh but a little over ½ ton in air, and have a breaking strain above 4 tons—equivalent to about 6¾ N.M. of its own length in air, and as much as 29 miles in water.†

However, when a sample of this cable had been under water for six months, Mr Willoughby Smith found that the contraction of the hemp had caused the core to protrude at several places.

In 1883 Captain Samuel Trott,‡ of the "Anglo" Company's S.S. "Minia," and Mr F. A. Hamilton strongly urged the claims of a light cable of their device. This, indeed, may be said to be the only light cable within recent years that has attracted serious attention.§

This was again a hemp-sheathed cable without any iron whatever.

It was constructed as follows :—The insulated conductor, of the ordinary type, had two separate and distinct hempen coverings. The first of these coverings to be wound round the core right-handedly and composed of strands laid up the reverse way. The second covering was wound left-handedly, and consisted of strands laid up from right to left. Fig. 121 shews one form of the Trott and Hamilton cable as here described.

---

* The close iron-sheathed cable has, in fact, been practically the form in use ever since the first submarine cable, with the exception of a few early important Atlantic and other deep-sea lines referred to in Part I. Bright and Clark's cable compound has been invariably employed subsequent to its introduction in 1862.

† This—the length of cable hanging vertically in water whose weight will cause the cable to break at the point of suspension—is very usually expressed as the "modulus of rupture" or the "modulus of tension." Strictly speaking, however, the latter should only stand for the weight it will bear, as a tensional strain, *without* breaking.

‡ Probably no man has had so extensive an experience as Captain Trott in repairing cables in deep water under more or less unfavourable conditions.

§ In 1874 Mr W. M. Bullivant obtained a patent (No. 1,159) for a very similar type, designed especially with much the same features.

**Unlaying and Laying up of Turns in a Cable.**—It was claimed for this arrangement that the contraction of either covering after immersion would not only resist any tendency to unlay in the other covering, but would also have the effect of tightening up the turns of both coverings, and so strengthen the cable generally.

As has already been pointed out, in the construction of an ordinary iron-sheathed cable provisions are made for avoiding any torsional twist in the wires when laying up. Moreover, if such a cable as fast as manu-factured was coiled direct on to a drum, and if it was possible to pay it out from the laying vessel to its destination from a drum, such a cable should always be free from torsion except when—whilst fixed at some point—it is drawn on to from another point.

As, however, any considerable length and weight of submarine cable requires of necessity to be coiled into tanks first in the factory and then

FIG. 121.—Trott and Hamilton's Light Cable.

from there into other tanks on board the laying vessel, the conditions are different.

In the first place, turns are put into the cable as the bight is coiled into the factory tank. These turns are taken out as the cable is uncoiled from the tank, but put in again on coiling into the tank of the laying vessel.*

As the cable is drawn up from the tank on board ship, in the course of paying it out to the bottom of the ocean, these turns again relieve them-selves. This action takes place almost entirely in the short length between the running coil and the hawse-pipe over the tank.

As the cable passes through the paying-out machinery, it becomes sub-jected to a strain varying according to the depth of water and the slack aimed at. This strain is at its maximum just outside the stern sheave, and

---

* This was all beautifully described and illustrated in Mr F. C. Webb's now almost classic paper on the " Laying and Repairing of Submarine Cables" (*Minutes of Proceed-ings Inst. C.E.*, 1858).

diminishes as the cable passes out and sinks to the bottom, at which point it arrives slack. The effect of this strain is to cause the cable to turn on its axis, or unlay, at the point at which it is at its maximum. This unlaying gradually lessens as the strain diminishes, until a point is reached where unlaying ceases and laying up again—or returning to its normal state— begins. The cable thus arrives at the bottom practically in the same state as it was when in the factory.

It has been contended by some writers—notably by Messrs Trott and Hamilton—that this is not the case, but the experience of these experts was more in regard to the *picking up* of cables. In this a very different state of affairs exists, particularly in the instance of old cables in which the outer cover- ing of yarns—laid in the reverse direction to the wires—has rotted away. The effect of picking up a cable under a considerable strain is to elongate the lay as the cable comes inboard over the bow sheave. This unlaying continues (if the strain be maintained) until the lay of the cable may be so much elongated as to cause the core to protrude in places, and the cable, where slack at the bottom, to form itself into complicated kinks and tangles which frequently break under even a moderate strain. Trouble from this cause frequently occurs after a length of cable has been picked up, and the ship stops to splice on new cable. Should she bring the lead at all vertical the cable at the bottom flies into kinks, and—on the least strain being subsequently applied—a break, or fault, follows.

As regards this unlaying tendency, however, whatever the circumstances may be, this should not become serious until the outside serving has rotted away. In other words, no such unlaying should occur with any form of cable when new—*i.e.*, on being laid for the first time—though some such objection may occasionally apply in the case of a shallow-water cable when relaid after some years.

It is contended by some that an outside covering of strong hemp cords, in taking all the working strains, tends to prevent this unlaying action ; and, if so, this places their employment at an additional advantage—besides those already mentioned.

Messrs Trott and Hamilton's hempen cable, without any iron wires whatever, was designed with a view to obviating the above difficulties, to which they drew special attention in the course of a paper introducing their cable to the profession generally.*

In the author's opinion, however, such objections do not apply until a

---

* "Submarine Telegraph Cables : Their Decay and Renewal," by Samuel Trott and F. A. Hamilton, *Jour. Soc. Tel. Engrs.*, 1884. This paper attracted some attention at the time, and incited an eminently practical discussion.

cable requires to be lifted, when it is free from the bottom, and the wires, perhaps, partially corroded ; and even then an iron-sheathed cable would, as a rule, be better capable of supporting its own weight without fracture than any unarmoured cable.

**Considerations regarding Hempen Cables.**—At first sight, in some ways, it must be admitted that a hempen cable without any iron appears to be quite the right thing for deep water.  At one time it was thought that on account of hemp being so light (in air) that some difficulty might be met with in inducing it to sink.  However, like jute, it absorbs water very readily.   There would, therefore, on this score be no difficulty in effecting the eventual submergence of a hemp-clad cable, though it would sink only slowly, and would be liable to be taken, for some length, out of its direct route owing to any prevailing surface and sub-surface currents.

Such a cable should, on the grounds of low specific gravity, be good for lifting in deep water ; though, as regards actual weight wet in air, this, by becoming readily "waterlogged," may be something considerable.

One of the great objections raised to hempen cables at first was that it was found to shrink so seriously in water as to strangle the core.   Due probably to improper construction, such cables were found also to unlay at quite an early stage of manufacture.*

By only employing hemp that has been thoroughly shrunk beforehand, and by careful and ingenious improvements in construction (based on extensive experience), Messrs Trott and Hamilton certainly succeeded in overcoming these objections.   The best and most costly (Italian) hemp was used : in fact generally, initial economy does not appear to have been one of the points claimed by the authors in favour of their cable, though ultimate economy was.

When new, this type of cable is probably as well able to support its own weight as an iron-sheathed cable.

However, though hemp alone was supposed at first to be practically indestructible in water, the strength of the hemp in hempen cables is not found to be at all sufficiently permanent—though certainly more so than when combined with iron—to be relied on as the sole medium of strength.

It is quite true that iron (even when galvanised, compounded and taped) tends to be prejudicially attacked—*i.e.*, corroded—by the carbonic acid dissolved in sea-water ; † and when once chemical action sets in here,

---

* Possibly both layers were applied in the same direction.

† The significance of the ends of the wires at a break becoming sometimes gradually reduced to needle points is not always certain; but it is usually, in most cases, considered to imply the existence of excessive chemical action, or excessive weakness—in the iron itself or in the galvanising—combined with a severe strain.

the hemp in an iron sheath rapidly decays.* Still, as hemp has nothing like the same actual strength to start with, and as—even when alone— it materially decays under many conditions, a hempen cable would not probably bear the same strain successfully, in the event of a repair some years after submergence, that an ordinary iron-sheathed cable would. In other words, hemp lasts better in salt-water when no iron is present: but as there is much less initial strength, the deterioration of a hempen line lying at the bottom of the sea is much more serious, as a rule—though often less in degree—than in the case of an ordinary iron-sheathed cable.

This is quite apart from the question as to whether a hempen cable would stand the wear and tear of being dragged along the bottom in the course of repairs. It is also apart from the question as to whether—on account of lightness—its presence would be shewn on the dynamometer sufficiently soon and with sufficient certainty.

Such cables would certainly require to be paid out with a great amount of slack (involving, therefore, an unusually great length in connecting given points telegraphically) to meet the drawback, mainly as regards raising, of a lower actual tensile strength and a greater area for skin friction.† This would naturally have the effect of inducing an increased tendency towards kinks.

Though, at first sight, a light and pliable cable might appear to more readily adapt itself and fit into the curvatures, or irregularities, of the. ocean bed, this is not so actually. Indeed, on the contrary, owing to its low specific gravity and comparatively great bulk, it naturally tends, whilst leaving the ship, to descend in too horizontal a line for good and per- manent laying in this respect. For the last additional reason, therefore, in order to obtain a fairly vertical line during paying out—similar to that in submerging an iron-sheathed cable—it would be necessary in the case of the hempen cable to drive the ship much slower, and to pay out a great deal more slack.

The question of the applicability in practice of a hempen cable depends, however, very largely on the nature of the bottom, as well as on the existence or non-existence of strong bottom, or sub-surface, currents. The hemp or canvas covering in an ordinary iron-sheathed cable only lasts for any material time in very still water, and is seldom, therefore, of much use

---

* Furthermore, the presence of iron in a cable can, on the whole, scarcely be favour- able from an electrical fault standpoint. Iron being a cathode, the tendency is for oxide of iron to be deposited.

† Moreover, the surface of the above cable being distinctly rough, the coefficient of friction is further increased on this account.

in quite shallow depths. Moving water introduces more decay by bringing a fresh supply of oxygen.* In shoal water, owing to the strong currents sometimes prevailing, the nature of the bottom is often found to be continually changing.

Hence, it will be seen that hempen cables could never be seriously considered for shallow water if only on account of the friction, decay, and abrasion they would be liable to experience, in the ordinary course—let alone icebergs. Comparatively warm teredo- and fish-ridden shallow waters would also put them outside the pale of practical politics; for a cable of this description would, indeed, prove a ready prey and acceptable meal to this class of enemy.

It is possible, however, that such cables might have a sphere of usefulness in places where it could be ascertained beyond doubt that no currents existed on the ocean bed. Indeed, under these circumstances—provided also that there was an entire absence of chemical action—a hempen cable would probably last longer than one that is iron-sheathed. Where, however, there is any chance of the line being shifted—and thus getting chafed over rocks by existing currents—a cable of this class would be quite out of the question.

It is believed, however, that this type has never been recommended for anything but absolutely deep water.

Under any circumstances, in the author's opinion, the "wringing action" referred to would be best defeated—if necessary—by some form of double sheathing of iron wires, each sheathing being extremely light, of opposite lays, and, moreover, of precisely the same length.

A length of about 50 N.M. of the Trott and Hamilton cable was made by the Telegraph Construction Company in 1889, and during the course of repairs inserted in deep water in the 1869 "Anglo" Atlantic between Brest and St Pierre. Again, in 1892 a similar length was put in, also in deep water. Both of these lengths were in circuit for some three or four years.

The more salient features regarding the question of such cables were ably dealt with by Mr James Graves in the course of a communication to the *Journal of the Society of Telegraph Engineers* by way of discussion on Messrs Trott and Hamilton's paper.†

---

* Thus, in small depths a cable should, really speaking, be actually *buried* in order to avoid moving matter.

In deep water, owing partly to the soft nature of the bottom, a cable usually buries itself—to a slight extent, at any rate.

† "On the Causes of Failure of Deep-Sea Cables," by James Graves (*Jour. Soc. Tel. Engrs.*, 1884).

Unless the possibility of recovering, and successfully repairing, hempen and other light cables can be made a greater certainty, it seems doubtful whether they will ever come into practical use, even for deep water—though very good in theory.

There can, however, be no question that if some form of light cable were devised which, whilst obviating the various objections—especially that of decay—applying to the ordinary iron and hemp combinations, really possessed the required strength, it would have a great future.

In the meanwhile it is generally considered more prudent to construct cables which—disregarding initial cost—can, at all events, be safely laid with probably greater lasting security ; and such as nowadays can, practically speaking, be repaired with absolute certainty.

**Light (Aluminium Sheathed) Cables.**—In 1893 Mr Edward Bright, M.Inst.C.E., and the author devised a species of light cable which en-

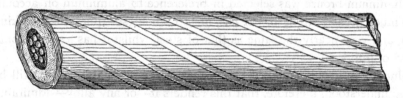

FIG. 122.—Bright's Aluminium-Bronze Cable.

deavoured to combine sufficient strength with lightness and freedom from corrosion, and such as might be found applicable for special cases.

The main feature about it was that for the ordinary iron sheathing wires of a deep-water cable was substituted aluminium bronze * " in the form of a riband laid upon, or surrounded by, flexible wrappings of hemp flax, jute, or other suitable substance." " In thus enormously reducing the weight and imparting freedom from corrosion, far greater durability is secured."

In this device (Fig. 122) the metal tape is applied spirally with an exceedingly long lay—varying with the type of cable—the length of which is (as in the ordinary wire sheathing) only limited by considerations of efficient binding.

Between the metal tape and the ordinary inner, tanned jute or hemp, serving, provision is made for a thin cotton tape (as a sort of "washer") previously soaked in Bright and Clark's silicated compound.† The latter is

---

* Aluminium when mixed with about 0.05 per cent. of silicon or tungstan is known as aluminium bronze.

† It may be here remarked that, for anti-boring insect purposes, in the preparation of this compound to its greatest advantage, a detail of some importance is the calcining of

applied with overlap, but one of the main features of novelty about this device is that the more or less thin and narrow metallic riband not only has no overlap, but there is a small space between the edges of each turn sufficient to allow of a very slight " play."

Over the metallic tape* another cotton tape may be applied of the same character as the former, and then—where required—a second metallic wrapping over that, applied in the opposite direction to the first, but similar in other respects.    The breadth of the metallic riband should, as a rule, be about $\frac{1}{4}$ inch with a spiral turn completed in 6 inches in the case of a cable or rope not exceeding two inches in diameter—increasing, of course, with larger cables.    The low specific gravity of aluminium and its alloys is an obvious point in its favour as against iron for the armour of submarine cables—provided that it will support an equal length of itself, and that it is equally durable.    It permits of a much greater length being carried by a given ship, besides taxing the machinery less, for any purposes to which (as rope generally) it may be applied.

Aluminium bronze was selected in preference to aluminium on account of its much greater strength.    It has, in fact, about six times the breaking strain, and very nearly the same as that of steel, whilst it has a weight only a third of iron.†

Aluminium is only equalled by copper as regards pliability.    It will be readily understood, therefore, that this renders it—or any alloy—admirably suited for the sheathing of a submarine cable.

Again, the riband form was preferred in this device on account of the further increased flexibility thereby ensured as compared with that obtained from wires of the same material under similar conditions.    It was on these grounds, moreover (as already stated), that the metal tape is not allowed to overlap—indeed, the edges do not even meet.    Again, such a taping could, under no circumstances, seriously damage the core under tension or pressure.

It has been suggested that there might be some difficulty in producing efficient joints in a sheath of this character, and that, in fact, there is no satisfactory method of jointing pieces of aluminium or aluminium alloys. A sufficient answer to this finds itself in the many miles of already existing aerial line composed of this material.

---

the flint down to the finest powder possible to ensure it properly mixing in a permanent manner with the pitch and tar.

* Provision was made for the aluminium sheathing taking the form of *wires*, if preferred.

† Thus, it is very largely used by the Post Office (as well as silicium bronze) for overhead telephone lines, especially in instances of very long spans.

Further, the durability of aluminium has been questioned (notwith-standing the above extensive system of aluminium lines), and a comparison has been drawn with aluminium torpedo boats which have shewn signs of somewhat early decay. In reply here, however, it may be pointed out that in the latter case *air* (in addition to water) is present, as an oxidising agent, in large quantities. This would not, of course, be so in the case of a submerged cable.

T.S. " Scotia" landing the " Eastern Extension" Company's Australia—New Zealand Cable, near Sydney, 1890.

# APPENDICES.

---

# APPENDIX I.

## BRITISH LEGAL STANDARD WIRE GAUGE.

| Gauge No. | Decimal Equivalents. Inch. | Metric Equivalents. Millimetres. | Gauge No. | Decimal Equivalents. Inch. | Metric Equivalents. Millimetres. |
|---|---|---|---|---|---|
| 7/0 | 0.500 | 12.700 | 23 | 0.024 | 0.610 |
| 6/0 | 0.464 | 11.785 | 24 | 0.022 | 0.559 |
| 5/0 | 0.432 | 10.973 | 25 | 0.020 | 0.508 |
| 4/0 | 0.400 | 10.160 | 26 | 0.018 | 0.457 |
| 3/0 | 0.372 | 9.449 | 27 | 0.0164 | 0.4166 |
| 2/0 | 0.348 | 8.839 | 28 | 0.0148 | 0.3759 |
| 0 | 0.324 | 8.229 | 29 | 0.0136 | 0.3454 |
| 1 | 0.300 | 7.620 | 30 | 0.0124 | 0.3150 |
| 2 | 0.276 | 7.010 | 31 | 0.0116 | 0.2946 |
| 3 | 0.252 | 6.401 | 32 | 0.0108 | 0.2743 |
| 4 | 0.232 | 5.893 | 33 | 0.0100 | 0.2540 |
| 5 | 0.212 | 5.385 | 34 | 0.0092 | 0.2337 |
| 6 | 0.192 | 4.877 | 35 | 0.0084 | 0 2134 |
| 7 | 0.176 | 4.470 | 36 | 0.0076 | 0.1930 |
| 8 | 0.160 | 4.064 | 37 | 0.0068 | 0.1727 |
| 9 | 0.144 | 3.658 | 38 | 0.0060 | 0.1524 |
| 10 | 0.128 | 3.251 | 39 | 0.0052 | 0.1321 |
| 11 | 0.116 | 2.946 | 40 | 0.0048 | 0.1219 |
| 12 | 0.104 | 2.642 | 41 | 0.0044 | 0.1118 |
| 13 | 0.092 | 2.337 | 42 | 0.0040 | 0.1016 |
| 14 | 0.080 | 2.032 | 43 | 0.0036 | 0.0914 |
| 15 | 0.072 | 1.829 | 44 | 0.0032 | 0.0813 |
| 16 | 0.064 | 1.626 | 45 | 0 0028 | 0.0711 |
| 17 | 0.056 | 1.422 | 46 | 0.0024 | 0.0610 |
| 18 | 0.048 | 1.219 | 47 | 0.0020 | 0.0508 |
| 19 | 0.040 | 1.016 | 48 | 0.0016 | 0.0406 |
| 20 | 0.036 | 0.914 | 49 | 0.0012 | 0.0305 |
| 21 | 0.032 | 0.813 | 50 | 0.0010 | 0.0254 |
| 22 | 0.028 | 0.711 | | | |

NOTE.—The above gauge was rendered legal by Government in 1884, conformably with a notice issued by the Board of Trade, which ran as follows :—"On and after 1st March, no other wire gauge can be used in trade in this country, that is, no contracts or dealings can be enforced legally which are made by any other sizes than those above given, as made by an Order in Council, dated 23rd August 1883."

# APPENDIX II.

---

ANGLO-AMERICAN TELEGRAPH CO.'S VALENTIA—HEART'S CONTENT
CABLE, 1894.

---

## SPECIFICATION.

THE following are the lengths and types of cable to be furnished by the contractors, under the foregoing contract, viz. :—

| Core $\frac{650}{400}$ Copper. Gutta-percha. | Lengths. N.M. |
|---|---|
| Type A.  Type B closed with 14 No. 1 (.300) galvanised   -   - | 4 |
| Type E.  10 No. 2 (.280) galvanised   -   -   -   - | 17 |
| Type B.  12 No. 6 (.200) galvanised   -   -   -   -   - | 347 |
| Type D.  18 No. 14 (.083) galvanised.   Each wire taped and compounded -   -   -   -   -   ;   -   -   - | 1,587 |
| Total   - | 1,955 |

CORE.

Conductor.    (A.) The conductor to consist of a central copper wire, .122 of an inch in diameter, surrounded by twelve copper wires, each .041 of an inch in diameter, the completed conductor to weigh 650 lbs. per N.M., or within 5 per cent. thereof, but the average weight per N.M. of the conductor shall not be less than that specified. The resistance per N.M. of the conductor at a temperature of 75° F. shall not be more than 1.9 ohms.

Insulator.    (B.) The conductor is to be insulated with three coatings of gutta-percha of improved inductive capacity, prepared according to Mr Willoughby Smith's system, alternating with three coatings of Chatterton's compound, and to weigh 400 lbs. per knot, or within 5 per cent. thereof, but the average weight per knot of the insulator shall not be less than that specified.  The resistance of the completed core to be not less than 150 megohms per N.M. after one minute's electrification, when tested at a temperature of 75° F., after twenty-four hours' immersion in water, fourteen days after manufacture, and the average inductive capacity per N.M. throughout the entire length is not to exceed .43 microfarads.

(c.) The core of all the types to be served with a good and sufficient serving of jute yarn, steeped in cutch or other preservative mixture, and applied wet, the yarn for the deep-sea type to be fine spun, of even diameter, and of good quality. Serving.

## Outer Coverings.

(d.) Type A. Type B to be served with tarred jute yarn, and again closed with fourteen galvanised BB iron wires, No. 1 B.W.G., equal to .300 of an inch when galvanised, or within 2½ per cent. thereof. The wire to be soft, and of good quality. Outer coverings.

(e.) Type E. The served core to be covered with ten galvanised BB iron wires, No. 2 B.W.G., equal to .280 of an inch when galvanised, or within 2½ per cent. thereof. The wire to bear a breaking strain of not less than 25 tons to the square inch, and to be of even diameter, soft, and of good quality.

(f.) Type B. The served core to be covered with twelve galvanised BB iron wires, No. 6 B.W.G., equal to .200 of an inch when galvanised, or within 2½ per cent. thereof. The wire to bear a breaking strain of not less than 30 tons to the square inch, with an elongation of not less than 10 per cent., and to stand not less than ten twists in a length of 6 inches.

(g.) Type D. The served core to be covered with eighteen galvanised homogeneous iron wires, each wire being well covered with a preservative compound and taped. The homogeneous wires to be No. 14 B.W.G., equal to .083 of an inch when galvanised, or within 2½ per cent. thereof, and to bear a breaking strain of not less than 85 tons per square inch, with an elongation of not less than 4 per cent. The wire to be capable of being bent round its own diameter three times, and unbent three times without breaking. The wire to be in bundles of not less than 2 cwt., and to have but one weld in each bundle.

(h.) Before being used for the sheathing of Types A, E, and B, the galvanised iron wire is to be heated in a kiln or oven just sufficiently to drive off all moisture, and whilst warm is to be dipped into a hot compound of coal tar and pitch mixed in approved proportions.

## Outside Serving.

(i.) Types A and E, manufactured as above, to be covered with two servings of jute yarn laid on spirally in opposite directions, alternately with two coatings of Bright and Clark's compound. Outside serving.

(j.) Types B and D, manufactured as above, to be covered with two of Johnson and Phillips' patent tapes, laid on spirally in opposite directions, alternately with two coatings of Bright and Clark's compound.

## General Clauses.

(k.) The cable when completed shall be coiled in suitable water-tight tanks, and be kept, as far as practicable, constantly under water. Cable to be kept under water.

(l.) The completed cable shall be coiled on board ship in water-tight tanks, and be kept as far as practicable under water until submerged. Tanks on board ship.

(m.) The electrical condition of the cable when shipped and also of the completed cable when laid shall be such as, having regard to its previous condition, and making due allowance for the mean actual temperature of the water, as shewn by the resistance of the conductor, to give no good grounds for believing that any fault exists in the insulator or conductor. Final electrical condition of cable.

# APPENDIX III.

## H.M. Post Office Telegraphs.

## Anglo-German Cable, 1891.

### SPECIFICATION.

1. *Conductors.*—Each conductor shall be formed of a strand of seven copper wires all of equal diameter, shall weigh 107 lbs. per N.M., and shall at a temperature of 75° F. have a resistance not higher than 11.65 ohms or lower than 11.18 ohms per N.M.

2. *Insulator or Dielectric.*—Each conductor shall be insulated by being covered with three alternate layers of Chatterton's compound and gutta-percha, beginning with a layer of the said compound, and no more compound shall be used than may be necessary to secure adhesion between the conductor and the layers of gutta-percha. The dielectric on each conductor shall weigh 150 lbs. per N.M., making the total weight of each conductor when covered with the dielectric 257 lbs. per N.M.

3. *Inductive Capacity.* — The inductive capacity of such insulated conductor (hereinafter called the core) shall not exceed .3333 microfarad per N.M.

4. *Labelling.*—Each coil of core before being placed in the temperature tank for testing shall be carefully labelled with the exact length of conductor and the exact weight of copper and dielectric respectively which it contains.

5. *Insulation Resistance.*—The insulation resistance of each coil of core shall be not less than 500 megohms per N.M. nor more than 1,800 megohms per N.M. after such coil shall have been kept in water maintained at a temperature of 75° F. for not less than twenty-four consecutive hours immediately preceding the test, and after electrification during one minute.

6. *Preservation.*—The core shall during the process of manufacture be carefully protected from sun and heat, and shall not be allowed to remain out of water.

7. *Joints.*—All joints shall be made by experienced workmen, and the contractors shall give timely notice to the Engineer-in-chief or other authorised officer of the Postmaster-General whenever a joint is about to be made, in order that he may test the same. The contractors shall allow time for a thorough testing of each and every

joint in the insulated trough by accumulation, and the leakage from any joint during one minute shall be not more than double that from an equal length of the perfect core.

8. *Serving.*—Each core shall during the process of stranding be wormed and served (save in the case of the core for the one-wire deep-sea cable, which shall be simply served) with best wet fully-tanned yarn, sufficient to receive the sheathing hereinafter described, and no loose threads shall in the process of sheathing be run through the closing machine. The said yarn shall be spun from the best Russian long-dressed hemp, and shall as regards tensile strength and number of twists be in accordance with the particulars given under the headings marked A. and B. in the Second Table. The core so served shall be kept in tanned water at ordinary temperature, and shall not be allowed to remain out of water except so far as may be necessary to feed the closing machine.

9. *Sheathing.*—The wire used for sheathing the served core shall be of homogeneous iron, well and smoothly galvanised, and shall be in accordance with the particulars contained in the First Table. The galvanising will be tested by taking samples from any coil or coils, and plunging them into a saturated solution of sulphate of copper at 60° F., and allowing them to remain in the solution for one minute, when they will be withdrawn and wiped clean. The galvanising shall admit of this process being four times performed with each sample without there being any sign of a reddish deposit of metallic copper on the wire. If after the examination of any particular quantity of iron wire, 10 per cent. of such wire does not meet all or any of the foregoing requirements, the whole of such quantity shall be rejected, and no such quantity or any part thereof shall on any account be again presented for examination and testing, and this stipulation shall be deemed to be and shall be treated as an essential condition of the contract. In the sheath of deep-sea cable, and in the inner sheath of shore-end cable, no weld in any one wire shall be within 6 feet of a weld in any other wire; and in the outer sheath of shore-end cable, no weld in any one wire of a strand shall be within 3 feet of a weld in any other wire of that strand. All welds made during the manufacture of the cable shall be re-galvanised. In laying up the cable the strand shall be spliced, and the ends at the butts bound down with spun yarn instead of being welded, the butts in any one wire of a strand being not less than 1 foot from those in any other wire of that strand.

10. *Compound and Serving—*

    (*a.*) *For Deep-sea Cable.*—The sheathed core shall be covered with three coatings of Bright and Clark's compound, and two servings of three-ply yarn, the said yarn being placed between the coatings of compound aforesaid, and being laid on in directions contrary to each other.

    (*b.*) *For Shore-end Cable.*—The inner sheath aforesaid shall be covered with two coatings of Bright and Clark's compound, a serving of yarn sufficient to take the outer sheath aforesaid being placed between the said two coatings of compound. The outer sheath shall be covered in all respects in the same manner as hereinbefore specified for the sheathed core of deep-sea cable.

The compound referred to in this paragraph shall consist of pitch 85 per cent., bitumen $12\frac{1}{2}$ per cent., and resin oil $2\frac{1}{2}$ per cent., and the yarn referred to shall be spun from the best Russian long-dressed hemp, and shall as regards tensile strength and number of twists be in accordance with the particulars given under the heading marked C. in the Second Table, and shall be saturated with gas-tar freed from acid and ammonia, but thoroughly dried after saturation, and before being used, so as to have no superfluous tar adhering.

11. *Measurement.*—A correct indicator shall be attached to the closing machine, and each N.M. of completed cable shall be marked or indicated in such manner as shall be agreed upon between the contractors and the Engineer-in-chief.

—————

FIRST TABLE (REFERRED TO IN PARAGRAPH 9 OF THE FOREGOING SPECIFICATION).

CABLE SHEATHING.

| Type of Cable. | Description of Cable Sheathing. | | | | | | | | | | |
|---|---|---|---|---|---|---|---|---|---|---|---|
| | Deep Sea. | | | Shore End. | | | | | | | |
| | Lay to be Left-handed. | | | Inner Sheath: Lay to be Left-handed. | | | Outer Sheath: Lay to be Left-handed. | | | | |
| | | | | | | | | | Strands: Lay to be Right-handed. | | |
| | Number of Wires. | Diameter of each Wire. | Length of Lay. | Number of Wires. | Diameter of each Wire. | Length of Lay. | Number of Strands. | Length of Lay. | Number of Wires in each Strand. | Diameter of each Wire. | Length of Lay. |
| | | Mils. | Ins. | | Mils. | Ins. | | Ins. | | Mils. | Ins. |
| 1-Wire - - | 10 | 280 | 13½ | 10 | 280 | 13½ | 12 | 21 | 3 | 220 | 5½ |
| 3-Wire | 11 | 280 | 14 | 11 | 280 | 14 | 12 | 21 | 3 | 220 | 5½ |
| 4-Wire - | 12 | 280 | 15 | 12 | 280 | 15 | 13 | 22 | 3 | 220 | 5½ |
| 6-Wire - - | 14 | 280 | 16½ | 14 | 280 | 16½ | 14 | 23 | 3 | 220 | 5½ |
| 7-Wire - - | 14 | 280 | 16½ | 14 | 280 | 16½ | 14 | 23 | 3 | 220 | 5½ |

The 280-mil. wire to have a *minimum* breaking weight of 3,500 lbs. and a *minimum* of ten twists in 6 inches.

The 220-mil. wire to have a *minimum* breaking weight of 2,300 lbs. and a *minimum* of twelve twists in 6 inches.

The diameter of any wire is not to differ from the standard diameter by more than 3 per cent. above or below.

SECOND TABLE (REFERRED TO IN PARAGRAPHS 8 AND 10 OF THE FOREGOING SPECIFICATION).

YARN.

| | A. Centre and Worming. (*Vide* Par. 8.) | | | | | B. Inner Serving. (*Vide* Par. 8.) | | | C. Outer Serving. (*Vide* Par. 10.) |
|---|---|---|---|---|---|---|---|---|---|
| Ply - - - - | 5 | 7 | 9 | 12 | 17 | Single | Single | Single | 3 |
| Weight, approximate, in lbs. per knot - - } | 25 | 24 | 28 | 40 | 112 | 13 | 16 | 25 | 10 |
| Twists, *maximum* number in 12 inches - - } | 6 | 6 | 6 | 6 | 4 | 11 | 10 | 9 | 6 |
| Breaking strain, *minimum*, in lbs. - - } | 175 | 168 | 196 | 280 | 784 | 90 | 112 | 175 | 70 |

The yarns indicated by the approximate weights given in this Table are those usually employed in the various types of cable, and their *minimum* breaking strains have been calculated on the basis that 1 N.M. of yarn weighing 1 lb. should have a *minimum* breaking strain of 7 lbs. In the event of yarns being used of weights differing from those mentioned in this Table, the same standard, namely, 7 lbs. per "knot-pound," shall be adopted in determining their *minimum* breaking strain.

For the purposes of the tests for tensile strength the samples of yarn taken shall be not less than 26 inches in length.

———

## CONDITIONS OF CONTRACT.

### 1. *Definitions.*

In the tender and the several schedules thereto—

(*a.*) The term "Engineer-in-chief" means the Engineer-in-chief of the Post Office.

(*b.*) The terms "nautical mile" and "knot" are synonymous, and are used to represent 2,029 lineal yards.

(*c.*) When used in a title indicating a type of cable the word "wire" is, for convenience, substituted for "conductor."

### 2. *Samples.*

The samples submitted with the tender shall, unless otherwise sanctioned by the

2 N

Engineer-in-chief, be taken as the standards by which the supply in bulk shall be governed.

### 3. *Free Access to Contractors' Works.*

The Engineer-in-chief and his agents shall at all reasonable times have free access to the contractors' works for the purpose of inspecting the process of manufacture in all its stages, and of examining and testing every portion of cable and the materials used in the manufacture thereof, and the contractors shall give every facility for such examination and testing.

### 4. *Power of Rejection.*

The Engineer-in-chief or his agents shall have power to reject any wire or other material used in the process of manufacture which shall appear to him or them to be of unsuitable description or of unsatisfactory quality.

### 5. *Contractors to provide Accommodation for Testing.*

For the testing of the insulated conductor during the whole process of manufacture, the contractors shall, at their own cost, provide the necessary batteries, and, if required, the necessary testing apparatus, together with a proper and separate room and leading-wires thereto and to the tank or tanks in which the cable shall be stored as hereinafter provided. The contractors shall also provide the apparatus necessary for making the various tests of iron wire and hemp yarn required by the specification.

### 6. *Storage.*

Each cable shall immediately on completion be passed into a tank or tanks of water, and shall be stored in the same tank or tanks and be kept therein under water until it is required by the Postmaster-General to be shipped or otherwise delivered, and during such time it shall be held by the contractors for the Postmaster-General under a wharfinger's warrant to be given by them to him for that purpose; but it shall nevertheless remain at the risk of the Postmaster-General as regards damage by fire. For such storage during any period within that stated no charge shall be made against the Postmaster-General. At no time during the occupation of a tank or tanks by cable under this contract shall any cable not the property of the Postmaster-General be placed in the said tank or tanks.

### 7. *Night or Sunday Work.*

Work under this contract shall not be carried on at night or on Sundays without the written consent of the Engineer-in-chief or other authorised officer of the Postmaster-General.

### 8. *Delivery.*

The cable shall, when required by the Postmaster-General, be shipped on board a vessel or barge to be placed by him at his cost alongside the contractors' works;

such vessel or barge, together with the labour required to coil down the cable therein, being also provided by and at the cost of the Postmaster-General; the necessary machinery, appliances, and articles required for shipping the said cable being provided by and at the sole expense of the contractors.

### 9. *Payment.*

Payment by the Postmaster-General will be made to the contractors within fourteen days after the completion of a cable to the satisfaction of the Engineer-in-chief, to be certified by writing under his hand.

### 10. *Non-fulfilment of Contract.*

If any cable shall be not completed to the satisfaction of the Engineer-in-chief, and not ready for delivery as aforesaid within the time named, the price payable in respect of any uncompleted cable shall be reduced by a sum equal to $\frac{1}{2}$ per cent. on that price for any complete day or fraction of a day which may elapse between the date on which the cable should have been completed and ready as aforesaid and the date on which the cable is actually completed and ready for delivery as aforesaid.

### 11. *Modification of Contract.*

The Postmaster-General may from time to time, by writing under the hand of the Engineer-in-chief addressed to the contractors, direct or prescribe any alteration or alterations to be made by the contractors in the manufacture of any cable with regard to the length thereof, the number of coatings of gutta-percha on the conductor, the ingredients of the compounds or other of the materials to be employed in such manufacture as aforesaid, or to the mode of such manufacture, and thereupon such alteration or alterations shall be executed by the contractors in the same manner as if such alteration or alterations had been directed or prescribed by this contract. Provided always, that if such alteration or alterations shall cause any increase or diminution in the cost of manufacturing the cable an equivalent allowance in respect thereof shall be accordingly added to or deducted from the price mentioned; and if any dispute shall arise between the Postmaster-General or the Engineer-in-chief and the contractors as to whether any such alteration or alterations have occasioned any increase or diminution of cost, or as to the amount of such increase or diminution of cost, then the same shall be settled by arbitration as hereinafter provided. Provided also, that as regards the length of the cable, the Postmaster-General shall not be at liberty to prescribe any diminution of the length thereof. And that if by reason of any alteration in the method of manufacture or of any increase of length some extended time beyond the date hereinbefore specified shall be required by the contractors for completing and delivering the cable, the contractors shall be allowed such extended time as the Engineer-in-chief may judge necessary for that purpose.

### 12. *Arbitration.*

All matters in difference between the contractors and the Postmaster-General which may arise under this contract shall be settled by arbitration in conformity with the provisions contained in the Common Law Procedure Act, 1854, in respect to the settlement of differences by arbitration.

### 13. *Members of Parliament.*

In pursuance of the Act 22 Geo. III., ch. 45, no Member of the House of Commons is to be admitted to any share or part of this contract, or to any benefit that may arise therefrom, according to the true intent and meaning of the said Act.

[PLATE XXIV.]

## APPENDIX IV.

**FORM A.**

### ELECTRICAL DATA.

| SECTION. *Where laid between.* | WHEN LAID. [Year.] | LENGTH. [N.M.] | CORE WEIGHTS. [Lbs. per N.M.] | | DIAMETERS, [Mils., Inches, or L.S.W.G.] | | AVERAGE SPECIFIC CONDUCTIVITY OF THE COPPER. [Per cent.] | ELECTRICAL VALUES WHEN LAID. Reduced to 75° Fah. [Per N.M.] | | | AVERAGE BOTTOM TEMPERATURE. [Fah.] | SIGNALS. | | | CONSTRUCTED AND LAID BY. [Contractors.] | WORKED BY. [Owners.] |
|---|---|---|---|---|---|---|---|---|---|---|---|---|---|---|---|---|
| | | | Conductor. [Copper.] | Dielectric. [G.P. or I.R.] | Conductor. [d] | Outside Core. [D.] | | Conductor Resistance. | Electro-Static Inductive Capacity. | Dielectric Resistance. | | Hand or Machine Transmission. | Receiving Instrument Employed. | Average Working Speed. [Letters per min.] | | |
| | | | | | | | | | | | | | | | | |

*Engineers* ....................

**FORM B.**

### MECHANICAL DATA.

| SECTION. *[Where laid between.]* | WHEN LAID. [Year.] | LENGTH. [N.M.] | CABLE TYPE. | CORE WEIGHTS. [Lbs. per N.M.] | | INNER SERVING. [Description.] | SHEATHING WIRES. | | | | | OUTER SERVING. [Description.] | Circumference. [Inches.] | COMPLETED CABLE. | | | Specific Gravity. | Breaking Strain. [Tons.] | | Modulus of Tension. [N.M.] |
|---|---|---|---|---|---|---|---|---|---|---|---|---|---|---|---|---|---|---|---|---|
| | | | | Conductor. | Dielectric. | | Number and Quality. | Diameter. [Mils. or L.S.W.G.] | Breaking Strain. [Tons per sq. in.] | Average Elongation. [Per cent.] | Number of Twists. [Average] in 6 in. | | | Weight. [Per N.M., in tons.] | | | | Calculated. | Actual. | |
| | | | | | | | | | | | | | | Dry. | Wet in Air. | In Sea Water. | | | | |

*Engineers* ....................

The material originally positioned here is too large for reproduction in this reissue. A PDF can be downloaded from the web address given on page iv of this book, by clicking on 'Resources Available'.

# PART III

## THE WORKING OF SUBMARINE TELEGRAPHS

# CONTENTS OF PART III.

# PART III.—WORKING.

## CHAPTER I.

### THEORY OF THE TRANSMISSION OF SIGNALS THROUGH CABLES.

Section 1.—Preliminary Remarks.

Section 2.—Propagation of an Electric Impulse in a Cylindrical Conductor : "Curbed" Signals : The Application of Condensers for Signalling Purposes—Alphabets—The 1869 Atlantic Cable taken as an example.
Dearlove's Transformer for Working Cables—Mechanical Analogy of Cable Working.

Section 3.—Signalling Speed : Absolute Velocity of Electricity—Data in Practice—Theoretical Calculations : Considerations involved : Latest Views—Further Practical Considerations and Comparisons.

### Section 1.—Preliminary Remarks.

It will be obvious that the commercial value of a submarine telegraph cable is dependent on the speed at which signals can be transmitted through it. This varies with the length of time a charge of electricity takes to produce its effect by the strength of current developed at the distant end—inversely according to the electro-static inductive capacity of the whole cable, and directly so to the conducting power of the conductor.

The number of submarine conductors between any two points on the globe being necessarily restricted by the great cost of establishment and maintenance, it very soon became necessary to discover the most advantageous dimensions of which to form the conductor and its insulator, and also the most suitable kind of transmitting instrument so as to work the cable at its maximum speed.

The solution of both these problems depended on the theory of electric propagation in a cylindrical conductor before the permanent state is established. In his celebrated "Treatise on the Mathematical Theory of the Galvanic Circuit," G. S. Ohm—as far back as in 1827—had worked out the corresponding differential equation, shewing it to be similar to the equation found by Fourier for the propagation of heat in a bar of unlimited length. But electrical telegraphy, which might have found useful applications for these data, only came into existence some ten years afterwards, and the labours of Ohm remained for a long time unnoticed.

In 1856, however, Professor William Thomson, F.R.S. (now Lord Kelvin), working on different lines to those of Ohm, discovered the direct analogy between the propagation of an electric impulse along a conductor of great length in proportion to sectional area, and the flow of heat through a bar of unlimited length. His results are nearest the truth when the manifestations of the variable state take place slowly; they are rigorously correct for bodies having great resistance, such as cotton threads, columns of oil, etc., and represent the phenomena of conduction in submarine cables with sufficient exactitude for all practical purposes.

Previously, Professor (afterwards Sir Charles) Wheatstone, F.R.S., and MM. Fizeau and Gounelle had instituted searching inquiries into the propagation velocity of electricity. The latter gentlemen conducted a number of elaborate researches on this subject from the year 1850, their last experiment in wave motion being made in 1863. The results obtained from the foregoing independent investigations revealed what were then considered grave discrepancies; * and some years later Professor Thomson pointed out that—unlike the velocity of light—no general calculation was admissible, and no general law could be laid down, for the speed of electrical transmission through all classes of conductors, independent of sectional area, length, and surroundings.† At the Paris Exhibition of 1881, the late Mr Robert Sabine

---

* Wheatstone had found that "electricity can travel at a rate *one and a half* times greater than that of light": in other words, "that the electric spark would go round our globe about twelve times in one *second*." According to Fizeau and Gounelle, "the beginning of an electric *current* may be transmitted along a copper wire at a velocity which is not greater than *three-fifths* the velocity of light."

† Indeed, in the course of correspondence on the subject with Professor (now Sir G. G.) Stokes, F.R.S., Professor Thomson pointed out that electricity has *no* velocity in the ordinary sense of the word.

For further particulars, see the "Mathematical and Physical Papers" of Sir William Thomson, LL.D., D.C.L., F.R.S. (Cambridge University Press, 1884); also 'Phil. Mag., 1856.

The truth of the above has been set forth recently in a very able manner by Professor W. E. Ayrton, F.R.S., during his lecture at the Imperial Institute on "Sixty Years of Submarine Telegraphy." Professor Ayrton exhibited a mechanical model illustrating the difference between the *sudden* opening of a door by a ball projected at it with a certain velocity, and the *gradual* opening of the door by the *gradual* increase of the pull at the

exhibited a system of ascertaining the contour and speed of electric waves passing through telegraph lines, by measuring the potential at different points along its length.   In the case of 304 N.M. of coiled cable, it was found that the wave travelled at 6,000 N.M. per second.

---

other end of a long piece of india-rubber, the latter representing the transmission of an electric signal.   For an account of Professor Ayrton's lecture, see the *Imperial Institute Journal* for July 1897 ; also *Nature*, 25th February 1897.

Anglo-American Telegraph Company's Station at Valentia, Ireland : Instrument Room.

SECTION 2.—PROPAGATION OF THE ELECTRIC IMPULSE IN A CYLIN-
DRICAL CONDUCTOR OF LIMITED LENGTH DURING THE VARIABLE
PERIOD.

Let A B (Fig. 1) represent a cable with its end B to earth, connected up
at A to a battery whose opposite pole is also to earth. In this cable we

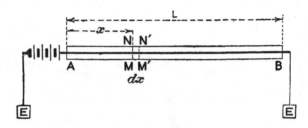

FIG. 1.

will consider a given volume of length $dx$ included between two sections
at right angles to the axis M N and M′ N′, and situated at distances A M $=x$
and A M′ $=x+dx$ from the point A. Expressing by

$P_0$ the potential at A,
P the potential at M at the time $t$,
$C_0$ the intensity, or strength, of the current at A at the same instant of time,
C    „    „    „    M    „    „
$C_1$    „    „    „    ·    B    ‵„    „
$\rho$ the conductor resistance per unit of length,
$k$ the electro-static capacity of the core per unit of length,
$r$ the resistance of the dielectric per unit of length,
L the length A B of the line

the potential and intensity, or strength, of current at the time $t$ in the
section M′ N′ will be $P+dP$ and $C+dC$.

The quantity of electricity $Cdt$ which spreads itself over the section M N
in the time $dt$ will divide into three parts; the first, which will proceed
along the conductor and spread over the section M′ N′, is represented by
$(C+dC)dt$; the second part, which traverses the insulation of the included

portion M N M′ N′, whose resistance is $\dfrac{r}{dx}$, will be equal to

$$\dfrac{\dfrac{P}{r}}{dx}\,dt$$

the third part will increase the electro-static charge in the included portion of cable, and can be expressed by $kdx\dfrac{dP}{dt}\,dt$.

We now have

$$Cdt = (C + dC)dt + \frac{Pdx}{r}\cdot dt + k\,\frac{dP}{dt}\,dx\,dt.$$

Reducing and remarking that by Ohm's law

$$C = -\frac{dP}{\rho dx} \qquad\qquad (1)$$

whence

$$dC = -\frac{1}{\rho}\,\frac{d^2P}{dx^2}\,dx$$

we get

$$\frac{1}{\rho}\,\frac{d^2P}{dx^2} - k\frac{dP}{dt} - \frac{1}{r}P = 0$$

or, assuming that

$$k\rho = a^2 \cdot \qquad\qquad (2)$$

and

$$\frac{\rho}{r} = \beta^2 \qquad\qquad (3)$$

$$\frac{d^2P}{dx^2} - a^2\,\frac{dP}{dt} - \beta^2 P = 0 \qquad\qquad (4)$$

The general integration of this equation has been given by Fourier, and is expressed thus—

$$\frac{P}{P_0} = \frac{e^{\beta(L-x)} - e^{-\beta(L-x)}}{c^{\beta L} - e^{-\beta L}} - 2\pi e^{-\frac{t}{kr}}\sum_{n=1}^{n=\infty}\frac{n}{n^2\pi^2 + \beta^2 L^2}e^{-\frac{n^2\pi^2}{a^2 L^2}t}\sin\frac{n\pi}{L}\cdot x$$

As a rule $\rho$ does not exceed 10 or 12 ohms, and $r$ reaches from 8,000 to 10,000 megohms; $\beta^2$ being therefore less than $\dfrac{1}{10^9}$. If we take $\beta^2 = 0$, which is the same thing as making $r = \infty$, or entirely neglecting loss of electricity through the insulation, the above integration is simplified and becomes

$$\frac{P}{P_0} = \frac{L-x}{L} - 2\sum_{n=1}^{n=\infty}\frac{1}{n\pi}e^{-\frac{n^2\pi^2}{a^2 L^2}t}\sin\frac{n\pi}{L}\,x \qquad\qquad (5)$$

Differentiating as regards $x$, and carrying the value obtained for $\dfrac{dP}{dx}$ into equation (1), we get for the current strength at the distance $x$

$$C = \frac{P_0}{\rho L}\left(1 + 2\sum_{n=1}^{n=\infty}e^{-\frac{n^2\pi^2}{a^2 L^2}t}\cos\frac{n\pi}{L}\,x\right)$$

At the end B of the conductor which is then to earth, $x = L$; $\sin n\pi$ being always *nil*, the potential is *nil*, and the current $C_1$ is

$$C_1 = \frac{P_0}{\rho L}\left(1 + 2\sum_{n=1}^{n=\infty} e - \frac{n^2\pi^2}{a^2L^2} t \cos n\pi\right) \qquad \qquad (6)$$

By giving to $n$ consecutive values 1, 2, 2, 3 ... the cosine acquires alternate values equal to $-1$ and $+1$. So that if we assume, for abbreviation, that

$$e - \frac{\pi^2}{a^2L^2} t = u \qquad \qquad (7)$$

equation (6) becomes

$$C_1 = \frac{P_0}{\rho L}\left\{1 - 2\left(u - u^4 + u^9 - u^{16} + u^{25} - \ldots\right)\right\} = F(t) \qquad (8)$$

With extremely small values of $t$, $u$ tends towards unity; at the limit the series $u - u^4 + u^9$ . . . equals $\frac{1}{2}$, and the current intensity is nothing. As the time interval increases, so $u$ diminishes, the series decreasing and the current increasing; but according to Sir William Thomson, the series only differs sensibly from its minimum value $\frac{1}{2}$ when

$$u > \tfrac{3}{4}$$

Using $\tau$ to express the time when this condition is attained, we have

$$e - \frac{\pi^2\tau}{a^2L^2} = \tfrac{3}{4} \qquad \qquad (9)$$

whence

$$\tau = \frac{a^2L^2}{\pi^2} \log_\epsilon \tfrac{4}{3} \qquad \qquad (10)$$

$\tau$ being expressed in seconds if $a$ and L are expressed in C.G.S. units,[*] or

$$\tau = \frac{k\rho L^2}{10^6} \times 0.02915 \text{ second} \qquad (11)$$

where $k$ stands for the electro-static capacity of the cable in microfarads per naut, $\rho$ the conductor resistance per naut in ohms, and L its length in nauts,

$$\tau = \frac{RK}{10^6} \times 0.02915 \text{ second} \qquad (12)$$

R representing the total conductor resistance in ohms, and K the total capacity of the line in microfarads.

---

[*] An ohm is equal to $10^9$ electro-magnetic absolute units of resistance; a microfarad is equal to $\frac{1}{10^{15}}$ electro-magnetic absolute units of capacity.

From this point onwards the series tends towards $o$, and $C_1$ increases up to its limit of value $\dfrac{P_0}{\rho L}$, which is only reached after an infinitely great interval of time.

**Thomson's Curve of Arrival.**—The curve I (Fig. 2) is the "arrival curve" * of a current in a cable, one end of which is to earth and the other end in connection with a constant source of electric potential. The times are counted on the axis of X, each division marked on this axis correspond-

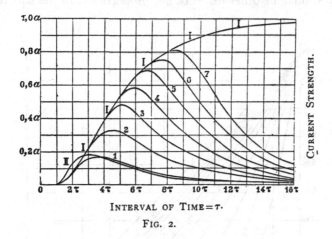

INTERVAL OF TIME $= \tau$.

FIG. 2.

ing to an interval of time equal to $\tau$.   The current intensities are counted on the axis of Y, the maximum intensity $\dfrac{P_0}{\rho L} = a$ being taken as unity and

---

* The above—commonly referred to as Thomson's "curve of arrival"—was first evolved by Sir William Thomson in 1855 in the course of a paper read before the Royal Society (see *Roy. Soc. Proceedings*).   This arrival curve in any given case is precisely and graphically reproduced in the signals received on the siphon recorder slip when the cable is being "worked," which, in fact, they are a true representation of under the existing conditions.

It will be seen, therefore, that for purposes of demonstration or experiment these signals are invaluable, and all the examples given in the following pages are nothing more nor less than "·recorder" signals.

A still more complete idea of what occurs may be obtained if the recorder at the sending, as well as at the receiving, end, be worked through—as is, indeed, usually done in practice.   Then the record on the two slips may be examined together.   It is necessary, however, to shunt the instrument at the sending end in order to keep the signals on the paper band.

In the "mirror," a short or long deflection to the left or right takes the place of a short or prolonged line above or below the zero line of the "recorder" slip.   As, however, the mirror system does not allow a record of the signals, it is not dwelt on in this chapter.   Moreover—partly for the above reason—the mirror system has been almost entirely supplanted by the "recorder."

divided into ten equal parts. We see that the curve does not quit the axis of X until an interval of time equal to τ has elapsed, and that the asymptote to the curve is a straight line parallel to the axis of X, at distance *a* from the starting point O. The current attains to about nine-tenths of its maximum strength after an interval of time 10 τ. As two similar cables, differing from each other only in respect to length, attain the same fraction of their maximum charge in the same interval of time, it follows practically from equation (10), that the times necessary for charging cables of different lengths but similar in other respects, are proportional to the squares of the lengths.

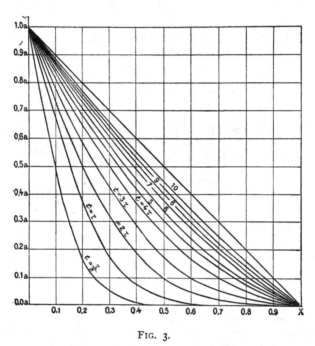

FIG. 3.

Equation (5) enables us to construct curves of potential in a cable at different times $\frac{\tau}{2}$, τ, 2τ, 3τ . . . of the variable period. In Fig. 3, O X represents the length of the cable, O Y the electro-motive force $P_0$ of the battery. The curves constantly approach the straight line X Y, but only coincide with it after an infinitely long interval of time.

Fig. 4 shews in the same manner the current intensities throughout the cable at the same instants of time. We see that in the middle of the cable the current strength is scarcely perceptible after a time-period $\frac{1}{2}$τ (shewn by the intersection of the ordinate 0.5 with the curve $t=\frac{1}{2}$), but that it rapidly increases from this moment. It attains its maximum value *a* after an interval of time between $3\tau$ and $4\tau$; all the subsequent curves passing through the same point shew that from this moment the current intensity loses in the first half of the cable, and gains in the other half.

If the cable is only connected to the battery at A (see Fig. 1) for a very short time $t_1$, and then at once put directly to earth, the charge of electricity will accumulate at both ends of the line. The current intensity at the B end can be determined for any instant by pre-supposing two states, the first due to the potential $P_0$ established at A at the moment from which the times are counted, and the second due to the potential $-P_0$

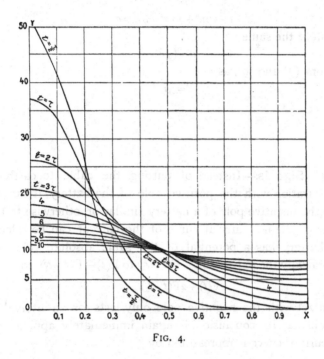

FIG. 4.

established at A after an interval $t-t_1$, and calculating the expression

$$C_1(t) = F(t) - F(t - t_1)$$

in which all values of the function F corresponding to negative time values will be considered as nullity.

Graphically, this comes to the same thing as taking the differences between the ordinates of the curve I (see Fig. 2) and those of a second identical curve supposed to be moved a distance $t_1$ towards the right. Curves I to 7, corresponding to contacts of respective duration $\tau$, $2\tau$, $3\tau$, ... $7\tau$, were traced in this manner. As a result we get a species of electric wave extending the whole length of the line.

If the duration of contact $\theta$ were infinitely short, the arrival curve of the current intensity would be represented by the equation

$$i = \theta \frac{dF}{dt} = \theta \frac{dF}{du} \frac{du}{dt}$$

or, going back to equations (7) and (8),

$$i = \theta \frac{2P_0}{\rho L} \frac{\pi_2}{a^2 L^2} (u - 4u^4 + 9u^9 - 16u^{16} + \ldots) \tag{13}$$

This intensity is represented by curve II (Fig. 2), and is maximum when $\frac{di}{dt} = o$, that is to say, when

$$u - 16u^4 + 81u^9 \ldots = o$$

which is sensibly the same as

$$u = (\tfrac{3}{4})^3$$

or, by equations (7) and (9), as

$$t = 3\tau.$$

"Curbed" Signals.—Instead of putting the cable to earth, after a contact of $t_1$ duration with the positive pole of the battery, it can be connected up to the negative pole of a battery similar or otherwise to the first. The ordinates of the new arrival curve of the current can be calculated if the second battery has a potential of the same absolute value $-P_0$, by adding algebraically the expression $-F(t - t_1)$ to $F(t) - F(t - t_1)$, so as to have

$$F(t) - 2F(t - t_1).$$

If the contact with the negative pole only lasts for an interval of time $t_2$, and the cable, at its conclusion, is again immediately applied to earth, the definite current curve is represented by

$$F(t) - 2F(t - t_1) + F(t - t_1 - t_2).$$

For example, if the first contact has lasted $4\tau$, and the second contact with a battery of $-P_0$ potential has lasted $3\tau$, and if the cable is immediately afterwards put to earth, curve 3 in Fig. 2 must be moved along upside down to the right as far as the abscisses $4\tau$, each ordinate of curve 4 being diminished by the length of the corresponding ordinate of curve 3. Thus, in Fig. 5 the ordinate $a\,d$ is equal to $a\,b$, less $b\,d$, which is equal to the ordinate $a\,c$. In this case the electric wave will be represented by the full line curve.

We proceed in a similar way when the second negative contact is followed by a third and positive one, and so on. The full line curves in

Figs. 6, 7, 8 represent the waves of arrival due to alternate contacts as indicated in the following table, the cable being put to earth again immediately afterwards :—

| Curves. | Duration of Contacts. | | | |
|---|---|---|---|---|
| | + | − | + | − |
| Fig. 6 | $4\tau$ | $4\tau$ | ... | ... |
| ,, 7 | $3\tau$ | $3\tau$ | $\tau$ | ... |
| ,, 8 | ... | $\tau$ | $4\tau$ | $2\tau$ |

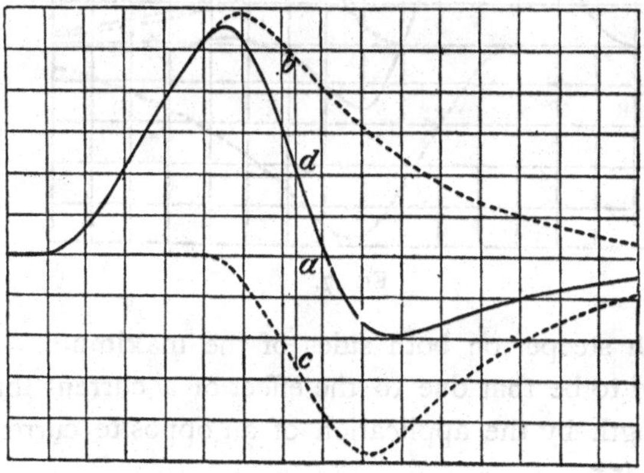

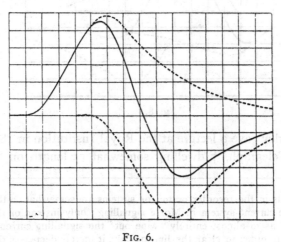

FIG. 5.

The cable may now be considered divided into sections, containing alternate positive and negative charges. These charges, after a time, run into one another, so that, the contacts being of suitable length, the electric wave will be much shorter than if there had been a single contact only. In this

FIG. 6.

way the cable is brought back much more rapidly to the neutral condition, after the charge due to the passage of the current, and the signals can therefore be made to succeed each other at much shorter intervals of time.

Signals obtained by a succession of usually about three to five very short alternate contacts are termed curbed signals in submarine telegraphy, the effect of the partially neutralising or "curbing" currents being to make the

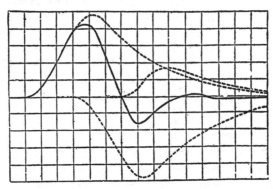

FIG. 7.

arrival curve much steeper on both sides of the maximum. A "curbed" signal may be said to be that due to the effect of a current that has been curbed in its strength by the application of an opposite current following immediately after it.*

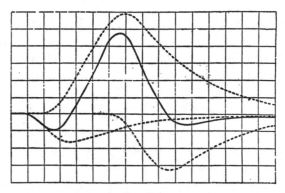

FIG. 8.

Mr C. F. Varley gave a lecture at the Royal Institution in 1867, entitled "The Atlantic Telegraph,"† dwelling somewhat at length on the general

---

* Curbed signals in the recorder or mirror systems are somewhat on the same principle as the double-current system in Morse signalling ; only instead of the secondary current being such as to almost entirely "wipe out" the signalling current after it has performed its duty in order to clear the line quickly, it merely decreases the volume of the working current—i.e., the current density—thus reducing the amount it does, and therefore increasing the rate at which signals can be effectively made to follow one another, besides rendering each more clear.

† *Proc. Roy. Inst.*, vol. v., p. 45.

principles underlying telegraphy through submarine cables.*   Again, in 1879, Mr James Graves read a most valuable paper before the Society of Telegraph Engineers on "Curbed Signals for Long Cables."   It may be remarked, in passing, that the system of curbing signals would be applicable for all cables possessing capacity, though less essential, of course, where the total retardation is low.   In practice it is only actually *necessary*, however, where high-speed working is aimed at.   For reasons of traffic, short cables are not, as a rule, worked at their maximum speed—*i.e.*, at that which their electrical values would afford if machine transmission devices were resorted to.

It is not usually worth while to curb any hand-sent signals, the quickest of which are not fast enough for this to become desirable.   In early days, however, and from time to time, various devices have been introduced for effecting the curbing of manually transmitted signals—to wit, those of Mr Edward Brailsford Bright (1858),† Professor William Thomson (1858), Sir

---

* This was some years after the papers of his brother, Mr S. A. Varley, on the same subject (before the Institution of Civil Engineers and the Society of Arts), referred to in Part I.   The latter gentleman called particular attention to the fact that the speed of signalling was as much dependent on the electrical resistance of the conductor as on electro-static capacity, and that the said resistance varied inversely as the sectional area. Professor William Thomson, F.R.S., had previously pointed to the real problems to be contended with—partly when criticising Mr Wildman Whitehouse's 1856 B.A. paper, which criticism was conducted in the correspondence columns of the *Athenæum*.

Professor Thomson here shewed that the rate of signalling would vary inversely as the *square* of the length, rather than merely with the length, as had been imagined by Mr Whitehouse in specifying the conductor for the first Atlantic cable.

He also pointed out here that with every cable there must be a certain time-period before the current begins to appear at the other end ; and that this time depends only on the cable—in fact on the product of the resistance of the conductor into the electro-static capacity, and practically not at all on the battery power.

For Thomson's complete theory, see *Proc. R. S.*

† This was the first suggestion of its kind.   Indeed, Mr E. B. Bright may be said to have been the originator of the principle (see his paper at the British Association meeting of that year).

Sir C. Bright's device consisted in an improvement on his brother's ; and was, perhaps, the most simple form ever published or patented.   This "compensating" or "curb" key was intended to meet the requirements of Atlantic telegraphy by clearing it promptly of each successive signalling current, so as to allow the next one to pass through in quicker succession.   The usual plan had been that of first sending a positive current of sufficient duration to produce the signal, and then wiping out the residual electricity left in the cable by a corresponding negative current.   It was thought, however, that it would be better to transmit in rapid succession a series of opposite currents of very small duration.   By this plan the cable is cleared much more rapidly, with a consequent increase in the working speed attainable.

Speaking generally, when depressing either lever of these keys, a clockwork train was set in motion, by the releasing of a cam.   This sent a series of weak reversals through the line.   The order of these would depend on which lever was depressed for the primary current.   Sometimes the curbing, or compensating, current appears to have been set up by the same E.M.F. as the primary current, though in force for a shorter period.   This was the case in Professor Thomson's 1858 patent.

Charles Bright (1860 and 1862), Mr Wildman Whitehouse (1860), and Messrs Thomson and Varley (1866). All of these consisted of keys with cam and train of clockwork wheel devices, forming a part of the transmitting key and set into action immediately after the depression of the lever. Such keys were termed "curb senders," or "curb keys." They were, however, never used to any extent in practice for the reason given; besides the fact of being before their time, inasmuch as at that period there was no demand for great rapidity in working. Moreover the appliances, in the shape of the cable and the operators, were not then prepared for innovations, or reforms, of this character.

**On the Use of Condensers\* for Signalling through Submarine Cables.**—By inserting a condenser at each end, between the cable and the sending and receiving instruments, so keeping the cable completely insulated, the signals are sharpened—becoming similar, in fact, to the above curbed signals;† and, moreover, earth currents through the cable are thereby obviated—*absolutely* so, if inserted at both ends.

In securing greater definition of signals by the use of condensers, the speed of working is also increased.‡ This is owing to the line being by this method sooner cleared of each successive signal, and therefore sooner ready to take the next, besides that the signals are thereby rendered more readable. In point of fact, in using condensers for cable signalling, time is not allowed for the sending condenser to be anything like fully charged before it is again discharged—only a small fraction of the total charge or current available (from the battery) being, in fact, used. It might naturally be supposed that the above would equally apply if the same manipulation of the key (by only momentary depressions) were adopted when applying a

---

\* The true definition of a condenser is said to be a storer, or accumulator of, electrical *energy*. This definition does not, however, tend to throw any light on the function performed by a condenser in the signalling circuit of a submarine cable.

† The application of condensers at each end of the cable has, amongst other features, so much the same effect as any "curbing" arrangements, that, in the opinion of many authorities, the manual keys and machine transmitters (as used nowadays for high-speed work) do not require to have curbing devices applied to them.

‡ Indeed, a curbing current (as effected in discharging the condensers by the key) serves the same purpose as the "wipe-out" current in the double-current Morse system. In both instances the primary current is fore-shortened, and the line is more or less cleared of any residual charge more quickly. It is not found possible, however, to produce this "wipe out" or curbing effect in any manual key employed for the mirror, or recorder, working at all suited to the nature of the signals, though automatic transmitters are, as will be seen, effectively fitted with supplementary curbing devices.

battery *direct* to the cable, but in practice the result is quite different. With a condenser, indeed, it is absolutely momentary impacts, or impulses, which are in operation.*

The system of inserting condensers at each end of the cable was first *patented* by Mr C. F. Varley in 1862 (No. 3,453), the condenser being termed an "induction plate" in those days. This method was not, however, put into practice until 1866, over the Atlantic cable of that year. The late Mr Willoughby Smith included it amongst the arrangements for his famous system, first applied on that expedition, for keeping up, contemporaneously, tests and communication between ship and shore whilst paying out. Mr Smith is said to have employed condensers at the *receiving* end quite independently of Mr Varley's patent. No condenser whatever was used at the sending end. The question of priority in this matter formed the subject of a somewhat animated debate on the occasion of Mr Willoughby Smith's paper at the Society of Telegraph Engineers in 1879, "On the Working of Long Submarine Cables." From this paper, it would, at any rate, appear that Mr W. Smith's arrangements were somewhat different to anything published in the aforesaid Varley patent.

In this patent, Mr Varley shewed, amongst other ingenious arrangements, a plan of shunting the induction plates, or condenser, by placing a high resistance in parallel with it, in order to increase the speed of working. It is not, however, a plan usually adopted nowadays, except on land lines, or short cables—usually for experimental purposes to vary the condenser effect—where the capacity is comparatively small, and where the siphon recorder is not employed.†

---

* The result is a much shorter period of rise than where no condensers are used : in the latter instance it is a case of steady current flow.

† The effect of placing a high resistance in parallel with the condenser at either (or both) the sending, or receiving, end of a cable is obviously that of lessening the condenser effect, by the shunting action involved. As has already been stated, the signalling, by siphon recorder, of long (ocean) cables is very considerably improved by the application of condensers at the ends ; and in such cases it is never usual to apply any shunt to the condensers, the required value for the condensers to give the best effect in proportion to the cable values being arrived at beforehand by experiment. It may sometimes happen, however, that for temporary or experimental work exactly the right capacity is not available with any set of condensers at hand, in which case a judicious shunting may be resorted to. This seldom applies in *ocean* cables, for which complete apparatus, based on preliminary experiments, is usually supplied. In certain instances, however—and especially as regards short lines—it might happen that though condensers were found desirable, there was not anything at hand of sufficiently low capacity value, consequently the condenser effect ("curbing," etc.) would be too great unless the signalling condensers were shunted in some way or another. A plan which is seldom resorted to on siphon

A speed, even up to twenty-five words a minute, was by means of condensers at the receiving end attained on the 1866 Atlantic, by manual transmission, working with the mirror instrument, with good mirror clerks at each end. It must be remembered that this speed involves the transmitting clerk manipulating the springs of the key in such a way as to make 375 contacts per minute, and the receiver has to distinguish, in the same space of time, the true from the false, in the 750 movements of the small beam of light.

Condensers were not usually interposed at the sending end of the cable till some time later—when the recorder came into use—though suggested in Mr Varley's patent.* This interposition of condensers is the plan

---

recorder circuits, it is more especially applicable to those cables worked by mirror or, still more, by Morse, which are of sufficient length for signalling condenser to be desirable —if only for curbing purposes. Some control and limit must be put on this curbing effect, inasmuch as in working with the mirror the signals must not be curbed to more than a certain extent, for high speed, where accomplished mirror clerks read the signals by their general character rather than by each separate movement of the spot. In those lines worked by the polarised-relay Morse system, which are of sufficient length, shunted signalling condensers are frequently used as above—partly on account of the retarding effect of the self-induction of the relay. The resistance used, as a rule, to shunt each set of condensers varies from 1,000 to 10,000 ohms. The same sort of shunting effect is sometimes brought about by reducing the capacity by placing another condenser in series with the primary signalling condenser. Thus, two similar condensers $c$ and $c_1$ when joined in series as shewn here—

will have half the capacity of one, owing to the distance between the outside extreme plates of each being thereby—in effect—doubled, without any increased surface, or area, being obtained for them, or for the inner plates.

The curbing of signals is, however, in the first instance, very little necessary on short lines, and is then only so when quite a high speed of signalling is involved, such as can alone be effected by machine transmission. This is seldom the case, partly owing to the fact that where the importance and traffic is such as demands it, there are usually, of necessity, several such cables to do the work—in case any one particular cable should break down. Thus—for the various reasons set forth—cables of less than a certain length seldom require to have signalling condensers applied to them, but if used they are very frequently supplied with adjustable shunt arrangements for varying conditions. Such arrangements are of course equally applicable at each end of the line, the effect being doubled by application at both ends.

* It need hardly be pointed out that a condenser is just as effective at the sending as at the receiving end of a cable, both in the curbing sense and in every other way, for any electrical effect must, of course, equally occur at each (either or both) end, there being no beginning or ending of an electrical circuit. Signalling condensers were only applied at the *receiving* end at first, owing to the fact that in those days of mirror signalling it was found that only a certain degree of "curbing" was desirable.

now universally adopted for signalling work over all long submarine lines. Some credit is also due to Mr J. C. Laws in the matter of suggestion. The effect, in the present day, of applying condensers at both ends of an Atlantic cable is practically to double the speed of signalling. This is partly due to the circumstance that the cable requires "clearing" after every contact; and this clearing is facilitated by any system of "curbing" such as is one of the features of the condenser. It is now thought that condensers at the *sending* end are especially desirable, supplemented, as a rule, by condensers in the receiving circuit.

It is generally advisable that the condensers for this purpose should be of comparatively low capacity, this value being usually a certain proportion of the total capacity of the cable. The object of interpolating condensers at either or both ends of the line is to obviate the difficulty of a comparatively long time being required for each signal, owing to the retardation of the line—partly due to the high capacity involved in a long cable, partly owing to its high resistance. A condenser of small capacity and practically no resistance can, however, be charged and discharged (or partially so, at any rate) almost instantaneously and much faster than such a cable could be even very partially. This high rate of charge and discharge of the condensers is not materially reduced when the said cable forms part of the inductive circuit, in which only partial charges and discharges—impacts, or impulses, in fact*—are an essential feature. Thus, to produce this effect of rapid charge and discharge most satisfactorily (by rapid primary and "curb" charges), it will be readily seen that the smaller the capacity of the condenser to be charged to the desired potential (sufficient to actuate the instrument in question to the required degree) the better—*i.e.*, the smaller the condenser, or condensers, the better, for then the smaller the quantity of electricity (Q) necessary to charge it to the required potential. Again, the smaller this quantity (Q) the quicker it is replaced or neutralised by a corresponding quantity of opposite sign, as effected by the curbing current following the primary ("working") current. Signalling condensers require, however, to have a certain amount of capacity, in order to keep up the size of the signals, as well as for maintaining their required (recognisable) character.

The manner in which the application of condensers to either or both

---

\* Briefly, the effect of condensers inserted in a cable circuit is that the period of potential rise is fore-shortened in such a way that *impulses* only are turned to account instead of a steady current flow. This is sometimes less accurately—though perhaps more clearly—defined by saying that the "*volume* of current" is checked by condensers and also by special curbing devices.

ends of a cable has the effect of "curbing" the signals may be gathered from Fig. 9.

This serves to illustrate the state of affairs when the positive pole is applied to line at the sending end. When, however, the key is allowed to fly up, a general discharge—or rather that which is ordinarily expressed as "discharge"—takes place. In effect it *is* a discharge; but actually, the following is more precisely the details of that which occurs:—

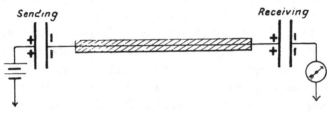

FIG. 9.

Electricity of the opposite sign comes up from the earth and restores complete neutrality by neutralising most of the previous (primary) charge throughout in such a manner as to act as a curb to the primary current. Thus it is not only perfectly correct to say that one of the functions of a condenser so placed is to have the same effect as an arranged for curbing current — following on each primary, or signalling,

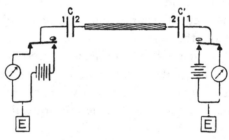

FIG. 10.

current—but, moreover, a precisely similar phenomenon is actually involved when such an electro-static system exists with condensers in circuit. The nature of the effect of applying a condenser at the receiving end is exactly the same as using one at the sending end, for what occurs at one end of a circuit (whether conductive or inductive) will correspondingly, and at the same moment, occur at the other, there being no beginning or end of a circuit.

If any change is effected by inserting a condenser at one end, double this change will result from additionally inserting one at the receiving end.

Supposing a battery communicates, at a given moment, a charge of positive electricity to coating No. 1 of the condenser C (Fig. 10), an equal quantity of neutral electricity being decomposed on coating No. 2 ; the negative electricity will be held by the charge of opposite sign on coating No. 1, and the positive portion set free will flow through the cable to accumulate on coating No. 2 of the condenser C'. Here, again, a quantity of neutral electricity is decomposed on No. 1 coating, equal to the positive charge on No. 2 coating, the negative portion being held by coating No. 2, and the positive portion set free to flow to earth through the receiving instrument. Directly the contact ceases, the positive electricity on coating No. 1 of condenser C returns to earth through E ; the negative electricity on coating No. 2 of the same condenser is set free, and recombines with the positive electricity in the cable, and on coating No. 2 of C'; lastly, the negative electricity on coating No. 1 of C', being in turn set free, flows to

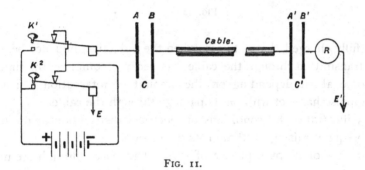

FIG. 11.

earth through E', and the cable recovers its neutral state ; or, as it has been expressed elsewhere* by Mr E. Raymond-Barker, a positive charge returns from earth, recombines with the negative charge hitherto held on C', and thus completes the normal equilibrium of the inductive system of cable and condensers.

Figs. 11 to 18 afford a practical and graphic analysis of the evolution and formation of siphon recorder signals on a long submarine cable worked with condensers.

In Fig. 11, $K^1$, $K^2$, represent a reversing key. A, B, and $A^1$, $B^1$, are plates of signalling condensers C and C' respectively, at two stations connected by cable.

Fig. 12 shews electrical distribution as produced by depression of $K^1$, which operation connects the + pole of the battery to plate A of C.

---

* "Lectures of Mr E. Raymond-Barker," the *Monthly Correspondent*, Madeira, May 1887.

Here the + pole of the battery being connected to A, thus throws on to it a *positive* charge which induces a corresponding *negative* charge on B which is connected to the cable, a *positive* charge being thrown on to A¹ at the other end of the cable.

This *positive* charge which now influences A¹ induces a *negative* one on B¹, whilst an equivalent positive charge goes to earth through R, thereby causing a deviation of the siphon—say, *above* the zero line, the siphon—however long the key may be kept depressed—falling back to zero when

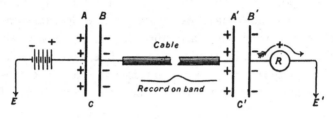

FIG. 12.

once C¹ is fully charged, that is to say, when the inductive influence between C and C¹, transmitted through the cable, has become complete, the time for this to come about depending on the amount of retardation which the induced impulse has met with on its passage through the cable.

Fig. 13 illustrates the conditions of electrical distribution as produced by the subsequent raising of K¹ after its depression.

This raising of K¹ puts plate A of C direct to earth through the upper

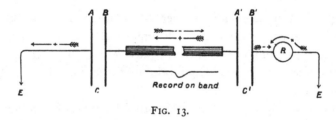

FIG. 13.

contacts of the key, thereby discharging from it the positive electricity which had been thrown upon it by the previous depression of K¹.

The + charge on A having thus passed to earth, the − charge on B of C recombines with the + charge on A′ of C′, whilst a + charge comes up, so to speak, from earth E¹ through R to recombine with and to neutralise the − charge which was on B¹ of C¹, and causes a deviation of the siphon similar in form to the former one in Fig. 12, but in the opposite direction, that is to say, the deviation is now *below* the zero line.

The siphon again falls back to zero, when the neutralisation, along the cable, betwixt C and $C^1$ has become complete, the time for this to come about depending, as before, upon the retardation in the cable, it being evident that all electrical pulsations, whether due to charges or to discharges, are equally affected by cable retardation.

Fig. 14 gives the electrical distribution after the depression of $K^2$ (in

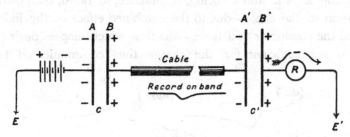

FIG. 14.

Fig. 11), which connects the negative pole of the battery to plate A of C; whilst in Fig. 15 we see the result of the subsequent raising of $K^2$, which discharges to earth the negative charge on A.

Fig. 16 illustrates the result of the depression and raising of $K^1$, or

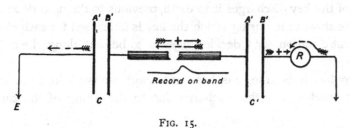

FIG. 15.

"dot" key, followed by the depression and raising of $K^2$, or "dash" key. In this example the keys have been kept down for some seconds of time, far longer, that is to say, than is necessary for charging the inductive system of condensers and cable.

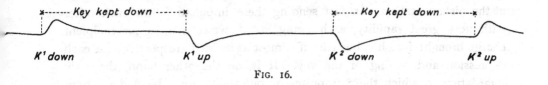

FIG. 16.

As a matter of fact, then, Fig. 16 gives the signal for the letter A ("dot" . . . "dash") as manipulated very slowly indeed.

Fig. 17 shews the letter A as transmitted not quite so slowly, and Fig. 18 the letter A at ordinary working speed.

Thus, it is evident that inordinately slow sending on a cable on which
condensers are used must produce unreadable signals, each charge deflection being followed up by a discharge deflection in the opposite direction ;
whilst the fact that properly formed signals can be produced only when the
rate of transmission is over a certain speed, is due to the manipulation of
the transmitting keys $K^1$ and $K^2$ being, in practice, so rapid, that, owing to
the retardation in the cable—due to the combined effect of the inductive
capacity and the conductor resistance—the time which elapses during each
depression is not sufficient for the charge thus communicated to the

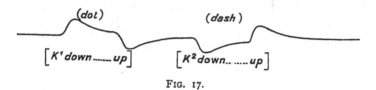

FIG. 17.

condenser A B to produce a complete effect upon the condenser A′ B′.    In
fact, the actual depression of each of the keys $K^1$, $K^2$, lasts for so short a time,
that the condenser A B can hardly be said to be properly charged, before
the raising of the key discharges it to earth, previous to the next depression.
However, the short time during which the key is depressed for each element
is of sufficient duration for a decided *impulse* to be sent along the cable to
the receiving end.

The impulse, it is true, has been cut short, in fact—in a manner—
reversed, or "curbed," by the discharge due to the lifting of the key, but

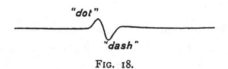

FIG. 18.

quite enough has gone forward for the production of a well-defined signal
at the distant station.    In fast sending these impulses follow each other
with very great rapidity, each being, as it were, due to an incipient
charge brought into life, but cut off almost at its birth, respectively at each
depression and raising of the key.    It is, on the other hand, the very
retardation to which these impulses or pulsations are subjected on their
passage through the cable, that serves to round them off, as it were,
and prevents their having the pointed and jagged appearance which is
the characteristic feature of recorder signals when not influenced by
retardation.

During the time required for charging the condenser C, in Fig. 10, its coating No. 1 remains at a constant potential $eo$ (Fig. 19), equal to the electro-motive force E of the sending battery, and represented by the straight line $ee_1$. The potential of No. 2 coating, in contact with the cable, decreases, on the contrary, from E to a certain value $E_1$, which theoretically can only be reached after an infinitely long interval, when the permanent state would be established ; the varying values of this potential during the same period of time would be represented by the ordinates of a curve such as $ee_2$, the asymptote to which would be a straight line parallel to the time axis $ox$ and situated at $E_1$ distance from it.    The insertion of condenser C into the circuit produces therefore, during the charging period, the same effect as an electro-motive force increasing from $o$ to $E - E_1$ of opposite name to that of the charging battery, and placed at the beginning of the cable.    As similar phenomena take place during the period of discharge, the ordinates of the arrival curve, immediately after contact is made or broken, are shortened in gaining proportion as the time interval increases.   We now see why the arrival curve is steeper when con-
densers are in circuit than when the cable is directly connected to the battery, and why it assimilates, on the other hand, to the curve obtained by applying a succession of alternating

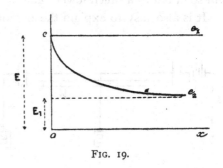

FIG. 19.

currents of diminishing potential immediately subsequent to the primary "working" current.

The increase of speed afforded by the use of condensers at both ends of the Brest—Saint-Pierre cable, of 1869, amounted, in fact, to 33 per cent.

By the use of condensers, although batteries of increased power become necessary, cables need only be charged, except at the extremities, to a potential actually lower than that which they would have if in direct communication with the battery.   For instance, if a 10-volt battery is required to work the receiving instruments through a cable of 1,000 micro-farads capacity, experience shews that by inserting condensers of 100 microfarads at each end, a battery of 25 or 30 volts is sufficient to ensure transmission of signals through the line.   Now it is easy to calculate approximately, in each of these two cases, the mean value of potential in a given section of the conductor.   For the sake of clearness, let us take the section commencing at the battery end and extending one-tenth of the total cable length.   If the battery is in direct contact with the cable and the latter to earth at the far end, the decreasing potentials—neglecting the usually comparatively small factor of loss through imperfect insulation—will be represented by a straight line whose exterior ordinates, for the length of line under consideration, are E and $\frac{9}{10}$ E, E representing the electro-motive force of the battery; the mean potential is therefore $\frac{95}{100}$ E. When the cable has a condenser at each end, the electro-motive force of the battery being increased to 3 E, if Q represents the charge taken up on each coating of the condenser C (Fig. 10), and $P_1$ the general potential of the cable at the end of the charging period, we have

$$Q = (3\ E - P_1)100 = P_1(1000 + 100)$$

whence

$$P_1 = \tfrac{1}{4} E$$

On the coating No. 2, and in the portions of cable nearest to it, the potential will fall during the first few instants after contact from 3 E to $\tfrac{1}{4}$ E; so the potential of the cable at points relatively close to the starting end will soon fall to a much lower value than in the preceding case.

It is also easy to explain the manner in which condensers act so as to render the receiving instruments very nearly insensible to the effects of earth currents. As these currents never develop very suddenly, except during magnetic storms, let us take as an instance an earth current having a potential of 100 volts—in fact, a current whose potential has increased from 0 to 100 volts in the space of 5 minutes or 300 seconds. Let it be assumed, also, that the resistance of the cable in which this current circulates is 7,500 ohms, its capacity K, 1,000 microfarads, and the separate capacity $k$ of the condensers, placed one at each end of the cable, 100 microfarads.

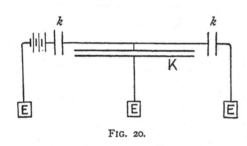

FIG. 20.

As we shall see further on, under these conditions $\tau$ is sensibly equal to 0.24 seconds, and for a double contact lasting $2.9\tau = 0.69$ seconds, the amplitude of the variation of the arriving current is 0.0269 of the limiting current strength.

E representing the electro-motive force applied to the first coating of one of the condensers (Fig. 20), and P the potential acquired by the cable, we have

$$k(E - P) = (K + k)P$$

The change in the condenser at the receiving station will therefore be

$$kP = \frac{k^2}{K + 2k} E$$

In the particular case under consideration, as the electro-motive force of the earth current varies by $\frac{1}{3}$ volt per second, the charge in the condenser at arrival increases in the same interval of time

$$\frac{100 \times 100}{1000 \times 200} \times \frac{1}{3} \times \frac{1}{106} = \frac{1}{360000} \text{ ampère}$$

The charge in the same condenser, with a 30-volt sending battery giving the same maximum intensity of arrival current as a 10-volt battery applied directly to the cable, varies per second

$$0.0269 \times \frac{10}{7500} = \frac{1}{30000} \text{ ampère}$$

The rate of charging in the second case will thus be about twelve times more rapid than in the first case.  With condensers, therefore, the signals will be as easily read as if the earth current had no existence—*i.e.*, they are practically unaffected thereby, especially as merely a strip of paper and an inferred zero is used for signalling purposes at the receiving station, in the place of the divided scale used for testing.

The quantity of electricity which enters a cable separated from the battery by condensers of variable capacity—the duration of the contacts with the battery remaining constant—increases with the capacity of the interposed condensers, without, however, exceeding the amount with which the cable would be charged under similar conditions, by direct connection to the battery.  This is easily understood if, using the same symbolic lettering as before, and designating by P the potential of the second coating of the condenser at any given instant during the charging period, and by $P^1$ the mean potential of the cable at the same moment, we notice that we always get

$$k(E - P) = KP^1$$

The limit of P being in fact the mean potential $P_1$ which would be acquired by the cable at the termination of an equal charging period, if in direct connection with the battery, $k(E - P)$ has also for limit $KP_1$, whatever the value of $k$.

These conclusions are confirmed by experiments recently made by M. Belz on the various underground and submarine lines connected up to the Marseilles office ; the results obtained with the 1880 Marseilles-Algiers cable being shewn graphically in Fig. 21.  The ordinates represent the entering charges compared with a charge obtained by lengthened contact with the battery, to which the value of 100 is given ; the condenser capacities being read off on the axis of x.  The contacts were produced with a Wheatstone automatic transmitter, their durations corresponding to speeds of 150 to 13 turns per minute, varying between 0.0166 and 0.115 seconds. The battery used consisted of 20 Callaud cells.

Each curve representing a charge has for its asymptote the straight line, parallel to the axis of x, corresponding to the charge acquired by the cable when connected direct to the battery for an equal duration of contact. With a 35-microfarad condenser, a capacity equal to one-fourth of that of the cable, the charge which enters the line after a contact of 0.05 second is only about 0.15 of the maximum charge, that is to say, of the charge which the cable without condensers would acquire after prolonged contact with a battery of $\frac{20 \times 15}{100} = 3$ Callaud cells. The strongest charge possible with a contact lasting 0.05 seconds, equal to 0.24 of the maximum, would be obtained without condensers after prolonged contact with 4.8 Callaud cells.

M. Belz also found that if the duration of contacts is short enough in proportion to the length of the line, the charges which enter—all else being equal—have the same value whether the far end of the line is insulated or to earth. The theory which we have just propounded gives a direct explanation of this fact. For the 1880 Marseilles-Algiers cable, $\tau$ being equal to 0.0208 second, the time 0.05 second, for instance, corresponds to $2.4\tau$ nearly. Now we have shewn that, at the expiration of the interval $\tau$, the strength of the arriving current at the far end is practically *nil*, and that it is very slight even at the end of a time $2.4\tau$. The quantity of electricity, therefore, which enters the cable at the starting end during such small intervals of time must therefore be sensibly the same, whatever the condition of the conductor may be at the far end.

The use of condensers offers a further advantage, as elsewhere explained, in prolonging the life of a defective cable, by keeping the fault at a permanent negative potential which prevents corrosion of the copper.

For the above reasons condensers are nowadays almost universally employed.

In the early days of submarine telegraphy, before the introduction of electrical condensers, it was a matter of vital importance to obtain some degree of definition and compactness in long-cable signals, the characteristic feature of which up to that time had been an undefined waviness, straggling in confused undulations. It was abundantly evident that an improvement of this nature, by increasing the speed at which signals could be safely transmitted, would greatly enhance the dividend-earning capacity of a cable.

For this reason much attention was paid to curbing devices, viz., arrangements for sending to line a succession of alternate currents of decreasing potential, resulting at the receiving end of the cable in a clearly defined signal.

Some of these devices were directly manual, that is to say, the sending key itself was made to transmit curbed signals, and these have been already referred to; whilst in others an electro-magnetic transmitter, actuated by hand-keys on a local circuit, was the curbing agent.   Later on—in 1875— the curb system was applied by Sir William Thomson and Professor Fleeming Jenkin to their "automatic curb-sender," which is governed in its action by the passage through the apparatus of a perforated paper band, after the manner of Wheatstone's automatic transmitter, though very different in detail.   This apparatus was, in fact, the first combination of the "curb key"—various patterns of which have already been briefly described —and the Wheatstone automatic transmitter.   It was only, however, used

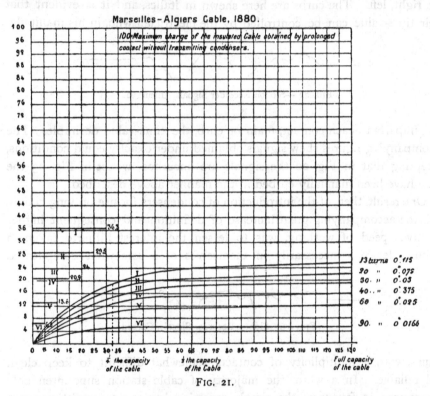

FIG. 21.

experimentally for a short time (on the Aden-Bombay cable), and was never considered to be quite a success for practical purposes.   Though exceedingly ingenious, it was rather too elaborate for regular work.   Moreover, like the *manual* curb key, it was really before its time, there being then very little demand for actual fast-speed work—such as may be attained on a cable by machine transmission.   Other names which have from time to time been associated with curb-signal devices, in connection with manual, or machine, transmission, are those of Siemens, Saunders, Muirhead, Gott, and Rymer-Jones.

It has also been pointed out by Mr Raymond-Barker that under certain

conditions—as, for instance, when a cable-hut or station is in communication with a telegraph ship through a great length of laid cable, *plus* a
considerable length of cable coiled in the ship's tanks, and when great
speed of transmission is not requisite or even possible—hand curbing can
be advantageously applied. The electrician would employ, as usual, an
ordinary reversing key, but would add a curbing contact after each main
contact. Thus the manipulation of the left and right finger-pieces of the
key for the transmission, *e.g.*, of the signal "understand" would be—left-
*right*, left-*right*, left-*right*, right-*left*, left-*right*, instead of merely left, left,
left, right, left. The curbs are here shewn in italics, and it is evident that
their time-value can be controlled by the electrician who, in his manipula-

FIG. 22.—Siphon Recorder Signals, Uncurbed.

tion, imparts to each contact, main or curb, the time-value desirable. The
accompanying figures shew signals obtained under exactly equal conditions,
excepting that in Fig. 22 the signals are uncurbed, whilst in Fig. 23 the
same have been manually curbed in the manner above described.

One result, then, of the introduction of condensers for use on long cables,
with the accompanying great increase of definition and consequent higher
effective speed of working, was to lessen the interest hitherto taken in
curbing devices — necessarily in themselves of a somewhat complicated

FIG. 23.—Siphon Recorder Signals, Curbed.

nature, with a multiplicity of contacts somewhat difficult to keep clean
and reliable. In a word the majority of cable-station superintendents
pronounced in favour of the simple reversing key as a transmitting agent
for signals, used in conjunction with condensers, as being a perfectly satisfactory system which extra curbing was thought to spoil.

Of late years, however, keen competition combined with a vast increase
of traffic over the various submarine cable systems has caused the matter of
curbed signals to be again taken up in conjunction with improved automatic
transmitters, these latter having been rendered necessary by the very high
traffic-carrying capacity of the two last laid Atlantic cables of 1894.

## ALPHABETS.

The alphabet used on submarine lines—if we except the printed characters on a few Hughes circuits between London and the Continent—is the International Morse Code.

On short cables (as in land-line telegraphy) this code is composed of "*dots*" and "*dashes*" rendered either visible, or merely audible, according as reading is carried on by sight or by sound.

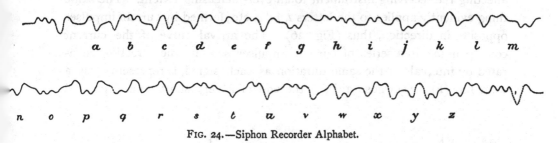

Fig. 24.—Siphon Recorder Alphabet.

On long submarine cables the same "*dots*" and "*dashes*" are represented by left and right deflections of a luminous image projected on to a scale from a mirror affixed to a magnetic needle, or by up and down deviations from a horizontal zero ink-line traced by a recording siphon. The first-mentioned system—the Morse pure and simple—requires a current in but

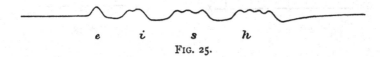

Fig. 25.

one direction; the two latter, like Wheatstone's single-needle apparatus, involve the use of reversals, or currents in opposite directions.

The reversal system, in which symbols corresponding respectively to dots and dashes are produced with positive and negative currents, may be

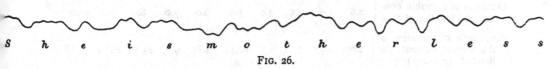

Fig. 26.

said to owe its origin to the point alphabet invented by Steinheil, although the latter was never officially adopted.

Alphabets involving the use of momentary contacts of constant and uniform duration and causing in the receiving instrument up and down, or right and left deviations from a zero position, enjoy the advantage of not requiring currents of prolonged application—*i.e.*, where the dash occupies

thrice the time of the dot. Nevertheless, after the completion of each
elementary signal, the cable by no means reverts at once to its former
electrical condition, and therefore time is to a certain extent taken up on
this account.

At the arrival end the signals necessarily depend in a large degree on
those which have preceded them. For instance, the letters *e, i, s, h* of the
Morse alphabet, formed as here (Fig. 25) by current impulses in one
direction only, give rise to increasing current intensities, consequently
affecting the receiving instrument to an ever-increasing extent. The same
thing evidently applies to the letters *t, m, o,* etc., formed by current impulses
opposite in direction, thus (Fig. 26). The arrival curve of the current
corresponding to a series of current impulses sent in one direction, separated by intervals of the same duration as each signal, is represented by a
wave line (Fig. 27) whose axis, at the end of a very short time, is parallel

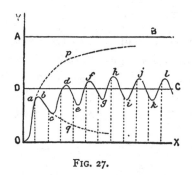

FIG. 27.

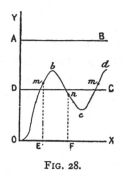

FIG. 28.

to the axis of X and at $\frac{1}{2}\dfrac{P_0}{\rho L}$ distance from it. The mean strength of the
arrival current thus rapidly becomes equal to half the intensity limit. Also
the amplitude of these undulations diminishes when the duration of the
contacts is shortened, as the accompanying table shews :—

| Duration of a double con-<br>tact, as a function of $\tau=1$ | 2.9 | 3.0 | 3.5 | 4.0 | 5.0 | 6.0 | 7.0 | 8.0 | 9.0 · | 10.0 |
|---|---|---|---|---|---|---|---|---|---|---|
| Amplitude of variation of<br>the arrival current, the<br>limit of current = 100 - | 2.69 | 2.97 | 4.52 | 6.31 | 10.42 | 14.85 | 19.67 | 24.42 | 29.11 | 33.68 |

If *m* E and *n* F (Fig. 28) represent the current strengths at which the
receiving instrument commences and ceases to work, its action will only be
regular when these limits fall on one side and the other of the line of
means D C. It is the difference of level between the points *m* and *n* which
constitutes the period of sensitiveness of the different instruments. If they

were both below D C, long signals would result separated by shortened intervals ; if both above D C, short signals would be given with long intervals. If the point *n* was below the hollows *c* . . . , a continuous signal would be received ; and if the point *m* was higher than the crests *b*, *d* , . . , nothing would be received.   In fact, the letters in which dots and dashes are combined are more or less impaired according to the nature of the currents which succeed them in the formation of signals.

Cable-telegraph operators, however, soon become accustomed to these effects, and read these signals with ease, even when the high speed of transmission together with the retardation of the cable has flattened out all definition of the separate elementary signals, and the elements have become merged into characters undulatory and irregular as far as distance of deviation from zero line is concerned, but of uniform characteristics and rate of formation, and therefore easily deciphered by an expert.   Take, for example, the word " Funchal," as given in Fig. 29.   Under such

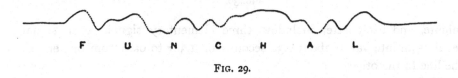

F      U      N      C      H      A      L

FIG. 29.

conditions it is of paramount importance not only that the speed of rapid transmission should be strictly uniform, but also that the time relation between current impacts and intervening spaces should be maintained constant.   In other words, a given letter ought always to present a uniform appearance in any given instance or conditions of signalling, and this is impossible without time uniformity of current contacts and spacing intervals, whatever may be the set ratio existing between the two.   A good recorder clerk does not, when reading, count the number of up and down deviations from the zero line.   When all the letters possess like characteristics his practised eye rapidly seizes each letter as a whole—short *words*, even, are read in this way—at a glance, just as the book reader quickly realises the sense of what he reads without closely analysing the formation of each letter or word.

On cables, therefore, where very high speeds are maintained, as on the recently laid Atlantic lines with heavy cores, some form of automatic transmitter is used which ensures the perfect uniformity of signals—transmitted at a speed too high to be efficiently maintained by hand.

APPLICATION OF THE FOREGOING TO THE 1869 ATLANTIC CABLE.

Let us take as an example the 1869 French Atlantic cable—the longest section in existence—whose length L is 2,584* N.M., capacity per naut $k=0.43$ microfarad, and copper resistance $r=2.93$ ohms per N.M. Formula (11) gives

$$\tau = \frac{0.43 \times 2.93 \times \overline{2.584}^2}{10^6} \times 0.029 = 0.245 \text{ second.}$$

The normal speed of transmission through this cable with mirror receiving instruments being about eleven or twelve words of five letters per

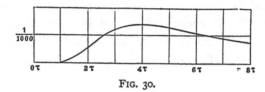

FIG. 30.

minute, and each letter including three elementary signals, each signal requires an interval of about 0.34 seconds or $1.38\tau$ to pass from one end of the line to the other.

If we take as 100 the intensity limit of arrival current from a battery of given potential, the intensities at different intervals of time counted from the commencement of contact will be represented by the numbers in the table on the following page.†

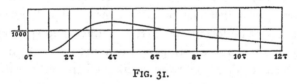

FIG. 31.

In this cable, then, half the value of the charge limit, at arrival, is only attained after an interval of $6\tau$, and to get nine-tenths of this charge, an interval of $13\tau$ is required. A receiving instrument, working under the influence of a current equal to 0.27 of the maximum, would only start after an interval of $4.12\tau$, or about 1.009 seconds. If the cable were disconnected from the battery at the starting end at the time $5\tau$ and put direct to earth, the arrival current intensity, at the time $6\tau$, would be $50.55770 - 0.01639 = 50.54131$; at $7\tau$, $60.41164 - 2.46081 = 57.95083$, and so on. If, at the

---

* Increased since to 2,717 N.M.
† Jenkin, "Electricity and Magnetism," 8th edition, p. 332.

time $7\tau$, the cable were again connected to the battery at the starting end, the arrival intensity, at the time $8\tau$, would be $68.42832 - 11.98582 + 0.01639 = 54.45889$. Lastly, if at the time $9\tau$, the cable were joined up to the

| $t$ in terms of $a$. | Strength of Current in Percentages. | $t$ in terms of $a$. | Strength of Current in Percentages. | $t$ in terms of $a$. | Strength of Current in Percentages. | $t$ in terms of $a$. | Strength of Current in Percentages. |
|---|---|---|---|---|---|---|---|
| .4 | .0000000271 | 1.1 | .04140636 | 3.5 | 18.48434 | 7.8 | 66.95995 |
| .5 | .0000005452 | 1.2 | .08927585 | 3.6 | 19.84366 | 8.0 | 68.42832 |
| .55 | .0000033639 | 1.3 | .1704802 | 3.7 | 21.21342 | 8.5 | 71.82887 |
| .60 | .000016714 | 1.4 | .2959955 | 3.8 | 22.59017 | 9.0 | 74.87172 |
| .62 | .000029252 | 1.5 | .476336 | 3.9 | 23.97071 | 9.5 | 77.59133 |
| .64 | .000049412 | 1.6 | .720788 | 4.0 | 25.35217 | 10.0 | 80.02000 |
| .66 | .000080817 | 1.7 | 1.036905 | 4.2 | 28.10757 | 10.5 | 82.18760 |
| .68 | .00012835 | 1.8 | 1.430252 | 4.4 | 30.83807 | 11.0 | 84.12139 |
| .70 | .00019845 | 1.9 | 1.904356 | 4.6 | 33.52902 | 12 | 87.38402 |
| .72 | .00029937 | 2.0 | 2.460812 | 4.8 | 36.16892 | 13 | 89.97752 |
| .74 | .00044152 | 2.1 | 3.09969 | 5.0 | 38.74814 | 14 | 92.03836 |
| .76 | .00063776 | 2.2 | 3.81846 | 5.2 | 41.26032 | 15 | 93.67565 |
| .78 | .00090371 | 2.3 | 4.61560 | 5.4 | 43.70028 | 16 | 94.97631 |
| .80 | .00125804 | 2.4 | 5.48661 | 5.6 | 46.06449 | 17 | 96.00951 |
| .82 | .00172272 | 2.5 | 6.42695 | 5.8 | 48.35070 | 18 | 96.83023 |
| .84 | .00232333 | 2.6 | 7.43163 | 6.0 | 50.55770 | 19 | 97.48215 |
| .86 | .00308919 | 2.7 | 8.49536 | 6.2 | 52.68501 | 20 | 98.00000 |
| .88 | .00405358 | 2.8 | 9.61264 | 6.4 | 54.73314 | 21 | 98.41134 |
| .90 | .00525387 | 2.9 | 10.77797 | 6.6 | 56.70294 | 22 | 98.73809 |
| .92 | .00673158 | 3.0 | 11.98582 | 6.8 | 58.9502 | 23 | 98.99763 |
| .94 | .00853247 | 3.1 | 13.23087 | 7.0 | 60.41164 | 24 | 99.20379 |
| .96 | .01070646 | 3.2 | 14.50800 | 7.2 | 62.15439 | 25 | 99.36754 |
| .98 | .01330764 | 3.3 | 15.81233 | 7.4 | 63.82523 | | |
| 1.00 | .01639420 | 3.4 | 17.13921 | 7.6 | 65.42636 | | |

negative pole of the same battery, the arrival current intensity, at the time $10\tau$, would be $80.02000 - 38.174814 + 11.98582 - 2 \times 0.01639 = 53.22490$, and so on.

If, at the starting end, a contact of $0.01\tau = 0.00245$ second is made with a 10-volt battery, or one of $0.1\tau = 0.0245$ second with a 1-volt battery, putting the same starting end of the cable to earth immediately after the contacts, the strengths of the arrival current will be represented by the numbers found in the two tables given below,* and by the curves in Figs. 30 and 31, the maximum intensity, corresponding to permanent contact with a 10-volt battery, being taken as equal to 10,000.

## TABLE I.

*Strength of Current at the Receiving Station after Contact with the Battery for a Period of*

| $\cdot\tau$ | 0.01$\tau$ | 0.1$\tau$ | Difference. |
|---|---|---|---|
| | And then cable put to earth. | | |
| | E = 10 volts. | E = 1 volt. | |
| 1.0 | 0.16 | 0.11 | 0.05 |
| 1.2 | 0.6 | 0.48 | 0.12 |
| 1.4 | 1.5 | 1.26 | 0.24 |
| 1.6 | 2.8 | 2.45 | 0.35 |
| 1.8 | 4.3 | 3.9 | 0.4 |
| 2.0 | 5.9 | 5.6 | 0.3 |
| 2.2 | 7.5 | 7.2 | 0.3 |
| 2.4 | 9.0 | 8.7 | 0.3 |
| 2.6 | 10.3 | 10.0 | 0.3 |
| 2.8 | 11.4 | 11.2 | 0.2 |
| 3.0 | 12.3 | 12.1 | 0.2 |
| 3.2 | 12.9 | 12.8 | 0.1 |
| 3.4 | 13.4 | 13.3 | 0.1 |
| 3.6 | 13.7 | 13.6 | 0.1 |
| 3.8 | 13.8 | 13.8 | 0.0 |
| 4.0 | 13.8 | 13.8† | 0.0 |
| 4.2 | 13.7 | | 0.0 |
| 4.4 | 13.6 | | 0.0 |
| 4.6 | 13.4 | | 0.0 |
| 4.8 | 13.1 | | 0.0 |
| 5.4 | 12.0 | | 0.0 |
| 5.9 | 11.0 | | 0.0 |
| 7.0 | 9.0 | | 0.0 |
| 8.1 | 7.0 | | 0.0 |
| 9.6 | 5.0 | | 0.0 |
| 12.8 | 3.0 | | 0.0 |
| 13.4 | 2.0 | | 0.0 |
| 16.5 | 1.0 | | 0.0 |
| 19.4 | 0.5 | | 0.0 |
| 23.5 | 0.2 | | 0.0 |
| 26.6 | 0.1 | | 0.0 |

* *Journal of the Society of Telegraph Engineers*, vol. viii., p. 101.
† Maximum at 3.84$a$.

TABLE II.

| | | | | | | | |
|---|---|---|---|---|---|---|---|
| $1\tau$ | 0.016 | $5\tau$ | 38.75 | $9\tau$ | 74.87 | $16\tau$ | 95.00 |
| $2\tau$ | 2.46 | $6\tau$ | 50.56 | $10\tau$ | 80.02 | $20\tau$ | 98.00 |
| $3\tau$ | 11.99 | $7\tau$ | 60.41 | $13\tau$ | 90.00 | $\infty$ | 100.00 |
| $4\tau$ | 25.35 | $8\tau$ | 68.43 | | | | |

The maximum intensity 0.001 384 is the same in both cases, as equation (13) shewed, the product $^{\theta}P_{0}$ remaining constant. This maximum is reached after a time $3.84\tau$, instead of $3\tau$ as given by formula (14). The signal appears, in the first case, at the time $1.26\tau$ with a current intensity of $\dfrac{0.81}{10000}$, and in the second case, at the time $1.30\tau$, that is, after an interval longer only by $0.04\tau$, with a current intensity of $\dfrac{0.82}{10000}$.

Four alternating contacts having the following durations,

1. Current +      -    $0.247\tau$
2.   ,,   −        $0.203\tau$
3.   ,,   +      -   $0.100\tau$
4.   ,,   −      -   $0.050\tau$
5. Cable to earth    -     -     -   $0.400\tau$

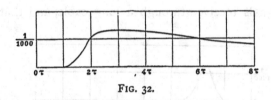

FIG. 32.

give a curbed signal, the intensity of current received at the arrival end being represented by the curve in Fig. 32. Fig. 33 shews the electric wave produced in the cable ; the first portion of the cable is divided, after a time $0.6\tau$, into sections containing alternating positive and negative charges of decreasing potential ; these charges combine together, discharging the cable so rapidly, that a new signal can be transmitted after an interval of 0.4 to $0.5\tau$, say nearly $0.3\tau$ before the first signal can make its appearance at the far end of the cable. On attempting to further reduce this interval of $0.4\tau$, the two sets of waves mix one with the other, and signals are no longer distinct. The time $1.1\tau$ may therefore be considered the limit of the speed with which signals can be transmitted, using mirror receiving instruments.

If the curbing system be used in conjunction with condensers inserted in the cable circuit, the arrival current intensity is represented by the ordinates of the curve (Fig. 34) which rises and falls much more abruptly than in the former case.

For the various reasons already shewn, condensers are universally employed at either or both ends for working a long length of cable.

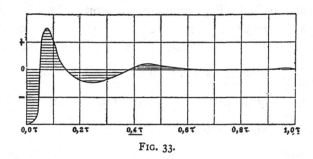

FIG. 33.

Some difference of opinion exists, however, as to the benefit derived from the use of additional curbing arrangements beyond that afforded by sending condensers. It is contended by several authorities—notably Mr T. J. Wilmot and Mr P. B. Delany—that extra curbs reduce too much the size of the signals received through a long cable, and that the desired

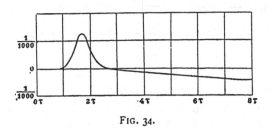

FIG. 34.

shape and character can be equally well secured by local means at the receiving end. It is generally admitted that though curbing transmitters should be an advantage—theoretically speaking—for comparatively short cables, these will always give excellent signals without a curb.

With a view to deciding whether a long cable works best with signalling condensers only, or when also employing special curbing apparatus, exhaustive experiments have been carried out under the following different conditions :—

1. Without either condenser or other curb.
2. With condensers, but without further curbing arrangements.
3. Without a condenser, but with a hand-key or machine curb.
4. With condensers and with auxiliary curb.

The last condition seems, as a rule, to give the best—or rather, the most reliable—results ; and especially so where duplex connections are in force, inasmuch as condensers may be said to be an essential for duplex working.

## DEARLOVE'S SYSTEM OF TRANSFORMERS IN PLACE OF CONDENSERS FOR WORKING CABLE CIRCUITS.

The following method of working cables patented (No. 8823, A.D. 1894) by Mr A. Dearlove, consists mainly in placing specially constructed trans-

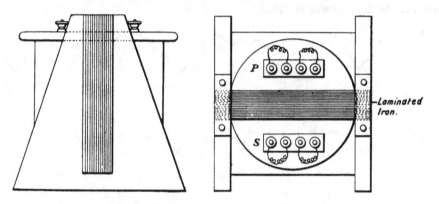

FIG. 35.—Dearlove's Cable Working Transformer (⅛ of actual size).

formers in the equivalent position of condensers as ordinarily used on cable circuits, both as regards the sending and receiving ends of the line.

The type of transformer employed is similar to those now generally supplied for electric lighting purposes, with laminated iron core and a *closed magnetic circuit*. The primary and secondary coils are wound with equal turns, and subdivided into four or more distinct and insulated circuits.

The transformer is represented in Fig. 35. It is equivalent in effect to 40 or 50 microfarads of capacity. Fig. 36 shews the circuit connections.

Cromwell Varley in 1862, and Mr G. K. Winter in 1873, suggested the use of induction coils for a similar purpose ; but the improvement claimed in this recent method is obtained by the adoption of the closed iron circuit, and the winding of the primary and secondary circuits—the ratio of the turns can be varied as is found most efficient for the circuit upon which it is used, but in no case must the primary circuit have a larger

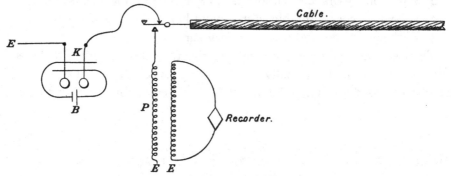

FIG. 36.—Connections of Dearlove's Transformer System.

number of turns than the secondary circuit. Thus no undue strain can possibly come upon the cable end under these conditions, provided that moderate battery power be used.

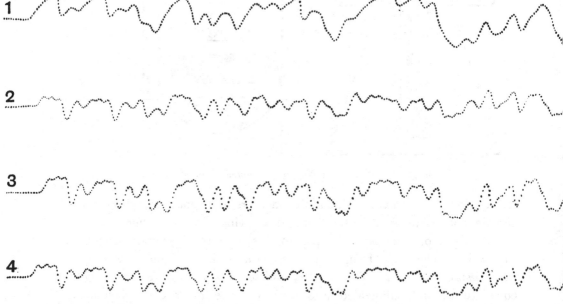

FIG. 37.—Signals obtained with Transformer, in place of Condensers.

The curbing action of the transformer is similar to that of the condenser, and the specimen signals given above shew this to be the case.

Particulars.—Line, 7,000 ohms. Cap., 330 mfds. Five cells. Eighteen words. Sending direct to Line with Automatic Transmitter.

REMARKS.

1. Signals with neither sending nor receiving condensers.   P = 200 ohms.   S = 200 ohms.
2.   „   no sending condensers, but receiving on transformer (slightly over-curbed).
3.   „   „   „   „   60 microfarads (insufficient curb).
4.   „   „   „   „   40   „

The amount of curb in the transformer can be regulated by the addition of no inductive resistance in the primary circuit, or by the introduction of a shunt, but it is seldom necessary to use this extra apparatus.

The particular transformer used has 5,000 turns in the primary circuit (P), and an equal number in the secondary circuit (S), both having a resistance of approximately 200 ohms.   The circuit being subdivided, the instrument is easily adapted to work either long or short cables, and can be recommended for use on cables up to 1,200 miles in length.

The best results are obtained on this system when the battery is applied direct to the cable as shewn in Fig. 36, but the transformer principle of working may be adapted to either the duplex or simplex methods of working.

There is nothing likely to get out of order or require attention in the instrument ; and the cost is less than a fourth of that of a condenser of equivalent capacity.

## MECHANICAL ANALOGY AND MODEL OF SUBMARINE CABLE WORKING.

In the course of a series of lectures at the Royal Institution,* Professor George Forbes, F.R.S., gave a hydro-mechanical analogy for the working of a submarine cable of certain length, and, therefore, possessing a certain factor of retardation, accompanied by a model † placed alongside of the electrical representation, both of which were in working order.

In Professor Forbes's model a thread (Fig. 38) is suspended in a viscous fluid.   A twisting force applied at the top represents E.M.F. ; the total twist in any length represents the charge in that portion of the cable ; the velocity of rotation at any point represents the current flowing at that part of the cable. A spring at top or bottom represents a condenser, whose capacity is measured by the twist (charge) given by unit twisting force (unit of E.M.F.).

---

* " Lectures on Alternating and Interrupted Electric Currents," by Professor George Forbes, F.R.S., M.Inst.C.E., delivered April 1895.
   † Exhibited a short time afterwards at the Royal Society soirée of that year.

By hanging a thread in a vertical tube filled with glycerine and water in suitable proportions, we have all the essentials for understanding the propagation of electric pressure through a submarine cable. But the measurement of the twisting force applied at the top is difficult. Any mechanism having momentum is objectionable, as this introduces the analogue of a heavy self-induction, which does not exist in the electrical problem. For this reason, Professor Forbes suspended the thread, and at each end vanes were supported which could be blown round by a constant pressure of air to represent the E.M.F. The spindle of the vanes is solid, and goes down into the liquid to the point where the thread representing the cable is attached. The vanes and air-jet represent the battery, and it

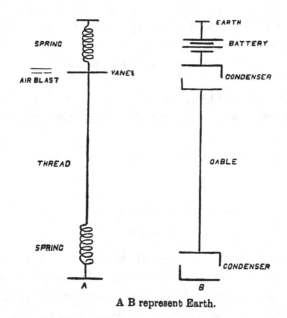

A B represent Earth.

FIG. 38.—Mechanical Analogy of Submarine Cable Working.

is found convenient to have two jets to rotate the vanes in opposite directions. These are actuated by two keys, + and −, which admit the air to one or other of the nozzles. If there be no sending condenser to be represented in the model, the vanes must be supported by a silk fibre without torsion. The vane spindle is held laterally by bearings.

Things being so arranged, we may assume that the air pressure used represents the E.M.F. of the sending battery, say 100 volts. If the cable which we wish to imitate has 800 microfarads capacity, the permanent charge on it would be $\frac{800}{10^6} \times 100 = 0.08$ coulomb. Fixing now the lower end of the thread, apply the air-jet and observe the twist. Suppose it is 10 revolutions. Then 1 revolution means 0.008 coulomb, and 1 revolution a second means 0.008 ampère. If now the resistance of the cable be 6,000 ohms, the permanent current would be $\frac{1}{60}$ ampère. This should be represented by $\frac{1}{60 \times .008} = \frac{1}{.48} = 2$ revolutions per second. Leave now the lower end free, and apply the air-jet at the top. If 2 revolutions per second be obtained, we have the right resistance in the model. If not, more water or glycerine should be added till the right resistance is secured.

It is impossible, however, to get mirror signals thus, as velocity indicates current, and the zero would move. If, however, we put a mirror at the lower end of the thread, and restrain its motion either by a spring or a magnet, it may be considered that we are putting in a receiving condenser, and the deflection indicates the potential at that point of the cable, just as a recorder does. We ought to add here, below the mirror, a vertical wire to represent the resistance of the recorder. If now, instead of having a torsionless fibre at the top, we attach the vanes to a spring, this is equivalent to a sending condenser, and we have a perfect representation of the signals received by the special cable which we have imitated.

The Commercial Cable Company : Station and Quarters at Waterville, Ireland.

### SECTION 3.—SPEED OF SIGNALLING.

The rate at which electrical impulses can be made to pass along any electrical conductor is—so far as we know at present—governed by two factors which form part of its constitution—*i.e.*, the conductor resistance of the line and its electro-static inductive capacity. Both of these influence the problem inversely, and to the same extent. This was first pointed out in a general way by Professor William Thomson (now Lord Kelvin) in the columns of the *Athenæum*, when confuting Mr Wildman Whitehouse's views on this subject in 1856; and was further, and more completely, dealt with by Mr S. A. Varley, at the Institution of Civil Engineers in 1858.* The independent views of Lord Kelvin and Mr Varley now take the form of a pretty generally admitted law (based on experimental proof), and is usually expressed

$$S = \frac{A}{KR}$$

Where S = the maximum speed of signalling on the line under consideration.

Where A = the constant—the multiple of K and R of a known cable giving a known speed.

Where K = total electro-static capacity of the cable under consideration.

Where R = total conductor resistance.

In fact, by dividing the KR of the known standard cable (= constant, A) by the KR of the cable under consideration, the relation of their respective KR (retardation) is arrived at; and therefore, the relation, also, of their respective signalling speeds.

This law is usually known as the KR law.† In the case of an aerial conductor, K naturally forms a very small factor; and under some circumstances may be almost neglected. But with an underground, or submarine, cable being, as a rule, considerable, it requires to be taken into equal account with the conductor resistance.

The product of K and R makes up, in fact, what is usually described as

---

* Mr Varley was, indeed, the first to put the whole matter on a perfectly clear basis in the course of his paper (at the above Institution) on " The Electrical Qualifications requisite in Long Submarine Telegraph Cables," as well as during one to the Society of Arts in the same year on " The Practical Bearing of the Theory of Electricity in Submarine Telegraphy."

† Electro-static capacity is very often expressed by F—originally standing for " farad," the unit of capacity; but inasmuch as this law is generally spoken of as the " KR law," K is, perhaps, preferable if only for the sake of uniformity.

the retardation of a cable in becoming charged throughout, which cannot, indeed, be constituted by either alone.  It also equally sets up a prolongation effect, in becoming entirely discharged, previous to transmitting another electrical impulse.*

Thus, in considering the type of conductor and thickness of dielectric which will give the required speed of signalling, or when calculating the maximum working speed afforded by a cable core of given dimensions, it becomes necessary to first arrive at the electrical resistance of the conductor and the electro-static capacity of the cable.  With these, and the same particulars regarding a cable giving a known speed, by the use of the above formula the corresponding signalling speed of the cable in question, or the core-type which must be employed to afford it, may be ascertained.

The speed of signalling through submarine cables is usually defined as the number of words which can be sent correctly from one end of the cable to the other in a single minute.  The number of letters contained in different words of the same language being exceedingly variable, the number of words transmitted in a given time will depend, to a great extent, on their length.  It has, in fact, happened during trials, that intentionally formed phrases consisting of very short words have occurred in succession over and over again, and a speed of transmission thus attained very different from the practical working speed of the cable.  This state of things is remedied by indicating the average number of letters of which the words are to be composed ; the number used formerly to be five, but has since risen to seven, as the result of the rules for counting words adopted by the International Telegraphic Conferences.†  To avoid all chance of misunderstandings, most managers and electricians of cable companies specify, as speed of transmission for traffic purposes, the number

---

* That which takes place in a cable when put into communication with an electric generator has been already dealt with, fairly fully, in the last section.  It suffices here to say that the retardation in any given cable is due to the static (holding back) effect of the induced charge outside the cable—a form of Leyden jar—on the primary charge in the conductor.  Similarly, when the battery contact is broken, and the conductor is put to earth, a cable discharges itself in the same, comparatively slow, manner on the above account.  In this case the phenomenon is, for distinction, described as prolongation.

† The fact that seven—or indeed nine—letters is found to be nearer the average than five in *cable* messages is explained by the circumstance that code words form a larger proportion of the traffic than in the case of land lines, in which five is still considered the proper standard number of letters to a word.

For the latter reason—associated with that of general uniformity—a word is still taken as representing five letters in all speed calculations for purposes of estimate and comparison.  The cable word is, in fact, only taken at seven to ten letters for commercial estimates and calculations in the direction of earning capacity during given working periods of the cable, and for the return on outlay.

of *letters* which can be sent through the cable per minute, for purposes of accuracy, and in order to be more comparable with each other.*

This second definition is, however, in one way no more concise than the first. The number of elementary signals forming the different letters of the Morse alphabet varies, in fact, from one to four. Thus it becomes necessary to state the average number of elementary signals of which the letters are to consist. This number is generally three, sometimes four, the intervals separating both letters and words being neglected.†

The first traces of electricity arrive at the remote end of the cable in an interval of time which is extremely short, though not infinitely so as would

---

* The traffic on some submarine lines consists of a very small proportion of code words or ciphers. On the other hand, the traffic on most of them is mainly so constituted, with words of ten letters and over.

† It may be suggested that speed of transmission would be more accurately defined by the number of elementary signals—"dots" and "dashes," more aptly termed recorder-line "ups" and "downs"—which can be sent through the cable per minute of time, all signals succeeding each other at equal intervals ; but this idea, though correct from a certain point of view, would scarcely answer in practice. It has already been shewn that signalling can be safely carried on at a speed at which all actual definition of recorder-line "ups" and "downs" will have disappeared, the value of each letter, or part of a letter, in terms of "ups" or "downs" being given, not in so many distinct waves in the ink-line, but by a more or less prolonged hanging-away of the siphon from the zero position. This being so, it would not be easy to correctly estimate the speed at which practicable signals can be produced on a given cable from the mere noting of the number of meaningless elementary signals received in uninterrupted succession in a given time. Such data would afford no such criterion of practical signalling as would be offered by the successful transmission within a given time of a certain number of standard words of seven letters, the words being chosen so as to give on an average, for the total number of words, a mean of about 3.1 elements per letter, which is the mean for the International Code.

The following list gives ten representative seven-letter words, of which the total mean number of elements per letter is 3.1—that is to say, equivalent to the mean number of elements per letter in the entire International Code Alphabet, *a* to *z* :—

LIST.

| | |
|---|---|
| Cockpit. | Forceps. |
| Bulrush. | Colony. |
| Redjowl. | Sobbing. |
| Cocoons. | Zadkiel. |
| Ploughs. | Follows. |

The use of these and other words of similar mean value in a trial of speed would give strictly reliable data in regard to the word capacity of a cable. In fact, such a test would be if anything somewhat too exacting, as, from the frequency with which the shorter letters occur in average words, the number of elements per letter in an average telegraphic message will be found to be something below the value 3.

A combination of the theoretically excellent element-unit with the more practically useful word-unit systems may be arrived at in the following manner.

The total mean number of elements per word in the list above given is 21.8, which

appear from equation (6)—in the first section of this chapter—which is not rigorously exact.   For this reason it becomes essential, on long lines,* in order to increase the nett returns of the cable, to use only the most sensitive receiving instruments, such as Sir William Thomson's mirror instrument or siphon recorder.   But the value of $C_1$ also depends on the potential $P_{o_1}$ and on the cable constants $r$, $k$, and L.   We shall now explain how the best results can be practically assured with a cable about to be laid between any two points, by a judicious choice of battery power on the one hand and a properly proportioned core on the other.†

We will first suppose that similar receiving apparatus is used at each

---

is equivalent to 3.1 elements per seven-word letter, the average for the International Telegraph Alphabet.

|  |  |
|---|---|
| Then elements per word = | 21.8 |
| Maximum space after each word equivalent to, say, 3 elements = | 3. |
| **Total elements per word** | **24.8** |

To arrive at a rigidly correct value for rate of speed of transmission through a cable, let us count the number of elements in the words transmitted in a given time, allowing also 3 elements for every word space.   The grand total divided by 24.8 will give the total number of standard words transmitted, which value, divided by time in minutes, will obviously give the number of standard seven-letter words per minute, spaces included.

For example, the lines—

> " I am the instrument of man's desire,
> To hold communion with his fellow-man
> In distant fields . . . in other climes afar,
> Swifter than flight of migratory bird—
> Nay, swift almost as speech from mouth to mouth . . . "

were sent through a cable in half a minute.

|  |  |
|---|---|
| Number of elements in words | 409 |
| Number of elements in intervening spaces at 3 elements per space | 105 |
| **Total elements** | **514** |

$$\frac{514}{24.8} = 20.7 \text{ standard words.}$$

$$\frac{20.7 \text{ words}}{0.5 \text{ minute}} = 41.4 \text{ standard seven-letter words per minute.}$$

\* On the basis that the working speed of a cable varies inversely—more or less—with the *square* of the length, it will be seen that any slight increase in the electrical constants beyond a certain figure becomes a serious matter—more and more so with further increase.   Hence, where a high speed of signalling is aimed at, it is most desirable in laying a long cable—if on this account alone—to avoid paying out more slack than is absolutely necessary for engineering reasons.

† At the outset it must be remembered, as a ruling principle, that it is more economical to reduce the retardation factor in the formula for speed (already given) by a low resistance (R) rather than by a low inductive capacity (K)—*i.e.*, by a large conductor with just sufficient insulation for electrical and mechanical purposes.   This is so in consideration of the relative cost of copper to gutta-percha or india-rubber, but the above principle must not be pushed too far, the limiting feature being fault liability.

end—mirror instruments, for instance—so that the same current intensity.
will be required to make them work.

Although the permanent state can never be established in theory, as $C_1$ only acquires its limiting value $a = \dfrac{P_0}{rL}$ with an infinite value for $t$, yet we have seen that in practice, after an interval of about $10\tau$, the arriving current increases but slowly, and may then be considered to have attained its maximum. Therefore, if we take two cables of similar construction, only differing as to their lengths L and L', the arriving current will acquire the same limiting value if the potentials $P_0$ and $P_0'$ are such that

$$\frac{P_0}{rL} = \frac{P_0'}{rL'}$$

or

$$\frac{P_0}{P_0'} = \frac{L}{L'}$$

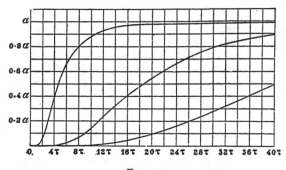

FIG. 39.

that is, if the electro-motive forces of the batteries employed are proportional to the lengths of the two lines.

If the receiving instruments work under the effect of a given fractional part, say $\frac{1}{80}$ of the current $\dfrac{P_0}{rL}$, the times $t = n\tau$ and $t' = n\tau'$ at which the intensity $\frac{1}{80}a$ is reached, will have the following ratio

$$\frac{\tau}{\tau'} = \frac{\dfrac{krL^2}{\pi^2}\log\epsilon\frac{4}{3}}{\dfrac{krL'^2}{\pi^2}\log\epsilon\frac{4}{3}}$$

consequently

$$\frac{S}{S'} = \frac{L^2}{L'^2}$$

The transmission speeds S and S' in two similar cables, under similar conditions, worked by batteries whose powers are proportional to their cable lengths, are, therefore, inversely proportional to the squares of the lengths of the two cables

$$\frac{S}{S'} = \frac{L'^2}{L^2}$$

If we consider, for instance, three cables whose lengths are as the numbers 1, 2, 3, connected to batteries of similar kind, the number of cells in each battery varying according to the same proportion, the arrival currents will have the same limiting intensity $a$, which in Fig. 39 is made equal to 100. If, again, the receiving instruments work with a current strength equal to 0.40$a$, we see by the figure that this strength will be reached in times which are sensibly to each other as the numbers 1, 4, and 9.

If two cables are of different lengths and also of different types, the same current intensity will be obtained when the electro-motive forces of the different batteries satisfy the proportion

$$\frac{P_0}{rL} = \frac{P_0'}{r'L'}$$

or

$$\frac{P_0}{P_0'} = \frac{rL}{r'L'} = \frac{R}{R'}$$

The electro-motive forces of the batteries should thus be proportional to the total resistances R and R′ of the two conductors.

The times $t = n\tau$ and $t' = n\tau'$ at which the same fractional part, $\frac{1}{80}a$ for instance, of the current intensity required to work the receiving instruments, is reached, will be to each other as

$$\frac{\tau}{\tau'} = \frac{\dfrac{krL^2}{\pi^2} \log \epsilon \frac{4}{3}}{\dfrac{k'r'L'^2}{\pi^2} \log \epsilon \frac{4}{3}}$$

whence

$$\frac{t}{t'} = \frac{k}{k'} \times \frac{r}{r'} \times \frac{L^2}{L'^2} = \frac{RK}{R'K'}$$

The transmission speeds $v$ and $v'$ are therefore

$$\frac{S}{S'} = \frac{k}{k'} \frac{r'}{r} \frac{L'^2}{L^2} = \frac{R'K'}{RK}$$

If the cores of the two cables are composed of the same substances, $d$ and D, $d'$ and D′ representing the respective diameters of conductors and cores, we have

$$\frac{S}{S'} = \frac{d^2}{d'^2} \frac{L'^2}{L^2} \frac{\log \dfrac{D}{d}}{\log \dfrac{D'}{d'}}$$

and, letting A stand for a constant, determined experimentally, which varies with the nature of the receiving instrument,

$$S = Ad^2 \frac{\log\frac{D}{d}}{L^2} \text{ signals per minute.}*$$

According to numerous experiments of Mr Willoughby Smith on cables with his improved gutta-percha core and with conductors containing 95 per cent. of pure copper, when mirror receiving instruments or siphon recorders are used,

$$A = 1297 \times 10^5$$

Therefore

$$S = \frac{1297 \times d^2 \times \log\frac{D}{d} \times 10^5}{L^2} \text{ signals per minute,}$$

D and $d$ being expressed in terms of the same, but any, unit, and L being expressed in N.M.

For a cable with a dielectric composed of another material, the constant would have to be multiplied by the inverse ratio of the capacities per N.M.† of the given cable and the cable with the aforesaid Willoughby Smith gutta-percha core, the diameters D and $d$ remaining the same. ‡

By applying this formula to the 1869 French Atlantic cable, we find that $S = 232$ signals or about 58 letters per minute.

---

* For purposes of ready calculation it is also well to remember that if a constant ratio be maintained between the weight (or diameter) of the conductor and insulator in the two cables, their relative speeds—provided precisely the same materials be used—will alone depend on their relative core dimensions and lengths. This is, of course, assuming the same conditions as regards bottom temperature, pressure, etc. ; or that any difference in this respect will be allowed for so far as is necessary.

† Throughout this book "N.M." or "naut" are used as abbreviations for "nautical mile." The reason for this choice in preference to "knot" has been explained in Parts I. and II.

‡ Willoughby Smith's "improved" gutta-percha, introduced in 1870, is taken as the standard dielectric throughout this chapter. Firstly, because it is the material of which the dielectric is composed in by far the greater length of cable at present in use at the bottom of the sea. Secondly, owing to it having a substantially lower capacity than any of the other "ordinary" gutta-percha so far produced, a core so insulated permits of a correspondingly higher speed. An india-rubber dielectric for the same reason permits of a still higher speed with corresponding dimensions, but on other accounts has only been employed on a comparatively small scale. Thirdly, the different forms of "ordinary' gutta-percha as manufactured by various contractors at one time or another with different available materials yield distinctly varying results as regards capacity—and therefore, as regards speed. On the other hand, the capacity of Willoughby Smith's gutta-percha being governed so much by the method of manufacture (referred to in the chapter on Gutta-percha, Part II.) naturally forms a much more uniform basis. The comparative specific inductive capacities of the above different materials are given in Part II., and the working speed with each bears the same ratio to one another under similar conditions and with dimensions common to all.

In a general way

$$S = \frac{A_1}{RK}$$

$A_1$ being a new constant, R and K expressing in ohms and microfarads respectively the total conductor resistance at the temperature of the waters in which the cable lies, and the total capacity of the cable.

**Data in Practice.**—According to Mr Willoughby Smith's experiments, $A_1$ would be equal to $108 \times 10^7$. On the other hand, Dr Alexander Muirhead in 1883, obtained through the Jay-Gould Atlantic cable of 1881, between Penzance and Nova Scotia, 7,288 letters in 81 minutes, including repetitions and comparisons, say about 90 letters or 270 elementary signals per minute. The length of this section of cable is 2,518 N.M. the conductor resistance R at 75° F. 8,320 ohms, and the capacity K of the cable 939 microfarads. It was constructed by Messrs Siemens Brothers, and, therefore, the dielectric is of their "ordinary" gutta-percha.* The temperature at the bottom of the Atlantic being about 37° F., the value of $A^1$ deduced from these data is

$$A^1 = 210 \times 10^7.$$

Dr Muirhead has also given the speed on the corresponding 1882 Jay-Gould cable as 18 (seven-letter) words per minute.

The Marseilles-Algiers cables of 1879 and 1880 give regularly during simplex working, with siphon recorders and automatic transmitters, 150 letters per minute, each letter requiring on an average four current impulses.

---

* Messrs Siemens Brothers devised special arrangements in the matter of the core of this and some other of their long Atlantic cables, with a view to increasing the attainable speed of signalling. The conductor of the cable at each end for some distance was of a larger type than in the middle so as to reduce the conductor resistance, the latter—from a manufacturing standpoint—being the more economical direction to work in for effecting this object than by reducing the capacity. The weight of dielectric is also not only less but, moreover, its *thickness* is less than in that part of the core in deep water (besides the material being somewhat different) where—on account of the increased pressure added to the greater difficulties and expense of repairs—extra precautions are necessary for ensuring a perfectly reliable insulation resistance.

Thus, in the first Jay-Gould "Atlantic" of 1881, the deep-sea portion was furnished with a core represented by 350 lbs. copper per N.M. to 300 lbs. gutta-percha : whilst that for the ends was constituted by $\frac{450 \text{ lbs. Cu}}{270 \text{ lbs. G.P.}}$ This device was in accordance with the notion that the rate of charge and discharge in an insulated conductor is mainly governed by its form at the generating end. It is not believed, however, that in practice any advantage was actually found ; moreover, the heavy conductor and small thickness of gutta-percha at the shoreward ends materially increased the risk of a decentralised conductor causing faults. This plan has, therefore, since been abandoned in more recent cables made by the above firm.

The resistance of these cables, at the temperature of the deep Mediterranean water (55° F.), being respectively 5,350 and 5,220 ohms, and their capacities from 136 to 129 microfarads, the approximate value of $A^1$

$$A^1 = 43 \times 10^7.$$

Modern fast-speed ocean telegraphy is represented by the two last Atlantic cables laid in 1894, one the property of the Commercial Cable Company and the other that of the Anglo-American Telegraph Company.

Both of these cables are entirely worked by machine transmission, full descriptions of the various systems and instruments of which will be found at the end of this book. The speed on the first-named cable, with a core composed of a copper conductor that weighs 500 lbs. per N.M., with a thickness of gutta-percha insulation represented by 320 lbs. per N.M., is said to be from 37 to 40 (five-letter) words per minute. On the second cable the ordinary working speed attained in practice (by automatic transmission) is as high as 47 (five-letter) words per minute, press work being satisfactorily carried out at a corresponding rate of 50 words per minute.*

Fig. 40 is a facsimile of signals received at Heart's Content station with automatic machine transmission from Valentia by means of Mr Herbert Taylor's latest automatic instrument, and with duplex connections in circuit.

Fig. 41 represents specimen code signals received at 247 letters (= 49.4 five-letter words) per minute, with Mr P. B. Delany's automatic sender.

Speeds of 250 letters and over have, indeed, been maintained in regular traffic over this cable for half-an-hour (and more) constant working. These speeds were carefully checked at each end by the company's officials.

This, the last Atlantic cable laid (one of five, more or less, working cables belonging to the " Anglo" Company†), may, certainly, be taken as the representative of .the most advanced stage of cable working in the present day.

It need hardly be said that both of the above lines are invariably worked duplex. Thus—in these days of "double-block," etc.—the speed is increased by 90 per cent., i.e., their working capacity is nearly doubled. The conductor of the "Commercial" cable is built up, according to Messrs Siemens' invariable custom with Atlantic cables, by a large central wire

---

* The *theoretical* speed of this cable, with a KR of $2.42 \times 10^6$, based on Mr Dearlove's " Tables" elsewhere alluded to, would be 48.8 (five-letter) words per minute.

† This company is by far the largest of the Atlantic companies, having requisitioned no less than eight Atlantic cables altogether, three of which, however, have fallen definitely out of use—i.e., those of 1858, 1865, and 1866. The "Anglo" Company comes only second to the "Eastern" as regards total mileage of cable.

[PLATE XXV.

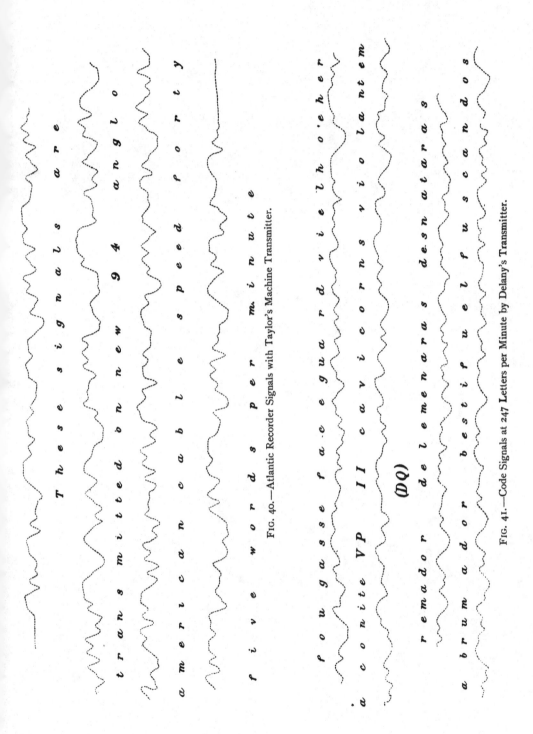

FIG. 40.—Atlantic Recorder Signals with Taylor's Machine Transmitter.

(DQ)

FIG. 41.—Code Signals at 247 Letters per Minute by Delany's Transmitter.

surrounded by a number of other wires of smaller diameter.* The advantage gained here over the *ordinary* strand conductor is that for given weights of conductor and insulating material, the area of the circle circumscribing the conductor is smaller and the thickness of the enveloping dielectric consequently greater than in an ordinary strand in which all the wires are of similar type. The result is that whereas the conductor resistance remains the same, the inductive capacity is smaller, and the speed of working proportionately increased.

In the "Anglo" cable, the same form of conductor was adopted,† and this conductor (650 lbs. to the N.M.) is by far the largest which has ever been made for telegraph purposes. A section view of the above in its gutta-percha insulating envelope is shewn in Fig. 42. We do not yet know in practice how to construct a core having the properties of Mr Oliver Heaviside's "distortionless circuit." In the absence of this, a step in the right direction was made in the design of the above latest Atlantic cable, by liberally increasing the size of the conductor.

FIG. 42.—Core of 1894 "Anglo" Atlantic.

The result of building a cable with this type of conductor is that the earning power per unit of cost of this cable is, compared with that of the 1873 "Anglo" Atlantic, as 1.55 is to unity. In other words, an increase of 55 per cent. in the earning power of a cable has been effected by automatic in place of hand sending—a somewhat remarkable feature!

With reference to the question of the high speeds now attained on Atlantic cables of the present day by machine transmission, it may be here remarked that a skilful clerk will decipher eighty or ninety ordinary messages per hour, and get through five hundred messages a day without making a serious error. These cablists who work at the "serpentine" signals, or "cabalistic zigzags," of the recorder, have been described as "the aristocrats of the telegraphic profession."

**Theoretical Calculations: Considerations involved.**—Let us now revert once more to the direct question of speed calculations. The considerable margin shewn by some of the preceding figures may be

---

* This may be looked upon as a modification of Messrs Bright and Clark's segmental conductor of 1863, described elsewhere; inasmuch as it also combines (though in a less marked manner) the electrical advantages of a solid wire with the mechanical advantages —in fact, necessities—of an ordinary equal-wire stranded conductor.

† The engineers to the Company being Messrs Clark, Forde, and Taylor; and the contractors, the Telegraph Construction and Maintenance Company.

accounted for by the fact that the speed of transmission is not quite inversely proportional to the square of the length, but appears to follow an intermediate law between the ratio of the length and that of the square of the length. Probably the propagation of the electric wave is in reality a more complex problem than it is supposed to be.* In the foregoing ordinary calculations for speed no account is taken of the reflex action of the current on itself, nor of self-induction;† neither has it been as yet possible to introduce definitely into the equations either the absorbent power of the dielectric or the resistance of the earth itself. It is uncertain, however, whether they enter into the problem in actual practice—*i.e.*, whether they actually affect the working speed.‡

Theoretically, if we consider the self-induction of the line and neglect on the other hand the conductor resistance, we find that the electric waves are propagated in the conductor in the same manner as vibrations in a perfectly elastic thread. The velocity of such waves, in an overhead copper

---

* Some, indeed, aver that—even for *practical* purposes—a modification in the speed formula is necessary in order to bring it within the required range of accuracy.

This view is largely founded on the circumstance that the *actual* speed obtained on a cable is usually in excess of that arrived at by calculation. This will only be noticeable where the cable in question is materially longer than that (of known values and affording a known result) which forms the basis for calculation.

Whether this is due to the *magnetic* induction coming into force in a long length it is impossible to say with any certainty at present.

† There can be no question that there is self-induction in a laid cable, and that its value is increased by the iron sheathing. Though no allowance is made for this in the artificial line adopted in practical duplex telegraphy, it is thought that the leading-up resistance tends to balance this.

‡ It is perhaps more than likely that they do *not*, when we remember that absorption through a gutta-percha or india-rubber dielectric is a somewhat slow acting process in relation to the time-period between the application of the battery and the receiving of the signal at the further end. In tending to clear the conductor, it might be argued that a great absorptive power, or electrification, would be a favourable condition. Again, it may also be viewed, with reason, that a highly absorptive (or polarising) dielectric tends to decrease the momentary effective electro-static capacity of the Leyden jar system ; but this latter would also give a corresponding lagging tendency to the signalling condensers. The effect of this last can only be overcome by additional curbing devices. In the ordinary insulation test on cables the expressions "high" and "low" (or "quick" and "slow") electrification are often used somewhat loosely with reference to the rate of fall in the observed deflection. These terms, however, are liable to tend to misunderstandings—indeed, to *contradictions*—when comparing plate condensers with cables. In the above remarks, therefore, *great* and *little* absorption are adopted by preference. In a cable these are signified during an insulation test in proportion to the percentage of difference in the deflection during any given period of time previous to the true resistance revealing itself by the ultimate permanent deflection due to leakage current alone.

Absorption may be regarded as something between capacity and insulation resistance. The term polarisation is less suitable on account of its very general use in quite different senses.

wire, is approximately $3 \times 10^{10}$ metres per second. The effect of conductor resistance is to alter the form of the electric waves, but their first effects are propagated with the same uniform velocity which they would have if the resistance was *nil*. In this sense only has the expression "absolute speed of electricity" any signification, and we then see that this velocity is comparable to that of light.

As the expression

$$S = \frac{A_1}{RK} = \frac{A}{rkL^2}$$

does not correspond to data furnished in practice, we shall substitute for it the following formula—

$$S = \frac{1}{rk}\left(\frac{A}{l} + \frac{A_1}{L_2}\right)$$

A and $A_1$ being two new constants which should be determined experimentally from a sufficient number of observations. Confining ourselves to the two particular cases referred to above, we find that

$$A = 826 \times 10^3$$

and

$$A_1 = 2769 \times 10^4$$

which gives

$$S = \frac{1}{rkL}\left(826 + \frac{27690}{L}\right) 10^3 \text{ signals per minute.}$$

Theoretically speaking, the magnetic induction introduced by the iron sheathing of an ordinary cable should tend to have a slightly beneficial effect on its working powers. This would be owing to a tendency on the part of the *magnetic* induction to neutralise a portion of the induced charge due to the electro-static capacity of the Leyden jar system. In actual practice, however, this is not supposed to come into the question of speed; and is, indeed, never allowed for in cable circuits, even when of very great length.[*]

---

[*] Practice is supported in this by the independent investigations of Lord Kelvin, F.R.S., the late Mr Charles Hockin, and Mr Oliver Heaviside, F.R.S.

If further evidence were necessary, it would be forthcoming in the fact that an artificial line constituted only by the total conductor resistance and the total electro-static capacity of the cable (*i.e.*, without any *electro-magnetic* capacity) exactly balances its electrical values. Moreover, the said "artificial" when signalled through reproduces precisely the same signals (as regards form and speed) as those obtained through the cable.

This statement requires to be qualified to some extent as regards extremely short cables worked at very high speeds. It is found in balancing these for duplex working, that less electro-static capacity is required than would be given by a perfect imitation of the electrical constants. This may be due to the effect of electro-magnetic inertia set up by the iron sheathing, under these conditions.

Again, there is the case of the alleged improvement in speed by separately brass-

Based on the preceding formulæ, Messrs Clark and Forde, in 1871, published a table of such a character as would enable any one to readily find without calculation the approximate attainable working speeds per minute through cores of different sizes for a cable of 1,000 N.M. In 1890 Mr Arthur Dearlove brought out a more extensive set of tables, and these together go to form a very useful little book of reference.* However, the various contracting parties naturally have their own special tables for this purpose, as well as to meet other similar objects, with constants based on their own material, etc.—which, indeed, require to be adjusted by change of materials available from time to time. Such tables usually also include (as a part of the story) the diameter ($d$) of the conductor, the external diameter (D) of the entire core, and the relation of the two, besides the electro-static capacity of the core. It would be of little use to present one of these speed tables, etc., as a specimen, for it would be only applicable at the time for the particular material it is based on and with the particular constants used. However, with the formula and the basis for constants, any table can readily be drawn up.

The working speed by the mirror or siphon recorder is limited by the rate of transmission. The practical limit of sending speed which a good operating clerk can keep up on these systems does not usually exceed 135 letters per minute. With automatic transmission, however, a speed of 600 elementary signals per minute has been reached on certain cables.† The length L of a cable being a given quantity, it seems useless therefore to diminish $r$ and $k$ below values whose product is less than

$$rk = \frac{1}{600L}\left(826 + \frac{27690}{L}\right)10^3$$

The electro-static capacity of vulcanised india-rubber cores being about 2 per cent. less than that of cores composed of Willoughby Smith's improved gutta-percha, the speed of transmission through vulcanised india-rubber core cables for equal dimensions is greater in the same proportion.

---

taping each core in a multiple-conductor cable, in place of a single taping for the whole when laid up with jute, or hemp, worming. This plan was first resorted to, with some success, by H.M. Postal Telegraph Department on the Bacton-Borkum cable of 1896, the application of the principle to *submarine* telegraphy being originally suggested by Messrs Dresing and Gulstad. With a view to partially neutralising electro-static induction by the magnetic induction between the taped cores, Messrs Felten and Guilleaume had independently adopted the above plan for underground land wires, as well as Mr Preece at a still earlier date.

* "Tables to Find the Working Speed of Cables," by Arthur Dearlove (E. and F. N. Spon).

† Indeed, practically speaking, the speed by machine working is only limited by the electrical equivalents of the line.

**Further Practical Considerations and Comparisons.**—During trials of short duration, very skilled operators have obtained with the mirror instrument a rather higher speed * than that afforded by the siphon recorder, but the practical working speed of the two forms of apparatus (if worked manually) is sensibly the same for reasons already given. By ordinary hand transmission this averages at twenty-five words per minute for more or less continuous working of several hours by efficient operative clerks, and thirty words per minute at a maximum for comparatively short periods of time. The Morse printing instrument, where the electrical values of the cable come into force, gives a speed of transmission one-fourteenth that of the siphon recorder. It can, indeed, only be used on cables below 400 miles in length,† and then only when the weights of copper and gutta-percha are not less than 107 and 140 lbs. per N.M. respectively.‡ In combination, however, with certain relays—notably that of Messrs Allan and Brown—it works at a much higher speed, and can then be used on cables considerably longer than 400 miles—up to something like double the length, indeed.

From what has transpired it will be seen that a clear distinction must always be borne in mind between the maximum speed of manipulating a key on the Morse and mirror (or siphon recorder) system and that which the cable will permit of. The Morse key lends itself to a slightly higher speed of working—in words per minute—than the double lever key of the siphon recorder (or mirror) system, notwithstanding the difference in the duration of contacts involved by dashes. But the length of cable which will "take" this speed is in the Morse system much more limited, even where "double current" and relays are employed. If a higher speed of transmission is adopted than the retardation of the line allows, the result is that the signals on any existing instrument§ get blurred into one another

---

* Sometimes up to thirty-five (five-letter) words per minute at a spurt.

† Unlike the case of mirror or siphon recorder hand transmission, first-class manipulators can work the Morse key at a rate as high as the fastest writer, i.e., about forty words per minute. Where such a speed as this is aimed at—and an "ordinary" speed of thirty words—the length of cable (with $\frac{107}{140}$ core) which can be worked by this system would not be more than about 200 miles, with an ordinary Siemens polarised relay.

‡ If ordinary Morse instruments were used on the Atlantic cables, it would take about five seconds from the time of battery contact until the relay at the distant end became sufficiently magnetised to record a firm signal, with the result that not much over one word per minute could be transmitted.

§ This all points to the circumstance that it is in the type of cable rather than in the signalling apparatus in which any radical change should be looked for, if we are to make any revolutionary increase in the speed of transmitting thought across the ocean by electrical conductors.

before arrival at the other end. Thus it is that on a long length of cable the speed at which the Morse system can be successfully worked is but a fraction of what the mirror or siphon recorder would take. On the other hand, it is *possible* to work any length of line at *some* speed—however slow —by the Morse system, provided the battery contacts are made in sufficiently slow succession and sufficiently long. The same applies to any instrument amenable to electro-telegraphy.

In many cases it might seem, at first sight, that a cable affording the maximum speed was wasted on short sections—especially where only second-rate clerks are employed, or where the traffic is such as only to necessitate the breaking in of each station for an hour or so per day. These lines, however, are almost invariably worked throughout their entire length —by joining up all the sections—at certain periods, and here the maximum speed is usually required.*

---

* As a rule, a coast cable of this description is laid mainly in order to bring the extreme points into communication. It is, indeed, usually only landed at the intermediate spots for the sake of subsidies, or guarantees of local traffic—at least sufficient to cover station expenses ; as well as to simplify working, testing, fault localisation, and repairs.

The Commercial Cable Company : Station and Quarters near Cape Canso, Nova Scotia.

# CHAPTER II.

## SIGNALLING APPARATUS.

## SECTION 1.—INTRODUCTORY REMARKS.

ON short submarine cables messages are interchanged by means of Morse or Hughes instruments, worked directly by the current through the line, or else with the help of relays.

When the cables exceed 500 miles or so in length, and it becomes necessary to use more sensitive apparatus, recourse is had either to the mirror receiving instrument, which only affords transient signals, or to the Thomson siphon recorder, which automatically inscribes the signals on a strip of paper, with a sensitiveness to current about equal to that of the reflecting instrument.*

---

* The heavy patent rights attached to the siphon recorder—involving a royalty of 10s. per N.M.—having expired in 1882, special patterns suited for short lengths (described further on) have since been gradually introduced in place of the Morse instrument. At the present time, the " Indo-European " and the Direct Spanish lines are the only circuits

When, as usually, the telegraph office is at a distance from the cable landing-place, communication between the two is established by means of underground lines laid in a trench or running through pipes. The connection between the shore-end and land-line conductors is made in the cable hut,* one or more lightning protectors being here inserted to protect the submarine cable from discharges of atmospheric electricity.†

Should the offices be too far removed from the landing-place to permit of the instruments being connected up to the cable through land lines, the messages are received and sent from the cable hut itself—suitably arranged for the purpose—traffic to and from the cable hut being carried on through overhead wires. In other cases the cable is joined up to the aerial line by means of relays placed in the cable hut or in the nearest office. Relays are also inserted at times between two sections of a submarine cable in order to avoid the necessity of retransmission.‡ Some of these instruments,§ such as Siemens' polarised relay, the relays of Stroh, D'Arlincourt, Froment, and others, are well known, and descriptions of them to be found in all works on general electro-telegraphy ; || others again, such as the Brown=Allan relay, belong more particularly to submarine telegraphy.

---

worked on the Morse system. The advantage of a record being obvious, the mirror instrument has similarly fallen out of use almost entirely, being only employed by the "Western and Brazilian" and the "West India and Panama" companies. In the author's opinion, however, the "mirror" should always have a sphere of usefulness during cable operations, if only on account of its extreme convenience in portability and simplification of adjustments. In speaking between ship and hut through a "shore end" a sounder could, in some instances, be still better turned to account. If made more sensitive, the latter might even be used for somewhat longer lengths.

* A device of the author's for ensuring this being permanently maintained was fully described in *The Electrician* of April 1897.

† Having regard to doubts as to the *nature* of the phenomena, some two or three guards of various descriptions are a wise precaution—for instance, one of Dr Lodge's, supplemented by a Saunders, Jamieson, or Bright protector, as well as by one of Siemens' plate guards. The latter (fully described by the author in vol. xix., p. 392, of the *Journal of the Institution of Electrical Engineers*) is perhaps best suited for an unfrequented hut, inasmuch as when one wire is fused another comes into the circuit automatically.

‡ This forms what is usually described as the Morse repeater, or translating, system, extensively employed in land telegraphs.

§ The above relays might become more generally useful if rendered efficiently acoustic. The delicate adjustment at present required to work the Morse coils would thus often be a matter of secondary importance. This would often be a great advantage for temporary operations where time is of the utmost importance, and where a special local battery may not always be available.

|| To wit, Culley's "Handbook of Practical Telegraphy," and Preece and Sivewright's "Text-Book of Telegraphy." Both of these are published by Messrs Longmans, Green, and Co. They also fully describe the Morse recorder and the Morse (or Vail) sounder.

We come lastly to duplex working—or sending two signals in opposite directions—at the same instant and through the same conductor. This method, worked out in theory by Dr Gintl of Vienna, as early as 1853, and rendered practicable by Mr J. B. Stearns with the use of condensers, in 1872, was successfully applied some years later to submarine cables of any length. The honour of this great achievement is principally due to Messrs Varley, De Sauty, Stearns, Muirhead, Taylor, Harwood, and Ailhaud.

Great Northern Telegraph Company : Amoy Station.

SECTION 2.—RELAYS AND SPECIAL METHODS FOR DISCHARGING
CABLES OF MODERATE LENGTH.

**Discharging Relays.** — Up to a distance of 150 miles the ordinary
Morse apparatus can be employed both on land and submarine lines.
Beyond this limit the rate of sending has to be considerably reduced, and
even then the signals soon commence to arrive in a mutilated condition,
due to the effects of alternately charging and discharging the cable.   The
state of things last described is remedied, in a great measure, by what is
known as the double-current Morse system,* which consists in following
up each successive signal by a weaker current impulse in the opposite
direction, in such a way as to facilitate the more complete discharge of the

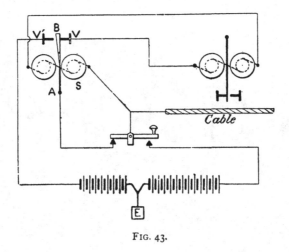

FIG. 43.

line.   The strength and direction of this reversed current should be so
regulated as to bring the conductor to the neutral condition, at the instant
when the ensuing signal commences.   For cables less than 300 miles in
length, the strength of the two currents should be in the proportion of
3 to 2.

Sometimes the line is simply put direct to earth for an instant after
each signal.

The discharging current can be sent into the line automatically by
means of the arrangement shewn in Fig. 43.   S is a Siemens polarised
relay whose resistance is about equal to that of the line combined with that
of the receiving instrument at the far end : the armature  the side on which

---

* The ordinary *single*-current Morse system is only capable of employment on *quite*
short cables.

makes connection with the discharging battery, is furnished with a spring intended to prolong the duration of contact on that side. The distance between the two screw contacts V and V is adjusted to give sufficient play to the armature A B. When the sending key is depressed the positive current from the line battery divides between the cable and the relay S, the armature A B making contact with V'. Directly the signal is completed and the key returns to normal position, the negative current from the discharging battery divides in the same way between the cable and the relay S, the positive charge remaining in the cable is neutralised and the armature returns into contact with the terminal V.

Several other arrangements effecting the same result in a still simpler manner have been devised in the last few years.

**Auxiliary Discharge Coils.**—Supposing that to the "line" terminal of a Morse sending key are joined up the cable and some coils whose joint resistance is about equal to that of the line, on depressing the key the portion of the current which enters the line is sensibly the same as if there were no resistance coils ; and when the key lifts, the line discharges itself almost entirely through the coils—before there is time for the key to resume its normal position.

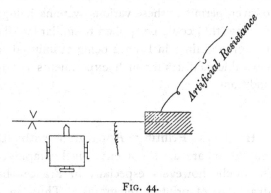

FIG. 44.

To avoid the weakening of the resultant current at the far end owing to its dividing between the receiving instrument and the artificial line, the latter may be connected to a spring blade, one end of which is fixed and the other end allowed a certain amount of play between two contacts in close juxtaposition. These contact pieces (Fig. 44) are situated slightly above and behind the lever of the sending key, and the blade is held down against the lower contact by a small spiral spring. A screw point on the key lever is adjusted so as to make contact with the blade directly the key begins to rise, the contact being retained until just before the key returns to its normal position.

This method, devised by Mons. E. Wünschendorff, M.I.E.E.,[*] in 1881,

---

[*] Engineer to the Paris French Government Telegraphs, and author of "Traité de Télégraphie Sous-Marine," a part of which work formed a basis for this volume.

when the first long subterranean lines in France were inaugurated, has given good results on the underground cable from Paris to Nancy (397 kilomètres or about 250 miles). Mons F Godfroy has lately obtained from it still better results by using single instead of double wound resistance coils, possessing, therefore, considerable self-induction. The extra current induced in the coils after each impulse, being of opposite direction in the receiving instrument to that of the discharge current from the cable, almost complete silence can be obtained with the telephone, thus indicating an almost perfect balance.

On their comparatively long lengths of subterranean lines the French Postal Telegraph Department have also experimented with, and adopted, several other ingenious applications of the Morse system with line discharging relays, pre-eminently those of Lacoine and Farjou, as well as the relays of Rambaud, D'Arlincourt, Willot, and what is known as the phono-signal system of M. Ader, in which two telephones are used, one for each current as received—after the manner of Bright's Bells. Space will not, however, permit of these various systems being described in detail, though, of course, they could be applied to similar lengths of submarine lines. These underground lines in France being of unusual length, they obviously afford favourable facilities for such experiments as a test also for submarine cable conditions.

**Hughes's Printing Apparatus with Discharge.** — The Hughes printing apparatus is not extensively employed in submarine telegraphy. The public, however—especially in France—have always appreciated the advantages of printed telegrams. Thus, in 1881, when the extensive system of French underground lines was first opened up, the telegraph administration of that country introduced a modification of this apparatus. On account of the similarity of working conditions between these lines and submarine cables, it may be of interest to indicate the arrangement which was then devised (with the assistance of Mr Borel), and which has been adopted by the French Telegraph Department.

With the automatic releasing trigger apparatus and with currents of same polarity, the discharge current produces, at the sending end, the effect of a current from the opposite end of the line, and prolongs the battery current at the receiving station. If the line is charged with several current impulses succeeding each other at very short intervals, the return current may acquire sufficient strength to print additional letters.

To obviate this defect, a steel cam was fitted on the cam axle in the centre of the available space between the two platinum contacts, and arranged so as to rub against and raise a flat spring at the required instant. This flat spring butts up against a set screw fitted to a second spring above, insulated from the rest of the apparatus by a plate of ivory or ebonite and in connection with a discharging battery. The form and position of the cam are such that contact with the discharging battery begins an instant after contact with the line battery finishes, being con-tinued until the line is. put in direct communication with earth through the framework of the electro-magnet. The compensating current lasts in this way about a quarter as long as the line current, and their relative intensities should be in about the same proportion.

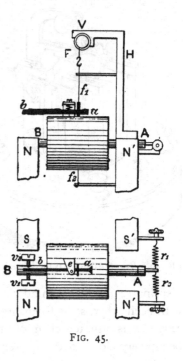

FIG. 45.

The set screw in the upper spring permits of the necessary adjustment.

Sometimes the discharging battery is replaced by a simple earth connection, in which case the discharge will be less complete unless the line is a very short one.

On the subterranean line between Paris and Nancy (397 kilomètres), this modi-fication of the Hughes apparatus works continuously at a speed of 105 to 110 chariot turns per minute. It is, in fact, capable of transmitting 160 letters per minute.

**Brown-Allan Relay**. — This apparatus (Fig. 45), introduced in 1878,* consists of a very light single cylindrical electro-magnet, suspended by a platinum wire F to a strong bracket H, so as to be perfectly mobile. The position of the electro-magnet may be accurately adjusted by turning the screw V and stretching to a greater or lesser extent the two silken threads $f_1$, $f_2$. The hollow core A B of the electro-magnet is placed in the centre of the vacant space between the poles N, S, N', S' of two powerful horse-shoe magnets, one of its ends A being attached to two opposing springs $r_1$ and $r_2$. A rod of soft iron $a\,b$, resting on the upper surface of

---

* The above invention—due to Mr Brown—is somewhat similar to an earlier apparatus of Mr Andrew Coleman. The Brown-Allan relay forms the subject of a patent (No. 1,757 of 1876) taken out in the name of George Allan and James Wallace Brown.

the electro-magnet, turns with but little friction round a vertical axis *e* attached to the reel, the small angle through which the rod deflects being limited by two stop screws $v_1$ and $v_2$. The resistance of the coil is usually 500 ohms, and on the passage of a current through the coil, the core cylinder becomes magnetised, causing its end A to approach the pole S, let us say. The rod *a b* follows this movement, but is almost immediately brought up against the stop $v_1$, where it closes a local circuit and leaves the coil to continue the movement alone. As soon as the latter commences to swing back to its normal position, the rod *a b* is carried back till it butts against the stop $v_2$ A general view of this suspended coil type of Brown-Allan relay — the form used in submarine telegraphy—is shewn in Fig. 46.

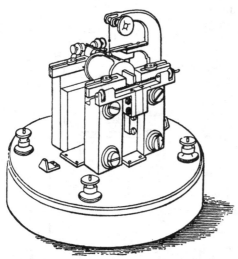

FIG. 46.—Brown-Allan Relay for Submarine Cables.

In later patterns of this relay,[*] the rod *a b* (Figs. 47 and 48) is mounted on the same axis *m n* as a second somewhat longer rod *c d*, a spring regulated by an adjusting screw *m* keeping the two rods in contact with each other, so as to ensure *a b* following the movements of *c d*. The axis *m n* is situated clear of the electro-magnet, the end *c* of the rod *c d* being very close to the extremity A of the core cylinder A B, whose every movement it follows. The rod *a b* is in this case also carried on by the first movement till it brings up against its stops $v_1$ and $v_2$, as before.

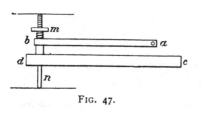

FIG. 47.

The connections are joined up as shewn in Fig. 49. In the receiving position the switch I is at contact No. 1, so as to allow the current from the cable to pass to earth through the coil M of the electro-magnet.

The contact of the armature *a b* against the stop $v_1$ closes the circuit of

---

* The above form is very similar to the Siemens polarised relay (invented about 1855), the main point of difference being that a jockey is included in the Brown-Allan instrument. This type is mostly—if not exclusively—used on land (aerial) lines; but sometimes also for quite short cables.

the local battery $p$ and the Morse receiver R. A small condenser C is sometimes inserted between the terminals of this receiver to diminish the sparking between the armature $a\,b$ and the stop $v_1$ caused by the "extra current."

To change to the sending position the switch I is moved to contact

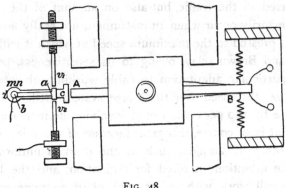

FIG. 48.

No. 3, which cuts the relay B out of circuit, substituting for it the current reverser K and the sending key D. The former of these is worked by the local battery $p_2$ when D is depressed; the armature L being pulled down by the electro-magnet M′, breaks the contact between the lever $f_1$; and the terminal $t$, at the same time leaving $f_2$ free to press up against $t$. The

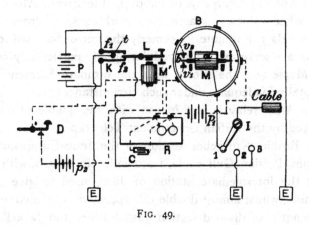

FIG. 49.

negative pole of the line battery P is thus put to earth, whilst the current from the positive pole flows into the cable by way of the armature L and No. 3 contact piece.

This relay (largely—in fact principally—used by the Eastern Telegraph Company) was the result of a great deal of experience and experiment. It supplied, at the time of its introduction, a long-felt want, more especiall in the case of cables exceeding 250 miles in length where Morse workin was desirable or unavoidable. Previously, numerous cables had bee worked on the Morse principle at the comparatively low rate of nine or te words a minute. This was much to the disadvantage of the companie which owned them, not only through the great and serious delays whic frequently occurred to the traffic, but also on account of the many error which must necessarily occur when an instrument not really adapted to th work required is pressed to the maximum speed at which it will respond.

The Allan and Brown relay, owing to its sensitiveness, peculiar con struction, and particular adaptation to cable working, obviates all thes evils. It increases the speed over the Morse system (with ordinary relay previously in use by 120 per cent., whilst involving no increase of trouble Its principal point is, of course, this great increase of speed it secures. Bu there are also other advantages, such as the marked uniformity of the signals, the little attention required for regulation, and the low battery power which it will work with on account of its extreme sensitiveness This relay, when once properly adjusted, will work for months withou requiring any further attention. It gives good results on lines of medium length. The "Eastern" Company's Porthcurno-Vigo section, 620 miles could not be worked practically on the Morse system with any ordinary relay, but a speed of twenty-six words per minute is obtainable with the " Allan and Brown."*

The Direct Spanish Company, in adopting the Brown-Allan relay or their Morse-worked cables, increased their working speed from twelve to twenty-seven words per minute. Formerly the mirror had had to be kep in readiness as a reserve and employed during any particularly busy time the ordinary Morse (i.e., the Morse with the ordinary Siemens or othe relay) being totally inadequate under such circumstances.

The Brown-Allan relay is also in use with the double-current Morse system employed by the Indian Government Telegraphs.

Again, the Bushire-Kurrachee section of the Indo-European Govern ment Telegraphs Persian Gulf cable, 1,050 miles in length, with metalli connection at the intermediate station of Jask, used to give sixty-five letters per minute, whilst almost double this speed was obtained on the two nearly equal lengths of divided section, Jask-Bushire and Jask-Kurrachee On working through with a Brown-Allan relay set up at Jask, the sending speed from Bushire to Kurrachee rose, during trials, to at least one hundred letters per minute.

---

* Now, however, the siphon recorder is in use for working this cable.

However, transmission speed trials direct from Malta to London, with Brown-Allan relays at Bona and Marseilles (maintaining the ordinary repeating installations at Paris and Beachy Head), only gave from forty-five to fifty letters per minute—much less than that obtained over the aerial line alone and through the cable alone, using in the latter case a mirror instrument or siphon recorder.

The secret of the success of the Brown-Allan relay lies in the fact that it contains what is technically known as a " moving zero." All other relays previously in use—such as the polarised relays of Siemens and Stroh—had "fixed" or "dead" zeros. No instrument can be worked with any reasonable speed, or success, through cables over 300 miles with the ordinary core for short lengths (under, say, 700 N.M.), which does not contain this quality of a "moving zero." The mirror, recorder, and Brown-Allan relay all include this element—hence their enhanced value and success.

On cables of 300 or 400 miles the jockey armature of the Brown-Allan relay can be attached with advantage to any other instrument, such as a Siemens, or other, relay, and this alteration will increase the speed certainly to the extent of 100 per cent. The Morse system even when supplemented by this relay would be quite unsuited, however, for ocean telegraphy—or, indeed, for submarine cables over 600 N.M. or so of ordinary core.

"Great Northern" Company's Telegraph Ship " H. C. Örsted ": Off Coast of Korea.

### SECTION 3.—THE MIRROR SYSTEM.

Cables cannot be satisfactorily worked on the Morse system—under any circumstances, with any existing relay or cable discharger—if the length in circuit be much over 700 miles—of the ordinary core for such a length. Thus when Professor Thomson (now Lord Kelvin) introduced in 1858 the mirror instrument—galvanometer as then called—it was a considerable advance in ocean telegraphy.[*]

**The Mirror "Speaker."**—The mirror receiving instrument (Fig. 50) differs from what is now known as Thomson's reflecting galvanometer in having one coil only and a system of magnetised needles attached (usually by means of light shellac) to the back of the mirror.   The latter, suspended

by a very short cocoon fibre, is enclosed in a brass tube which is inserted through the centre of the coil as in dead-beat galvanometers. A stout semi-cylindrical magnet completely covers the upper half of the coil, and can be revolved to a limited extent round a vertical axis by a screw placed at the top. The needle can be brought to any convenient normal position by slightly turning the controlling magnet in the required direction.

FIG. 50.—Mirror Signalling Instrument.

The entire apparatus is supported by a stand on which are fixed four terminals.   These terminals connect with two or more separate circuits of 1,000 and 2,000 ohms, or any other required resistance, severally wound on the reel.   By joining up to the terminals in different ways the resistance can be obtained nearest to that which gives the maximum effect theoretically.

The rays of light from the lamp, after passing through a long narrow

---

* A great point of advantage in both the mirror and siphon recorder instruments (as compared with the Morse instruments) is the fact of their action being dependent on *changes* of current-strength only as measured by distance from the zero line—*i.e.*, by a potential curve ; and the fact that, therefore, it is not necessary to wait for the entire clearing of the line before a fresh signal can be sent out.

slit in a screen, are focussed by a lens, and are reflected back from the mirror on to a strip of paper supported on a stand or scale.*  The luminous image or "spot" may be finally adjusted by the controlling magnet so that as long as the instrument is at rest, the spot will remain in the centre of the scale.

It has become a very general practice—in the first instance with Messrs Siemens Brothers—to fill the brass tube containing the mirror with quite dilute glycerine.†  Vibratory oscillations are considerably reduced by this means without in any way spoiling the signals.  This plan is especially applicable to a cable subject to strong earth currents, or in the case of a faulty line.

Again, within the last few years Mr Walter Judd, of the Eastern Extension Telegraph Company, has introduced a still further improvement. According to Mr Judd's device, a rod of soft iron is inserted into the mirror tube, furnished with a shoulder piece at one end to prevent it actually touching the mirror by entering too far along the tube.  The core becomes magnetised by every current passing through the coils, thus producing greater amplitude in the signals.  Again, owing to the fact that it does not give up or reverse its magnetism as rapidly as the signalling currents pass through the coils, the signals are steadied and rounded off—a considerable advantage on short lines, as well as in the instance of a defective cable, or one subject to strong earth currents.  By adjusting the length of core within the tube the signals can be brought to any required size.

This last device has been rendered still more effectual by winding the soft iron core with fine silk-covered wire, connected up in series with the mirror coils.  The above transformation of the core into an electro-magnet is due to Mr John Rymer-Jones, of Silvertown.  This gentleman has lately developed an improved form of instrument especially designed for use on board ship.  The latter was fully described by him at the time in the columns of the *Electrical Review* (vol. xl., p. 39).  Here, too, Mr Rymer-Jones shews

---

* It is very usual, however, to adopt what is commonly known as Jacob's transparent scale, the principle of which is described elsewhere.  By this means it is no longer necessary to darken the room.  The method is, in fact, even more applicable to constant signalling purposes than to periodic testing purposes.  Another plan is to reflect the spot on to the receiver's writing pad.  In ordinary work, however, the message has to be dictated, as received, to another clerk.  To obviate this, the author has suggested that the receiver, though unable to write the message properly, whilst reading the movements of the spot at a high speed, would soon be able to key the letters of a typewriter.

† The details of this plan have recently been considerably simplified and improved by Mr Frank Jacob on behalf of Messrs Siemens Brothers and Co.  For further particulars see the *Electrical Review* of 24th December 1896.

very clearly the requirements for sensibility, dead-beatness, etc.,* under the varying conditions in which his improved apparatus can be used by different adjustments for each case.†

The universal (suspended coil) galvanometer of Mr H. W. Sullivan‡ can also be turned to good account for signalling—especially where dead-beatness forms one of the requirements, as in ship to shore communication on a cable with a fault in it.§

The semi-cylindrical controlling magnet of the ordinary mirror instrument is sometimes replaced by two bar magnets (Fig. 51) which pass in through the coils right up to the brass tube containing the mirror presenting poles of opposite names. Two toothed wheels, *a* and *b*, engaging in racks on the magnets enable the instrument to be regulated with the greatest nicety.

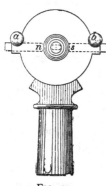

FIG. 51.

A common practice with the Silvertown Company, in connection with the cables in which they have been concerned, is to employ and provide what is known as a water resistance.‖ This instrument (Fig. 52) is inserted in the receiving circuit between the cable and the receiving instrument, or between the latter and the earth, thereby introducing a certain resistance in such a way as to diminish the size of the signals to the required degree. The value of this resistance may be varied by raising or lowering the

---

* In Mr Rymer-Jones's device, the front of the tube is additionally closed by a transparent cap. Thus the movement of the mirror is still further damped by the confined air. The beneficial effect of the above is naturally much more marked in the instance of sharp signals when there is little or no capacity in the line circuit to produce retardation. It is, therefore, peculiarly applicable for working on quite a short length of cable, or for practice purposes through an "artificial."

† The two extreme conditions under which the instrument can be used enables the ship after cutting a cable, in the course of repair work, to speak in either direction—*i.e.*, through a long section or on practically short circuit, if close in shore. Thus, the necessity for a Morse instrument during cable operations is entirely obviated here.

‡ For a full description of this instrument reference is made to the *Electrician* and the *Electrical Review* of 22nd March 1895.

§ Again, it is believed that Mr Thomas Clark (chief electrician to the Telegraph Construction Company) and Mr T. E. Weatherall have devised another marine galvanometer recently, which, additionally, meets these ends as in the above case.

‖ This apparatus is a cheap and handy substitute for a set of adjustable resistance coils, though less reliable. In the course of some experiments made by the author, it was found that—speaking roughly—the resistance of hot fresh-water was as much greater than cold fresh-water as the latter was in comparison with salt-water of the same temperature. This is, of course, no less natural--on the score of the extinction of animal life under a given rise of temperature—than that distilled water has a still higher specific electrical resistance.

insulated loop conductor L, thus interposing a greater or less area of water between the contact pieces A, B, and the respective terminals C, D below. With fresh-water the whole length of each tube represents many thousand ohms resistance. Unfortunately the contact pieces get dirty and involve too high a minimum resistance, as well as a very *variable* quantity. Moreover, the water is liable to leak. Both of these objections have, however, been to a great extent, if not entirely, overcome in recent improved forms.

Where very little capacity or resistance is present the retardation is so small that the signals become very shaky. They may be steadied to a great extent by the interposition of artificial retardation (resistance and capacity) at either or both ends of the circuit. As a rule, however, where the ordinary form of mirror instrument is in use, the lowest length of cable, with the usual core, which can be effectively worked on this system is about 100 miles ; and this only with a very stiff suspension, besides a good deal of artificial resistance and capacity. Hitherto, therefore, in the case of lengths below this, the Morse system has been adopted.

As has already been shewn, the latter introduces many objections in the way of delays, misadventures, etc., especially if the adjustment of the relay is not properly understood at either end.

It is not improbable, however, that when modified in the complete manner suggested by Mr Rymer-Jones, the "mirror" might be made to work satisfactorily on quite short lengths. If so, it would be peculiarly well adapted for taking the place of the Morse entirely during cable operations for signalling purposes between ship and shore.

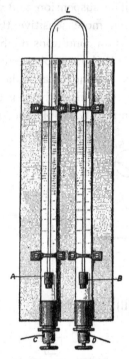

FIG. 52.—Water Resistance Apparatus.

**Suspended Coil Mirror.**—A mirror instrument based on the suspended coil principle has been found a capital substitute for the ordinary, suspended magnet, mirror on occasions. In this, the suspended coil is free to move in the field—*i.e.*, between the poles—of a powerful permanent magnet. Mr E. Raymond-Barker experimented with such an instrument on the occasion of a fault on the Brazilian Submarine Company's system, his idea being to be able to work the cable notwithstanding the fault by using such an instrument as would be sympathetic to firmly transmitted signals, but which—unlike the ordinary mirror instrument with exceedingly light suspension—would be beyond the influence of sudden vibrations arising from the fault. The suspension consisted of a long-range mirror on a 500-ohm coil, kept in position by fibres and weights, as in the siphon recorder hereafter described.

This instrument has an action which greatly tends to smooth off jerky vibrations; whilst, at the same time, readily responding to battery impulses. It is a capital "all-round" form of mirror instrument, as one adjustment suffices for signals from cables of very unequal lengths.

Notwithstanding the great comparative weight of coil and mirror, and the general stiffness and rigidity imparted to the signals by the lower bifilar suspension and weights, the instrument as a signalling apparatus is even more sensitive than the ordinary suspended magnet mirror—under certain conditions doubly as sensitive, in fact.

It should not be supposed, however, that this latter is any special point, except under particular circumstances. The ordinary mirror instrument is quite sufficiently sensitive for all ordinary cable signalling, besides being an ideally simple and portable apparatus.

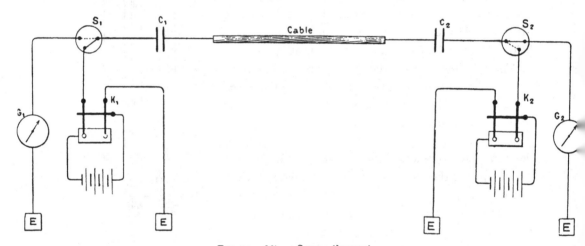

FIG. 53.—Mirror System Connection.

### STATION INSTALLATION.

When the line is worked simplex, each station sending and receiving alternately, the general scheme of installation is represented by Fig. 53. By depressing one knob or the other of the two-lever key M, the operator sends a positive or negative charge into the neighbouring condenser C, producing a deflection to one side or the other on the mirror receiver at the other end of the line. He works the switch S each time he wishes to change from sending to receiving, or *vice versâ*. In the figure, station 1 is in the sending position, and station 2 ready to receive.

**Transmitting Keys.**—During the period of mirror signalling on long cables previous to the introduction of the siphon recorder, various forms of transmitting keys have from time to time been introduced, all of them being on the double tappet system, as in the case of needle instrument working. Latterly these have been very much improved on in details of construction, various devices having been incorporated therein for discharging the cable, by instantaneously and automatically putting it to earth after the transmission of each signal. These improvements were

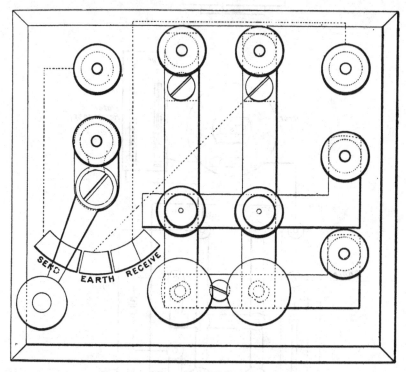

FIG. 54.—Mirror Key and Switch (scale = ⅔).

mainly introduced since the adoption of the siphon recorder in place of the mirror instrument on most of the long lines, the same form of key being of course applicable to either case. The switch for changing from sending to receiving and *vice versâ* is very usually nowadays incorporated with the key. Fig. 54 represents a good substantial and reliable species of combined key and switch, as designed by Mr W. A. Price for the Silvertown Company.* Fig. 55 shews the simplest form of connection

---

* Even by experienced clerks, attendance to hand switches is liable to be forgotten. Some years ago, Mr R. K. Gray, M.Inst.C.E., designed an ingenious double-lever

at a station in connection with a short cable (*without* condensers)—having a small amount of traffic—worked by the mirror system. The connections are sometimes arranged in such a way that the current is also conveyed through the instrument at the sending end for the purpose of assurance that there is a complete circuit. This is a good plan in the case of temporary operations, such as during, and just after, the original laying

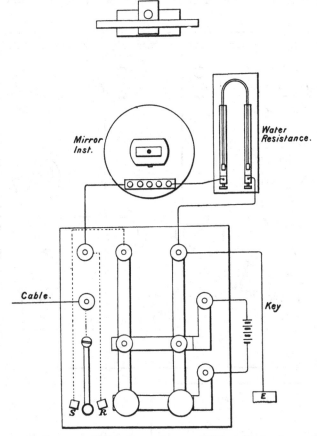

FIG. 55.—Simple Connections at a "Mirror Station."

of the cable, or in connection with repairing operations. A shunt to the mirror at the sending station requires to be then thrown into circuit. By this means, moreover, some estimate may be arrived at as to the character

---

signalling key without any hand switch. The latter was replaced by an automatic switch arrangement underneath the levers of the key, the movement of which influenced them. Unfortunately it was found that the object aimed at here was not always realised in practice.

of the signals at the further end, the outgoing signals acting as a rough guide.

To obtain concise signals much depends on well-timed manipulation at the sending end, and duration of contacts suited to the length of the line and the electrical qualities of the core. Too short current impulses may be insufficient to properly charge the cable, and the signals will suffer at the distant end in consequence; too long contacts, on the other hand, may develop counter signals difficult to distinguish from the real ones.

To understand this last effect, suppose one of the key levers to be held down till both condensers and the cable are completely charged; the spot will then return to, and remain at, zero. The moment the lever is let go the discharge from $c_1$, $c_2$ (see Fig. 53) and the cable will cause an instantaneous reversal current, which will produce a second deflection at the receiving end opposite in direction to the first. Each depression of one key lever thus giving two opposite signals, correct reading would be impossible. On the other hand, if the duration of contact is very short, the galvanometer needle $G_2$, after deflecting, will immediately return to zero without passing this point, the discharge current merely accelerating the return movement. If, however, the key lever is kept down a little too long, the needle having already started returning to zero, the discharge, when it occurs, will cause the needle to continue on its travel and give more or less of a deflection on the other side of zero.

**Earth Connection.**—In the case of submarine cables under ordinary circumstances, the sheathing wires of the cable itself is the proper earth connection to adopt. Several wires should be connected together and made use of for this purpose in the manner described elsewhere.

When, however, very sensitive instruments, like mirror receivers, are installed in telegraph offices where other conductors are brought in (especially if they are overhead wires, such as require powerful batteries), special precautions must be taken with regard to the earth connections. What, in fact, happens is, that a portion of the charge in the land lines— small it is true—instead of dissipating in the earth, leaves its own earth connections only to re-enter by those of the submarine cable, where it interferes with the mirrors. A case of this kind occurred, in 1875, at both ends of the Marseilles-Algiers 1871 cable. At the Marseilles end the mischief was easily remedied by soldering the sheathing wires of the underground cable (which was about $3\frac{1}{2}$ miles long, and used as an earth for the mirror instruments only) to the sheathing of the submarine cable in the hut. At Algiers, where the ground is drier, it was only possible to prevent diffusion of the electricity from the different earth connections

(special plates, water-pipes, etc.), by using as an earth wire for the submarine line, the interior conductor of an armoured cable which went
direct to the sea, at about 1½ miles from the station, and was joined up to
a large plate of copper immersed in the water,* after the manner suggested
by Cromwell Varley.

Mutual induction between two electric circuits (cables, or otherwise),
closely located and running in a more or less parallel line, is another
source of trouble which requires to be considered and guarded against.

Quite recently, in the course of a paper † read before the Institution of
Electrical Engineers, Mr A. P. Trotter ‡ shewed—by a practical example at
Cape Town—how seriously an electrical tramway line could influence the
working of a submarine cable whose shore end was sufficiently near. On
this occasion the present writer suggested that with care it might always be
possible to balance these inductive effects in the compensating circuit of a
duplex system, though its accurate maintenance would be a matter of some
difficulty.

Neighbouring electric light mains are similarly liable to be a source of
annoyance—especially if conveying high potential currents : and it is for
this reason that their whereabouts are restricted by the Post Office land-
line system, particularly where the telephone is in question.

## System of Checking Transmission.

When it is desired to keep a check, at the sending end, on the messages
transmitted, double-lever keys are sometimes used of a special form designed
by Mr H. A. C. Saunders. They consist of two Morse keys in connection
with each other through their pivots, and furnished with steel spring blades
at the rear end, these springs being insulated from the levers to which they
are attached by thin sheets of ebonite and working between two stop
contacts.

The connections are joined up as in Fig. 56, and the working is
as follows :—On, say, depressing the knob $k'$, the positive current from
the line battery P charges the condenser C, and a current in the same

---

* It was possibly with a view to obviating the above objections that as long ago as
1856 Mr Wildman Whitehouse and Mr Samuel Statham appear to have taken out a
joint patent (specification No. 1,745 of that year) for using a wire as a "return" instead
of the earth, either in a single circuit or for several circuits combined.

† *Jour. I.E.E.*, vol. xxvi., p. 501.

‡ Lately appointed Electrical Adviser to the Cape Government.

direction, coming from the first half of the battery $p$, passes through the Siemens form of Morse receiver M with duplicate polarised armature (Fig. 57). Depressing the knob $k^2$ would send a negative charge into the condenser C, and a similar current through the receiver M. In the first instance, one of the two armatures of the receiver would be attracted, and a blue mark printed on the paper, the movement of the second armature producing the same effect in the second case. In this way two parallel rows of dots or short marks are printed on the strip of paper, the marks in one row corresponding to dots, and those in the second row to

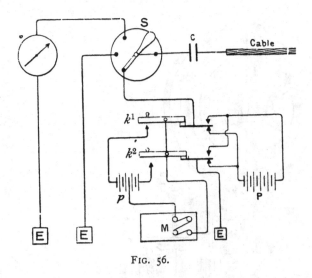

FIG. 56.

dashes, of the Morse alphabet. These symbols read quite as easily as the characters of the Morse code—

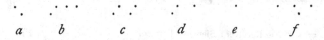

The commutator S has three different positions, the intermediate contact being connected directly to earth, so as to enable the condensers to discharge themselves rapidly when changing from the sending to the receiving position.

In the cases of the two Atlantic cables between Penzance and Canso, laid in 1881 and 1882, belonging to the American Telegraph and Cable Company, each condenser C has a capacity of 130 microfarads. They consist of 13 boxes of 10 microfarads each, the several boxes being subdivided into 5 sections of 2 microfarads each which can be joined up together in different ways according to requirements. The line battery P consists of 10 Siemens and Halske cells, of considerable internal resistance, but giving a very constant electro-motive force.*

**Record of Mirror Signals as Received.**—Attempts have also been made to utilise the radiometer of Professor (now Sir William) Crookes, F.R.S., for recording signals produced by the mirror receiver. An instrument based on this principle was shewn in the French Commercial Section at the International Electrical Exhibition at Vienna, in 1883. The moving parts of the radiometer (Fig. 58) merely include,

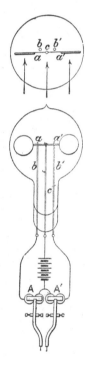

Fig. 57.—Morse Receiver: Siemens' Pattern.                    Fig. 58.

in this instance, two vanes of which the front faces are both coated with lamp-black. Behind the aluminium bar $a\,a'$ connecting the two vanes, are the ends of two platinum wires $b, b'$, which pass out through the glass at the lower end and connect to the coils of two Morse electro-magnets A, A'; the other ends of these coils are joined up to the negative pole of a battery

---

* "Die Einrichtung der Küstenstationen langen Unterseekabel," Dr von Tobler, *Elektro-technische Zeitschrift*, 1884, p. 76.

whose positive pole communicates with the metallic pivot $c$ on which the vanes of the radiometer oscillate. The apparatus is placed in front of a mirror receiver in such a way that the spot, when at rest, appears in the centre of the space separating the two vanes. When the spot deflects to the right, for instance, the vane $a'$ being repelled, makes contact with the wire $b'$, and closes the circuit of the electro-magnet $A'$, whose armature is accordingly attracted. If the styles attached to the armature are brought close enough together to print the marks on the same strip of paper, the signals can be read without difficulty by means of the Steinheil alphabet.

This apparatus, however, has never been turned to account in the practical working of our submarine cable systems.

T.S. "Silvertown" Landing the Cable at Fernando de Noronha, 1892.

### SECTION 4.—SIPHON RECORDER WORK.

The siphon* recorder, commonly known by the more simple title "recorder," was first invented by Sir William Thomson in the year 1867, but since then it has received numerous improvements.

As already stated in Part I., the siphon recorder has gradually supplanted the mirror instrument on all the existing long cables where the mirror had been in force, excepting in those of the "West India and Panama" Company, and in some of the sections of the "Western and Brazilian" Company. Practically speaking, equally sensitive to weak currents (though requiring more skill in technical knowledge and attention), it has the enormous advantage of yielding a record of the signals as received—and also as *sent*, if desired. Again, being much less trying to the sight, it is a more satisfactory instrument for working.

Though at a very slight disadvantage as regards speed of working (under given common conditions), the feature of a record being obtained was in itself a sufficient point of advantage to bring about the gradual replacement of the mirror instrument by the siphon recorder as soon as the latter had been invented. Inasmuch as the Post Office and other Government inland telegraph systems have given up the use of records or recording instruments for their systems, the great advantage of a record for cable working may not appear very obvious. The case of submarine telegraphy effected by competing companies is, however, entirely different to that of land telegraphs managed by the State. Thus: (1.) In submarine telegraphy there is an advantage in any "slip" method, in that errors are less liable to occur, as the clerk has thereby a chance of referring back if in doubt about any word, or words. (2.) Errors are a more serious matter for the companies' cables than for the State-controlled lines—indeed, the Government do not—in a complete sense—undertake the *onus* of them. It is, therefore, much more important that errors should be kept at a minimum in submarine cable system, and the message sent out absolutely correct to start with. (3.) It is more essential to be able to trace errors, inasmuch as the competing routes undertake to obtain repetitions free of charge, provided that the error is found to rest with them.

The siphon recorder possesses a still further advantage over the mirror system in that—with the usual manual transmission, at any rate — it requires only one ordinary clerk at the receiving end, instead of a first-rate mirror clerk and a writer.

---

* Oblivious of its Greek derivation, this word is often erroneously spelt *syphon*—especially on grocers' and chemists' labels to certain vessels containing mineral water.

On the other hand, the initial cost of the siphon recorder is, of course, very much greater than that of the mirror instrument—the proportion being about 10 to 1, in fact.* Still, since the royalty has been taken off the siphon recorder costs very much less than it used to, being about £75 (with vibrator)—or about £45, with direct writer, for short cables—instead of some £600 including royalty and sole manufacturer's charges.

**The Instrument.** — In its latest form the recorder consists of an exceedingly light coil of wire, delicately suspended between the two poles of a powerful electro-magnet, and capable of turning on a vertical axis. The coil is so arranged that, in its normal position, the plane of the coils

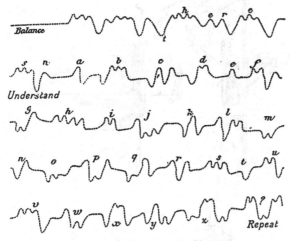

FIG. 59.—Siphon Recorder Signals.

of wire is parallel to the north and south line of the magnet. On a current circulating in the coil, the plane of the wires tends to take up a position at right angles to the north and south line, the coil turning one way or the other, according to the direction of the current.†

The motions of the coil are transmitted by silken threads to an extremely thin glass siphon, mobile on a horizontal axis; one end of this

---

* The mirror instrument is sold at about £5, and the double-current Morse at some £18.

† There are at least seven text-books which describe the siphon recorder, but which, when treating of the suspended coil, remain content with the statement that a current through it causes it to be "impelled across the magnetic field in one direction or the other, according to the polarity and strength of the current passing through it," or words to that effect—as if this movement were some phenomenon too complicated for brief explanation. Once, however, it is realised that a current-traversed coil is practically a magnet, the *rationale* of its deflections becomes self-evident.

siphon dips into a reservoir of ink, and the other end approaches to within a very small distance of a strip of paper driven through the instrument at a uniform rate. The ink being in permanent communication with a small electro-static machine, and the paper with earth, the ink is attracted to the paper, and issues from the capillary tube of the siphon with a rapid succession of tiny spurts. A continuous dotted line is thus inscribed on the strip of paper, which line is straight when the coil is at rest in its normal position, and undulating when the coil oscillates from one side to the other under the influence of positive and negative currents passing through it (Fig. 59). The deviations above and below an imaginary

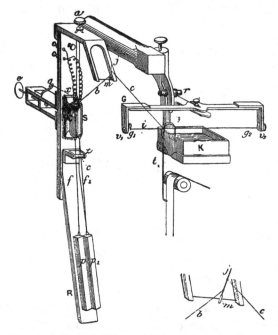

FIG. 60.—Suspended Coil and Siphon attached to Frame.

axial line corresponding respectively to the dots and dashes of the Morse code—*i.e.*, the signals, which exactly represent the arrival curves of the currents—are read very easily when the summits are well defined. We see by the figure, which is a facsimile of signals received on the cable from Pernambuco to St Vincent (1,844 N.M.), that when several current throws of same name succeed each other, they arrive so rapidly that the coil has not time to return, in the intervals, to its position of stable equilibrium, so that, as we have already shewn, the signals are produced by the first portions only of the electric undulations sent through the cable.

Let us now proceed to examine in detail the various essential parts of the apparatus.

**Coil and Electro-Magnet.**—The coil s (Fig. 60) is formed of several hundred turns of silk-covered copper wire of 0.003 inch diameter, and is wound to a resistance of about 500 ohms. The successive convolutions of the wire, wound on right-handed, are glued one to each other, sufficient rigidity being thus obtained as to enable a frame to be dispensed with. To render the instrument available for duplex working on the differential principle a second wire is often joined to the first, the ends of the two circuits being soldered to small spirals whose free ends are connected to four terminals (Fig. 61) insulated from the framework of the instrument by a sheet of ebonite N. The electrical resistance of each circuit is about 250 ohms. When working simplex, or by *Bridge*

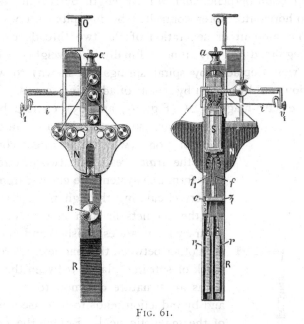

FIG. 61.

duplex, two of the terminals are connected together with a wire, so that the current will pass through the two circuits successively and in the same direction.

The coil is suspended by a fine silken thread whose length can be regulated by means of the screw-nut $a$. Two threads, $f$ and $f_1$, secured to the lower corners of the coil (see Fig. 60), pass under a bracket $z$, resting against the roller $c$, and are kept stretched by two weights $p$ and $p_1$ weighing 25 grammes each, and sliding in an inclined plane R. The bracket $z$ is movable to a certain extent up or down the upright frame to which it can be clamped by a screw-nut $n$ (Fig. 61). This method of suspension brings the coil back to a given azimuth, after the passage of a current; by raising the bracket $z$, the oscillation period of the coil is diminished, and *vice versâ*.

In more recent patterns, the weights $p$ and $p_1$ (see Fig. 60) have been done away with, the coil being, in some cases, also attached to its frame from the bottom, besides being suspended from the top.   The French Government Telegraph Department several years ago applied a modified form of siphon recorder to their Marseilles-Algiers cable, and in this the coil was maintained in position as follows. The two threads fastened to the lower corners of the coil, instead of being engaged in two grooves on a horizontal cylinder, adjustable by hand (Fig. 62), are pulled apart in a direction perpendicular to their length by two hooks attached to horizontal fibres controlled by fine-pitch screws (Fig. 63).

FIG. 62.

In this way the amount of separation of the two threads can be most conveniently regulated at any time.   Finally the weights attached to these threads were replaced by spiral springs (as shewn) to which any degree of tension could be given by means of adjusting screws.

Two large electro-magnets M (Figs. 64 and 65), excited by a local battery, are supported by a semi-cylindrical piece of iron L, which is in direct contact with both the armatures.   The two electro-magnets thus form one system, the current from the local battery circulating through the coils surrounding the magnets in such a way that poles of contrary name are established and placed facing each other between the magnets.   A rectangular block of soft iron placed between the two poles forms an armature common to the two poles, and by induction tends to increase the intensity of the magnetic field.   Round the rectangular block is suspended the coil S, which is thus

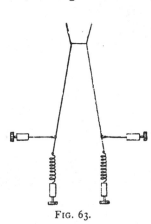

FIG. 63.

placed in a magnetic field of great intensity.   By means of the screws V and $V^1$, the two coils M surrounding the electro-magnet can be moved nearer to, or further from, the suspended coil S.   The battery which excites the electro-magnets is divided into two half-batteries which can be connected up in the circuit separately or combined by means of the commutator H (Fig. 66).   A short block of soft iron C', invariably fixed to one of the uprights of the framework, is placed as a core in the centre of the suspended coil S, so as to further increase the intensity of the magnetic field.

A sort of door W (see Fig. 64) gives ready access to the ebonite plate N in case of need, and enables all the delicate parts of the mechanism to be

[PLATE XXVI.

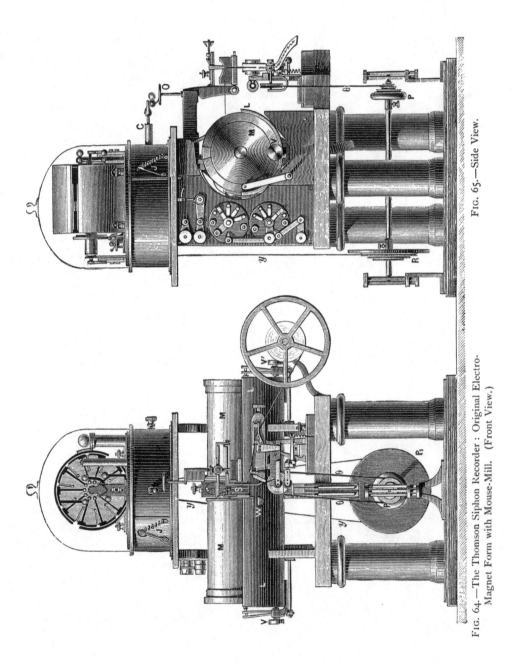

FIG. 65.—Side View.

FIG. 64.—The Thomson Siphon Recorder: Original Electro-Magnet Form with Mouse-Mill. (Front View.)

[To face p. 608.

inspected and repaired when necessary. Fig. 67 shews another pattern of a long cable siphon recorder in which the electro-magnet has been substituted by a strong compound permanent magnet.

**Siphon Record of Signals.** — The siphon, shewn natural size in Fig. 68, is a glass tube of very small diameter, bent twice at right angles near one of its ends, and to an angle of 130° at about ¼ inch from the opposite end. This tube being very fragile, it is well to know how to replace it without having recourse to the makers. To do this, select a glass tube of about 1 inch in dia-

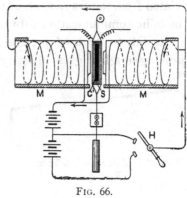

FIG. 66.

meter, with sides about ⅛ inch thick. The central portion must now be warmed for a length of about 1 inch in the flame of a gas-jet or spirit-lamp, turning the tube round and round during the whole operation. When the glass is sufficiently softened, it can be drawn out to the required diameter, and when cold it is cut into lengths of about 4 inches.

To make a siphon from one of these pieces, hold it in one hand, applying the flame of a lucifer match with the other hand to the point where the bend is required. The free end will then droop by its own weight till the necessary curvature is obtained. The longer leg can be cut to the proper length with scissors, and the end rubbed on an emery stone so as to get a fine smooth end which can be brought into close proximity to the paper without risk of tearing.

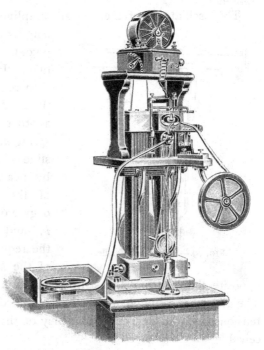

FIG. 67.—Thomson Siphon Recorder, with Permanent Magnet.

The siphon *t* thus formed is fastened with a little white wax to the cradle Q (Figs. 60 and 69) of aluminium, which is itself secured to a platinum wire $i$. This wire is stretched between two springs $g_1$ and $g_2$, and can be given a slight torsion by turning the mill-headed clips $v_1$ and $v_2$. The upper end of the siphon dips into the ink-well K, which is insulated from the framework of the apparatus by a rod of ebonite, and also in metallic communication with the disc O (Figs. 64 and 65).

The ink to be used preferably is a blue aniline liquid, obtained by dissolving in a half glass of water as much of the crystals as can be held on the point of a penknife blade. This solution—of a beautiful dark blue colour, and perfectly fluid—does not thicken or form deposits, and can be produced in very small quantities.

FIG. 68.—Glass Siphon of Thomson Recorder.

The motions of the coil S are conveyed to the siphon $t$ by two silken threads $b$ and $c$ (Fig. 60), one of which connects an upper corner of the coil and the centre of the lever $j$, and the other joins the upper end of this same lever to a point D on the aluminium saddle of the siphon Q (Fig. 69).

The oscillations of the coil are amplified in the ratio of the relative lengths of the lever arms of $j$, which is pivoted at $m$, and is called the multiplier. In rear of the suspended coil—with regard to the point to which the thread $b$ fastens—is attached (Fig. 60) another silk thread $x$ stretched by a spring $q$, this spring being carried in a slide which can be moved to and fro by means of the screw $e$. The tension of the spring $q$ is adjusted to the degree of torsion on the platinum wire $i$; and the screw $e$, being turned in the required direction, brings the point of the siphon to its normal position opposite the centre of the strip when no current is circulating in the coil.

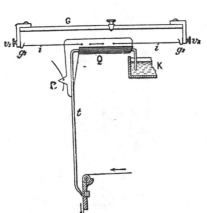

FIG. 69.—Siphon and Cradle.

The play of the siphon under the combined influence of the two opposing tensions and the deflecting tendency of the coil will now be easily conceivable.

The frame G which serves to stretch the platinum wire $i$ is capable of a slight vertical displacement up or down. It can be clamped by means of the set screw $r$ so as to bring the point of the siphon opposite the required point on the strip of paper. When the instrument is to be idle for some hours, it is well to withdraw the upper end of the siphon from the ink-well, to avoid any chance of stoppage in the tube. To do this, loosen the clamping screw $r$, and carefully raise the frame G; the slot in which the clamp travels during this operation is so shaped that the tension of the thread $c$ is not appreciably increased. Should the ink have accidentally dried in the interior of the siphon, the tube will have to be cleaned by soaking in sulphuric acid.

The strip of paper, or tape, which is unwound from the reel R (Fig. 70), is stretched by passing under a spring $\alpha$, passes over the guide roller $\beta$, and

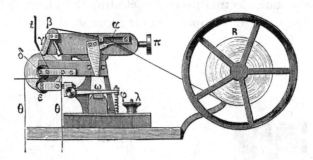

FIG. 70.—Paper Strip and Drawing-off Apparatus.

then descends over a projecting plate $\gamma$ immediately facing the point of the siphon $t$. It is drawn onwards between two rollers $\delta$ and $\epsilon$, the upper one receiving its rotary motion from the magneto-electric machine by means of the cord $\theta$ and multiplying pulleys P, and the lower one being pressed against its companion $\delta$ by the action of the spring $\phi$ on the bent lever $\omega$.

The tape is adjusted to the proper distance from the siphon by turning the mill-headed screw $\pi$, and the tension of the bent lever spring is regulated by the nut $\lambda$.

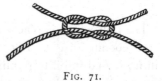

FIG. 71.

Whip-cord is the best line to employ for transmitting the motion, and when the two ends are tied with a reef-knot as in Fig. 71, the cord passes easily round the pulley without stretching. A slight elongation, however, of the cord will not matter, as the axle of the lower pulleys P works in a slide, the weight of the pulleys being supported by the bight of the cord itself.*

As dry paper would have too great electrical resistance, the paper tape

---

* In more recent forms of siphon recorder, the whip-cord has been replaced by leather boot-lace, or by an india-rubber thong of circular section.

used is soaked in a solution of 2 parts of nitrate of ammonia in 100 parts of water. The paper never quite dries, being kept slightly moist by the deliquescence of the salt which it contains. Sometimes it is found sufficient to keep the rolls, for some time before use, over vessels containing water, the paper taking up, in this way, the required amount of moisture.

**The Electric Mill.**—This piece of mechanism, commonly known as the "mouse-mill," does duty both as an electro-magnetic motor and also as a generator of static electricity on the principle of the Varley machine. The motor serves to draw the paper strip forward in front of the siphon ; the generator of static electricity serves to electrify the ink which is conveyed from the well $K$, through the siphon, to the paper strip.

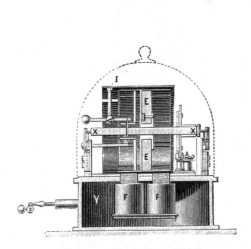

FIG. 72.—The "Mouse-Mill."

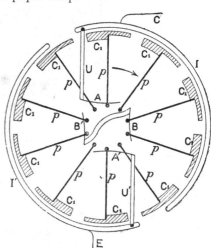

FIG. 73.—Side View of "Mouse-Mill," shewing Electrical Connections.

The mill, which is placed at the top of the complete instrument, consists of an ebonite disc $E$ (Fig. 72) whose axis $X$ revolves on roller bearings half immersed in oil to minimise friction. Round the periphery of the disc (Fig. 73) are placed at equal intervals pieces of soft iron $C_1$ with their outer surfaces parallel to the circumference of the disc, insulated from each other but connected to ten metallic rods $p$. These radial rods terminate in rounded knobs, arranged symmetrically round and at equal distances from the centre of rotation, which come successively into contact with four gilt-steel springs $A$, $A'$, $B$, $B'$. The springs $A$ and $A'$ are connected by copper rods $U$ and $U'$ to two fixed pieces of iron $I$ and $I'$ semi-cylindrical in shape, which almost completely surround the moving parts of the mill and form the inducers ; the first of these is in communication with a conductor $C$, the second with the earth through the frame of the instrument. The springs $B$

and B' are in metallic connection with each other, but insulated from the remainder of the apparatus.

The principle of Varley's machine tells us that if an electric charge, however feeble, is given to one of the pieces $C_1$, and if the axis X of the mill receives continuous rotary motion, the charge on the inducer I will increase very quickly. It is even necessary to cover the pieces $C_1$ and the inducers with paraffin, to prevent the sparking which would soon occur from one to the other.

The axis X is rotated automatically by the electro-magnetic machine. To effect this a horse-shoe electro-magnet F is placed in a receptacle V underneath the mill, the magnet being excited by a local battery whose circuit is alternately closed and broken. The height to which the two cores of the electro-magnet F project is dependent on the length of the pieces $C_1$. Directly the circuit of the local battery is closed the magnet F attracts the piece $C_1$ nearest to it; the current is inter-

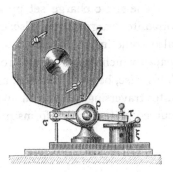

rupted when the armature $C_1$ has passed the poles of the electro-magnet, but the motion still continues for an instant or so, in virtue of the acquired momentum. As the next arma-ture approaches the magnet, the circuit is again closed, this second armature is in its turn attracted, and the movement of rotation continues indefinitely.

To effect the automatic make-and-break of the circuit at the exact instant required,

FIG. 74.

a disc Z (Fig. 74) having the form of a regular decagon is placed behind the inducers and arranged to rotate with the axis X. Below this disc is a lever $\sigma\tau$, pivoted on an axis $\rho$; the end of the lever $\sigma$ immediately below the centre of the disc carries a tiny roller which bears against the rim of the disc, the opposite end of the lever $\tau$ is traversed by a screw, the platinum point of which rests on a small drum $\xi$. The disc Z, being carried round with the rotary motion of X, retains the lever $\sigma\tau$ in the position shewn in the figure so long as the roller $\sigma$ is in contact with the central portion of one of its ten sides, the battery circuit being closed meanwhile. When one of the angles of the disc passes over the roller, the latter is depressed, causing the opposite end of the lever to rise and break the local circuit at $\tau$. Therefore, assuming the disc Z to be keyed to the axis X in such a manner that the lever $\sigma\tau$ will start tilting at the exact instant when one of the armatures $C_1$ commences to leave the poles of the electro-magnet, the breaks in the current will occur just as described above. The disc, carried forward by the momentum of the moving parts, will allow the roller to rise against the next flat side, and the lever $\sigma\tau$ will regain its normal position. The local circuit will now be closed at $\tau$, and the same action recurring as each side of the disc passes the roller, the rotary motion will be maintained.

As the sparking which occurs between the end $\tau$ of the lever $\sigma\tau$ and the plate $\xi$, in course of time disintegrates the surface of the latter, arrangements are made by which the plate can be moved by means of a small handle so as to present a fresh surface to contact with $t$ when necessary.

The rotary motion of the axis X is transmitted by a cord $y$ (see Figs. 64 and 65) to the pulley $P_1$ mounted on the same shaft as the pulley P which draws the strip of paper through the apparatus.

By means of the commutator J graduated resistances can be inserted in the circuit of the electro-magnet F, and the speed of the mill varied accordingly. The least supplementary resistance corresponds of course to the highest velocity.

The static charge set up on the conductor C induces a corresponding opposite charge on the disc, or plate, O, fixed at a proper distance. It is also sometimes communicated by *conduction*,* or with a strip of damp paper which hangs down from the conductor C and rests on the plate O. This disc, being connected, as we know, to the ink-well K, the electricity, after traversing the siphon and the strip of paper, goes to earth through the curved plate $\gamma$ which forms part of the metallic framework.

The Vibrator.—The passage of the ink through the siphon is attended with great difficulty whenever the degree of atmospheric moisture exceeds a certain limit, owing to the loss of static electricity through the air and surrounding parts of the instrument. Thus, in 1884, Mr G. F. Pescod, of the Central and South American Telegraph Company, devised a method for establishing and maintaining vibration in the siphon by means of electro-magnets.† In adjusting a recorder suspension he had noticed—on accidentally shaking the table on which the recorder was placed—that the signals were greatly enlarged, and that the definition was much improved when the recorder was subject to the vibratory movement thus produced.

---

* The former plan is, however, usually preferred, for by this means adjustments can be readily made by inserting any required resistance between C and O.

† Sir William Thomson had previously designed something of the sort, but it does not appear to have been used in practice, though a reference thereto was made in his siphon recorder original patent.

This observation suggested to Mr Pescod to attach a fibre connecting the armature of an ordinary electric make-and-break system with the fibre suspending the recorder coil, so that the latter should be kept constantly in a state of vibration. It was thought—and rightly so—that this would to a great extent overcome the inertia of the coil, thereby leaving the received current little to do beyond effecting the extra movement necessary for making the signals.

It is unnecessary to point out the enormous value of this simple application, various modifications of which have since been adopted at every "recorder" telegraph station throughout the world.

FIG. 75.—Siphon Recorder Signals *without* Vibration.

Fig. 75 is the facsimile of a piece of recorder slip received on a cable with a fault in it * when working *without* vibrator,† whilst Fig. 76 is an illustration of the character of signals received *with* this apparatus (of Mr Pescod's) applied. From a comparison of these the advantage will be obvious.

In 1885 Mr Pescod published in the *Journal of the Society of Telegraph Engineers* ‡ a few lines with reference to this invention, after it had been in use for nearly a year on the "Central and South American" Company's

FIG. 76.—Siphon Recorder Signals *with* Vibration.

cables, and had proved of special value in working a section whose "KR" was $9.2 \times 10^6$, as well as on one of $11.4 \times 10^6$.§

In 1886 Mr Charles Cuttriss (electrician to the Commercial Cable Company) devised a similar plan. Here, a small style, drawn from

---

* The application of this device under such circumstances is peculiarly to the point, for by its means the sensitiveness is so far increased as to obviate the necessity of an increase in the battery power.

† And also without any electrification of the ink, the siphon being drawn directly over the surface of the slip, thereby introducing considerable friction.

‡ Vol. xiv., p. 343.

§ One of this company's officials, Mr F. P. Walker, subsequently suggested, instead of attaching a fibre between the armature and the recorder suspension, that the armature should be made to play directly on the fibre suspending the coil. This is found to be a convenient plan. It is then necessary that the fibre should be protected against abrasion, or else toughened.

very fine iron wire, is placed in the end of the siphon, and separated from the armature of an electro-magnet by the thickness of the paper only. A special vibrator determines the excitation of the magnet at regular and rapid intervals. The style vibrating in unison with the armature of the electro-magnet withdraws a small quantity of ink from the tube at every impulse, tracing a line on the paper formed of fine dots very close together.*

FIG. 77.—Muirhead's Improved Siphon Recorder.

**Later Patterns of the Siphon Recorder.**—When applying the siphon recorder to their cables between Marseilles and Algiers, the French Postal

---

* Since then several other forms of vibratory devices have been introduced—notably those of Dickenson and Ash.

The description of a recent type, as supplied by Messrs Siemens Brothers, is as follows :—A light iron reed is attached to the siphon-cradle and an electro-magnet is fitted on the frame, or bridge, with its poles opposite the reed. The electro-magnet is

Telegraphs adopted a modified form of this instrument. The electrical (mouse-mill) method of unwinding the paper was abolished, and the ordinary clockwork arrangement (as in the Morse recording instrument) substituted for it.* To avoid shaking the siphon suspension, this drawing-off apparatus was placed on a separate table. Moreover, the curved plate against which the point of the siphon presses—usually known as "the saddle"—was made movable in three different directions, so as to render it easily adjustable to the required position. These special arrangements were novelties at the time, it is believed, as well as those previously mentioned in connection with the suspension attachments.†

Perhaps the latest type of Thomson siphon recorder for working long cables is that designed by Dr Alexander Muirhead,‡ and of which Fig. 77 gives a general view. In this pattern the component parts of the instrument are so arranged that the various adjustments of it can be more readily effected than in the original form.

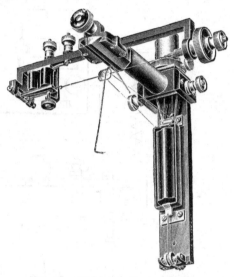

The most novel feature in Dr Muirhead's instrument is in the suspension piece, of which Fig. 78 is a general view and Fig. 79 a diagram of the connections. Here the coil is connected by means of two fibres to the aluminium cradle

FIG. 78.—Muirhead's Suspension Piece.

to which the glass siphon is cemented. The torsion of the stretched, phosphor-bronze, strip which carries the aluminium cradle and that of the two fibres together, give the directive force to the suspended coil.

The apparatus § employed by Dr Muirhead for imparting the necessary

energised intermittently by currents from a primary battery controlled by an automatic "make-and-break." The apparatus is provided with adjustments by which the rate of vibration can be made synchronous with the natural rate of the siphon.

* In still more recent patterns—as is shewn a few lines on—the electric motor system has been again returned to, though now as a separate machine for this purpose alone, the mouse-mill with its electrifying provisions being altogether a thing of the past.

† Again, this was probably the first instance of a direct writing recorder, as used on similar short lengths of cable, recent forms of which are described further on.

‡ See patent specification No. 20,793 of 1893.

§ Based on a patent taken out in 1893 in the names of Alexander Muirhead and Robert Henry Edgar. See specification No. 3,413 of that year.

vibration to the siphon consists of a small electro-magnet fixed at one end of the bridge-piece carrying the siphon suspension, whose armature is attached to the stretched wire on which the siphon rides. A vibratory current is passed through the coil of this electro-magnet from a make-and-break apparatus constructed like that of an ordinary electric trembling bell. The pair of coils on the bridge-piece which carries the siphon are connected in series with the interruptor, a battery and a sliding resistance.

To get the interruptor and siphon in tune, the weight C (Fig. 79) is

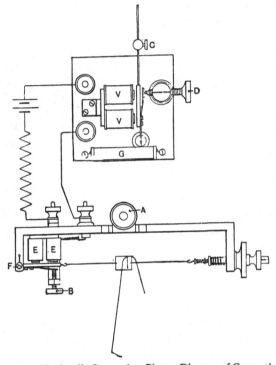

FIG. 79.—Muirhead's Suspension Piece : Diagram of Connections.

first placed in a suitable position, and the final adjustment is made with the screw D. A coil G is fitted on the interruptor to absorb the spark at the point of make-and-break. In this device the clockwork for drawing the paper is replaced by a modern form of electro-motor placed at the top of the instrument, as shewn in the figure.* The motor is energised by a battery consisting of Thomson trays (or gravity cells) equivalent to a total E.M.F. of 4 volts.

---

* Ordinary motors of small type have several distinct advantages over clockwork for this purpose. To wit, the obviation of winding, and, therefore, of periodic jars to the apparatus. All siphon recorders as now made and supplied by Messrs Muirhead and Co. are furnished with such motors.

Mr Frank Jacob has designed for Messrs Siemens Brothers a somewhat similar suspension piece. The make-and-break of the electro-magnet, however, in this case involves the vibration of the siphon only instead of the entire cradle, as has been already shewn.

Mr Charles Cuttriss, of the Commercial Cable Company, has also devised a modification of the Thomson siphon recorder in a very convenient and efficient form, and this instrument—a general view of which may be found at the end of this section (p. 630)—is in use on the Commercial Company's Atlantic system.

### Installation of the Apparatus and its Connections at Sending and Receiving Stations.—As the power of checking messages at the

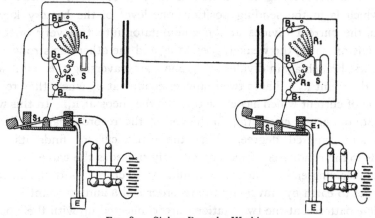

FIG. 80.—Siphon Recorder Working.

stations from which they are despatched is an important point in the practical working of cables, the current which actuates the receiving instrument at the distant station is in " recorder " working almost invariably made to first traverse the apparatus at the sending end. On account, however, of the great delicacy of the siphon recorder, only a very small portion of the line current is permitted to pass through this instrument at the home station. An ordinary slide resistance R (Fig. 80), forming a shunt to the coil S of the recorder, is placed on the left of the apparatus, but only put into circuit when *sending*. The maximum resistance of this adjustment is about 8 ohms, as a rule.*

---

* By this plan, proof is obtained of the signals sent out by the transmitting clerk, and thus in case of error its origin can be traced. Moreover, by this means assurance is

A second set of resistances $R_2$, all of which are above 500 ohms, forms a second shunt to the coil S, its principal object being to rapidly deaden the oscillations of the coil. This shunt enables the currents which are induced in the coil (when the latter turns in the magnetic field set up by the large electro-magnets) to circulate in a closed circuit, and to oppose the movements of the coil, in accordance with Lenz's law.

A specially designed commutator, consisting of a brass lever pivoting on a horizontal axis, is placed between the receiving instrument, the earth, and the battery; it is worked each time when changing from the sending to the receiving position, or *vice versâ*. The centre contact $E_1$ enables the cable, in the first place, to discharge itself for an instant almost directly to earth, the current having only to pass through the low resistance rheostat $R_1$. With this object in view, the heights of the different contacts are so arranged that the lever, in tilting, touches $E_1$ before finally quitting S.

The connections being established as in Fig. 80, if, at the left-hand station, which is in the sending position, one lever of the battery key is depressed, the current divides at the commutator into two very unequal portions, one of which, the weaker, goes to $B_1$, and the other, the stronger, to $B_3$. The first branch again divides at $B_1$, one part traversing the coil S, and the other the shunt $R_2$. These two branches reunite at $B_2$, where they rejoin the portion of current which has come through the rheostat $R_1$. In this way the strength of current circulating in the coil of the recorder can be diminished to any required degree, whilst the entire current finds its way afterwards to the condenser which immediately precedes the cable.

At the receiving end, when the commutator lever is raised, the current can only get to earth by traversing the recorder coil S and its shunt $R_2$.

The line batteries at the two stations are connected up with their poles inverse to each other, so that signals can be read the same way, no matter from which end they come.

Sometimes the small portion of the current which is conveyed through the coil of the recorder at the sending end goes directly from the coil to earth. In this case the connections are arranged as in Fig. 81. At the sending station the circuit is broken at contact No. 1 and closed at contacts Nos. 2 and 3; at the receiving end, on the contrary, the circuit is closed at 1 and broken at 2 and 3. An auxiliary resistance of 5,000 ohms or more,

---

secured of there being a complete circuit, and of the instrument at the further end being earthed ready for receiving. It may be mentioned, however, that if a cable be very faulty, even though the further end be insulated, signals may be noticeable at the sending end.

To effect the same result in *duplex* working special arrangements require to be made.

according to circumstances, is inserted between the terminal $B_2$ and the earth. The current, at the start, divides into two parts which reach $B_1$ and $B_3$ simultaneously; at $B_1$ a further division takes place, the larger portion going to the condenser, the remainder traversing the coil S—and, in case of need, its shunt $R_2$—to rejoin at $B_2$ the current which has come through the rheostat $R_1$: both the latter currents now escape to earth through the resistance $R_3$. At the receiving end the current traverses the recorder coil S and the rheostat $R_2$ which forms the shunt, passing to earth through contact No. 1.

Callaud, Minotto, or Siemens-Halske cells are generally used for the line batteries.* For the local circuit it is better to employ the large surface tray batteries on the Daniell principle (specially designed for this purpose by Sir William Thomson), the internal resistance of which does not exceed 0.1 ohm per cell. Three of such cells are generally sufficient to work the mouse-mill, and but nine are required to give the necessary intensity to the magnetic field surrounding the coil S of the recorder.

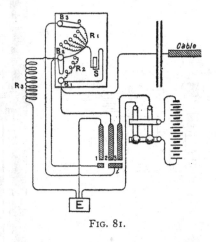

FIG. 81.

The number of cells used for signalling naturally varies with the length of line. Between Marseilles and Malta, a distance of 834 miles, four to five cells are enough. On the Malta-Alexandria section, 927 miles in length, eight to ten cells are found necessary.

**Permanent Magnet Recorders.**—In 1884 a new form of Thomson siphon recorder was introduced which rendered it suitable for working

---

* Of late often replaced by dry cells of extremely low resistance and great staying power ; or preferably by accumulators, as the former sometimes fail—especially in hot climates.

Where so small a portion of the charge is employed for doing work, increase in E.M.F. will produce very little increase in speed. On the other hand, reducing the battery resistance is of paramount value, as a greater rush of current into the line at the first moment is thereby secured. It is on this account that the above forms of battery are employed where high speeds are aimed at through a given cable.

The insertion of condensers for signalling purposes at the receiving end of a cable, and still more when, as is usual, they are placed at both ends, involves a somewhat higher battery power.

cables of *moderate* length, as well as on long cables. In this modification, the electro-magnets M (see Fig. 66) are replaced by permanent magnets, the mouse-mill being completely suppressed, and the rheostats $R_1$ and $R_2$ as well as the two local batteries dispensed with. A general idea of these simplified instruments, with their connections, is given in Fig. 82. Here the siphon has to be of somewhat larger diameter so that the ink may circulate by its own weight, without being carried on to the paper by the flow of electricity as in the more complete apparatus. The strip of paper

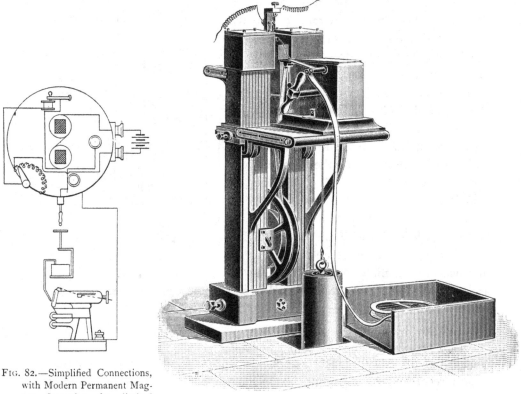

FIG. 82.—Simplified Connections, with Modern Permanent Magnet Recorder for limited lengths.

FIG. 83.—Thomson Siphon Recorder : Simplified Permanent Magnet Pattern.

is drawn past the siphon by clockwork. The general appearance of this instrument is shewn in Fig. 83, also, in section, by Fig. 84. This form of recorder (with permanent magnets) is, with all its modifications, by degrees entirely replacing the old electro-magnet form.

Again, Fig. 85 serves to shew a recent type of permanent magnet recorder adopted by the French Postal Telegraphs for their Government cables, with the side adjustment to suspension piece as already referred to.

**Direct Writing Recorders.**—On short cables, where it is not necessary to vibrate the siphon (on account of the greater strength of effective current), and in which the rate of signalling adopted is not near the limit at which the cable can be worked, modifications of the Thomson recorder in the form of direct writing siphon recorders are used. A convenient pattern of such an instrument, designed by Dr Muirhead, is shewn in Fig. 86. The siphon in this case is attached to an aluminium frame fastened to the

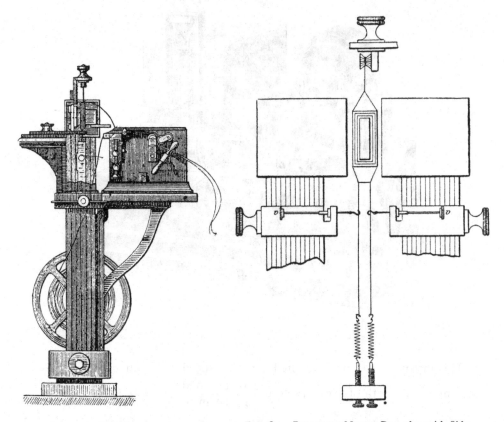

FIG. 84.

FIG. 85.—Permanent Magnet Recorder, with Side
Adjustment to Suspension Piece.

coil itself which is suspended bifilarly top and bottom; the lower fibres can be adjusted laterally, and there is also an adjustment for varying the tension on the fibres themselves.

The platform over which the paper passes beneath the marking end of the siphon is in this instrument provided with adjustments in two directions, so that the paper may be raised up into light contact with the siphon and levelled to suit the movements of the siphon. The direct writer being free from "jars," as much as 120 words per minute—and even more—have been obtained on this form of siphon recorder, the signals being, moreover, well defined.*

In this device the paper is drawn by means of a small electro-motor. The motor is conveniently placed at the side and provided with double gearing acting directly on the paper draw-off, the object of this double gearing being to extend the range of speed at which the paper is required to travel.

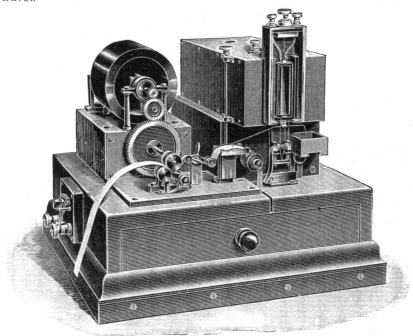

FIG. 86.—Muirhead's Direct Writing Recorder, for Short Cables.

This instrument is well adapted for high-speed work on cables up to 1,000 miles in length of ordinary core, $\dfrac{107 \text{ lbs. conductor}}{140 \text{ lbs. insulator}}$ per N.M.

Another improved form of short cable siphon recorder, with clockwork movement, as made by White of Glasgow (the maker of all Lord Kelvin's inventions), is shewn in Fig. 87. This instrument is provided with a comparatively small, but strong, permanent magnet (in horse-shoe, instead of long straight bar magnet, form), and is conveniently fitted with "sending" and "receiving" shunts. The whole apparatus has a very compact appearance.

* On an emergency the speed with this instrument has approached something like 200 words a minute.

**The Smith Switch, or Commutator.**—The lever commutator in the Thomson recorder is sometimes replaced by an apparatus designed by Mr

FIG. 87.—White's Direct Writing Siphon Recorder.

Benjamin Smith, giving precisely the same connections but in a neater manner. Fig. 88 shews the internal arrangements of this, and Fig. 89 an

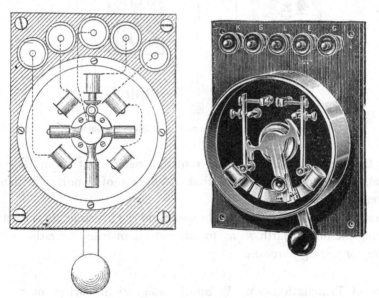

FIGS. 88 and 89.—Benjamin Smith's Recorder Switch Key.

outside view. It consists principally of four cross arms which butt against stops fitted with contact pieces when rotated horizontally by a lever or

handle mounted on the same vertical axis. The connections are established as in Fig. 90. In the position shewn, the line is in its normal
condition, directly to earth through the commutator. To place the contacts
for sending, the lever is drawn to the left, and it must be moved over to a
corresponding position on the right when changing to the relieving position.

The condenser, in this installation, is not situated between the line and
the receiver, but is placed between the receiver and the earth ; thus only the
signals received come through the condenser. The result of this arrange-

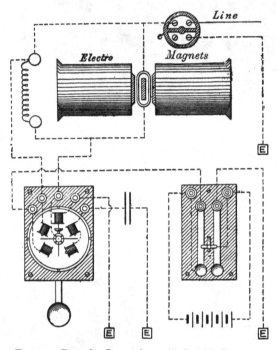

Fig. 90.—Recorder Connections with Smith's Commutator.

ment is that the cable and condenser remain constantly at about the same
potential, and experience proves that the work of sending is thereby
facilitated.

The small commutator between the cable and the instruments is intended
for putting the line to earth so as to cut out all instruments either during
storms or for testing purposes.

**Manual Translation.**—Mr B. Smith designed his switch more especially for purposes of ready manual translation at an intermediate station.
Manual translation used sometimes to be known as the " human relay
system." Even in the early days of land telegraphy, before the Govern-

PLATE XXVII.

ELLIOTT BROS.
LONDON

PLATE XXVII.]

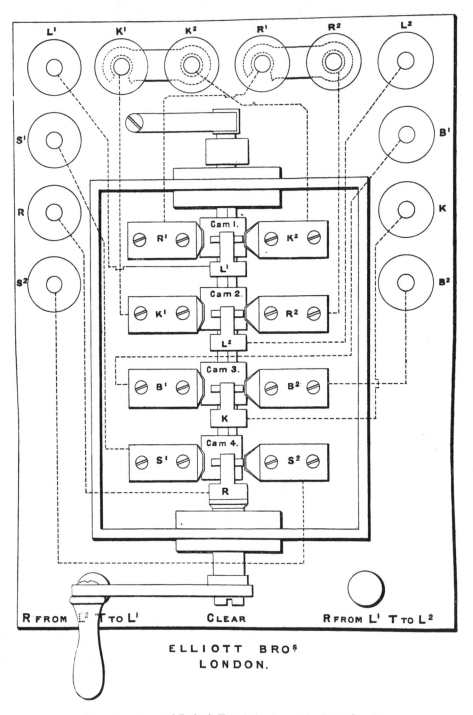

FIG. 91.—Raymond-Barker's Translation Switch for Cable Circuits.

[To face p. 627.

ment took over the telegraphs, systems of manual translation were in vogue in the absence of automatic relays. It was, in fact, very soon discovered that an immense saving of time and labour—not to speak of a reduced chance of error—occurred when the clerk at an intermediate station sent off the message to the required station direct from the Morse slip, instead of writing out the message and then handing it over to another clerk to retransmit it on another circuit. Mr Smith set his mind to organising a similar complete and regular system of human translation, or repetition, for submarine cable circuits where an electrc-mechanical relay is impracticable. It was mainly in order to meet these requirements (in a ready form) for the Eastern Telegraph Company's system of cables that Mr Smith's switch was designed, as above described.

Mr E. Raymond-Barker also devised a commutator, or switch, to meet the same purpose, and this is very largely used by various other companies, being an exceedingly efficient and compact affair. Fig. 91 represents a general plan of this instrument.

The various conditions, as regards connections, under which the switch can be turned to account, in the case of a repeating or translating (intermediate) station on the one hand, and in that of a terminal station on the other, are clearly shewn in Figs. 92 and 93 (see Plate XXVIII. overleaf).*

As already stated in connection with the mirror system, various cable transmitting keys have been devised from time to time. The improvements on the original reversing key (similar in principle to that used in prehistoric needle-signalling days) consists mainly of greater excellence of workmanship, more reliable contacts, and a more substantial form generally. Subsequent modifications combine the sending and receiving switch with the transmitting key, thus making the key form a part of the circuit at all times. Such combined keys are invariably used now, on account of the convenience of the switch being so ready at hand for the operator. There are several keys of this pattern, all naturally similar in principle. It will suffice if we content ourselves with a full description of one of them.

**Dickenson's Key.**—This is the device of Mr W. Dickenson, of the Anglo-American Telegraph Company, and unites the functions of a current reversing key and commutator or switch. The commutator handle has to be worked every time when changing from sending to receiving, or *vice versâ*.

The reversing arrangement consists of two ordinary key levers $t_1$ and

---

* It may be remarked that the intermediate station repeats—or "keys"—the message on through the other cable to the next station, except when he receives a signal (usually "I Q.") which indicates that it is for his station.

$t_2$ (Fig. 94) entirely independent of each other. Their upper contacts are connected together and also to terminal No. 2 ; the lower contacts, or those against which the levers butt when depressed, are connected to each other in a similar way, as well as to terminal No. 3.

The commutator is placed between the levers, and consists of an ebonite eccentric secured to a horizontal axis turned by a crank handle $v$. Above the eccentric is fitted a metallic contact piece communicating with terminal No. 1. Three springs, insulated one from the other and placed two on the right and one on the left, press against this contact piece when they are not forced outwards by the projecting part of the eccentric. The spring on the left connects to terminal No. 5, the front spring on the right to terminal No. 6, and the spring in rear to the pivot of the lever $t_2$, the pivot of the lever $t_1$ being connected to No. 4 terminal. In the position shewn by the figure, the two right-hand springs are forced away to the right by the eccentric, and consequently insulated ; the left spring, on the contrary, is pressing against the contact piece. By turning over the handle $v$ 180°, the left-hand spring is insulated by the eccentric and the other two make contact. During the rotation of the commutator there comes a time when all three springs are making contact simultaneously, but this only lasts for a moment if the handle is worked quickly. This arrangement, as we shall see further on, facilitates the discharge of the cable.

FIG. 94.—Dickenson's Transmitting Key.

The entire apparatus—except the commutator handle and the tappets—is enclosed in a round brass box with a glass top.

The connections are made as in Fig. 95, where $U_1$ and $U_2$ represent two variable resistances containing four coils each ; the resistances of the coils in $U_1$ are respectively 1, 1.5, 2, and 3 ohms ; in $U_2$ each coil has an equal resistance of 500 ohms. The latter rheostat, as we have already explained, serves to deaden the oscillations of the coil s of the recorder, the least effect in this direction being obtained with one coil, and the greatest effect with all four coils in circuit. The current at the sending end—or a portion of it, at least—is made to traverse the recorder coil s for facilitating the checking of messages, as previously narrated.

The commutator being in the receiving position (which is the one shewn in the figures), if a charge of electricity arrives through the cable at the condenser C, a charge of similar sign is set free from the opposite coating of the condenser. This charge divides into two parts at the rheostat $U_2$, one part going to $n$ through the coils of $U_2$ which are in circuit, and the other

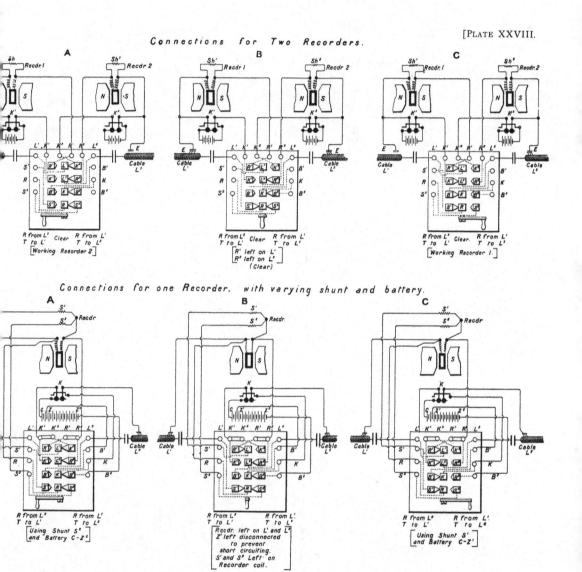

Connections for Two Recorders. [PLATE XXVIII.

Connections for one Recorder, with varying shunt and battery.

FIGS. 92 and 93.—Alternative Arrangements with Raymond-Barker's Manual Translation Switch.

[ *To face p.* 628.

after traversing the coil s, rejoins the first part at *n*. As the resistance of the coil s is at the best equal to only one of the resistance coils in $U_2$, at least half the total quantity of electricity passes in the form of current through the recorder. The two reunited currents now go to terminal No. 1, thence through the contact piece, left-hand spring, and terminal No. 5 to the mirror receiver $G_0$, and so to earth. This mirror receiver $G_0$ is only intended to serve as a stand-by in case of accident to the recorder; under ordinary circumstances it is kept out of circuit by means of a plug inserted between terminals 4 and 5.

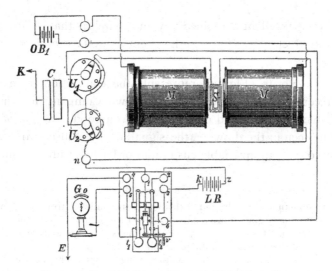

FIG. 95.—Recorder Connections with Dickenson's Key.

On changing to the sending position by turning over the handle *v*, we see that, at the instant when the three springs are simultaneously pressing on the contact piece, the cable is in communication with the earth through the condenser C, and any portion of the charge still remaining in the line can readily escape. Now suppose the knob of the lever $t_2$ be depressed, the negative current from the line battery L B, leaving the pole *z*, passes through terminal No. 2, the upper contact of the lever $t_2$, the upper contact of $t_1$ through terminal No. 4, and so to earth. The positive current, leaving the pole *k*, passes through terminal No. 3, the lower contact of $t_2$, through this lever to the rear right-hand spring, and so to the contact piece against which the two springs on the right are pressing. Here the electricity divides into two portions; one part being conveyed through terminal No. 1 to *n*, where a further division takes place, one portion traversing the coils of the rheostat $U_2$ which are in circuit, and the other part going through the recorder coil s to rejoin its fellow-current at the axis of the rheostat $U_2$; the second part of the principal current on leaving the contact piece, passes through the front right-hand spring to terminal No. 6, whence to the rheostat $U_2$, and finally to the axis of $U_2$. The now reunited current flows with its full strength into the condenser C, immediately preceding the cable. The resistance in circuit of $U_1$ being always very slight, the greater portion of the current traverses this rheostat from the first, so that the current passing through the recorder is of no greater intensity than that received from the distant end.

The circulation of current produced by depressing the other lever $t_1$ is precisely similar but opposite in direction.

O $B_1$ is a local battery which serves to excite the electro-magnets M M.

Another very excellent combined key and switch is that of Messrs Peeling and Davis, which serves for recording sent as well as received messages.

The Plate opposite (Fig. 96) represents the connections at a recorder station as recently furnished by the Silvertown Company for the South American Cable Company, and may be taken as a good example of others.

As may be seen in the drawing, the signalling condensers are arranged in *parallel*, according to invariable custom, in order to secure the sum of the capacities of each.*

---

* Trouble is sometimes experienced with these—especially in tropical climes. The insulation has occasionally gone down from about 800 to about 150 ohms per N.M., due to a rise of temperature from 65° to 75° F. By the same temperature increase, the capacity has also materially increased on occasions.

It would seem as though a warm temperature brings about a chemical action in the wax which alters its specific qualities as regards insulation and inductive resistance.

The Siphon Recorder : Cuttriss Pattern.

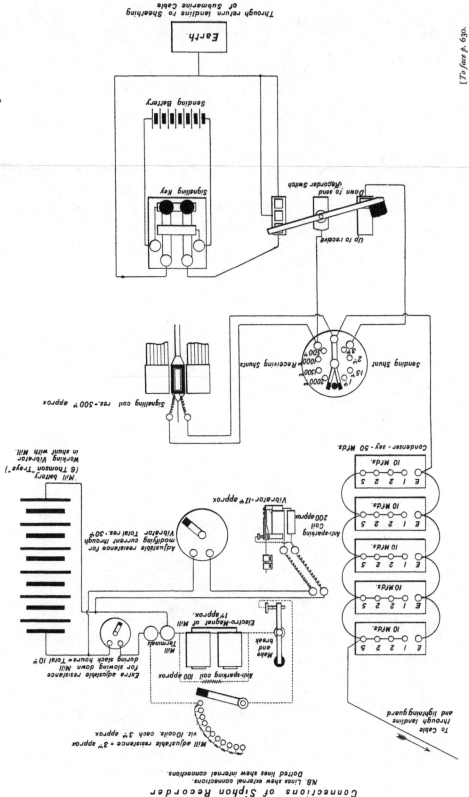

[PLATE XXIX.

FIG. 96.—Siphon Recorder Apparatus installed at the Stations of the South American Cable Company.

Through return landline to Shearthing of Submarine Cable

Earth.

Sending Battery

Signalling Key

Recorder Switch
Down to send
Up to receive

Receiving Shunts
Sending Shunt
500
1000
1500
2000
2°
15°
1°

Signalling coil res.·500ᵂ approx

Condenser · say · 50 Mfds.

10 Mfds.    10 Mfds.    10 Mfds.    10 Mfds.    10 Mfds.
E 1 2 5    5 2 1 E    E 1 2 5    5 2 1 E    E 1 2 5

Mill battery
(6 Thomson "Trays")
Working Vibrator
in shunt with Mill.

Vibrator ·1″ approx

Adjustable resistance for
modifying current through
Vibrator  Total res.·30ᵂ

Anti-sparking
Coil
200 approx

Extra adjustable resistance
for slowing down Mill
during slack hours = Total 10ᵂ

Mill
Terminals

Electro-Magnet of Mill
1″ approx.

Make
and
break

Anti-sparking coil  100 approx

Mill adjustable resistance · 3ᵂ approx
viz. 10 coils, each 3ᵂ approx

Connections of Siphon Recorder
NB. Lines shew external connections.
Dotted lines shew internal connections.

To Cable
Through landline
and lightning guard

## Section 5.—Other Similar Apparatus.

**Lauritzen's Undulator.**—This apparatus, which has certain points of similarity with the siphon recorder, is chiefly used on the submarine lines of the Great Northern Telegraph Company.* It consists principally of four straight electro-magnets placed at the four corners of a square (Figs. 97 and 98), the coils of the magnets being joined up in one continuous circuit. The direction of winding is the same for the coils situated in the same diagonal, and opposite to that of the coils in the other diagonal. The poles, for instance, at the upper ends of the cores, with the same current, will be north for the diagonal N, N, and south for the diagonal S, S.

In the space between the four electro-magnets are four very light steel magnets, $aa'$, $bb'$, $cc'$, $dd'$, joined at their central portions to a rod $mn$, turning between two pivots with the least possible friction. The poles $a$, $c$, $b'd'$ of the permanent magnets are of similar name, and opposite to the

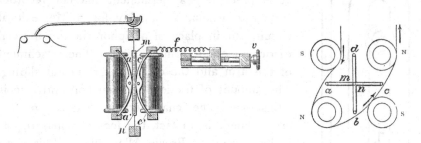

Figs. 97 and 98.—Lauritzen's Undulator.

poles $a'$, $c'$, $b$ and $d$. A spring $f$, whose tension is regulated by a screw $v$, tends to retain the moving parts of the apparatus in a given azimuth.

When a current passes through the coils of the four electro-magnets, the combined action of the cores on the permanent magnets tends to turn the moving parts in a given direction depending on that of the current. Thus, if the rod $mn$ is connected to a bent tube of small diameter, one end of which dips into a reservoir of ink, the other end lightly pressing

---

* Previous to the appearance of Lauritzen's "undulator," this company had been in the habit of employing Wheatstone's receiver for working their cables. This latter is a high-speed Morse instrument of elaborate and highly sensitive form, being much lighter in construction than the ordinary Morse, and having a relay with very delicate tongues. In conjunction with the Wheatstone transmitting machine, this instrument can be worked at an exceedingly high speed. With the above automatic transmitter it is exclusively used in the United Kingdom Post Office system for press telegraph work.

against a strip of paper drawn along by clockwork, the signals traced by the point of the tube on the paper will be similar to the indications of the recorder.

The total resistance of the four exciting coils is about 1,000 ohms. The instrument is so exceedingly sensitive, that readable signals are obtained through a circuit of 30,000 ohms resistance with a single Leclanché cell.

On lines of medium length (350 to 700 miles), where the undulator is chiefly used, it gives a working speed of 400 letters per minute (or about seventy words per minute) simplex.*

**Siemens' Permanent Magnet Relay.**—Within recent years Messrs Siemens Brothers have devised a very excellent modification of the siphon recorder well adapted for working cables up to say 700 N.M. in length of ordinary $\frac{107}{140}$ core.

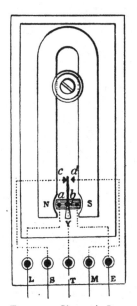

It is practically a permanent magnet recorder turned to account as a relay for a Morse local circuit, but in place of a siphon the moving coil carries a metallic arm (Fig. 99). The attachment of this arm and the coil is not rigid but sliding. The amount of friction between coil and arm is adjustable. The characteristic feature in this arrangement is, in fact, the jockey armature—as is the case in the Brown-Allan relay. This combination, however, permits of a rather higher speed than the latter on a given cable, or the same speed through a somewhat greater length of the same type.

It is used on the short sections of the "Commercial" Company's Atlantic system, and also on some of the cables belonging to the French Compagnie Française des Câbles Télé-graphiques in the West Indies and elsewhere.

FIG. 99.—Siemens' Permanent Magnet Relay.

---

* In connection with this company's system should be mentioned the "dextrineur" of Mr Mygind, a machine employed for pasting down the slip on a handy sheet of paper as fast as it runs from the instrument. It is used at most of the "Great Northern" stations (as well as by the "Anglo" Company), and is especially useful in cases of retransmission, where the telegram is read direct from the slip, and the delay and risk of errors occasioned by the copying of the telegram—previous to retransmission—thus avoided. The receiving

**The Ader Recorder.**—Just as this book goes to the press, an ingenious apparatus has been devised by M. Ader, making use of the principles of photography. Here the coil of the siphon recorder is superseded by a long fine vertical wire through which the current passes. As the wire is drawn from side to side by the poles of the magnet (in the field of which it is stretched) a ray of light from a lamp throws its wavering shadow on a travelling strip of photographic paper—automatically developed *en passant* —in the form of an undulating white line (on a black ground), representing the signals of the message.

This apparatus is claimed to be more sensitive than any preceding instrument, whether siphon recorder or mirror. It has been tried on the French Atlantic cable between Brest and St Pierre, the result being 600 signals per minute instead of 400. It is reported that with this system New York and Paris are shortly to be put into *direct* communication.

It may, of course, turn out that the sinuous line on the sensitised paper is more troublesome to read than the siphon-recorder line, but this is scarcely likely. The absence of the friction between the siphon and the paper, together with the small inertia of the moving parts, are advantages that ought to be brought to account—if only on the score of legibility. Again, its simple construction as compared with the siphon recorder, and the absence of a vibrator (which frequently gives trouble and requires close attention in order to maintain a regular rate of vibration and sensibility), are additional points in favour of M. Ader's instrument.

Altogether, there seems little doubt that the idea of recording messages by photographic means will be largely adopted in the future—at any rate for working very long cables.

Further, it is somewhat surprising that the application of photography to the deflections produced on "dead-beat" reflecting galvanometers and on the mirror "speaker" has not yet been turned to account in practice as a means of avoiding the fatigue experienced by the eye when following the rapid

---

clerk, by means of the "dextrineur," can receive messages faster, moreover, having no loose tape to attend to. Where the "dextrineur" is in use, immediate manual translation (as read from the slip) is of course impossible. This apparatus is partly intended for the purpose of preserving all the slip in each message separately—for possible checking or reference—in a convenient form. Thus, when a message arrives, it is—as fast as it comes —gummed on to a sheet, then written in English words by a clerk, this word copy being delivered to the person it is directed to, the original gummed slip (as received) being at the same time filed, instead of the ordinary word copy and loose slip. In the case of an intermediate station, the clerk translates—*i.e.*, repeats on—from the slip ready gummed on to conveniently formed sheets.

movements of a bright light.*   It is certain that the much smaller movements of a spot of light—such as would sufficiently serve if recorded by
photography—would enable a much higher speed to be attained than with
the comparatively large amplitude required for visual telegraphy.   Indeed,
the extreme sensibility which reflecting galvanometers are capable of
justifies us in anticipating by this system an even higher speed than that
already claimed for the Ader recorder.   With reference to this latter,
again, those who have experienced the difficulties of deciphering recorder
slip, following the trail of a nervous, sinuous, line, will greet the photographic
method cheerfully, as one of great promise.

---

* The application of photography to the purposes of recording mirror signals was
actually suggested by Mr H. G. Cheeseman as much as twenty years ago ; and devices
based on this principle have since been patented several times by various people.

Western and Brazilian Telegraph Company : Landing Place at Rio de Janeiro.

# CHAPTER III.

## DUPLEX TELEGRAPHY.*

SECTION 1.—History—Differential Principle : Wheatstone Bridge Principle—Varley's Artificial Cable : Stearns' Method—De Sauty's Method.

SECTION 2.—Modern Practice : Muirhead's Inductive Resistance : Muirhead and Taylor's Method : "Double-Block"—Practical Examples and Installations : Duplex Directions for Short Circuits : Duplex Directions for Long Circuits—Comparison of the Principal Duplex Systems — Other Methods : Benjamin Smith's : Harwood's : Jacob's : Ailhaud's—Comparison of Varley's and Muirhead's Artificial Line—Similes of Duplex Telegraphy.
Quadruplex and Multiplex Experiments on Cables.

### SECTION 1.—HISTORICAL SKETCH.

DUPLEX telegraphy is constituted by the simultaneous transmission of messages in both directions from each end of the conducting line.

The principle of duplex working is to render the receiving instruments insensible to the currents sent into the cable at their own end, whilst they faithfully record all signals coming from the distant station. Up to the present this problem has been solved mainly in two ways, one or the other of which embrace all the special methods suggested by different inventors. Firstly, the differential system, due, in principle, to Dr Gintl, of Vienna, who in 1853 first shewed in a practical way how it was possible to send two messages in *opposite* directions, on the same wire at the same time. In the following year (1854) this system was perfected by Herr C. Frischen, of Hanover,† and adopted in the same year by Messrs Siemens

---

* An entire chapter is devoted to this subject (besides being referred to somewhat fully in Part I.) on account of its vast importance—*i.e.*, its immense bearing on the earning capacity of a given line. But for the duplex system, as now practised, we should require many thousand more miles of cable to carry our messages here, there, and everywhere.

† In this year (1854) R. S. Newall included in a patent a method of telegraphy such as would be now described as "duplex" telegraphy, and was at that time spoken of as a double-speaking system. It is referred to in Part I. of this book. This claim was "for the arranging or combining of electric telegraph apparatus in such a manner as to render it possible to telegraph simultaneously in opposite directions between two stations, using one line wire and the earth as the means of communication." This formed the first *English* patent for duplex telegraphy. The second English patent was that of the

and Halske, partly in virtue of a somewhat similar patent of Dr Werner
Siemens. Secondly, the Wheatstone bridge system, the idea of which was
vaguely put forward in 1858 by Mr Farmer, an American, and clearly
demonstrated in 1863 by M. Maron, of Berlin,* who, however, never
brought the matter to a practical conclusion.†

The differential system entails the use of receivers containing two coils,
the wires of which are wound in opposite directions. Dr Gintl's method
was to join up one of these coils to line, and the other to a suitably adjusted
rheostat. Using a double contact key, he was able to send currents from
two distinct batteries, through the two circuits simultaneously, the effects of
the two currents on the receiver at the sending end mutually cancelling
one another.

Messrs Frischen and Siemens‡ also wound the two coils of their
receivers opposite ways, but they made them of exactly equal resistance,
and placed them under identical conditions with regard to their influence
on the magnetised needle. The two coils were joined up at one end to the
same battery P (Fig. 100), the other end of one coil being connected to line,
and the further end of the second coil to a resistance R, equal to that of the
line. The battery current from P thus divides into two equal parts whose
effects on the needle of the receiver G mutually cancel each other. The
current from the distant station, on the contrary, goes to earth either by
passing successively, and in the same direction, through the two coil circuits,

---

brothers Bright (referred to elsewhere), and the third that of Mr W. H. Preece (1855),
this latter being a differential system. Mr Preece has always been closely identified with
duplex telegraphy : in 1879 he gave a series of Society of Arts Cantor Lectures, which
included one on this subject. This formed the most complete account of duplex telegraphy in existence, put in such a way—as is his wont—that the uninitiated could readily
follow him. See *Jour. Soc. Arts*, vol. xxvii., p. 991.

* M. Maron used a doubly wound coil with two separate batteries and a double-
pointed key.

† Besides the differential and Wheatstone bridge systems of duplex telegraphy, there
was the patent of 1855 (No. 2,103) taken out by Sir Charles Bright and Mr Edward
Bright, which included a system of what may be termed "simple circuit duplex," whereby
cross-communication may be maintained when the apparatus is ready for operation in
either simultaneously. This method involved a system of relays throwing into circuit
Bright's acoustic telegraph bell instruments or "phonetic apparatus" as it was then
described. It was employed successfully on the "Magnetic" Company's lines between
London and Liverpool until the time when the gradual failing of all underground
lines in this country would not any longer permit of duplex working. On the extensive
aerial line system being established, it was found that duplex telegraphy was no longer
necessary to meet the then prevailing traffic.

‡ Messrs Siemens and Halske had united with Mr Frischen over their respective duplex
patents so as to work them jointly. They established a duplex circuit from Hanover to
Göttingen in the year of their patent (1854), and between then and 1856 erected duplex
apparatus at some ninety telegraph stations.

and thence through the resistance R, when the key D is up ; or else, when D is down—by far the greater part of the current at all events—passes through one coil only and the battery P. Its action on the magnetised needle is thus sensibly similar in both cases, and the receiver works precisely the same whether its own station is sending or not. The differential

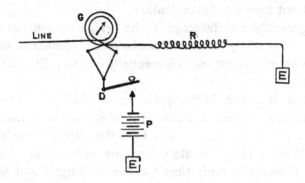

FIG. 100.—Duplex Telegraphy on the Differential Principle.

system of duplex telegraphy may be likened to a "tug of war." When both sides are equally strong in the latter, no matter what power is used to pull one way, if an equal power be exerted to pull the other, there will be a neutral result, or zero indication. So it is in a "differential" instrument,

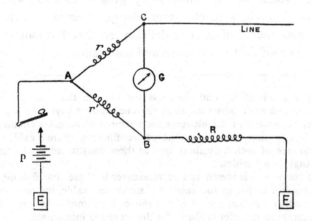

FIG. 101.—Duplex Telegraphy on the Wheatstone Bridge Principle.

whether it be a galvanometer or a relay, the object of which is to divide the current equally but oppositely, and thereby to produce no effect, however strong the currents sent through it may be.

The rheostat and other means of final adjustments are employed for balancing, in this as in the bridge system about to be described.

In the Wheatstone bridge system (Fig. 101),* the line forms the fourth arm of a bridge whose distant angle is to earth. The resistances R, r, and r' being adjusted to give a balance, the points B and C are at the same potential, and the current from the battery P cannot traverse the diagonal B C. It is evident that if the receiver G is placed in this diagonal, it will not be affected by the current of the sending battery, but will be able to reproduce all signals sent out from the distant station.

Speaking generally, an advantage in the bridge system lies in the fact that *any* kind of signalling instrument can be used; whereas in the "differential" system an instrument with differentially wound coils is, of necessity, involved.

The practical difficulties in the application of either one or the other of the above methods are caused by a sharp throw, of short duration, communicated to the needle of the receiver each time the circuit of the battery P is broken and closed. The alternate deflections, opposite in direction, are due to a flow of electricity every time the line is charged and discharged. They have been practically entirely eliminated so far as regards the duplexing of overhead wires by the special application of condensers to a duplex circuit as first instituted by Mr J. B. Stearns, of America. By this means the flow of electricity due to the line charging and discharging is exactly balanced by the opposing rush caused by the charge entering and leaving the condenser.†

Submarine cables, having considerably greater electro-static capacity than aerial lines,‡ retain a much larger charge; besides which, they are worked with instruments of greater delicacy, so that the balance in their case is much more difficult to establish and maintain.

---

* For the sake of simplicity and clearness the keys in the first few (theoretical) diagrams of this chapter are represented as single-lever contact keys. In actual practice, the signalling key is, however, a double-lever arrangement on all cable circuits unless the Morse system is in operation—which is probably never the case where a cable is duplexed, except in the instance of such busy short lines as those constituted by the Anglo-Continental and the Anglo-Irish cables.

Similarly the receiver G is shewn as a galvanometer in these first few diagrams. It need hardly be remarked that in the case of a submarine cable, the receiver would be almost invariably either a siphon recorder or a mirror instrument—as a rule the former. Thus, in later diagrams a recorder is shewn for the receiving instrument.

† Mr Stearns' form of artificial cable—embodied in his patent for duplex telegraphy of 1872—was, in actual fact, precisely the same as Varley's plan of representing a cable by alternate resistance coils and condenser (tinfoil and paraffin paper) for the purposes of reproduction at the sending end in order to obtain an idea of what the signals would be like at the receiving end under given conditions as regards battery power, etc.

‡ Indeed, owing to the distance from the ground at which aerial land telegraph wires are usually erected, overhead lines have, practically speaking, no electro-static capacity excepting in very wet weather.

Thus, though a considerable number of land lines had been duplexed,* it was not until after Mr Stearns conceived the plan (by his patent of 1872) of inserting condensers in the artificial line, so as to represent the capacity of the cable—in addition to the resistance by resistance coils, as in land-line duplexing—that any material length of cable was duplexed.

In duplexing the short sections of one of the "Anglo" Atlantic cables in 1873, and the main section some time afterwards (in 1878), Mr Stearns adopted the differential system, as previously adopted by the P.O. Telegraph Department.† This was effected by the siphon recorder being furnished with two separate differentially wound coils, each wound doubly according to anti-self-induction principles. Practically all the other submarine cables which have been duplexed—almost entirely by Messrs Muirhead—have been duplexed on a Wheatstone bridge basis.

As stated before, Mr Stearns' arrangement was in effect a precise reproduction of a plan of Mr C. F. Varley (patented in 1862)‡ for repro-

FIG. 102.—Varley's Artificial Line.

ducing at any station what occurs in the cable in question—at any rate, as regards the manner of building up an artificial line by what is termed a

---

* In America they had worked their land lines on the duplex system, without any sort of condensers, as early as 1867, to be followed by the European Continent. Germany was, however, undoubtedly the first country to employ any form of duplex telegraphy, the differential system of Frischen and Siemens being adopted the very year—1854—in which it was invented, from Hanover to Göttingen, and gradually on a large scale. The duplex telegraphy was not taken up till several years later by the United Kingdom. This was gradually after the purchase of the land telegraphs by the State in 1870. Previously the "Electric" Company had employed the Wheatstone automatic system for press work ever since it came out in 1864; and they did not require to work their lines by a duplex method, with the traffic of that time, for any other circuits.

† When the duplex system was taken up by the Postal Telegraph authorities, the plan adopted throughout (as now) was the differential method, on account of its simplicity, and owing to the fact of a form of differential instrument—invented by Cromwell Varley about 1850—being in very general use in the United Kingdom Government Telegraphs. The method of Mr Eden was the first employed in England. However, no efficient or continuous duplexing even of land lines occurred until Mr Stearns made known his method, in 1872, for the application of condensers to the compensating circuit to represent the capacity of the line—small though it be in the case of aerial wires.

‡ See specification No. 3,453 of that year—a "master patent"—the second claim being for "employing a 'test circuit' formed by 'induction plates' and resistance, so adjusted to each other as to produce an artificial line, possessing the same amount of retardation as the cable itself."

"step by step" device,* as shewn here in Fig. 102, and again in Fig. 103. In Mr Stearns' arrangements, the two circuits of the recorder coil had each a resistance of about 250 ohms. The artificial line which balanced the cable (Fig. 104) consists of a metal riband having very considerable specific resistance wound round another thick cord and afterwards covered with the thinnest possible layer of an insulating substance possessing great

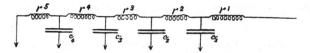

FIG. 103.—Stearns' Artificial Lines.

inductive capacity. The balance is regulated by varying the resistances $r$ and $r'$, a resistance being inserted between the condenser C and the cable, in case of need. The strength of the arrival current through the receiver can be diminished by connecting terminals 1 and 2 with a shunt.

In duplex telegraphy by differential systems, the battery forms part of the circuit when the key is down and a signal is being sent; but as soon as

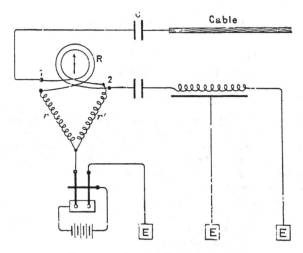

FIG. 104.—Stearns' Duplex Method.

the signal ceases, the battery is no longer a portion of the circuit. All batteries—even those used for duplex cable signalling—possess *some*, though very little, internal resistance. This item is sometimes provided

* Mr C. F. Varley's patented artificial cable was beautifully described by him in the Royal Institution lecture of 1867, previously referred to. It is, however, a question whether the views there stated did not originate with his brother, Mr S. A. Varley.

for by inserting between the key and the earth a resistance coil equal to the internal resistance of the battery, so that whether the key be in use or not the total resistance shall remain unchanged. The current must, in fact, go

> "To earth through batt'ry when the sending key's depress'd,
> And thro' equivalent resistance when at rest."

This was actually effected by Mr·Stearns in applying his system in the early days. It would, however never now be put into practice, as regards submarine cables—with the double-block Wheatstone bridge system—for the battery resistance, besides being exceedingly low, does not come in here.

Mr Stearns met with very considerable difficulty in obtaining anything like a permanent balance, according to the principles of duplex telegraphy, on the short length he first experimented with, and still more on the main section.* This difficulty was no doubt due to the large number of points of contact (and corresponding terminals) between the condenser and the resistance coils involved in Varley's artificial cable, as adopted by Stearns, thus introducing the liability of variation to a large extent, though— in theory—the finer the subdivisions,† the more complete the repro- duction of the state of affairs in a cable.‡ In point of fact, a really accurate balance could not possibly be maintained by this method, the result being that the working speed was considerably reduced each way, in order that the disturbed signals on an imperfectly balanced line might be understood, and the number of words carried each, way by the cable on 'duplex" was nothing like the theoretical double number of that on "simplex." §

---

* No doubt the only reason that an electrically long cable is often found to be more difficult to accurately "balance" than a short one is that in a long cable a more highly sensitive signalling instrument (siphon recorder, or mirror) is employed than on quite a short line where the Morse or some such instrument may be in vogue. Thus, in the first case it is necessary that a more complete and absolute balance be maintained (as well as attained), the instrument in question being actuated by slight *alterations* of the current only. But for the above fact, theoretically speaking, the longer the cable the more easy the balance, owing to the increased retardation of the line more completely deadening any defects in this respect.

† In order to obtain the maximum capacity effect from a number of condensers, regard must always be had for the fact that the sum of the capacities of each is only secured when they are joined up in parallel.

‡ This has, indeed, actually been urged as an advantage for this species of "artificial" over that of Muirhead, where the resistance and capacity are formed in one. It must, however, be remembered that the latter offers extremely fine additional methods of adjustment : thus, any such claim appears to be scarcely borne out in practice.

§ In 1881, in the course of a Joint Report to the Submarine Telegraph Company on Stearns' Duplex Method, Mr C. F. Varley and Professor W. E. Ayrton pronounced it to be, practically speaking, unworkable. It should be remarked, however, that at that time the system formed the subject of a lawsuit—Stearns *v.* Submarine Telegraph Company.

The late Mr C. V. de Sauty is not unusually accredited with having accomplished the first duplexing of any material length of cable. This was in 1873, on the submarine line between Lisbon and Gibraltar, 365 N.M. Mr De Sauty employed the bridge method, using, like Stearns, the Varley artificial cable (the only form then known), as shewn in Fig. 105—*i.e.*, a series of resistances shunted by condensers at alternating points, thus constituting an imitation line having both resistance and capacity, the product of which should be the same in the "artificial" as in the real cable to be balanced.* This first piece of duplexing by Mr De Sauty, though of much interest and utility experimentally, was not looked upon as a complete practical success in a permanent working sense. Indeed, an artificial line of this nature, if it

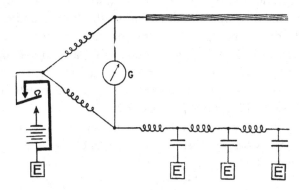

FIG. 105.—De Sauty's Method.

is to be an absolutely perfect equivalent to the cable, would require a positively unlimited number of subdivisions. This condition was fully realised by Messrs Muirhead and by Mr Herbert Taylor. Thus, in 1875 the difficulty was overcome in a most ingenious manner by a joint patent (No. 684 of that year) standing in the names of Herbert Arnaud Taylor and Alexander Muirhead. This was for an entirely novel form of artificial line, combining throughout, in one, the function of resistance and capacity—as in a cable—instead of separately and alternately.† The result of this device

---

* A full description of Mr De Sauty's duplexing arrangements on this cable will be found in vol. ii., p. 138, of the *Journal of the Society of Telegraph Engineers*, 1873.

† As a matter of fact, in the above patent, plumbago was specified (by Mr Taylor, the originator) for the conductor and plates ; whereas tinfoil, previously and independently tried by Mr John Muirhead, has since been adopted in universal practice. Dr A. Muirhead first devised the gridiron form for the conductor strip in place of a continuous length, and in substitution for ordinary plates condensers. The late Mr Robert Sabine is also said to have been responsible for some gridiron type of condenser.

was that duplex cable working from that time became possible in practice ; and the system has since been applied successively and successfully to almost all the cables now in operation at the bottom of the sea, thereby increasing their working capacity, in the present day, by some 90 per cent.— *i.e.*, very nearly doubling it. The duplex system is, in fact, equally effectual when applied to cables as in its application to aerial wires.

## SECTION 2.—MODERN PRACTICE.

We will now proceed to describe the salient features of duplexing submarine lines as now more commonly practised.

**Muirhead's Method.**—The artificial line which Messrs Muirhead and Taylor suitably termed "inductive resistance" (as combining, in one, resistance and capacity) consists of alternate sheets of paraffined paper and tinfoil. First comes a sheet of paraffined paper over which is placed a sheet of tinfoil cut so as to give a number of long narrow zigzag strips (Fig. 106). The tinfoil is then covered with a second sheet of paraffined paper over which is a complete sheet of tinfoil, which is in turn covered with insulating substance, then a second sheet of tinfoil cut out as before, and so on. The strips of tinfoil are joined up together, so that the electric current can circulate through them from end to end : they form, in fact, the conductor of the artificial line, and correspondingly represent the conductor of the cable.

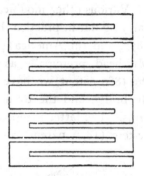

FIG. 106. — Muirhead's Inductive Resistance.

The complete sheets of tinfoil are also connected together, and are in communication with the earth : they represent the sheathing of the cable of which the sheets of paraffined paper form the dielectric. By cutting the sheets of tinfoil into strips of varying width, the resistance of the line can be sensibly adjusted to its capacity per unit of length in each particular case.

Fig. 107 serves to more clearly shew the theory of Muirhead's artificial line above described, and also to illustrate the system of elements of the series, a certain number of which are connected up to form what are termed "parts" or "sections," in which all the exterior complete sheets of tinfoil are joined together in such a way as to present practically no resistance. Perhaps, however, Fig. 108 makes the principle of this artificial line still more explicit, as it preserves the gridiron—zigzag—form of the conductor (tinfoil) plate, with the outer, earth, tinfoil plate surrounding it, the space between them forming the dielectric.

According to the instructions in the patent, the cut-out sheets of tinfoil may be occasionally replaced by strips of blotting paper, in the pulp of which 50 per cent. of its own weight of finely powdered black-lead has been incorporated. The tinfoil, however, gives the best results, owing to the difficulty of getting good electrical contacts with black-lead (plumbago) paper.

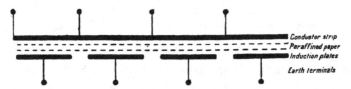

FIG. 107.—Muirhead's Artificial Cable.

Various other modifications and improvements as regards methods of carrying out in practice the duplexing of cables were later—in 1875, 1876, 1877, and (in America) 1888—patented by Messrs John and Alexander Muirhead, principally as regards means of final adjustment. Thus, in the 1880 patent, we have a rheostat of very low resistance at the apex of the bridge, and the total abolition of high resistance in the proportional arms. The latter is replaced here by what is known as the "double-block" system (1876 patent) as constituted by condensers in each proportional

FIG. 108.

part. However, the main feature of the Muirhead system (which rendered the duplexing of long cables really practicable) was the novel form of artificial cable of the original 1875 patent as above described.

Again, Messrs A. and J. Muirhead have constructed artificial lines with silk or cotton covered copper wire plunged into molten paraffin and afterwards covered with strips of gilded copper wound on spirally. The central copper wire—the size of which is calculated according to the required resistance—represents the conductor of the artificial line ; the outer sheathing is in communication with the earth. This kind of inductive resistance is better suited as an artificial (balancing) representation of aerial wires than tor cable work.

As a general rule the product of the total resistance R into the total capacity K of the artificial line should be equal to the product of the cable resistance R′ into its capacity K′,

$$RK = R'K'$$

Where the proportional arms of the bridge are equal, Dr Muirhead takes

$$R = R'$$

and consequently

$$K = K'$$

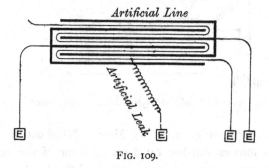

FIG. 109.

Inasmuch as no artificial line can ever exactly reproduce the conditions existing in a submarine cable—the several sections of which may be of varying construction, and laid at very unequal depths in water differing considerably in temperature—Messrs Muirhead found it necessary to add certain arrangements, by means of which the first portions of the artificial

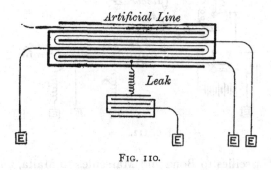

FIG. 110.

lines can be more accurately balanced against the first few miles of cable to be duplexed. Resistance coils are inserted, on the one hand, between the receiver and the lines, real and artificial: supplementary condensers, on the other hand, being introduced to increase the capacity, either of the receiver itself, or of those portions of the circuit with which it is in immediate contact.

When a cable is faulty, Messrs Muirhead insert in the artificial line escape circuits or "leaks," through adjustable resistances (Fig. 109). Sometimes, these "leaks," instead of going directly to earth, terminate in condensers of small capacity (Fig. 110).

**Practical Installations and Examples**.—Fig. 111 shews the theoretical

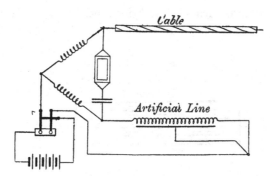

FIG. 111.—Muirhead and Taylor's First Duplex System.

arrangement of connections adopted by Messrs Muirhead and Taylor in the duplex installations established by them on some of the early cables of the Eastern Telegraph Company, and on that of the Direct United States Cable Company.

The lines through which trials of this method were first made were the

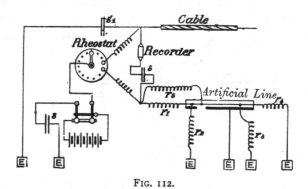

FIG. 112.

two cables from Marseilles to Bona and Marseilles to Malta, *via* Bona ; the former in 1875, and the latter in 1876. We have for these two cables :—

|                          | Marseilles-Bona Section. |   | Marseilles-Malta Section. |
|--------------------------|:------------------------:|---|:-------------------------:|
| Length in N.M.           | 447.66                   | - | 848                       |
| Copper resistance        | 5,210 ω                  |   | 9,632 ω                   |
| Electro-static capacity  | 1²8.7 φ                  | - | 238.24 φ                  |
| Insulation per N.M.      | 3,124 Ω                  | - | 2,113 Ω                   |

The arrangement of apparatus and connections was as shewn in Fig. 112. In the duplex installation of these cables Messrs Muirhead and Taylor, balancing the cable with their form of artificial line on the Wheatstone bridge system, placed a condenser in the bridge containing the receiving instrument, with a view to blocking off any possible earth currents. Balance was generally obtained under the following conditions:—

| CABLES. | PROPORTIONAL ARMS. | | Capacity of the Receiving Condenser. | ARTIFICIAL LINE. | |
| --- | --- | --- | --- | --- | --- |
| | Upper. | Lower. | | Resistance. | Capacity. |
| Marseilles-Bona | 1,000 ω | 1,000 ω | 40 φ | 5,035 ω | 97.4 φ |
| Marseilles-Malta = | 2,000 | 2,025 | 40 | 17,000 to 18,000 | 190 to 230 |

On the Marseilles-Bona cable, 8 Leclanché cells were used for signalling when successfully balanced by duplex apparatus, equivalent to 11.8 Daniell cells; and on the cable from Marseilles to Malta, 22 Leclanché cells, equivalent to 32.6 Daniell cells.

The working speed through both these lines with the recorder and skilled operators was, almost at the outset of duplex work being effected, as much as twenty-five English words per minute each way.

At Aden, on the Aden-Bombay section, 1,817 miles in length, balance was obtained by making $s$ (see Fig. 112) = 1.23 microfarad, $r_1 = 0$, $r_2 = 0$, $r_3 = 210,000$ ohms (this last resistance being applied at a point 250 miles along the artificial line counting from the bottom), $r_4 = \infty$, and $r_5 = \infty$. The total resistance of the artificial line was 11,827 ohms (about three-quarters of that of the real cable), its capacity 656 microfarads; the resistances in the proportional arms were 2,000 and 3,000 ohms respectively, the first resistance being close to the cable, and the second one near the artificial line. At the Bombay end, $\rho_1 = 2,005$ ohms, $\rho_2 = 2,035$, $r_1 = 60$, $r_2 = 120$, $r_3 = \infty$, $r_4 = 175,000$, and $r_5 = \infty$.

The cable from Ballinskelligs Bay (Ireland) to Torbay (Nova Scotia), belonging to the Direct United States Cable Company, was the first ocean cable to which the system of duplex working was applied. This was established with complete success by Messrs Muirhead and Taylor in 1878.*

---

* As already stated, one of the short sections of the "Anglo" Company's cables had been experimentally duplexed by Mr J. B. Stearns in 1873; but not sufficient success was met with to warrant a long trans-Atlantic section of this company's cable system being "duplexed" till five years later.

The length of the cable is 2,423 N.M., its conductor resistance 7,315 Siemens units,* and its capacity 987.6 microfarads.

At Ballinskelligs balance was obtained by making $r_1 = 22$, $r_2 = 1,400$, $r_4 = 1,600$ S.U. ( = Siemens units), $r_5 = \infty$, $s_1 = 0.06\phi$, $s = 60$ microfarads. The proportional arms were $\rho_1 = 2,000$ ohms, $\rho_2 = 2,036$ ohms, the higher resistance being on the side of the artificial line.

At Torbay, the balance corresponds to $r_1 = 0$, $r_2 = 50$, $r_3 = 5,000$ at a distance representing 1,600 N.M. of the real line, $r_4 = \infty$, $r_5 = 90,000$ S.U., $s_1 = 4.37\phi$, $s = 60$ microfarads, $\rho_1 = 2,000$ ohms, $\rho_2 = 2,010$ ohms.†

The ratio of the capacities of each artificial line at Ballinskelligs and Torbay to that of the cable was $\frac{4}{5}$. The working speed through this line was

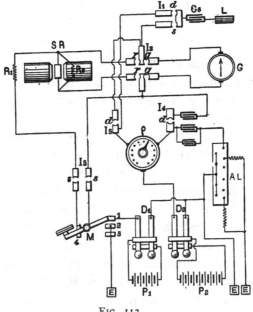

FIG. 113.

found to be about 100 letters per minute each way, or very nearly twice the speed with simplex working. The receiver was a mirror galvanometer. The adjustment of the duplex is more difficult with this instrument than with the recorder for two reasons. In the first place, the mirror being more sensitive than the recorder, and the spot always moving along the scale in a straight line, the lateral displacement of the zero may easily cause confusion in the signals ; in the recorder any displacement of zero takes place

* The Siemens unit is equivalent to 0.9434 legal ohm.
† A. Muirhead, *The Telegraphic Journal and Electrical Review*, 1879.

transversely to the paper, and the legibility of the signals is but slightly affected. Secondly, as the mirror signals are only transient, it is necessary for them to be more clearly defined than those of the recorder, which can be deciphered at leisure. The mirror instrument has since been replaced by the siphon recorder for working this cable, the duplex system, as originally applied, yielding—for the above reasons—still more satisfactory results.

The practical duplex installation at the majority of the stations of the Eastern Telegraph Company is shewn in Fig. 113, where A L represents the Muirhead artificial line, S R the siphon recorder, and $\rho$ a slide resistance box. A simple system of commutators enables the changes to be made from simplex to duplex working, or from recorder to mirror receiving, and *vice versâ*. To receive on the recorder, the holes marked *r* in the commutator $I_2$ are plugged, and those marked *g* unplugged. The reverse operation is performed in order to receive on the mirror instrument.

During simplex working, the holes marked *s* in commutators $I_1$, $I_3$, $I_4$, $I_5$

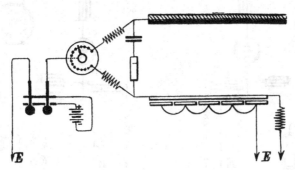

FIG. 114.—Muirhead's Duplex System, shewing "Artificial" Connections.

are plugged, and those marked *d* left open. The key $D_1$ is used for sending, the lever of the reversing switch M connecting contacts 1 and 4. For instance, the left-hand knob of D, being depressed, the positive current goes straight to earth. The negative current, passing through contact No. 1, divides into two portions at the lever M. One part, passing through $I_3$, $I_2$, the recorder S R and its shunt $R_2$, $I_2$, and $I_1$, arrives at the condenser $C_2$; the other part passes through $I_3$, $R_1$, and rejoins the first portion.

For duplex working the holes *s* are unplugged, and those marked *d* plugged at $I_1$, $I_3$, $I_4$, $I_5$. The key $D_2$ is in communication with a battery $P_2$ of somewhat greater power than $P_1$, and is employed for sending. The switch M is not then in use.

The manner in which the principle of the Muirhead and Taylor artificial line is carried out in practice will be readily followed from Fig. 114, shewing the working connections.

**Muirhead's " Double-Block."**—What is now known as the " double-block " system of duplex formed the salient feature in the 1876 patent of Messrs J. and A. Muirhead. This is constituted (Fig. 115) by having condensers in each proportional arm, for the double purpose of warding or " blocking " off earth currents and—in place of high resistance—acting as a reliable " block " to the current so as to ensure a sufficiency passing (across the bridge) through the recorder when signals are to be received. Messrs Muirhead and Taylor at first so inserted condensers in the bridge arms in conjunction with a resistance of materially less value than they had previously been in the custom of using. Now, however, Dr Muirhead almost invariably adopts what may be termed double-block " pure and simple "—*i.e.*, condensers alone in the proportional parts without any resistance there whatever—thus doing away with all retardation in the

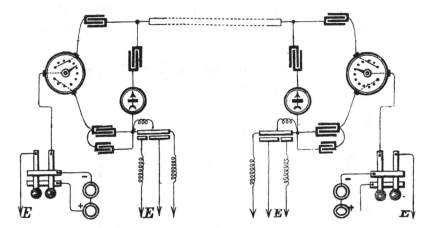

FIG. 115.—Muirhead's " Double-Block " Duplex.

two branches which previously was present on account of the dual presence of capacity and resistance.

The main point of novelty in the Muirhead 1880 (American) patent was the introduction of an adjustable low resistance box (in the convenient form of a rheostat) at the apex of the bridge arms. The introduction here of this little instrument, besides forming the main bone of contention in a patent case involving some £45,000,* has also proved of great use for purposes of final and accurate adjustment. Each coil in the resistance represents a small fraction of an ohm, so that it will be obvious that not only must it be capable of offering facilities for very fine adjustments, but

---

* The cardinal points of this case were very fully gone into by the author in the course of an article in *Engineering* of 28th December 1894.

[PLATE XXV.]

PLATE XXX.]

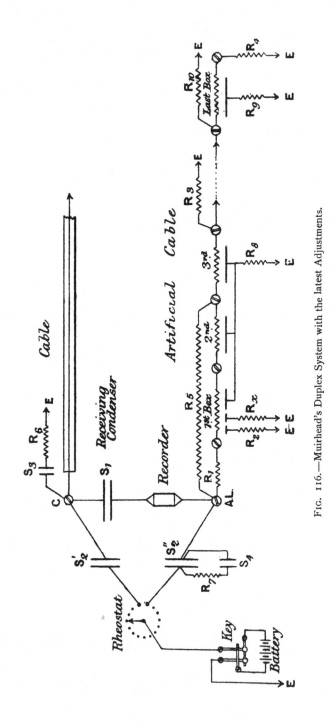

FIG. 116.—Muirhead's Duplex System with the latest Adjustments.

also that it does not do so at the expense of the working speed, inasmuch as being of such a low value it cannot be said to introduce any fresh resistance. Besides this instrument there are several other nice adjustments which may be finally inserted when necessary. These take the form of resistance (in shunt and otherwise), and also that of capacity, shunted or not, as the case may be. Thus, the discharge from a condenser can be retarded by resistance coils used in connection with small subdivided condensers. It may also be retarded by a coil of wire wound round a soft iron core placed between one side of the condenser and the earth, by giving the discharge current some work to do in magnetising the soft iron core. The retardation so established may be regulated by the length, or diameter, of the core. The first of these methods is usually the plan adopted for final adjustments; and several of such adjustments are included in the illustrations here given of "up-to-date" duplex installations.* Fig. 116 gives a

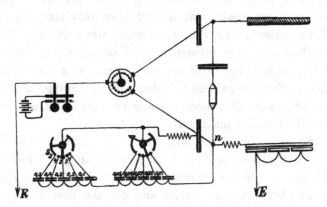

FIG. 117.—Muirhead's Modern Duplexing Apparatus, with Slide Condensers.

fair diagrammatic idea of the duplex arrangements of the present day (for ocean cables) on the Muirhead complete double-block system with all final adjustments, shewing the various boxes of Muirhead artificial line.

Again, Fig. 117 also illustrates a modern duplex installation (absolute double-block with Muirhead "artificial"), and clearly shews the manner of adjustable slide condensers of fractional values as used nowadays for final capacity corrections. In the present day the effect of the application of

---

* The other means of prolonging the discharge from a condenser consists (1) of an apparatus made up of a bobbin of silk-covered wire wound with a sheet of tinfoil between each layer of wire, and allowing the discharge to traverse the wire, while the tinfoil sheets are connected together and joined to the earth ; or (2) by arranging one set of the sheets of a condenser in the form of a long riband, possessing resistance. The first of these (1) constitutes, in fact, Varley's artificial line as employed by Stearns for his main compensating circuit, and the second (2) that of Muirhead.

duplex apparatus by a Wheatstone bridge arrangement on the double-block system with Muirhead inductive-resistance artificial line is to increase the working, and therefore earning, capacity of the cable by over 90 per cent.—*i.e.*, very nearly doubling it—where before by the previous method of duplexing in vogue a 30 per cent. increase of speed over simplex was the most that could be secured.

The plate on the opposite page (Fig. 118) gives an idea of the state of things in actual practice at the stations at each end of a cable worked by duplex, on the Muirhead system.

**Duplex Directions for Short Circuits.**—On relay Morse circuits, by Muirhead's Wheatstone bridge duplex, the easiest method of perfecting the balance is as follows :—Insert a mirror or other sensitive galvanometer in place of the relay; get a rough balance for dots or short contacts (made preferably with a cable or reversing key) when the duplex connections are on at the opposite station, and when that station's switch is at "send." Then, after noting the "false zero" when the key is at rest, hold down the key and plug resistance in or out of the resistance coils at the end of the artificial line until the same zero, as at first observed, is obtained. Then perfect the balance for "dots" by "pumping" the keys and altering the rheostat and adjustment (1) and perhaps (2) as well, until the spot remains motionless—or practically so.

After making any change of adjustments (1) and (2) always alter the rheostat until the best possible balance is obtained. Thus, suppose the balance cannot be obtained satisfactorily with the rheostat only, plug in or out capacity by .01 mf. at a time, and get the best effect with the rheostat. If this is not efficacious, increase or diminish the shunt by 1,000 ohms, and vary the rheostat and condenser as before.

On cables over 300 miles in length further adjustments are very often required, when the recorder or mirror is the instrument used, and are shewn in Fig. 116, viz., (3) a small set of resistance coils (one amounting to 100 ohms in the aggregate is sufficient) is inserted between the first line terminal of the artificial line and the terminal marked A L of the bridge. On cables over 300 miles in length it is not recommended to leave the bridge in circuit permanently, but only to employ it to facilitate the preliminary balancing of the cable, two sets of condensers being substituted for the two arms of the bridge, as in Fig. 116, and explained below. Other adjustments are sometimes found necessary to perfect the balance on long circuits, the principal of which are the following :—(4) and (5) Two sets of resistance coils, of higher resistance than the last (3), inserted respectively between the first and second capacity terminals of the artificial cable and the earth ;

[PLATE XXXI.

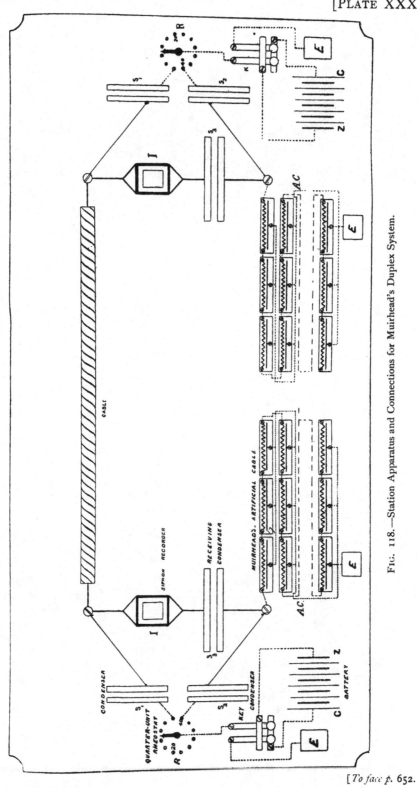

Fig. 118.—Station Apparatus and Connections for Muirhead's Duplex System.

[To face p. 652.

(6) one or more leakage circuits, coils of wire of from 5,000 to 100,000 ohms resistance, inserted between certain of the line terminals at the far end of the artificial cable and the earth. Fig. 116 shews all these adjustments in position, but it is seldom that more than four of them are required.

The balances at Penzance station of the Western Union Telegraph Company, Lisbon and Pernambuco stations of the Brazilian Submarine Telegraph Company, are given below as typical cases :—

### PENZANCE, CORNWALL.

| | |
|---|---|
| $S_1 = 80$ microfarads. | $R_1 = 25$ ohms. |
| $S_2' = 80$ ,, | $R_2 = 30$ ,, |
| $S_2'' = 80$ ,. | $R_3 = 100,000$ ohms on ninth box. |
| $S_3 = 4.402$ ,, | $R_5 = 22,000$ on first five boxes. |
| $S_4 = .504$ ,, | |

### LISBON (No. 2 LISBON-MADEIRA CABLE).

| | |
|---|---|
| $S_2' = 40$ microfarads. | $R_1 = 11.5$ ohms. |
| $S_2'' = 38$ ,, | $R_9 = 2,000$ ,, |
| $S_4 = 3$ ,, | |

### PERNAMBUCO (St VINCENT CABLE).

| | |
|---|---|
| $S_1 = 70$ microfarads. | $R_3 = 180,000$ ohms between ninth and tenth box. |
| $S_2' = 70$ ,, | $R_7 = 640$ ohms. |
| $S_2'' = 66$ ,, | $R_8 = 15$ ,, |
| $S_4 = 4.67$ ,, | $R_{10} = 3,200$ ,, between nineteenth and twentieth box. |
| $R_1 = 18.8$ ohms. | |
| $R_2 = 600$ ,, | |

**Duplex Directions for Long Circuits.**—In all cases of long cables the readiest method of obtaining balance will be found to be as follows :— Starting with the adjustable condenser $S_3$ and the whole of the shunt box $R_5$ to the first four boxes in circuit, keeping the knob of the rheostat in one hand, send a rather rapid succession of "dots" by means of the key, and turn the knob backwards and forwards until the best balance procurable is obtained. If a satisfactory result is not so obtained, insert 05 mf. in $S_3$, increase or diminish it by .01 or .02 mf. at a time, after each alteration obtaining the best balance possible by means of the rheostat. Notice, of course, whether such alteration of the condenser improves matters or the reverse, and go on in the direction of improvement till the most suitable capacity is found. Should the last adjustment not be effectual, alter the resistance of the shunt by 5,000 or 10,000 ohms, and repeat the adjustment with the subdivided condenser and the rheostat as before. When getting the balance near, alter the shunt by smaller amounts, say 1,000 at a time. Should there still remain a "jar" of the mirror, or blurred line on the recorder slip, while signalling on one key, which is unaffected by any variation of the condenser $S_3$ and the shunt $R_5$, then insert the resistance box $R_2$ between the first capacity or earth terminal of the first box of the artificial line and the earth. If on sending reversals a sharp wave is produced, which cannot be eliminated by means of any of these adjustments, a leakage circuit $R_3$ will have to be applied at some point of the far end of the artificial line, the position to be found by trials. The condenser $S_3$ can be dispensed with in a great many cases by inserting resistance $R_1$, between the beginning of the artificial line and the bridge Sometimes on very long cables another set of resistance $R_x$ is required, either alone or in conjunction with $R_2$, placed within the second earth or capacity terminal of the artificial cable and the earth, in order to eliminate the "jar." From day to day it will be found necessary to alter the shunt $R_5$, or resistances $R_1$ and $R_x$, according to the variation of temperature of the artificial line and the underground or land line lead to the cable.

On long cables, after obtaining the nearest balance by means of the bridge and the various adjustments above described, the condenser $S_2$ is removed, and two sets of condensers $S'_2$ and $S''_2$ are inserted in place of the arms of the bridge, one between the terminal $a$ of the rheostat and C, and the other between the terminal $b$ and A L. To adjust these two sets of condensers, it is necessary to have a finely subdivided condenser $S_4$ in connection with $S''_2$, and the mode of proceeding in order to re-establish the balance is the same as above described, the condenser $S_4$ taking the place of the condenser $S_3$. The subdivided condenser $S_4$ is, of course, in parallel with $S''_2$, thereby increasing the total capacity (to the value of $S_4$) by increasing the area of the plates to that extent.

The best mode of proceeding to correct for slight variations in the balance is as follows :—First get nearest result by means of the rheostat and condenser $S_4$ while sending rapidly on one key. If then, on sending reversals, a sharp wave or jar is produced, alter the resistance $R_1$, or the shunt $S_5$, if used, and get nearest result by means of the subdivided condenser $S_4$, with the assistance of the rheostat. Proceed thus in the direction of improvement. Should there be a sharp return wave, or flick, after doing all that can be done by means of these adjustments, then the resistances between the first and second earth terminals and the earth will have to be altered, and perhaps the position of the "leak."

When the line is made up of a mixture of land wires on poles, and underground or submarine cable, the "balancing" becomes more complicated. On a mixed line of cable (submarine or subterranean) and land wire, in those cases where they are to be worked together as one circuit duplexing is rather a troublesome matter, and special arrangements require to be made. Thus, when the cable is at the distant portion of the circuit, its discharge being prolonged, will not be observed till after that from the line portion. If this mixed line also begins with a short cable or underground wires, there will be an immediate discharge from that, independent of that from the land wire, and this will be again followed by that from the distant cable. It will be seen, therefore, that such a circuit requires a complicated compensation to reproduce the static effects. The short cable, or underground line, will require a small condenser arranged for immediate discharge, then a larger one must be used for the distant slow discharge, which discharge is also retarded.

**Comparison of the Principal Duplex Systems.** — It should be remarked, however, in conclusion, that in practice every one who is in the habit of duplexing cables has their own particular methods of effecting a balance—quite at variance with one another very often—and these methods are continually changing from time to time. We have, however, given one instance of the course which may be adopted as applied to Muirhead's method.

The "double-block" system—by Wheatstone bridge—of Muirhead and Taylor (above described) has almost entirely superseded all other duplex methods, whether using the Muirhead or Varley form of artificial line. Since the practical application of duplex telegraphy to cables, ocean or otherwise, some 80,000 miles of cable have been duplexed—indeed, about half of the total mileage now working. Out of this, about 75,000 N.M. have been duplexed by Muirhead's system—i.e., using the Muirhead and Taylor form of artificial line, and also adopting the bridge method.* Indeed, since a joint agreement was come to between Mr Stearns and Messrs Muirhead and Taylor in 1878, all the duplexing of cables has been done by the latter gentlemen, till quite recently.

What the actual point of superiority is in Muirhead's artificial cable over that of Varley—as adopted by Stearns—cannot, of course, be said with any certainty. It would, however, appear, at the outset, to be a more faithful imitation ; for in a cable it is not a case of alternate resistance and

---

* Just as Stearns' method (of Varley's artificial and differential principle) is invariably adopted for overhead land-line duplex work. It is obviously better here, where the capacity to be inserted in the artificial line is a variable and, at the most, an infinitely low value as compared with the resistance required. There being so little capacity, comparatively speaking, the advantages of Muirhead's "inductive resistance" does not apply in this case. The capacity of an aerial line is roughly about one microfarad for every 100 miles, under ordinary circumstances and in average instances.

capacity, but of the two combined throughout the length of the conductor. One thing is clear, and this is that, in practice, Muirhead's method enables a balance to be more readily obtained—and, what is more to the point, more assuredly *maintained*—that is to say, as far as concerns a submarine, or subterranean, line, possessing material capacity as well as resistance.

It would be both interesting and instructive to compare the effect of balancing the same cable by the two methods. It is believed, however, that this has never been done. The greater difficulty of establishing a permanently complete balance on the Stearns system certainly tends to affect the speed of working, for with an imperfect balance the signalling speed has to be kept within certain limits, owing to the disturbing influence on the received signals thereby incurred.

The trouble experienced in setting up and maintaining an absolutely complete balance in Stearns' method by an ordinary Varley artificial line is partly due, very probably, to the large number of contact pieces at each junction of capacity and resistance,\* which naturally tend to be a source of leakage and variability seriously affecting the prospects of a complete and permanent balance where such sensitive instruments are in question as those used for ocean telegraphy.†

**Other Methods.**—Besides the Stearns and Muirhead systems of duplex telegraphy as applied to cables, there are several other forms of the differential and bridge methods, employing either Varley's artificial cable or the Taylor-Muirhead artificial. Instances of these are the systems of Harwood, Benjamin Smith, and Jacob in this country ; and of M. Ailhaud in France.

Mr Benjamin Smith's method of duplexing (devised in 1876) makes use of the Wheatstone bridge principle, and employs a Varley artificial cable for the compensating circuit. The novel feature of this plan consists in the insertion of a high resistance in front of, and in series with, the cable. Mr Smith, in the course of his experience in the attempted duplexing of

---

\* Though, theoretically speaking—as already stated—the greater the number of subdivisions the nearer the approach to the true imitation of a cable in which each section (of minimum dimensions) constitutes both resistance and capacity.

† In the same way, a cable with a fault in it will not work duplex (though, unless a very bad fault, it works better simplex) on account of the balance being thereby upset. Faults being practically always variable, cannot be efficiently allowed for in the adjustments--that is to say, if the defect form an important item in the prevailing conditions.

It may be mentioned incidentally that a fault being close to the sending end of a simplex circuit is, as a rule, a favourable condition, unless it be so close that the battery power has to be kept very low in order to avoid breaking down the fault.

cables with varying types of conductors and varying thicknesses for the dielectric, conceived the idea that by so inserting a high resistance he would render his cable more easy to balance on account of the extra retardation thereby introduced. This would have the effect of causing the current charges to be materially slower: or rather, to put it more accurately, it would thereby take a longer time for any upsetting of an absolute balance to affect the signals—so much so that the period might be avoided altogether; and, moreover, when such a change *did* take place, it would not be so serious.

In Harwood's method (as devised in 1879 on the bridge principle with

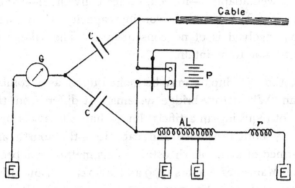

FIG. 119.—Harwood's Duplex.

Muirhead's "artificial" line) the battery P, Fig. 119, is in the bridge which usually contains the receiving instrument G; and reciprocally, condensers of about equal capacity to those usually inserted in the line for simplex working, are placed in the proportional arms. It may be remarked, however, that where the transmitting key is placed in the bridge instead of at the apex (or fork) of the proportional arms—in fact, when the sending key and receiving instrument are reversed in their positions as above—it is usually thought advisable to insert a condenser in the bridge along with the key.

A Muirhead artificial line A L, supplemented by an extra adjustable resistance,* forms the third arm of the bridge, the cable being the fourth arm.

This plan gives good results on lines of medium length. It has been in use on the cable from Porthcurno to Lisbon, the length of which is 850 miles, the copper resistance 8,050 ohms, and the capacity about 250 micro-

---

* This supplementary resistance forms one of the extra means of balance adjustment, and (as may be seen from the various recent diagrams) is common to all modern methods in practice for duplexing, at any rate, short lengths.

farads. The artificial line used in duplexing this cable is enclosed, at Porthcurno, in nine boxes, its resistance being 5,819 ohms, and its capacity 192 microfarads. The supplementary resistance at the end of the artificial line is of 3,000 ohms, and the condensers in the comparison arms of the bridge have each a capacity of about 40 microfarads. The sending battery is composed of ten Minotto cells, and the working speed is about 140 letters per minute each way.

It will be obvious from a study of the connections that in this method it is most important that the battery be efficiently insulated.* This in itself is a point against its adoption under certain circumstances. It is, moreover, rendered somewhat imperfect owing to the extra retardation it introduces.† Nevertheless, it is well adapted—indeed, rather convenient—for cables that are worked at a speed far below their working capacity, where, therefore, the extra retardation involved is of no consequence. The balance being less sensitive, is less trouble to maintain.

Ailhaud's method of duplex may be looked upon as a combination of the differential and Wheatstone bridge systems. It differs from the preceding methods in not requiring an artificial line—that is to say, a line possessing simultaneously both resistance and capacity—the fourth arm of the bridge being formed of a simple rheostat. This method has been adopted for duplexing the Marseilles-Algiers 1879 and 1880 cables, and other French Government lines, with complete success. An advantage claimed for it is the high degree of variation as regards scope and nature of adjustment.

**Comparison of Varley and Muirhead Artificial Lines.**—However, so far as English practice is concerned in the present day, almost all of those of our submarine cables which are duplexed employ the Wheatstone bridge principle with the Muirhead-Taylor form of artificial line for the compensating circuit—constituting, in fact, what is commonly known as Muirhead's method. In most instances the duplexing of these cables have been carried out, or superintended, by Dr Muirhead himself.

Nevertheless, there are those, even now, who argue that the Varley

---

* Whatever is in the bridge must be thoroughly well insulated—in fact, its two ends must be equi-potential. A battery naturally requires more care to insulate than an instrument of any sort.

† By Harwood's arrangement there is retardation just near the battery, at the *sending* end (through the compensating circuit) instead of at the receiving end. This has the effect of rendering it an insensitive method by, as it were, reducing the strength of the current at the outset. Harwood's plan is, on the other hand, one which necessarily implies easy balancing. It is, however, suited for certain cases only—chiefly those of short circuits.

artificial line is preferable for balancing purposes* on account of the greater scope for nice adjustment, both as regards resistance and capacity, constituted by these alternations of separate resistance and capacity, which can thus each be dealt with separately—under any special change of conditions, for instance.† There is no doubt that vast improvements have been made in the practical application of the Varley "artificial" to cable-duplex work.

Thus, during the last five years the Silvertown Company have successfully duplexed a number of cables—for the Western and Brazilian Company, for the Central and South American Company, and for others—on the Wheatstone bridge principle, but employing (like Mr Benjamin Smith) the Varley artificial line for compensation.

Previously—in 1888—Messrs Siemens Brothers had duplexed the Pouyer-Quertier Atlantic cable by Jacob's method of that year, using a Varley artificial line. Messrs Siemens have also applied duplex apparatus to some of the shorter (multiple-core) sections of the "Commercial" and other Atlantic cables by means of Jacob's system of multiplex telegraphy with multiple-core cables. This system (patented 1882) consists, in fact, of balancing one insulated conductor against the other; thus, where a multiple cable is in question, obviating the use of any special compensating apparatus.

It is also considered by some that the Varley "artificial" is preferable to Muirhead's for submarine cable duplex work (as well as for land-line duplex) on the score of the independence of the resistance and capacity components for the following reason. It is stated that the tinfoil in the combined inductive resistance is very much influenced as regards the resistance factor by changes of temperature such as may be made in the varying and various climates that the apparatus may be installed in. This feature

---

* On the other hand, the leading up resistance $R_1$ (Fig. 116) in the connections of a Muirhead artificial line tends to balance any self-induction in a cable. The coils in a Varley "artificial" being necessarily wound non-inductively, this is not so here. In a coiled cable the self-induction is considerable, of course; but even in the linear conductor of a cable, *some* self-induction is involved.

† In point of fact, the main difficulty to be met with in the practice of duplexing cables lies in the insulation of the condensers, which is liable to vary—and, indeed, under some circumstances, to gradually fail altogether—owing, for instance, to climatic conditions. This state of things may be brought about by any chemical change in the wax of the paraffined paper, causing it to gradually decompose, thus altering its specific qualities as regards insulation and inductive resistance. A fall in insulation from 800 megohms to 150 megohms (per N.M.) has been known to occur by a rise of temperature from 65° to 75° F. only, the inductive capacity being at the same time materially increased. Actual mechanical shrinkage of the dielectric is another source of trouble which requires to be guarded against with reference to the condensers. It will be readily understood that any change in the value of the condensers (and thus of the compensating circuit) thus brought about has the effect of entirely upsetting the balance.

requires, therefore, to be carefully guarded against or compensated for, in
order to maintain an efficient balance; whereas in the Varley "artificial,"
the resistance being provided for separately, can be made up from a material
—like German silver, or any similar alloy—which has a very low temperature coefficient, electrically speaking. On the other hand, it must be
remembered that the tinfoil of Muirhead's "artificial" is buried in a mass,
and that even supposing it is so affected the apparatus is rendered complete by a rheostat for final adjustment according to surrounding conditions
at any given moment. Be this as it may, the objection raised does not
appear to apply in practice, as there is no actual evidence of inefficiency due
to the above cause.

## Similes.

Various similes have been given to bring home the principles of duplex

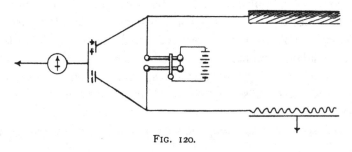

FIG. 120.

telegraphy. Perhaps the best is that of the "tug of war," by which, if equal
forces are applied in opposite directions, no result ensues. In the case of

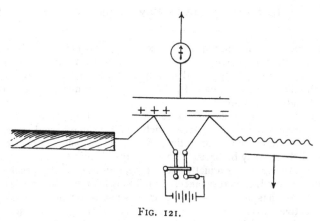

FIG. 121.

the connections in Harwood's bridge method, the bridle simile has been
aptly used on the same basis. This is suggested by Figs. 120 and 121.

## QUADRUPLEX EXPERIMENTS.

In 1878 Dr Alexander Muirhead, in collaboration with Messrs J. A. Briggs and G. K. Winter, conducted a series of experiments* in the direction of the application of quadruplex telegraphy to submarine cable working.† The results attained were not, however, of a sufficiently satisfactory nature to warrant the system being adopted practically in the few cases where it would be of special value.‡ Moreover, as long as the signalling instruments for working long submarine cables are of so delicate a character as the recorder—influenced by *changes* of current—there does not seem much prospect of quadruplex telegraphy being applicable here on any extensive scale. On very short cables, land-line "quadruplex" may be applied; the difficulties increasing gradually with increase of capacity and consequent increase of retardation. It may be further added that attempts have been made experimentally even to instal "sextuplex" and "octuplex" telegraphy on submarine cable circuits—*i.e.*, by which three or four messages might be simultaneously transmitted in both directions through the same conductor. These, of course, met with still less success.

---

* On the basis of their patent, No. 4,590 of that year.

† The term "quadruplex" is given to those arrangements which enable two signals to be sent in one direction and two others in the opposite direction at the same instant of time. Quadruplex telegraphy is, in fact, constituted by the combination of duplex telegraphy with what is known as "diplex" (or "biplex") telegraphy, which latter consists of the simultaneous sending of two signals in the same direction : this is generally dependent on the use of two relays of such a nature (*i.e.*, one "polarised," the other unpolarised) that one works solely under the influence of currents of given strength and varying direction, whilst the other is alone acted on by currents of varying strength, but in one direction only, the same compensating arrangements being effective as those employed in ordinary duplex telegraphy. Diplex telegraphy, indeed, is set up by two messages being sent through a cable in the same direction simultaneously, and when this cable is also duplexed, quadruplex telegraphy is then established. Van Rysselberghe's anti-induction system is an instance of this ; as well as, more recently, the phonopore of Mr Langdon-Davies alluded to towards the end of this book, and the plan of M. Picard.

‡ It is found that on short cables with a heavy traffic, such as those between this country and the Continent, the Wheatstone automatic with Wheatstone repeater (Morse system) get through the work with all the required expedition.

# CHAPTER IV.

## AUTOMATIC MACHINE TRANSMISSION.

General Remarks—Belz-Brahic System—Taylor's Automatic Transmitter—Delany's System—Taylor and Dearlove's Automatic Curb Transmitter—Wilmot's Transmitter: Cuttriss': Muirhead's Curb Transmitter: Muirhead and Saunders's Curb Transmitter: Price's Electrical Contact Apparatus for Transmitters—Advantages of Mechanical Transmissions—Application to Long Cables for High Speed Working.

THE problem of how to increase the working and earning capacity of a given submarine cable came to be considered principally in connection with trans-Atlantic telegraphy owing to an ever-increasing pressure of traffic, accompanied by a natural unwillingness to incur the heavy initial expense of further duplicate cables.

The rate of working by ordinary manual transmission is practically limited on a long cable—assuming for the moment the core to admit of an unrestricted signalling speed—by the rate which the transmitting clerk can maintain for a given period of time. For, inasmuch as to an experienced recorder clerk, recorder signals are just as easy to read as ordinary writing, the speed of reception is only limited by the rate of writing; and as the different portions of the tape can be "received" by several different clerks, this rate of reception may be looked upon as unlimited when compared with the rate of transmission, to which, at any rate, it puts no check. In the days of pure and simple manual transmission, however, the plan of splitting up the tape into various hands was never practised, and the speed of transmission was therefore not infrequently limited by the speed of reception owing to irregularities in the sending due to the personal element of various operators coming into force.*

**Belz-Brahic System.**—Messrs Belz and Brahic, in 1879, were, it is believed, the first to conceive the idea that if the Wheatstone automatic trans-

---

* This, again, is assuming that the core be of such a type as not to put a lower limit on the working speed.

PLATE LXXIII.

PLATE XXXII.]

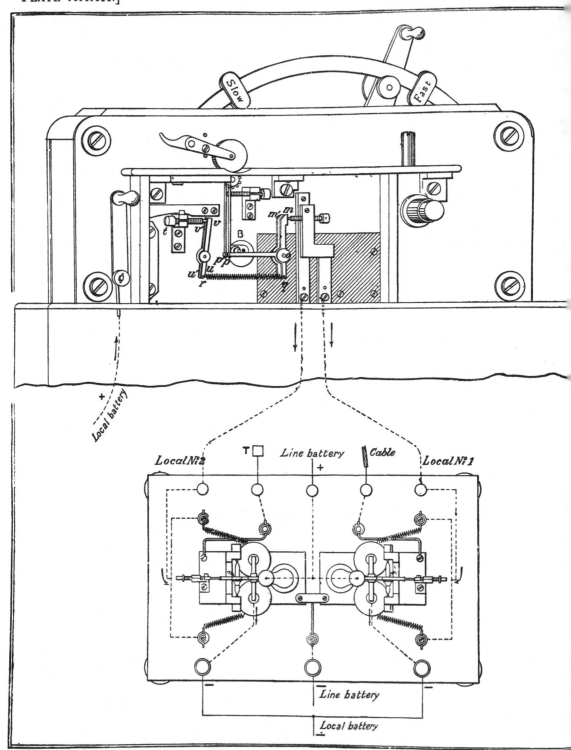

FIG. 122.—Belz-Brahic Automatic Transmission Machine.

[*To face p.* 663.

mitter,* as employed on aerial lines,† could be modified in certain directions in such a way as to render it suitable for working cables, the above difficulties would, at any rate, be partially met. To put this idea into practice they adopted, in a most ingenious manner, the perforated strip transmitter of the above-mentioned automatic apparatus, and added a Froment relay arranged as a current reverser.

The balance beam B of the modified transmitter (see Fig. 122 on opposite page) is of metal instead of ebonite, and works by means of one pin only. The reverser, with the levers and rods connected to it, is suppressed and replaced by two bent levers $p o m$, $p' o' m'$, mounted on the same pivot, having to do with the needles at one end, and with two screw stops insulated from each other at the other end. Two special springs $q r$, $q' r'$, tend to keep the bent levers constantly pressing against the screw stops. These springs are attached to the lower ends of the levers $u v$, $u' v'$, so that their tension can be regulated by means of a screw $t$ acting on the upper ends of the levers. The whole of the mechanism is of very solid construction, and capable of working for months without being touched.

The apparatus being merely required to send dots, the perforator (Fig. 123) was simplified. It has three punches only, one for the positive current impulses corresponding to dots, the second for the negative currents giving the dashes of the Morse alphabet, and the third for drawing the strip of paper along. Fig. 124 gives an elevation and plan of the punching apparatus as it appears with its cover on.

Fig. 122 (opposite) shews in plan the connections of the Froment relay arranged as a current reverser. The positive pole of a local battery is joined up to the framework of the automatic transmitter, the current being admitted—by the make-and-break motion of the apparatus—to one pair of relay coils, or to the other as required. The two coils of

---

* So beautifully perfected in detail by the mechanical genius and skill of Mr Augustus Stroh, this instrument is sometimes worked (simplex) at a rate of 600 words per minute on the Press circuits where it is used.

In ordinary practice, however, the speed is more often about half the above. It is found that when a certain rate is exceeded, it becomes impossible for the clerical staff to deal with the slip as it positively *pours* out. Moreover, very often, the delays due to repetitions are so much greater at the higher speeds, that a more moderate rate gets through the work quicker in the end. This being so—fascinating as the 1,000 words a minute pointed to by inventors may be—in practice it is rather a wild notion at present. In view of various undertakings and prophecies, these remarks apply to "up to date" and future cable telegraphy as much as to inland telegraph systems, so long as the present general methods are in use.

† The late Mr Robert Sabine is said to have experimented with the Wheatstone transmitter for working the cable between Calais and Fano (Denmark) at quite an early date.

each pair are joined up in parallel to lessen the resistance of each circuit. One pole of the line battery is connected up to the two upper screw stops and the other pole to the two lower stops between which the moving pallets

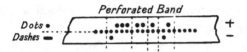

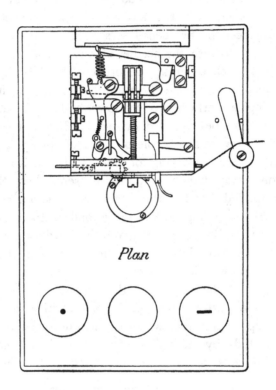

*Plan*

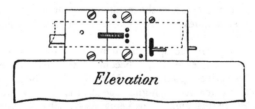

*Elevation*

FIG. 123.—Paper Perforator.

of the relay work. The framework of one of these pallets is joined up to the cable, and that of the other to the earth. According as one pallet or the other is attracted by the current of the local battery, so is the positive or negative pole of the line battery put to earth, and a negative or positive current sent into the cable.

This reversing relay, actuated by the automatic transmitter, is, in fact, merely an automatic cable key, being a double-action instrument with two tongs as in the Muirhead-Winter quadruplex apparatus. The French Government Telegraphs employ this Belz-Brahic automatic system in working their cables between Marseilles and Algiers.

The details of installation at a station at one extremity of the line are shewn in Fig. 125 (see Plate XXXIII., overleaf), and may be taken as a fair sample of a short cable recorder duplex system with automatic trans-mission. Here, going from right to left, we find:—

The resistance R situated immediately before the cable.

The recorder with coil differentially wound.

The train of clockwork for unwinding the strip of paper.

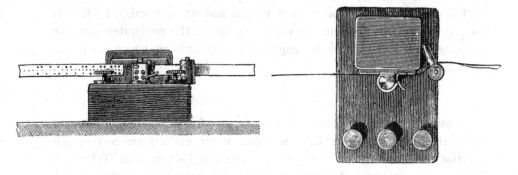

FIG. 124.—Punching Apparatus: Elevation and Plan.

The proportional arms of the bridge.

The sending condenser A.

The condenser B, with the adjustable set of resistances for controlling it.

The set of resistances $\beta$, forming the third arm of the bridge.

The condenser C and the set of resistances $\gamma$.

The condenser D and the set of resistances $\delta$.

A two-lever key for hand sending when the automatic transmitter is not in use.

Two four-way commutators for changing over from one key to the other.

A second double-lever key, termed the duplex key, to be used in case of need, for hand sending.

The automatic transmitter.

Lastly, the reversing relay as already described.

The wheel of the mechanical transmitter which impels the paper strip works regularly at the rate of thirty turns per minute. This wheel having twenty teeth, one turn gives twenty current impulses or spaces ; now twenty impulses go to one average French word of five letters, including the spaces between letters and words. The working speed over the Algiers cables, under these conditions, averages therefore thirty words per minute, or 1,800 per hour for simplex, and somewhere about 3,500 for duplex working through each cable.

It should be remarked, however, that this speed is merely limited by the type of core and by the limit of requirements. This, or any other, form of mechanical transmitter is in itself capable of working up to very much higher speeds—say 1,000 words a minute, at the least. Thus, where automatic transmission is in force, the working rate is only limited by the construction (and consequent degree of retardation) of the cable, together with the necessities of the particular case—i.e., whether or no a large, and correspondingly costly, core is warranted by the estimated traffic receipts.*

The service is carried on at each station and on each cable by four or five punching clerks and one key clerk who inserts the perforated strips in the transmitter and cuts off in lengths the strip arriving, which he passes on to the writers.

**Taylor's Automatic Cable Transmitter.**—After MM. Belz and Brahic, Mr Herbert Taylor, M.Inst.C.E., was one of the earliest workers in this direction. In 1888† he designed for the Anglo-American Telegraph Company an automatic transmitter, which was very successfully worked. Since then he has given further attention to the subject, and has perfected an instrument, a description of which follows.

This later form of automatic transmitter, designed and patented by Mr Taylor, has distinct and important improvements to which reference will be made, as may be gathered from a study and short explanation of Fig. 126 (Plate XXXIV.).

---

* In ordinary manual transmission, however, probably no operative clerk could send for any length of time, if at all, even at this (thirty word) rate—much less at fifty words (=about 750 current impulses) a minute, which is the speed attained, with the automatic transmitter, on the latest Atlantic cable, whose core was purposely designed to permit of such work being rendered practicable.

† At the end of this same year, Mr Julius Timm, of the Great Northern Telegraph Company, patented a modification (see specification 18,966[88]) of Wheatstone's transmitter suitable for working cables. This was first used on the Hongkong-Shanghai section of the above system in 1889, and was fully described in the *Telegraphic Journal and Electrical Review* the following year.

[PLATE XXXIII.

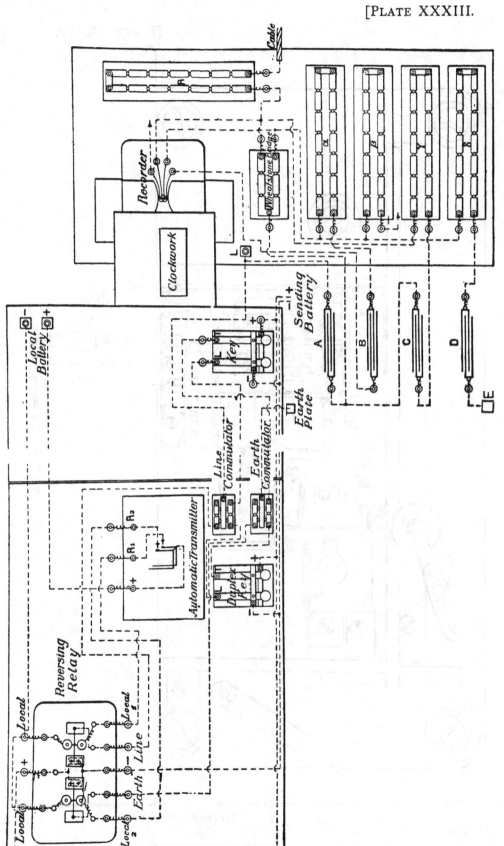

FIG. 125.—Early Automatic Submarine Telegraphy by the French Government Telegraphs: Diagram of Connections on the Marseilles—Algiers Cable (Duplexed).

[PLATE XXXIV.

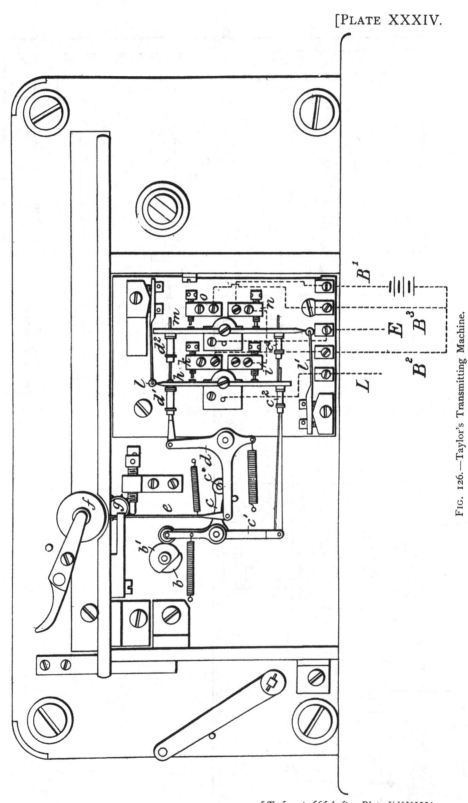

FIG. 126.—Taylor's Transmitting Machine.

As in the case of the Belz-Brahic apparatus, this instrument is of the Wheatstone type, but adapted to transmit the cable code of signals.

The two levers $m$ and $h$ may be regarded as representing the two spring blades of the usual hand-sending key. In this instrument the two levers are controlled in their movements by two needles, and the movements of the needles are in turn governed by a paper tape which is punched in accordance with the message to be transmitted.

The cams $b$, $b'$ control respectively the motion of the contact levers $m$, $h$. One of these cams at the proper time allows either contact lever to move, and thus to put line and battery in connection; the other cam, at a regulated period after this operation, connects the line with earth or its equivalent.

These two cams $b$, $b'$ act through a bell crank lever $c$, and they are fixed upon an axis which is rotated continuously by the clockwork; moreover, as constructed they are adjustable relatively to each other. By this combination the comparative duration of the battery and earth contact with line, to produce a given signal, may be adjusted, until the most suitable arrangement is found for the working of the circuit in connection with the instrument. In some degree this regulation is of the nature of a "curb," as by diminishing or increasing the period of the earth contact with line relative to that of the battery contact with line, so is the curbing effect on the signal diminished or increased.

The adjustments and the action of the transmitter are simple, and the speed of working is only limited by the electrical constants of the line.

The speed obtained with this form of "auto" when sending Press messages on the 1894 "Anglo" cable was as high as fifty five-letter words per minute.

The connections and circuit need no explanation, being clearly shewn in the figure.

**The Delany Automatic System.**—Mr Patrick B. Delany, whose name is well known in connection with a method of synchronous multiplex telegraphy—adopted by the Post Office—for land lines, has of late years devoted his energies almost exclusively to ocean cable telegraphy. This has given eminently satisfactory results in the course of trials on Atlantic cables subsequent to its invention in 1893.[*]

The principal features of the system are an electro-magnetic perforating machine and a new way of making contact through the perforated tape.

---

[*] See English patent specifications Nos. 21,630 and 23,687 of that year.

The perforator (Fig. 127) comprises a key having three light levers similar to the lightest form of cable keys. The buttons of these levers are grouped in clover form—dot, space, and dash.

An electro-magnet C operates the dot punch, another B the dash punch,* and a third A the simple, but very effective, mechanism for spacing—performing the function of the space key in the ordinary Wheatstone puncher, but, if anything, more accurately, and without making a centre row of holes. The only holes in the tape are, in fact, for the signals. The perforating (dot and dash) electro-magnets are again illustrated in further

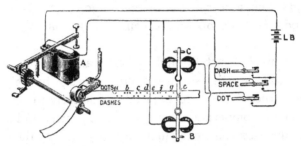

FIG. 127.—Paper Perforating Apparatus in Delany's Automatic Telegraphy.

detail by Fig. 128, from which it will be seen that a small punch is attached to the extreme end of each of the levers.

By tapping the key levers lightly with one finger messages are perforated with a minimum amount of labour and at a speed about equal to the rate effected by a Wheatstone puncher. Both the dot and the dash magnets are in series with the space magnet ; and in series with the space magnet alone

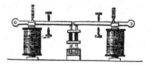

FIG. 128.—Electro-Magnetic Punchers.

is a fourth electro-magnet, which, by a novel application of a simple friction operating on the hub of the paper reel, turns the latter a little each time the space key is put down, so that slack tape is provided in advance for the punches. The action of this unwinding device is not arbitrary as to the intermittent turning of the reel. It does not turn it a definite amount

---

* In thus speaking familiarly of the "dot" and "dash" punch, it must be remembered that there is in this system when applied to siphon recorder, or mirror, working no difference in the *duration* of the contacts, but only in the current so established, one producing a positive, and the other a negative, current.

each time. Owing to the constantly diminishing diameter of the reel such an arrangement would not answer. The unwinder ceases to turn the reel after two or three convolutions of tape have been unwound. The friction put upon the flange of the reel, by the unwound tape banking up against the frame in which the reel is set, prevents the reel from responding further to the action of the "unwinder" until some of the slack tape has been used up. This constantly supplied slack tape prevents any tripping of the spacing mechanism while punching is proceeding.

Should a punch fail to perform its work properly, the failure is instantly detected by the operator's ear.

All the magnets are worked from a common battery L B, about 20 volts being used. As it is open circuit work, the drain on the battery is not great.

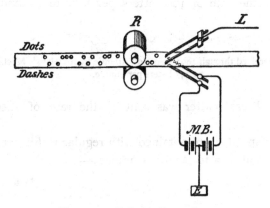

FIG. 129.—Delany's Transmitter.

In his transmitter, Mr Delany has also made an entirely new departure. Instead of retaining the Wheatstone rocking beam plan (or some of its modifications), or the equivalent step-by-step plan, by means of the clockworked rollers R (Fig. 129), he draws the tape between contact brushes, the upper set pressing downward on the top of the tape, whilst the lower ones press upward on the lower side of the tape. These contact brushes, in circuit with the transmitting battery M B, by means of the line L, comprise five or six steel wires each, and—under considerable pressure—meet through the perforations, thus making absolutely perfect electrical contact. When the paper intervenes, of course the contact is most effectually broken, and the edges of the holes in the tape have the effect of keeping the contact brushes bright and clean. There can be no failure or "sticking"—cleaning, burnishing, or adjusting is here rendered unnecessary. There is practically no mechanism beyond the paper pulling machinery. Any division of time desired may be made between marking and spacing—*i.e.*, between charging and discharging the cable; for, unlike the rocking beam, or step-by-step, principle, the duration of impulses as compared with the length of the space, or discharge, depends solely on the size of the holes in the strip. The apportionment of time at present is about two-thirds in favour of charging the cable to one-third for discharge.

By the Delany system the contacts of the automatic transmitter operate electro-magnetic transmitters which, in turn, send the signal impulses into the cable.

With this system about 300 code messages of regular traffic were sent consecutively at an average speed of 88 letters per minute over the Direct United States cable, Ballinskelligs, Ireland, to Halifax; 35 code messages, regular traffic, were sent at an average speed of 95 letters per minute, some of them being at the rate of 100 letters per minute, without an "R.Q."

FIG. 130.—Signals received through 1894 "Anglo" Atlantic from Delany's Machine Transmitter at 247 letters per minute.

(repeat signal). Press matter was sent at the rate of 120 letters per minute.

The following speeds were obtained with regular traffic over the Anglo-American cables, Valentia to Heart's Content :—

| Cables. | | | | | | | Letters per Minute. |
|---|---|---|---|---|---|---|---|
| 1873 | - | - | - | - | - | - | 153 |
| 1874 | | - | - | - | - | - | 153 |
| 1880 | - | - | - | - | - | | 133 |
| 1894 | - | | - | - | - | - | 251 |

These speeds were carefully timed by officials of the company at both ends of the cables, and were actually transmitted letters with extra spaces between words, computed on the basis of seven letters to a word, and 3.7 impulses to the letter.*

Fig. 130 is a specimen of signals over the Anglo-American '94 cable at 247 letters per minute by the Delany system.

The Delany automatic transmitter is capable of recording on prepared chemical paper perfect signals at the rate of as much as 10,000 letters per minute.

---

* It should be remembered that while it has been the custom to compute cable speeds, as in land lines, on the basis of five letters to the word, cable code messages average seven to ten letters per word ; and in taking credit for extra spaces between words, the latter rule should be followed to get at exact results.

**Taylor and Dearlove's Automatic Curb Transmitter.**—Fig. 131 (Plate XXXV., overleaf) illustrates the general construction of an instrument patented by Messrs H. A. Taylor and Arthur Dearlove in 1894 (specification No. 20,807). It combines, with minor improvements, the plain automatic sender of Mr H. A. Taylor, as previously described, with an additional automatic curbing apparatus, which provides that each signalling current sent into line is immediately followed by another current of opposite sign and shorter duration.*

To effect this operation two levers $u$, $v$, constituting a reversing key, are employed. Their motion is governed directly by the action of an adjustable snail-shaped cam $t$, fixed upon the same axis as the cams $b$, $b'$. The respective poles of the line battery are connected to the levers $u$, $v$, and, during the passage of each signal, the lever $q$ is rocked by the action of the snail cam, and a reversal of the battery takes place during the passage of a signal—sooner or later according to the adjustment of the screw $s$.

The whole of the mechanism connected with the "curbing." is mounted on a sliding platform $r$, and is so arranged that any desired degree of "curb" can be obtained by moving forward or backward this platform, thus causing the rocking lever $q$ to engage with the snail cam at an earlier or later period of the revolution, as the case may be.

To shew clearly the enormous range of adjustment possible with this instrument, and the effect of various degrees of "curbing," the slips (Fig. 132, Plate XXXVI.) obtained upon an artificial line, are reproduced further on.

In connection with this method of working, Messrs Taylor and Dearlove arrange for a permanent record of the messages transmitted. This is carried out by means of two relays $R^1$, $R^2$, in conjunction with a "Herring" or "Steinheil" receiver H worked by a local battery; and the apparatus is so connected, that unless the signal actually passes to line, no record is obtained. The polarised relays are connected *between the battery* and the contact levers of the reversing key, and are of unusually low resistance— about $2\frac{1}{2}$ ohms. They, therefore, introduce no retarding effect upon the charge entering either condensers or line.

---

* The application to hand transmitting keys of "curbing" mechanisms, though frequently tried, has never been practically successful.

The necessary mechanical additions introduced considerable complication, and they were all more or less defective in construction. Thus, the use of keys so fitted resulted in small actual advantage. Improvements in working were looked for, and more easily obtained, by a proper disposition of battery and condenser power.

For various reasons, the addition of curbing gear to automatic transmitters was more necessary, and this has been successfully effected. The curbing apparatus is made to work with machine-like regularity: its range of adjustment is large ; and of course it may or may not be used in combination with the ordinary sending condensers.

The connections of these relays are made in such a manner that one relay $R_1$ receives, and is actuated by, the currents which signal "dashes," and the relay $R_2$ receives, and is actuated by, the currents which signal "dots." The curbing currents by reason of the position of the relays have no effect upon them, except to pull the tongues over harder on to what is called the "dead stop"; neither does the discharge from the cable pass through the coils of either relay. In the local circuit of these relays is the "Herring," which records the outgoing signals, and unless the signal actually passes to line no record is possible; the avoidance of errors and the effectiveness of the "check" is by these arrangements greatly increased.

The three cams controlling the movements of the needles, contact levers, and curb in Taylor's and Taylor and Dearlove's transmitters are shewn in Fig. 131 (as well as in Fig. 126), $b$, $b'$, and $t$. The clockwork is stopped by moving a lever, which, at the same time, by a switching arrangement, transfers line and earth to a hand-key for calls, corrections, etc., as is necessary or desirable.

The compactness and the actual working of this latter instrument is remarkably good. A considerable increase in the speed of transmission is obtained under ordinary circumstances, but especially is this noticeable when it is used upon long circuits with a high "KR."

The slips (as referred to above) and the conditions under which they were obtained are given on the folding sheet opposite.

These specimens of signals obtained under the varying conditions here noted prove, beyond all doubt, the effectiveness and free range of the curbing apparatus provided in this automatic transmitter of Messrs Taylor and Dearlove. For instance, in slip 1 no sending condensers are used, but the curb is adjusted suitably. Here the definition and uniformity of the signals are perfect, nor is this result to be obtained by other means, such as a variation of the battery or condenser power.

**Other Transmitters.**—Besides the automatic transmitting instruments and systems which we have here described, there are also several others which, it is believed, have similarly proved highly satisfactory. For instance, Mr T. J. Wilmot in 1890 brought out an automatic instrument, being— like the Belz-Brahic, the early Taylor "automatic," and others—a neat modification of Wheatstone's adapted to recorder, for cable work, instead of Morse signals, by alterations in the form of the levers. Then there is the transmitter of Mr Charles Cuttriss (electrician to the Commercial Cable Company in New York), the great feature of which is that the prepared transmitting paper takes indentations instead of holes from the punching

[PLATE XXXV.

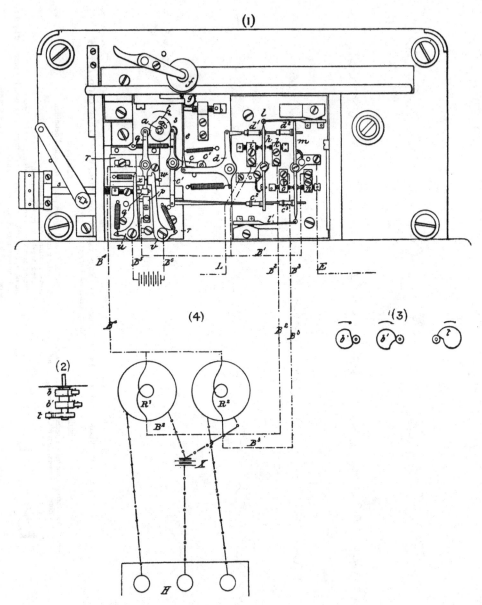

FIG. 131.—Taylor and Dearlove's Automatic Curb Transmitter.

[To face p. 672.

*N.B.*—These signals were obtained on a line of 7,000 ohms and 330 mfds. or $KR = 2.31 \times 10^4$.   Six Leclanché cells.   Speed, twenty-eight, five-letter, words per minute.

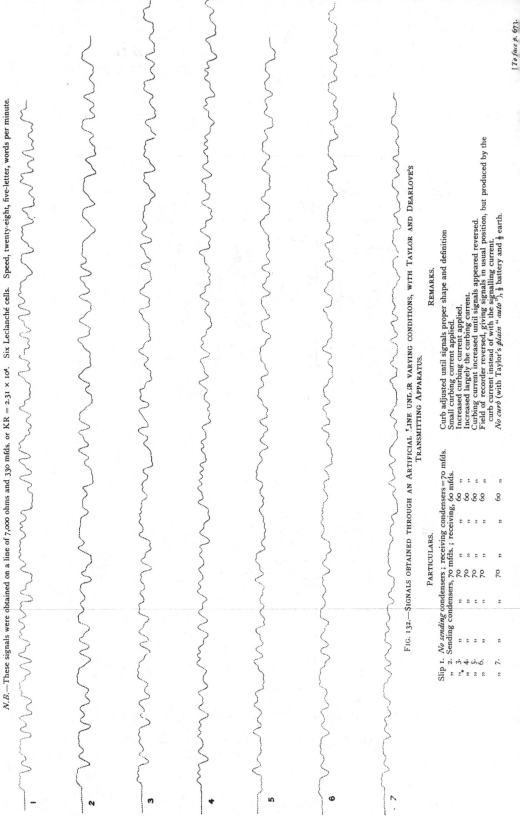

FIG. 132.—SIGNALS OBTAINED THROUGH AN ARTIFICIAL LINE UNDER VARYING CONDITIONS, WITH TAYLOR AND DEARLOVE'S TRANSMITTING APPARATUS.

PARTICULARS.

Slip 1. *No sending* condensers ; receiving condensers = 70 mfds.
,, 2. Sending condensers, 70 mfds. ; receiving, 60 mfds.
,, 3.   ,,   ,, 70   ,,   ,,   ,, 60   ,,
,, 4.   ,,   ,, 70   ,,   ,,   ,, 60   ,,
,, 5.   ,,   ,, 70   ,,   ,,   ,, 60   ,,
,, 6.   ,,   ,, 70   ,,   ,,   ,, 60   ,,
,, 7.   ,,   ,, 70   ,,   ,,   ,, 60   ,,

REMARKS.

Curb adjusted until signals proper shape and definition
Small curbing current applied.
Increased curbing current applied.
Increased largely the curbing current.
Curbing current increased until signals appeared reversed.
Field of recorder reversed, giving signals in usual position, but produced by the curb current instead of with the signalling current.
*No curb* (with Taylor's *plain* "auto"), $\frac{2}{3}$ battery and $\frac{1}{3}$ earth.

[*To face p.* 673.

The material originally positioned here is too large for reproduction in this reissue. A PDF can be downloaded from the web address given on page iv of this book, by clicking on 'Resources Available'.

apparatus. Both of these mechanical transmitters have been, and are, largely used on the Commercial Company's cables.*

FIG. 133.—Wilmot's Automatic Transmitter.

In 1893 Dr Alexander Muirhead and Mr H. A. C. Saunders joined hands over another curb transmitter (in the form of an automatic key)

FIG. 134.—The Cuttriss Transmitter.

worked on a somewhat different principle, but this apparatus requires extremely regular transmission of a special kind, and there is some difficulty in ensuring this.

* Mr Wilmot, the above company's superintendent at Waterville, was one of the very first workers in the field of automatic telegraphy. His apparatus was soon recognised as being a successful solution to its application for cables. The Wilmot transmitter is not furnished with any curbing device, the inventor being one of those who consider that by using condensers all the required curbing effect is afforded. A general view of this instrument is given in Fig. 133, and the like representation of that of Mr Cuttriss in Fig. 134.

Again, in 1894, Dr Muirhead devised an ingenious automatic sender which also includes arrangements for curbing each signalling current. A great feature in this instrument (already largely used by the "Eastern" and allied companies) is that it perfectly ensures the attainment of a correct relation for the length of battery and length of earth contacts. It also affords means of altering the duration of contacts without stopping the apparatus, according to varying conditions of the line. This device is, in fact, very different in principle from the Wheatstone transmitter, and is, therefore, unlike some of the other cable automatics in this respect. It is said to increase the working speed of a cable by as much as 40 per cent.

In 1892 Mr W. A. Price invented an altogether novel method of effecting electrical contacts especially applicable to automatic transmitters for working cables. This was intended to meet the difficulties experienced in the proper contact making and breaking through the punched holes of the slip of the various modifications of Wheatstone's transmitting instrument. In his patented invention, Mr Price proposed to ensure good electrical contact by means of streams and jets of mercury from an Archimedean screw apparatus in connection with an Archimedean pump.

Unfortunately space would not permit of all these being described in detail. A sufficient idea of some of them has, however, been presented to the reader. In all of these, as in Wheatstone's transmitter, clockwork is employed for drawing off the paper, in the usual way, adjustable to almost any desired speed. The point of novelty in each, as apart from Wheatstone's apparatus for land-line circuits, is (1) the various more certain methods of ensuring a sufficiently good electrical contact where a high retardation figure is involved as is the case in long submarine cables ; and in some instances (2) the application of curbing arrangements as suitable for long cables of high retardation.

**Advantages of Automatic Working.**—From the introductory and subsequent remarks it will be seen that the advantages of machine (automatic) transmission over ordinary manual transmission may be summed up as follows :—

1. Higher signalling speed, here limited only by the dimensions of the core.* 2. Greater uniformity or regularity, by obviating varying, or bad,

---

* Very few operators can keep up a speed of cable transmission much over 120 letters per minute—*i.e.*, 20 words—for any length of time, though even 35 words per minute can be worked up to at a stretch for a very short time. With automatic sending, however, the speed is only limited (1) by what the dimensions of the core will permit of for readable signals ; and (2) by what the receiving instrument will take. This, with the siphon recorder as at present used, is at least 100 words per minute. Thus, an increase

clerk-sent signals. 3. Improved definition with a given speed. 4. Smaller number of errors.

**Its Application in Practice.**—The application of automatic transmission has been, however, so far almost entirely confined to long ocean cables, such as those across the Atlantic, to some of the longer and busier sections of the "Eastern" Company, and by "through working" to the "Great Northern" Company's European system. But the French Government Telegraphs were, it is believed, the first to make use of the earliest cable automatic—that of Belz and Brahic—and to apply it to their Marseilles-Algiers cables. In some special cases automatic transmission has been, and is, employed on short cables; but, as a rule, it is not—only for want of sufficient traffic to warrant its necessity.

In Atlantic cables, automatic working has become imperative, and is almost exclusively adopted, in order to keep pace with the continually increasing traffic without adding to the facilities of the system by further duplications. In short cables, however, duplications are less costly items; and are often absolutely essential in any case, for purposes of telegraphic security. Thus, a short cable is seldom pressed for traffic in the same way that Atlantic and other long cables are—except in the instance of the Anglo-Continental and Anglo-Irish cables. The latter adopt the automatic system very largely for Press work, etc.

---

of as much as 35 per cent. may, within reasonable limits, be looked for in the working speed and earning power per unit of cost of a cable by the application of a system of machine transmission in place of hand sending.

Cable-Laying up the Amazon River: Breves Station.

# CHAPTER V.

## RECENT DEVELOPMENTS.

## SECTION 1.—WORKING-THROUGH EXPERIMENTS.

MR P. B. DELANY has recently devised a system for operating cables by relay and sounder. This cannot, of course, compete with the siphon recorder, or mirror, systems from a speed point of view. It can merely be looked upon as a decided improvement in this way on the ordinary Morse, Brown-Allan relay, systems, for short and moderate lengths : it is, however, unsuited for long and very busy circuits, where it is essential to get the very utmost work that is possible out of a given cable.

Mr Delany has also drawn out a system for repeating from land lines into cables. In the course of some experiments he has successfully transmitted automatic signals from London to Heart's Content with regular traffic at 178 letters per minute, two repeaters being used—one at Haverfordwest, in Wales ; the other at Valentia.

As compared with the mirror system, for short length cables (within certain limits) the Morse system, besides having the advantage of a record, actually permits of a slightly more rapid transmission, owing to the difference between the handling of a single lever and the depression of two tappets. Notwithstanding the increased duration of contact involved by a "dash," Morse working is absolutely the faster under these conditions. When, however, the length gets beyond this (with the ordinary limits of core), the Morse instrument becomes slower in its action—even with an increase of battery power—and the mirror answers more readily to the reduced current, being a more sensitive instrument than any relay applied to the Morse.

For a high speed of signalling on short cables—though the Morse is preferable for the above reasons to the *mirror*—it would seem that some form of direct-writing siphon recorder should be adopted, especially if the exigencies of the case are such as to warrant a comparatively heavy core for machine transmission. Such an instrument can now be made as cheaply as a Morse with its necessary relay; and with the same advantages of record, the speed would undoubtedly be brought up to a higher pitch where necessary.

It would appear as though working land lines in direct communication with cables could be more reasonably and advantageously effected on the siphon recorder system than on that of the Morse*—condensers being employed, of course, to ward off earth currents.

Quite recently experiments have been made by Dr Muirhead on behalf of the Indian Government Telegraph Department through long lengths of cable and land line with a view to testing whether they could be worked direct, without any intervening apparatus, on a universal siphon recorder system. The results of these experiments were, it is believed, only partially satisfactory. The siphon was too much affected by induction from neighbouring land lines. The system worked well, and at a fair speed, when the *cable* was at the sending end of the circuit, but not when the transmission took place from the other terminus—owing to the current being at the outset subjected to inductive influences as above.

Some interesting experiments in *long*-distance telegraphy on the Morse system, as regards land lines, were carried out last year in Australia. In the course of these, the longest stretch of line was spoken over that has ever been known to be—*i.e.*, 7,314 miles, almost entirely round the Australian Continent, from Cape York, Queensland, to Derby, Western Australia. This was effected by means of automatic repeaters. There were no less than fifteen intermediate stations, all several hundred miles apart. When, therefore, it is taken into consideration the number of armatures that had to be attracted and released on the make-and-break of each signal, together with the many local circuits in operation, the results of these experiments may be said to have been eminently satisfactory.

**Muirhead's Universal Transmitter.**—A simple automatic transmitter has lately been designed by Dr Muirhead, which can be used equally well

---

* Indeed, in the opinion of many, a universal Morse system—*i.e.*, cables joined direct to Morse-worked land lines—would render the work slower than by retransmission.

for Morse or recorder working. The perforated paper is passed through the transmitter (illustrated by Figs. 135 and 136) in the same way as in the well-known Wheatstone instrument. In this case, however, it is not essential to use a spur wheel for drawing the paper through; nor is it necessary to perforate the centre holes in the oiled paper, though preferably done in order to guide the paper should there be any irregularity in the line of perforations.

For Morse working a special perforator is employed. The dots are punched square by means of the left-hand lever, the paper being moved on $\frac{1}{10}$ inch: the dashes are punched with the right-hand lever, making an oblong hole, and the paper moved on $\frac{2}{10}$ inch. Two levers A, A are placed

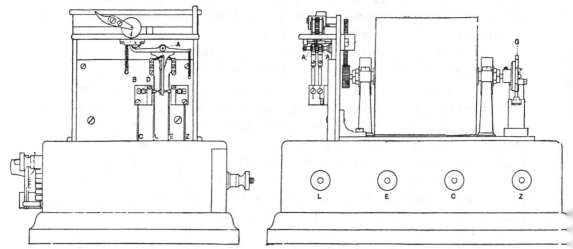

FIG. 135.—Muirhead's Automatic (Universal) Transmitter, for Morse or Recorder Working. (Front View.)　　　FIG. 136.—Muirhead's Universal Transmitter. (End View.)

laterally, having nose-shaped ends trailing against the paper: one or other of them falls into the holes made in the paper as it passes through the transmitter, and makes a short or long contact as the case may be. This contact is made with B or D at the other end of the lever pivoted at A. The connection is broken by the paper coming into contact with the nose-shaped end of the lever, which forces it out of the hole. The currents sent into the line are made by means of a local transmitter, which enables the man in charge of the instrument to tell by the sound how the signals are passing out.

An instrument of this description was tried with success on a land line of about 300 miles in length with a relay at the distant end transmitting the signals through 500 N.M. of submarine cable having a resistance of about 4,000 ohms and a capacity of 170 microfarads. The signals were received on an Allan and Brown suspended coil (cable) relay and recorded on an ordinary Morse instrument, with a speed of 45 words per minute. For recorder working the same transmitter may be used in a similar way, but an ordinary cable perforator would in this case be employed. Both contacts being of the same duration, the only difference is in the external connection.

**Cable Relays.**—With a view to effecting automatic, instead of manual, translation between lengths of submarine line, the cable relay may be said to have been the dream of the cable manager for years, and one that has seriously occupied the minds of many an electrician. The late Mr C. V. de Sauty, Mr Walter Judd, Mr Charles Cuttriss, and Mr E. Raymond-Barker have especially turned their attention to the subject, but so far no absolutely complete success has been met with in practice.*

The first idea that occurs to the experimenter is naturally a relay with fixed contacts; but, as Mr Cuttriss has expressed it, the "sticktion" here involved must always be fatal to success.

Both Mr De Sauty and Mr Raymond-Barker have endeavoured to turn the suspended coil to account as a cable relay. Some years ago the last-named devised an apparatus consisting of a suspended coil in a magnetic field with a projecting tongue to close two local circuits after the manner of Bright's Bells. So far, however, it would seem that cable relays are not within the sphere of practical teletics—not, at any rate, when applied to the end of a long cable involving a varying zero at the relay in proportion to the strength of the more or less exhausted current.† If this were to be made a theoretical success, the apparatus would necessarily be so delicate and complicated that it could never be left to itself. It would, in fact, require the attendance of a skilled and experienced person. This being so, the human relay, or manual translation, method is the obvious remedy. In a word, though it is possible that the cable relay may be applicable to short circuits, in the case of really long cables it does not appear as though it would be possible to obtain impulses sufficiently clear to operate any dead beat relay, such as would be required to actuate signalling levers.

If, however, the cable relay could be made a complete success for those circuits which employ the siphon recorder, it might be turned to account in doing away with the siphon and the friction against the paper which takes

---

* Since the above was written, it is believed that Messrs Herbert Taylor and Arthur Dearlove have obtained good results with a cable relay of their device. See patent specification No. 11,482[95].

† As an *alarum* relay, however, this instrument would work anything from a bell to a mid-day gun, or time ball.

place at every vibration. If this could be effected, the result would be that of obtaining a sharper and more defined movement, reproducible upon a simpler, and less sensitive, local instrument *—especially in the instance of a long cable, where a gain of something like 20 per cent. in speed might be realised as the result of the application of a really efficient relay.

**Automatic Punching.**—In connection with the subject of repeating systems from one line to another, it may be remarked that quite recently Mr Arthur Dearlove has devised an ingenious method of punching the Wheatstone transmitter tape at a distance.† The punching instrument is constructed so as to be capable of being operated by feeble currents—*i.e.,* electric impulses—automatically transmitted through a land line or cable. The tape so prepared does duty then in connection with a transmitter at the end of the cable through which it is desired to retransmit the message. This system works very well at a rate up to thirty words, or more, per minute, and it would seem as though it might have a considerable sphere of utility for working through a series of cables with intermediate repeating stations, and when they could not be successfully worked through direct in a continuous length from the extreme ends only.

This apparatus appears, at any rate, to well meet some of the difficulties encountered, as mentioned above, in cable-relay construction and working.

---

* By watching the movement of the siphon cradle when the siphon is off, this gain in definition becomes apparent.

† Patent specification No. 9,167 of 1895.

Cable-Laying up the Amazon River : Station at Parintins (Villa Bella).

## SECTION 2.—PHENOMENA IN LONG-DISTANCE CABLE TELEGRAPHY.

The conductor resistance throttles the flow of current just as a partial stoppage does in a speaking tube. Any throttling causes a sudden recoil or back thrust, which will hinder the next current, whether of air or electricity. This can be remedied by providing a leak, so that the accumulated air or electricity, instead of being thrust back, escapes as waste. This is what occurs at a large fault, and explains the splendid signals often obtained through a faulty cable. The current which arrives at the far end is naturally weak, but a succession of such currents can follow with great rapidity because they are not hindered by recoil currents—*i.e.*, currents which are echoed back by the CR, and which actually flow out at the sending end. Thus, theoretically speaking, an increased speed of signalling should be obtained from a cable having a number of leaks judiciously arranged,* using such an electro-motive force that the current reaching the distant end would still be strong enough for producing sufficient influence on the receiving apparatus.

Several electricians have experimented in this direction (with various forms of artificial leaks,† condensers, induction coils, etc., interposed at different points along the line) both for purposes of high-speed telegraphy and long-distance telephony.

Monsieur Godfroy, Professor M. I. Pupin, Sc.D., and others have tackled this question in different ways within recent years. Dr Pupin proposed once more to increase the working speed of long cables by dividing them into sections and connecting them up by condenser repeaters.

Though condensers at each end of a cable have a highly beneficial effect in cutting off earth currents, and in further improving the definition by curbing the signals, their interposition at random would probably be prejudicial rather than otherwise. It must be remembered that the dielectric of all condensers other than air condensers are possessed of considerable viscosity, thereby introducing material absorption and thus setting up a lag effect of some degree or other.

No doubt need be entertained that success will finally crown the efforts-

---

* Professor G. F. Fitzgerald, F.R.S., has very clearly defined in further detail the effect of leakage on wave propagation through telegraph circuits in *The Electrician* for 24th May 1894.

† The idea of establishing leaks of a certain resistance between the conductor of a cable and the earth is now a matter of comparatively ancient history. The efficacy thereof for improving the signals on a long cable was first pointed to by the late Professor Fleeming Jenkin, F.R.S., somewhere in the sixties.

already made regarding these two problems, and that ocean telephony will not be the last which electrical science holds in reserve as a benefaction for the years to come.

So far as concerns any further substantial increase in the speed attainable for submarine telegraphy, it seems pretty evident that if this is to be effected it will be by an entire revolution in the form of conductor, dielectric and completed cable, rather than in the signalling apparatus. The latter has probably reached its limits of sensitiveness—already extremely high— and any further increased sensibility of the instruments is likely to be at the expense of steadiness, and would tend to bring them within the range of influence of other surrounding forces.

By way of improving the present means of signalling upon cables, Mr Oliver Heaviside, F.R.S.—who treated the subject mathematically in the *Philosophical Magazine* in 1879—has prominently advocated the introduction of both " leak " circuits and self-induction into cable lines.

To practically effect some of the above ideas, electrical engineers have devised new forms of cores accompanied by devices which, to a certain extent, realise the theoretical advantages put forward.

**Taylor and Dearlove's Leak Cable.**—In the specification of a patent obtained by Messrs H. A. Taylor and Arthur Dearlove—No. 13,136 of 1894

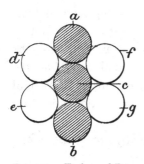

FIG. 137.—Taylor and Dearlove's Leak Circuit Cable.

—we find the construction of a core described, which provides for the introduction of a leak circuit and self-induction being set up in the cable.

This object is obtained by including two or more conductors insulated from each other in the same strand. In order that the covered and bare wires when laid up round the central wire should form a symmetrical strand, the weight of metal in the covered wires is so reduced that their diameter when covered is the same as that of the bare wires.

The central wire is of iron or copper, and is insulated from the surrounding wires, two or more of which are themselves insulated.

The wires $a$, $b$, and $c$ (Fig. 137) are insulated with cotton. It follows that the grouping of the wires for signalling purposes may be made as thought most desirable, each group consisting of two contiguous bare wires and one covered one.

When the central wire is of iron, the self-induction of the circuit is increased, and in consequence the electro-static capacity of the core diminished.

Some experiments were recently made by Mr Arthur Dearlove with variable "leaks" on two cables,* respectively 1,120 N.M. and 1,066 N.M. in length, of the same CR per N.M. The "leak" was inserted at the mid-station, or approximately in the centre of the circuit. This position Mr Heaviside recommends as the best when only one leak circuit is inserted.

The total CR resistance of the circuit was 18,345 ohms, and the leaks at the central point were varied from 10,000 ohms to 1,000 ohms. The conditions of the line during the work were as usual—*i.e.*, both sending and receiving condensers were in circuit, and a battery power of 50 cells was used with automatic transmission.

As the "leak" resistance was varied from time to time, the same message at the same speed was transmitted, and the resulting signals on the siphon recorder carefully compared. The general conclusion of the experiment was that with 5,000 ohms to earth, the signals were more clearly defined than either under ordinary conditions or with lower resistance in the leak; but such increased definition did not admit of a higher speed of transmission, though the beneficial effect of the leak was distinctly noticeable.

A more extended trial of this "leak" principle has been since made with' an artificial cable, which lends itself peculiarly well to an experiment of this nature.

A number of leaks were inserted at intervals along the line, and their position as well as their resistance was varied to find the most suitable and advantageous result. The effect of the "leaks" on the received signals was, as in the previously mentioned experiment, improved definition, in the arrival signal, which was in character similar to that obtained when curb sending is employed.

The improvement noted was not sufficient to encourage the idea that, with the present receiving instruments and method of working, largely increased working speed is to be looked for by the adoption of the "leak." With higher speeds of transmission than at present ordinarily practised in cable work, the "leak" principle may prove more efficient.

---

* Zanzibar to Seychelles and Seychelles to Mauritius.

Again, later, another experiment was made with a special form of leak, according to the joint patent of Mr H. A. Taylor and Mr A. Dearlove, already described and illustrated by Fig. 137. This, in place of an ordinary fault, took the form of a shunt of high resistance between the main circuit and the earth. The conductor resistance of the main circuit being some 20,000 ω, the best results were obtained when the leak—placed about half-way along—was composed of about 5,000 ω. Automatic sending was adopted, and a marked improvement in definition was noticeable. But this was not sufficient to make up for the reduced amplitude of the signals, due to fall of potential and loss of current at the leaks.

**Independent Expert Opinions.**—Mr F. Alexander Taylor, of the Eastern Telegraph Company, and Mr H. W. Sullivan, M.I.E.E., have independently come to similar conclusions, which they have made known quite recently in the columns of *The Electrician*. Again, Mr Rymer-Jones, M.I.E.E. (chief electrician on the Silvertown Company's cable expeditions), has made a series of experiments with leaks interposed in an artificial cable, his results—pointing the same way—being set forth in the *Electrical Review* a short while back.

The only conclusion arrived at in regard to the application of leaks on the ordinary cable circuits of to-day, is that when suitably disposed along the line, in obviating the choking effect of retardation, they secure increased definition for the signals, though not sufficiently to permit of any substantial increase in the working speed—even with the battery power raised within reason.

The Telegraph Ship " Faraday."

## Section 3.—New Proposed Methods for Rapid Cable-Signalling and Long-Distance Cable Telephony.

At various times and in various ways different inventors have from time to time proposed schemes for abandoning the use of the earth as a return circuit,* employing a complete metallic circuit instead. These schemes have sometimes taken the form of a cable enclosing two separate insulated conductors, and sometimes that of two distinct cables laid together.

The first plan has already been adopted for telephony where the induction involved by an earth return (in the case of other cables similarly using an "earth return" are alongside) is an important item of retardation. It has of late years been also proposed to apply it to ocean telegraphy.

As compared with the speed of signalling on land lines, the speed attainable on long submarine cables is—as has been shewn in previous pages in this book—very slow indeed. The phenomenon of retardation (predicted by Faraday as the necessary consequence of the electro-static charging of the dielectric) intervenes to limit the rate of working. Consider what this implies. It is not that the velocity of electricity itself is changed: the effect is wholly different. On first making contact with a battery at one end of a cable, the potential at that end is raised, and at once a flow begins all along the cable. With the rapidity of light—185,000 miles a second, or so—a current is set up throughout the conductor; but most of the current that rushes in at the transmitting end during the first instant, though it flows at this rate, does not reach the other end, being, so to speak, arrested on its way in order to charge the surface of the surrounding gutta-percha. The current in the distant parts does not gain its proper strength until the nearer parts have become charged; and, when the transmitter ceases to send in any more current, the accumulated charges go on discharging, thus prolonging the signal. Hence this lateral accumulation, due to the capacity of the cable, retards the signals, and blurs them together —like the echoes in a badly-acoustic hall—if they succeed one another too quickly. In the oldest Atlantic cables the speed of signalling was only six to eight words per minute. By the use of curb-senders and automatic transmitters this speed was raised considerably. In the newest Atlantic cables, in spite of the much greater weight of copper employed in them, and the use of these automatic devices at the trans-

---

* As has previously been mentioned, the late Mr Samuel Statham and Mr Wildman Whitehouse positively secured a joint patent for this in 1856.

mitting and receiving ends, the speed of signalling has still only reached some 40 to 50 words a minute, whilst the possibility of telephoning through such cables is as remote as ever.　Many years ago it was suggested that a condenser inserted in the circuit at either or both ends of the cable would, amongst other things, exercise a curbing effect, and tend to make the signals shorter and sharper; indeed condensers are so used now at the termini of all long cables.　It was also suggested by Varley that a further aid might be given by using at each end of the line an inductive shunt, consisting of an electro-magnet placed in derivation from the end of the cable to earth.　The use of such inductive shunts has been found advantageous on long lines, and has lately been revived by Mr Godfroy at the terminus of a cable.　But the fact remains that as long as these lateral tendencies to the retention of the charges exist all along the length of the cable, any improvements which merely affect the apparatus at the ends of the cable will fail in providing a real remedy.　Since the capacity and resistance that cause retardation are distributed all along the cable, the remedy must also be a *distributed* remedy.　The compensation cannot be effected by devices at the *ends* of the cable only: compensating devices must be applied at the middle, or, preferably, at several intermediate points also.

**Silvanus Thompson's Proposed Cable.**—A suggestion to face this problem was made by Professor Silvanus P. Thompson, F.R.S., in 1891,[*] and took the form of proposals for a wholly new departure.　He proposed to construct an entirely novel species of cable in which inductive shunts to compensate the tendency to lateral discharge are introduced at intermediate points.　One method of carrying out this suggestion is to construct the cable with two copper wire cores, as a going and a return wire, enclosed side by side within one sheath, the two being connected together at one or more intermediate points by an inductive shunt.　Fig. 138 will roughly illustrate the fundamental idea.　The two main conductors of copper A A′ and B B′, each in its proper sheath of gutta-percha, are continuous all the way through the cable from end to end.　But at one or more intermediate points—as represented in the figure—the cable is made a little thicker so as also to enclose a third wire which is joined as a shunt from a point *a* on one conductor to a point *b* on the other conductor.　This third wire should be one of high specific resistance,[†] and should be made magneto-inductive either by coiling it around

---

[*] Based on patent specification No. 22,304 of that year.

[†] Moreover, it runs in the insulation between the cores from one conductor for 100 miles or more before it connects up to the other.

an iron wire as a core, or by using an iron wire—or a wire overspun with iron.*
This is in fact a simple and ingenious way of introducing a Varley shunt in
the middle of a cable. As already pointed out, it has long been known to
telegraph engineers that a slight leak in the middle of a long cable
materially assists the speed of signalling, as it enables the retarding charges
to dissipate, and acts in the same way as lessening the electro-static
capacity of the system. But a mere leak is liable to degenerate, as it
implies a fault in the insulation which may at any time develop into a
total breakdown. Dr Thompson's device is like a leak that cannot so
break down, being just as well insulated as the rest of the cable. But it is
something more than a leak, for, being made inductive, it actually sets up
compensating tendencies just where they are wanted. On this plan a very
long cable would be made up of a number of alternate sections of two-wire
cable and of three-wire cable jointed together, so as to provide inductive
shunts at intervals along the length. Figs. 139 and 140 illustrate

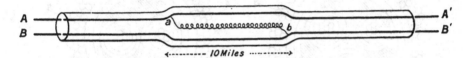

FIG. 138.—Silvanus Thompson's Proposed Cable, with Inductive Shunts or Leaks.

diagrammatically sections of the two-wire and of the three-wire portions of
such a cable. The main conductors A and B are continuous right through
the cable. The auxiliary, or neutralising, third conductor C constitutes the
inductive shunt.

A good deal of evidence exists, both in the experience of the cable
stations and in experiments made in the laboratory upon artificial cables,
that shunts at intermediate points do have the beneficial effects which
this new construction is designed to secure. It is thought by many that
the path to success lies in the direction thus indicated. A new cable
constructed on this plan would not cost twice as much as an ordinary

---

* This, again, renders it a comparatively non-inviting path for the current. Thus, a
current travelling along it will be further delayed by being compelled by a natural law to
magnetise the wire around it. To take a *simile*, if the water which fills a six-inch pipe at
good pressure tries to turn aside through a one-inch pipe crammed with tiny water-wheels,
it stands to reason that the small bypath will not be an inviting one. So in the cable the
main current ignores the thin wire, and passes on to do its duty at the opposite instrument. On the other hand, that bugbear the clinging current, being left behind by the
main portion of the current, leisurely saunters along the thin wire and into the second
wire, where it mingles with the returning current and is out of the way of the rapidly
following signal.

single cable, especially if the cost of laying—which is a large proportion of the whole cost—were taken into account ; and, if properly proportioned, it is contended that it would have something like ten times the carrying capacity.    To make a single cable across the Atlantic do the work of the ten working Atlantic cables of the present day would be no mean achieve-ment, and this does not seem to be at all impracticable in the light of present knowledge.    On the contrary it seems probable that a much greater increase in the speed of signalling is quite within the range of engineering possibilities.

This elaborate and novel scheme was first brought forward by Professor Thompson in the course of a paper on " Ocean Telephony," read in 1893 at the International Electrical Congress of Chicago, in which he was advocating the then much-talked-of trans-Atlantic telephone line.

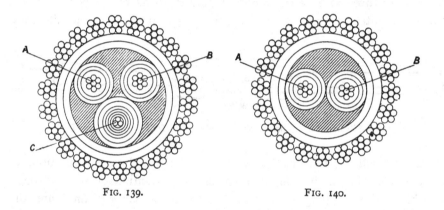

FIG. 139.                              FIG. 140.

This was not on account of these special leak devices being any more applicable to telephony than to telegraphy.*    They are theoretically well adapted for overcoming the difficulties present (in the form of retardation) in any long line—telegraph or telephone—of substantial capacity and of substantial resistance.    It was thought, however, at the time that such a device was especially in demand to render long-distance telephony feasible.

Nevertheless, this ingenious scheme of Professor Thompson's can scarcely be said at present to have emerged from the experimental stage, so far as regards *submarine* telegraphy.    Theoretically, it is perfect, and should at any rate be applicable to *subterranean* cables possessing considerable retardation.

If Professor Thompson's devices can eventually be turned to practical

---

* Except in so far that retardation forms a more serious evil in the problems of telephony.

account for submarine cables, as much as for subterranean lines, this will mean—as already shewn—a great advance in the working speed attainable for submarine telegraphy with a core of given dimensions and cost.

As regards the application to telephony over long cables—crossing the Atlantic, for instance—it may be remarked that the necessity of transmitting actual speech across the ocean must eventually declare itself in the same unmistakable terms as did the necessity of transmitting thought fifty years ago. Where, however, submarine work is in question, such a device obviously introduces practical mechanical difficulties in a long length of cable which it is hoped may be overcome in the course of further modifications. Whether such inductive shunts when fitted could be relied upon with a cable that undergoes as much handling and actual strain as a submarine cable before it is safely deposited at the bottom of the sea is a question; and whether subsequently they would interfere with the proper carrying out of repairs—or repairs interfere with them—is also matter for consideration.

Be this as it may, the above device certainly contains one element of advantage over leaks set up by a connection between the copper conductor and iron sheathing wires, for here—by Dr Thompson's plan—we have a good non-corrodable joint under the protection of the insulating material.

**Preece's Cable.**—As illustrated in Fig. 141, Mr W. H. Preece, F.R.S. (engineer-in-chief to H.M. Post Office), has given us another solution of the problem of high-speed telegraphy—but more especially for long-distance telephony—in his proposed multiple-conductor cable (see patent specification No. 13,894 of 1892).

Here, it will be seen, there are two semicircular conductors in each core so as to provide a complete metallic circuit. These are covered with brown paper, by way of insulation. They are laid up with their flat sides pressing closely against one another, with the intervening sheet of brown paper. The twin conductor thus formed is insulated with gutta-percha in the usual way. It is claimed that as the conductors are made to approach each other, the electro-magnetic induction increases at a greater rate than does the electro-static induction, until at a certain point the one actually neutralises the other.* When introducing this type of cable at the British

---

* It was on this same principle—as well as to prevent mutual induction from wire to wire—that in the last Anglo-German cable (of 1896), the Post Office authorities had each core separately enveloped in brass tape, instead of tape being applied outside the whole as had been their practice, heretofore.

Mr P. C. Dresing had independently made a similar suggestion; and this principle

Association Meeting of 1896, Mr Preece stated that he had little fear of being unable to speak between England and Germany.*

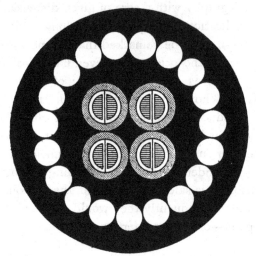

FIG. 141.—Preece's Proposed Multiple-Conductor Cable for Long-distance Telephony and High-speed Telegraphy.

**Smith and Granville's Cable.**—Mr Willoughby S. Smith and Mr W. P. Granville have sought to secure the same ends, in an ingenious manner, by a multiple-core cable (Fig. 142), with a tubular air-space in the middle, so as to reduce to a minimum the inductive capacity between the various conductors.‡

FIG. 142. † — Smith and Granville's Air-Space Core.

Some miles of the above have been manufactured for G.P.O. purposes; but whether the principle could be turned to account for *submarine* work is doubtful, owing to the water-pressure at considerable depths. There is also the increased difficulty in construction which the air-space would, in the first instance, appear to entail. Again, paying out or picking up over a drum in deep water would tend to flatten the core,

---

has been recognised for some time past by Messrs Felten and Guilleaume and others in the construction of aerial telephone cables.

* "Electrical Disturbances in Submarine Cables," by W. H. Preece, C.B., F.R.S., read before Section A of the British Association. B.A. Meeting at Liverpool, 1896.

† From the above illustration it will be seen that this form of cable is literally in the nature of a hollow tube, which, however, is stopped up at intervals by partitions that divide the air-space into water-tight compartments. Here the four separate gutta-percha insulated conductors are of such a section and so symmetrically twisted together as to form this longitudinal air-space, the whole being finally covered outside in a thick casing of gutta-percha, thus forming one large substantial core. The conductors are laid in a spiral with one turn in every eight inches of the cable.

‡ See patent specification No. 8,573 of 1895.

and in so pressing the various conductors closer together, the capacity would be raised once more, even if the thin dielectric round each were not damaged. In laying a cable of this class in warm climates this risk would, necessarily, be increased. On water getting in at a fault, the whole section between partition walls—whatever the distance be—would require to be cut out in the course of repairs. Moreover, if the four cores adhere firmly to the outer gutta-percha covering, jointing would probably not be very easily effected. These, then, form some of the difficulties to be contended with in regard to this ingenious type of multiple-core line.*

* However, on 30th June of this year (1897), H. M. telegraph ship "Monarch" laid a cable between Beaulieu, in Hampshire, and Gurnard Bay, Isle of Wight, a distance of 1.94 N.M., in connection with a new telephone line then in course of construction between Southampton and Newport. This cable was manufactured by the Telegraph Construction and Maintenance Company, in accordance with the above plan. Short lengths of the "air-space" core were subjected to a hydraulic pressure of 700 lbs. per square inch (equivalent to 260 fathoms) for a considerable time without any sign of damage or alteration in shape, and the electric tests shew results which are highly satisfactory.

The advantages of this form of cable will become apparent by a reference to the following illustrations and figures :—

| Diagrams shewing Actual Size of Cores. | | Total Weights. | | $KR$ per Naut, Closed Circuit. |
|---|---|---|---|---|
| Cores of the London-Paris telephone cable. | | 640 lb. copper wire. 1200 lb. gutta-percha. | Copper resistance, 7.48 ohms per naut. Inductive capacity between diagonal wires, .1385 microfarad per naut. | 2.0719 |
| Willoughby Smith and W. P. Granville's new air-space core. | | 940 lb. copper wire. 940 lb. gutta-percha. | Copper resistance, 5.163 ohms per naut. Inductive capacity between diagonal wires, .098 microfarad per naut. | 1.012 |
| Cores required to give copper resistance and inductive capacity approximately similar to that of the new air-space core. | | 940 lb. copper wire. 5100 lb. gutta-percha. | Copper resistance, 5.163 ohms per naut. Inductive capacity between diagonal wires, .108 microfarad per naut. | 1.115 |

Tests taken since laying the cable give excellent telephonic results, with perfect freedom from "cross-talk" and external induction. The inductive capacity of the laid

**Barr and Phillips' Cable.**—Mr Mark Barr and Mr C. E. S. Phillips have introduced to public notice yet another ingenious solution to the problem of combining high efficiency with long-distance telegraphy and telephony. They propose to employ what is practically a double conductor; but the return circuit for each wire, or group of wires, would, in this case, be the earth again.

The principle of this arrangement is, primarily, that of having the helpful leakage between two conductors at the centre of the cable instead of putting leaks between the core and sheath. In this device the inventors rightly recognise that, if leakage is obtained by the use of paper or other poor insulation, the impervious quality of the cable is destroyed.* And on the other hand, if the leakage is obtained by putting connections from core to sheath, the resistance of the leaks is too low unless coils are used.

In the Barr and Phillips cable the mode of signalling is such that the electro-static lines of force, extending from the multiple-conductor core to

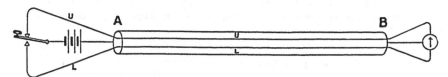

FIG. 143.—Barr and Phillips' Proposed Plan for High-speed Submarine Telegraphy
and Long-distance Telephony.

the sheath, undergo very little fluctuation during working. Hence the compound (or double) core may be insulated from the sheath with material of comparatively high specific inductive capacity such as gutta-percha. And it is suggested that no more gutta-percha need be used than that required for mechanical imperviousness.

Secondly, as will be seen from the diagram, the object is to signal in such a way that a rapid discharge may take place around the sending end

---

cable is ·095 microfarad per N.M. between the diagonal wires, but the length, 1·94 N.M., is too short to shew the advantage, if any, to be obtained by the close proximity of the conductors without water intervening.

If the vexed question of leakage should ever be settled in favour of the leak, then the air-space will prove of value as a conduit for containing the inserted resistances.

* With reference to a paper by Professor E. J. Houston and Dr A. E. Kennedy, this fact was especially dwelt on recently by the author. See "Problems of Ocean Telegraphy," by Charles Bright, F.R.S.E., *The Electrician*, 30th April 1897.

Mr H. Kingsford has further pointed out that in adopting a low insulation dielectric it would be necessary to embody with the cable at least one well-insulated wire to enable localisation to be effected in the case of a break. For further particulars, see *Journal Inst. Elec. Engineers*, vol. xix., Part 891, also *The Electrical Review*, 28th May 1897.

of the cable at each key movement, thus accelerating the speed of transmission aside from any improvement due to leakage. On beginning to
signal, the key (Fig. 143) is in contact with the upper conductor U, and a
steady current is flowing in that conductor from A to B. It may be added
that the lower conducting strand is negative to the upper.

The upper contact is arranged to follow the key when the latter is
depressed until connection is made with the lower contact. At this
moment a rapid discharge takes place around the end A of the cable. On
further depressing the key—the movement being continuous, of course—the
upper contact ceases to follow, and only the lower contact touches the key.
The current is now flowing in the lower strand L and in the same direction
as before through the cable—from A to B.

Between L and U (which, by the way, represent two *groups* of wires
alternately placed) is an insulating material of comparatively low insulation resistance and low specific inductive capacity, such as paper. The
receiving instrument is a differential galvanometer.

It will be seen that in this cable there is no wide departure necessary
in the process of manufacture, and that the cost depends upon the same
elements as in the existing cables.

It is claimed that the speed is faster with this arrangement than with
any other disposition of the same weight of copper and gutta-percha.
Moreover, it is said to be possible to duplex this cable.

In conclusion, it is but fair to suppose that we may be on the eve of
developments of a fundamental and radical character in the methods of
submarine electrical communication, more especially, and suitably, as
regards the structure of the cable.

**Langdon-Davies' Phonopore.**—Long-distance telephony through submarine cables—and the telephone in combination with the telegraph—
are forcing themselves on the attention of inventors. Thus, within recent
years experiments have been made with Mr C. Langdon-Davies' phonopore system on submarine cables. If the cable be above a certain length,
however, capacity stands in the way of success in that direction.

This instrument, invented and patented in 1884-85, will be found fully
described elsewhere. It certainly forms a most valuable adjunct to land-
line systems for purposes of diplex telegraphy—the line being *du*plexed
also, if required—and it is perhaps surprising that it has not been yet turned
to further account. On aerial lines, it has proved to be capable of working

through 500 miles and over. It is already doing good work on the Great Western, Midland, Great Eastern, and Brighton railways. If applied in connection with ordinary duplex telegraphy, the combined systems effect no less than 180 (twenty or thirty worded) messages an hour!

**Munro and Bright's Telephone Recorder.**—Mr John Munro and the author have considered the possibility of a recording telephone, it having occurred to the former that if facilities for recording could be given to the telephone a remarkably sensitive form of recording instrument would be obtained. Accordingly Mr Munro and the author experimented with what may be termed a "telephone recorder" with a view to its use on land lines and cables either as an acoustic or writing instrument at will. The telephone is of the Gower-Bell (powerful magnet) type. The ordinary telephonic current in passing through its coils modifies the magnetism of the soft iron pole piece in such a way as to attract the thin iron strip or disc (placed over or between the poles of the magnet), thus—through suitable multiplying gear—actuating the siphon pen by a multiplying lever, and causing it to record its movements in the usual way on the slip. The iron strip must have resilience in it—a spring in fact—in order to bring it back to zero after the signal current has passed and had its effect.

T.S. "Silvertown" in Smyth's Channel, Straits of Magellan, on her Homeward Voyage after Laying the Central and South American Company's Duplicate Cables, 1891.

# "WIRELESS" TELEGRAPHY.

## MARCONI'S SYSTEM.

OF late years there have been various suggestions * in the direction of telegraphy without any conductor between the points of communication—sometimes under the pseudonym "signalling through space." † These suggestions have been based on experiments of a more or less practical character, carried out by Mr W. H. Preece, C.B., F.R.S., and others ; but success has only been met with over comparatively short distances. The most marked advance in wireless telegraphy is that which has been recently made by Signor Guglielmo Marconi, who, with the co-operation of Mr Preece and the Post Office authorities, has succeeded in annihilating space to the extent of nearly nine miles. It is thought, therefore, that an account of Signor Marconi's system as developed since last year would be of interest in connection with our subject—if only as being the very latest thing in Inductive Telegraphy.

It will be seen here that we have not to do with a new "force"—as has been suggested by some of the non-technical journals—but a well-devised combination of facts already known to the scientific world.

Marconi's wireless telegraph is based on the principle of turning Hertzian waves to account in transmitting them through the luminiferous ether by means of electric sparks. It is now recognised that the universe is filled with a homogeneous elastic yet continuous medium which transmits heat, light, electricity, and other forms of energy from one point to another. The discovery of the real existence of this "ether" is, as Mr Preece has remarked,‡ one of the great events of the Victorian era. Of its nature,

---

* Most of which are enumerated towards the end of Part I.

† In most of these, though there is no continuous cable from point to point, there is a corresponding length laterally on each bank or shore, between which the induction takes place, and which, indeed, may be said to form part of the transmitter and receiver respectively : thus, then, the expression "telegraphy without wires" is scarcely applicable, strictly speaking.

‡ "Signalling through Space without Wires," Royal Institution Discourse, Friday, 4th June 1897.

however, we are still ignorant. On the other hand, we know that any disturbance of the ether must originate with some disturbance of matter. One of the greatest scientific achievements of our generation is the magnificent generalisation of Clerk Maxwell that all these disturbances are of precisely the same kind, and that they differ only in degree as to length and frequency. Light is an electro-magnetic phenomenon, and electricity in its progress through space follows the law of optics. In 1888, the late Professor Heinrich Hertz proved experimentally the actuality of electric waves, whilst Clerk Maxwell had foretold their existence some thirty-five years past. Hertz shewed that it was possible to produce electric waves such as would penetrate doors, walls, etc., and excite a spark in a resonator tuned to the same frequency as the transmitter. But this was not done over a greater distance than a few yards. Marconi has succeeded in making the resonator work a Morse recording instrument and print a telegraphic message—as before stated—at a distance of nearly nine miles. To pass from a general statement to a more or less detailed account, we first come to Mr Marconi's method of transmission.

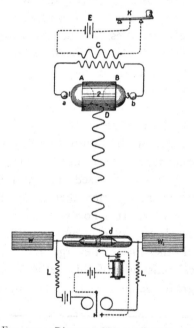

FIG. 144.—Diagram of Marconi's Telegraph.

**The Transmitter.**—This is practically one of Professor Righli's form of Hertz radiators with certain important additions to meet prevailing circumstances. These are referred to further on.

In Fig. 144—which gives a general idea of the connections by Marconi's system—there are two spheres, A and B, of *solid* brass, four inches in diameter. These are fixed in an oil-tight case D of insulating material ; and it will be seen that a hemisphere of each is exposed, the other being immersed in vaseline oil. Two small spheres, *a* and *b*, are fixed close to the large spheres, each being connected to one end of the secondary circuit in the induction coil C, the primary circuit of which is excited by a battery E thrown in and out of circuit by the Morse key K. Now, whenever the key K is depressed, sparks pass between 1, 2, and 3, and since the system A B contains capacity and electric inertia, oscillations are set up in it of extreme rapidity. The line of propagation is D*d*, and the frequency of oscillation is probably about 250 millions per second.

The distance at which effects are produced with such rapid oscillations depends chiefly on the energy in the discharge that passes. In the Post Office experiments, a six-inch spark coil sufficed through anything up to four miles, but for greater distances a more powerful coil—one emitting sparks 20 inches long—was used. This distance increases with the diameter of the spheres A and B ; and, moreover, it is nearly doubled if they are solid instead of hollow.

It is easy to see how the generation of these waves in the ether can be controlled by making or breaking the electric current in the induction coil, which yields the spark ; but how are the waves caused to shew these signals at the distant end ?

**The Receiver.**—Here lies the crux—aye, and probably the success—of Marconi's system. It is not the Hertzian resonator, pure and simple, but a combination of this instrument with a much more sensitive detector of electric waves discovered by M. Edouard Branly in 1890. This gentleman found that loose metal dust coheres when such waves pass through it, and by cohering it becomes a better conductor of electricity.* M. Branly's instrument—the origin of the "coherer"†—consisted of a glass tube con-

---

* Somewhat the same principle was turned to account by Mr S. A. Varley in his lightning guard of many years back.

† This expression was first used by Professor Oliver Lodge, F.R.S., who in 1894—whilst discoursing at the Royal Institution on "The Work of Hertz"—shewed very similar apparatus, and suggested it as a basis for telegraphy without any continuous conducting wire. For his "coherer," Dr Lodge employed iron or brass filings in quite a high vacuum—far higher than that of Marconi.

The term "coherer" is liable to lead to confusion owing to having a subjective as well as objective sense. In a Physical Society paper of May 1894, Mr Rollo Appleyard described some experiments with solid rods of gutta-percha mixed with brass filings, where a similar phenomenon was observed. This, therefore, he designated a "sensitive dielectric," which term has the advantage of avoiding the above-mentioned ambiguity ; but metallic filings could scarcely be termed a *dielectric*.

taining metallic filings,* connected to an external circuit fused through its ends. If the filings (or dust) form a part of a circuit with a Morse instrument and battery, the arrangement can be so adjusted that the "Morse" will give a signal whenever an etheric wave passes through the filings.

Such in principle is the telegraph of Marconi. Let us turn to details once more. The relay (see Fig. 144) consists of a small glass tube four centimetres long, into which two silver pole-pieces are tightly fitted, separated from each other by about half a millimetre—a thin space which is filled up by a mixture of fine nickel and silver filings, to which is added a trace of mercury.† The tube is exhausted to a vacuum of four mm. and sealed. A single cell ("local") lies in wait in the coherer circuit to force the current through, when the resistance in the filings is broken down by the electric waves. This feeble current (of what may be termed the primary circuit)—insufficient in itself to actuate a telegraphic instrument *directly*—influences a delicate relay, which, in turn, throws a more powerful battery into the secondary circuit where the heavier work has to be done. In its normal condition the metallic powder is virtually an insulator. The particles then lie in higgledy-piggledy fashion, and only lightly touch each other. But when electric waves sweep down upon them, they become "polarised" and order is at once installed. Being then more or less subject to pressure,‡ they "cohere"—electrical contact ensues, and a current passes. The electrical resistance of Marconi's relay—*i.e.*, the resistance of the thin disc of loose powder—is practically infinite when in its normal, disordered, condition. This resistance drops to as low a figure as five ohms when the absorption of the electric waves is sufficiently intense. Thus, we see the so-called coherer is capable of being rapidly converted from the condition of an insulator to that of a conductor.

In his 1890 investigations, Mons. Branly found that the high resistance of the "coherer" could be restored by mechanical tapping. In the language of Dr Lodge, Marconi "*de*coheres" the metallic filings by an ingenious automatic trembler. He causes the local, or secondary, current to very rapidly vibrate a small hammer-head against the glass tube. This it does effectually, and in so doing makes a sound by which any telegraph operator could read Morse signals, independently of the Morse

---

* Copper, aluminium, and iron.

† The actual proportions said to be affected by Signor Marconi are 96 per cent. nickel to 4 per cent. silver, with but a *soupçon* of mercury.

‡ This would naturally result just as it would if a fleet of ships partly in "column" (end on) were ordered into "line" (broadside on) where there was only a limited amount of line space.

ink-writer worked in series with it from the same current—*i.e.*, that of the secondary circuit.

**General Working.**—The exhausted tube has two wings (see W and W, in Fig. 144), which, by their size, may be made to tune the receiver to the transmitter by varying the capacity of the apparatus.* Choking coils L and L (see Fig. 144) are made use of to prevent the energy escaping.

Oscillations set up in the transmitter fall upon the receiver tuned in sympathy with it, coherence follows, local currents are established, and signals made.

In open clear spaces within sight of each other nothing more is wanted ; but when obstacles intervene and great distances are in question, height is needed. Tall masts (with metal cones on the top), kites and balloons, have

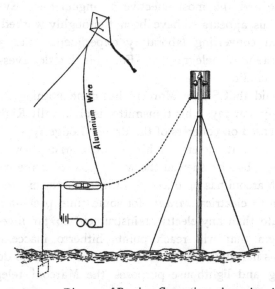

FIG. 145.—Marconi's System : Diagram of Receiver Connections when using either Kite or Pole.

been employed to " clear intervening obstacles," and to give capacity. The wings W and W (Fig. 144) are then removed : one pole is connected to earth (Fig. 145), and the other extended up to the top of the mast, or else fastened to a balloon or kite—by means of a wire, as shewn in the figure. The wire and kite (or balloon), covered with tinfoil, here becomes the wing, in which

---

* This may be effected by increasing the length of the wings W and W of the receiver in Fig. 144. The proper length may be found beforehand experimentally close to the transmitter, inasmuch as distance has no bearing in the matter ; and this, it will be understood, is the only reasonable plan in practice.

case one wing of the transmitter should also be put to earth, and the other correspondingly raised by means of a conducting pole, if not the apparatus itself.*

By Marconi's system, it is said to be easy to transmit many messages in any direction at the same time. It is only necessary to tune the transmitters and receivers to the same frequency or "note." †

**General Review.**—It is difficult to detect what exactly constitutes the improved results attained by Marconi as against anything that had previously been attained—60 yards as a maximum—in 1894, by Dr Lodge.‡ Is it in the particular mixture employed for the filings of the coherer? It is quite clear, at any rate, that Signor Marconi has gone through a great deal of very careful experimental work to discover the best materials for use and the most effective arrangements. Every detail of Marconi's apparatus appears to have been thoroughly worked out, and he has succeeded in converting laboratory experiments into a novel and yet practical system of telegraphy. For this he deserves—like other inventors—every credit.

It has been said that Signor Marconi has done nothing new. He has not discovered any new rays: his transmitter is practically Righli's radiator, and his receiver based on coherers of the Branly-Lodge type. However, as Mr Preece reminds us at the end of his lecture, Christopher Columbus did not invent the egg, but he shewed the world how to make it stand on its end. Similarly, Marconi has produced from means—the existence of which have been known to electrical *savants* for some time past—a new electric eye more delicate than any electrical instrument so far invented, and a telegraphic system that will reach points hitherto inaccessible by the ordinary means as narrated in this volume. There can be no doubt, in fact, that for shipping and lighthouse purposes the Marconi telegraphy will

---

* In raising the line of action by means of the elevated transmitter and receiver as above, a marked increase in the range of effective transmission is secured, though it is not so easy to explain how this comes about. At first sight it might appear to be due to the waves being thereby enabled to reach the receiver without passing through hills or other solid obstructions, which no doubt have a greater resistance than the air. But the advantage of thus raising the sphere of operations has been experienced without the existence of hills—*i.e.*, in cases of a level surface of water between transmitter and receiver.

† With reference to the above, it may be questioned in practice, if the receivers were in close juxtaposition, whether they would not interfere with one another electrically.

‡ This gentleman, however, pointed out that his apparatus—crude as it was at that time—would probably work through a distance of half a mile.

prove invaluable ; * and to Signor Marconi nothing but praise is due for the ingenuity and perseverance with which he has developed a successful practical application of some of the latest discoveries in electrical science— like others who have revolutionised commerce and added to the wealth of nations by facilitating interchange of intelligence.

Here we have a remarkable example of the disinterested manner in which the Post Office authorities really take up an invention that seems promising. It should be remembered that Mr Preece had for some time been experimenting in the direction of Inductive Telegraphy on lines of his own, but when Marconi's system was brought to his notice he immediately recognised its superiority—as he stated, indeed, in the course of his recent Royal Institution lecture (previously alluded to), from which the material for this section is very largely drawn.†

For further information on the subject the reader is especially referred to the columns of the *Electrical Review.*‡

### TESLA'S RESEARCHES.

Mr Nikola Tesla is said to be at work on another form of wireless telegraphy, the idea of which has been in his mind for a long time. It may be remembered that when in England several years ago he prophesied that telegraphic messages would yet be sent through the earth to Australia— certainly a simplification of the Pacific cable question !—and not only messages, but electric power, if need be. Mr Tesla now claims to have made good his words to the extent of sending electric signals through 20 miles of ground without any conducting wire.

According to his own account he seems to have found the problem rather easy to solve ; but he will not vouchsafe any information as to how

---

* A great deal of sensational nonsense has been written in popular journals and newspapers about Marconi waves penetrating metal as easily as anything else, notwithstanding the time-honoured doctrines of Faraday. They may, however, be regarded in the light of "florid imaginations" pointing to some "striking picture"—in this case that of men-of-war, gunpowder magazines, or forts, being blown up though many miles out of sight—but with these we have nothing to do here, beyond referring to them with ridicule.

† "Signalling through Space without Wires," by W. H. Preece, C.B., F.R.S., reproduced in *The Electrician.*

‡ Since going to press Signor Marconi's complete specification (No. 120,369 of 1896) has been published, the patent having been accepted on 2nd July of the present year (1897). *The Electrician* of 17th September, whilst giving a very full abstract of this, also reproduces some excellent photographs of Dr Lodge's somewhat similar apparatus for the same purpose, shewn at the Oxford B. A. Meeting of 1894.

he does it until his work is complete. All we know is that he employs vibratory currents of electricity, and waves, or oscillations, set up by them in the luminiferous ether. This latter is coming to be regarded not only as an inexhaustible store of power, but a universal *vehicle* of power—a kind of all-round, omnipresent, gearing.

Other wonders may yet be in store for us, or the coming generations, but in the meanwhile the cable manufacturer need have no misgivings—neither need the telegraph engineer.

West African Telegraph Company : Loanda Cable Hut.*

---

\* The interest, if not the satisfaction, with which the reader may now close this volume would certainly have been increased if only photography had been equal to the task of reproducing the prevailing smell at this benighted and picturesque spot !

# APPENDIX.

---

## "RECORDER" SIGNALS UNDER VARYING CONDITIONS.

### THE LETTER "A" UNCURBED AND CURBED.

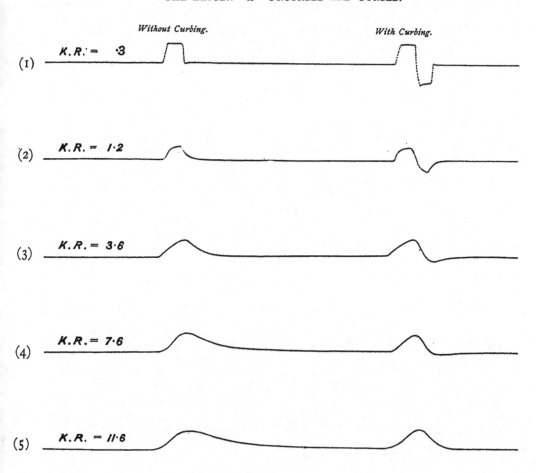

Actual Signals obtained through Muirhead's (Inductive-Resistance) Artificial Cable, under different conditions.

N.B.—*Case* (1) *has practically no retardation, and Case* (5) *has proportionately a great deal.*

# INDEX.

INDEX.

# The India Rubber, Gutta Percha, and Telegraph Works Company Ltd.

### *HEAD OFFICES:*
### 100-106 CANNON STREET, LONDON.
### 97 BOULEVARD SEBASTOPOL, PARIS.

# ELECTRICAL ENGINEERS.

# CABLES
## Submarine, Subterranean, and Aerial.

# DYNAMOS
# INSTRUMENTS, BATTERIES,
### AND
# ALL ELECTRICAL APPARATUS.

### *WORKS:*
## SILVERTOWN, LONDON, E.
## PERSAN-BEAUMONT, FRANCE.

Telegraphic Addresses:

"GRAYSILVER," LONDON.     "INDIA-RUBBER," PERSAN.

# SIEMENS BROTHERS & CO. LIMITED,

### Electrical and Telegraph Engineers.

---

## SUBMARINE, SUBFLUVIAL, SUBTERRANEAN, AND AERIAL CABLES.

### IRON POSTS, INSULATORS, INSTRUMENTS, BATTERIES,

#### And all Appurtenances for Telegraph and Cable Stations.

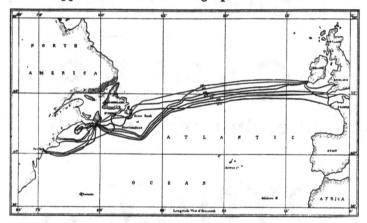

**Seven out of the Eleven Living Atlantic Cables were made and laid by Siemens Brothers & Co. Limited.**

### ELECTRIC LIGHTING AND TRANSMISSION OF POWER, DYNAMOS, MOTORS, ALTERNATORS, ELECTRIC RAILWAYS AND TRAMWAYS, CENTRAL STATIONS.

---

### Head Office—12 QUEEN ANNE'S GATE, LONDON S.W.

#### BRANCHES:

21 GRAINGER STREET WEST, NEWCASTLE-ON-TYNE.
261 WEST GEORGE STREET, GLASGOW.
65 PITT STREET, SYDNEY, N.S.W.
46 and 48 MARKET STREET, MELBOURNE.

#### Works—WOOLWICH, KENT.

Cable Address—"SIEMENS, LONDON."   Codes—"A.B.C," "A1," "Engineering,"
"MOREING & M'CUTCHEON."

# SOME
# SUBMARINE TELEGRAPH CABLES
## Manufactured by us.

**1857.**  **1897.**

**DEEP SEA CABLE.**

Australia to Tasmania.
Balearic Islands and Barcelona.
British Burmah.
Persian Gulf, 1,651 miles.
Shore End of Atlantic.
For Behring Sea.
England and Germany.
Persian Gulf.
Tasmania to Australia.
Bornholm to Libau.
Sweden to Russia.
England and Norway.
Salcombe to Brest.
Marseilles to Bona.
Bona to Malta.
North German Cable.
French Atlantic Cable, 1,086 miles.
Montevidean Cables.
Falmouth and Lisbon.
Alexandria, Candia, Greece, and Italy.
England and Denmark.
Sweden to Gottland.

England and France.
Italy and Turkey.
Scotland and Ireland.
Cook's Straits, N.Z.
England and Belgium.
Constantinople to Odessa.
France to Denmark.
Italian Cables.
Spanish Cables.
Denmark and Sweden.
United States of America, West Indies.
La Guayra, Venezuela to Curacao.
San Domingo to Curacao.
Puerto Plata to St Nicholas Mole, Hayti.
Aquadores, Cuba, to Cay Manera.
Halifax, Nova Scotia, to Bermudas.
Port-au-Prince to St Nicholas Mole, Hayti.
Martinique to Surinam.
Bahamas to Florida.

# W. T. HENLEY'S TELEGRAPH WORKS CO.
## LIMITED,
### Indiarubber and Gutta-Percha Manufacturers.
#### Cables and Wires of EVERY DESCRIPTION.

Warehouse and Offices—27 MARTIN'S LANE, LONDON, E.C.
## WORKS—
## NORTH WOOLWICH, LONDON, E.
### Australia—
## CROMWELL BUILDINGS, BOURKE & ELIZABETH STREETS, MELBOURNE.

*Telegraphic Address—"HENLEY'S WORKS, LONDON."*       *Telephone No., 1734.*

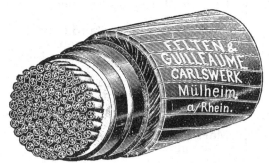
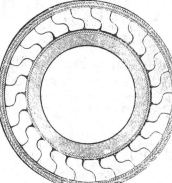

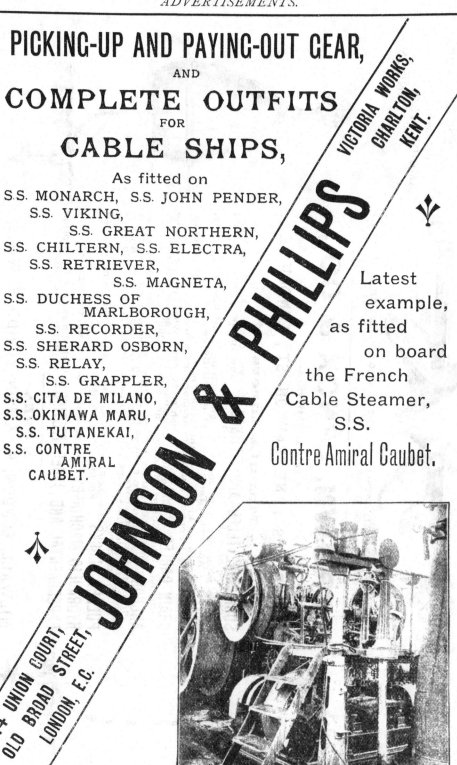

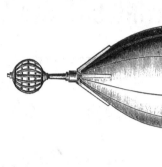

# By CHARLES BRIGHT, F.R.S.E.
### A.M.INST.C.E., M.I.MECH.E., M.I.E.E.

## I.

# SCIENCE AND ENGINEERING DURING THE VICTORIAN ERA.

## II.

# THE EVOLUTION OF THE TELEGRAPH.

## III.

# UNDERGROUND CABLES.

## IV.

# TWENTY-FIVE YEARS' SUBMARINE TELEGRAPHY.

### IN PREPARATION.

## V.

# SUBMARINE SURVEY.

## VI.

# COAST COMMUNICATION AND "WIRELESS" TELEGRAPHY.

Printed in the United States
By Bookmasters